SCIENTIFIC METHODS AND
CULTURAL HERITAGE

SCIENTIFIC METHODS AND CULTURAL HERITAGE

An introduction to the application of materials science to archaeometry and conservation science

Gilberto Artioli
Università degli Studi di Padova

with contributions from

I. Angelini, F. Berna, M. Bicchieri, M. Brustolon, G. Chiari, A. Kaplan,
B. Lavédrine, J. Mazurek, L. Pel, M. Schilling, D. Stulik,
M.R. Valluzzi, J. Wouters

OXFORD
UNIVERSITY PRESS

Great Clarendon Street, Oxford, OX2 6DP,
United Kingdom

Oxford University Press is a department of the University of Oxford.
It furthers the University's objective of excellence in research, scholarship,
and education by publishing worldwide. Oxford is a registered trade mark of
Oxford University Press in the UK and in certain other countries

© Gilberto Artioli 2010

The moral rights of the author have been asserted

First published in 2010
Reprinted 2012

All rights reserved. No part of this publication may be reproduced, stored in
a retrieval system, or transmitted, in any form or by any means, without the
prior permission in writing of Oxford University Press, or as expressly permitted
by law, by licence or under terms agreed with the appropriate reprographics
rights organization. Enquiries concerning reproduction outside the scope of the
above should be sent to the Rights Department, Oxford University Press, at the
address above

You must not circulate this work in any other form
and you must impose this same condition on any acquirer

British Library Cataloguing in Publication Data
Data available

Library of Congress Cataloging in Publication Data
Data available

ISBN 978-0-19-954826-2

a Lety

...ἀληθείη δὲ παρέστω
σοὶ καὶ ἐμοί, πάντων χρῆμα δικαιότατον.
Μίμνερμος

Preface

The book is intended to be an introduction to materials science for non-science majors working with materials and related problems connected with cultural heritage. The rather challenging experience of teaching an introductory course on "Technical analyses and methods" to archaeologists in the past years actually turned out to be a stimulus to find novel and simple ways to convey scientific concepts to students with little or no background in physics, mathematics, chemistry, biology, and Earth sciences. The manuscript naturally started with the reorganization of the notes prepared for the course, and it was purposely planned as an updated textbook concerning techniques, methods and applications of scientific methods to cultural heritage, written using a minimal amount of technical terms and formulae. However, being a scientist who is dealing on a daily basis with archaeologists, curators, conservation scientists, and many other diverse people involved with the investigation of the cultural heritage, I soon realized how complex the field is, and how rapidly it is developing. The actors in the field span from pure humanists, some of them unfortunately lacking any clue to the potential of scientific investigations, to pure scientists, well acquainted with one or more techniques, but for the most part fairly ignorant about cultural heritage problems. In the middle there is an increasing army of researchers, caught by two fires, struggling with a complex multidisciplinary field that many consider to be a routine service to archaeology and art conservation, but that actually needs a great deal of engagement, ingenuity, team work, and an amazing level of critical assessment.

Although the information delivered is meant to be accurate and up to date according to my experience and understanding of the field, it is important to be aware that the investigation techniques in materials science are being continuously developed, optimized and improved, thus needing a continuous transfer of knowledge between science, archaeometry, and conservation. I hope that the selected rather basic level of descriptions appeals both to non-science majors, giving them an introductory broad overview of the analytical contribution to cultural heritage problems, and also to science majors in disciplines that are not directly involved with cultural heritage, showing them alternative applications of the scientific methods that they learned or that they are learning during their education and training. None of the topics is treated at a more than introductory level, not even those that are close to my personal research. Rather, the attempt is to show the potentials of the multitude of available techniques, and anyone interested in specific instruments or methods is referred to the cited technical literature. The unusual style of the book and all the drawbacks and pitfalls are of course to be blamed on my personal choices and idiosyncrasies.

One of the major issues raised at the earliest stages of the project, when I was trying to define the topics covered by the book, concerns the very definition of this area of research, and how it relates to commonly recognized disciplines such as archaeometry and conservation science. After long and often animated discussions with colleagues, including archaeologists, curators, conservators, and colleague scientists I decided that *"application of materials science to archaeometry and conservation science"* is broad enough to cover just about anything scientific that is happening in the field related to cultural heritage, and this largely includes the more conservative definitions of archaeometry and conservation science. After all, the very same scientific techniques and methodologies developed in materials science are adopted by conservation scientists and by archaeologists when in need of analytical data. They just need to be inserted in a coherent framework of investigation and to be appropriately approached and used. Every analysis is unique, because it invariably involves unique objects, materials, and questions.

Los Angeles, August 2007
Padova, July 2009

Acknowledgements

I would like to acknowledge the Getty Conservation Institute for granting me a Guest Scholarship in 2007 and allowing me to start this project. I warmly thank Giacomo and all the staff at the GCI for useful discussions and tips, and for the great time I had in LA. Jerry Podany kindly attempted to broaden my restricted view of archaeometry, I hope he succeeded somehow... I am also grateful to friends who patiently introduced me to archaeological problems and logics of inquiring, at least as far as a non-archaeologist can go: Raffaele C. De Marinis, Paolo Bellintani, Marica Venturino.

All collaborators who contributed to the volume with very interesting and up-to-date technical sections are acknowledged. The book substantially improved thanks to their expertise and specialized knowledge.

Mrs Barbara Vianello invaluably helped in organizing my files, the images, the copyright permissions and, in critical moments, also my mind and my time. Matteo Parisatto and Michele Secco prepared and/or modified some of the graphics. Stefano Castelli skilfully managed slide scanning and photographic work, even at short notice. Judy Santos, Valerie Greathouse, Tom Shreves and Annette Pedrosian @ GCI; and Emanuela Danieletto and Serenella Boesso @ UNIPD readily and cheerfully satisfied my pressing and obsessive reference search. A special hug to Judy and Emanuela!

A number of friends and colleagues checked part of the chapters and helped me to spot and correct several inconsistencies: Ivana Angelini, Beril Bicer-Simsir, David Bourgarit, Fiorenza Cella, Tiziano Cerulli, Giacomo Chiari, Brian Egloff, Alessandro Guastoni, Marco Franzini, Marco Martini, Gianmario Molin, Vincenzo Nociti, Sandro Recchia, Alberta Silvestri. I thank them for their time and effort, and of course I take all the blame for the final outcome.

Table of Contents

List of Contributors   xv

1. Introduction   1
  1.1 Analytical methods applied to cultural heritage:
     problems and requirements   5
  1.2 The view of the artist: materials   7
  1.3 The view of the archaeologist: interpreting the past   9
  1.4 The view of the curator and the conservator:
     conservation issues   12
  1.5 The view of the scientist: optimizing the techniques   14

2. Overview of the analytical techniques   16
  2.1 Available techniques: a problem of energy,
     space, time, and cost   17
        Box 2.a State-of-the-art science at large facilities   21
    2.1.1 Spectroscopy between physics and chemistry   29
        Box 2.b XRF X-ray fluorescence spectroscopy   34
        Box 2.c OES Optical emission spectroscopy   37
        Box 2.d AAS Atomic absorption spectroscopy   40
        Box 2.e IR-RS Infrared and Raman spectroscopies   42
        Box 2.f UV–Vis Ultraviolet and visible spectroscopies   46
    2.1.2 Mineralogy and petrology: the Earth Sciences beyond
        diffraction and optical microscopy   47
        Box 2.g XRD X-ray diffraction   50
        Box 2.h ND Neutron diffraction   52
        Box 2.i ED Electron diffraction   53
        Box 2.j TG-DTA-DSC Thermal analyses   55
    2.1.3 Through the eyes of science: imaging techniques
        at different scales   57
        Box 2.k OM Optical microscopy   64
        Box 2.l SEM-TEM Electron microscopy   66
        Box 2.m X-rays and neutron radiography   69
        Box 2.n CT X-ray and neutron computed tomography   71
        Box 2.o Reflectometry in painting diagnostics   75
    2.1.4 Imaging the past: visual reconstruction and analysis   79
        Box 2.p Wide area survey and remote sensing   84
        Box 2.q Virtual reality in cultural heritage   90

2.2	Sampling problems: invasive, micro-invasive, and non-invasive methods	94
2.3	A question of boundaries: detection limits and limiting questions	101
2.4	How to match problems, materials, and methods	106
	2.4.1 One material, multiple answers	111
	2.4.2 Should I sacrifice the sample or the significance of the measurement?	113
	2.4.3 Optimization of the techniques	114
	2.4.4 Cost-effective information content: developing analytical strategies and risk assessment	118
	Box 2.r Portable non-invasive XRD/XRF instrument: a new way of looking at objects [G. Chiari]	119
	2.4.5 Data treatment, statistics, and presentation of the results: making science clear	125
	Box 2.s Accuracy and precision in scientific measurements	128
2.5	Time and dating: an overview	130
	2.5.1 Which time scale?	131
	2.5.2 Chemical time scales: the assessment of long term behaviour of cultural artefacts [B. Lavédrine]	136
	2.5.2.1 Durability and changes	137
	2.5.2.2 Artificial ageing methods	140
	2.5.3 Relative methods: fundamentals	157
	2.5.3.1 Palynology	158
	2.5.3.2 Obsidian and quartz hydration	159
	2.5.3.3 Lichenometry	161
	2.5.3.4 Chemical changes in materials	162
	2.5.3.5 Aminoacid racemization method	163
	2.5.3.6 Geomagnetic polarity reversals and archaeomagnetism	164
	2.5.4 Absolute methods: fundamentals	166
	2.5.4.1 Dendrochronology	167
	2.5.4.2 Methods based on radioactive decay processes	170
	Box 2.t MS Mass spectrometry	178
	Box 2.u ^{14}C calibration procedures	181
	Box 2.v Stable isotopes	184
	2.5.4.3 Methods based on electron trap accumulation	188
	Box 2.w TL and OSL luminescence	189
	Box 2.x EPR Electron paramagnetic resonance spectroscopy [M. Brustolon]	192
	2.5.5 Advantages and pitfalls	196
3.	**Materials and case studies: how to meet the needs**	**199**
	3.0.1 Pyrotechnology	204
3.1	Structural materials I: lithics, rocks, stones, structural clay products, ceramics	209
	3.1.1 Lithics, rocks, stones	210

Box 3.a NAA-PGAA-NRCA Neutron-based analysis	214
Box 3.b Rock art characterization and preservation	226
3.1.2 Structural clay products, ceramics	229
3.1.2.1 Chemical and mineralogical composition	232
3.1.2.2 The physical-chemistry of the firing process	234
3.1.2.3 Physical properties and classification	238
3.1.2.4 Characterization methods and interpretation	239
3.2 Structural materials II: cements, mortars, and other binders	242
3.2.1 Lime-based materials	242
3.2.1.1 Dating of mortars	250
3.2.2 Gypsum-based materials	250
3.2.3 Clinker-based materials	251
Box 3.c On-site investigation of masonry [M.R. Valluzzi]	253
Box 3.d Degradation and conservation of binders. A tale of pores and water	259
3.3 Pigments	266
Box 3.e The measurement of colour	274
3.4 Glass and faience	278
3.4.1 The nature and composition of vitreous materials	283
3.4.2 Source materials and manufacturing techniques	289
3.4.3 Glass alteration and degradation processes	295
3.4.4 Analytical techniques for glass studies	298
Box 3.f XPS X-ray photoelectron spectroscopy	299
Box 3.g AES – EELS Electron spectroscopies	301
Box 3.h XAS X-ray absorption spectroscopy	303
3.5 Metals	305
3.5.1 Metal science	307
Box 3.i PIXE Proton induced X-ray emission and ion beam analysis	313
Box 3.j Metal corrosion	317
3.5.2 Ore, mines, smelting	320
3.5.2.1 Provenance	326
3.5.2.2 Diffusion vs multiple invention	330
3.5.2.3 The investigation of metallurgical sites	331
Box 3.k Smelting slags: the key to ancient metals extraction	334
3.5.3 Characterization of metal objects	339
3.5.3.1 Chemical composition	339
3.5.3.2 Physical properties and microstructure	340
3.5.3.3 Surface analysis	342
Box 3.l Metallography and crystallographic texture analysis	343
3.6 Gems	348
3.7 Organic materials	355
3.7.1 Bones and ivory	356
3.7.1.1 Bone material: major components and hierarchical organization [F. Berna]	357

3.7.1.2 Mineralized collagen fibril [F. Berna]   359
3.7.1.3 Ivory   360
Box 3.m Bone alteration and diagenesis [F. Berna]   364
3.7.2 Amber and resins [I. Angelini]   367
    3.7.2.1 Resins   370
    3.7.2.2 Amber and copal   372
    3.7.2.3 Analytical methods for the study of resins and amber   381
3.7.3 Paper [M. Bicchieri]   384
    3.7.3.1 Paper degradation   384
    3.7.3.2 Paper analysis: destructive   388
    3.7.3.3 Paper analysis: non-destructive   391
3.7.4 Fibres and textiles   395
3.7.5 Natural organic dyes and pigments in art [J. Wouters]   399
    3.7.5.1 HPLC-PDA as a protocol for the analysis of natural dyes   402
    3.7.5.2 Aspects of natural organic dye and pigment analysis in practice   403
Box 3.n NMR Nuclear magnetic resonance [L. Pel]   408
Box 3.o HPLC High performance liquid chromatography [J. Wouters]   410
Box 3.p GC/MS Principles of gas chromatography/mass spectrometry [M. Schilling]   413
Box 3.q The use of antibodies for molecular recognition [J. Mazurek]   415

3.8 An example of complex composite materials and processes: Photography [D. Stulik and A. Kaplan]   419
  3.8.1 Visual and microscopic methods of identification of photographs   421
  3.8.2 Identification of photographs using non-contact and non-destructive analytical methods and procedures   423
  3.8.3 X-ray Fluorescence Spectroscopy   423
  3.8.4 Quantitative XRF analysis of photographs   424
  3.8.5 Fourier Transform Infrared Spectrometry   426
  3.8.6 Portable laboratory for research of photographs   428
  3.8.7 Analysis of photographs using invasive methods of analysis   430
  3.8.8 The future of scientific and analytical techniques in research of photographs   431

4. Present and future trends: analytical strategies and problems   **434**

Reference list   **438**
Permission list   **507**
Image acknowledgements   **515**
Index   **519**

List of Contributors

I. Angelini
Università di Padova, Padova

G. Artioli
Università di Padova, Padova

F. Berna
Boston University, Boston

M. Bicchieri
Istituto Centrale per il Restauro e la Conservazione del Patrimonio Archivistico e Librario, Roma

M. Brustolon
Università di Padova, Padova

G. Chiari
Getty Conservation Institute, Los Angeles

A. Kaplan
Getty Conservation Institute, Los Angeles

B. Lavédrine
Centre de Recherche sur la Conservation des Collections, Paris

J. Mazurek
Getty Conservation Institute, Los Angeles

L. Pel
Eindhoven University of Technology, Eindhoven

M. Schilling
Getty Conservation Institute, Los Angeles

D. Stulik
Getty Conservation Institute, Los Angeles

M.R. Valluzzi
Università di Padova, Padova

J. Wouters
Zwijndrecht, Belgium

Introduction

1

1.1 Analytical methods applied to cultural heritage: problems and requirements   5
1.2 The view of the artist: materials   7
1.3 The view of the archaeologist: interpreting the past   9
1.4 The view of the curator and the conservator: conservation issues   12
1.5 The view of the scientist: optimizing the techniques   14

A scientific discipline came into existence in the last part of the 20th century that is broadly defined as the application of scientific principles and methods to the characterization of materials that are related to cultural heritage. The field is so vast that there is no general agreement on the extent or even the definition of this discipline. As a materials scientist involved with the technical analyses of art and archaeological materials, at different times I find myself redefined as an archaeometrist, a conservation scientist or simply as an "odd" kind of scientist. Before tackling the technical details of the analyses, I believe it is important to realize and explore the different levels of understanding and the variety of implications involved with the technical analyses of cultural heritage materials.

A widely used term for the discipline is **archaeometry**, which has been in use since the founding of the journal *Archaeometry* in Oxford in 1958. Before that Christopher Hawkes suggested using "archaeological science" (Tite 1991) to describe the discipline concerning the quantitative characterization of ancient objects and processes. Archaeometry is nowadays largely used as an alternative term for "archaeological science", or "science in archaeology". However, it is mostly accepted that the archaeometric applications to archaeology, as wide and far ranging as they are (Brothwell and Pollard 2001), are distinct from those related to **conservation science**, which is more closely related to curatorship, art history, museum conservation practice, and restoration (Winter 2005). The general aim is to quantitatively help in solving problems related to understanding the nature of ancient and art materials and processes, the interpretation of these materials with respect to human cultural history, and to aid in preservation and conservation of the objects for the future. The boundaries of archaeometry and conservation science therefore, from the point of view of the scientific analyses, are often blurred and intimately mixed. After all, the same instruments are used for analytical investigations, and for the most part the analytical protocols are independent of the age of the objects.

I personally tend to transcend the original definition of archaeometry as being limited to archaeological applications, and favour the re-definition of archaeometry as the broad application of materials science within the present trend of integration between archaeology and conservation (Agnew and Bridgland 2006). The term archaeometry will therefore be used throughout the book in this broadened acceptation. However, it should be clear that this personal opinion is far from being generally accepted, and both the terms archaeometry and conservation science nowadays have their own lives and often slightly different fields of application. From the point of view of materials science, it is also hard to confine the cultural heritage of the past to archaeological objects and art objects alone, as many recent treatises seem to imply (Leute 1987, Creagh and Bradley 2000), since many cultural objects are quite hard to define: Jerry Podany made me fully aware of how difficult it is to define things, and I fully agree that we should also take into account fairly recent materials that quickly became archaeological (early plastics, 20th century reinforced concrete, the engines of the industrial revolution... all in need of proper investigations and conservation), art works that are rapidly becoming archaeological, and archaeological artefacts that are treated as art, whereas the majority of them will certainly never be seen as such.

The application of scientific methods to archaeometry and conservation science requires a remarkable degree of interaction between specialists dealing directly with the different aspects of cultural heritage materials, such as archaeologists, curators, conservation scientists, art historians, and artists on one side, and scientists of different disciplines on the other (Fig. 1.1). In fact, at present there are very few analysts who have a complete training and background in the arts or humanities, mainly because of the high level of physico-chemical knowledge that is required to carry out the technical analyses and to interpret the results. On the other hand, most professionals involved with cultural heritage have a purely humanistic or artistic background, but often with little or no training in scientific principles and techniques. The traditional academic separation between humanistic and scientific curricula in higher education has hindered the broad exchange of competences between art and archaeology, and science. Unfortunately this is the case in most countries, independent of geographical, historical or political developments. Only a few laboratories associated with museums and art institutions, or institutions devoted to technical conservation, often national research institutes, have developed efficient archaeometrical and analytical tools. At the same time they maintain a continuous close contact with cultural heritage problems and issues. Frequently research on cultural heritage is carried out on an occasional basis by scientists, mostly located in universities, who develop an interest in rather specific problems. They happen to work on the project for a limited amount of time, and then return to their main field. Continuity and the build-up of knowledge is often a problem. The need for an improved dialogue between disciplines is a recurring theme. Almost every scientist working in the field has strong opinions about this, and many of them openly ask for action (Tite 1991, Brothwell and Pollard 2001, Maggetti 2006).

Within this context, archaeometry and conservation science can be perceived as bridges between humanistic, art and conservation disciplines on the one hand, and hard sciences on the other. This interface between disciplines is a fertile field where methodologies, analytical techniques, and especially the ideas developed in each specific scientific area may be enhanced, optimized, and ultimately applied with some ingenuity to diverse interdisciplinary environments.

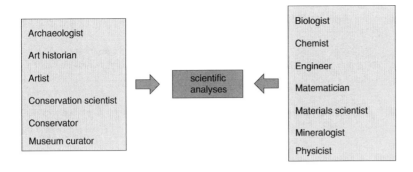

Fig. 1.1. Science for the cultural heritage at the crossroad of a number of different professions and disciplines.

Within the natural and physical sciences many techniques have been developed to solve specific problems of concern to just a few disciplines. At times this makes it difficult to match the problem being investigated with the scientists who have the appropriate background and competence to tackle it. For example NMR, vibrational spectroscopies and chromatographic techniques are commonly used by chemists, whereas surface probes (i.e. XPS, EELS, Auger, AFM, etc.) and electron microscopy are common tools of physicists; optical microscopy and X-ray diffraction are traditionally used by mineralogists and petrologists, and radiography and tomography are everyday tools of materials scientists, engineers, and medical physicists. It is not at all uncommon for the uninformed archaeologist or art historian to approach a "scientist" for the analysis of a specific material, without a critical assessment of the appropriate technique. If the "scientist" does not possess a broad overview of the existing techniques and analytical procedures and an appropriate understanding of the problems involved, he or she will inevitably use the technique that is readily available in his or her laboratory, or the one that yields the quickest and cheapest answer. This random process very often results in the wrong technique being applied and in inappropriate or useless information being produced (Fig 1.2(a)).

Moreover, many scientists are more interested in the details related to the optimization and the use of a specific technique, such as the lowering of standard deviations or probing smaller areas with the beams, rather than in the general cultural problem related to the objects being investigated. In the words of Pollard: "to carry out superb chemistry in support of trivial or meaningless archaeology" (Pollard *et al.* 2007)....This attitude is unfortunately quite widespread, and it is well captured in Fig. 1.2(a), inspired by the famous Gary Larson cartoon "Early microscope". Figures 1.2(b) and 1.2(c) depict a different attitude, where scientific results provide unexpected results: the outcome of an investigation critically depend on the mutual collaborative effort made between the scientific and the humanistic sides at every stage of the project.

It should also be said that many scientists are fascinated and challenged by the possibility of analysing art or archaeological objects because they are beautiful and because they intrinsically remind us of the very human processes that produced them, be it artistic creation or cultural and historical developments.

Fig. 1.2. Science and archaeometry. The complex cultural exchange and relationship between scientists and archaeologists or conservators must determine the selection of the analytical techniques, methodology, and data interpretation. Ultimately it gives the analytical answer: (a) expected and/or trivial, (b) unexpected and incorrect, or (c) unexpected and correct. A great number of other possibilities can be envisaged. (Courtesy of L. Artioli)

The analysis of objects related to cultural heritage adds an exciting flavour to the scientific analysis, so that a number of chemists are very willing to analyse paint pigments in place of smelly organic precipitates, or many mineralogists and petrologists are happy to take a look at archaeological pottery fragments rather than play with dirty soils and crushed rocks. Sometimes scientists seem to think that ancient materials are simpler than modern ones. They soon find out that the opposite is true: cultural heritage materials can be extremely challenging to analyse, and need to be looked at with special attention and, often, specially developed protocols.

Archaeometry is at present located in the broad, undefined, and rapidly changing area between two extremes: on one hand are the problems and materials pertaining to cultural heritage, often in search of appropriate characterization techniques and methods, and on the other hand scientists familiar with technical developments and protocols, often in search of appropriate materials and applications.

It is therefore important to appreciate that each analytical technique offers specific advantages and limitations, and that a broad view and understanding of the available techniques is mandatory in order to take advantage of the full spectrum of modern analytical tools. Modern archaeometry is a discipline that is growing rapidly both in its scope and maturity. The application of analytical techniques deriving from the natural and physical sciences is, of course, not new. Caley (1951) cites almost a hundred publications describing scientific investigations of antiquities or art works published before 1875, mainly analyses of archaeological metals. A number of classical books adequately cover the methodological developments and the basic applications up to the late 1990s (see, for example, Brothwell and Higgs 1969, Tite 1972, Goffer 1980, Leute 1987, Bowman 1991, Pollard and Heron 1996, Lambert 1997). However, what is new in archaeometry is the increasing quality of the science involved and the growing level of the interaction between analytical scientists and archaeologists or conservation specialists concerning the problems of data interpretation. The rapid development of large scale scientific facilities in the last decades, such as pulsed neutron and synchrotron radiation sources, makes the gap between the available analytical potential and the actual cultural heritage applications even more striking.

It is encouraging that in recent years scientists from a variety of disciplines have shown an increased appreciation for the experimental and scientific problems involved with art and archaeology. Furthermore, many educational institutions have started to promote undergraduate, graduate and higher curricula courses in conservation and archaeology with a definite increased interest in several technical aspects of archaeometry. These trends should eventually contribute to the maturity of the field, and to the creation of operating personnel who have an adequate perception of both the humanistic and scientific aspects of the problems involved.

It is of course impossible to describe in a single book or even in a series of books all the possible methods by which materials can be analysed. Recent volumes have attempted to address specific technical issues, each one adopting a different strategy and selecting areas of application from the point of view of the techniques, or of the materials to be analysed (see, for example, Ciliberto and Spoto 2000, Creagh and Bradley 2000, Henderson 2000, Janssens and Van Grieken 2004, Martini *et al*. 2004, Bradley and Creagh 2006, 2007, Pollard *et al*. 2007). Indeed most of the volumes, although presenting excellent material, have a multi-authored character and they clearly show the fragmentation and heterogeneity of this research area, defined as "a conceptually large research field populated by a relatively small numbers of researchers using techniques borrowed from elsewhere" (Winter 2005).

This volume complements the existing literature by presenting an overview of the modern techniques, though focusing on the information content that can be extracted using different experimental strategies rather than on the technicalities themselves. Sometimes the experimental parameters and conditions are illustrated because they are important and they ought to be taken into account when planning an analytical project. This may well mark the difference between the production of random analytical results and the insertion of well-focused and well-thought out data into a broader research plan.

1.1 Analytical methods applied to cultural heritage: problems and requirements

Archaeometry is concerned with every application that employs chemico-physical characterization methods to understand the nature and the changes in time of art and archaeological materials. In doing so, archaeometry uses all available techniques and methods developed within the most diverse and specialized scientific disciplines, from geophysics to molecular biology (Tite 1991) or, taken alphabetically, "from aerology to zymurgy" in the words of Goffer (2007). The scientific tools, and therefore archaeometry, must be appropriately used to serve two broad areas of application: conservation science and archaeology (Fig. 1.3). The two areas tend to be distinct in scope and methodology, though there is substantial overlap in terms of cultural background. Most objects may, at the same time, be considered as art from the curatorial and aesthetic points of view and as archaeological objects from the historical and cultural points of view. Archaeometry can and should provide analytical techniques and scientific support to serve all possible perspectives.

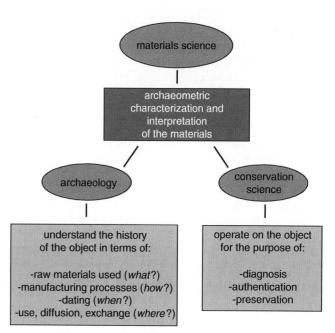

Fig. 1.3. Schematic areas of application of archaeometric analyses.

The object of the investigation must necessarily be a physical material, either artistic or archaeological in nature. No matter how fuzzy or disputable the definition of the object is, no matter what are its size and complexity, from a rock to a whole archaeological site, from a metal fragment to an oil painting, the object is universally linked to human activity. Therefore, to understanding the nature of the object is to penetrate not only its intimate physico-chemical nature, but to unveil in part the human process that produced it, whether artistic creation or cultural development. Archaeometry is ultimately investigating not the object itself, nor the assembly of materials composing the object, but the people behind it. The conservation and the archaeological investigations often involve similar characterization techniques, but they differ substantially in the way that problems are defined and in their final scopes.

At a time when conservation is rapidly developing the concept of cultural heritage as derived from the traditional Western intellectual tradition and is steadily moving towards a more universal understanding of the world heritage (Bouchenaki 2003, 2004), the practice of **conservation** of artefacts, objects, and works of art is nonetheless deeply concerned with the detailed physico-chemical characterization of the materials, in order to understand their past chemical evolution, foresee their possible chemical evolution in the future, and indicate the best means of restoring them to their pristine state, preserve their present state, and even carefully describe the objects' history and modification through time. The relationship between science and conservation is certainly an old one, and it has often been troublesome, because the final restoration strategy is partly dependent on the ever-changing ideas of art sociology, philosophy, and criticism (Price et al. 1996). Nonetheless conservation must be firmly bound to the solid understanding of the physico-chemical and structural aspects of the objects and the processes involved. The role of science is to provide information and quantitative data upon which curatorship and conservation strategies are based. Furthermore, many pieces of information derived from the analyses are used to support authentication issues. Authentication is an important field of application linked to both curatorship and art history (Craddock 2007).

As a start, **archaeology** employs archaeometrical techniques in order to understand the surviving material culture of past societies. The objects are analysed in order to reconstruct (1) the procurement and processing of raw materials, (2) the production technology, including the manufacturing and decorating techniques, and (3) the use, distribution, trade or exchange of the objects. The investigation assumes that a better knowledge of the production-and-use chain ("chaîne opératoire" in the words of Tite, in Brothwell and Pollard 2001) of the artefacts helps us gain an understanding of the people associated with them. Eventually the analyses will supply clues for the interpretation of how a new technology was discovered or chosen, and why it was adopted at a particular time and place. The answers will ultimately derive from a holistic approach, taking into account that the production, distribution and use of a particular type of artefact are necessarily dependent on environmental, technological, economic, social, political and ideological contexts and practices (Diamond 1999, Renfrew and Bahn 2008). Archaeology, in fact, ought to encompass and combine the data supplied by a large number of disciplines, each contributing to the reconstruction of the human puzzle.

1.2 The view of the artist: materials

Artists are seldom interested in a scientific understanding of the detailed physical properties of the materials they use for artistic creation, although this was truer in the past than at present. An artist's interest in the material(s) that he or she uses is generally purely aesthetical and phenomenological, and focuses on whether the material can be employed to bring the artist's ideas into shape. The archetypal image of Michelangelo standing in the quarry carefully and painstakingly selecting the marble blocks to be used for his sculptures is an example of the deep and complex relationship that exists in art between the artist and matter. The limbs of Christ emerging from the stone in the Pietà Rondanini are a tangible example of the artist's miraculous creation. In most cases the artist's attitude towards materials is based on intuition, artistic ingenuity, and personal experience rather than on a scientific knowledge of the material's properties. The result can be catastrophic, especially when the artist departs from traditional wisdom and recipes and adventures into discovering new materials or mixtures, actually experimenting with new creative techniques or art forms. The history of art is loaded with examples of masterpieces made with non-durable or easily degradable materials, the result of poor personal experience, limited availability of raw materials, the choice of cheap materials, or even the precise and sometimes fatal choice of the artist. Leonardo's experiments with a variety of pigments, or the instability of some of Mark Rothko's recent paintings (Cranmer 1987), illustrate the point well.

Few artists are personally concerned in an informed way with the possible deterioration of the structural or aesthetic properties of their works, so that scientific analyses and physico-chemical data of material properties are thus happily ignored. As a consequence, curators' headaches, endless art criticism, the expansion of the restorers' job market, and possibly significant scientific investigations are produced. In only a few cases, where repetitive production techniques are used, for example in the case of art ceramics, details of the raw materials and transformation processes are analysed and often routinely controlled in order to optimize the artist's or the craft worker's time and to reduce the number of shapeless fragments produced. The astonishing colour and expression of the terracottas of Guido Mazzoni (Fig. 1.4), the delicate glazes of Luca della Robbia (Fig. 3.51), or the vivid colours of Islamic lustre-ware (Fig. 1.5) are all examples of carefully mastered ceramic production methods that employ extremely controlled and highly technological manufacturing techniques.

There are cases in which the artist accurately and consciously selects the materials to be used in their work because of their durability and workability,

Fig. 1.4. Head of an old man. Guido Mazzoni, ca. 1480. Galleria Estense, Modena, Italy (Inv. 4178; photo P. Terzi). (Bonsanti and Piccinini 2009)

Fig. 1.5. Bowl lustre-ware. Tell-Minis style, Syria, ca. AD 1075–1100. Royal Ontario Museum, Toronto, Canada.

or because of specific aesthetic properties. Michelangelo's well-thought out use of a high pozzolan content in the plaster support of the Last Judgment fresco in the Sistine Chapel (Michelangelo: La Cappella Sistina 1999) is an appropriate example of selecting a material based on physical properties and durability. Yves Klein's obsession with blue and his careful selection of the specific components of his International Klein Blue paint, artificial ultramarine in a polyvinyl acetate medium (Mancusi-Ungaro 1982), is an extreme example of a material selected for its final appearance and exclusivity. The architectural counterpart of this extreme may be found in Frank Geary's choice of titanium for the shining and long lasting metal coverage of the Disney Hall in Los Angeles. This is a perfect fusion, as they say in LA, of artistic and technological intents requiring a highly developed understanding of material properties and a knowledgeable control of the production process.

1.3 The view of the archaeologist: interpreting the past

The function of technical analyses is to aid archaeology to unravel the human past. Independent of the conceptual foundations of modern archaeological theory and practice, emphasizing processual, post-processual or interpretive modelling (Renfrew and Bahn 2008), there is no doubt that scientific investigations and proper interpretation of observations plays an increasing role in archaeology. Archaeological scientists and archaeological theorists ought to join forces to use information derived from material culture to understand past societies (Jones 2004). Scientific methods can be used by archaeologists at all stages of their work, from project planning and site surveying, to object characterization and interpretation. Most scientific disciplines can actively contribute to archaeology, as each archaeological excavation is a unique, irreproducible event. Any information lost is lost forever, and archaeologist ought to be very well aware of the intrinsically interdisciplinary nature of their work and of the need to interact in depth with a multitude of specialists before, during and after an excavation.

Since the first general compilation of scientific applications to archaeology (Laming 1952) up to the comprehensive publication of Brothwell and Higgs (1969), the focus of archaeometry has been on the physico-chemical analysis of artefacts in order to define the nature of the materials, the manufacturing processes, and the trade routes, in addition to the obvious and almost obsessive requests for dating excavation materials and their contexts. Historically, Henderson (2000) traces the development of scientific archaeology through several stages:

1) an early phase defined as "early stirring", i.e. a phase of ground breaking attempts to face the problems of ancient technology and materials compositions based on little or no previous knowledge and experience;
2) a long phase of "invention", during which archaeology explored the potential of techniques and instrumentation borrowed from a variety of scientific disciplines, sometimes adapting them to its specific needs;
3) a subsequent "primary phase of archaeological science research", exemplified by the establishment Oxford laboratory and by Renfrew's innovative work in linking scientific data and human past behaviour and, finally,
4) the present mature phase of "holistic archaeological science", hopefully witnessing a full interaction between all scientific disciplines linked to archaeology, and a fruitful integration of a variety of scientific data into archaeological contexts and interpretation, thus enriching and expanding our knowledge the past.

Many different disciplines have recently come into close contact with archaeology. Biological and biomolecular studies, including genetic techniques, yield clues on the palaeobiology of ancient humans, their health, pathology, diet and nutrition. Palaeoenvironmental and palaeoecological information help in situating past civilizations into the contexts of broad geographical and climatological landscape. Prospecting techniques and remote sensing have greatly advanced large scale site surveying (Figs 1.6 and 1.7). Microbeam and penetrating beam techniques, especially those based on synchrotron sources and spallation neutron sources, have opened entirely new possibilities in physico-chemical analyses. Advanced statistical methods, large databases, visualization programmes, and fast computers are of fundamental help in data handling, interpretation, and modelling. Many of these recent developments are briefly reviewed in the excellent manual edited by Brothwell and Pollard (2001).

In fact, the available techniques are so diverse and numerous that the inexperienced archaeologist may find it difficult even to know which technique

Fig 1.6. Aerial photograph of the small and large villages in the Terramara of S. Rosa (Poviglio, Reggio Emilia Italy). The picture taken from above clearly shows the boundaries of the area that was subsequently excavated. (Courtesy of M. Cremaschi, University of Milano)

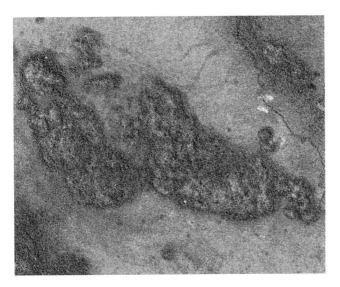

Fig 1.7. High resolution false-colour image taken from the IKONOS satellite of the Guatemalan lowlands. Advanced filtering of colour plants help in identifying Maya archaeological sites. (From Server and Irwin 2003; image available at the NASA Marshall Space Flight Center website: http://www.nasa.gov/centers/marshall/multimedia/photos/2006/photos06-018.html)
Original colour picture reproduced as plate 1.

should be sought and applied, let alone have a detailed knowledge of how each technique works. This volume attempts to introduce the criteria upon which the techniques should be chosen in the planning phase and the amount of information they can provide, and to convey the idea that archaeologists should undertake team-work as much as possible with appropriate scientists, within the limits of time and cost of their project. It is quite clear that answers to complex questions can hardly be derived from one technique alone, especially when the problems involve parameters such as time, chemical changes, population dynamics, and the like. Therefore archaeologists should gather as much information as possible on all aspects of their investigations, given that useful data can be obtained only by the efficient and clever use of suitable methods and techniques, the thoughtful interpretation of hard data, and continuous interaction with specialists during data analysis. The best way to avoid the misuse of scientific techniques (Fig. 1.2(a)) is to select techniques that are appropriately based on well chosen samples and well posed questions.

Too often scientific data have been used to prove or disprove highly prejudicial assumptions, and data tables have been added to Appendices in articles just to make the content look more "scientific". One would hope that those days are over, though understandably it is often extremely difficult to contextualize scientific observations, to interpret them with an unbiased attitude, and to propose far-reaching and conclusive discussions. Sometimes this ideal process is difficult even within the boundaries of hard physical sciences, where technical questions are often more precisely outlined. It is of course much harder when the field is broader, undefined, and the concept of materiality (Gosden 1994) is intimately mixed with notions of local cultural tradition and identity (Graves-Brown *et al.* 1996). The metaphoric and symbolic values of the past material culture will be at the heart of discussions between science-based and theoretical archaeologists for quite a while (Miller 1994, Tilley 1999, Jones 2004). However, there should be a strong tendency for archaeologists to understand the properties of the materials and the thermodynamic and kinetic basis of technological processes; these are fundamental to any subsequent interpretation (Killick 2004).

1.4 The view of the curator and the conservator: conservation issues

Starting from the basics, even the simple cataloguing of museum objects may greatly profit from scientific investigation to refine the descriptions of the objects. It is often disturbing to read catalogue entries such as "wooden box with metal hinges and ivory inlay", where with a little effort it might read "wooden box (willow, i.e. *Salix sp.*) with bronze hinges (90% copper–10% tin alloy) and walrus ivory inlay" (Oddy 1984). Not only is this description more rigorous and satisfying, but the information content is far larger, and it greatly helps researchers to retrieve this information from the catalogue or the database for further study and comparison purposes.

However, it is obvious that that simple museum cataloguing and conservation is only a small part of modern curatorship. Whether we see a work of art as a physical object, or we attach aesthetic and metaphysical values to it, there is a strong legacy to understanding its nature, its making and its history, and to preserve it for the future because heritage conservation is an integral part of civil society. The objectives of conservation briefly described above, i.e.

the characterization of the physical condition of the object, is actually just a part of the conservatorial tasks that nowadays include managerial abilities and wide-ranging cultural and social attitudes, given the rapid changes brought about by recent globalization, population mobility and market economies (Avrami *et al.* 2000). Starting from a view of cultural heritage that is limited to masterpieces and prehistoric and historic monuments, the concept nowadays may well include everyday objects that are significant to specific societies or groups: vernacular architecture, entire archaeological sites and areas, and even natural and cultural landscapes. The challenges of understanding and preserving cultural heritage materials are therefore multiplied and the skills required of curators are commensurately expanded. Conservation science may therefore currently be just the technical portion of the widest field of curatorship and conservation. The traditional view of a curator is that of a professional formed by the fundamental texts dealing with the historical and philosophical issues of the conservation of cultural heritage (Price *et al.* 1996). Among the competencies expected from curators who have to manage and decide on the extent and impact of interventions there should be a good understanding of the behaviour of the constituent materials and the production techniques of the objects, of the available techniques to be used to evaluate the mechanisms of changes and degradation, and a knowledge of the most appropriate products and techniques to be applied during conservation. The detailed list of educational skills and training thought to be needed by a mid-career conservator could be daunting (GCI-AIC Report 2002). Again, the issues involved are so far-reaching that collaboration with scientists and team-work is an obvious answer.

Most of the problems involved in conservation can be rationalized as a function of time. The issues to be considered are the reconstruction of the materials and the processes used at the time the object was produced (past), the understanding of the changes that have occurred since then (present), and the intervention to stop or slow down the processes of decay and alteration (future). Scientifically speaking, the problem is to relate the **kinetics of chemical processes** to the human time-scale (Section 2.5.2). To control the chemical kinetics, modern conservation management includes control of the environmental variables surrounding the objects, such as temperature, humidity, light, etc. and their continuous monitoring. An example of the rapidly changing needs of modern society for specific conservation issues is the preservation of images and information on recently developed media. We are all aware of the difficult issues related to the restoration and preservation of old parchments and paper books (Section 3.7.3) and many are aware of the growing need to understand and solve problems related to the preservation of rapidly degrading photographic collections, whose materials were developed only after the first decades of the 1800s ((Section 3.8) Lavédrine 2003, 2009). Few, however, are fully aware of the rapid degradation of recent media, such as magnetic tapes, optical discs and solid state memories, which have been used only for decades (Rothenberg 1995). The very rapid switch from analogue to digital coding and the continual development of new information supports, together with the present massive production of data, has created enormous problems of data preservation and data access, many of the core issues are still to be solved.

1.5 The view of the scientist: optimizing the techniques

Of course the issues briefly discussed for art, archaeology and conservation ought to be fully appreciated by the scientists in charge of the technical and analytical operations. Useful scientific results are rarely produced without a close and continuous interaction with the group investigating the object and proposing the problems. The interaction should be: **before** the analysis in order to plan for appropriate sampling techniques, analytical methodologies, and time/cost evaluations; **during** the analysis, to check that the analysed parameters match the problem posed, and **after** the analysis in the data treatment and interpretation stages (Ciliberto and Spoto 2000). If the collaboration is successful, very often new problems or unexpected interpretations are generated by brainstorming and open inquisitive minds. The rate of scientific success and progress may indeed be gauged by the number of alternative solutions stemming from the investigation. In fact one may well be suspicious of results that just confirm preset assumptions...

Two fundamental contributions should be introduced by the scientists involved in the technical operations: (1) the proposal of alternative or complementary techniques to extract the maximum amount of information from the analysis, and (2) optimization of the experimental methodologies to the specific needs of the problem. Given the huge variety of materials and shapes composing objects to be analysed, the standard experimental setting and protocols of many scientific techniques are unsuitable for obtaining appropriate results. This is why very often standard analyses produce totally inadequate results in terms of sensitivity, spatial resolution, counting statistics and so on. In common practice, very seldom does one obtain the required information straightforwardly. The dedicated scientist has a number of different strategies in order to improve the results considerably, for example by optimizing the source, by changing the geometry of the measurement, or by using more sensitive detectors. This is a winning strategy and it is the one that is absolutely required when using very flexible instrumentations at large scale facilities. Some of these implementations will be described in the chapters devoted to techniques. All this, of course, requires time and money, and it only happens if a constructive and long lasting interaction exists between the analytical group and that requesting the analyses.

In brief, it is not universally appreciated that many techniques that are sample-invasive if performed accordingly to standard or widely used protocols, i.e. those that may be familiar and easily implemented in many laboratories, may also be used in a totally non-invasive or micro-invasive way. These alternative protocols are often more acceptable and convenient for precious archaeological or art samples, though they may not be implemented in all laboratories, simply because the required instrumentation is unavailable. Several examples, including X-ray diffraction, IR spectroscopy, and tomography are discussed in Section 2.4.3, and infinite possibilities exist for development. It is expected that the complex problems related to the cultural heritage will be investigated by an ever increasing array of instrumental techniques and optimized methodologies. One of the roles of the scientists is precisely that of optimizing such analytical methodologies to meet the requirements of archaeometry, and to make the technique available to a wide range of applications for the cultural heritage.

Further reading

Agnew N. and Bridgland J. (2006) Of the past, for the future: Integrating archaeology and conservation. The Getty Conservation Institute, Los Angeles.

Brothwell D.R. and Pollard A.M. (2001) Handbook of archaeological sciences. John Wiley & Sons, Chichester.

Price N.S., Kirby Talley Jr M. and Melucco Vaccaro A. (eds) (1996) Historical and philosophical issues in the conservation of cultural heritage. The Getty Conservation Institute, Los Angeles.

Renfrew C. and Bahn P. (2008) Archaeology: Theories, methods and practice. 5th Edition. Thames & Hudson, London.

2 Overview of the analytical techniques

2.1 Available techniques: a problem of energy, space, time, and cost   17
2.2 Sampling problems: invasive, micro-invasive, and non-invasive methods   94
2.3 A question of boundaries: detection limits and limiting questions   101
2.4 How to match problems, materials, and methods   106
2.5 Time and dating: an overview   130

Given the wide range of materials and problems involved in archaeometry, as defined in Chapter 1, and taking into account the need for continuous interaction between the people managing the cultural heritage and the scientists, it is important that both parties are aware of (1) the full spectrum of techniques that are available for scientific investigation and analysis, (2) their advantages and limitations, and (3) the kind of information that they can provide.

In this chapter we will explore how scientific measurements, often involving absorption or emission of radiation in different parts of the electromagnetic spectrum, are not only powerful tools for the characterization of matter, but are also the mean of understanding and quantifying a variety of phenomena that are important for modern everyday life and past human achievements, from the perception of colour to the heat loss from buildings, from pyrotechnologies used in the technological transformation of natural materials to the behaviour of bones after burial.

Of course technical details and specific operational procedures of each technique are complex and they will have to be acquired by contacting experts of the particular technique and by perusing technical manuals. However, the presented introductory material should convey the general notion that the same material can be investigated by different techniques, each technique working at **specific energy** or energies, with specific **spatial scale** and resolution, and with widely varying **limits of detection** and sensitivity. Therefore each analytical technique and even each mode of operation within a single technique may produce very different results and information. These analytical complexities should be viewed not as a problem, but rather as a set of incredibly powerful tools for acquiring complementary data towards the solution of archaeometric problems, both in archaeology and in conservation issues. Once the archaeometric questions are defined, then following specific protocols of intervention and project planning (Section 2.4), the specific decisions involved in the archaeometric analysis of art and archaeological materials are:

- sampling,
- choice of the analytical technique and methods,
- time and costs of the analyses,
- data handling and interpretation.

These issues will be discussed in detail, because they encompass all the parameters that should be taken into account when planning scientific investigations.

2.1 Available techniques: a problem of energy, space, time, and cost

A few principles are necessary to understand the operations and the limits of the available analytical techniques. First, most experimental techniques use a probe to investigate the material of interest, be it a natural mineral or a synthetic compound, present in different states: gaseous, liquid, and solid, both amorphous and crystalline. The probe is usually a beam of electromagnetic radiation (Fig. 2.1), though incident particle beams (electrons, neutrons and protons) are also used. The fundamental relation proposed by De Broglie in 1924, the wave–particle duality principle of quantum mechanics, ensures that the two treatments are equivalent and that we may treat particle beams as waves for all analytical applications. It should be recognized that modern analytical techniques invariably probe the chemico-physical nature of the atoms, so that an understanding of the basics of physical chemistry is mandatory to assess the potential and the shortcomings of the techniques. The choice and evaluation of each technique must encompass the energy range involved, the volume of sample probed, and the time and costs of the analyses. Each of these parameters offers a different perspective of the available techniques, and the final choice may actually be a combination of different methods, or a compromise between different approaches. In any case the choice must be made on well-defined and well-thought parameters.

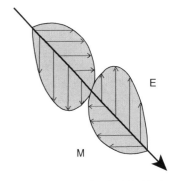

Fig. 2.1. Oscillating electric (E = blue) and magnetic (M = red) fields for a propagating electromagnetic wave. (Modified from B. M. Tissue, Chemistry Hypermedia website, original image available at: http://www.chem.vt.edu/chem-ed/light/em-rad.html)

Energy

The one and only fundamental formula that any non-scientist ought to learn is the relationship between energy and wavelength for photons ($E = h\nu = hc/\lambda$, where E = energy, ν = frequency, λ = wavelength, h = Plank's constant, c = speed of light in vacuum). This formula enables curious non-scientists to sweep through the full spectrum of electromagnetic radiation (Fig. 2.2), and to quickly understand why beams having different wavelengths carry a different amount of energy and therefore interact with very different quantum energy levels of matter. High-energy radiation bands (i.e. γ- and X-rays) have short wavelengths and high frequencies. They interact with quantized levels of the atoms with comparably high energy, such as nuclear levels or electrons located in orbitals close to the nucleus (core electrons). Medium energy radiation bands (i.e. UV, Vis) largely interact with electrons located in the outermost orbitals of the atoms (valence electrons). Low energy radiation bands (i.e. microwaves and radio waves) have long wavelengths and low frequencies. They interact with very low energy levels and quantum jumps, such as electrons located between atoms forming the chemical bonds, or they may excite the low frequency modes related to atomic and molecular diffusion and movements. As an example, everybody, of course, is familiar with the increased thermal motion (and the consequent increase in temperature) of the water molecules in food that is induced by microwave heating.

Each wave can be uniquely characterized by its wavelength, frequency, or energy. It is just a matter of convenience and tradition that chemists in

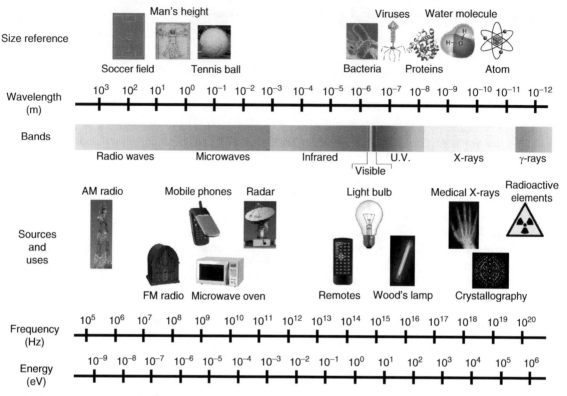

Fig. 2.2. Visual representation of the electromagnetic spectrum, with wavelength, frequency and energy scales. (Modified from the LBL educational website, original image available at: http://www.lbl.gov/MicroWorlds/ALSTool/EMSpec/EMSpec2.html)

vibrational spectroscopy prefer to use frequencies (generally in units of cm^{-1} or Hertz), mineralogists using X-rays talk about wavelengths (in Å, Ångstrom or 10^{-10} m), and physicists in electron microscopy define the electron beam with energy (in eV, electron-volts). The wavelength ranges of different radiation bands are reported in Table 2.1. By using the fundamental equation given above, anyone can easily convert values between wavelength, frequency, or energy units of the electromagnetic waves. When dealing with particles, the conversion is conceptually the same, although one must also take into consideration the mass of the particle involved. Any introductory textbook of physics or physical chemistry will help in understanding these details.

Any material bombarded by radiation or particles interacts with the incident beam, emitting particles and scattered radiation that are analysed in order to acquire information on the chemical or physical state of the atoms or molecules composing it. As an example Fig. 2.3 shows the physical absorption and emission processes that are simultaneously activated by an accelerated electron beam impinging on a thin layer of material. In general the instrument used to carry out the measurement is defined based on the nature of the incident beam, the process actually measured, and the type of radiation or particle

Table 2.1. Frequencies and wavelengths of the different bands of the electromagnetic spectrum, with the processes they activate at the atomic level

Electromagnetic band	Frequency (Hertz)	Wavelength (m)	Energy levels excited
γ-rays (gamma-rays)	$>10^{20}$	$<10^{-12}$	nuclear levels
X-rays	10^{20}–10^{17}	10^{-12}–10^{-9}	core electrons (inner e⁻)
UV, ultraviolet	10^{17}–7.5×10^{14}	10^{-9}–4.0×10^{-7}	valence electrons (outer e⁻)
Vis, visible	7.5×10^{14}–4.0×10^{14}	4.0×10^{-7}–7.5×10^{-7}	
NIR, near-infrared	4.0×10^{14}–1.2×10^{14}	7.5×10^{-7}–2.5×10^{-6}	molecular vibrations
IR, infrared	1.2×10^{14}–1.2×10^{13}	2.5×10^{-6}–2.5×10^{-5}	
Microwaves	1.2×10^{13}–3.0×10^{11}	2.5×10^{-5}–10^{-3}	molecular rotations, librations, phonons
Radio waves	$3.0 \times 10^{11}>$	$<10^{-3}$	nuclear resonance, diffusion

detected (Table 2.2). This provides a preliminary overview of the instrumental techniques based on the physical process that they are designed to detect and measure. The fact that each instrument employs a specific kind of electromagnetic radiation or particle to probe matter, and measures only a narrow selection of the re-emitted radiation or particles, optimizes the technical design of each laboratory instrument for rather specific experiments, and is usually limited to one kind of measurement. That is why in subsequent chapters we will refer to instruments that are available in scientific laboratories as spectrometers, diffractometers, microscopes, and so on, depending on their specific design.

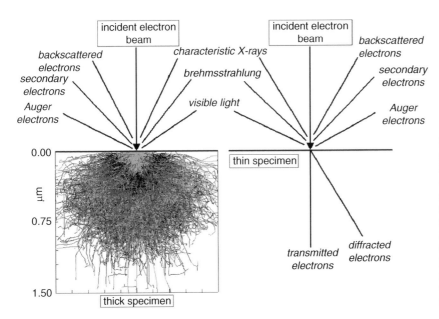

Fig. 2.3. Scattered radiation and secondary particles produced by an accelerated electron beam on a thick (left) and a thin (right) layer of material. Every time an incident beam of radiation or a particle interacts with matter, a number of simultaneous processes are excited: each experiment is designed to specifically measure one or more signals emitted from the sample. The simulation within the bulk sample shows the penetration of electrons into the sample (i.e. the probed sample volume): it is from this interaction volume that the scattered signal is emitted.

Table 2.2. Definition of some of the most commonly employed instruments on the basis of the nature of the probe, the process excited, and the radiation or particles detected. Acronyms of the most diffused analytical techniques are explained. Acronyms of techniques not listed here may be interpreted at http://www.chemie.de/tools/acronym.php3?language=e

Analytical technique	Incident probe	Interaction	Detected signal
X-ray diffraction, XRD	X-rays	elastic scattering	X-rays
Neutron diffraction, ND	neutrons	elastic scattering	neutrons
Electron diffraction, ED	electrons	elastic scattering	electrons
X-ray fluorescence, XRF	X-rays	ionization-emission	X-rays
Proton induced X-ray fluorescence PIXE	protons	ionization-emission	X-rays
Scanning electron microscopy-energy dispersive spectroscopy SEM-EDS	electrons	ionization- emission	X-rays
Electron probe micro-analysis, EPMA	electrons	ionization-emission	X-rays
Atomic absorption spectroscopy, AAS	visible	absorption	visible
Optical emission spectroscopy, OES	temperature	ionization-emission	visible
Neutron activation analysis, NAA	neutrons	nuclear excitation-emission	delayed γ-rays
Prompt gamma activation analysis, PGAA	protons	nuclear excitation-emission	prompt γ-rays
Visible spectroscopy, Vis	visible	absorption	visible
Ultraviolet spectroscopy, UV	ultraviolet	absorption	ultraviolet
Infrared spectroscopy, IR	infrared	absorption	infrared
Raman spectroscopy, RS	infrared	anelastic scattering	infrared
Mössbauer spectroscopy, MöS	γ-rays	nuclear resonance	γ-rays
X-ray absorption spectroscopy, XAS	X-rays	absorption	X-rays
X-ray photoelectron spectroscopy, XPS	X-rays	ionization-emission	electrons
Auger electron spectroscopy, AES	electrons	ionization-emission	electrons
Electron energy loss spectroscopy, EELS	electrons	absorption-energy transfer	electrons
Inelastic neutron scattering, INS	neutrons	collective excitations	neutrons
Thermoluminescence, TL	temperature	electron trap annealing	visible
Optically stimulated luminescence, OSL	visible	electron trap annealing	visible
Electron paramagnetic resonance, EPR	microwave	absorption resonance under EM field	microwave

It is important however to understand that currently there is a class of large scale sources of radiation or particles (Box 2.a) that cannot be classified as specific instruments, insofar as they provide very intense beams over a wide energy spectrum, and very different kinds of instruments can be designed and installed at these facilities to exploit the available probe beams.

> **Box 2.a State-of-the-art science at large facilities: advanced tools for modern investigations**
>
> A very important piece to be added to the analytical mosaic is that represented by so-called **large-scale facilities**. They are large research centres, generally developed at international level and located in the proximity of research laboratories to take advantage of existing technological infrastructures, which are based on very powerful and intense sources of radiation or particles of the kind that are not available at independent institutions because of cost, technical complexity or organizational power. Aside from a number of accelerator facilities mainly used for high energy physics, there are several types of large-scales facilities that are useful for materials characterization. They can be described as **neutron sources** producing continuous or pulsed neutron beams (steady-state nuclear reactors and pulsed spallation sources, Fig. 2.a.1), as **synchrotron radiation sources** producing highly brilliant photon beams in the energy range from γ-rays to IR (Fig. 2.a.2), and as **particle accelerators** producing collimated energetic beams of particles (protons, ions) that are useful for a variety of spectroscopic techniques, such as AMS, PIXE, or PGAA among others (Boxes 2.t, 3.i, 3.a).
>
> Starting information on the nature and properties of each source can be found in basic textbooks (for example Carrondo and Jeffrey 1987, Baruchel *et al.* 1993, Monaco and Artioli 2002, Mobilio and Vlaic 2003, Wenk 2006), and updated lists of currently active sources are to be found on-line for both synchrotrons (http://www.esrf.eu/UsersAndScience/Links/Synchrotrons) and neutron sources (http://www.ill.fr/Info/links.html).
>
> The first time that one uses a large-scale facility one may find it difficult, because it requires a certain knowledge of the properties and potential of each source in order to obtain reliable and useful data. Not only does each neutron and synchrotron source have different characteristics, but virtually each beamline at the facility has special
>
>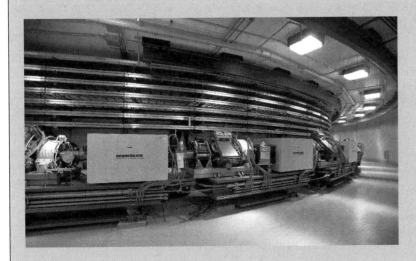
>
> **Fig. 2.a.1.** Equipment installation inside the ring tunnel of the Spallation Neutron Source, Oak Ridge National Laboratory, US. (Courtesy of ORNL, image available at: http://www.physorg.com/news65118008.html)
>
> *(cont.)*

Box 2.a (*Continued*)

Fig. 2.a.2. Nocturnal view of the ESRF synchrotron ring, Grenoble, France. (Courtesy of P. Ginter/ESRF, image available at: http://www.esrf.eu/Infrastructure/Computing/Networks/Site).

optics and instrumentation so that the final beam at the user end is different and optimized in terms of energy range and resolution, beam size and divergence, photon or particle flux and brilliance, etc. The beamline at which the experiment has to be carried out must be appropriately selected, and for inexperienced users the beamline selection and experiment preparation must be carefully done in collaboration with beamline scientists. The flow chart of an experiment usually implies: the planning of the experimental details with beamline scientists; the formal application of beamtime allocation (commonly meeting two deadlines per year); the screening of the scientific merits and instrumental feasibility of the experiment by appropriate selection panels; allocation of beamtime; travelling to the facility for the allocated period; carrying out the data collection and treatment and, finally, reporting the outcome of the experimental session to the facility. The whole process of beamtime request, allocation and carrying out the experiment may take from a few months to one year, so that experiments must be planned well in advance. Long-term involvement and collaboration between users' groups and beamline scientists commonly facilitates the process and the rate of success. It is also important to know that most large-scale facilities have national or international financial support for users' programmes, so that the travel and the logistic expenses of the users are often partially or totally covered by the facilities themselves.

The experimental technical complexities, the length of the application process, and the distance from the facility are often seen as unnecessary annoyances by the users. This is not the case, of course, as experimental data collected at state-of-the-art sources are frequently orders of magnitudes better than those collected in conventional laboratories in terms of sensitivity, spatial and energy resolution, peak to background ratio, etc. Furthermore, many experiments performed with neutron and synchrotron sources are simply impossible to do with laboratory instruments, so that these sources should be seen as totally new probes to characterize materials at space and energy windows and scales that were previously unavailable.

It is not possible to list in this volume the infinite possibilities offered by existing and developing new facilities. Some reviews, although far from being exhaustive, may provide glimpses of what kinds of experiments are possible on cultural heritage materials at large-scale sources (Kockelmann *et al.* 2000, 2006; Dooryhee *et al.* 2006, Creagh 2007). In general, experiments at neutron sources, especially at spallation sources, tend to exploit the neutron penetration into thick objects, the wide momentum transfer of the pulsed sources, and the intrinsic nuclear interaction between neutrons and matter. At synchrotrons, experiments employing highly brilliant and collimated micro-beams of X-rays are often employed to obtain diffraction, spectroscopic and imaging data with unprecedented space and energy resolution. Indeed the possibility of simultaneously obtaining high quality data by several techniques on µg-quantities of material is considerably extending the information that we can obtain on micro-samples. The flexibility of the instrumental configurations at synchrotron and neutron sources is also a bonus parameter that makes large-scale facilities invaluable for almost any new experiment that one may think of.

For in-depth information

Baruchel J. *et al.* (1993) Neutron and synchrotron radiation for condensed matter studies. HERCULES, Editions de Physique–Springer-Verlag.

Creagh D. (2007) Synchrotron radiation and its use in art, archaeometry, and cultural heritage studies. In: Bradley D. and Creagh D. (eds) (2007) Physical techniques in the study of art, archaeology and cultural heritage. Vol. 2. Elsevier, Amsterdam.

Dooryhée E., Menu M. and Susini J. (eds) (2006) Synchrotron radiation in art and archaeology. Special Issue of Appl. Phys. A, **83**, no. 2.

If there is no energy exchange between the probing beam and the material, then the radiation is defined as being scattered elastically, and the process is described as **diffraction** (i.e. wave interference). If the interaction involves an exchange of energy between the incident radiation and the atoms in the materials, i.e. we have absorption and re-emission of radiation at defined frequencies, the process is anelastic and is treated as **spectroscopy**. When radiation or particles interact with the atoms in the sample, the absorbed energy induces the excitation of specific energy levels, starting from nuclear levels excited at high energies and progressively probing electron and molecular levels excited at lower energies (Table 2.1). Spectroscopic techniques can therefore also be classified on the basis of the energy levels that they can probe (Table 2.5); this is referred to in technical books as the energy transfer. The excited physical process and the specific energy transfer make each technique unique in the information they provide.

In general, any quantitative measurement of the intensity–energy relationship in any part of the electromagnetic spectrum is called **spectrophotometry**. The specific case of measurements in the visible part of the spectrum is called **colorimetry**, because it is intimately related to our perception of colour (Box 3.e).

Both diffraction and spectroscopic processes are physically involved in **imaging** techniques, i.e. those techniques that produce a visual image or a map of the space distribution of heterogeneities in the sample. A few experimental techniques dealing with spatial distribution of matter at the atomic level

Table 2.3. The most widespread techniques used in archaeometry, based on the type of information they provide

Information	Spectroscopy	Diffraction	Imaging
Chemical	XRF, IR, RS, XPS, PIXE, AAS, OES, NAA, NMR, UV, Vis		SEM-EDS
Physical	AES, XPS, XAS, EELS		radiography, tomography, reflectrography, SEM
Mineralogical	IR, RS	XRD, ND, ED	optical microscopy
Age	MS, TL, OSL, EPR		

are based on the direct measurement of inter-atomic forces or tunnelling contact energies (atomic force microscopy AFM, scanning tunnelling microscopy (STM), etc.), these techniques are employed in the characterization of materials, but they have yet to find archaeometric applications.

A number of other important analytical techniques are not based on the probe–matter interaction described above, namely **gas and liquid chromatography** (GCy, LCy), **mass spectrometry** (MS) and **thermal analyses** (TGA, DTA, DSC). All of these have been successfully applied in archaeometry and they will be described in detail.

In general, the resulting information provided by measurements is one of the following:

- *the chemical nature of the material*, i.e. the presence (qualitative chemical analysis) and relative or absolute proportions (quantitative chemical analysis) of the chemical elements in the sample;
- *the physical status of the material*, i.e. the measure of the physical, textural and specific parameters of the object that may be useful to interpret its use, manufacturing, physical properties and integrity;
- *the mineralogical nature of the material*, i.e. the state of aggregation of the chemical elements into crystalline or amorphous phases. This information is crucial to understanding the history of the object, the raw components, its manufacturing processes and its alteration state.
- *the age*: dating techniques provide relative or absolute values of the age of the sample.

Table 2.3 lists some of the most used techniques in archaeometry based on the type of information provided.

Space

There is another general aspect that should be clear when approaching scientific measurements: the scale at which the measurement takes place. Humans perceive dimensions through the personal experience of the senses, largely from seeing, touching, and travelling, mostly in the 10^{-3}–10^4 m range, i.e. from the smallest things visible by eye (fraction of millimetres) to the farthest object we can see and reach (tens of kilometres). We can see the Moon, the Sun and the stars, but we can hardly perceive their distance without instrumental measurement. Science on the contrary, through instrumental probing, has access to the whole range of 44 powers of ten presently known in nature, from the size of the Universe (10^{25} m) to the sub-quark region of fundamental particles (10^{-18} m) (Morrison and Morrison 1994).

The intrinsic nature of natural and synthetic materials is highly heterogeneous. We may perceive this heterogeneity at different levels, from the atomic to the macroscopic (Table 2.4):

- **Atomic** or **nanoscopic** level (heterogeneities at the 10^{-10}–10^{-9} m scale). This is related to single atoms, atoms clusters, or single molecules. It is essentially measured by analytical chemistry techniques or atomic/molecular microscopy.
- **Microscopic** level (heterogeneities at the 10^{-8}–10^{-5} m scale). This is related to the long range ordering of atoms into the crystalline state. It is essentially measured by crystallographic techniques, especially diffraction.
- **Mesoscopic** level (heterogeneities at the 10^{-6}–10^{-3} m scale). This is related to the shape and orientation of crystals in single-phase or multiple-phase crystalline aggregates. It is essentially measured by electron and optical microscopy techniques or by techniques that are sensitive to crystal orientation (texture analysis).
- **Macroscopic** level (heterogeneities above 10^{-3} m scale). This is related to the macroscopic properties of the materials, from millimetre sized specimens to samples as large as whole buildings, archaeological sites, or geographical areas. Most imaging techniques and all of the techniques measuring the physical properties of the materials commonly deal with matter at the macroscopic level.

The physical properties that we want to measure may be related to different scales, for example colour is generated by the absorption processes of light by specific localized electron levels, and it is necessary to investigate the material at the atomic or sub-atomic level to interpret the fundamental nature of the process. Colour may also be generated by light interference caused by structural heterogeneities at the light's wavelength scale (4.0×10^{-7}–7.5×10^{-7} m) such as in opal, and in this case we need to image or characterize the material at the microscopic level. Metal hardness or ceramic mechanical resistance is related to the type, size, and shape of their crystalline or amorphous domains. To understand the mechanical properties of the materials one needs to characterize the structural and textural features at the mesoscopic level.

All human observations carried out without specialized instruments are performed at the macroscopic level. This is the case, for example, in visual

Table 2.4. Representation of matter at different scales (Fig. 2.4.8 © The J. Paul Getty Trust 1991. All rights reserved)

Atomic (nanoscopic) scale 10^{-10}–10^{-9} m

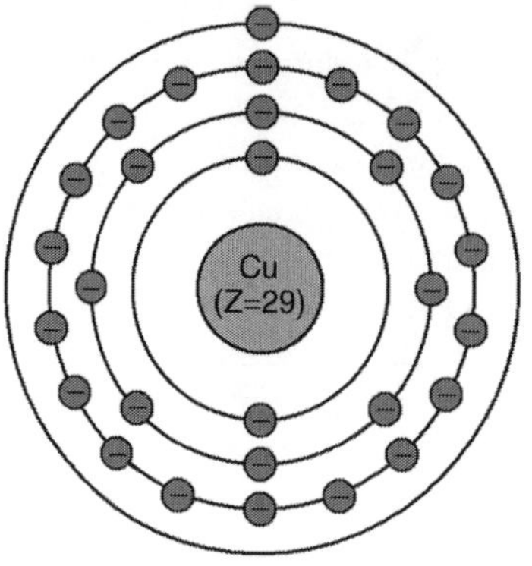

Fig. 2.4.1

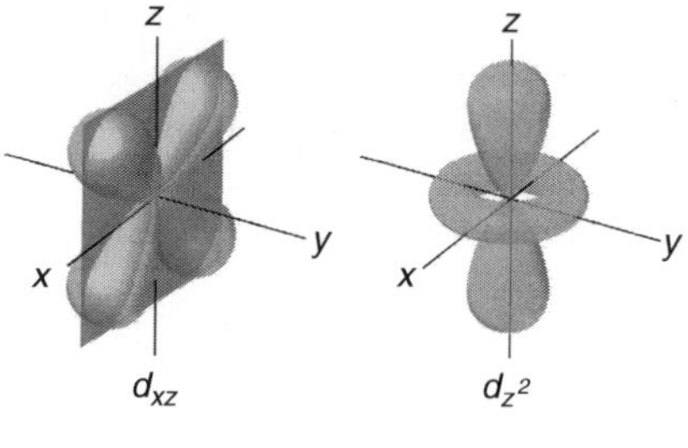

Fig. 2.4.2

Microscopic scale 10^{-6} m

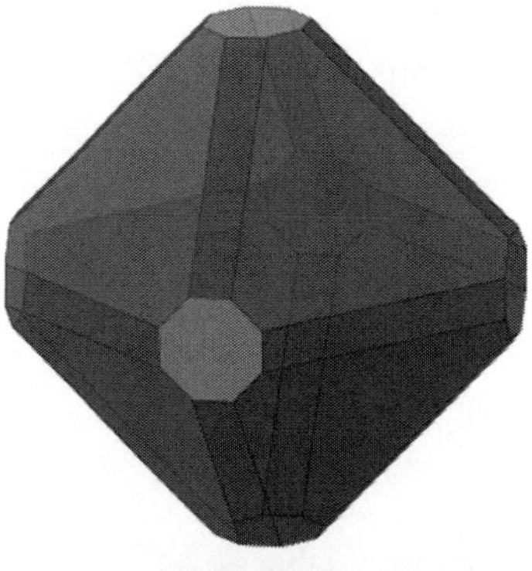

Fig. 2.4.3

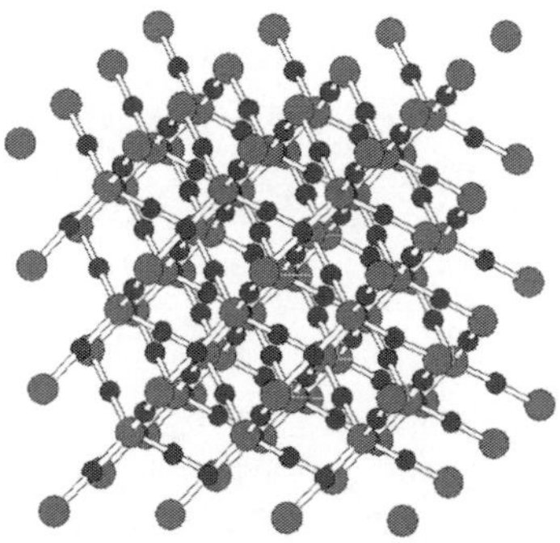

Fig. 2.4.4

Mesoscopic scale 10^{-3} m

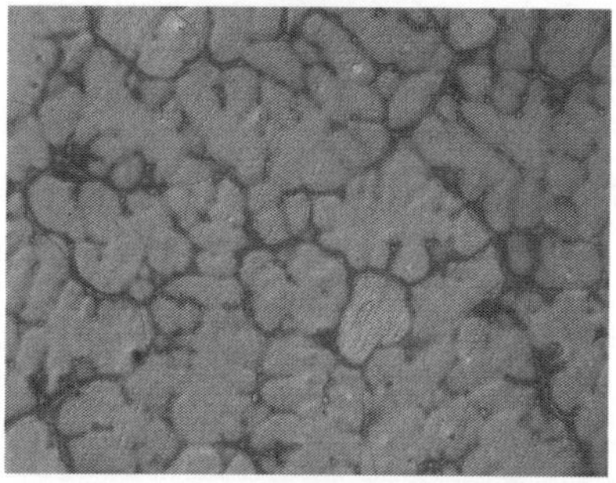

Fig. 2.4.5

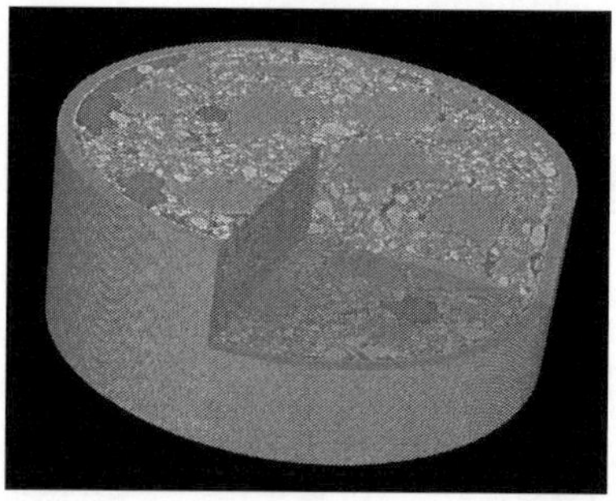

Fig. 2.4.6

Macroscopic scale $>10^{0}$ m

Fig. 2.4.7

Fig. 2.4.8

or photographic inspections of museum objects, buildings, and sites. Even at the macroscopic level, however, we should distinguish between **qualitative** observations concerning the external appearance of materials (i.e. shape, surface roughness, colour, etc.), which are normally confined to large scale heterogeneities at the surface of the object, and the **quantitative** measurement of the physico-chemical properties of the object, which requires instrumentation appropriate to the task and let us penetrate beyond the surface of the object. Reflectographic or radiographic analyses of paintings, mummy tomography, and radar site surveys are examples of attempts to penetrate into the object to understand its intimate nature at a large scale, though complete interpretation of the material generally requires the understanding and the characterization of heterogeneities and properties within the object at a much finer scale. Experimental techniques do exactly this: they allow us to penetrate the material and investigate its components down to the atomic level or below.

The choice of the scale at which the object is being investigated must be based on at least two principal concepts:

1) **Probed volume**. Each technique specifically probes a certain volume of the material investigated. This volume depends on the size and collimation of the incident beam, on the penetration of the beam into the material, which in turn depends on the energy of the beam and the chemical nature and density of the material, and on the extraction volume of the detected signal. Sometimes it is possible to change (most often to decrease) the standard volume probed by a technique under routine conditions, by changing the source of radiation, the geometry of the measurement, or the adopted detectors. Independently of the technique, any change of the instrumental configuration has direct effects on the data collection time, and on the resolution and significance of the measurements.

2) **Representativity**. The volume sampled and/or probed must be representative of the material and of the properties to be measured. The representativity of the sample is of course related to the scale of the heterogeneities present in the sample, and whether we want to measure the average values or the spatial distributions of the experimental parameters. This concept is critical in determining the quality of the information embodied in the measured data and its statistical significance. There are fundamental scientific and archaeometric considerations to be taken into account during the selection of the scale and the location of the sample to be analysed, especially for the analysis of complex, composite, and highly heterogeneous materials. In the words of Oddy (1984): "Too often university or commercial scientists will take unnecessarily large samples, or take samples (from alteration layers) which are unrepresentative." The recent diffusion of micro-beam techniques and portable surface probes makes the issue of great actuality (Box 2.r), and it will be discussed in detail in relationship with the sampling problems.

Time and cost

Performing scientific analyses involves a substantial amount of human and hardware resources. Sample preparation, instrument calibration, data collection, data treatment and interpretation must be carried out by personnel with highly specialized training and experience. Scientific instrumentation is rather expensive, especially if non-commercial and specially developed instruments are to be used. All this contributes to the cost of the analyses.

The vast majority of experimental investigations is carried out on the instruments available in small-scale scientific laboratories, located at universities, research centres, industrial development laboratories, or in a variety of other institutions that are more closely related to cultural heritage, such as museums, testing or conservation centres. Most of these laboratories have been created for specific purposes, and their focus is to provide technical know-how and instrumental tools to solve problems related to specific classes of materials or processes. However, the financial and manpower costs involved with experimental techniques are generally high, even at laboratories relying on substantial investments and running budgets. Therefore in general severe choices have to be made as to the number, type, and technical configuration of the instruments to be installed, let alone the managing and upgrade costs. The consequences for everybody who is planning any type of experimental characterization are that (1) experimental investigations are costly, so that a specific budget should be inserted into the general conservation or research projects from the start, and (2) each experimental laboratory has a limited set of instruments available for the analyses, a problem that can be bypassed by collaboration between different laboratories or even by creating networks of laboratories entertaining continuous interactions and shared protocols. The latter strategy, for example, has been adopted by a series of recent EU actions related to cultural heritage (Boutaine 2006).

However, nowadays only a small part of the analyses performed are planned and carried out within the framework of long-term projects, collaborations between institutions, or large-scale actions. The vast majority of archaeometric investigations is actually carried out by personal contacts and short-term contracts. This has profound consequences on: (1) the **quality** of the data, because the scientific laboratory carrying out the measurements in most cases is not fully devoted to archaeometric analyses, the personnel are not used to treating cultural heritage samples (how many measurements are carried out daily by untrained students!), the protocols adopted and calibration procedures are not finalized for specific archaeometric problems and, ultimately, archaeometry is not a priority for the laboratory, but rather just an additional application of the existing know-how; and (2) the length of **time** taken for the analysis, which may be unnecessarily long because the analytical procedures are not optimized, and non-dedicated personnel carry out the measurements on dead or low-priority machine time. These informal collaborations may have positive effects on the (3) **cost** of the analyses that, as they are performed as a side activity of the laboratory, do not require specific personnel or instrumentation costs, in addition to the usual running costs of the analytical instruments.

There is, of course, in all scientific measurements an intrinsic omnipresent compromise between quality, time and cost that must always be carefully taken into account and accepted for every scientific measurement. It implies that high quality data, that is data having superior precision, lower detection limits, better statistics, etc. must necessarily be obtained at the expenses of longer analytical time (i.e. more careful sample preparation and data treatment, longer measurement time, etc.) and higher costs. However, this is even more so in the case of archaeometric measurements, where the compromise must be defined and confronted right at the start of the project planning: Which technique should be adopted? At which laboratory should the measurement be carried out? Should routine technology or state-of-the-art high technology be used? Of course the time required and the cost of the analyses critically follow from the decisions taken.

In taking the decision, is important to realize that the techniques should match the problems. There are always a number of analytical answers to each problem, and it is useful to assess in advance the information content deriving from the possible analytical solutions.

Strategies are far ranging. Sometimes it may prove useful to test several techniques on the same sample to obtain reliable assessment parameters when there is no previous experience of the specific problem or material, before adopting a particular technique. It may be worth checking whether the data to be measured can be compared with existing databases and/or literature. It may be decided to characterize the materials at different levels, for example to use a cheap and quick technique for rapid screening of a large number of samples, and then to use a more sophisticated technique for in-depth characterization of a selected number of samples. On the one hand, there is no limit to the amount of time and money we can spend on scientific analyses, if we really want to understand the nature of the materials and of the problems in detail (see Box 2.a to enter the world of large scale facilities, also defined as super-microscopes). On the other hand, there is little reason to spend any amount of time or money on scientific analyses if the archaeometric problems are not well or appropriately defined.

2.1.1 Spectroscopy between physics and chemistry

Spectroscopy is the study of the interaction between electromagnetic radiation (or particles) and matter. **Spectrometry** is the experimental measurement of these interactions. The instrument that performs such measurements is a spectrometer, and the experimental data concerning the interaction is referred to as a spectrogram, or a spectrum. The theory and practice of spectroscopy require a substantial knowledge of the physical properties of the probe, of the mechanisms of the radiation–matter interaction, and of the chemical structure of the atoms and molecules comprising the matter.

We generally define spectroscopic processes as all interactions of radiation with matter involving a quantized **exchange of energy** (energy transfer). In all instances we consider the transfer of **energy quanta** from the incident photons (or particles) to the electrons or the nuclei of the atoms, which depends on the frequency of the incident beam, and involves the basic phenomena of energy absorption and emission (Barry *et al.* 2000, 2001).

The **absorption** of radiation is based on the excitation from the **ground state** to an **excited state.** The change in energy level is defined as a transition between states. We measure the frequency and the magnitude of the absorbed radiation in order to get information on the energy levels of the atoms (electrons, nuclei) to be investigated. The experimental techniques measuring the energy-dependence of the absorption of radiation from matter are the **absorption spectroscopies.**

The **emission** of radiation from the atoms on decay from the excited state to the ground state (also called relaxation, often involving transitions to intermediate energy levels) is also a powerful source of information on the intrinsic energy states of the atoms in the material. This is due to the coupling of the absorbed radiation to internal atomic processes, which causes re-emission of radiation at a different frequency. The frequency shift, the intensity, and the direction (or momentum) of the emitted radiation are related to the chemical nature and environment of the atoms and are measured by a number of **emission spectroscopic techniques.**

With reference to the visible part of the electromagnetic spectrum, Fig. 2.5 shows that chemical elements emit radiation at specific wavelengths during relaxation from an excited state, for example that produced by thermal energy, and they absorb radiation at the same wavelengths during excitation by an incident radiation beam. This is the famous Kirchhoff and Bunsen experiment that in 1859 led the way to the quantum theory of atoms and opened the possibility of the spectrochemical analysis of matter, including emission and absorption spectroscopies. There is of course a direct relationship between the wavelength (and frequency) of the absorbed/emitted radiation and the energy levels involved in the change. The general relationship once again involves the fundamental Plank's constant (h): $\Delta E = h\nu = hc/\lambda$, where ΔE = energy difference between the starting and the final energy levels, ν and λ = frequency and wavelength of the absorbed/emitted photon, and c = speed of light in vacuum. The existence of this direct relationship allows

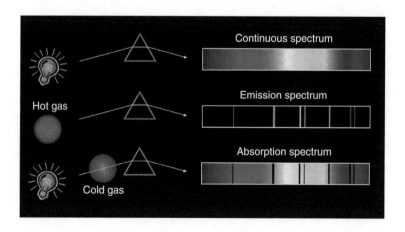

Fig. 2.5. Mechanisms of production of the continuous spectrum, the emission spectrum, and the absorption spectrum. The experimental presence of emission and absorption lines at the same wavelengths was fundamental to linking the electronic structure and the optical properties of atoms. (Modified from the Astronomy and Astrophysics website of the Columbia University, original image available at: http://www.astro.columbia.edu/~archung/labs/fall2001/lec04_fall01.html)
Original colour picture reproduced as plate 2.

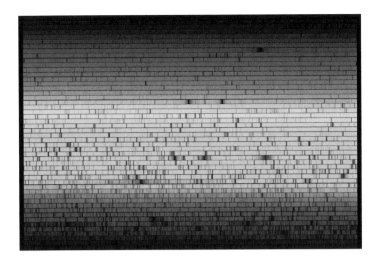

Fig. 2.6. Dark absorption lines in the continuum visible emission spectrum of the Sun's atmosphere. (Courtesy of N.A. Sharp, NOAO/NSO/Kitt Peak FTS/AURA/NSF; image available at: http://www.noao.edu/image_gallery/html/im0600.html)
Original colour picture reproduced as plate 3.

us to measure experimentally the wavelength of the absorbed/emitted radiation using spectroscopic techniques, and straightforwardly to derive physical or chemical information relating to the atomic or molecular nature of the material being investigated. The detailed analysis of the visible absorption spectra even allows us to recognize the elements present in the atmosphere of distant celestial bodies, as shown by the dark absorption lines present in the continuous emission spectrum of the Sun (Fig. 2.6).

In every experiment, three basic parameters of the incident/emitted radiation are to be measured, known, or assumed: the **energy** (or frequency, or wavelength), the **direction**, and the **intensity** of the beam. The energy and the direction carry the information on the nature and structure of the atoms in the samples at different levels, as discussed below, whereas the intensity of the measured beam is related to the number of interacting particles, atoms, or molecules, and is therefore the basis for quantifying the components of the system.

For example, quantification of chemical species by absorption spectroscopies relies on Lambert–Beer's law (Fig. 2.7), which states that the absorbance ($A = \log(I_0/I)$, where I_0 is the intensity of the incident radiation and I is the intensity of the radiation after passing through a layer t of matter) for a given frequency is $A = act$, where a is the molar absorption coefficient of the substance analysed, and c is the concentration of the absorber in the sample. This universal relationship allows us to measure the concentration of absorbers (i.e. atoms, molecules, electrons, nuclei, etc.) in the sample by knowing from the literature the fundamental properties of matter (in this case the molar absorption coefficient of the absorber, a or absorptivity), by experimentally fixing the thickness of the sample (t), and by measuring the number of photons (i.e. the intensity) in the incident and transmitted beams. The same process is at the root of the absorption contrast that is visible in the images produced by an electromagnetic beam passing through a sample (absorption imaging by radiography, tomography, Box 2.m, 2.n).

Fig. 2.7. Graphical representation of the attenuation of the incident beam through a sample during the absorption process. It is used to quantify the concentration of the absorber by means of the Lambert-Beer's law (Modified from Wikipedia: Beer–Lambert law, original image available at: http://en.wikipedia.org/wiki/Beer-Lambert_law)

A direct consequence of quantum absorption by matter is the activation of a number of processes within the material by the absorbed radiation. The Grotthus–Draper's law, for example, also called the principle of photochemical activation, states that only the light that is absorbed by a system can bring about a photochemical change. Besides being at the base of the photosynthetic reaction in plants, it will be shown that this also causes all processes of photo-induced degradation in artwork (Section 2.5.2.2).

When secondary radiation is re-emitted the process is called fluorescence, phosphorescence or luminescence, depending on the physical mechanism.

Independently of its energy, the re-emitted radiation is called **fluorescence** if the excitation and relaxation processes are for most practical purposes simultaneous and they do not involve intermediate states, although physically there is a short fluorescence relaxation lifetime of the order of a few nanoseconds, which is the time the system resides in the excited state, and which follows an exponential decay process into the ground state. The name of the process is derived from the mineral fluorite, which produces bright blue, red or green visible fluorescence when irradiated with UV radiation.

Fluorescence emitted in the visible range (**luminescence**) can be activated by thermal energy excitation (thermoluminescence) or by light excitation (optoluminescence). Both physical processes are widely used for dating purposes (Section 2.5.4.3, Box 2.w). Other familiar luminescence phenomena are the northern and southern polar lights (Aurora Borealis) produced by the solar wind ions entering the Earth's magnetosphere, and the bioluminescence in the tail of fireflies and glow worms or in the special organs (photophores) of abyssal fish. Interestingly, the mechanism of bioluminescence in fireflies is based on the enzyme luciferase interacting with appropriate proteins, such as the luciferin pigment in the case of the European firefly (*Lampyris noctiluca*) or blood cells in the case of forensic applications, which adopts genetically modified luciferase for detection of blood on crime scenes.

Another common luminescence phenomenon known as Čerenkov's radiation is the pale blue light emitted in the pool of nuclear reactors, by electrons having energy higher than light travelling through water (approx. 0.75c).

The re-emission is called **phosphorescence** if the relaxation process is delayed by intermediate non-radiative states and the excited state is maintained for much longer times. Again, everybody is familiar with phosphorescent objects such as children's toys and decorations glowing in the dark for hours after being activated by thermal or radiation exposure for a sufficient time.

In spectroscopy, we can use incident radiation with frequencies covering the whole electromagnetic spectrum, from radio waves to gamma rays. The basic scheme for the interpretation of the absorption/emission of radiation between energy levels is the same in all cases, but the physical nature of the energy levels investigated depends on the energy range that we are using as the probe (Tables 2.1 and 2.5).

Some of the high energy levels such as the nuclear and the core electron ones are intimately linked to the quantum structure of the atom, and they are relatively unaffected by the chemical bonds that the atom forms with the neighbouring atoms. These levels provide information on the nature of the chemical element being investigated, i.e. its **atomic number** (Z) and its position in the **periodic table** (an enjoyable perusal of the periodic chart can be made at http://www.webelements.com/). Several spectroscopic techniques such as NAA, XRF, AAS, OES and others, which probe these high-energy levels of the atoms are therefore commonly used to obtain information on the **elemental chemistry** of the material, i.e. to identify and quantify the elements present in the sample, either in absolute quantities (compared with analytical standards) or in relative proportions. These techniques are commonly developed, optimized, and applied in the field of analytical chemistry and they differ greatly in the volume of analysed material, in the Z-range of the accessible elements, and in the detection limits.

Some of the energy levels at lower energy are related to the outer electrons of the atoms, either in the valence band, in the conduction band, or in localized inter atomic orbitals. These levels provide information on the **electronic configuration** of the atom (i.e. its valence state), on the **chemical bonds** that the atom forms with neighbouring atoms, and on the local **coordination geometry** of the atom, i.e. the inter-atomic distances and angles. Some techniques such as UV, Vis, XPS, AES, and in part EELS directly probe the electronic states, others such as XAS measure the valence states and local geometry indirectly by the effects they have on the absorption edges relative to core-electrons ionization. Many of these techniques are commonly applied in the field of solid-state physics and semiconductor technology, and in the understanding of the ionization state and short-range crystal chemistry of the elements in solid compounds, both crystals and glasses.

Some of the spectroscopic techniques probe the vibrations of couples or groups of atoms in clusters or molecules, and they are therefore extremely useful in identifying and characterizing groups of bound atoms in inorganic systems or organic molecules. These spectroscopies (IR, RS) are referred to as **vibrational spectroscopies** because they directly measure the quantum vibrational states (or **vibrational modes**) of molecules. In fact the atoms linked by chemical bonds in molecular systems vibrate at frequencies in the range 10^{12}–10^{14} Hz, which corresponds to the infrared part of the electromagnetic spectrum (Fig. 2.2). This means that the oscillating electrical field of the incident infrared radiation interacts with fluctuations in the dipole moment of the molecule. If the frequency of the radiation matches the vibrational frequency of the molecule then radiation will be absorbed, causing a change in the amplitude of molecular vibration (molecular excitation) and absorption peaks in the incident beam (IR absorption spectroscopy). Alternatively, there might be an anelastic exchange of energy between the incident radiation beam and the vibrating atomic group, so that the scattered radiation has an increased or decreased energy with respect to the incident beam, the difference being measured as a frequency shift (Raman scattering).

The frequency of vibration is controlled by the bonding forces between the atoms, like vibrating masses connected by springs. Each time an atom is displaced with respect to the others in the molecule, the attraction force between the atoms (i.e. the chemical bond) acts like the restoring force in a spring and tends to restore the equilibrium state. In doing so the molecule undergoes a complex motion that can be described by the vibrational components of the molecule (stretching, bending, rotation, etc.; see Fig. 2.8), the frequency of each component being proportional to the inter-atomic force in the bond (i.e. the greater the strength of the bond, the greater the frequency of vibration). As in the mechanics of springs (Hooke's law), if the attractive force between the atoms is proportional to the displacement of the atom from the equilibrium position, then the system can be described to be in the so-called harmonic model: the atoms vibrate in phase with the same frequency (albeit with different amplitudes) and the frequency of each vibration can be calculated from basic principles, that is bond lengths, bond angles, and bond force constants. Each vibration of the molecule in which there is in-phase motion of the atoms is called a **normal vibration** mode. In an atomic system or for molecule containing N atoms, there are $3N - 6$ vibrational modes ($3N - 5$ in a diatomic molecule).

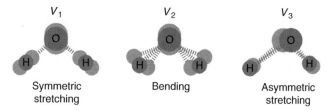

Fig. 2.8. Vibrational modes of a simple non-linear three atom molecule such as water. Each of the normal vibration modes can be excited and produces a specific absorption peak in the infrared region of the spectrum.

Table 2.5. Classification of techniques based on the energy band used as a probe, the excited process and the kind of information that they yield

Excited process	Elemental chemistry information	Valence state, atomic coordination, or chemical bond information	Vibrational states and molecular information
Nuclear resonance		MöS, NMR	
Nuclear capture	NAA, PGAA		
Core electrons transitions	EPMA, XRF AAS, OES SEM-EDS	XAS	
Valence electron transitions		XPS, AES, EELS UV, Vis	
Vibrational states			IR, RS

Box 2.b XRF X-ray fluorescence spectroscopy

X-ray fluorescence spectroscopy (**XRF**) is one of the most used techniques for the elemental analysis of materials (Parsons 1997, Mantler and Schreiner 2000). The essentials of the technique are fairly simple: core electrons of the atoms are expelled by high energy primary X-rays, the unstable ionized atom then relaxes to the ground state by a series of electron jumps into lower orbitals to fill the vacancies, thereby emitting fluorescence photons in the X-ray region, which correspond to the quantum structure of the atom (Fig. 2.b.1). The fluorescence emission, also called secondary X-rays, represents the "characteristic" spectrum of the atom, and it may be readily used to identify and quantify the chemical elements. The calculated X-ray emission lines for all elements in the periodic table are an

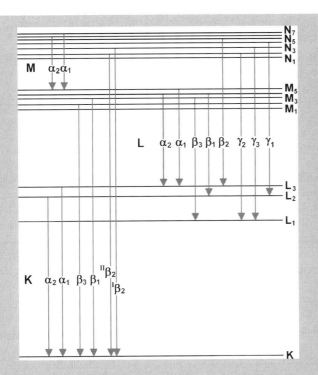

Fig. 2.b.1. Graphical representation of the energy levels of electrons of the K, L, M, N levels in an atom, and the Kα, Kβ, Lα, Lβ, Lγ and Mα electron transitions corresponding to the characteristic X-ray emission lines.

essential tool for the interpretation of all measured spectra. They are, for example, listed in the precious *X-ray Data Booklet* (http://xdb.lbl.gov/). Although the X-ray fluorescence process can be activated by ionizing the atoms with γ-rays produced by radioactive elements (for example ^{57}Co, ^{109}Cd, ^{125}I or ^{241}Am), sufficiently energetic electrons produced in electron microscopes (Box 2.l) or accelerated protons (PIXE, Box 3.i), by far the most widespread incident probe is an X-ray beam produced by laboratory tubes operated at a high voltage (typically in the range 20–60 kV). The range of elements that can efficiently be analysed with laboratory instruments depends on the energy of the primary X-rays, so that different regions of the periodic table are probed using X-ray tubes with different anodes, such as Cu, Mo, Pd, Rh, Ag, Au or W, each of which produce a characteristic radiation and different energy. Of course the selected anode must be of an element not contained in the sample to be analysed, because of the overlap with the tube characteristic lines. As an example, a W anode (59.3 keV) may excite the K lines of elements with atomic numbers 15–55 and the L lines of elements with atomic numbers 65–90 (Fig. 2.b.2). The K lines of the elements with atomic numbers of less than Na ($Z = 11$) are easily absorbed by even a few centimetres of air, and a few micrometres in the sample bulk, and they cannot routinely be analysed. If the incident X-ray beam is produced by a synchrotron ring, then the XRF technique is called **SRIXE** (synchrotron radiation induced X-ray emission), and it offers a few advantages with respect to the same experiments performed with laboratory instrumentation, namely a much more collimated probe, a very low intrinsic background, and the possibility of energy-tuning the source to selectively excite specific fluorescence lines (Adams *et al.* 1998, Janssen 2004).

The experimental setup for the measurement of the XRF spectra is composed of an X-ray or γ-ray source used to irradiate the sample, a sample stage and holder, and a system for detecting the fluorescent X-rays. Typical XRF spectra show the intensity of the fluorescence X-rays emitted by the sample (commonly in counts/unit time) as a function of energy (in eV), as shown in Fig. 2.b.3. After appropriate calibration and correction for matrix self-absorption effects and detector efficiency, the integrated areas of the fluorescence peaks are converted into relative or absolute concentrations of the analysed chemical elements.

(*cont.*)

Box 2.b (*Continued*)

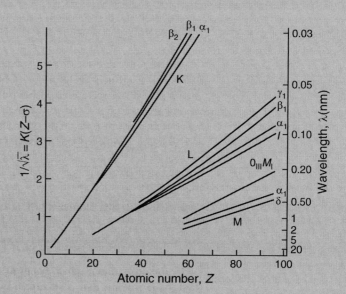

Fig. 2.b.2. Moseley diagrams for the K, L and M series. (Modified from Jenkins R., 1999)

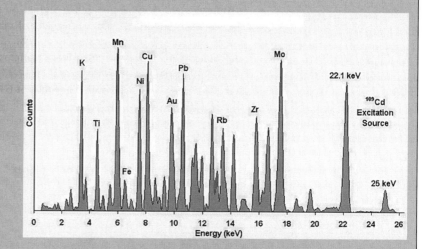

Fig. 2.b.3. High energy XRF spectrum obtained with a radioactive 109Cd source of 22.1 keV.

There are two main types of detector systems: those based on solid-state detectors and multi-channel analysers that are capable of directly measuring the energy of the fluorescent X-rays (the system is called **EDS**: energy dispersive spectrometry) and those based on a diffraction grating, typically a single crystal analyser, which measures the energy of the fluorescent X-ray indirectly by diffracting the different wavelengths of the collimated fluorescent signal at different angles (the system is called **WDS**: wavelength dispersive spectrometry). The WDS spectrometers have lower counting rates but higher energy resolution (typically 5–10 eV) and therefore they can discriminate overlapping fluorescence lines more efficiently, whereas the EDS spectrometers have higher counting rates but sensibly lower energy resolution (typically 150–200 eV). As a consequence, ED-XRF systems are often used for fast low resolution measurements, for example in portable apparatuses used for field measurements (Longoni *et al.* 1998, Bronk *et al.* 2001), whereas WD-XRF systems in the laboratory provide for high-resolution measurements with detection limits down to the ppm level for ideal samples. They require longer counting times and/or a larger amount of sample. EDS spectrometers are often associated with electron microscopes (SEM, TEM, Box 2.l) to provide complementary chemical measurements in addition to imaging and diffraction, whereas several WDS spectrometers are mainly associated with fine-focused electron beams in so-called electron probe micro-analysers (**EPMA**). The latter type of instrument, also called electron micro-probes, commonly offer a good analytical compromise between element detection limits (in the 1000 ppm range), small probed areas and measurement flexibility.

The potentially high sensitivity for most elements and the relative ease of operation make XRF a widespread analytical technique in archaeometry (Ferretti 2000). New generation instruments based on EDS are commercially available both for laboratory and portable field measurements. However, it should be clear that for XRF measurements performed in non-ideal conditions (i.e. a sample in vacuum, flat analysed surface, low sample attenuation, etc.), where the measurement is performed directly on a material without proper sampling and preparation, obtaining reliable quantitative results may be very hard, because of the surface geometry, possible surface patinae and alterations, irregular shape of the probed sample volume, non-optimal geometry of detection, etc.

Microbeam instruments exist that perform two-dimensional scans over limited areas of the sample surface using a focused beam, thus obtaining chemical information for each point and producing data that is useful for chemical imaging and mapping (Section 2.1.4). The technique is called scanning XRF (**SXRF**, Scott 2001).

One extension of XRF, called total reflection XRF (**TRXRF**), allows measurement of elements in the sample in ultra trace concentrations (µg/g in solids and ng/g in liquids) by using angles of incidence on the specimen surface below the critical angle, typically $< 0.1°$ (Klockenkämper 1997). In these conditions the primary beam is totally reflected, it penetrates into the substrate for only a few nm, and there is an optimal interaction between the primary beam and the sample. Absorption and matrix effects on the fluorescence signal can largely be neglected, and the sensitivity of the measurement is greatly enhanced. The TRXRF has been successfully applied in the chemical analysis of paint pigments, surface varnishes and manuscript illuminations (Vandenabeele and Moens 2005).

For in-depth information

Janssen K. (2004) X-ray based methods of analysis. In: Janssens K. and Van Grieken R. (ed.) (2004) Non destructive microanalysis of cultural heritage materials. Comprehensive analytical chemistry series. Vol. XLII. Elsevier, Amsterdam.

Jenkins R., Gould R.W., Gedcke D. (1981) Quantitative X-ray spectrometry. Marcel Dekker, New York.

Kramar U. (2000) X-ray fluorescence spectrometers. In: Lindon J.C., Tranter G.E. and Holmes J.L. (eds) Encyclopedia of spectroscopy and spectrometry, Vol. 3, Academic Press, London. pp. 2467–2477.

Box 2.c OES Optical emission spectroscopy

Optical emission spectroscopy, also called **atomic emission spectroscopy** (AES) is a spectroscopic technique that examines the wavelengths of photons emitted by atoms during their transition from an excited state to a lower energy state. Each element emits a characteristic set of discrete wavelengths according to its electronic structure (Fig. 2.c.1) and by measuring these wavelengths the elemental composition of the sample can be determined.

Figure 2.c.2. shows that each element has a characteristic optical emission spectrum, and therefore the measurement of the frequency and intensity of the lines allows qualitative and quantitative assessment of the chemical

(cont.)

Box 2.c (*Continued*)

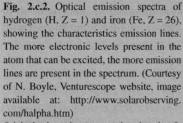

Fig. 2.c.1. Graphical representation of the electron transitions in a hydrogen atom, corresponding to the characteristic emission lines. The Lyman series are emitted in the UV region, the Balmer series in the visible region (Fig. 2.c.2), and the Paschen, Brackett, and Pfund series in the IR region.

Fig. 2.c.2. Optical emission spectra of hydrogen (H, Z = 1) and iron (Fe, Z = 26), showing the characteristics emission lines. The more electronic levels present in the atom that can be excited, the more emission lines are present in the spectrum. (Courtesy of N. Boyle, Venturescope website, image available at: http://www.solarobserving.com/halpha.htm)
Original colour pictures reproduced as plate 8.

composition of the sample. The OES experimental setup is composed of a system of atomic excitation, which is part of the emission source, an optical system for the dispersion of the different wavelengths, and some device that measures the emitted lines in the region encompassing the near-UV, the Vis and the near IR regions. The atomic emission source must provide for sample vaporization, dissociation and excitation of all the lines of interest for the elements in the sample. In the early instruments this was produced by combustion (flame-OES) or by an electric current discharge between two electrodes (spark-OES). Such instruments are now considered to be obsolete in terms of efficiency of excitation, the amount of sample used, and excitation temperatures. Most modern instruments rely on excitation sources based on inductively coupled plasma (ICP), although a number of other sources are available such as direct current plasma (DCP), microwave induced plasma (MIP) or capacitively coupled microwave plasma (CCMP). Bonchin *et al.* (2000) provide an exhaustive description of the emission sources. In fact the ICP sources on modern OES instruments are so widespread that very often, and mistakenly, the source is identified with the whole measurement instrument, which is then referred to as ICP in place of the correct full acronym, **ICP-OES**. One should always remember that the ICP source may actually be coupled to another measurement device, such as a mass spectrometer (Box 2.t).

Beyond the atomic emission source, the OES spectrometer must disperse the optical emission spectrum into its components, and this operation is performed using a monochromator or a polychromator (Fig. 2.c.3). Monochromators operate in sequential mode, insofar as they must scan the spectrum and measure one individual line at a time, they allow for slow measurements, but all the wavelengths in the spectrum can be observed. Polychromators select a number of wavelengths for simultaneous measurement through a slit system, and therefore they provide for faster data collection of a fixed number of elements, at the expense of the flexibility to observe all possible lines in the emission spectrum. In both spectrometers the detectors measuring the incoming photon flux can be photomultiplier tubes or solid state devices.

The basic assumption in OES spectroscopy measurements is that the emitted photon flux is proportional to the concentration of atoms in the sample. Therefore the stability of the excitation source and the sample introduction processes are critical parameters of the experiment. The samples are fed by an argon gas stream into the excitation plasma source (typically an ICP) as aerosols through nebulizers of different kinds, though torches and electrothermal vaporizers are also used. Solid samples therefore are commonly totally dissolved in acids before the analysis, though laser ablation techniques may be used to extract minimum quantities of materials directly from the solid sample into the plasma source. Since the material to be analysed is transformed into an atomic stream that is destroyed and lost after the analysis, the ICP-OES method is considered invasive and it requires a minimum amount of material to be sampled and sacrificed. However modern instruments provide low background noise and very low detection limits (1–100 ppb depending on the element and instrumental configuration), so that the technique is now widely used for trace element analysis in provenance studies, especially of ceramics and metals that may be sampled and easily dissolved (Tykot 2004).

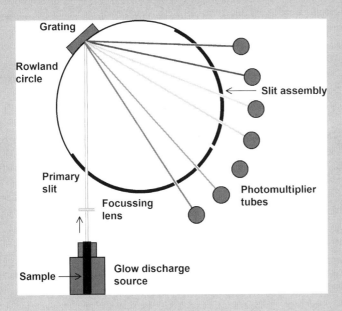

Fig. 2.c.3. Paschen–Runge type polychromator used in OES spectrometers to simultaneously focus individual emission lines to element-specific detectors.

(*cont.*)

Box 2.c (*Continued*)

A recent extremely interesting development in the field of emission spectroscopy is the coupled use of a powerful Nd/YAG laser microbeam to generate a hot plasma directly from the surface of the object, and then to analyse the emission of the plasma during cooling using an optical spectrometer, either a scanning monochromator or a non-scanning polychromator. In the latter case, if a spectrometer is used that is able to acquire all possible wavelengths from the near infrared to the deep ultraviolet, all elements can be simultaneously analysed. The technique is called laser-induced breakdown spectroscopy (**LIBS,** Miziolek *et al.* 2006) or, less frequently, laser-induced plasma spectroscopy (LIPS). The LIBS technique can employ portable instrumentation, the optical detection system may use fibre optics to make the setup more flexible in cultural heritage applications, and it requires a very small amount of material. The laser-produced hole in the sample is of the order of 100–500 μm, depending on the power of the laser and on the material. LIBS is therefore extremely promising as a non-contact micro-invasive analytical technique. Detection limits are in the range 1–100 ppm depending on the element and the matrix effects.

For in-depth information

Boumans P.W.J.M. (1987) Inductively coupled plasma emission spectroscopy. Part 1. Chemical Analysis Series, Vol. 90. John Wiley & Sons, New York.

Montaser A. and Golightly D.W. (eds) (1992) Inductively coupled plasmas in analytical atomic spectrometry. VCH, Weinheim.

Box 2.d AAS Atomic absorption spectroscopy

Atomic absorption spectroscopy is also a technique for the quantitative chemical analysis of samples in solutions based on optical spectroscopy. It relies on the proportionality between the absorption of specific wavelengths in the optical spectrum and the concentration of an element in the atomic vapour probed by the incident beam (Fig. 2.7).

The instrument is composed of an atomizer that transfers the atoms of the condensed sample to the gas phase using different processes, including flame excitation, electric discharges, laser radiation, electron bombardment, and inductively coupled plasmas. The technique is therefore destructive, because the original sample must be dissolved, nebulized, and dissociated into isolated atoms to be analysed. High-temperature flame (flame-AAS, or FAAS) and electrothermal heating (electrothermal-AAS, or ETAAS) atomizers are the most widely used techniques for atomic gas production.

The incident beam is provided by a source of light, commonly a hollow-cathode lamp or an electrode-less discharge lamp, emitting a beam that probes the layer of gas produced by the atomizer. The lamp must contain the same element that is to be analysed in the sample, so that the excited atoms in the lamp emit their characteristic spectrum in the UV and visible regions, which are necessary to activate the absorption process in the analyte atoms present in the sample gas. AAS instruments require a specific lamp for each element to be measured. Lamps are available for over 65 elements, though some lamps with alloy cathodes may be used for several metals. The characteristic lines of the element are dispersed and isolated from the emission spectrum exiting the sample chamber by a monochromator system, usually a diffraction grating–slit system, and their intensity is recorded on the detector, commonly a photomultiplier tube or charge-coupled device (CCD) (Fig. 2.d.1). If the flow rate of the sample from the atomizer is uniform and the concentration of the analyte atoms in the solution is constant, then the attenuation of the incident beam through the sample is proportional

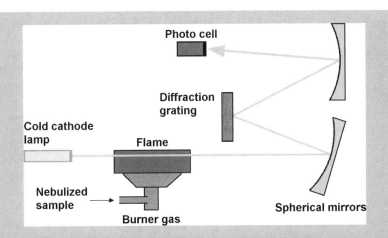

Fig. 2.d.1. Schematic representation of an AAS instrument. The light source and the flame atomizer may be replaced by different devices, such as brighter discharge lamps or electrothermal atomizers, though the measurement scheme remains unaltered. (Modified from Chromatography-online .org website, original image available at: http://www.chromatography-online. org/GC-Tandem/Atomic-Spectroscopy/ Flame-AA/rs33.html)

to the concentration of measured element (Beer's law, Section 2.1.1) and quantitative measurements are possible by external calibration of the attenuation curves as a function of concentration. As is often the case with spectroscopic methods, the best results are obtained when the analyte element in the calibrating solutions are in the same concentration range as in the sample to be analysed. Several methods of background correction and compensation for source drift are available (Hill and Fisher 2000).

Flame-AAS spectrometers commonly reach a lower detection limit in the range 1–10 ppm for most elements, whereas instrument with electrothermal atomizers, such as graphite furnace–AAS spectrometers may reach detection limits as low as 1–10 ppb for many metals. The high sensitivity and the relatively simple instrumental configuration and measurement protocols, together with the limited costs of the spectrometers have made AAS one of the main analytical techniques of choice for archaeometric analyses in the past (Hughes *et al.* 1976), especially in

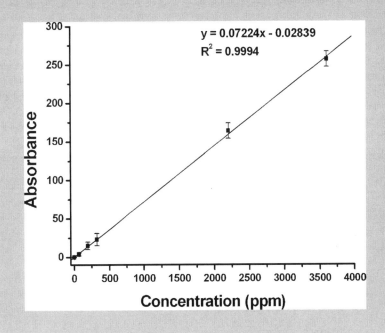

Fig. 2.d.2. Example of a calibration curve obtained by measuring several solutions with known concentrations of Cr. Measurements of unknown samples that have Cr concentration in the same range can be interpreted by the use of such curves.

(*cont.*)

Box 2.d (*Continued*)

the 1980s, when the method rapidly became very popular and largely replaced earlier flame-OES spectrometers. The major disadvantage of the method is the time-consuming limitation to the analysis of one element at a time (Fig. 2.d.2). Even the possibility of analysing a few elements at the same time, as many as six in modern spectrometers, does not critically improve the practice of AAS spectrometry, which is gradually being replaced by the more convenient, albeit more expensive, ICP-OES spectrometry for trace element analyses in provenance studies.

In the past AAS has been widely used for the measurement of major, minor and trace elements in ceramics and metals (review in Pollard *et al.* 2007, Chapter 3), the results being mostly comparable to ICP-OES data for major elements, but showing some consistent discrepancy for minor elements between the two techniques (Hatcher *et al.* 1995).

For in-depth information

Ebdon L., Evans E.H., Fisher A. and Hill S.J. (1998) An introduction to analytical atomic spectrometry. John Wiley & Sons, Chichester.

Haswell S.J. (ed.) (1991) Atomic absorption spectrometry. Theory, design and application. Analytical Spectroscopy Library. Elsevier, Amsterdam.

Box 2.e IR-RS Infrared and Raman spectroscopies

Infrared radiation does not possess enough energy to excite transitions in the electronic states, but it is of the right energy to interact with the vibrational levels of molecules and bound groups of atoms (Table 2.1). The absorption of infrared radiation during the excitation of transitions between vibrational levels (**IR absorption spectroscopy**) or the anelastic exchange of energy between the incident beam and the vibrating atomic group (**Raman scattering**) provide for powerful methods to probe the vibrational states of the molecules. Each molecule or group of atoms has specific vibrational properties that depend on the nature of the atoms, the inter-atomic chemical bonds, and the dynamic group properties (rotations, etc.), therefore an experimental measurement of the vibrational states yields for each molecule a specific spectrum in the infrared region that can be used to identify the compound (Figs 2.e.1 and 2.e.2).

For a molecule to absorb IR, the vibrations or rotations within a molecule must cause a net change in the dipole moment of the electric charge (Section 2.1.1), so that the symmetry of each specific vibration mode determines whether that mode is present in the spectrum (i.e. the mode is **active**) or not (i.e. the mode is **inactive**). The set of conditions allowing for the activation of the modes is called the **selection rules** and can be rigorously derived using group theory of representations. For example if the mode is symmetric, i.e. it has a centre of symmetry such as in a linear molecule, then the dipole will not be changed during the increased amplitude of vibration of the molecule, and the mode will not be infrared active.

In Raman scattering the mechanism is different, insofar as the incident radiation actually deforms the atomic charge and it induces an instantaneous dipole moment in the molecule. Depending on the polarizability of the molecule, if the electron charge deformation corresponds to a possible vibrational state of the molecule, the mode is Raman active and a Raman scattering peak will be observed. Because of these different mechanisms and selection rules, IR and RS spectroscopies measure different vibrational modes and therefore provide complementary information on the analysed material.

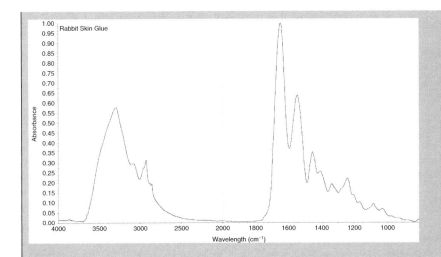

Fig. 2.e.1. FTIR spectum of rabbit skin glue. The intensity and frequency of the absorption peaks uniquely identify the compound. (Courtesy of H. Khanjian, GCI)

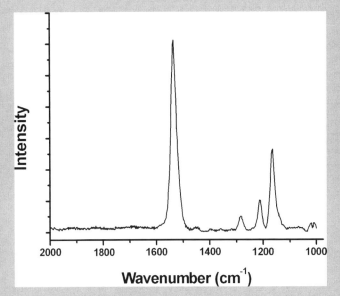

Fig. 2.e.2. Raman scattering spectrum of saffron (crocetin, carotenoid dicarboxylic acid, C20H24O4), an organic yellow pigment extracted from the stigma of the crocus flower largely used in antiquity. (Modified from the UCL Raman Spectroscopic Library of Natural and Synthetic Pigments website, original image available at: http://www.chem.ucl.ac.uk/resources/raman/pigfiles/saffron.html)

The vibrational spectroscopies cannot be used to analyse single atoms and elemental compositions, but rather they are used to identify molecules in organic materials and bound atom clusters in inorganic materials, for example structural groups in crystalline phases such as the carbonate group $[CO_3]^{-2}$, water molecules $[H_2O]$, hydroxyl groups $[OH]^-$, silicate groups $[SiO_4]^{-4}$, and many others. Both techniques are extremely sensitive to very diluted concentrations of the excited molecules, though absolute quantification requires complex calibration procedures, and they are mostly used in qualitative or semi-quantitative analyses. In most cases IR and Raman spectra allow the same straightforward identification of the molecules that are present in the sample, though in principle one technique can be used to observe modes that are inactive in the other one. Isolated water is a common example of a molecule that is not Raman active. This means that we have to use IR to observe small quantity of waters in the sample; it also means that we have to use RS to observe the signal of other molecules dispersed in water or the

(*cont.*)

Box 2.e (*Continued*)

signal of OH-containing molecules such as textile fibres, because the water/hydroxyl signal totally dominates the IR spectrum. On the other hand, sometimes IR spectroscopy is preferred when the Raman signal is obscured by highly fluorescent coatings or impurities in material.

The IR absorption measurements are mostly performed in transmission geometry, by measuring the attenuation of the incident beam as a function of frequency (v), commonly expressed in wavenumbers (that is $1/\lambda$ in cm^{-1}). The infrared spectral range is usually subdivided into three regions: the near IR, which is the closest to visible ($\lambda = 750–2500$ nm, $v = 13300–4000\, cm^{-1}$), the middle IR ($\lambda = 2.5–20\,\mu m$, $v = 4000–500\, cm^{-1}$), and the far IR, which is the closest to microwaves ($\lambda = 20–1000\,\mu m$, $v = 500–10\, cm^{-1}$). In dispersive geometry, the instrumentation is composed of an infrared radiation source, commonly a rare earth oxide cylinder heated at $1500\,°C$ (called a Nernst glower), the sample chamber, a monochromator (usually based on an optical grating) accessing one or more of the spectral regions, and a detection system. The sample can be gas, liquid or solid, in the latter case it is a pellet prepared by pressurizing the powdered sample in a KBr matrix, or oil-dispersed crystals held between two KCl plates. The beam from the source is split into two separate beams, one passing through the sample and the other passing through a reference substance, usually the solvent or the oil present with the sample. The two beams are then directly read at the detector and compared. The reference signal is used to cancel out the effect of the solvent in the sample (null principle).

Most modern instruments have a Michelson interferometer (Fig. 2.e.3) in place of the monochromator. The beam from the source, before passing through the sample, is split into two beams by a half mirror where half the beam is transmitted and half is reflected. The two beams are back-reflected by two perpendicular mirrors (one fixed and one moveable) and then focused on the detector. If no sample is present, as the moveable mirror is displaced by fractions of the wavelength, a cosine wave pattern of constructive and destructive interference occurs in the signal. If the sample is present, then the interference pattern is modulated by the absorbance in the sample. A simple Fourier transform of the signal allows the mathematical reconstruction of the absorbance pattern, as if it were recorded in dispersive mode. This method, called Fourier transform IR spectroscopy (FTIR) is universally adopted in modern instruments because of the improved resolution, better signal-to-noise ratio, and rapidity of measurements.

Fig. 2.e.3. Scheme of a Michelson interferometer generally used in FTIR absorption spectroscopy. (Modified from Encyclopaedia Britannica website, original image available at: http://www.britannica.com/EBchecked/topic-art/340440/17906/The-Michelson-interferometer-consists-of-a-half-transparent-mirror-oriented)

Another very useful experimental development allows the measurement of the absorption spectra in reflectance geometry in place of transmission geometry, and the reflectance process is exploited in two techniques (Fringeli 2000). The first is called **diffuse reflectance Fourier transform spectroscopy (DRIFT)**, and it permits direct measurements of IR spectra on the surface of the objects, or the collection of data on the surface of powdered samples with a minimal quantity of samples, thus minimizing the invasivity of the measurements. The other is called **attenuated total reflectance (ATR)** and it also enables samples to be examined directly in the solid or liquid state without further preparation. ATR uses the property of total internal reflection called the **evanescent wave**. A beam of infrared radiation is passed through the ATR crystal (generally germanium, zinc selenide or diamond) so that it reflects at least once off the internal surface in contact with the sample. This reflection forms an evanescent wave that extends into the sample, typically by a few micrometres. The beam is then collected by a detector as it exits the crystal. Liquid samples are directly poured on the crystal, whereas solid samples are crushed and pressed against the crystal surface.

The Raman spectrometers are composed of a light source, often a helium/neon laser, an optical sample illumination system positioned on a high-precision optical bench, and a spectrophotometer detecting the signal scattered at 90° from the incident beam. The spectrophotometer has a prism or a grating device that allows one to record the intensity of the scattering as a function of the frequency shift from the non-interacting elastic radiation (Rayleigh scattering). Similarly to **FTIR** instruments, most modern Raman spectrometers are based on interferometer systems and Fourier transform-based signal treatment to improve the quality of the measured spectra (Fourier transform Raman spectroscopy, **FTRS**).

Special applications are used to increase the surface sensitivity of the technique (surface enhanced Raman spectroscopy, or **SERS**) to analyse very thin layers of materials, such as glazes or alteration layers. Finally, the recent developments in coupling both IR and Raman spectroscopies with optical microscopes (Turrell and Corset 1996, Smith and Clark 2004) make the techniques very powerful to identify and study micro-samples or micro-components of heterogeneous systems, such as paint layers or soil micro-stratigraphies. It is now possible to collect good quality Raman microscopy spectra on samples with dimensions of 1–10 µm, making the sampling tolerable in most practical cases.

Raman spectroscopy is used to analyse a variety of materials (Vandenabeele 2004, Smith 2006), though it turns out to be particularly useful in the analysis of pigments, especially because of the easy access to the low wavenumber region ($<500 \, cm^{-1}$), which is the far infrared region accessible only with difficulty by common IR spectrometers. Extensive databases of Raman spectra have been developed for identification of pigment phases; for example a very useful on-line resource is the collection of spectra available at the UCL Department of Chemistry website (www.chem.ucl.ac.uk/resources/raman/index.html). The identification potential of compounds by RS may be substantially enhanced by the combined use of LIBS (Box 2.c), which provides the elemental composition of the sample (Bruder *et al.* 2007). The use of a Nd/YAG laser (1064 nm) as a source makes RS very important for a variety of cultural heritage applications, especially for the study of naturally fluorescent biomaterials such as horn, hoof, tortoise shell, and ivory (Edwards 2000, Edwards and Munshi 2005).

IR spectroscopy also is a primary tool for identification of art and archaeological materials, because of the small amount of material needed for the analysis of almost any substance. It is especially used in the identification of organic compounds, but it is also employed with success in the identification of inorganic phases and minerals. Similarly to RS, extensive databases exist with the reference IR spectra of most common compounds. (See the extensive reference list in Derrick *et al.* 1999).

Beyond identification, the speed of the measurements and their sensitivity make IR spectroscopy a very useful tool to monitor the advancement of chemical reactions. The high sensitivity of IR to OH groups makes the technique particularly convenient for the study of the hydration process of archaeological and natural glass. A specific example is the application to the study of the surface hydration layer in obsidian (Stevenson *et al.* 2001), which is

(*cont.*)

Box 2.e (*Continued*)

used as a dating technique (Section 2.5.3.2). Amber identification and provenance studies (Section 3.7.2.3) in the past were also mainly based on IR characterization (Beck 1986).

IR and Raman spectroscopic mapping of surfaces, for example paintings or manuscripts, are also very useful complementary techniques for imaging (Section 2.1.4). They are commonly performed using fibre optics technology (**FORS**: *fibre optics reflectance spectroscopy*, Bacci 2000; beware, sometimes the same acronym is also used for *fibre-optics Raman spectroscopy*, Edwards 2004) and mapping pixel by pixel the area of interest in the IR spectral range. This provides enhanced sensitivity for visualizing the distribution of molecules and pigments on the investigated surface, even when imaging in the visible region does not provide information. IR and Raman imaging are especially useful for visualizing discoloured inks on manuscripts or preparatory drawings underlying the paint layers.

For in-depth information

Colthup N.B., Daly L.H. and Wiberly S.E. (1990) Introduction to infrared and Raman spectroscopies. Academic Press, Boston.

Derrick M.R., Stulik D. and Landry J.M. (1999) Infrared spectroscopy in conservation science. Scientific Tools for Conservation. The Getty Conservation Institute, Los Angeles.

Edwards H.G.M. and Chalmers J.M. (eds) (2005) Raman spectroscopy in archaeology and art history. The Royal Society of Chemistry, Cambridge.

Box 2.f UV–Vis Ultraviolet and visible spectroscopies

Optical spectroscopy is related to the transitions of electrons between the outermost energy levels of the atoms, whose energies are in the range from near infrared, through visible, to ultraviolet (Table 2.1). From the application point of view, UV and Vis spectroscopies are extremely important in cultural heritage because they deal with the measurement of colour (Hunt 1998), and therefore they allow quantification of the visible effects related to the production, use, and degradation of the materials used for colouring: ink, pigments, paints, dyes, etc. From the physico-chemical point of view, these spectroscopies give us insight into the electronic **mechanisms of colour** (Section 2.1.4). It is at this stage important to remark that: (1) colour is produced by the selective absorption or re-emission of light, (2) there are a number of different physico-chemical mechanisms causing colour (Nassau 1983), and (3) colour can be measured by spectroscopic methods (Bacci 2004).

Routinely, the measurement of the UV (200–400 nm), Vis (400–750 nm), and near IR (750–2500 nm) absorption of selected wavelengths by the sample is performed by means of transmission spectrophotometry, similarly to AAS and IR spectroscopies. In the single beam arrangement, a light from a suitable source covering the spectral range is used, typically deuterium or xenon arcs, tungsten filament lamps, quartz halogen lamps, or tunable lasers. A monochromator, either a prism or a diffraction grating, is positioned between the source and the sample and allows selection of the wavelength of the incident beam passing through the sample in a glass or plastic cuvette. A detector, commonly a photomultiplier tube or a charge-coupled device, is then employed to record the photon intensity as a function of the wavelength (Fig. 2.f.1). In the dual beam arrangement a beam splitter positioned after the monochromator produces two beams, one passing through the sample and the other through the reference cell. As in IR spectroscopy, this configuration eliminates the problems related to source fluctuations.

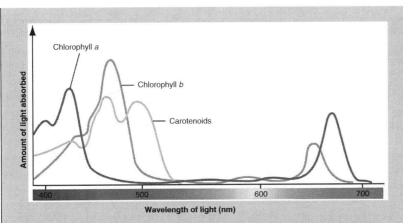

Fig. 2.f.1. Vis absorption spectrum of light from molecular compounds contained in plant leaves. Both types of chlorophyll (a and b) strongly absorb in the blue and red regions, thus reflecting the intermediate region that causes the green colour of plants. Carotenoids strongly absorb only in the blue region, thus shifting the colour of the leaves to yellow, orange, and red in autumn, when the chlorophyll in leaves disappears.
Original colour picture reproduced as plate 9.

However, due to the frequent need in cultural heritage to measure the absorption spectra in situ, for example on a wall or canvas paintings, the transmission measurements are obviously impossible, and therefore techniques based on surface reflectance have been developed, based on fibre optics and digital recording technologies.

For in-depth information

Clark B.J., Frost T. and Russell M.A. (eds) (1993) Techniques in visible and ultraviolet spectroscopy. Practical absorption spectroscopy, Vol. 3. Chapman & Hall, London.

2.1.2 Mineralogy and petrology: the Earth Sciences beyond diffraction and optical microscopy

Besides chemical analysis, the traditional experimental tools for the characterization of minerals and rocks, are **optical microscopy** (OM: Box 2.k; Bloss 1999) and **X-ray diffraction** using both single crystal and powder specimens (XRD: Box 2.g; Giacovazzo 2002). They are both relatively inexpensive and rapid techniques that allow crystalline phases in complex polyphasic materials to be identified. Modern mineralogy and petrology are of course not limited just to optical microscopy or X-ray diffraction in the characterization of natural materials, but rather they take full advantage of the vast array of spectroscopic and microscopic techniques developed in other scientific fields. It should be noted that both optical microscopy and diffraction are largely confined to Earth sciences laboratories, with the notable exception of reflected-light optical microscopy, which has become a routine tool of metallurgists (Box 3.k; Scott 1991). This is certainly not the place to discuss the historical and cultural reasons for the present general situation that, among other effects, implies a substantially reduced amount of mineralogy and crystallography being taught at higher education levels almost everywhere, but it is important to understand why so many mineralogists and petrologists have recently been involved with archaeometry: Earth sciences laboratories are the natural places where complex materials such as archaeological pottery, ancient metallurgical slags, or building stones and mortars can be studied.

When sample fragments can be sacrificed for preparing a polished thin section that is transparent to light, optical microscopy is an invaluable tool that allows one, in a matter of minutes, to identify all the main crystalline phases, their mutual relationship, texture, morphology and size, and often allows an almost complete interpretation of the nature and history of the sample, including post-use alteration processes. Optical microscopy may be considered a cheap preliminary tool to obtain a number of preliminary pieces of information on complex samples (identification and location of crystal phases, alteration state, textural information, etc.). The investigation of the material can then be continued and expanded by other techniques, for example by point chemical analyses (SEM-EDS, EPMA, PIXE), often carried out on the very same thin section that was prepared for optical microscopy. The main limitation of the technique is that grains and domains that are smaller that the resolution obtainable by visible radiation (approx. 0.1–0.5 µm) are not observable. Electron microscopy (Box 2.1) should be used to visualize smaller objects to higher magnifications.

Diffraction is also an extremely powerful tool because it is the most straightforward instrumental technique for identifying crystalline phases. Since diffraction is the only technique for probing the long range order of atoms in the sample, in most cases the large amount of information contained in the Bragg peaks allows the crystalline phases to be identified, even when spectroscopic methods, confined to the local atomic environment, may produce ambiguous results. X-ray diffraction (XRD) is performed using relatively cheap instruments that are available in most laboratories. It is therefore an accessible, fast and inexpensive technique. Moreover, with modern generation diffractometers, XRD may often be performed in a non-invasive mode (Section 2.2).

The combination of traditional mineralogical and petrological tools with the variety of spectroscopic and microscopic techniques employed in the physicochemical characterization of materials is essential in the modern investigation of complex archaeometric materials. These are often produced by thermal transformation of natural minerals and rocks, where numerous crystalline and amorphous phases are often present at the same time, and the chemical composition easily encompasses most elements of the periodic table, including major, minor, trace and ultra-trace elements. Knowledge of the nature and properties of multi-elemental, polyphasic natural materials is often necessary to interpret properly any archaeometric material derived from them. Present trends seems to be that many mineralogists and petrologists are moving away from the traditional fields of optical microscopy and X-ray diffraction, or at least they are adding more and more techniques to their toolbox. This is positive and at times creates new and fertile areas of research. However, too often simple, powerful, and traditional techniques such as OM and XRD are neglected and ignored, especially when mineralogists and petrologists are not part of archaeometric or conservation working teams. The abandonment of these techniques or their incompetent use, a trend linked to the decrease in the teaching of mineralogy, petrology, and crystallography in academic curricula, could indeed become a shortcoming for archaeometry in the future.

One area of investigations where the Earth Sciences as a whole play an important role is that of the **provenancing** of materials. The concept of provenance

(Wilson and Pollard 2001, Tykot 2004) nowadays has a rather precise meaning, which is "to measure well-defined chemical or isotopic parameters that can be used as **tracers** of the object or of the materials used to manufacture it". The problem of reconstructing ancient production, exchange, and trade systems is of course of primary importance in prehistorical archaeology, where no historical written sources are available, though it is also vital in the investigation of recent history, when direct information is missing, incomplete, or even plainly incorrect. Of course the concept of chemical and isotopic tracers, both stable and radiogenic, is deeply entrenched with the interpretation of the origin and evolution of igneous and metamorphic rocks and meteorites. It has become an everyday tool in many geological disciplines, so that it is fundamental in any Earth Sciences curriculum (see for example Cox *et al.* 1979, Faure 1998, Banner 2004). The large increase in the application of chemical analyses in archaeometry after the 1960s, which is what Pollard and Heron (2000) call the "golden age" of archaeological chemistry, indeed witnessed a substantial effort to define and measure the **chemical fingerprints** appropriate to most kinds of materials. Despite many successful applications, it is quite clear that fingerprinting of materials is to be investigated with caution and a great deal of know-how, and that a number of conditions and assumptions must be satisfied (modified from Wilson and Pollard 2001, and Tykot 2004):

1) measurable chemical or isotopic parameters must be transmitted unchanged (or reliably related through a predictive model) from the raw geological source to the finite object during the manufacturing process, and post-depositional processes should be absent or appropriately measured and modelled;
2) all relevant sources must be known, and measured fingerprinting parameters (analytically and statistically comparable to the observed data) must be available for all of them. A close match of a suite of measured parameters to a source is by no mean evidence of provenance, unless all other possible sources are measured and discriminated;
3) the intra-source parameter variation (source homogeneity) must be less than the observed inter-source parameter variation (source discrimination);
4) mixing of different sources does not occur during the processing of materials or, if it does occur, it must be properly modelled;
5) discriminating models must be developed using sound statistical treatments, and interpreted in the frame of known human behaviour;
6) discrimination should be based on as many parameters as possible, and all parameters must be measured with sufficient analytical precision to enable statistical discrimination.

It is clear that a complete provenance investigation requires at least competence in the distribution and properties of the geological materials (Earth Sciences), in the analytical details of the measurements (analytical and physical chemistry), in the historical background and ancient manufacturing processes (historical and experimental archaeology), in modern analogue manufacturing processes (chemical engineering, industrial chemistry), in the statistical treatment and modelling of the data (mathematics and statistics), and possibly in many other disciplines. Needles to say, the above assumptions and the full array of competencies needed are hardly met in practice, and this obviously calls for much stronger communication and collaboration between the researchers. The detailed advantages and pitfalls of the commonly used tracing methodologies for provenancing the different classes of materials will be dealt with in Chapter 3. However, it is here stressed that points (1)–(4) of the above list require a high degree of geological competence, especially concerning the distribution of the mineral resources in the territory, the evaluation of their importance and exploitability in past times, and especially in the understanding of the behaviour of the measured chemical and isotopic tracers in the frame of geological processes. Without an appropriate contribution from the Earth Sciences any measurement and fingerprint modelling is bound to be highly empirical and qualitative. In the words of Shackley (1998) concerning stone materials, "…archaeometric data in the laboratory is only good as the geoprospection that produced the source standards", and "sampling is as important as the precision and accuracy of instrumental results". Taking into account that, excluding organics (or maybe including them…), virtually all archaeological materials are made of geomaterials, the statement may well be extended beyond stones.

Box 2.g XRD X-ray diffraction

X-ray diffraction (**XRD**) from the solid state is a fundamental technique for the characterization of synthetic and natural materials. Since the discovery of Bragg-type X-ray diffraction from periodic crystal lattices, the technique has proven an essential tool for: (1) the identification of crystalline compounds, (2) the quantitative analysis of polyphasic mixtures, (3) the study of the long-range atomic structure of crystals, including detailed charge density studies, and (4) the physical analysis of crystalline aggregates, including orientation texture, crystallite size distribution, and lattice micro-strain effects.

During the first half the last century such routine applications were developed into standard analysis procedures, and during the last decades they became so widely used that data are now available for most known inorganic compounds and minerals, to the extent that electronic databases are now accessible for automatic identification and rapid retrieval of crystallographic and structural information of crystalline substances. The routine identification or quantification of crystalline phases by diffraction is by far the most widespread application in archaeometry, although non-destructive analysis of a material's texture has recently been proven to be a powerful tool especially in metal investigations (Box 3.k). The focus of modern crystallographic research on inorganic solids is the interpretation of the physical and chemical properties of materials in terms of their ideal or defective atomic structure, and the transfer of the acquired crystal chemical knowledge to the engineering of solid state compounds with novel technological properties.

The last twenty years have also witnessed the development of dedicated second and third generation synchrotron radiation sources in the region of hard X-rays (Box 2.a). The availability of brilliant, polarized, and collimated synchrotron radiation beams with a wavelength that is tunable over a wide spectral range has made a number of interesting advances possible in materials characterization. State-of-the-art investigations of inorganic compounds and minerals by synchrotron X-ray diffraction include in situ dynamic diffraction of kinetic processes and phase transformations, structural characterization of compounds at ultra-high pressures, and the use of resonant scattering effects for element-selective structural analysis. None of these advanced applications has yet found direct archaeometric applications, with only high temperature diffraction being sometimes used as a novel tool for in situ triggering of the reactions taking place during the firing of old ceramics, with the aim of reconstructing the original firing temperatures (Section 3.1).

XRD data collection and analysis is routinely performed in any laboratory dealing with inorganic or mineral compounds. Samples are commonly in the form of small single crystals (with sizes in the range 0.1–1.0 mm), polycrystalline aggregates (powders, i.e. assemblages of 10^3–10^6 crystallites with size around 1 μm), or oriented specimens (fibres, thin films, polished surfaces). Therefore the experimental techniques and data analysis methods to be applied may vary, depending on the physical state and the chemical composition of the sample besides, of course, the nature of the requested information. Typical laboratory applications in inorganic chemistry, solid state physics and mineralogy may involve: (1) the identification of unknown crystalline phases, (2) the quantification of phase abundance in polycrystalline mixtures, (3) crystal structure determination and refinement, (4) physical analysis of the sample in terms of crystallite size, lattice micro-strain, or orientation texture of the crystal domains.

In archaeometry, XRD easily represents the cheapest and most reliable technique for identifying crystalline phases both in natural and man-made materials such as metals, metal slags, ceramics, soils, building stones, pigments, plasters, etc.

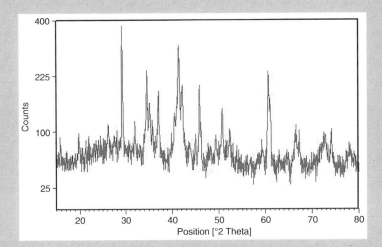

Fig. 2.g.1. Typical X-ray diffraction powder pattern obtained by a laboratory 2-circle diffractometer on a poorly diffracting sample. The visible Bragg's diffraction peaks are used to identify and quantify the crystalline phases in the sample.

The experimental apparatuses commonly involve a sealed-tube or rotating anode X-ray source, a 2-circles (for powders) or a 4-circles (for single crystals) automated diffractometer for sample orientation with respect to the incident beam, and one or more detector banks for detection of the diffracted signal. The data collected are generally intensity profiles of the diffracted X-rays as a function of the scattering angle (Fig. 2.g.1), which show the Bragg-diffracted peaks characteristic of each crystalline phase. The angular position, the integrated intensity, and the peak profile shape of the diffracted signal are important for the subsequent analysis based on Bragg's Law ($2d \sin \theta = n\lambda$, where d is the distance between lattice planes in the crystal, θ is half the diffraction angle between the incident and diffracted beams, and λ is the wavelength of the incoming radiation), graphically shown in Fig. 2.g.2.

For in-depth information

Giacovazzo C. (ed.) (2002) Fundamentals of crystallography. IUCr Texts on Crystallography, no. 7. Oxford University Press, Oxford.

(cont.)

Box 2.g (*Continued*)

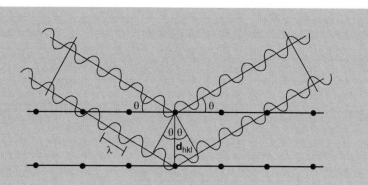

Fig. 2.g.2. Graphical representation of the interference process by waves produced by the ordered arrangement of atoms in a crystal.

Box 2.h ND Neutron diffraction

Neutrons can be produced by nuclear reactors, through fission of heavy nuclei, or by pulsed sources, through spallation of nuclei bombarded by protons. Neutron scattering in cultural heritage investigations has long been confined to chemical studies, especially by the use of neutron activation analysis (**NAA**, Box 3.a), which is also the basic process used in imaging by autoradiography (Box 2.m) and, more recently, by prompt-gamma activation analysis (**PGAA**) and resonance capture analysis (**NRCA**). However, materials scientists and engineers have long exploited the peculiar aspects of the neutron–matter interaction for the analysis of materials (Carrondo and Jeffrey 1987, Baruchel *et al.* 1993). Neutrons, being heavy uncharged particles, interact primarily with the atomic nuclei so that the character of the scattering process is entirely different from that of a photon–matter interaction. There is no need to enter into the physical details, but it is important to note that neutron beams can penetrate thick portions of matter, independently of its chemical nature, and that they are therefore an ideal probe for the **non-invasive** analysis of undisturbed art and archaeological objects (Kockelmann *et al.* 2006a). Keeping in mind the different nature of absorption for particles and photons (Section 2.2), X-rays usually penetrate a few tens or hundred of micrometres, depending on the energy used and the density of the material, and electrons are totally absorbed within hundreds of nanometres from the surface of the sample (see Table 2.8). Neutrons, on the other hand, can easily penetrate several or even tens of centimetres in most materials, including metals and rocks, with the notable exception of a few very absorbing elements that have high neutron cross-sections (H, B, Li, In, Cd, Gd).

The fantastic penetration of neutron beams has been widely exploited in: (1) the imaging of large objects by neutron radiography and tomography, which is described elsewhere (Section 2.1.4) and (2) the non-destructive analysis of crystalline materials by neutron diffraction (**ND**). Taking into account the nature of the neutron scattering process, the diffraction of neutron beams described as waves is a process that is conceptually similar to that of electromagnetic radiation, as described in Box 2.g for X-rays. Similarly, it is described by Bragg's law, is performed by similar instruments, and provides comparable information, namely the chemical and crystallographic nature of the crystalline components. ND is therefore very valuable in the identification of crystal phases within thick objects, the results being largely unaffected by surface alteration, coating layers or even thick containers (Fig. 2.h.1).

Furthermore, the use of modern time-of-flight (**TOF**) methods on spallation sources (Kockelmann *et al.* 2006b) poses several advantages with respect to experiments with thermal neutrons produced by reactor sources. Figure 2.h.2 shows the typical arrangement of detector banks in an ND experiment using TOF geometry. Using such a geometry the sample, the beam and the detectors are fixed and it is possible to perform phase identification and quantification, strain analysis, depth profiling and full texture analysis in relatively short time. These experiments can be extremely useful to (1) evaluate hidden features in thick objects, (2) assess the depth and degree of alteration

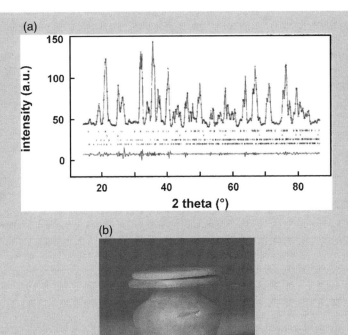

Fig. 2.h.1. ND powder patter (a) obtained at ILL on Egyptian cosmetics preserved inside a sealed calcite vase from the Louvre collection (b). The neutron data allow non-invasive identification of the component phases (galena, cerussite, laurionite, phosgenite) without opening the container. (From Martinetto *et al.* 2003)

and corrosion of surface layers, and (3) perform non-invasive investigations for the reconstruction and interpretation of the manufacturing processes of metal objects (Box 3.k). The rapidity of the measurement may be necessary to minimize the radioactivity induced in the object by the neutron flux.

For in-depth information

Alexandrov Yu.A. (1998) Fundamental properties of the neutron. Oxford University Press, Oxford.

Bacon G.E. (1975) Neutron diffraction. Clarendon Press, Oxford.

Lovesey S.W. (1989) Theory of neutron scattering from condensed matter. Oxford University Press, Oxford.

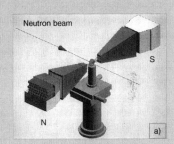

Fig. 2.h.2. Typical arrangement of neutron detector banks in a time-of-flight experiments exploiting the pulsed neutron beam from a spallation source (ENGIN-X beamline, ISIS, UK). (From Siano *et al.* 2003)

Box 2.i ED Electron diffraction

Electrons are fundamental particles that can be described as waves, according to the De Broglie wave–particle duality principle, and as such they are diffracted by periodic crystal lattices, as described for X-rays and neutrons. Since electrons are charged particles they interact via Coulomb forces with the combined electrostatic potential formed by the electron cloud and the atomic nucleus. The practical consequences of the strong electron–matter interaction are that electrons are quickly adsorbed by matter and therefore electron diffraction (**ED**) should be performed on

(*cont.*)

Box 2.i (*Continued*)

Fig. 2.i.1. Single crystal ED pattern of a small airborne particle of a phyllosilicate mineral (kaolinite) obtained on a FEI Tecnai F20 G2 high resolution TEM instrument.

crystals of the order of few tens of nanometres. We can imagine that one single crystallite extracted from a polycrystalline aggregate (composed of millions of grains) used as a powder sample for XRD or ND may indeed be used as a suitable single crystal for electron diffraction.

ED is usually performed in a transmission electron microscope (**TEM**, Box 2.l) where the electrons pass through a thin layer of the material. The resulting diffraction pattern is then observed on a fluorescent screen, recorded on photographic film or using a CCD camera positioned below the sample. Figure 2.i.1 shows the diffraction pattern of a single particle of a phyllosilicate mineral from airborne dust. For the appropriate characterization of complex crystal phases several diffraction planes must be recorded, and are often combined with the chemical composition measured by ED-XRF (Box 2.b) on the same crystal at the time of the diffraction experiment. In fact, many modern TEM instruments are equipped with energy dispersive spectrometers in order to perform imaging, diffraction and chemical analysis at the same time, so that they can be used to identify the shape, the dimensions, the crystalline nature and the chemistry of very minute particles. The micro-sampling that is necessary to obtain TEM samples makes the technique virtually non-invasive, provided that the extracted particles are representative of the sample (Section 2.2).

There is another very interesting application of ED performed in the backscattering mode, which is increasingly used in the analysis of textured materials. These are materials that show a preferred orientation of the crystalline components caused by the production, working or usage processes of objects. The technique is called electron backscatter diffraction (**EBSD**), also known as backscatter Kikuchi diffraction (**BKD**) and it is based on the measurement of the diffraction pattern formed by the electrons backscattered towards a screen or a CCD camera positioned at an angle from the sample surface. EBSD is usually performed in a scanning electron microscope (**SEM**, Box 2.l) where some of the electrons from the incident beam interact with the lattice planes of the crystal, satisfy Bragg's law, and are directed towards the detector where they form the Kikuchi lines corresponding to the planes of the crystal (Fig. 2.i.2). From the analysis of the line orientation, it is possible for each image to identify the crystal phase and the orientation of the crystal relative to the sample's reference system. Since the measurements are performed in a scanning microscope, there will be an array of point measurements, and a map can be produced where all points with the same orientation are assigned as belonging to the same crystal. The final map will show the shape and orientation (random or preferred) of all crystals in the sample, and can be used to interpret the so-called crystallographic **texture** of the sample. This information may also be obtained by other methods, such as optical microscopy and X-ray or neutron

(a) 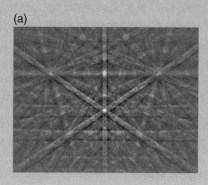 (b)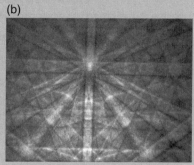

Fig. 2.i.2. Intersecting Kikuchi lines in a single crystal EBSD pattern of (a) a silicon crystal (Image courtesy of L. Peruzzo, CNR, University of Padova) and of (b) a copper crystal. (Image courtesy of R. Keller, NIST)

diffraction, and it is very useful in helping to determine the manufacturing process of the sample, especially metals (Box 3.l).

Although very powerful, high resolution ED (**HR-ED**) and EBSD are to be considered rather labour intensive, so that in the field of cultural heritage they can hardly be considered to be routine techniques for phase characterization. However, it should be noticed that ED is one of the very few techniques that allow measurements on extremely small samples, which may go undetected by most other techniques. A number of examples are already available where HR-ED has been used to characterize in detail the nature and significance of crystalline nanoparticles in ancient materials, mostly small crystals and metallic clusters in prehistoric and medieval glass (Fig. 2.i.3, the archaeometric problems and significance are described in detail in Section 3.4). It is expected that the nanoscience revolution at present dominating many fields of materials science will soon produce results and basic knowledge that will be directly transferable to applications related to cultural heritage. Then electron microscopes may well become invaluable instruments of a nano-archaeometry revolution.

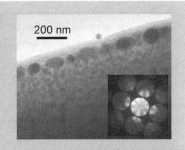

Fig. 2.i.3. Transmission electron image showing the shape, distribution, and nanometric size of the copper particles in a medieval glazed lustre ware. The inset image shows the convergent beam electron diffraction (CBED) pattern obtained on a single particle, which allows direct and unambiguous identification of the exsolved crystal phase. (From Pérez-Arantegui *et al.* 2001)

For in-depth information

Cowley J.M. (1992) Electron diffraction techniques. IUCr Monographs on Crystallography, Vols 1 and 2. Oxford University Press, Oxford.

Schwartz A.J., Kumar M. and Adams B.L. (eds) (2000) Electron backscatter diffraction in materials science. Kluwer Academic/Plenum Publishers, New York.

Spence J.H. and DeWitt B.S. (2003) High-resolution electron microscopy. Monographs on the Physics and Chemistry of Materials, Oxford University Press, Oxford.

Box 2.j TG-DTA-DSC Thermal analyses

Thermal analysis is a branch of materials science where some properties of materials are studied as they change with temperature. In principle all techniques, including scattering, diffraction, spectroscopy and physical measurements can be used to monitor physico-chemical changes in matter as a function of temperature, i.e. almost all measurements can be made temperature-dependent. Nevertheless, at present the term thermal analysis is usually reserved for a narrow range of techniques that are expressly designed to measure changes in the physical properties of materials with temperature, namely **thermogravimetry** (**TG,** also called differential TG or **DTG,** or thermogravimetric analysis **TGA**), **differential thermal analysis** (**DTA**) and **differential scanning calorimetry** (**DSC**).

Thermogravimetry measures the changes in sample weight driven by the change in temperature (weight loss curve). Instrumental parameters to be defined are the atmosphere surrounding the sample (air, reducing or oxidizing conditions, vacuum), the maximum temperature reached (commonly about 1000°C), and the heating rate. Sometimes the gas produced during the high temperature reactions are fluxed to an FT-IR or an MS analyser, in order to characterize their chemical or molecular composition (this is called **evolved gas analysis**). As many weight loss curves look similar, the weight and temperature measurements are performed with superior precision, and a **derivative weight loss curve** can be used to the temperature or the temperature range at which the weight loss process occurs. The technique yields two kinds of information: (1) the response of the material to temperature changes, which is

(*cont.*)

Box 2.j (*Continued*)

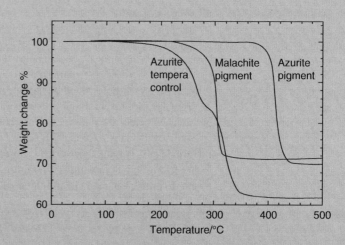

Fig. 2.j.1. Weight loss curves obtained for reference purposes by TGA on azurite and malachite pigments. (Odlyha *et al.* 2000)

crucial to understanding its behaviour during manufacturing by ancient pyrotechnology, modern industrial thermal treatment or use by fire. This in turn is (2) closely related to the material's chemical, structural and textural nature. The TG/DTG curve is therefore a fingerprint of the material (Fig. 2.j.1), even when the signal is not or is hardly interpretable based on scattering or spectroscopic data. Typical examples are clay minerals and organic polymers.

Differential thermal analysis measures the temperature difference between the material under study and an inert reference material (typically Al_2O_3) during heating under identical conditions. The differential temperature is generally measured by recording the electrical current between two thermocouples, one embedded in the sample and the other embedded in the reference material. The differential temperature (ΔT) is generally measured by recording the electrical current between two thermocouples, one embedded in the sample and the other embedded in the reference material. The temperature difference ΔT is then plotted against time or temperature (DTA curve or thermogram). Any reaction in the sample, either **exothermic** (i.e. releasing heat) or **endothermic** (i.e. absorbing heat), can be detected as a variation in T relative to the inert reference. The area of a DTA peak is directly related to the **enthalpy** change during the reaction. Therefore, a DTA curve provides data on all transformations that occur during heating, such as structure decompositions, phase transformations, glass transitions, crystallization, melting, sublimation, etc. Furthermore, as in TG measurements, the DTA curve can be used as a fingerprint for identification purposes.

Differential scanning calorimetry measures as a function of temperature the difference in the amount of heat required to increase the temperature of the sample and of the reference material, which has a well-defined heat capacity. All transitions and reactions occurring in the sample at high T involve energy changes or heat capacity changes, so that more or less heat (depending whether the reaction is endothermic or exothermic) will need to flow to it than to the reference to maintain both at the same temperature. DSC can be used to measure a number of characteristic properties of the sample, very similarly to DTA. Specifically, the ability to determine transition temperatures and enthalpies makes DSC an invaluable tool in determining the **phase diagrams** for materials.

Because of the sensitivity in detecting subtle phase transitions and also reactions in amorphous, organic, and polymeric materials, DSC and DTA have been largely used in the measurement of the properties of materials used in conservation. In fact, thermoanalytical techniques provide a great deal of information on the actual physicochemical state of precious objects, the changes in thermal stability on the ageing of samples prepared according to traditional recipes, and the impact or damage caused by museum environments on cultural objects (Odlyha 2000). The changes taking place in collagen-based materials such as leather and parchments (Fig. 2.j.2, Chahine 2000) and bones (Nielsen-Marsh 2000, Lozano *et al.* 2003) are particularly important.

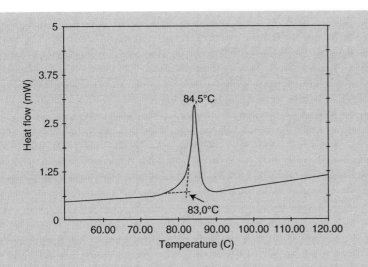

Fig. 2.j.2. DSC curves of leather (Chahine 2000). The experimental measure of the enthalphy of denaturation of collagen in leather and parchment is important to understand the present state and hydrothermal stability of artefacts during ageing.

In archaeometry thermal analysis has sometimes been employed to characterize the nature of materials, such as amber (Ragazzi *et al*. 2003), and to understand better the processes involved in ancient pyrotechnology, especially pottery (Kingery 1974) and glass (Heide *et al*. 2000).

For in-depth information

Brown M.E. (2001) Introduction to thermal analysis. Techniques and applications, 2nd Edition Academic Press, New York.

Gabbott P. (2007) Principles and applications of thermal analysis. Blackwell Publishing.

Haines P.J. (2001) Principles of thermal analysis and calorimetry. RSC Paperbacks. Royal Society of Chemistry.

2.1.3 Through the eyes of science: imaging techniques at different scales

Modern **imaging science** (Barrett and Myers 2004) is concerned with the generation, collection, duplication, analysis, modification, and visualization of images. Simple browsing of the word "imaging" in Wikipedia yields intriguing and far reaching subfields such as "digital imaging", "medical imaging", "radar imaging and survey", "chemical imaging", "document imaging and storage", "imaging software", and so on. Each of these areas, and many more, have profound implications related to cultural heritage. Since the invention of photography, the possibility of storing visual information on film and, more recently, in digital format has greatly enhanced education, communication, conservation, and analysis of the cultural heritage. Archaeology, museum management and display, conservation science, and art criticism have always exploited as much as possible the possibility of **reproducing** objects and sites for a variety of reasons including cataloguing, communication, increasing public access, education, and marketing. The visual **recording** of irreproducible situations and events, such as the different steps of archaeological excavations, or the original location of objects subsequently stored in museums and collections, is of course an invaluable complement to written reports and descriptions. The discussion whether an image is a more faithful, reliable, or complete record than a subjective drawing and textual description could be far reaching. In fact, to this day many archaeologists still prefer line drawings of objects, which enhance details and contours, to photographic or 3D images, which they claim are often confusing, misleading, and artificial. Many disciplines, including architecture and archaeology, hardly recognize digital and computer-generated pictures as scholarly. On a personal note, I truly love old and modern drawings and prints of collection items simply because they are beautiful and they carry a great deal of information on the attitude, culture, and spirit of the recorder, though there is no doubt that digital and scientific images are far less subjective, and the apparent confusion they create is due mainly to the huge amount of information content that the images themselves contain, and that must be properly decoded and interpreted. As they say "A picture is worth a thousand words". I couldn't agree more, though this brings us directly to the deep meaning of **visualization,** defined as "the complex technical process of creating images and graphics to communicate a message". As discussed later, we are to be greatly concerned with the reliability of the imaged data, with the difference between the data and the models used to represent them, and with the purpose of the whole visual operation.

Given than since prehistory *homo sapiens* has always attempted to communicate both abstract and concrete ideas through visual means, and there is no ultimate boundary between visualization, imagery, and art, it may be safe to state that we are here mostly concerned with the use of graphic tools such as images, diagrams, and animations, nowadays invariably produced through **computer graphics**, for the primary purpose of reporting and communicating scientific information concerning the cultural heritage (MacDonald 2006).

The key is to select in each case the image techniques and scales that are more appropriate to report the required **information content**. As briefly discussed above, with scientific instruments we can visualize the object of interest on totally different scales. Each visualization technique sizes the object at a different scale level, from the atomic view to the human macroscopic perspective, and images at different scales produced by different instruments contain very different kind of information, in much the same way that geographical maps at different scales yield information of totally different natures: would you use your city map to find the relative positions of Chicago and Beijing or would you use the continental map of Africa to find the location of the nearest market square? Let us therefore try to rationalize the most useful and commonly used techniques for imaging the cultural heritage in terms of their scale and information content. Concerning the scale we may say that: (1) at the **macroscopic** level the image shows the whole object and/or its surrounding environment (Box 2.p). The most common applications are:

- Large area survey. For example that is: the use of instruments to map the relative location of objects before, during, and after archaeological excavations, the elaboration of thematic data for paleogeographical or palaeoclimatic reconstruction, site management, etc.
- Diagnostics of architectural buildings. The structural and thermographic mapping of buildings of any kind for conservation and stabilization purposes.
- Object imaging. The visualization of the present shape, integrity, and state of the object. The image is produced for cataloguing and inventory purposes, to list the objects found in a particular site or collection, for comparison purposes, for preliminary visual analysis of the object prior to conservative intervention, for educational purposes and divulgation, even for reproduction and marketing, including the computer-aided production of three-dimensional replicas.

Whereas we say that (2) at the **mesoscopic**, **microscopic,** or **atomic** levels, the images show increasingly smaller portions of the object that are generally used during scientific analyses for visualizing specific information for:

- Conservation. To have a complete knowledge of the object's parts, their alteration state, and the nature and distribution of the materials composing it, in order to select the areas of intervention and plan the preservation process.
- Study and interpretation of the object. The image shows very selected portions of the object to emphasize and illustrate the investigations, aiming to understand of the object's history, manufacturing, usage, and evolution. Each investigation must necessarily go through several steps of imaging and data recording at different scales, with or without sampling.

We have already introduced the concept that all matter is heterogeneous, at one scale or another, and it is precisely this inhomogeneity that creates contrast in the image. What is apparently homogeneous at one level may show contrast and heterogeneities at smaller scales, and all materials are intrinsically discontinuous at the sub-microscopic and atomic level. Many instrumental techniques, starting with optical microscopy, were specifically developed to observed fine features of materials that were unobservable at the macroscopic level. The whole history of optics, with the introduction of glass lenses to improve human vision in the form of glasses, telescopes, and optical microscopes focuses on this concept. One possibility for a start to the discrimination of modern imaging techniques is to realize that some of them simply **probe the surface** or the external layers of the observed object (surface techniques), whereas other techniques allow **the internal structure** of the material to be probed (bulk techniques). The latter are invariably based on penetrating probes, such as an electron, X-ray, or neutron beam passing through the sample. By realizing that all imaging techniques actually record a signal emitted or re-emitted from the object after being hit by the source probe (Table 2.2), we can easily understand that the penetration of the observation largely depends on the energy of the probe and the nature of the material. For example, thick layers of glass or water are transparent to light, so visible radiation can be used to probe these materials at depth, whereas even thin foils of metal are opaque and completely absorb light, so that visible light can be used to image only the surface of metals. The same concept is applicable to all bands of the electromagnetic spectrum, and to all materials; this will be enlarged upon later in the chapter. Table 2.6 reports the most used surface or bulk imaging techniques classified according to the scale at which they probe matter.

It is useful to appreciate that some techniques (i.e. tomography, reflectometry, remote sensing) are performed on the untreated sample, and this might be of great advantage when studying cultural heritage materials (Section 2.2). Other techniques need partial treatment of the sample (i.e. SEM requires coverage of the sample surface by a conductive carbon or metal layer for electron dispersion), or even substantial sample manipulation in order to get the appropriate sample thickness for optimal beam transmission (i.e. TEM, OM) or optimal polishing (EBSD). **Invasiveness** must therefore be taken into account during the choice of the technique appropriate for the analysis.

The other fundamental parameter to be considered is time, that is whether the image is take in one shot, as in photographic images, or whether the image is built up pixel by pixel by the recording of the secondary emission related to the physical and/or chemical nature of the material, and measured by spectroscopic techniques at the desired energy. These images, based on **pixel mapping** can be produced by almost any spectroscopic or scattering technique using

Table 2.6. General classification of surface and bulk imaging techniques based on the scale of the observation

	Surface imaging	Bulk imaging
Atomic scale	■ Atomic force microscopy ■ Scanning tunnelling microscopy	■ High Resolution Transmission Electron Microscopy, HR-TEM (Box 2.l) ■ Depth profiling, (see text)
Microscopic scale	■ Scanning Electron Microscopy, SEM (Box 2.l) ■ Micro-chemical mapping (see text)	■ Phase contrast X-ray radiography (Box 2.m) ■ Synchrotron radiation micro-tomography (Box 2.n)
Mesoscopic scale	■ Electron backscattering diffraction, EBSD (Box 2.i) ■ X-ray texture analysis, XRTA (Box 3.k)	■ Optical Microscopy, OM (Box 2.k) ■ X-ray topography ■ Neutron Texture Analysis, NTA (Box 3.k)
Macroscopic scale	■ Photography ■ Reflectography / Reflectometry (Box 2.o) ■ Chemical mapping (see text) ■ Laser scanning ■ Multi-spectral image acquisition ■ Remote imaging (Box 2.p)	■ X-Ray/Neutron radiography (Box 2.m) ■ X-Ray/Neutron tomography (Box 2.n)

a narrow probe scanning the sample, and instrumentally synchronizing the scanning beam with the spectrometer analysing the selected signal. If we are interested in the two-dimensional distribution of selected chemical elements on the surface, appropriate **chemical maps** may be obtained by recording at each point the characteristic X-ray fluorescence signal from the core electrons (Table 2.1). This may be performed by XRF, PIXE or, SEM-EDS, and in principle by any technique equipped with an EDS or WDS spectrometer (Box 2.b) and a scanning probe that is energetic enough to excite the X-ray fluorescence process. Other techniques do not measure in the X-ray region, but rather they attempt to exploit the IR, Vis, or UV radiation bands to extract complementary information on painted surfaces. These techniques are commonly referred to as **reflectometry** (Box 2.o). All reflectometric and surface X-ray fluorescence techniques are totally non-invasive, excluding the rare instances where colour centres are activated by the incident probe, and the sample may undergo changes in colour. Other mapping techniques are based on the physical extraction of a small number of atoms from the surface, which are then analysed by mass spectrometry or optical-emission spectrometry. These techniques commonly employ a small laser beam to atomize the surface (LA-MS Box 2.t, LIBS Box 2.c) or an ion beam (SIMS Box 2.t) and they are effectively micro-destructive, though the holes produced by the laser or ion beams are not detected by the human eye and they are visible only at high magnification by electron microscopy (SEM). Figures 2.9 and 2.10 show elemental maps obtained by PIXE on ink-written paper and painted papyrus, respectively. There is a substantial amount of information derived from the careful comparison of the elemental maps with the conventional images obtained by visible light, because the distribution of specific elements

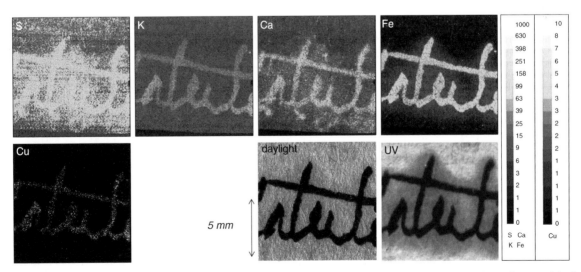

Fig. 2.9. PIXE elemental maps of an 18th century manuscript compared with the visible and ultraviolet light images. The UV image and the S, Ca and Fe maps clearly show the halo due to diffusion of some of the ink components out of the writing. (From Remazeilles *et al.* 2001)

 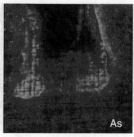 

Fig. 2.10. Micro-PIXE elemental maps of pigments on a papyrus of the "Book of the Dead". The distribution of the elements and the correlation with the visible drawing makes it possible to identify the nature of the pigments and to discriminate the elements, such as As, deriving from conservation treatments. (From Olsson *et al.* 2001)

is related to the nature of the original or re-painted pigments, or to the chemical composition of the ink and the diffusion/alteration processes in the paper layer.

When the probe is an extremely collimated and bright microbeam of radiation produced by a synchrotron (Box 2.a), then we may actually talk about **micro-mapping**, with spatial resolutions being effectively in the sub-micrometre range. The next generation of synchrotron instruments are aiming for nanoscale resolution.

Of course the signal selected for recording must match the specific problem to be investigated, for example in place of chemical probes we can use diffraction and obtain **crystal phase maps** (see, for example, Dooryhée *et al.* 2005), or techniques that can discriminate electron energies, (such as XPS or XAS) and obtain **maps of the oxidation state** of the ions, and so on. Of course pixel mapping requires more sophisticated and specialized instruments than the routine bulk laboratory measurements, so that the time and costs required for the analyses are substantially affected. The specific analysis of painted layers by Vis, IR, and UV radiation reflectometry is discussed in detail in Box 2.o, because of its widespread use and importance in the analysis and interpretation of paintings and frescoes.

In principle, 2D scanning can be further extended in the three dimensions by the use of non-invasive tomographic techniques. Although in its infancy, chemical and phase mapping in 3D is growing fast and in the future will probably be a key technique in the understanding of the processes active in undisturbed cultural heritage samples. The combination of synchrotron radiation absorption micro-tomography with phase (micro-XRD) and chemically-sensitive (micro-XAS, micro-XRF) techniques, or the combination of TOF neutron diffraction and PGAA or NCRA analyses are leading the way in this direction. Although many of these analyses are performed on the same samples in sequence (see, for example, Kasztovszky *et al.* 2007), there are no fundamental reasons against them being performed rapidly and simultaneously at the same experimental station.

Another very specific application of penetrating techniques is the possibility of measuring chemical parameters progressively in the material. The method is broadly defined as **depth profiling** and it allows one to assess the penetration of specific elements within the sample surface, producing chemical diffusion curves in the material. Since diffusion is essentially controlled by temperature and time, if we know the environmental conditions of our sample the penetration of an element is a function of time. This is the physical basis for many of the relative methods for dating based on chemical diffusion (Section 2.5.3), such as bone and flint fluorination, obsidian and quartz hydration, etc. Depth profiling can be performed in two ways. The first is micro-destructive, and consists in ablating a micro-hole (of the order of 100 μm) in the sample by laser or ion beams, and progressively extracting atoms or ions that are being analysed by different means, commonly mass spectrometry or other spectroscopic techniques (XPS, LIBS, etc.). This is the basis for LA-MS and SIMS (Box 2.t). The net result is a chemical profile in time that is proportional to the depth of the hole. The second method is non-destructive, and it implies energetic ion beams (Box 3.i) that have a specific interaction with the species to be detected. Depending on the interaction, two techniques are employed, the first is based on a nuclear reaction between the accelerated ions and the chemical diffusive species in the sample (nuclear reaction analysis, NRA), and the second measures the elastic recoil energy of the diffused species (energy recoil detection, ERD). Both techniques produce concentration profiles of the analysed element as a function of depth; in particular they are used for hydrogen detection and quantification (Ziegler *et al.* 1978, Langford 1992).

Table 2.7 attempts to take invasiveness and time into account during the selection of the appropriate technique

Table 2.7. Classification of imaging techniques based on beam penetration, scan mode and invasiveness

Penetration	Scanning mode	Invasiveness	Examples of the techniques
Surface techniques	One shot recording	Invasive non-invasive	• Reflected light optical microscopy • Photography
	Pixel scan	Invasive	• Scanning electron microscopy • Chemical mapping • Texture mapping by EBSD
		Non-invasive	• Atomic force microscopy • 2D, 3D laser scanning • Reflectometry • Optical coherence tomography • Multi-spectral image acquisition • Remote sensing
Bulk techniques	One shot recording	Invasive	• Transmitted light optical microscopy • Transmission electron microscopy
		Non-invasive	• X-Ray/Neutron radiography • X-Ray/Neutron tomography
	Pixel scan	Invasive	• X-Ray topography
		Non-invasive	• Texture mapping by neutron diffraction • Phase mapping by diffraction tomography

Box 2.k OM Optical microscopy

Optical microscopy (**OM**), performed by the use of a **polarized light microscope** (Fig. 2.k.1), is one of the oldest and most widespread tools for the mineralogical and petrographic investigation of minerals and rocks, although the fact that it is mostly applied to geologic samples is derived from purely historical reasons and it is generally applicable to all solid samples. Visible light must pass through a thin section of the specimen (**transmitted-light TL-OM,** or polarized-light OM), or it must be reflected by the polished surface of a thick sample (**reflected-light RL-OM**). All compounds with valence electrons confined in closed or localized orbitals (ionic and covalent bonds, molecular compounds) absorb light in very narrow bands, so that the sample is largely transparent to light and the analysis can be performed in transmission mode. Most rock-forming minerals such as silicates, carbonates, halides, etc. behave in this way. When delocalized electrons are present in the compound (metallic bond) most visible wavelengths are absorbed and the analysis must be performed in reflecting mode. The latter technique is commonly applied to metals and oxides and it is traditionally used in the investigation of mineral ore deposits and in the metallurgical characterization of metals and alloys (**metallographic analysis** by reflected-light OM, Box 3.m).

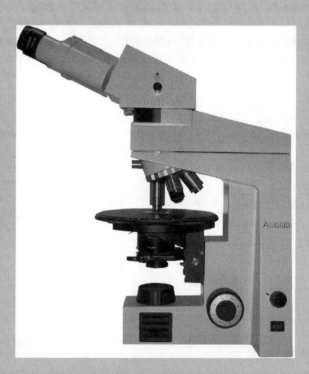

Fig. 2.k.1. Photograph of a laboratory (Axiolab, Carl Zeiss) polarizing microscope for OM observations. (Courtesy of C. Mazzoli, University of Padova)

The OM technique is always invasive because it requires a thin section or a polished and etched surface of the sample. Although it is cheap, it is always statistically meaningful because it allows investigation of centimetre-sized portions of the material, and if based on robust crystallographic knowledge it provides rapid access to a wealth of information: phase identification; grain size, shape and distribution; presence of defects such as cleavage planes, twin boundaries, and mechanical deformations; the nature and distribution of inclusions; textural relationships and genetic history of the crystal phases; the presence of amorphous components; and much more. Practical observations made under the microscope encompass: the visual estimation of crystal size and distribution (the smallest feature that can be observed is limited by the wavelength of light and is of the order of 0.5 µm) absorption colours and pleochroism by plane-polarized light, interference colours and birefringence by the use of cross-polarizers, measurement of the relative difference in refraction index between adjacent phases using Becke lines, absolute measurement of the refraction index by means of reference oils, and evaluation of the crystal symmetry by measuring the allowed vibration directions of light in the crystals. Operational details can be found in most mineralogy textbooks. A special technique call **dark field microscopy** employs patch and direct illumination stop blocks coupled by a lens condenser, so that only the light scattered by defects and boundaries is observed above the dark background. The technique is used to evidence inclusions and defects in homogeneous matrices.

Although OM can be used in the investigation of every type of material, the two most common applications are in the characterization of ceramics (Fig. 2.k.2) and metals (Fig. 2.k.3). In both cases OM provides a rapid and cheap analytical technique for the preliminary investigation of a statistically significant portion of the sample, and yields a complete qualitative overview of the phase composition and texture. Quantification of the observations requires a great deal more work, generally involving digital processing of the OM images, chemical maps (for example by SEM-EDS, Box 2.l) or point chemical analysis of selected phases (for example by EPMA, Box 2.b). Modern microscopes can easily be

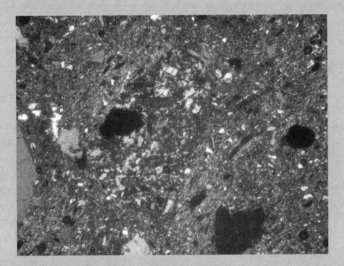

Fig. 2.k.2. Cross-polarized transmitted-light OM image of a ceramic shard under the microscope (horizontal field of view is 4 mm). The size and distribution of the temper grains, mostly polycrystalline rock fragments, within the fine grained matrix can be rapidly estimated. The latter is composed of small isolated minerals embedded in the glassy reddish paste formed by clay firing. (Courtesy of L. Maritan, University of Padova)
Original colour picture reproduced as plate 10.

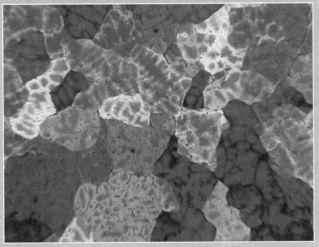

Fig. 2.k.3. Cross-polarized reflected-light OM image of a thermally recrystallized copper sample showing clear evidence of the original dendrites formed during casting.
Original colour picture reproduced as plate 11.

(*cont.*)

Box 2.k (*Continued*)

equipped with CCD cameras, so that **digital images** are systematically recorded and analysed (Section 2.1.4). Many qualitative observations of the past, especially concerning the statistical analysis of the size, shape and space distribution of specific phases may be made substantially more quantitative by the use of specific analytical software.

For in-depth information

Bloss F.D. (1961) An introduction to the methods of optical crystallography. Holt, Rinehart and Winston, New York.

Box 2.l SEM-TEM Electron microscopy

Electron microscopes are instruments that are specifically developed to obtain images of matter at high **magnification**. Since they employ accelerated electrons in place of light, they can reach much higher magnification than optical microscopes. Typically, **scanning electron microscopes** (**SEM**) employ electron beams accelerated in the range 10^2–10^4 eV, corresponding to wavelengths of the order of 0.123–0.012 nm, and therefore may reach magnifications over 100 000 times, compared with magnification factors for the best optical microscopes of about 1000. **Transmission electron microscopes** (**TEM**) employ electron beams accelerated in the range 10^4–10^5 eV, corresponding to wavelengths of 12.2–3.7 pm, and they reach magnification factors of over 10^7. Working at such high magnification and taking into account that wavelengths of 1 pm are about a hundred times smaller than inter-atomic distances, the best TEM instruments have a resolving power that is sufficient to discriminate atoms.

The electron microscope uses electrostatic and electromagnetic lenses to control the direction and divergence of the electron beam, in much the same way that glass lenses are used in optical microscopes to define the image in the focal plane. In addition to the energy of the beam, the basic difference between SEM and TEM instruments is the geometry of operation.

In the scanning microscope the electron beam hits the surface of the sample in a raster scan, and the detector (positioned at an angle above the sample) is synchronized with the beam and collects at each point the signal resulting from interactions of the electron beam with atoms at or near the surface of the sample (Fig. 2.3). The experimentally observed image may be formed by **backscattered electrons** (i.e. high energy electrons of the incident beam backscattered from the surface), **secondary electrons** (i.e. low energy electrons (< 50 eV) emitted by the atoms in the sample following ionization processes), **characteristic X-rays** (i.e. fluorescence X-ray produced by the ionization processes), **light** (i.e. visible photons emitted in non-metallic materials by an electron–hole recombination, the process is called **cathodoluminescence**) and others. All these types of signal require specialized detectors for their detection, and they are not usually all present on a single instrument.

Low energy secondary electrons originate within a few nanometres from the sample surface. The contrast in the signal depends on the number of secondary electrons reaching the detector. The number of electrons emitted from the surface depends on its inclination to the beam. If the beam enters the sample perpendicular to the surface, the number of detected electrons is much higher than the number emitted from an inclined surfaces on the opposite side from the detector (dark surfaces), and much lower than the number of electrons emitted from the slopes facing the detector (bright surfaces). As the angle of incidence increases, the escape distance of one side of the beam will decrease, and more secondary electrons will be emitted and detected. Thus steep surfaces and edges tend to be brighter than flat surfaces, which results in images with a well-defined, three-dimensional appearance (Fig. 2.l.1). Backscattered electrons are also commonly used in SEM instruments to produce images of the topography of the

Fig. 2.1.1. Secondary electrons SEM image of pollen grains. Images as such allow fantastic magnification and definition of the surface topography of the materials. (Courtesy of L. Howard, Electron Microscope Facility, Dartmouth College; image available at: http://remf.dartmouth.edu/imagesindex.html)

surface, because the contrast is mainly derived from the angle between the surface morphology, the incident beam and the detector. The backscattering process is commonly enhanced by gold coating the surface in poorly scattering samples, for example biological materials. Since heavy elements (high Z) backscatter electrons more strongly than light elements (low Z), they appear brighter in the image. Therefore backscattered electrons are used to detect the contrast between areas with different chemical compositions when the image contrast is not dominated by the surface morphology, for example on fine grained fracture surfaces or on flat polished surfaces (Fig. 2.1.2, Fig. 2.15, Fig. 2.16). The chemical information on the different phases is considerably enhanced if an EDS spectrometer is also available on the same SEM instrument (**SEM-EDS**). In this case, besides the backscattered and secondary electrons, it is also possible to record at each raster point the intensity of the characteristic X-rays lines of specific elements,

Fig. 2.1.2 Backscattered electrons SEM image of a 100 μm fragment of a protohistoric glass (Artioli *et al.* 2008). The image shows the complex heterogeneous texture of the material, containing a number of zoned augitic clinopyroxene crystals embedded in the silicate glass matrix. The large and the small white circles are metal and sulphide droplets unmixed from the silicate glass at high temperature (Courtesy of A. Polla, University of Padova).

(*cont.*)

Box 2.1 (*Continued*)

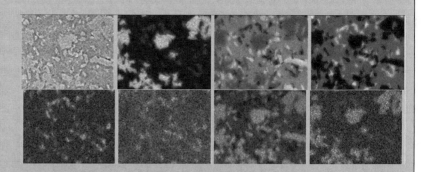

Fig. 2.1.3. Backscattered electrons image (upper left) of a polyphasic sample containing several non-stoichiometric silicate and oxide phases. The maps of the elemental distribution greatly help in defining the partitioning of the elements among the different phases. The images are listed in order of increasing atomic number: Mg < Al < Si < Ca < Ti < Cr < Fe.

in order to produce maps of elemental distribution (Fig. 2.1.3). The maps are extremely useful in interpreting element distributions and genetic processes in complex polyphasic materials.

In transmission electron microscopes the detector is positioned below the sample, so that the electron beam passes through the thin sample and produces a high resolution image of a small portion of the material. Nanometre-sized particles, and even their quasi-atomic structure may be imaged this way (Fig. 2.1.4). Furthermore, besides high resolution imaging, TEM instruments can be equipped with EDS spectrometers for chemical analysis of the imaged particles, similar to SEM-EDS, and further they can be operated in diffraction mode to produce crystallographic information (ED, Box 2.i).

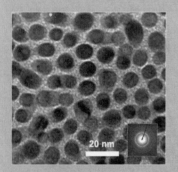

Fig. 2.1.4. HREM image of a copper-coloured lustre from a 16th century ceramic fragment from Gubbio, Italy. The nanometric copper metal particles (average dimensions 8 nm) are homogeneously spaced in the matrix with a minimum distances between the particle surfaces of about 2 nm. The inserted image shows the selected area electron diffraction pattern (SAED) of the particles. (Courtesy of B. G. Brunetti, University of Perugia)

Preparation of the samples for SEM is fairly trivial, and usually only a metal coating (for imaging) or a graphite coating (for chemical analysis) is required. The preparation of samples for TEM is a little more laborious, insofar as it requires preparation of very thin layers of materials, usually produced by grinding or ion thinning, deposited on copper/graphite nets. In all instances the samples have to be measured in a vacuum, as the air molecules (N_2, O_2, H_2O, Ar, CO_2 in order of relative abundance) would scatter and absorb the electrons. The measurement in a vacuum is generally not a problem for anhydrous materials, where the major problem usually is the perturbation of the sample by the energy deposited by the electron beam, though it may cause changes in hydrous systems. In these cases, the sample may be partially kept under a low-pressure of water (up to 2.5 kPa) in modern environmental scanning electron microscopes (**ESEM**), which allows hydrated samples to be measured in a wet atmosphere.

For in-depth information

Egerton R.F. (2005) Physical principles of electron microscopy: an introduction to TEM, SEM, and AEM. Springer, New York.

Goodhew P.J., Humphreys F.J. (1988) Electron microscopy and analysis. Taylor & Francis, London.

Kiely C.J. (ed.) (1999) Electron microscopy and analysis. Proceedings of the Institute of Physics Electron Microscopy and Analysis Group Conference, University of Sheffield, 24–27 August 1999. Institute of Physics Conference Series. Institute of Physics Publishing, London.

Box 2.m X-rays and neutron radiography

Radiography started in 1895 with the discovery of X-rays, also referred to as Röntgen rays after the German physicist (Wilhelm Conrad Röntgen) who first described their properties. Their very first application was the imaging of thick objects, before realizing their electromagnetic nature. Besides scientific applications, from the start X-ray radiography had a huge impact on society and medicine because of the possibility of imaging projections of the human body.

Projectional radiography is conceptually very simple, as it allows the visualization of the density distribution within an object by differential absorption. Experimentally it is performed by recording on a linear or an area detector (photographic film, image plates and fluorescent screens, CCD cameras) the image of the X-rays produced by a source, normally an X-ray tube, and passing through an object. The instrumentation geometry varies depending on the source of radiation, the size of the imaged object, the required resolution and the type of detector. The most common geometry is cone beam geometry used in laboratory sources (point source, 40–50° divergent beam, direct imaging of the photons) though, depending on the application, it varies from high-resolution parallel-beam micro-radiography at synchrotron sources, to high energy radiography of large objects using X-ray sources developed for industrial components. Casali (2006) nicely summarizes in a tutorial way the advantages and disadvantages of the available instrumental configuration and their use in cultural heritage.

In digital X-ray radiography the **absorption contrast** producing the image is related to the differential thickness of the imaged object or to the inhomogeneous distribution of matter within the object (i.e. of electron density) or both. Accordingly, absorption is described by the Lambert–Beer's formula $I = I_o \exp(-\mu t)$ (Sect. 2.1.1, Fig. 2.7), where I is the intensity of the transmitted radiation, I_o is the intensity of the incident radiation, t is the thickness of the material, and μ is the linear absorption coefficient (in units of cm^{-1}), which for X-rays is dependent on the energy of the radiation used and on the mean atomic number of the atoms in the imaged material. Experimental optimization of the technique requires the selection of the energy appropriate to the thickness and the nature of the material, control of the angular divergence of the source, and a definition of the source to sample (d_{ss}) and the sample to detector (d_{sd}) distances. For a given divergence of the X-ray beam, the **magnification** factor (M) of the image is related to the ratio between the two distances ($M = [d_{ss} + d_{sd}]/ d_{ss}$). The greater the sample-to-detector distance, the higher is the spatial resolution in the image, at the expense of intensity, counting statistics, and data collection time.

X-ray radiography is a relatively cheap, non-destructive and powerful technique to image the internal features of thick objects, and as such it has long been used as a diagnostic tool in cultural heritage, mainly to analyse archaeological objects (Fig. 2.m.1) and ancient mummified bodies (Fig. 2.m.2). Many averagely-absorbing materials can be imaged with laboratory X-rays, including ceramics, soft organics, paper, wood, and thin layers or rocks and light metals. Paintings and other art works are routinely analysed with museum X-ray

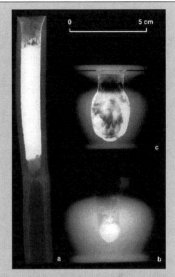

Fig. 2.m.1. X-ray radiography of different make-up receptacles from the Egyptian collections of the Louvre Museum. The white areas show the distribution of the X-ray absorbing lead powders present in the make-up: (a) reed case, still full of makeup. (b) alabaster recipient with a fabric lid. (c) alabaster recipient and cover. It contains a small amount of make-up attached on the inner wall. (From Martinetto *et al.* 1999; photo courtesy of C2RMF)

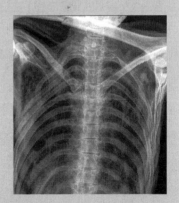

Fig. 2.m.2. X-ray radiograph of the chest of the Alpine Iceman performed during the 2001 re-analysis of the body. The image clearly shows the arrowhead stuck deep under the shoulder. (Museo Archeologico dell'Alto Adige, Bolzano, Italy, http://www.iceman.it)

(cont.)

Box 2.m (*Continued*)

sources. If thicker and/or denser materials are to be analysed, then industrial type high energy X-ray sources are used. With these sources, which are commercially developed to image small industrial components, it is possible to penetrate several centimetres of copper and bronze, and therefore they are appropriate for many archaeological objects.

In order to image larger and denser objects, such as full size marble and bronze statues it is necessary to use **neutron radiography**. Unlike for X-rays, neutrons are attenuated by some light materials (i.e. H, B, Li) and therefore are totally useless for hydrogen-loaded materials such as the human body, whereas they penetrate many heavy materials because of their low nuclear cross-section (Box 2.h). X-rays and neutrons are therefore two fairly complementary sources for materials imaging (Lehmann *et al.* 2007). Fig. 2.m.3 shows a Roman iron dagger imaged by neutrons and X-rays, which allow the detection of different internal features: the neutron image in the centre shows high absorption in the fractures, which are filled by some kind of organic substance that was probably inserted during cleaning and consolidation of the object; the X-ray image on the right shows high absorption in the handle area, clearly due to a thicker internal layer of uncorroded metal. A much clearer vision of the internal part of the handle can be extracted through the tomographic reconstruction (Box 2.n, Fig. 2.n.5).

Finally, a very peculiar form of radiography (**autoradiography**), which is mostly used for painting analysis is performed using neutron beams as irradiation sources for the activation of secondary processes (Taft *et al.* 1992). The nucleus captures a neutron and turns into a metastable radioactive isotope that will decay with its characteristic half-life emitting an energy quantum in the range of γ-rays (nuclear processes, Table 2.1). The probability of capture

Fig. 2.m.3. A dagger from excavations near the Roman Vindonissa camp (photo left) was inspected with neutrons (middle) and X-ray (right). (From Lehmann *et al.* 2007)

depends on the activation cross-section that is specific to every isotope. During irradiation, the painting is fixed on a support in front of the opening of a neutron guide at a small angle (< 5°) to the neutrons' direction, so that their main free path within the paint layer is maximized. The support is then scanned with a velocity of a few cm/s, allowing for a uniform activation of the total area of the panel. Owing to the short irradiation time, on average only about

0.4% of the atoms become radioactive and the method is considered non-invasive. Induced radioactivity commonly falls below safety levels within days. These secondary γ-emissions are detected at different times and form in the 2D detector images corresponding to the distribution of the specific radioactive isotopes. Table 2.m.1 lists some of the most common active isotopes employed in autoradiographic imaging. Since each pigment has a specific chemistry, the elemental distribution corresponds to the use of specific colours in the picture. Neutron autoradiography is capable of revealing different paint layers buried during the creation of the painting. In many cases, the individual brushstroke applied by the artist is made visible as well as changes and corrections introduced during the painting process.

Table 2.m.1. List of the most common radioactive isotopes used in neutron autoradiography (modified from Laurenze-Landsberg et al. 2003 and Leonardi 2005)

Chemical element	Associated pigment	Isotope	Half-life
Manganese	umber, ochre, brown colours	^{54}Mn	2.6 h
Copper	malachite, azurite, verdigris	^{64}Cu	13 h
		^{66}Cu	5.1 m
Sodium	canvas, ultramarine	^{24}Na	15 h
Arsenic	smalt, realgar	^{76}As	26.5 h
Antimony	Naples yellow	^{122}Sb	2.7 d
		^{124}Sb	60 d
Phosphorus	bone black	^{32}P	14.3 d
Mercury	vermilion	^{203}Hg	47 d
Cobalt	smalt	^{60}Co	5.3 y

For in-depth information

Ainsworth M.W., Brealey J., Haverkamp-Begemann E., Meyers P., Groen K., Cotter M.J., Van Zelst L. and Sayre E.V. (1982) Art and autoradiography: insights into the genesis of paintings by Rembrandt, Van Dyck and Vermeer. The Metropolitan Museum of Art, New York.

Domanus J.C. and Bayon G. (1992) Practical neutron radiography. Commission of the European Communities, Euratom Neutron Radiography Working Group. Springer, Berlin.

Lang J. and Middleton A. (2005) Radiography of cultural material. Butterworth-Heinemann.

Box 2.n CT X-ray and neutron computed tomography

Computed tomography (CT) is the most used method in medicine and materials science to obtain the image of the whole or partial internal portion of a bulky object (Baruchel et al. 2000). CT is actually a mathematical treatment where the internal 3D distribution of matter in a body is reconstructed and visualized starting from a number of 2D radiographic projections (Box 2.m). In principle the computed reconstruction can be applied to data that is different from the absorption images, as in the case of ultrasonic and capacitance tomography or nuclear magnetic resonance imaging. However in the field of materials science and **non-destructive testing**, absorption imaging using X-rays, γ-rays or neutrons is by far the most convenient and most adopted technique.

2D image radiographs can be obtained using different instrumental setups, mostly in transmission geometry. Exactly as in radiography, the type of source, the source-to-object and the object-to-detector distances, and the size

(*cont.*)

Box 2.n (*Continued*)

and resolution of the detector determine the geometry and the final resolution of the reconstructed image (Baruchel *et al.* 2000, Casali 2006). The vast majority of laboratory instruments employ a point source, a cone geometry and an area detector. While the speed of the data collection is crucial for medical CT scans, where the imaged body part is fixed and several sources move around the object for scanning (Fig. 2.n.1), non-destructive tomographic testing in materials science is normally performed by a fixed source and detector geometry, and by discrete rotation of the object around one axis (Fig. 2.n.2). The larger the number of available 2D projections, the higher the geometrical and positional definition of the features producing the physical absorption contrast (Fig. 2.n.3).

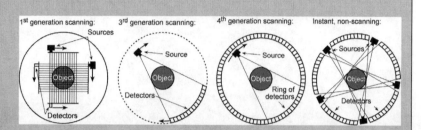

Fig. 2.n.1. Measurement geometries used for medical X-ray CT scanners of 1st, 3rd and 4th generation. Also shown is the instant configuration in which all ray-sum measurements are carried out simultaneously. (Modified from Johansen 2005)

The instrumental configuration must meet the needs of the investigation. The two basic constraints for the selection of the instrumental components are the **size of the imaged object** and the final required **resolution** of the voxels (i.e. volume pixels) in the reconstructed 3D tomographic image. Of course they are not independent, since it is by no means viable to scan very large objects at the maximum possible resolution (nowadays well below 1 µm). Besides practical considerations, such as the homogeneity of the source, the area coverage of the detectors, and the dimensions of the smallest sensitive element in the detector, the size of the acquired images would be computationally untreatable by available software and computers. Typical reconstructed datasets of $2048 \times 2048 \times 2048$ voxels for medium sized objects (of the order of $5 \times 5 \times 5 \, cm^3$) involve images with voxel resolutions of 120 µm, comparable to that of medical images (typically of $60 \times 60 \times 1000 \, \mu m^3$). At the two extremes, we can image metre-sized objects with resolution above the millimetre, or we can image

Fig. 2.n.2. Diagram of the instrumental setup for tomographic experiments in materials science. (Modified from NDT Resource Centre website, original image available at: http://www.ndt-ed.org/EducationResources/CommunityCollege/Radiography/AdvancedTechniques/computedtomography.htm)

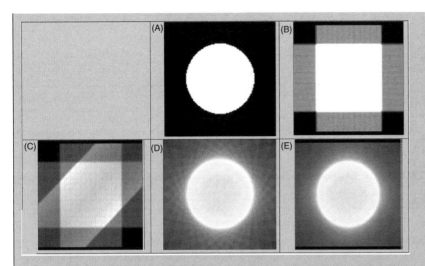

Fig. 2.n.3. Virtual reconstruction of a round object (A) using: (B) two projections at 0°, 90°; (C) three projections at 0°, 45°, 90°; (D) thirty projections at every 12°; and (E) a thousand projections at every 0.36°. The increasing definition and reliability of the reconstructed image is evident. (Courtesy of P. Bleuet, ESRF)

millimetre-sized objects with resolution below a micrometre. Three applications to archaeological materials are shown as examples. The first (Fig. 2.n.4) shows a heavily corroded and altered Bronze Age copper ingot imaged with an industrial-type high energy X-ray source at the Getty Conservation Institute. The tomography (voxel resolution around 40 μm) allows a distinction of the alteration layers, and the non-destructive analysis of the pristine copper core. The second one shows a section (Fig. 2.n.5) of the tomographic reconstruction of the handle of the Roman sword radiographed in Fig. 2.m.3. The third one (Fig. 2.n.6) shows the high-resolution

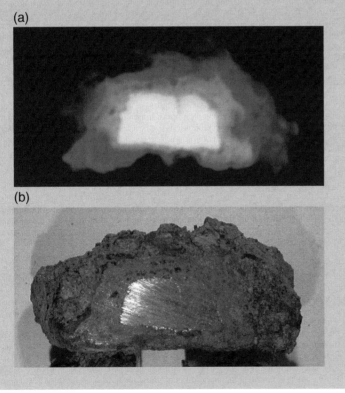

Fig. 2.n.4. Heavily altered and corroded protohistoric copper ingots. The tomography: (a) allows perfect non-invasive visualization and characterization of the unaltered copper core, of the cuprite layer, and of the malachite-brochantite-soil surface layer. (b) shows the photographic image a section cut from a similar ingot for comparison.

Original colour pictures reproduced as plate 12.

(cont.)

Box 2.n (*Continued*)

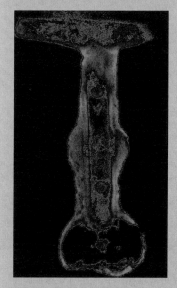

Fig. 2.n.5. False-colour section of the tomographic reconstruction of the handle of the Roman sword whose neutron and X-ray radiographs are shown in Fig. 2.m.3. (Courtesy of E.H. Lehmann, PSI)

Fig. 2.n.6. Example of a tomographic image of a Bronze Age faience fragment from Lavagnone, Brescia, Italy, showing a virtual section with voids and crystal inclusion. The glass phase is perfectly distinguishable by the absorption contrast, and its volume and distribution can be estimated. The tomographic data were collected at the SYRMEP beamline at the ELETTRA synchrotron, Trieste, Italy. (Artioli *et al.* 2008).

micro-tomographic image of a portion of a Bronze Age faience bead collected using synchrotron radiation at the ELETTRA Syrmep beamline. The image (voxel resolution around 4 μm) allows evaluation of the shape and distribution of the quartz grains and the voids within the glass matrix, thus providing insights into the type and manufacturing techniques of the early European faience materials.

Microtomography using synchrotron radiation can at present achieve amazing submicron resolution (about 0.1 μm). Furthermore the visualization of internal boundaries and contours may be substantially enhanced using **phase contrast imaging** (Cloetens *et al.* 1999, 2000). The technique, also defined as **holotomography**, is based on the inhomogeneous refractive index and/or thickness distributions in the sample. Thanks to the spatial coherence in the synchrotron beam, especially on long experimental beamlines, Fresnel diffraction turns local variations in phase into changes in intensity, and thus into an image. This requires that the detector is placed at some distance from the sample, and that images have to be recorded at more than one distance from the sample. Their processing provides quantitative phase maps that are combined for the final 3D tomographic reconstruction. The technique is potentially extremely useful to detect sub-micrometre sized features in undisturbed materials, for example soils, alteration layers, or inclusions buried under the surface of the objects.

Neutron tomography works pretty much the same way as the described photon tomography, taking into account the properties of the neutrons (Box 2.h). Because of the low intensity of the neutron flux, micro-beams are not possible, so that the best spatial resolution available in neutron radiography and tomography at present is around 0.1 mm. However, the excellent penetration of neutron beams provides for great sources to image thick and dense objects, and their characteristic sensitivity for light elements makes them appropriate to detect organics or water in inorganic materials (Richards *et al.* 2004, Lehmann *et al.* 2007, Kardjilov *et al.* 2007).

Another tomographic technique of recent introduction but of considerable interest is **optical coherence tomography** (Targowski *et al.* 2008). This is a relatively new, but nevertheless well-established, technique that offers the possibility for non-invasive in situ characterization of the inner structure of semi-transparent objects that weakly absorb and scatter light. It is non-contact and perfectly safe for objects as long as the exposure to the light source is limited to a certain level. It is limited to light-transparent materials, therefore the technique may be used for example to investigate at micrometre resolution the stratigraphy of painting layers, the deterioration of stained glass, and the internal structure of ceramic glazes.

The **non-invasivity** is the most important feature of all tomographic techniques and it is routinely exploited in the preliminary study of archaeological and art objects. Tomography is used to understand the nature and internal distribution of the materials and the fine features within the objects, for example it is used to select appropriate sampling areas for subsequent analyses or as a diagnostic technique in conservation. A very instructive case study is the tomographic investigation of the so-called **Antikythera Mechanism** (Freeth *et al*. 2006). The impressive 3D reconstruction of the gear, based on high resolution tomographic and photographic images, allowed the researchers to make sense of an ensemble of 82 puzzling metallic fragments dating back to the 2nd century BC. Confirming and extending previous hypotheses of the mechanism functions, the recent investigation proposes that ancient Greek technology was far more advanced than is commonly believed, and that the object was a very sophisticated astronomical calculator preceding the known astronomical clocks of the 14th century by a millennium of apparent technological gap (Marchant 2006).

For in-depth information

Casali F. (2006) X-ray and neutron digital radiography and computed tomography for cultural heritage. In: Bradley D. and Creagh D. (eds) Physical techniques in the study of art, archaeology and cultural heritage. Vol. 1. Elsevier, Amsterdam. pp. 41–124.

Kalender W.A. (2000) Computed tomography. MCD Verlag, Munich.

Kak A.C. and Slaney M. (1988) Principles of computerized tomographic imaging. IEEE Press, New York.

Box 2.o Reflectometry in painting diagnostics

Paintings are important and widespread works of art that need a considerable amount of scientific knowledge concerning both their historical knowledge and conservation issues. The study should in principle characterize completely the **present state of the art work**, in order to (1) understand and interpret its origin, (2) record its past history and modifications, (3) identify the active changes in the materials and their mechanisms and (4) define future interventions and preservation conditions.

In the investigation of paintings, we are dealing with materials that are heterogeneous at the nano-scale, applied differently in different parts of the work, and are very often undergoing different evolution and modifications depending on local conditions. A painting is essentially composed of: the support, the preparatory ground, the preparatory sketch (usually made by charcoal, graphite, black bone, hematite, etc.), an "imprimitura" transparent coverage of the drawing that is the support of the overlaying multiple painting layers and, finally, a covering protecting varnish (Leonardi 2005). Figure 2.o.1 shows a typical sequence of layers and materials composing the physical structure of a painting. The wall paintings and frescoes have a similar structure, given that the base support is the wall, and that the water-based colours are entrapped by chemical changes (carbonation) in the preparatory lime layer. There are two basic strategies for investigating a painted layer: **in situ analysis** using probes that analyse the layers to a certain depth, and **ex situ analysis** of micro-samples extracted through the different layers (Fig. 2.17). In situ non-invasive analysis is the obvious choice in most instances, and direct sampling is performed only when ambiguities remain from previous investigations.

A number of techniques are employed to image and analyse the paintings at different scales and depths. At one time or another, all of the techniques described in this volume have been used to investigate framed or wall paintings, though some are most often used because they have proven to be most effective from the point of view of: (a) **probed layer** (i.e. depth of penetration), (b) **speed of operation** (one shot mode or scan mode Table 2.7), (c) **portability and instrumental flexibility** (Box 2.r).

(*cont.*)

Box 2.o (*Continued*)

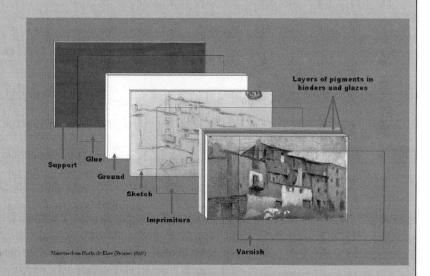

Fig. 2.o.1. Physical layers of different materials composing a painting. (From Leonardi 2005)

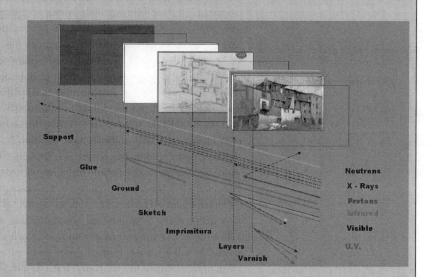

Fig. 2.o.2. Schematic representation of the different penetration depth and probed layers of techniques employing various radiation and particle sources. (From Leonardi 2005)

Figure 2.o.2 shows the penetration depth of different probes in a standard supported painting. It is important to understand that each technique probes different layers and provides unique and complementary information on the object.

- **Neutrons** pass right through the object, no matter how thick it is, and in doing so they activate detectable delayed radioactivity (Box 2.m). The use of neutron radiography and autoradiography is thus possible to produce element-selective images (Table 2.m.1) of extended areas or of the whole painting. It is used to map underlying layers of paint caused, for example, by the artist's corrections, re-used canvases or restorations.

- **High energy X-rays** may also be used to produce radiographies of the object. Image contrast is produced depending on the thickness and nature of the painting's support and paint layers. A few pigments containing heavy elements (Pb, Hg, Sb) create appreciable contrast, though in general X-ray radiography is best used for assessment of the state of the wooden frame and structural support (Fig. 2.o.3(b)).
- **Protons** have a very low penetration and are mainly used to produce secondary X-ray (PIXE) maps of elemental distributions in the surface layers. Since several chemical elements can be detected at each point of the scan, their quantitative evaluation provides for fairly straightforward identification of the pigment's composition. Modern portable XRF instruments can in principle produce very similar elemental

Fig. 2.o.3. Image of the painting "Christ carrying the cross" taken by (a) visible light, (b) X-ray radiation in transmission, (c) UV reflectometry and (d) IR reflectometry. (Musée d'art de Joliette, Quebec, Canada (© CCI, 2009).

(*cont.*)

Box 2.o (*Continued*)

maps, though at lower sensibility and precision, and portable LIBS probes are also very promising instrumentations.
- **Infrared radiation (IR)** is mostly backscattered by the gypsum or calcite ground layer, unless there is a highly absorbing material such as the graphitic carbon of the preparatory sketch. In this case the underlying drawing is highly visible in the IR reflectometric image. Quantitative analysis of the IR absorption bands (Box 2.e) is also very useful for identifying pigments and organic materials used as pigment binders, glues and restoration materials (Fig. 2.o.3(d)).
- **Visible radiation (Vis)** images commonly visualize the 2D distribution of the surface pigments absorbing in the light region. These are the most familiar pictures of art works (Fig. 2.o.3(a)), though careful measure and calibration of colour, for example during removal and cleaning of the outer varnish layer, is by no means a trivial task (Bacci 2004). Sometimes the use of monochromatic light or white light used in grazing incidence may help in the characterization of the surface layers.
- **Ultraviolet radiation (UV)** produces characteristic luminescence from surface polymer components, mainly used as coating varnishes and pigment binding agents. The light produced by UV irradiation depends on the polymerization process and thus on the age of the varnish, so that recent surface retouches and repainting are easily detectable (Fig. 2.o.3(c)).

There is a substantial amount of complementary information to be gained if all the techniques are applied to the same art object. However, the parameters constraining the investigation strategy may vary depending on the specificity of the art work or the costs. For example it is obvious that wall paintings and, very often, museum masterpieces cannot be moved to a large neutron or synchrotron facility. Curators are, for good reasons, increasingly reluctant to allow any kind of sampling or transportation of the objects. Neutron, proton, and synchrotron analyses therefore are ruled out for wall paintings and for objects that cannot be transported to reactors or accelerators. However some large museums have their own proton accelerators (i.e. the AGLAE facility at the Louvre) and several have internal laboratories with high energy X-ray sources for radiographic examinations (i.e. the Getty Museum in Los Angeles).

Inevitably however, the non-invasive in situ techniques are the ones that are most frequently adopted: the measurements can be performed at the storage location of the art work, thus minimizing handling and transportation risks, they require no sampling, and they provide plenty of useful information, as discussed before. UV-VIS-IR spectroscopies performed in reflectance mode using fibre optics technology are therefore widely used in the analysis and characterization of painted surfaces, including wall paintings (Bacci 2004, Hain *et al.* 2003, Mairinger 2004). These reflectometric techniques are increasingly used in parallel with LIBS and fibre-optics Raman spectroscopy (Edwards and de Faria 2004, Bruder *et al.* 2007) for a powerful combination of portable probes that are capable of chemical mapping and pigment identification. Illuminated manuscripts, parchments, painted statues, and many other art works may be studied with the same non-invasive techniques.

For in-depth information

Bacci M. (2000) UV-VIS-NIR, FT-IR, FORS spectroscopies. In: Ciliberto E. and Spoto G. (eds) (2000) Modern analytical methods in art and archaeometry. John Wiley & Sons, New York. Ch. 12.

Kortüm G. (1969) Reflectance spectroscopy. Springer-Verlag, Berlin-New York.

2.1.4 Imaging the past: visual reconstruction and analysis

The techniques for collecting digital image data, both two-dimensional and three-dimensional, that are suitable for analysis by powerful computer programs are nowadays widespread tools for everybody working in cultural heritage. One can laser scan objects up to the size of buildings and sites, or get the internal pictures of metre-sized objects by tomography in order to have the data for any subsequent **virtual rendering** of the scanned objects. Most data at present end up in enjoyable animations of the object, the building, or the site rotating and shifting under different light conditions on the screen while viewers are comfortably seated in front of the TV set or a laptop computer screen. The overall field of the application of **information technology** to cultural heritage is rapidly developing, so that we are talking about virtual archaeology (Barceló *et al.* 2000) and the fourth dimension in cultural heritage (Stadtarchäologie Wien 2004). However, we should cautiously distinguish between virtual 3D rendering of real data, modelling of the data using **virtual reconstruction** and **virtual reality**, and the use of advanced information technology for research and management of the cultural heritage.

Taking for granted that virtual aids are important tools for the research and management in the cultural heritage, I see at least three points that need to be properly addressed and clarified: (1) authenticity versus virtual reality. Are we looking at real objects or models? (2) access and standardization of large databases, which involves the integration of different kind of data, and (3) the use of virtual reconstruction in active research.

Authenticity. From the point of view of the end-user, one of the major issues in the digital world is in distinguishing the real data from the interpretation. Photographic images, laser scans of surfaces and 3D tomographic images are measured data that can be nicely visualized using advanced computer graphics and rendering tools. There is a whole field of research dealing with the production and treatment of these experimental data, which are designed to be stored in digital form for future reference and long-lasting preservation, rather than to promote **virtual access** to museum, collections, buildings and sites (Lahanier 2004, MacDonald 2006). There are practical problems related to light sources, colour accuracy, spatial resolution and image definition, relief detection and measurement, data storage, and so on. Assuming that satisfactory experimental data are available, then the problems to be faced during visualization are mostly those of realism (rendering, illumination, texture) and the realistic models used for graphical representation. Visualization in this context can be defined as the process of creating a *geometric representation* of the regularity present in the experimental data set: joining points with lines, fitting surfaces to lines, or "solidifying" connected surfaces (Gershon 1994). In this sense, we realize that taking a 2D digital photograph and producing a visual representation of the same surface from a laser scan or spectroscopic data are entirely different processes. In the second case, we are already dealing with mathematical models of the data and while, on one hand, the visualized reconstructed picture usually contains information that is invisible in the simple photographic image, on the other hand this is a **model** of reality, showing a conceptual interpretation, or one of the possible theoretical projections of the model (Barceló 2001).

Figures. 2.11–2.13 show an ideal and exciting case of virtual reconstruction: the modelling of the full set of teeth of a Middle Miocene hominoid from Thailand (Chaimanee *et al.* 2003) based on advanced synchrotron 3D tomographic data. The figures show the photographs of the real findings (**2D-images**: Fig. 2.11), the 3D rendering of the tomographic scans performed on the fragments by synchrotron radiation tomography (**3D visualization**: Fig. 2.12), and the final reconstruction of the denture by virtual modelling of the missing teeth and parts based on the available information (**virtual reconstruction**: Fig. 2.13). In this near-ideal case study, a virtual reconstruction has been masterfully used to create a reliable model of the full denture, where the starting objects and the virtual part are mixed but totally distinguishable thanks to the different rendering colours. In other instances, it may be much more difficult to separate the reconstructed parts from the real evidence. Of course, we all know from intuition that if the hard evidence is abundant and well preserved,

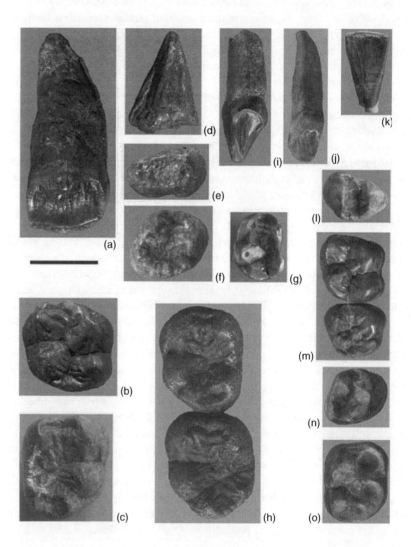

Fig 2.11. Photographs of the fragments of teeth of a Middle Miocene hominoid from Thailand. (From Chaimanee *et al.* 2003)

2.1 *Available techniques* 81

Fig. 2.12. 3D computerized tomography scans of teeth fragments shown in Fig. 2.11. The tomographic data were collected at beamline ID-19 of ESRF, Grenoble. (Courtesy of P. Tafforeau, ESRF)

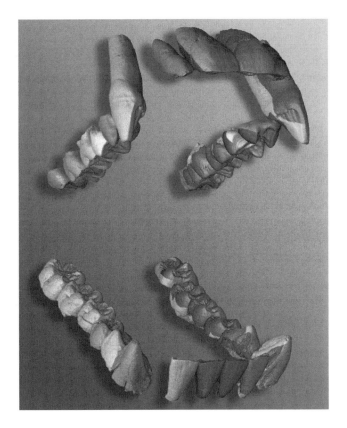

Fig. 2.13. Fully reconstructed denture of the hominoid starting from the virtual 3D CT scans shown in Fig. 2.12 (pale grey), and modelling the missing teeth on the basis of the available information. (Courtesy of P. Tafforeau, ESRF)

then the final reconstruction is reliable, whereas if the model is based on scarce and poorly preserved evidence it will be highly subjective and possibly unreliable. This is a part of the normal deductive process in science. However, in the specific case of a virtual reconstruction there is an additional pitfall nested in the realism used to technically render the images (see, for example, Van Gool et al. 2004): the more realistic they are, the more difficult it is to discriminate the model from the reality. The issue is the subject of lively debates (Barceló 2002, Kantner 2000), and it ought to be seriously considered when defining the philosophy and aims of the virtual reconstruction process.

Databases. The step beyond reconstruction is **virtual reality** (VR), defined as an environment where the human operator is transported into an interactive environment. When VR comes into play, the models are further complicated by the multiple information layers that are attached to the images, because the objects should be described by more than the simple shape or appearance, and additional input should be provided by computer-based information and databases. The aim of using advanced information technology in cultural heritage is not that of producing nice images, but rather that of providing powerful tools for "visualizing and understanding the complex, detailed, dynamic, multisensory history that is our shared cultural patrimony" (Sanders 2001). Virtual reality used in cultural heritage must quickly add robust information content in order to be a satisfactory tool for the current needs of all activities involved in cultural heritage: communication, education, research, and management. IT people are trying to move on, developing tools such as augmented reality environments, explanatory dynamic environments, knowledge management, and so on. The future outlook is related to the so-called **virtual cultural heritage** (Frischer et al. 2002). The process is clearly connected to ever-increasing computer power and more sophisticated software tools, though they must necessarily be based on reliable, flexible, and friendly databases.

The crucial importance of extended experimental datasets is sometimes not fully appreciated. There is a general belief that, exactly as we see in TV fiction for murder investigations, one needs just a tiny amount of evidence (a chemical analysis, a molecular spectrum, a diffraction pattern, a few isotope ratios,...) to solve any case, no matter how complicated the problem is. As seen on television, you push the button and the computer always cranks the right solution out, even presenting it in pretty images and colours. Few people realize the huge amount of data that must be available to allow appropriate comparison, checking and interpretation of every single experimental number, assuming that the algorithms and deduction procedures are well determined and correct. The analogy between forensic science and archaeometry is actually appropriate: scarce evidence, multiparameter space, and few clues. In addition, in most cases the databases that we need for comparison are simply non-existent. The vast majority of the art works and the archaeological finds in museums and collection has never been properly analysed. Many of the (few) existing analyses have been obtained with non-standardized or obsolete techniques and the published data are often not comparable to those we may presently collect using many of the techniques described in this book. Almost every laboratory routinely working in cultural heritage has its own limited dataset of reference compounds of the materials commonly analysed. Interlaboratory calibrations are uncommon, expensive, and labour intensive. Experimental databases and calibration protocols in archaeometry and conservation science are to be a major issue if we seek a future virtual reality environment that is also "real", not only "virtual".

Research. Scientists are curious people, and to them the most appealing side of an object (a building, a site, a landscape) is not the measured visible part, but the missing part. In practical terms this means understanding the complete shape and function of incomplete or puzzling artefacts, interpreting past manufacturing processes, inserting the object into its living context, and exploring the minds of the people who developed and used the object itself. Virtual reality is a great tool for such a challenging job though, as discussed, it must be firmly based on reliable data and models. Several areas of scientific investigations have begun to take full advantage of information technology at several stages of the process.

At the level of data acquisition, for example, archaeological excavations extensively use referenced systems and GIS-based technologies (Box 2.p) for integrating geographical, geological, archaeological, and geomorphological data. Not only is the information content of the excavation record increased for all subsequent analyses, but the data are readily usable and compatible for a number of large scale applications such as ancient landscape reconstruction, urban planning, excavation management, etc. Museums mainly use advanced information technology for enhancing the views of heritage, for rationalizing inventories and improving virtual access to the collections, and to assess the state of the stored items for conservation planning (MacDonalds 2006). Digitized collections databases will undoubtedly also prove extremely useful to cultural heritage management and research programmes, though to be really useful they have to be cross-linked with multi-level information, and sooner or later they have to include the time evolution (of the object, of the sites, of the cultures) among the accessible models. To adopt and extend a concept about GIS put forward a while ago by Lock and Harris (1997), we might say that "...the final stage is the development of an integrated management system in which virtual reality becomes just one component within an information system with modelling, decision support, and analytical capabilities".

As far as specific conservation issues are concerned, the image data obtained by several techniques (reflectography, radiography, tomography) are of course fundamental to all diagnostic investigations. They allow preliminary assessment of the degraded parts of the object in need of intervention, the distribution of matter at the surface or inside the object, and the locations where samples are eventually to be taken for subsequent analyses of the materials. There are also a few rather innovative and specialized applications of virtual reconstruction techniques to conservation problems: the jigsaw puzzle-like computer-aided recombination of fragmented objects such as shattered pottery or detached frescoes, and the virtual modelling of the original appearance of faded colours in paintings. Both applications are described in detail in Box 2.q.

Box 2.p Wide area survey and remote sensing

In Section 2.1 we crudely grouped together all objects and measurements of matter above the millimetre as "macroscopic", from the point of view of materials science. Of course most of the perceived physical reality in the cultural heritage is in this class, and it is obvious that different techniques are to be used for different large-scale applications. Following the line of thinking already introduced, we may try to explain the use and application of techniques at the macroscopic level on the basis of the size of the object. In fact, at the centimetre to metre size (10^{-2}–10^0 m) common to art and archaeological objects, we still use most of the techniques that have been introduced for materials analysis: 3D digitization technologies for surface mapping (Pieraccini *et al.* 2001), reflectography, radiography or tomography to penetrate within the object (Boxes 2.m, 2.n, 2.o). As we move up in size, not only is the object of the investigation larger, but the questions and the focus of the investigations are different, though in the end we are still concerned with measuring the heterogeneous distribution of matter within the body.

When we are dealing with objects in the size range of tens to hundreds of metres (10^1–10^2 m), we are interested in archaeological structures and sites (graves, walls, fireplaces, etc.) or architectural buildings (houses, temples, streets, channels, etc.). The main problems at this level are: (a) I do not know where the objects are. Can we find them? (i.e. **intrasite survey and mapping**) or (b) I do know where each object is, but what is its extension and conservation state? (i.e. **structure diagnostics**). At an even larger scale (> 10^2 m) we get into the field of wide area prospection (i.e. **intersite survey and mapping**). This is used not only for classical archaeological prospection (Renfrew and Bahn 2008, Chapter 3; Gaffney 2008), but also for urban mapping, land organization and exploitation, palaeo-landscape reconstruction, interpretation of regional resources and trades, cultural heritage management and safeguard, maps of archaeological risk for modern urbanistics, and so on. Problems as such are very timely and urgent, though they are quite beyond materials science and the scope of this volume. However, a few techniques introduced in the last twenty years or so are rapidly becoming fundamental in many fields related to cultural heritage. This brief description is meant as a general introduction to their application potential.

At the level of **intrasite survey**, a number of consolidated **geophysical techniques** are widely used in the ground survey of archaeological sites (Leute 1987, Chapter 3 of Renfrew and Bahn 2008). They are **electrical resistance** surveying, **magnetic field gradient** surveying and **ground-penetrating radar** (**GPR**) surveying. Acoustic, seismic, and chemical prospection are much less used (Craddock and Hughes 1985). Leute (1987) provides a basic introduction to the archaeological applications of standard resistivity and magnetometric measurements: electrical resistance methods are especially useful in detecting underground stone structures or sand/gravel filled pits (high-resistivity positive anomalies) and trenches containing organically enriched fill, clays, and high salinity soils (negative anomalies). Local deviations of the Earth's magnetic field measured by magnetometry and caused by the archaeological record are due to the remnant magnetization and the magnetic susceptibility of buried materials. Both mechanisms are dependent on the presence of iron, commonly in the form of iron oxides in metal objects, soils, ceramic shards, and fire structures. Besides Fe-containing metals, most of the observed underground features with increased magnetic field values are referred to buried features containing fired materials (hearths, fire-altered soils, bricks, furnaces) because the remnant magnetization originates during cooling of the iron oxide phases through the Curie temperature, which is in the range 565–675 °C for common Fe-oxides.

Ground-penetrating radar (Conyers 2005) employs short polarized radar pulses (i.e. electromagnetic radiation in the microwave UHF–VHF region, Table 2.1) to image the subsurface by detecting the reflected signals from buried structures and interfaces (Fig. 2.p.1). When the emitted wave hits a boundary related to a change in the dielectric constant, the reflected return signal varies, exactly as in seismology. GPR can be used in a variety of media, including rock, soil, ice, fresh water and architectural structures. The architectural investigations carried out at the amazing Basilica di Collemaggio, L'Aquila, Italy by Ranalli and co-workers (2004) is a fairly complete example of the potentials of GPR in architectural engineering. It can detect objects, changes in material, voids and fractures. The depth range of GPR is limited by the electrical conductivity of the ground, and by the frequency used. Higher

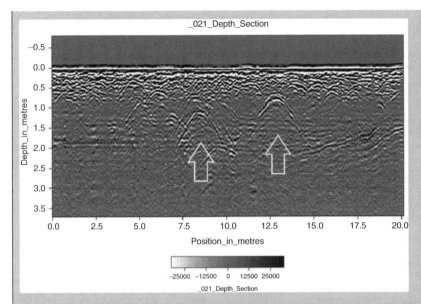

Fig. 2.p.1. GPR depth section (profile) collected at a historic cemetery in Alabama, USA. Arrows indicate very distinct reflections, probably associated with human burials. "Hyperbolic" reflections, appearing as an inverted U, are typically associated with discrete objects. The many smaller reflections near the surface are likely to be caused by tree roots. Because of the nature of the site, subsurface testing was not conducted, but these interpretations are supported by surface indications (grave markers and depressions) and with line-to-line patterning within the GPR data. (Image available at: http://en.wikipedia.org/wiki/Image:LINE21.jpg)

frequencies penetrate less, but yield better resolution. Optimal depth penetration is achieved in dry sandy soils or compact dry materials such as igneous or sedimentary rocks, and concrete where the depth of penetration could be up to 15 m. In moist and/or clay-laden soils and soils with high electrical conductivity, penetration is sometimes only a few centimetres. Recent developments in GPR include 3D tomographic data acquisition (Forte and Pipan 2008), and optimized signal-to-noise ratio by the use of multi-offset data to image complex environments (Booth et al. 2008, Gaffney 2008).

With regard to historic buildings and architectural structures, several techniques are available for the diagnosis of the structure stability and health at different scales (Binda 2005). Some of them are described elsewhere in the volume (Box 3.b), and only two basic survey techniques are recalled here: **photogrammetry** and **thermographic testing**.

Photogrammetry in a sense is the first remote sensing technology ever developed, being as old as modern photography itself. In photogrammetry the geometric properties of objects are determined from photographic images (McGlone et al. 2004). The 2D geometrical information contained in each image (distances and angles) are used to reconstruct the shape and the metric of the object by projective geometry, triangulation, stereoscopy, bundle adjustment, and a number of more sophisticated algorithms. The use of multiple images is warranted by a set of common points. Photogrammetry and Light Detection and Ranging (**LiDAR,** also called Laser Detection and Ranging **LaDAR**) data complement each other, insofar as photogrammetry is more accurate in the x- and y-direction while LiDAR is more accurate in the z-direction. A 3D visualization can be created by georeferencing the aerial photos and LiDAR data in the same reference frame, orthorectifying the aerial photos, and then projecting the images on top of the LiDAR grid. Photogrammetry has been widely used in the rapid measurement of buildings and in the topographic mapping of areas and large archaeological sites. Extreme and fascinating applications include the merging of live actions and computer-created images in movie post-production and, more to the point, the virtual reconstruction of disappeared art and archaeological artefacts from the available 2D images. The best example is the virtual reconstruction of the Bamiyan, Afghanistan, Buddha statues destroyed by the Taliban militia in 2001: the ETH group employed advanced photogrammetric techniques to produce visual reconstructions of the statues to guide the restoration (www.photogrammetry.ethz.ch/research/bamiyan/index.html).

(cont.)

Box 2.p (*Continued*)

Fig. 2.p.2. The large statue of the Buddha in Bamiyan, Afghanistan, before the destruction in 2001 (left), and its photogrammetric reconstruction by the ETH group (right) (Gruen *et al.* 2002). (Courtesy of F. Remondino, ETH Zurich)

Thermographic testing (or infra-red thermography IRT), although mostly known for breast medical screening, is especially important in the diagnostics of buildings and large art works (Maldague 1993, 2001; Ludwig 2004; Avdelidis and Moropoulou 2004). In industry it is mainly used for the visualization of the heat flow and energy leaks out of modern buildings. Although weakly penetrating (a few centimetres) the technique is easy to use and yields infrared images of thermal anomalies in walls and architectural structures that are related to internal discontinuities in the structure or in the nature and properties of the constituting materials, or to surface features such as restorations, alterations, salt deposits, and biotic infestations. Thermographic images are therefore extremely useful to model water and heat diffusion in the structures, and to diagnose detachments and deteriorations in plasters and frescoes. In some instances infrared thermal imaging has also been used in a survey of kilometre-sized areas, in order to detect and map large-scale structures (Scollar *et al.* 1990).

However the most used techniques for wide area survey are **aerial photography** (or better **aerophotogrammetry**), and **satellite imagery**. The advantages of gaining a good aerial view of the ground had been long appreciated by archaeologists as a means of obtaining an overview of the site context. Views of archaeological sites from the sky are always breathtaking (Gerster and Trumpler 2005). Early investigators attempted to gain birds-eye views of sites using hot air balloons, scaffolds or even cameras attached to kites, though of course it was the airplane and its military use during the First and Second World Wars that put the archaeologists in a position to use the technique to discover and record archaeological sites. **Aerial photography**, if used for quantitative mapping, is just a long-distance extension of plane or stereoscopic photogrammetry, so that the processing and treatment of the images is performed using photogrammetric algorithms. However aerial photography is largely used for wide-area surveys to search for clues indicating the presence and location of archaeological sites, which

Fig. 2.p.3. Thermographic image of a church wall (Santa Maria Incoronata, Martinengo, Bergamo, Italy) showing, through the heat flow, architectural structures hidden in the chapel wall. (Courtesy of N. G. Ludwig, University of Milano)

are more visible from the air than from the ground. Some of them are only visible from the sky, for example the notorious geoglyphs located in the Pampa de Jumana, near the towns of Nazca and Palpa, Peru, which are also the subject of an extended photogrammetry-based reconstruction programme (Sauerbier *et al.* 2006; see also: www.photogrammetry.ethz.ch/research/peru/index.html). A number of subtle physical effects make the archaeological structures detectable in the images, such as shadow marks (cast on the ground when the Sun is low on the horizon), cropmarks (due to differential growth of the crops in the presence of buried features and soil anomalies), frostmarks (due to the abnormal accumulation of snow and frost along land depressions), and soil marks (due to the different colour of anthropogenically affected soils). Sometimes particular climatic conditions such as heavy rains or drought make the soil marks more evident (Fig. 1.6).

Satellite imagery may be the ultimate resource for global surveying, as artificial satellites in orbit sweep the whole of the Earth's surface and systematically record images at different wavelengths. Depending on the type of

(*cont.*)

Box 2.p (*Continued*)

sensors and the orbit's altitude from the Earth's surface, each satellite yields images at different spectral and geometric resolution. The spectral resolution depends on the recorded energy band and on the bit-depth of the sensors. Geometric resolution (expressed in **ground sample distance**, or GSD) reflects the satellite's ability to image an area effectively in a single pixel. Typical Landsat Thematic Mapper (landsat.org) GSD is about 30 m (Fig 2.p.4, left), where each pixel contains the average information of a $30 \times 30\,m^2$ area on the ground. This is largely insufficient to detect small archaeological structures. Other satellites, including for example IKONOS (Fig. 2.p.4, right), CORONA, and especially QuickBird produced images at much higher resolution, down to a GSD of less than 1 m (Table 2.p.1). The last commercial satellite (GeoEye 1, www.geoeye.com), launched in 2008 has an amazing resolution in the range 0.4–0.5 m, i.e. one single pixel of a Landsat image is comparable to 3600 pixels of a GeoEye image. As Fig 2.p.4 shows, images at metre resolutions are sufficient to extract useful archaeological information (see for example Sever and Irwin 2003, Challis *et al.* 2002–2004, Goossens *et al.* 2006). When the resolution is

Fig. 2.p.4. Comparison of two false-colour space-based images captured by two commercial Earth-observation satellites (Landsat TM, left, and IKONOS, right) depicting the ancient ruins of Tikal, a Maya city deep in the Guatemalan rainforest. The Landsat imaging system has a nominal resolution of 30 metres, while the IKONOS can capture a nominal resolution as close as 1 metre, a scale at which individual pyramids, pathways and small structures become apparent. (Image available at the NASA Marshall Space Flight Center website: http://www.nasa.gov/centers/marshall/multimedia/photos/2006/photos06-018.html)

Table 2.p.1. Spatial and spectral resolution of satellite images available from historical launches (modified from Challis *et al.* 2002–2004)

Sensor	Spatial resolution (m)	Spectral range (μm)	Cost (US$) per scene	Scene size (km)
AVHRR	1100	5 bands, 0.58–12.5	100	3000×1500
Landsat MSS	80	4 bands, 0.5–1.1	425	170×185
Landsat TM	30 (120 m band 6)	7 bands, 0.45–2.35	1500	70×185
Landsat ETM +	15	Pan (band 8), 0.52–0.90	1500	170×185
SPOT XS	20	3 bands, 0.5–0.89	1250	60×60
SPOT PAN	10	1 band, 0.51–0.73	1250	60×60
IKONOS	1	4 bands, 0.45–0.88	1210	11×11
GeoEye-1	0.41	4 bands, 0.45–0.90	2888	15.2×15.2
QuickBird Pan	0.61–0.72	1 band 0.45–0.90	6120	16.5×16.5
QuickBird MS	2.44–2.88	4 bands 0.45–0.90	6800	16.5×16.5
CORONA	2 +	Single photographic image	18	17×232

not enough, the satellite imagery is sometimes supplemented with aerial photography, which has a higher GSD resolution, though in general it is more expensive per square metre, and the multi-spectral character of the satellite is exploited and combined with detailed topographical information to produce thematic maps of the territory. Almost twenty years ago, Cox (1992), for example, showed how carefully interpreted Landsat TM satellite images could significantly contribute to extending archaeological knowledge about the wetland areas of Cumbria, UK and Sever and Irwin (2003) lead the way for a number of studies (Saturno *et al.* 2007, Garrison *et al.* 2008) that successfully located a number of Maya sites, roadways, and canals in the Guatemalan forest, thus contributing to the reconstruction and interpretation of Mayan land use. Fowler (2002) recently concluded that "satellite imagery, although not a substitute for conventional aerial photography, represents a complementary source of information when prospecting for archaeological features. In a regional context, low resolution multispectral imagery can be used for the prospection for areas of high archaeological potential through the use of image processing and modelling techniques and, together with medium resolution imagery, can be used to prepare base maps of regions for which up to date mapping is not available. High-resolution imagery, together with conventional aerial photographs, can be used subsequently to detect and map archaeological features."

The combination of digital datasets of different natures, such as satellite images, aerial photographs and manually recorded reference or survey points on the ground (by **global positioning systems**, or **GPS**), requires that the images are spatially synchronized using a common reference system (**georeferencing**, Hill 2006). This combination of raster and vectorial images is performed by **geographical information systems (GIS)** (see for example: Wheatley and Gillings 2002, DeMers 2004). A GIS system is any advanced information system that is capable of integrating, storing, editing, analysing, sharing, and displaying **geographically referenced information**. In general GIS applications are tools that allow users to create interactive queries, analyse spatial information, edit data, maps, and present the results of all these operations. The manipulation and use of digital geographic data through GIS software is nowadays routine in modern archaeology and conservation science (Karman and Knowles 2002). The introduction and rapid diffusion is undoubtedly related to the historical and practical importance of the use of maps and plans in everyday practice, besides the very attractive power to visualize spatial information at any scale. In the words of Lock and Harris (1997) "while many other computer-based and quantitative methods remain conceptually remote to the majority of archaeologists, even elementary GIS analysis and especially output in the form of maps is of immediate relevance". However, caution is needed because one should always distinguish between data, tools, and models. GIS is a powerful tool, but needs precise and carefully collected data in order to deliver useful and reliable models. This is even more so when GIS systems are used to map landscape and site variations through time (Siart *et al.* 2008). Hot issues in the field include the use of four-dimensional GIS (3D plus time), object-oriented GIS, the relationship between GIS and 'Virtual Reality' technologies, and the integration of GIS with distributed systems and the Internet (Lock 2000).

In summary, wide area surveys at any scale and location of surface and underground structures is fundamental for the knowledge, the conservation and proper management of cultural heritage. The time has come to fully exploit the available tools for advanced data integration and analysis.

For in-depth information

Campbell J.D. (2008) Introduction to remote sensing. 4th Edition. The Guildford Press, New York.

Lock G. (ed) (2000) Beyond the map: Archaeology and spatial technologies. IOS Press, Amsterdam.

Parcak S.H. (2009) Satellite remote sensing for archaeology. Routledge, Taylor and Francis Group, Abingdon, UK–New York, US.

Wiseman J.R. and El-Baz F. (eds) (2007) Remote sensing and archaeology. Springer-Verlag, Berlin-New York.

Witten A.J. (2006) Handbook of geophysics and archaeology. Equinox Publishing Ltd. London.

Box 2.q Virtual reality in cultural heritage

In recent years, there has been fervent activity in the acquisition, processing, storage, transmission, representation, analysis and presentation of cultural-heritage-related image data. International efforts in these areas have produced promising results from both high-visibility projects and an ever-growing number of local initiatives. The purposes of individual projects are highly varied and may include material analysis (dating and provenance determination), discovery and interpretation of ancient technologies, examination of details (tool marks, brush strokes, craquelure), enhanced knowledge of conservation materials and processes, and cultural heritage dissemination and presentation. In all cases, digital signal processing is the foundation upon which these applications must rely in order to deliver both optimal and highly repeatable results (Barni *et al*. 2008).

Major issues concerning the application of information technology to cultural heritage problems include data acquisition and processing (MacDonald 2006, Remondino *et al*. 2008), virtual reconstruction and modelling (Barceló 2001, 2002) and the insertion of virtual reality techniques and methods into research and knowledge management (Davenport and Prusak 1998). While theoretical issues are briefly discussed in Section 2.1.4, we present below a few outstanding examples of recent achievements made by information technology into active research in archaeology and conservation. The selected examples are meant to give just a glimpse of the infinite possibilities that computer technology is offering as new tools stem from the direct interplay between mathematical modelling of images and complex applications.

The first case study concerns the **computer assisted restoration** of the Mantegna frescoes of the Ovetari Chapel in the Chiesa degli Eremitani in Padova, Italy (Fig. 2.q.1). On March 11 1944, the church was destroyed by allied bombing along with the inestimable frescoes by Andrea Mantegna and his school. In the following 60 years, several attempts were made to restore the recovered fragments by traditional methods, with little success. Then in 1994 the Soprintendenza ai Beni Storici ed Artistici di Venezia launched an advanced restoration project (Cazzato *et al*. 2006, http://www.progettomantegna.it), which started with the digital classification of the 72 500 fragments and then proceeded through a collaboration with the University of Padova with the development of innovative and efficient pattern recognition algorithms to map the original position and orientation of the fragments in the wall fresco. The theoretical and practical challenges were enormous, given that the available fragments covered merely about 10% of the original area, and that the only usable images of the whole representation were low resolution black and white photographs taken in 1900 and 1920. Virtual localization of the fragile fragments meant accurate planning of the final restoration and realistic evaluation of how much of the fresco would be visible upon completion, without the need for extensive and dangerous manipulation. The advanced mathematical and computational techniques developed for the task are reported in technical papers (Fornasier and Toniolo 2005, Fornasier 2006) and included solutions to problems such as the fuzzy information of the fragment's boundaries, the mismatch between the grey-converted shades of the colour digital images and the greyscale of the fresco's photographs, the lack of a priori knowledge of the image orientation, and many other problems. The results are remarkable (Fig. 2.q.2) and not only provided crucial information for the actual restoration, which was concluded in 2007, but also showed the way for future and even more challenging tasks, such as the reconstruction of the Giotto's and Cimabue's frescoes in Assisi, which were destroyed by the 1997 earthquake.

Automatic reassembling of fractured objects

Similar objectives are faced by a number of recent applications involving automatic systems that are capable of aiding archaeologists in the painstaking job of reconstructing shattered artefacts, such as ceramics and statues. When the totality of the fragments are recovered from the excavation and they all belong to one object (for example a pot from a grave) the task is relatively easy. A more realistic situation involves partial recovery of fragments from an unknown number of different objects (for example a layer of discarded broken pots from a pottery workshop).

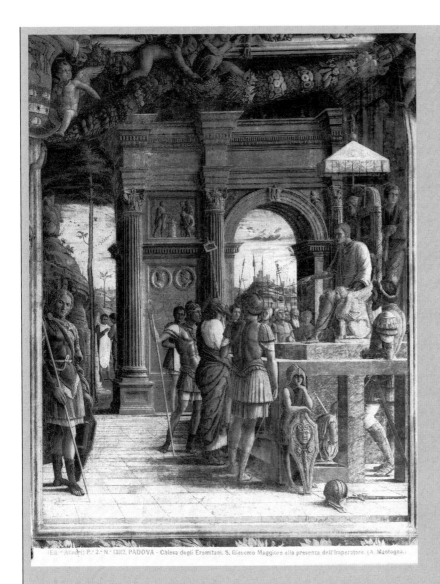

Fig. 2.q.1. Andrea Mantegna: "San Giacomo in front of Erode", Cappella Ovetari, Eremitani Church, Padova (1452). 1900 photograph taken before the destruction in 1944. (© Archivi Alinari, Firenze)

Ideally, therefore, the automatic system should simultaneously solve an unknown number of multiple puzzles where all the puzzle pieces are mixed together in an unorganized pile and each puzzle may be missing an unknown number of pieces (Willis and Cooper 2008). This apparently unsolvable problem has attracted a lot of active research in the recent past, partly triggered by the availability of high-definition 3D **shape acquisition systems** (i.e. laser scan, tomography, Table 2.7) for the rapid and precise measurement of the fragment's morphology. The models needed for solving these problems are new and challenging, and most involve 3D space, an area that is largely unexplored by the signal processing community. The mathematical and computational details of the algorithms developed to perform the task (Fig. 2.q.3 and 2.q.4) are well beyond the scope of this volume, and the state of the art is nicely described by Willis and Cooper in the cited review (2008). Here it is sufficient to note that most of the developed algorithms are based on pure geometry-matching strategies, and there is ample room for improvement by inserting

(*cont.*)

Box 2.q (*Continued*)

Fig. 2.q.2. Computer-based reconstruction using efficient pattern matching techniques (Fornasier 2006).

additional matching constrains such as 2D images painted on the surface of 3D objects (i.e. drawings, graffiti, and decorations of ceramics) or an overall symmetry of the object, such as the 17 crystallographic plane space groups (for tiles or mosaic floors) or the axi-rotational symmetry (for wheel pots). However, even if the developed available automatic systems have not yet become simple everyday tools for archaeologists, restorers and researchers, they are already at the stage of providing interesting results, as shown by the outcome of the Forma Urbis Romae project at the University of Stanford (Koller *et al.* 2006, http://formaurbis.stanford.edu), where expert users have discovered 20 new matches for the reconstruction problem of the 1186 pieces that have been studied for hundreds of years (Koller and Levoy 2006).

A similar problem involves the laser scanning and virtual rendering of thousands of prehistoric clay tablets (Section 3.1.2) incised with cuneiform inscriptions (Anderson and Levoy 2002, Kumar *et al.* 2003). The large-scale project (The Cuneiform Digital Library Initiative: http://cdli.ucla.edu/index.html) has the multiple functions of saving a fragile clay archive that is slowly degrading, and at the same time making the inscriptions virtually accessible for every scholar.

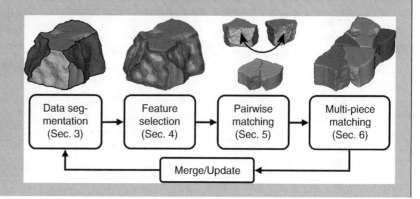

Fig. 2.q.3. The generic process of object reconstruction. It consists of four steps (from left to right): (a) classify the boundaries of interest for matching, (b) compute a shape model for these boundaries in terms of features, (c) search for correct pairwise matches of the computed features, and (d) implement a merge and update procedure that merges pairwise matches into larger multipiece configurations (Huang *et al.* 2006).
Original colour picture reproduced as plate 13.

Colour fading

Another totally different application of information technology is related to the investigation of colour fading. The change and disappearance of paint colour on art objects is a rather complex process that is affected by the type and physical state of the pigment, the type of paint application, the storage conditions and, of course, the lighting conditions. Fading due to light exposure is a serious concern for the long-term preservation of coloured art objects, books, archival materials, watercolours and woodblock prints, textiles and photographs, and all the objects that have been observed to be prone to fading damage (**light-sensitive materials,** see Section 2.5.2.2). The problem is generally faced in designing appropriate lighting for galleries, in the definition of acceptable light exposure doses for specific kind of sensitive material, and in exhibition rotation schedules for sensitive objects. However most curatorial prescriptions and strategies are based on the empirical and somehow subjective observation of past colour changes on the same kind of material. The introduction of digital imaging with careful calibration and representation of colour (Lossau and Liebetruth 2000, Berns 2001) has made the measurement of changes over time more quantitative, though it is the direct measurement of lightfastness by the use of a non-invasive **micro-fading tester** (Whitmore 2002) that has really made possible the link between light exposure and fading for each type of light-sensitive material. The instrument, initially introduced to measure the light sensitivity of materials for exhibition management and planning, can actually provide the spectral curves of colour fading versus the time for controlled light doses. By using these calibration curves and defining the lighting exposure conditions of the pigmented material, we can virtually extrapolate the changes *into the future*: we can simulate the fading of the colour of the art work during an exhibition period (Morris and Whitmore 2007) or *in the past*, i.e. reconstruct the original colour of faded objects. Virtual reality will then show the evolution of the image colours in time.

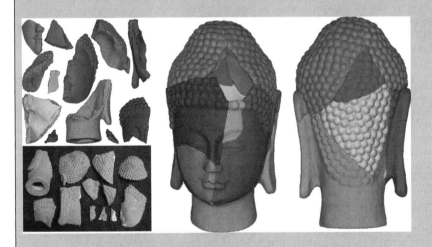

Fig. 2.q.4. Reassembling a fractured head model (Huang *et al*. 2006).
Original colour picture reproduced as plate 14.

For in-depth information

Barceló J.A., Forte M. and Sanders D.H. (2000) Virtual reality in archaeology. British Archaeological Reports, International Series no. 843. ArcheoPress, Oxford.

Willis A.R. and Cooper D.B. (2008) Computational reconstruction of ancient artifacts. Signal Processing Magazine 25, 65–83.

2.2 Sampling problems: invasive, micro-invasive, and non-invasive methods

Sampling is a key issue in determining the success of a scientific analysis. No matter how sophisticated a technique is and how optimized is the analytical protocol, if the samples are inappropriate or inadequate, the results will be meaningless. Among scientists, this concept is often summarized in the phrase "garbage in, garbage out". All possible care should be taken in selecting the number, the volume, and the position of the samples to be measured in order for them to be significant. Each of these parameters needs specific discussion.

Furthermore, sampling as the physical action of extracting a part of the material for the analysis that is often to be completely destroyed during the process, is somehow contradictory with the very same philosophical concept of conservation (Pallot-Frossard *et al.* 2007). It should be generally appreciated that in the investigation of cultural heritage any unnecessary damage must be avoided. Sampling practices such as those witnessed in the past on many museum objects (Fig. 2.14) are of course totally unacceptable by modern standards because of the requirements of the analytical techniques. Sometimes sampling is not possible at all because of the uniqueness, the fragile state, or the aesthetics of the object.

Possible solutions reside in micro-sampling or in the use of non-invasive techniques. **Micro-sampling** is often possible because of the high sensitivity of modern analytical techniques. For example, accelerator-based or laser-ablation mass spectrometry and several of the analytical techniques based on micro-beams require orders of magnitude less of the sample (in volume or mass) than was required in the past. X-ray diffraction can be performed in the laboratory on a few microgrammes of inorganic powder material. State-of-the-art IR techniques based on diffuse reflectance or through the microscope may

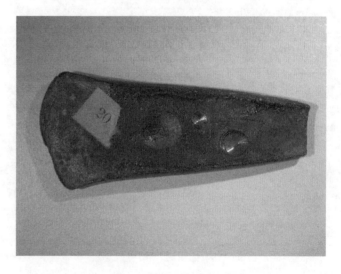

Fig. 2.14. Example of extensive sampling performed in the last century for chemical analyses on a prehistoric copper axe. Invasive sampling as such is of course totally unacceptable by modern standards. Museo Gaetano Chierici di Paletnologia, Remedello, Italy.

operate on single grains or a milligramme-level amount of powder. However micro-sampling, besides being as least non-destructive as possible, must also be representative, and representativity needs the object size and homogeneity to be carefully assessed. The scale of the sample's heterogeneity and the amount of probed sample critically affect the possibility of comparing data obtained by different techniques. For example, it may be problematic to compare the results of bulk analytical techniques such as XRF, AAS or OES to elemental data obtained by modern micro-beam techniques such as PIXE or EPMA.

Non-invasive techniques are all those techniques that provide some kind of examination of the sample without extraction and consumption of material. Depending on the particle or the energy of the beam and the physical process detected, some non-invasive techniques only probe the surface of the sample, whereas others induce and detect a signal at depth within the material and can effectively be considered to be bulk techniques. However, one should be rather cautious in using surface non-invasive techniques, which usually probe the very outer layers of the materials. The problem is not related to the technique, for which the penetration depth of analysis is well known or measurable, but rather to the material itself, which is often altered or modified at the surface so that the analytical results may prove totally misleading with respect to the true composition of the object. Typical examples are glasses (Fig. 2.15) that are frequently depleted of alkali elements at the surface, and metals that often have compositional changes at the surface with respect to the bulk due to manufacturing techniques (tin enrichment of bronzes) or to the electrochemical re-precipitation during burial (copper enrichment of brasses and bronzes).

When incident beam techniques are employed to probe materials at depth, it is important to take into account: (a) the penetration of the beam into the material (**penetration depth**), which depends on the energy and angle of incidence

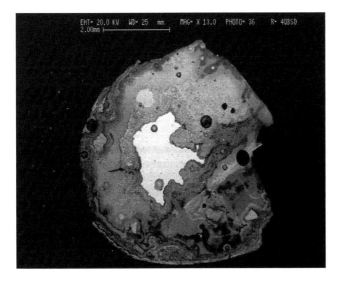

Fig. 2.15. Backscattered electron SEM image of a sectioned Bronze Age glass bead, showing a core of unaltered glass and a thick outer layer of alkali-depleted glass. Surface non-invasive analysis of this specific object would produce highly misleading compositional results. (From Artioli 2004)

Table 2.8. Mean penetration depth of different electromagnetic or particle beams into materials of different nature and composition

Material (ρ in g/cm³)	Soft X-rays (1 keV)	Hard X-rays (30 keV)	Thermal neutrons (25.3 meV)	Protons (2 MeV)	Electrons (30 keV)
Fe (7.87)	0.1 μm	1.6 mm	8.3 mm	19.2 μm	2.1 μm
Quartz (2.64)	1.2 μm	4.9 mm	3.5 cm	39.2 μm	6.2 μm
Bone (1.80)	1.3 μm	2.7 mm	2.8 cm	46.5 μm	9.1 μm
Polypropylene (0.90)	5.9 μm	4.7 cm	1.5 mm	74.6 μm	18.3 μm
Wood (0.59)	5.2 μm	6.1 cm	5.2 cm	144.7 μm	27.9 μm

of the incoming beam, and on the linear absorption coefficient of the material (i.e. its elemental composition), (b) the escape depth of the re-emitted signal, which is often re-absorbed by the material itself before reaching the surface for detection, and (c) the geometry of the detection system. Such parameters can be theoretically calculated or experimentally measured and calibrated. In general, techniques employing electrons and electromagnetic radiation softer than X-rays as incident beams sample a very thin layer of material. Techniques employing neutrons and high energy electromagnetic radiation (X-rays and γ-rays) are intrinsically sampling thicker layers of materials, although the signal produced at depth cannot necessarily be detected. Proton beams penetrate little into the material, and the penetration depth depends on the energy of the beam. Table 2.8 provides indicative numbers for the penetration depth of typical beams in different materials. Technically, the penetration depth is defined as the distance within the material at which the power of the beam is attenuated to $1/e = 0.37$ of the incident value or, in case of protons, the distance at which the particles loose their kinetic energy and slow to rest (**stopping power**). Perusal of the table shows that, depending on the material and the type of interaction, the penetration of the beam in the material involves submicrometre- to tens of centimetre-thick layers.

Techniques employing laser beams that ablate the surface of the material and extract atoms at depth for the analysis (for example LA-MS, LA-XPS, LIBS, etc.) are very useful because they can effectively remove the alteration or contamination layers present at the surface and probe the internal part of the material. Again these techniques can be considered micro-invasive because the hole produced for the analysis is of the order of few tens of micrometres and it is in most cases invisible without examination by electron microscopy.

Furthermore it is important to realize that many techniques that are normally invasive or even destructive when used in routine modes of operation, may become partially or totally non-invasive if optimized for the specific material to be analysed (Section 2.4.3).

Even if the technique itself is non-invasive, sometimes the object needs to be moved from its storage location if it is to be analysed by special instruments, and this may sometimes pose some risk of damage during manipulation and transport. To avoid this risk, and also to analyse large objects that cannot be moved (architectural structures, statues, wall paintings, in situ archaeological layers, etc.) a certain number of **portable** techniques have been developed, by which it is possible to acquire information directly on the objects (**in situ techniques**). Portable XRF, Raman and even XRD are rapidly becoming very popular because of their ease of use and portability, however the results are very often extremely qualitative because of the surface contamination and geometry limitations. Extreme care should be taken in their application and especially in the interpretation of the results (Chiari 2008, and Box 2.r).

Number of samples

For the identification and characterization of homogeneous materials, the number of samples is simply the minimum required to make sure that the results are statistically significant and reproducible. If the volume of each sample is larger than the intrinsic dimension of the heterogeneities in the material, then the sample may be considered homogeneous, and a few samples are enough to ensure that all measurements are comparable. One measurement alone is always to be taken as suspect, because there are an infinite number of unexpected features (for example inclusions, or alteration layers) that may invalidate the sampling and the analytical results. These may only become evident after a comparative series of analyses on the same or on similar objects.

For the characterization of macroscopically heterogeneous or composite materials, of course the sampling procedures need to be multiplied by the number of single homogeneous parts. An extreme and difficult case is exemplified by diagnosis in the field of cultural heritage conservation, where typically large objects (vases, large statues, monuments, or even cultural heritage sites) need to be investigated. These projects often require tens, hundreds or thousands of samples to be taken and measured. In all cases the number and volume of the samples must be carefully gauged against the size and heterogeneity of the object. If preliminary knowledge of the object and the material is lacking, then the sampling procedure may take place in successive steps. First a sampling campaign is planned to survey and map the size and distribution of the material's heterogeneities, the integrity, and the individual components of the object. This is often performed with the aid of surface or bulk imaging techniques (Section 2.1.3) and with the in situ use of portable instruments. Then a series of samples are taken, representative of the different materials comprising the object, or of the different alteration states of the material present in different parts of the object. The extracted sample can be anywhere from micrometre-sized in small composite homogeneous objects (jewels, beads), to millimetre-sized in large heterogeneous objects (pottery, metals), to centimetre-sized in objects that have undergone different degrees of alteration in different parts (building stones, wall paintings). After a first series of tests performed by a number of different techniques, sampling can be reasonably focused on specific areas of the object, and limited to the minimum amount required by the technique selected on the basis of the test results.

Volume of the sample

In the case of invasive sampling, the volume of the sample to be extracted by mechanical operations (chipping, drilling, etc.) depends on the instrumental

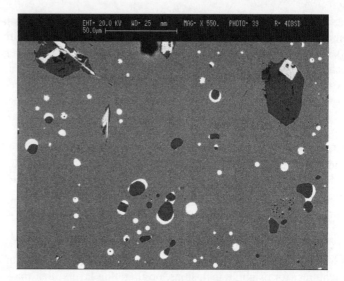

Fig. 2.16. Segregations caused by unmixing in the copper matrix of a prehistoric tool. Backscattered electrons SEM image. The 1–10 μm size segregations are mostly lead metal (white) and copper sulphides of digenite-chalcocite composition (dark grey). Scale bar is 50 μm. (Courtesy of I. Angelini, University of Padova)

requirements of the technique to be used and on the heterogeneities of the object: the larger the dimensions of the heterogeneities, the larger needs to be the field of observation of the measurement, and therefore the volume of the sample. Every material has inclusions and compositional heterogeneities at different levels.

Metals and alloys typically show inclusions and segregations at the 1–100 μm level, even when they appear homogeneous at the macroscopic scale (Fig. 2.16). This means that samples much smaller than 0.1–0.5 mm tend to be insignificant, because of the risk of analysing inclusion-free samples, or inclusion-only samples. Both of them may not be representative of the actual mean bulk composition of the metal. This also means that techniques using incident beams of the order of 10 μm or less (**micro-beam techniques**, i.e. TEM, EPMA, etc.) may have problems in analysing a representative portion of the bulk material. In the case of micro-beam techniques, a large number of point analyses will be necessary to attain a statistically significant composition of the material, or else specific protocols of measurement must be adopted, which artificially increase the volume probed by the micro-beam, for example by relative random movement of the beam and the sample during the measurement. Techniques using larger incident beams routinely produce reliable average analysis of the material.

Ceramics frequently have millimetre-sized grog and temper inclusions. Large heterogeneities of this kind are also common in binders such as concrete and plaster, in structural materials such as bricks and stone masonry, and in many other materials. A representative specimen in this case needs to be of the order of centimetres, and this is the size of the samples normally used for the preparation of thick or thin sections for optical microscopic analyses. The large polished samples used for optical microscopy can subsequently be analysed by micro-beam techniques (SEM-EDS, EPMA, etc.) to map the elemental distribution or to have the individual small-volume analysis of each component, including matrix, inclusions, segregations, alterations, etc.

If a mean bulk chemical analysis of a heterogeneous material is required, for example by NAA, MS, AAS, or OES, then a large volume of material (at least one order of magnitude larger than the largest inclusion) needs to be prepared and homogenized. This is commonly performed by powderizing the material and by thoroughly randomizing the particles, or by total dissolution and digestion of the sample by acid attack. In the first case, the homogeneous sample of the material to be measured is a pressed pellet of the randomized powder (XRF, NAA, IR), sometimes diluted by a binder composed of a material that is transparent to the incident radiation to reduce self-absorption and preferred orientation (i.e. H_3BO_3, cellulose or graphite for XRF, KBr for IR, etc.). In the second case the liquid solution is subsequently atomized, ionized, and often converted into a high temperature plasma by various methods (Bonchin et al. 2000) before being introduced into the measurement chamber (AAS, OES, MS).

Mortars and plasters commonly have millimetre-sized inert inclusions, and samples of 2–3 centimetres are usually representative of the material. Concrete, on the other hand, is made with much large aggregate particles, up to several centimetres. To avoid sampling of the aggregate alone, concrete samples up to 5–10 centimetres must be taken.

Paint layers on canvas, wood or walls are often sub-millimetre-sized, and they are often sandwiched between underlying preparatory layers and outer preservation varnishes. The samples need to be as least invasive as possible, but they need to go through the full stratigraphy of the material to offer a complete picture of the layer sequence. The samples are therefore commonly millimetre-sized, though the analyses have to be performed with micro-beams (with a size comparable to the layer thickness) to obtain the composition of each distinct layer (Fig. 2.17). Of course alternative non-invasive techniques are available for the analysis of paintings (Box 2.o).

Position of the sample

If sampling is possible, and the volume of each specimen has been defined based on instrumental requirements, then the final sampling positions on the object must be defined. The most important issues concerning the sampling positions are **alteration** and **representativity**.

The **surface** part of an archaeological object or an art work is invariably different from the **core**. For the most part this is due to the chemical processes occurring over time (alteration, degradation, abrasion, etching, precipitation, corrosion, etc.), though very often there is a man-made covering layer that from the start was expressly meant to change the aesthetics and/or the function of the material (varnishes and waxes on paintings, glazes on ceramics, clay slips and linings on various kinds of objects, resins and tar on wood, etc.) and, finally, there may be layers that were deposited long after the use of the object for treatment, restoration, or conservation. In many cases, for example when dealing with surface corrosion layers of ancient metals, it may be very difficult to discriminate between man-made and alteration layers. Thin patinas originally made to change the appearance of the object may be deeply embedded in the alteration layer, so that extreme caution must be used both in

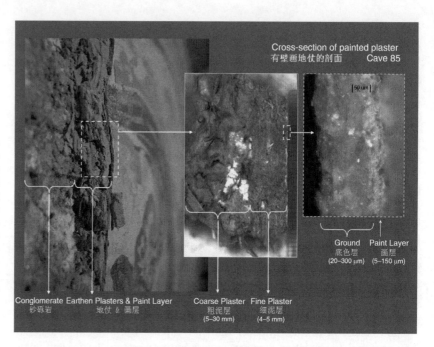

Fig 2.17. Wall painting stratigraphy of a Mogao wall painting. The image illustrates the paint layers in Cave 85. The paintings were executed as fine line drawings in red and black ink on a ground preparation covering two base layers of earthen plaster, and then filled in with mineral pigments and washes of organic colourants. The elemental analysis of each layer must be obtained using microbeams smaller than the individual layer thickness, i.e. of the order of 10–20 μm. (Courtesy of L. Wong, GCI Mogao Project; image available at: http://www.getty.edu/conservation/field_projects/mogao/mogao_images.html#mogao29)

cleaning the object and in sampling for analyses. In conservation diagnostics it is of course preferable to have samples that are representative of both the pristine and the degraded part of the object, in order to define the first and last steps of the degradation process and its mechanisms. Imaging techniques (microscopy, radiography, tomography) used as preliminary mapping tools of the heterogeneities in the material can be very important in defining the sampling strategy.

Representativity is related to the heterogeneity of the material. To plan a rational sampling strategy, we must know the number and positions of the different parts of the object, and the distribution of matter in each part, including surface layers. If this is known from preliminary imaging, and the minimum number of samples has been selected based on instrumental requirements (reproducibility), then the sampling areas are confined to the unaltered portions that are most easily reachable from the outside. Sizable sampling of course is usually performed in the areas that are the most hidden from sight, and if possible it is carried out before cleaning and restoration, to avoid pristine information being lost and extraneous conservation material being added to the system.

In summary, we may state that sampling is not a stand-alone operation. It should be preceded by historical studies and careful evaluation of the object, both from the point of view of the conservator and from that of the materials scientist, and by non-invasive mapping and imaging to confirm altered or restored parts and exclude unidentified ones. When non-destructive tests do not yield enough information, final sampling is performed to access sufficient material for in-depth analyses.

2.3 A question of boundaries: detection limits and limiting questions

The obvious starting question in materials science is: What is the sample made of? (Table 2.10). This implies that by far the most common investigations are qualitative and quantitative chemical analyses, performed through very different techniques (Table 2.3). The final aims of chemical analyses are clear to anyone: (1) to identify all elements present in the sample, (2) to quantify the relative and absolute amounts of each element, and (3) to evaluate the precision and accuracy of the results (Box 2.s). Actually, these tasks are not as trivial as it may seem, though they are routinely performed and the results are analysed and interpreted. To start with identification and quantification: it is simply not possible to identify and measure all the elements of the periodic table present in a sample, unless it is a pure compound with certified purity and stoichiometry. In real life most samples are made of a small number of elements that make up most of the compound (**major elements**), and a certain number of elements that make up almost all the rest of the compound (**minor elements**). There is no absolute limit for an element to be a major or a minor one. Common practice suggests that minor elements are present in amounts of a few wt% at most, and that they must not necessarily be present in the compound, that is the chemical compound does not change in nature if a minor component is missing, whereas if a major component is missing then we are dealing with a different compound. The sum of minor and major elements is above 99 wt% of the material. The elements that are present below the 0.1 wt% level (1000 ppm) are arbitrarily considered to be **trace elements**. Please be aware that there is no codification of the term. In copper and bronze archaeometallurgy, for example, the dozen or so meaningful elements may have widely varying concentrations below 20.0 wt %, so that the unusual scheme developed by Waterbolk and Butler (1965) to escape a linear concentration scale is still often used in the archaeometallurgical literature. They observed that alloy or inclusion elements may be present in ancient copper in concentrations varying from several percent to below 1 ppm, and that artefacts may be profitably classified with classes defined on a geometrical scale, in much the same way that many rocks or meteorites are geochemically classified on logarithmic scales. Waterbolk and Butler therefore proposed five classes of concentration (VH = very high = 7.5–23.7 wt%, H = high = 0.75–7.5 wt%, M = medium = 75–750 ppm, L = low = 7.5–75 ppm and VL = very low = 0.75–7.5 ppm) that turn out to be very successful for metal type discrimination and still encounter estimators and a variety of applications, though they have little analytical meaning because they cross the detection limit of many techniques.

In principle, chemical analyses may be carried out with high precision, for example it is possible with modern mass spectrometry to detect elemental levels below 1 ppt (10^{-10} wt%). Using such extreme analytical conditions many geological or archaeometric samples would show most of the elements of the periodic table to be present in trace or ultra-trace levels. Even the simple task of identifying all the elements present in the sample therefore would be not trivial, and it would require an infinite amount of time and money. Thus, when we talk about chemical analysis, we usually refer to the detection and quantification of major and minor elements. Trace elements are looked for only to answer very specific questions, such as in provenancing or materials contamination and treatment.

Concerning major element chemistry, one of the common problems is to understand if and how the obtained results can be compared with the results of other laboratories and with those available in the scientific literature. Extreme care should be taken when comparisons are made between the results obtained by different techniques or obtained by the same technique with different instruments, for example adopting different experimental set-ups, or at different times, given that the instrumental details rapidly develop and improve with time. For example, comparison of bulk elemental analysis performed by different methods is often very difficult because the selected set of analysed elements and the detection level critically depend on instrumental and operating choices.

Many techniques of quantitative chemical analysis only analyse a limited set of elements, numbering a dozen or so. If the material is well known and it has been analysed before in the laboratory in charge of the investigation, most likely the selected elements correspond to the actual composition, and no major or minor element remains undetected. However, if the material is not commonly investigated in the analytical laboratory, then element selection may be a problem because of the instrumentation, the calibration standards, or simply the defined analytical protocol. It is unfortunate that so many incomplete analyses find their way into the published literature. Missing elements may often be spotted in non-normalized analyses, which clearly show lack of components or analytical problems, but if the compositional results are normalized to 100 wt% (beware of normalized data!), then the information on the elements not analysed and possibly present is lost forever, and the results are therefore biased and misleading.

For example for metal analysis it is common to measure the 4th raw transition elements ($24 < Z < 30$, i.e. Cr, Mn, Fe, Co, Ni, Cu, Zn), and those elements usually found as alloying or impurity components in historical metals (Section 3.5), that is As, Ag, Sn, Sb, Au, Hg, Pb, Bi. On top of the metallic elements, it may be useful to check the presence and amount of some non-metallic elements often derived from the ore minerals or the smelting process, such as the chalcophile elements (S, Se, Te), P, C, and O, the last two of course being important in steel and partially oxidized metals. In principle, this makes the total number of elements to be measured to be around 20. In practice, this is true only if the analysis is performed by modern MS techniques, and most other techniques have limitations, either due to measuring time or reachable energetic range, etc. so that at best a dozen elements are commonly quantified.

Prehistoric and historic pre-20th century glass is almost invariably silica-based (SiO_2), with a few alkali and alkali–lime elements used to lower the melting point and increase durability (Section 3.4). In principle six elements (O, Si, Na, Mg, K, Ca) make up most of the glass, and any additional component is included as an impurity in the starting materials or added intentionally to modify the optical properties, such as an opacifier or colorants. In the case of analysis of glass, the six base elements need to be analysed to characterize the bulk composition, and all other elements are left to the skill of the analyst, who should perform a pre-scan of the element present, or to the specific requests of whoever is submitting the piece for analysis.

Instruments that are routinely used for metals use different measuring parameters and calibration procedures to those used to analyse silicate minerals or glass. Unless the instrument is completely and properly re-calibrated, there will very likely be missing or badly quantified elements. This is one of the reasons to call for continuous collaboration and interaction between users and operators in analytical laboratories, so that optimal and appropriate protocols are used in the measurements. Given the effort, time, and costs required to keep the analytical instruments running and properly calibrated, it is unlikely that a one-time sample would be analysed following the appropriate protocol, unless performed in a laboratory that specialized in that very kind of material. If a metal is chemically analysed on a SEM-EDS instrument that is routinely calibrated for silicate glass (or a glass sample is measured on an instrument calibrated for metals), you may be fairly confident that part of the above-discussed elements are missed, misidentified or badly quantified.

Assuming that appropriate procedures are used to quantify the major elements, the question arises as to whether the selected technique is appropriate for detecting all minor and trace components. The **smallest concentration** that we can measure with a particular technique is defined as three times the signal corresponding to the instrumental noise level (or background, i.e. the signal detected when no sample is measured), and it is called the **detection limit (DL),** or limit of detection (LoD). In fact this is the point that defines whether the element or compound is **present or not**. The commonly reported **estimated standard deviation** (σ, or **e.s.d.**, i.e. the root-mean-square deviation of the values from their mean) of a measured value contains information on the variance of repeated measurements, and therefore intrinsically on the instrumental noise level. If the σ has been correctly evaluated, then an element is present when its measured value is at least 3σ.

Each instrument has specific detection limits for different elements, depending on the physical processes involved, the instrumental components and the concentration of the element in the sample. Approximate values of DL for the most diffused analytical techniques are listed in Table 2.9.

Sometimes the relative ratios between elements, in place of absolute quantities, are of importance to the problem. Measuring ratios is generally easier and safer than quantifying the absolute elemental concentrations. This strategy is commonly adopted in the measurement of isotopic abundances.

Having realized the potential of each analytical instrument, the next step is to use each instrument to answer the archaeometric questions, appropriately translated into analytical terms (Table 2.10). The translation of a humanistic problem into a series of rational questions and ultimately into a flow chart of scientific experiments designed to answer them is by no means evident or automatic. It may be relatively easy to apply known techniques to problems that have been faced and rationalized before, and that are described more or

Table 2.9. Indicative limits of detection of elements for the most used chemical analysis techniques

Technique	Detectable elements (Z)	Detection limits (ppm) [10 000 ppm = 1%]
SEM-EDS	normal 11–92 (windowless 6–92)	1000–5000
EPMA(WDS)	11–92	100–1000
XRF	11–92	10–100
PIXE	11–92	1–100
NAA	9–83	0.00001–0.1 (see Fig. 3.a.1)
PGAA	1–92	1–100 (see Fig. 3.a.2)
ICP-OES	1–92	0.001–0.1
LIBS	1–92	1–30
AAS	11–92	1–10
MS	1–92 (including isotopes)	0.001–0.01

Table 2.10. Specific questions to be addressed when planning a scientific investigation. Acronyms of the techniques are interpreted in Table 2.2

Question addressed	General methodology	Commonly used techniques
Is chemical element Z present in my sample? i.e. *What is the chemical nature of the sample?*	• qualitative chemical analysis	XRF, OES, LIBS, AAS, EPMA, PIXE, NAA, SEM-EDS, MS
What is the (relative/absolute) concentration of element Z in the sample?	• quantitative chemical analysis	
How is element Z distributed in the sample? i.e. *How are the different material components distributed?*	• spatially resolved chemical analysis • optical/electron microscopy coupled with chemical analysis	XRF, EPMA, LIBS, SEM-EDS, PIXE
What is the crystal chemistry and crystallography of element Z, and how is element Z chemically bonded to other atoms in the sample? i.e. *What crystal phases are present?*	• short range chemical spectroscopy • vibrational spectroscopy • crystal chemical analysis • phase analysis / crystallographic analysis	MöS, XAS, NMR, IR, RS, UV-Vis, XRD
What is the detailed electronic structure of element Z? i.e. *What is the valence and the chemical bonding?*	• atomic/electron spectroscopies	Auger, EELS, XPS
Is there a relationship between the measurable parameters and the physical properties of the material? i.e. *Why is the object made this way?*	• materials properties analysis • structural crystallography (average structure) • defect/texture/strain analysis (real structure)	porosity measurements residual stress deformation mechanical resistance texture analysis structural phase analysis

less in detail in the literature. Science always relies on the shoulders of giants, or at least proceeds on available footprints, so that the advancement process is very often limited to the systematically repeated application of a well-developed protocol, or to the optimization of the methodology to slightly different materials. However, sometimes sparks of ingenuity and favourable conditions lead to completely new experiments and to novel insights into the problems. Interdisciplinary contexts and areas at the boundary between established fields of research are notoriously very fertile in stimulating novel approaches, and cultural heritage studies intrinsically have such an interdisciplinary character.

Let us go through the neodymium example, though most probably it is still unknown to the majority of the readers. All scientists know that Nd is very rare and it is normally present at the trace level in natural materials (1–20 ppm), and that it should therefore be measured with techniques that have a very low DL (for example XRF, NAA, or MS). Nonetheless for the most part the scientists would not know in which materials to look for it, and why. Geochemists and petrologist also know about the Sm–Nd systematics in rocks (De Paolo 1988, Banner 2004, Dickin 2005); they are aware of the radioactive decay evolution of the system (i.e. the two elements are joined in a parent–daughter relationship by the alpha-decay of ^{147}Sm to ^{143}Nd with a half life of 1.06×10^{11} years). Therefore they have a clue that different minerals produced at different times from different sources have variable quantities of Nd and, above all, they carry a different isotopic signature, measured as the ^{143}Nd/^{144}Nd ratio. Thus, the important quantity to be measured is the Nd isotope ratio, which is measurable only by MS. However, all this does not mean much to an archaeologist unless we add the crucial information that the Sm and Nd contents and isotope ratios in silicate minerals are characteristic parameters of their genetic processes, and therefore of their provenance. The curious archaeologist at this stage is tempted by the investigations of dozens of silicate-containing or silicate-derived materials (glass, ceramics, flints), disregarding the fact that once the measurements on the objects have been made, there are hardly any appropriate local or regional systematics of Sm and Nd in rocks with which to compare the results obtained. Of course many readers may have realized that this example is not fictitious at all: using Nd as a potential tracer for the provenancing of raw materials in archaeology is a reality and it is a very promising method indeed, albeit the application is just at the beginning (Brady and Coleman 2000, Degryse and Schneider 2008). The neodymium story is a successful story of a reasonably well-developed method that has crossed disciplinary boundaries, and is making an impact in a field that was certainly not planned from the start. This happened because: (1) the problem was clearly defined; (2) new tools for provenancing were necessary and actively sought after, and (3) researchers with inquisitive and open minds created a bridge between geology, geochemistry and archaeology.

However, the story also raises some puzzling issues. Are there enough environments where such creative developments may occur? Curiosity and knowledge do not seem to be at the top of the list of priorities in many research groups, often being overwhelmed by fund-raising competition and publication pressure. At times the technical competence and the specialized scholarship necessary to produce sensible advances requires so much effort that few researchers can afford the time, cost and manpower to explore new areas. Academic and educational curricula are rapidly shrinking, for several compelling reasons, truly interdisciplinary courses are hardly organized, and when they are the level of teaching is at best introductory. How many archaeologists have sufficient background in isotope geochemistry? And how many physicists or geologists know the historical and conservation problems embodied in the object that they are analysing? The lack of communication (at best) or even disregard or disinterest (at worst) between the two communities has been repeatedly remarked upon (Shackley 1998, Wilson and Pollard 2001), but unfortunately it is still the norm. In the specific case of neodymium systematics, who is going to take up the task of building the reference database of rock isotope values that will serve as the interpretive frame for historical movement of materials? Do we have structures for this?

Questions without measurements are mere curiosity. Measurements without systematics, databases and interpretive models are mostly useless. Models without experiments and applications are intellectual exercises. It is unfortunately a fact that questions, experiments, and models are produced by different, often hardly-communicating groups of researchers. Real advances occur when forces are synchronized and synergy's at work.

2.4 How to match problems, materials, and methods

It is assumed that any technique that is selected for the analysis of cultural heritage materials must (a) be totally non-invasive or, at least, use the least amount of material required to produce meaningful results; (b) produce useful, precise (i.e. reproducible), and accurate (i.e. calibrated) results; and (c) be routinely accessible to the user community.

Having established this, the choice of the appropriate analytical technique(s) to be used for a particular problem is sometimes difficult, and in order to provide useful and reliable data several aspects must be considered:

- **focusing the problem**: it is not useful to analyze an object or a sample if no questions have been asked. Within the frame of the general problem, one may want to start with simple questions (What material is the object made of?, Is the material altered?, etc.) so that all parties involved know what they are talking about. Given the interdisciplinary character of many investigations and the totally differing backgrounds of the researchers involved, this is not trivial, and it should be taken seriously. Then one can move step-by-step to more complex issues based on the available information or the observed preliminary data (Table 2.10)

- **understanding the principles** of each available technique, including:
  - the level and completeness of the *information* provided (Section 2.1)
  - the *amount of sample* required and the *damage* produced (Section 2.2)
  - the level of *accuracy and precision* needed (Box 2.s)
  - the *detection limits* of the technique (Section 2.3);
- **the availability and cost of the analyses:** deeper levels of understanding require more costly and time-consuming techniques;
- **the possibility of interaction with the analysts:** this may be crucial in order to understand and interpret the analytical results, the potentials and limitations properly;
- **the quantity of the sample available** to be analysed: the possibility of extracting a certain amount of material for invasive analyses, or the selection of non-invasive analytical techniques;
- **a comparison of the results with available literature data and databases**. In order to insert the data into a broader context, it is necessary to make use of extensive databases and to make sure that the adopted technique is standardized against generally adopted protocols in order to have calibrated results.

The ability of the archaeometrist is to match the original problem with the appropriate techniques, and to translate the problems that triggered the investigations from their cultural heritage context to a level of terminology that is understandable by scientists and technicians. There are a number of straightforward technical questions to be posed to the scientist (Table 2.10) in order to obtain useful measurements. As we increase the complexity of the questions (and of the measurements) in order to gather more detailed information, i.e. as we move down through the list of questions in Table 2.10, there is an associated considerable increase in time, manpower and cost required by the analyses.

The more complex the problem is, the higher are the requirements for the analysis. In many cases one technique is not sufficient to yield all the necessary information and to solve the problem unambiguously. Very often combined analyses yielding complementary information are required to solve complex analytical puzzles. On the one hand novel analytical techniques, if introduced in a collaborative and receptive environment, may greatly help in solving old problems and stimulate new questions. For example Ambers and Freestone (2005) enlighteningly show how the relatively recent introduction of Raman spectroscopy in the Department of Scientific Research of the British Museum in London, a laboratory already well acquainted and experienced with several scientific techniques, considerably helped in extending the laboratory's potential in several ways. On the other hand, there are countless examples and anecdotes concerning the superficial and uninformative interactions between conservators or archaeologist and scientists, based on badly defined problems and questions. For example, an 1890 photograph analysed by modern SEM-EDS techniques produced no useful results, because the microscopist, unfamiliar with photographic material and unaware of the introduction of baryta-based paper in 1885, spent hours of precious machine time in investigating the size, morphology, and composition of the baryte (barium sulphate) grains intermixed with cellulose. The information on baryte was fairly useless to the curator of the photographic collection stimulating the investigations, who was actually interested in the presence of minor elements eventually yielding clues to the original photographic process.

Another example: a small fragment of marble from a very famous renaissance masterpiece was smuggled by the restorers in charge of the cleaning of a statue to a large-scale synchrotron facility, where it was analysed by state-of-the-art diffraction instrumentation. The totally useless result, much as depicted in Fig. 1.2 (a), showed the marble to be composed of pure calcite (calcium carbonate), information already available to everybody acquainted with marble composition, and in any case easily obtainable with much cheaper techniques. The real issue is in the microstructure and fabric of the calcite grains, in the carbon and oxygen isotopic composition, and in the presence of other trace chemical elements: This is the information used to provenance the marble, to authenticate marble statues, and to understand whether alteration processes are currently acting on the surface.

A synopsis of the most common techniques used in the investigation of cultural heritage is given in Table 2.11. The table also shows the frequency of use of the listed techniques in laboratories engaged in cultural heritage investigations within the LabsTech network that participated in a 2005 survey (Boutaine 2006). Besides digital and classical film photography, which were of course still much used at the time of the survey, by far the most cited and diffused techniques were, and presumably still are, optical microscopy (OP) in reflected and transmitted light, scanning electron microscopy (SEM), powder diffractometry (XRD), and infrared spectroscopy (IR). The survey clearly shows how low-cost information-rich techniques dominate the scene. The techniques listed in Table 2.11 were selected on the basis of information content and complementarity, rather than frequency of use, and this is reflected in the range of popularity resulting from the survey. Sometimes a cheap analytical solution is sufficient to solve the problems, at least in the first approximation. At other times more exotic and expensive techniques must be approached. A well managed and constantly wired network of laboratories is possibly a great solution for cultural heritage problems, offering competent advice in a bottom-up sequence.

The preceding discussion relies on the assumption that the material scientist is confronted with a well-defined general problem, and that only the details must be sorted out, in order to clarify the scopes and modes of the measurements. The detailed problems may depend on the cultural heritage context, but they may be grouped into a few broad objectives:

1) **nature and state of the material components** (to understand the physical reality, the function, the behaviour);
2) **interpretation of the history** (reconstruction of the origin, the manufacturing processes, the diffusion, the past and future time evolution);
3) **authenticity and age** (of the object, building, site, etc.).

Table 2.11. Summary of information obtained on different materials by some of the most used complementary techniques. The numbers below the technique acronym indicates the frequency of use in laboratories involved in cultural heritage studies (Boutaine 2006)

Material	XRD	IR-RS	XRF-AAS-OES-NAA	EPMA-PIXE	MS	SEM-EDS	Radiography/ Tomography	Reflected/ transmitted light- OM	Other techniques
	49	60-25	30-23-23	24-17	16	76	42	87-75	
Ceramics	Major and minor phase identification	Identification of some phases	Bulk chemistry (major elements)	Point chemical analysis (major elements)	Trace elements Isotope analysis (Pb)	Elemental mapping Reaction rims	Internal images	Major phase identification Inclusions, Temper, Texture Alteration	TL-OSL
Stones, Rocks, Lithics	Major and minor phase identification Alteration layers	Identification of some phases	Bulk chemistry (major elements)	Point chemical analysis (major elements)	Trace elements Isotope analysis (Pb)	Elemental mapping Reaction rims	Internal images	Major phase identification Inclusions, Texture, Alteration	OSL
Bone, Ivory	Apatite identification	Identification of species	Relative age (F,N,U content)		Trace elements Stable isotopes		Internal structures		EPR
Amber		Amber provenance			Molecular composition		Imaging of plant/animal inclusions	Imaging of plant/ animal inclusions	Py-GC TGA-DTG
Metals	Major and minor phase identification Texture analysis		Bulk chemistry (major elements)	Point chemical analysis (major elements)	Trace elements Isotope analysis (Pb,Cu,Zn,Ag,Sn,Au)	Elemental mapping Inclusions Segregations	Internal images	Metallography	LIBS
Metals, Slags, Ores	Major and minor phase identification	Identification of some phases	Bulk chemistry (major elements)	Point chemical analysis (major elements)	Trace elements Isotope analysis (Pb,Cu,Zn,Ag,Sn,Au)	Elemental mapping Reaction rims Inclusions Segregations		Major phase identification Metallography	Texture analysis
Glass, Glazes, Faiences	Crystalline phase identification		Bulk chemistry (major elements)	Point chemical analysis (major elements)	Trace elements Stable isotopes	Elemental mapping Reaction rims Inclusions Segregations	Distribution of inclusions Glass-crystal texture		TEM XAS XPS
Organics		Molecular identification			Trace elements Stable isotopes				NMR HR-LC GC-MS
Pigments	Phase identification	Phase identification	Bulk chemistry (major elements)	Point chemical analysis (major elements)	Trace elements Isotope analysis (Pb, Fe, Sn,...)	Elemental mapping			LIBS XAS XPS

Of course everybody involved with the cultural heritage immediately appreciates the implications of such issues for the object(s) with which he/she is dealing, at any scale and of any nature. And every materials scientist immediately appreciate the potentials of the technique(s) that he/she masters for the measurements, at any scale and energy.

The broad issues listed above include virtually all possible objectives that we may imagine to be related to cultural heritage, whether we are dealing with the restoration of a 20th century concrete building, the interpretation of a painting, or a survey of an archaeological site. Of course these are only useful for defining the context in which we are moving, and they should be broken down into specific questions in order to be operative, but they should always be kept in mind during the investigation, for the measurements to be meaningful.

The first objective, to gain insights into the **nature and state** of the material components, can be easily and straightforwardly translated into the technical questions needed for the measurements and listed in Table 2.10. These are the questions that the technician operating the instrument who is unaware of the problems in the background should be asked. Is there vermilion in this pigment? Is there a preparatory drawing underneath the painting? Is this wooden object made of cypress wood? Is there a burial chamber under that mound? Is this copper or bronze? Why is this stained glass discoloured? What is this efflorescence on the wall? These are all straightforward questions that can be related to experimental tests and instrumental measurements.

However, the experimental details related to the nature and state of the material must be evaluated within a more general interpretation of the **history** of the object in order to be meaningful. The archaeologist, the museum curator, the conservation scientist, the art historian, the restorer,... all want the object to be a piece of their story. Why was vermilion used by that particular artist? Was vermilion a colour on his usual palette? Are all similar objects made of cypress wood? How many of the known mounds of the area contain a chamber, and can we get to the chamber without fully destroying the mound? Is the non-copper content of the metal intentional or casual alloying? How can we stop the discolouring process of the glass window? Can we avoid the production of efflorescent crystallization that produces micro-cracking in this plaster? Such questions transcend the essential materials data measured during the characterization of the object, and make it the starting point for a historical tale involving artistic production and human life. In conservation science, this means extrapolating the data obtained today on the object back into its past, to understand the artistic and material history of the art work, check its stability and durability, and plan for its future. In archaeological terms this means moving beyond the "What is left?" question attempting to characterize the material's evidence, and tackling the typical "Where, when and how?" questions that attempt to collocate the object into the variety of human experiences (Renfrew and Bahn 2008). The further question "Why was it made?" is probably the most interesting, though it leaps straight into socio-anthropological issues.

Decoding the history of the material and comparing the measured results with those available for similar or coeval objects yields the basic information necessary to insert the work into a relative or absolute **time sequence**.

The aesthetic, stylistic and chemical evolution of the material in time provides reference scales that can be used to interpret technological and artistic changes throughout human history (Section 2.5). The same knowledge is used to recognize forged antiquities and art works. In fact, **authenticity** tests are based on laboratory data that expose artistic, stylistic, and physico-chemical features that are inconsistent with the supposed age of the artefact (Jones *et al.* 1990, Craddock and Bowman 1991, Oddy 2004). Extensive scientific analysis of the certified works of an artist may also help in the attribution task, to the point that connoisseurship increasingly leads the way to **technical art history** (Ainsworth 2005).

2.4.1 One material, multiple answers

As shown, each technique yields specific information depending on the principles of interaction, size and energy of the beam, and mode of operation. The same artefact may be investigated using different probes and different scales. Thus, in principle, a variety of **complementary information** may be collected and integrated to obtain a complete assessment of the problems faced, the only limitations being time and cost. Because of these limitations, appropriate **investigative strategies** must be developed by evaluating problems, priorities and risks and measuring them against available resources. The analytical strategies should be based on such planning.

When analytical choices are to be made, one should first assess what kind of information is required. Is it physical, chemical, or mineralogical? (Table 2.3). Then one should proceed by evaluating the techniques that may yield similar information, but using different probing processes. Mineralogical information (i.e. phase identification and composition) may be obtained by diffraction, Raman spectroscopy, optical and electron microscopy, etc. Chemical information may be obtained by a number of different spectroscopies, albeit with different resolution, limits of detection and sample requirements. All these details should be adequately planned and discussed with the scientists involved, fully appreciating the fact that by using different techniques we may obtain different answers for the same material.

Two issues are critical when comparing measurements performed by different laboratories and/or different techniques. The first concerns the reliability and reproducibility of the measured data, in terms of precision (Box 2.s) and the limits of detection (Section 2.3). The measurement of samples with known composition, which has usually already been measured on other instruments, is the key to evaluating whether the selected technique can answer the analytical questions within the accepted error and detection limits.

Secondly, even for techniques that provide very precise results, the comparison of the data obtained by different instruments and techniques is not straightforward. We are dealing with the absolute values resulting from an experiment, and the accuracy of the results (Box 2.s) is estimated by careful measurement of certified standards using the same instrumental conditions used for the samples. The measured data are commonly re-calibrated using empirical or semi-empirical corrections in order to comply with the expected standard values. The problem of analysing and comparing **data from different sources** is important, especially when extended databases and multiple data sets are being treated and analysed. The most common problems are: (1) how to produce measurements that are comparable to the literature; (2) how to assess the validity and the consistency of the data in a database or in the literature; and (3) how to combine data from different sources (i.e. different instruments) in order to have homogeneous data, for example to perform statistical analysis. These are very difficult problems. In general, making repeated measurements on standard samples is the only reasonable way to assess the accuracy and reliability of the data. Disregarding the precision of the measurement, which depends on the specific instrument and analytical protocol, the absolute observed value resulting from independent experiments ideally should be the same within 3 e.s.d.s (3σ). This of course is hardly ever the real case, especially when dealing with complex and heterogeneous materials such as those related to cultural heritage, where sampling problems, alteration, contamination and all sorts of practical issues affect the measurements. Simple re-grinding and re-preparation of a specimen of a heterogeneous material such as an archaeological ceramic or a mosaic tessera very often produces large differences in results, even for major elements or phase amounts. Errors and discrepancies of up to a few percent may be observed between values that should in principle be perfectly identical. The analytical project must therefore contemplate a series of measurements that are specifically planned to define the variation of the parameters in the material under investigation, the influence of methodological variables in results, and the importance of the variance in observations with respect to the problem.

The reliability of the measured data is a fundamental point for discussion during the interpretation of the results, and it becomes a critical issue when dealing with **extensive analytical projects**. This point is well illustrated by the outcome of two broad past projects, discussed in detail by Pollard *et al.* (Pollard *et al.* 2007, Chapter 3.5), which made a large use of emission or absorption spectroscopies: (a) the analysis of Greek and Cypriot pottery by the Research Laboratory for Archaeology and the History of Art (Oxford University, UK) and the Fitch Laboratory (British School at Athens, Greece), mainly based on early flame-OES data; and (b) the chemical analyses of copper and bronze European objects carried out first in the UK by the Ancient Mining and Metallurgy Committee of the Royal Anthropological Society under the guidance of Vere Gordon Childe, followed by the large German SAM programme (Junghans *et al.* 1968–1974). The latter included more than 20 000 metal analyses and both databases mainly included optical spectrometric data.

When dealing with such extended databases, the issue of data consistency is at the core of any advanced statistical treatment and, furthermore, the technical developments in the analytical techniques since these projects were undertaken have been so great that modern analytical data are hardly comparable to those present in the database. If one wishes to compare recently collected data with the existing dataset, one must at least make sure that a comparable subset of data is used, and seriously consider the possibility of modifying the values and their e.s.d.s (of the data, of the dataset, or of both) in order to make them statistically comparable and cross-calibrated as much as possible through common reference standards.

2.4.2 Should I sacrifice the sample or the significance of the measurement?

Most modern analytical techniques may perform reliable analyses on a quantity of sample in the range 50–100 mg. To have an intuitive reference, 100 mg of lead ($\rho = 11.34$ g/cm^3) means a cube with sides of about 2.1 mm, whereas the same mass is equivalent to a cube of bone ($\rho = 1.80$ g/cm^3) with sides of about 3.8 mm or a cube of amber ($\rho = 1.05$ g/cm^3) with sides of about 4.6 mm ($\rho = 1.05$ g/cm^3). Many high-sensitivity bulk techniques may indeed analyse a quantity of materials that is lower by one or two orders of magnitude, and microbeam techniques perform the analyses on volumes of the order of 10^{-3}–10^{-6} mm^3 (which is equivalent to cubes with sides 10–100 µm). However, working with small samples brings into question the risk of poor significance and representativity, as discussed above at length, and furthermore the measurement time required to gain enough counting statistics with an extremely small quantity of sample may be unacceptably long. As a general rule, good statistics and reliable data are proportional to the quantity of sample and the intensity of the incident beam in the first approximation. Therefore the choice of technique used for the analysis must be carefully assessed, taking into account the minimum amount of material to be sacrificed to obtain **meaningful results**.

The issue is particularly important when considering absolute dating, especially using radiocarbon techniques. The uncertainties in the measurements are propagated through the calibration curves (Box 2.u) and poor measurements of the C isotope ratios result in unacceptable errors in the calibrated absolute dates. The uncertainties in the mass-spectrometric measurements are inversely proportional to the amount of sample available for the analysis. The situation in radiocarbon dating is largely acceptable with the advent of modern accelerated mass spectrometry (AMS), which requires very small amounts of material (of a few mg) to produce statistically significant values, but the problem of having a large enough sample to produce statistically meaningful results is common to all techniques.

It is virtually impossible to compile a table listing the minimum amounts of material necessary for the different techniques, such as that compiled for detection limits, because the quantity required may be extremely variable even using the same technique and different instrumental conditions. Furthermore, the issue of sample quantity or volume is generally not very important in analytical laboratories, because geologist and mineralogist commonly have plenty of rocks available, chemists may synthesize their compounds at pleasure, and engineers count on mass-produced industrial components. Sometimes plenty of samples are also available for archaeometric measurements, for example pottery shards or metal smelting slags. However, a more frequent situation is that even micro-sampling may sometimes be considered to be too invasive. Scientists and technicians must therefore change and optimize the routine protocols of analysis to make measurements on much smaller samples, and nonetheless obtain reliable results.

For example, in conventional X-ray powder diffractometry (Box 2.g) employing tube sources, satisfactory measurements are normally performed using about 500 mm^3 of powdered material, generally packed in square metal holders with cavities of about $22 \times 22 \times 1$ mm^3, or discs of diameter 27 mm and thickness 2.4 mm. Considering a normal packing coefficient of the powder (about 0.6) and a material with average density (for example quartz, SiO_2, $\rho = 2.64$ g/cm^3) this means that in normal laboratories about 0.8 g of quartz are necessary for routine XRD measurements, that is a cube of about 6.7 mm in size of solid material. This is way too much material for normal sampling of archaeological or art objects. Of course some diffraction laboratories specialize in small quantities of materials, for example by taking advantage of parallel-beam, focusing and grazing incidence geometries, and special sample preparations, such as very thin dispersions on zero-background plates. Under these instrumental conditions the quantity of sample can be decreased by 4–5 orders of magnitude, down to tens of microgrammes (Dapiaggi et al. 2007), though very few laboratories may perform diffraction measurements in such demanding instrumental configurations.

The XRD example was looked at in detail, just to show that even techniques that are considered routine may be optimized and adapted for special requirements, so that working with a very small amount of material, such as that resulting from micro-sampling (Section 2.2), does not necessarily mean obtaining lower precision and uncertainties in the results. Reliable and meaningful results may be obtained on micro-samples in many laboratories, but it certainly requires a good deal of manpower and dedication. The key is, once more, in developing longstanding collaborations and frequent exchanges of views and know-how between the partners.

A substantial number of dedicated laboratories with optimized instruments, technical competence, and expertise in the analysis of cultural heritage materials are now available, and they are continuously increasing in number, thanks to focused national and international programmes devoted to developing research infrastructures for cultural heritage (Boutaine 2006). Furthermore, it should not be forgotten that superior quality data on an extremely small amount of sample may be collected at large-scale facilities (Box 2.a), where increasing attention is paid to the application of advanced techniques to cultural heritage applications.

2.4.3 Optimization of the techniques

In the previous section the idea of optimizing the technical details of instrumentation and measurement protocols expressly to take into account the requirements of cultural heritage samples has been introduced; it is related to the volume of material analysed. The volume of the sample is one of the most important parameters, because it directly affects the sampling procedure and, in many instances, the whole analytical strategy. However, depending on the nature of the sample and the instrument involved, the sensitivity and/or the efficiency of the technique may be increased: (a) by changing the source (in the extreme case by using intense radiation sources); (b) by changing the geometry of the measurement (for example by switching from transmission into reflection geometry); or (c) by increasing the efficiency of the detectors. In fact detection is probably the single most limiting factor in most experiments. This is certainly the case at large facilities where detectors are notoriously the weak link in the experimental chain.

Although infinite possibilities exist to improve and optimize the measurements, citing a couple of practical cases may be useful in order to clarify what may be done. In the case of FTIR spectroscopy, for example (Box 2.e), which is widely used for identification of organics and pigments, most samples are commonly measured in transmission mode, though they may also be measured in reflectance geometry, in ATR or DRIFT modes. In the specific case of amber, FTIR has long been used to identify the material and recognize the characteristic signature of Baltic amber (Beck 1986), though experiments performed in traditional FTIR transmission mode commonly require about 20 mg of material (Spragg 2000). This quantity is not always easily extracted from important museum pieces, because the sampling area may be clearly visible, so that a method has been developed using reflectance mode and requiring much less sample, in the range 0.1–0.3 mg, which is about 200 times less than in transmission mode. Figure 2.18 shows that IR curves measured with different geometries may be comparable and useful for identification and provenance purposes (Angelini and Bellintani 2005).

As discussed in Section 2.1, all techniques using incident radiation need special care in the selection of the energy employed in the experiment. The energy and the flux of the beam influence the activated process, as well as the thickness of the sample that is passed through, and the spatial and energy resolution of the results. In **densitometric** measurements, for example, careful calibration of the X-ray beams is needed for absolute measurement of the inorganic mass

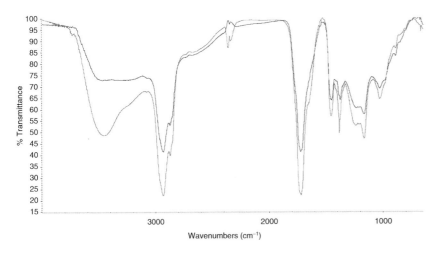

Fig. 2.18. FTIR spectra of Baltic amber. The two spectra were collected in transmission (blue curve) and in diffuse reflectance mode (red curve). Despite the use in DRIFT of a much smaller amount of sample, the two curves are perfectly comparable and useful for identification purposes. (Modified from Angelini and Bellintani 2005)
Original colour picture reproduced as plate 4.

in bones (Fig. 2.19). The technique is commonly used in the medical diagnosis to evaluate the significant decreases in bone density (osteoporosis), by a special two-beam technique called **dual-energy X-ray absorptiometry** (**DEXA**). The dual-energy spectrum from an X-ray source is obtained either by means of alternating pulses of low and high kV that are applied to the X-ray tube (the low- and high-energy spectra are then measured separately), or through the application of a constant potential to the X-ray source while using a K-edge filter to separate the energy spectrum into two narrow energy bands. In the latter case an energy-discriminating detector with a dual-channel analyser counts the resultant photons. The use of two energies allows bone mineral to be assessed independently of soft-tissue heterogeneities. Single measurements may be turned into two- or three-dimensional scans using DEXA scanners. Densitometric techniques similar to those used in the medical measurements are used to investigate archaeological skeletons (Farquharson and Brickley 2000).

In transmission imaging techniques, which include radiography and tomography, the choice of radiation is also critical. Thick objects may be imaged by neutrons and high-energy X-rays, though even for low-absorbing materials, such as the faience shown in Fig. 2.n.5, the X-ray beam must be optimized so as to enhance the absorption contrast between the glassy matrix (containing mostly Si and O, but with minor quantities of Na, K, Ca, Mg and Cu) and the crystalline quartz inclusions (SiO_2), which are lacking the alkali components. In the specific case, the fine-scale heterogeneities in the material require the experiment to be optimized not only for the transmission/absorption contrast, but also for the final resolution of the image, which must be adequate to show the micrometre-sized texture of the faience. Collimated synchrotron radiation is necessary for such a tomographic experiment, and the intrinsic coherence of the beam may furthermore be used to enhance the discontinuities and the boundaries of the grains by operating in phase contrast mode (Baruchel *et al.* 2000). This is just a simple example of how tomographic imaging requires well thought out optimization of experimental geometry and instrument choices.

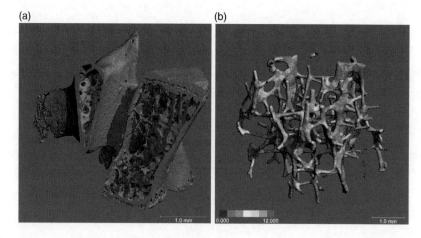

Fig. 2.19. The amount and distribution of the inorganic part (apatite) in bone may be imaged and quantified using radiographic and tomographic techniques. Image (a) is the sectioned and segmented tomographic scan of a mouse vertebra. Image (b) is the processed image of the internal trabecular part of a bone, where the colour coding corresponds to the local thickness of the apatitic structure. (Courtesy of SCANCO Medical AG)
Original colour picture reproduced as plate 5.

Concerning X-ray powder diffraction, in Section 2.4.2 it is discussed how alternative and specialized instrumentation may critically decrease the amount of powder needed for the measurement. We may expand it further, and show that nowadays it is fairly easy to perform X-ray powder diffraction directly on an object in a totally non-invasive way, so that not even micro-sampling is required. Unfortunately this implies the use of different experimental geometries, different X-ray sources and/or different detectors, and many laboratories are not equipped for such non-standard analyses. Sometimes they are not even aware of such alternative analytical protocols. Measurements of sub-milligramme amounts of powdered sample are possible using improved sources (rotating anodes, micro-sources or synchrotrons), or using a focusing technique to increase the flux in a cylindrical sample (Debye geometry), and by using efficient linear detectors. Direct measurements of unground micro-samples are possible using the transmission geometries and two-dimensional detectors (CCD, image plates, image intensifiers) commonly used in single crystal X-ray diffraction (Monaco and Artioli 2002). This requires integration of the powder spectrum over the whole explored reciprocal space (Figs 2.20 and 2.21). Alternatively, measurement of the XRPD pattern directly on the object may be performed either using parallel-beam geometries on laboratory tube sources (with the use of curved mirrors made by multilayer technology), or penetrating sources, such as high-energy synchrotron X-rays (Fig. 2.22) or neutrons (Kockelmann *et al.* 2006a). It should be mentioned that advanced powder diffraction is not limited to identification and quantitative information of the crystal phases, though they are certainly the most common applications. Rather, if properly measured and analysed the diffraction data may provide very useful information on the size of the crystallites (Martinetto *et al.* 1999, Ungar *et al.* 2002), on their orientation (see Box 3.k), and even on the high temperature processes, for example reconstructing the firing processes of ceramics by direct measurement of the high temperature phase changes.

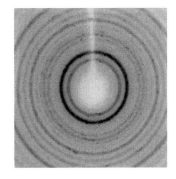

Fig. 2.20. Diffraction image collected on a protohistoric faience bead in a totally non-invasive mode, using a single crystal XRD diffractometer and a CCD detector. (From Artioli 2004)

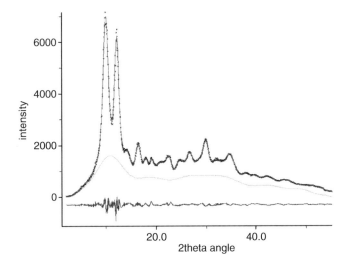

Fig. 2.21. Diffraction spectrum obtained by integration of the Debye rings present in the CCD image of Fig. 2.20 (From Artioli 2004). The well-modelled profile shows the quartz and cristobalite phases present as crystalline inclusions in the faience.

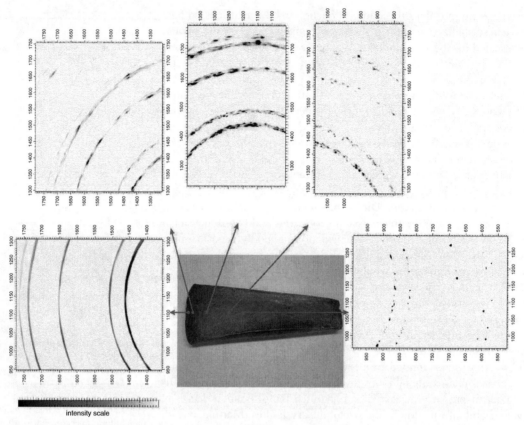

Fig. 2.22. Diffraction rings collected on the Iceman copper axe using highly penetrating synchrotron radiation. The experiment was performed at ESRF beamline ID15B (From Artioli *et al.* 2003. Museo Archeologico dell'Alto Adige, Bolzano, Italy, http://www.iceman.it.)

In short, measurements on cultural heritage materials should be performed in specialized labs, preferably in those that have previous experience of analysis of the same kind of objects or materials, and paying special attention to measurement conditions and protocols. This would ensure a higher probability of getting useful results with minimum offence to the specimen.

2.4.4 Cost-effective information content: developing analytical strategies and risk assessment

The objectives of the scientific analyses of cultural heritage materials reflect the variety of disciplines involved. Adopting the same scale-based approach that was introduced earlier, we may start with the **objects**. In archaeometry the materials studies are aimed at the reconstruction of production, distribution and use of the archaeological objects, besides being the base for subsequent interpretation of the reconstructed life cycle of the object and of its users (Tite

Box 2.r Portable non-invasive XRD/XRF instrument: a new way of looking at objects [G. Chiari]

In studying "Cultural Heritage" materials conservation scientists try to leave the objects intact as much as possible. One can send beams of various kinds and analyse how they are modified by the interaction with the artwork without having to take samples. In the last few decades tremendous progress has been made in miniaturizing instruments for all analytical techniques, often making them portable.

Analyses are usually defined as "non-invasive" when no sample is required; "semi-destructive" when a sample needs to be taken from the object but is left intact for further study, and "destructive" when the sample is consumed in the analytical process. In addition to these very general descriptions one should quantify many other factors, such as the relevance of the information retrieved (the need for and impact of such information, e.g. the possibility of saving an object which is in a poor state of conservation), the risk involved in the transport of the object to the lab, the intrinsic value of the object, the impact that even taking a nano-sample may have on the object, etc. Economics may be another important factor in deciding what type of analysis should be done: insurance companies charge large sums to cover the risk of transportation of art to the laboratory. For all these reasons "non-invasive" analytical techniques, especially those using instruments that are "portable" (NIP), have had a large impact on the conservation science field in recent years. In addition to the strictly technical aspects, these portable instruments have altered in a profound way how we can best analyse art objects. In fact, with NIP one can obtain results almost in real time. As a consequence, the relationship between conservation scientists and their "partners", conservators, museum curators, art historians, site managers etc., has changed. Questions can often be answered while the whole team is present and actively discussing the object, generating new brainstorming that includes just-obtained results. This produces conclusions that were not possible before the introduction of NIP.

The miniaturization of computers, excitation sources and detectors has made possible the advent of many NIP instruments. The most obvious example is hand-held X-Ray fluorescence (XRF). Capable of being brought everywhere and used on the whole object (without surface preparation), a portable XRF gives us the elemental composition (at least in part) of the spot we are looking at. It does not, however, give quantitative results (something to be kept in mind). It does not give the map of the elements distribution unless a very large number of point analyses are done. In general, XRF does not penetrate much beyond the surface. Most importantly, it does not tell how the elements are combined to form molecules and crystals. In spite of these limitations, the portable XRF produced a real revolution in the field for two reasons: (1) it is *non-invasive*; (2) it has become *portable* (NIP: the unit goes to the objects and not vice versa).

An instrument performing both X-ray diffraction (XRD) and XRF simultaneously could overcome most of the limitations of just using XRF. Named DUETTO, it has been recently manufactured by **inXitu** (psarrazin@inxitu.com) using a design that was jointly developed with the Getty Conservation Institute, Los Angeles. XRD and XRF on the same spot (documented in high definition by a digital camera, which also allows for adjusting the focusing plane) are collected by an energy-dispersive CCD. The tube selected has a copper anode, giving priority to the quality of the XRD data. Battery-operated, the head weights about 8 kg and is truly portable. Photons with CuKα energy are considered to be diffracted and all the others are interpreted as fluorescence.

In a reasonably short space of time (5–10 minutes) a rough but interpretable XRD pattern can be obtained. Better patterns, almost comparable to a bench diffractometer, need longer exposures. Having XRF data on the same spot greatly helps the interpretation of the XRD, especially for very quick analyses that are necessarily of poor quality. It is always a good strategy not to rely on one type of analysis only.

The positioning of the instrument is critical. Mounted on an XYZ translation system, it can be positioned on the flat surface of the object with a tolerance of 0.02 mm with the help of a laser line reflected by the surface that is forced, by varying the distance of head to object, to match a crosshair while being monitored by the digital camera. Obviously, since the goal is to measure objects of various sorts without sampling, the coarse texture of some surfaces may be sufficient to reduce the resolution of the XRD giving broad peaks.

(cont.)

Box 2.r (*Continued*)

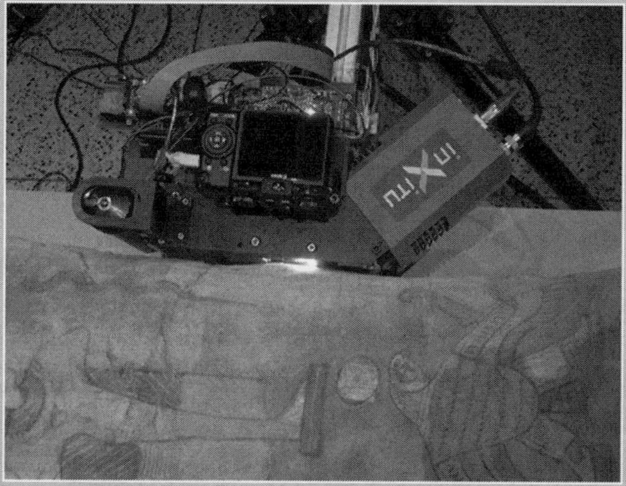

Fig. 2.r.1. The combined XRD/XRF DUETTO instrument without its cover to show part of the interior, as positioned to analyse a red shroud Fayum mummy from the Getty Museum. (Courtesy of G. Chiari, GCI)

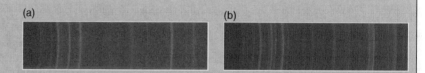

Fig. 2.r.2. (a) CdS pure pigment and (b) commercial Cadmium Orange.

Figure 2.r.1 shows the instrument without its cover to expose part of the interior, positioned to analyse a red shroud Fayum mummy from the Getty Museum. The X-ray tube on the left, the camera in the centre and the detector on the right are clearly visible. The distance between the object's surface (focusing plane) and the hardware is 2 mm so that no physical contact is needed. On the top part the generator is visible, so that the high voltage wires are enclosed in the head of the instrument, reducing risk for the operator.

In Fig. 2.r.2 the analysis of cadmium orange (chemically CdS doped with a small quantity of CdSe) is shown. The pattern (a) refers to the pure pigment while the part (b) refers to a commercial product (Cadmium Orange in a tube of oil paint). In these "raw data" it can easily be seen that the pattern (b) contains more lines, and that these are dotted, indicating a substance made of coarse grains.

From the raw data of Fig. 2.r.2, using proprietary software one can extract a conventional 2D XRD diffractogram as shown in Figure 2.r.3. It is possible using software to perform the integration in strips of various thicknesses, which greatly influences the resolution, obviously at the expense of intensity. It is also possible to use software to eliminate high intensity spots due to accidental large grains. All the pixels of different energy are interpreted as XRF as shown in Fig. 2.r.4.

Using a proper sample older, the instrument can also work as a normal powder diffractometer for those cases in which a small sample of material is available, still combining XRD and XRF. The grains may be coarse since a special sound driven device takes care of providing a random distribution of the orientation.

The miniaturization and complete portability of the XRD/XRF instrument described here obviously is achieved at the expense of higher performance. The major drawbacks are: for XRD the range of 2θ observable with the Cu tube is 20–50°. This is because the angle of the incident beam cannot be less than 10° for geometric reasons and the dimensions of the tube; and the area of the detector only allows an upper angle of 50° 2θ. This may be a

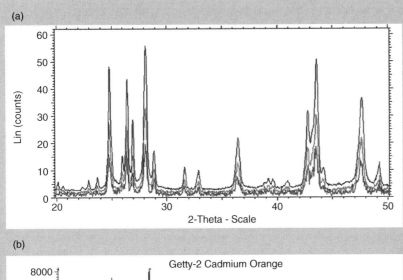

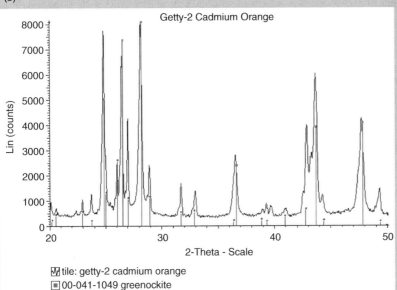

Fig. 2.r.3. (a) The patterns collected after 5 minutes, 50 minutes and 8.3 hours are shown. In spite of a low peak/background ratio, the 5-minute pattern was interpretable, especially using the chemical information provided by XRF. (b) XRD pattern of Cadmium Orange with 8.3 hours exposure.

problem for substances like clay minerals that have their most characteristic lines at low angles. Often, though, the diffraction peaks in the above-mentioned range are sufficient to identify the substance. For example palygorskite (principal line at $2\theta = 8.54$) was the first candidate in an automatic search even without the presence of the principal line.

As mentioned earlier, the positioning of the head with respect to the object is critical. A large roughness of the surface may cause a shift in the lines due to the offset of the focusing plane. In these cases, perhaps, the addition of an internal standard, such as silicon or alumina on the surface to be analysed, would permit this error to be corrected by software. The powder can be easily brushed away after the analysis is performed.

(*cont.*)

Box 2.r (*Continued*)

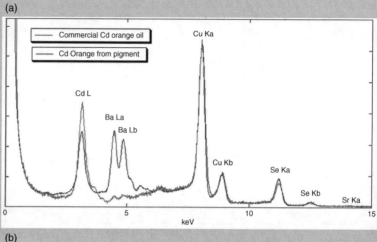

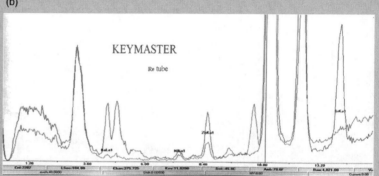

Fig. 2.r.4 (a, b). XRF patterns of Cadmium Orange and commercial product as measured by DUETTO compared with a specialized ortable XRF (Bruker). From the XRF data one can see that barium is present in the commercial oil paint but not in the pure pigment. From the XRD the barium can be better characterized as barite ($BaSO_4$). The selenium peaks are clearly identifiable in the XRF. For comparison we reported the portion of spectrum common to DUETTO and one of the more commonly used portable XRF instruments.

The use of a Cu tube is not optimum for XRF, since many important K lines are not excited for heavier atoms. Therefore, identification has often to rely upon L and M lines, which present larger superposition.

In conclusion, this new instrument will help to further limit the need to take samples. Sampling will still be needed when the characteristics of the object itself requires it. In the cases mostly concerned with quantitative data (provenance or dating), samples will always be needed.

One cannot expect a small portable instrument to yield the same results as a well-calibrated bench instrument. Therefore the data obtained should always be considered for what they are: a valuable way of getting valid information in a short time without having to move the object and without damaging it in any way. The sampling can be greatly facilitated by grouping together regions or the object on the basis of their composition (analytical imaging). In this way the number of samples can be greatly reduced and their relevance increased. One should, however, be ready to accept that a few miniaturized and selected samples may be taken in order to answer the questions to which these techniques do not provide answers.

For in-depth information

Chiari G. (2008) Saving art in situ. Nature 453 (8), 159.

2001). In conservation science the examination of artefacts is devoted to evaluating the condition and value of a prospective purchase, to quantifying the type and amount of conservation required, and to evaluating whether it is safe for the artefact to be loaned and/or exhibited (Oddy 1984).

Then, moving away from individual objects and taking a broader view at monuments, buildings, and sites, we may recall that "...archaeological sites are easily destroyed once they have been excavated, because they become exposed to the elements, to visitors and, often, to looters. It is becoming disturbingly clear that cultural patrimony is under siege, as magnificent monuments—from Angkor in Cambodia to Machu Picchu in Peru—crumble. Modern archaeologists must increasingly consider how to preserve sites they have unearthed or how to examine sites without touching them" (El-Baz 1997). Fortunately, in recent years not only the level of attention is increasing, but also a more comprehensive approach to the conservation of cultural heritage sites is being made (Agnew 1997, Agnew and Bridgland 2006). Moving away from a limited materials-based approach, a holistic view of site conservation attempts to focus on the values related to the conservation process itself. There is a rising effort to integrate and coordinate the investigation and conservation efforts on all scales, from objects to heritage town centres to large regional cultural paths. Within this broad view, the strategies of **analytical study** must be part of a rigorous approach to knowledge acquisition, conservation planning, and management of the material cultural heritage resources. Besides general cultural significance, the values attached to the material and the intangible cultural heritage include the definition of cultural identity (Graves-Brown *et al.* 1996), social and national bonds, economic profit, and political messages (Avrami *et al.* 2000, De la Torre 2002). The dynamics between values and the benefits of knowledge and conservation is complex, and in the decision processes there is an increasing need to interface the overall planning stage (***what*** to study and conserve, and ***why***) with the technical details (***how*** to study and conserve).

This conceptual framework is somehow well defined in the field of conservation, where the **diagnosis** (i.e. the investigation and analysis of the cause or nature of the degradation of the system) is expected to set the experimental ground for the **prognosis** (i.e. the prospected evolution of the system as anticipated from the models and the evidence from previous cases). In principle these two steps yield enough information for the conservation strategy to be decided: the key is the correct assessment of the **rate of the chemical reactions** causing the alteration and degradation processes (Section 2.5.2). By accelerated testing of the materials (Feller 1994), historical analysis of known occurrences of the process (Brimblecombe 2005), and the development of adequate experimental damage functions (Benarie 1991, Delalieux *et al.* 2002), the rate and mechanisms of deterioration should be defined. This sets a time frame for the intervention and, if satisfactory solutions are unavailable, intermediary or temporary treatments may be selected. For example, reburial of excavated sites is a popular solution when the site does not allow proper conservation and management. The extrapolation of the time rate of degradation should take into account the uncertainty in the measured or estimated parameters, and combine them in a worst-possible scenario. Despite such a rigorous **risk assessment**, there is unfortunately always space for unanticipated consequences, side-effects, and deadly synergies between multiple causes of degradation (Agnew 2003), so that there is ample space for analysis and tests before, during and even after the conservative intervention. Given the limited lifetime of any material used for treatments (Section 2.5.2), the impossibility of a complete reversibility of the process, and the complex interaction between causes and intervention procedures, the idea of a successful "definitive one-time solution seems, in fact, to be somewhat of a myth" (Agnew 2003). Continuous monitoring and changes in the intervention strategies may be required. With regard to the costs of conservation, the economic and cultural values assessed are often hardly comparable. Many politicians and economists nowadays seem to rely on the concept of cultural tourism as the added value to foster conservation and valorization (Cowen 1998, Frey 2000). Though tourism may be a determinant for the future of archaeological sites, the social, cultural and economic impacts should be compatible with the objectives of long-term preservation. The total risk is also increased in proportion to the exposure (Cunliffe 2006):

$$Risk = Vulnerability \times Exposure \times Hazard$$

Even the concept of cultural tourism therefore should be thoroughly considered in the light of the interdependence between cultural values attached to heritage (i.e. its cultural importance), risk assessment, and its economic potential, so that a "**heritage ecosystem approach**" may be more appropriate (Greffe 2004).

Despite the numerous problems, however, when dealing with the conservation of cultural heritage at least the conceptual frame supporting the decision-making process seems to be sufficiently defined. On the other hand, the investigation process of the archaeological heritage is much fuzzier, and it gets more undefined the more we adventure into the remote past. The archaeological excavation itself paradoxically gains cumulative knowledge by destroying the site and removing the objects. The idea and the tools for totally non-invasive archaeological investigation are yet to come, though at least modern geophysical, geochemical and remote survey techniques (Box 2.p) greatly help in confining the areas that should be physically excavated, both at the regional and site levels, thus reducing the time, costs and invasiveness of surface surveys and extensive random test-pits.

During the excavation, good and complete recording is essential to store information that would otherwise be irremediably lost. As in conservation, thorough risk assessment should be performed of what is worth bringing to light (and conserving), what is worth leaving underground for future investigations, and what is worth being lost during the excavation process. Unfortunately, in many cases, time-frame, limited budget or current expectations seem to have a higher priority in the decision process than proper risk assessment.

The increasing diffusion of advanced GIS (Box 2.p) and virtual reality techniques (Box 2.q) during the excavation and classification of the material evidence may slowly introduce the direct mapping of the information in four dimensions (3D space and time). Ideally all information retrieved by archaeologists should be stored in easily accessible shared databases for immediate retrieval and analysis, and extensively integrated with the archaeometric results. Poor recording, non-standard format of the stored data, limited exchange of information, and only intra-disciplinary diffusion of knowledge, are some of the common problems in the field. The chronic scarcity of funding resources must be resolved through collaboration, creative thinking and long-term commitment.

2.4.5 Data treatment, statistics, and presentation of the results: making science clear

The question as to whether nature is truly mathematical has been bugging scientists even before Galileo's statement quoting:

> La filosofia è scritta in questo grandissimo libro che continuamente ci sta aperto innanzi a gli occhi (io dico l'universo), ma non si può intendere se prima non s'impara a intender la lingua, e conoscer i caratteri, ne' quali è scritto. Egli è scritto in lingua matematica, e i caratteri son triangoli, cerchi, ed altre figure geometriche, senza i quali mezi è impossibile a intenderne umanamente parola; senza questi è un aggirarsi vanamente per un oscuro laberinto. (Galileo Galilei, *Il Saggiatore*, 1623).

Though, provocatively, there is ground for challenging the role and use of mathematics in science (please read the amazing paper "*The pernicious influence of mathematics on science*" by Schwartz 1962), it may be safely stated that mathematics helps us understand the world around us by revealing patterns and order in the observations performed in the diverse disciplines. Scientists use mathematics not only to organize and manipulate data, but to rationalize and describe the relationships in the physical world. "Working physicists adopt a quantitative approach to almost all investigations, and physics tends to be taken as a model for how any successful science should be formulated" (Davies 1992). In the last part of the twentieth century many disciplines related to science struggled, and some are still struggling, to become truly quantitative and archaeology has certainly followed this main trend. This is clearly shown by the number of fundamental textbooks addressing the use of mathematics in the field, starting with the seminal books by Orton (1980) and Shennan (1988), up to the more recent and specialized volumes (Baxter 1994, 2003, Orton 2000). In fact, when one looks at these widely used textbooks one soon realizes that of all mathematical disciplines the only one that is really emphasized in archaeology is statistics. Further, even if there is a small core of highly competent mathematicians in the field, "the bulk of the people practising the discipline do not even understand what they [the mathematicians] are talking about, never mind its implications" (Shennan 1988). Now, this could also be stated for a number of different research fields, and things in archaeology are slowly changing (note for example the seven chapters dedicated to statistics in Brothwell and Pollard 2001), in particular thanks to interdisciplinary collaborations and the introduction of hopefully appropriate statistical training in academic curricula (see Drennan 2001, for a discussion of the issue). Two points should be mentioned at this stage.

1) In the first place statistics is a huge field, and in most cases introductory courses provide at best a rather soft understanding of the basic concepts of statistical methods. Since many students do not have sufficient mathematical background, and many teachers are not sufficiently acquainted with the intrinsic problems of archaeological data, many concepts are taught in an intuitive way and many methods are presented as black boxes. Clearly, there is a long way to go in order to introduce sufficient statistical culture to the generation of forming archaeologists, especially because archaeological data really require fully developed statistical treatments and analysis, well beyond the basic tests of normal distribution and the commonly used cluster-type analysis. Rather, the full array of the most advanced tools of **multivariate analysis** (principal component analysis, discriminant analysis, correspondence analysis, etc. (Baxter 1994, Baxter and Buck 2000)) and beyond (Bayesian approaches, genetic algorithms, neural networks, fuzzy logic, etc. (Buck *et al.* 1996, Ripley 1996, Baxter 2004)) can, and should, be efficiently used to try and create order out of apparently untreatable and chaotic data. The issue is especially relevant in classification problems and provenancing studies. Datasets with non-normal or biased distributions are the norm in archaeological and cultural heritage records, therefore tailor made data treatments and interpretations have to be developed in each specific case, with careful control kept of the assumptions and of the simplifications made. Some strategies may have a statistical meaning, but they may be completely unphysical or they may have little to do with our understanding of past human behaviour. Thus a great deal of critical assessment is required: one should always remind oneself that research is guided by paradigms, theories, and appropriate questions. Most of the statistical analyses involved require sophisticated software, luckily available as free programs thanks to the GNU project (see the **R software package**: http://www.r-project.org/), and a good deal of know-how and computational effort. Though statistics may be applied to anything in cultural heritage, from sampling of just one object to statistical sampling of a large number of objects (Orton 2000, Simon and Utz 2007), pottery analysis seem to be an area where most statistical methods have been tested and their validity checked (for example Rice 1987, Sinopoli 1991, Orton *et al.* 1993), mainly because of the complex nature of the assemblages, the (fortunately) large numbers of populations, and the challenging questions to be answered, about raw materials, manufacturing techniques, provenancing and diffusion, etc. Advanced statistical analysis may be applied to any kind of analytical data, once we take into proper account the nature of the data, their physical significance, and the aims of the study. Out of the huge literature devoted to pottery, recent examples encompass application of complex multivariate analysis to NAA (Glas-

cock *et al.* 2004) and ICP-OES compositional data (Fermo *et al.* 2004), and tentative applications of artificial neural networks to LIBS (Ramil *et al.* 2008) and ICP-OES compositional data (Lopez-Molinero *et al.* 2000). Studies involving the bulk chemical composition represent by far the most common type of pottery investigation. It is to be expected that, if morphological, optical, mineralogical, textural, surface decoration and other data are added to the basic chemical information, pottery analysis may rapidly develop into a highly computerized kind of study (Gilboa *et al.* 2004).

2) Secondly, statistics is just a small part of mathematics, and of course countless other mathematical sub-disciplines fruitfully contribute to the quantitative investigation and management of cultural heritage. Just to mention a few cases brilliantly discussed by Main (1991) in its far-sighted review, applied mathematics and **computer science** have radically changed all computational aspects, **sampling theory** is widely used both in the field and in the management of museum collections, **graph theory** is at the core of many indexing and classification software programs, **pattern recognition** is evolving into 3D reconstruction and modelling, **multi-dimensional scaling** and **image analysis** are pushing past qualitative measurements such as optical microscopy and aerial photographs into the quantitative realm. Of course all aspects of theoretical and applied mathematics and informatics contribute directly or indirectly to the research concerning cultural heritage, and it is obviously impossible to get enough expertise in all areas. However a strong mathematical background, certainly much stronger than that presently offered in most curricula, might certainly help to take better advantage of the advanced theoretical and experimental tools that are continuously being developed. In a complex world where an answer is not simply true or false, a fair understanding of the data treatment and interpretation is mandatory.

The introduction of computers as everyday tools and the astonishing development in information technology in the last thirty years has had a profound impact on all sciences. Conservation and archaeology were not immune to this revolution. All analytical instruments nowadays are computer controlled, and all data, including survey and excavation data, are stored, treated and analysed on laptops almost in real-time. Digital imaging, 3D-reconstructions, laser scanning, GIS and virtual reality (Section 2.1.4, Box 2.p, 2.q) are profoundly changing the way that archaeological objects, art works, buildings and sites are perceived, modelled and visualized. Mathematical algorithms embedded in the software and the power of the hardware have changed for ever the way that we collect data, analyse them, and present the results. We can do old things better and faster, though we may also do things that were unthinkable or inaccessible in the past, for example full 3D tomographic reconstructions would simply be unfeasible without powerful computers. There is just one caveat: beautiful images are not the reality, they are just models (Section 2.1.4). In the words of Drennan (2001), with which I partly agree:

> …much of the difference between the analyses we are able to perform now and those we were able to carry out 30 years ago has to do with appealing graphical output and user convenience. Dramatic increases

in computing power have, to a very large extent, been sopped up by graphical user interfaces.

Baxter (2008) also seems to think that little conceptual advancement has occurred in the last 30 years or so, besides improvements in the computational aspects.

Having scrupulously and painstakingly keypunched the cards of my early Fortran-based computer programs, at a time when one missing or mistyped card meant days of wasted human and computer time (the computer curmudgeon attitude, in Drennan's words), I am probably allowed to remark that nowadays it is much easier to run programs a hundred times, with little or no check of the input and even less thought about the interpretation of the results (the modern student attitude). This should nonetheless not prevent us using the present cheap computer bonanza as much as possible, however paying due attention to the theoretical and computational aspects of data processing. At all stages of the investigation we should have the required skills in order to be in control of the process and of the results (not vice versa!), and to do so we need to understand: (1) the theory and practice behind the analytical technique (the measurement); (2) the algorithms and principles used in the data treatment (the analysis); (3) the model that we are applying to solve the posed problem (the interpretation) and (4) the way we present the results (the visualization).

Let him cast the first stone, he that has the complete grasp and control of each step of the process! . . . unless of course he/she is an active part of a strongly collaborating interdisciplinary group.

Box 2.s Accuracy and precision in scientific measurements

Researchers not acquainted with the basic principles of scientific measurements should at least make an effort to understand the meaning of accuracy and precision, because these are fundamental concepts in data treatment and interpretation. Both parameters are needed to assess the validity of a measurement, so that confidence can be attached to any general inference or deduction derived from the data.

Accuracy is a measure of how close the analytical results are to the true value. It can be considered dependent on system errors, i.e. on the adopted analytical measuring device. Accuracy is checked and estimated by internal or external standards, so that **calibration procedures** are critical to obtain absolute data, which can be compared with reference values and data measured by other techniques or instrumentation. In principle, accuracy should be examined over an experimental range that extends beyond the range of the analysed samples.

Precision is a measure of the spread of the data from a specific series of experiments, and it is an estimate of how reproducible the measurements are. It can be considered to be dependent on random errors. The coefficient of variation of the data is better estimated on a large number of data points. Depending on the analytical problem, multiple measurements can be performed on the same sample with the same instrument (analytical reproducibility of the instrument and of the measurement protocol), on different samples taken from the same specimen (homogeneity of the sample), on the same sample prepared for the analysis with different protocols (reproducibility of the sample preparation procedures, absence of contamination, absence of human intervention), and on the same sample measured with different instruments and/or different laboratories (inter laboratory precision).

A graphical representation of the concepts of accuracy and precision is given in Fig. 2.s.1. In terms of measured data, let's say for example absolute dates performed by some dating technique on an archaeological

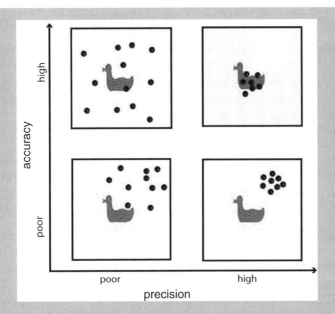

Fig. 2.s.1. Graphical representation of accuracy and precision. The two extreme cases are shown in the upper left, in which the measurements are instrumentally poorly reproducible, but the mean value is very close to the true value (poor precision, high accuracy), and in the lower right, in which the measurements are very reproducible, but the instrument is badly calibrated and the mean value deviates substantially from the true value (high precision, poor accuracy).

sample having actual age 2500 BC, it would be easy to recognize (and discard) a measurement performed with poor precision and poor accuracy (lower left image), because it reads something like 3500 BC (±1000 years), which is a poorly reliable (i.e. high e.s.d.) and implausible mean value that is clearly offset from any expected or calibrated value. By the same token it is also easy to recognize (and accept) a measurement performed with high accuracy and high precision (upper left image): it would read something like 2500 BC (±100 years), a clear sign of very reproducible and well calibrated measurements. Estimated standard deviations (e.s.d. or σ) are the root-mean-square deviations of the values from their mean (see Section 2.3).

However, in real life it could be much more difficult to assess the reliability of the other two measurements. In fact a poorly precise but highly accurate measurement (upper left image) might read as 2500 BC (±1000 years); the mean value is actually correct and well calibrated, but it has an unacceptably high estimated standard deviation (e.s.d.) possibly due to sample contamination, sample preparation or measurement instability. Are we prepared to accept this value despite the large estimated e.s.d.s ? On what ground?

Even more tricky might be to spot the case depicted in the lower right image: a highly precise but poorly accurate measurement. The mean value might read as 3500 BC (±100 years). The lower e.s.d. and good reproducibility of the measurements are actually deceptive, and without any estimation of the accuracy provided by standard calibration or inter laboratory comparison it might be very difficult to discard the measurement. These are the cases where measurements are not to be taken at face value and critical discussion of the data is highly welcome.

With regard to dating, uncalibrated radiocarbon dates are a good example of precise but inaccurate dates, as they need dendrochronological calibration (Box 2.u) to become accurate. At the opposite end, thermoluminescence dates frequently have a limited experimental precision, though they are for the most part fairly accurate.

For in-depth information

Frost T. (2000) Quantitative analysis. In: Lindon J.C., Tranter G.E. and Holmes J.L. (eds) Encyclopedia of spectroscopy and spectrometry. Academic Press, London. Vol. 3. pp. 1931–1936.

Shennan S. (1988) Quantifying archaeology. Edinburgh University Press, Edinburgh, and Academic Press, San Diego. 2nd Edition Reprinted (1997).

2.5 Time and dating: an overview

One of the major concerns of the researchers in a number of different disciplines (archaeology, geology, palaeoanthropology, palaeontology, etc.) is how to place the object of study in the time scale. What is the age of the specimen? Is the almost obsessive question routinely asked to the scientist in charge, partly because of the importance of the time information in the reconstruction of the past, and partly because dating is one of the very first successful achievements of archaeometry (Aitken 1990, Taylor and Aitken 1997, Garrison 2001, Hedges 2001).

When talking about time scales, there is hardly a unifying common sense, because every individual has a perception of time related to their own personal experience. Electronic transitions are phenomena occurring with characteristic time intervals in the picosecond range (10^{-12} s), so that chemists and physicists are well acquainted with sub-second time scales. Solid state transitions in materials undergoing phase transformations have kinetics in the range of seconds, hours, and days; accordingly mineralogists, materials scientists, and engineers commonly operate in the minute–hour time scale of the laboratory. Experiments lasting more than a few hours are often allowed only in unmanned mode, because of working timetables and student scarcity. The **lifetime** direct experience of human beings generally spans a few decades, and we have to rely on older people's accounting or on written records to expand our knowledge of the past. Historians examine written records, which in some places are not older than a few hundred years (i.e. Australia, South America), and at most they are a few thousand years old (i.e. Mesopotamia, Near East), and therefore they cover just a small fraction of the history of mankind. The study of prehistory (i.e. the period between the appearance of hominids, however they may be defined [approximately 3–4 millions years ago], and the appearance of written records [approximately 3000 years ago]) makes it necessary for the paleoethnologist and the palaeoanthropologist to rely on scientific methods to put events into a relative or absolute **time sequence** (Renfrew and Bahn 2008). Geologists and palaeontologists commonly deal with even longer time intervals, since they are concerned with processes covering the whole history of the Earth, including continental drift and the evolution of life (Hawkesworth and van Calsteren 1992).

Archaeologists of course are in great need of understanding the temporal location of the artefacts, the sites, the layers they are excavating, but they are not the only ones. The dating game (Zimmer 2001) is an intriguing task involving anthropology, forensic science, art history and a countless variety of other disciplines. The procedures and the logics used to identify false artworks are not dissimilar to the ones used for dating. In fact many of the dating techniques are being shared by different breeds of scientists, especially at central experimental facilities, all engaged in the game of inserting the object of their study into the time line.

1) Whatever the scope of the investigation, when dealing with time one always ought to be aware of a few general concepts. We may consider time on an **absolute scale** (absolute dating of the object/culture), or on

a **relative scale** (relative dating of the object/culture with respect to younger and older objects/cultures). Also, when dealing with the past material record, one should be aware that cultures evolve and technical abilities go forward at different paces and times in different places. Taking a critical approach to age and dating is fundamental to placing the evolution of natural processes and human civilizations in the proper time frame (Diamond 1999, Gould 1990).

2) The available dating techniques are based on entirely different physical processes, so that they have specific limits of accuracy and precision. **Each dating method** is therefore based on a specific analytical technique and, due to physical limitations, it may be applied only to a **well-defined time interval**.

3) Each material has a defined **lifetime**, depending on its physico-chemical characteristics. If on the one hand we wish to understand how the artefact was produced and used, on the other hand we should also be concerned with its chemical lifespan, which involves understanding its evolution in time, including its alteration, degradation, conservation, and a number of other implications (Section 2.5.2). In short, we have to be concerned with the **thermodynamic** stability (or metastability, or instability) of the object in its environment, and with the **kinetics** of the inevitable changes acting upon it.

4) The fading and disappearance of cultural heritage is not only the major concern of conservation science, which should be involved with the future durability and preservation of the tangible heritage, but it also has profound implications in archaeology. First because what is excavated is only a small part of the material's evidence of the past, and second because, since different materials degrade at different rates, the more we go back in time the more the archaeological record is biased towards durable materials (Caple 2001).

5) When applying dating techniques to the archaeological record, security of context is a crucial issue. We must be confident that the measured object is consistent with the layer that we wish to date, and must exclude any contamination.

2.5.1 Which time scale?

Time scales can be absolute or relative. Absolute time scales are measured in seconds (s), following the international SI system. However, depending on the specific application, our experience, and frame of mind we usually employ a number of other units including minutes, hours, days, months, years, lustrums, decades, thousands of years (ky), millions of years (My) and billions of years (Gy). There is usually not much need to go beyond about 13 Gy, which is the assumed time of the Big Bang and the coming into existence of our Universe.

Absolute dating techniques are supposed to insert the object or the event of interest in a timeline that spans the whole or part of the age of the Earth, assumed to be about 4.5 Gy. Starting from the geological time scale (Fig. 2.23), if we are interested in human evolution we have to narrow our search down

132 Overview of the analytical techniques

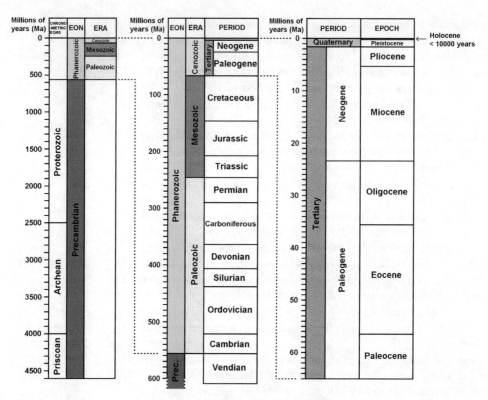

Fig. 2.23. Geologic time scale. The eons, eras, periods and epochs are generally delimited by major changes in the Earth's history, such as mass extinctions, climate changes, recognized changes in plate tectonics regimes or life evolution (see for example Harland *et al.* 1990). (Modified from the Geoscience website of the University of Calgary, original image available at: http://www.geo.ucalgary.ca/~macrae/timescale/timescale.html).

to the last part of the Cenozoic Era, that is the Pliocene (the last Epoch of the Tertiary Period) starting at about 5.3 My and then focus on the Pleistocene (the first Epoch of the Quaternary Period) starting at about 10 ky ago, and on the Holocene, which is the Epoch of the Quaternary in which we live. It is impressive that the whole prehistory and history of mankind, which is the whole object of archaeology and palaeoanthropology, lies in the very last time bit of Earth's history (Fig. 2.24).

In order to assess absolute dates, each technique experimentally measures a specific physical property of the material: it can be a chemical quantity, an isotopic ratio, a defect density or any measurable property that varies continuously with time. The variation may be linear or it may follow more complex functions, but in any case the property versus time dependency has to be well known and calibrated (Section 2.5.3). If the rigorous time variation is not known, or it is modified by local parameters, then the dating methods at best provide only relative information.

All dating methods (Table 2.12, Wintle 1996, Hedges 2001, Schwarcz 2002), whether absolute or relative, have their own time scale and interval of application. One of the most challenging tasks is the one of cross-referencing and

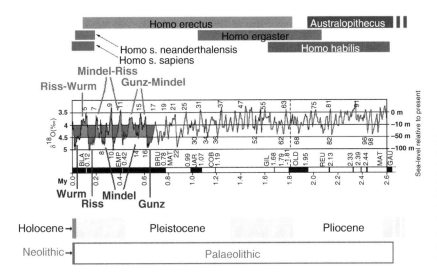

Fig. 2.24. Close-up of the most recent part of the geologic time scale, corresponding to the last 2.5 My or so, i.e. the Holocene, the Pleistocene, and the last part of the Pliocene Epochs. The geologic time scale is compared with the archaeological scale, the palaeoanthropological scale, the temperature scale derived from oxygen isotopes, the sea level changes, the most recent glacial periods and the geomagnetic reversals.

cross-calibrating different methods and scales. Several practical situations can be identified. For example, we may want to fix a number of absolute points in our relative sequence. This is one of the most common practices in archaeology, i.e. the absolute dating of a few key layers within an excavation stratigraphy by radiocarbon techniques, in order to insert it into a chronological model. Sometimes sequences from different sites are available and it is desirable to link them chronologically. This may be attained by the use of reference layers (guide or mark layers), which are well distinguished from the others in the sequence (by composition, chemical or isotopic markers, pollens, serially and culturally defined objects, etc.) and are present in all or some of the available stratigraphies. This is crucial if we wish to expand the interpretation of our layer sequence within a wider geographical context, for example by linking the excavated layers to the available sequences for environmental reconstruction (paleoclimate, temperature, soil erosion, vegetation and agriculture, etc.), each one made using different physical parameters and time resolution (Rapp and Hill 1998, Edwards 2001).

The **cross-referencing of time scales**, which is the combined use of independently obtained sequences, each one based on specific physical, chemical, biological, or geological parameters, is one of the exciting challenges of dating (Hedges 2001). The sequences and the data that are being accumulated in several areas serve as a strong reference framework for the interpretation of the past. Within the archaeological stratigraphy itself there is a wealth of information that needs to be interconnected and interpreted: the composition and texture of the excavated layers (studied by micromorphology, soil chemistry, mineralogy, palynologists, etc.) and the objects found (material culture) in the first place. Then, to get a broader picture, we should obtain information on past climate, temperature, landscape, environment, vegetation, population, etc. Much of this information provides independent curves and sequences that need to be cross-linked, matched, and interpreted.

Table 2.12. Synopsis of the most used dating techniques

	Method	Physical basis	Time interval
Relative dating	Archaeological seriation	Typology/Frequency sequences of objects	0–Palaeolithic
	Geological stratigraphy Geomorphology Paleoclimatology	Sedimentary and geochemical geochronology	Holocene–Pleistocene
	Chemical stratigraphy	Chemical variations	Any
	Palynology	Variation in the frequency of pollen palynomorphs	
	Geomagnetic polarity	Sequence of inversions of magnetic pole	780 ky–200 My
	Lichenometry	Growth of lichens on rock surfaces	100 y–10 ky
	Obsidian hydration	Surface hydration of natural volcanic glass	1–500 ky
	Aminoacid racemization	Time to reach racemic equilibrium of L/D molecules	1–300 ky
Absolute dating	Dendrochronology	Sequence of tree rings	0–10 ky
	Archaeomagnetism	Magnetic pole wandering	0–3 ky
	Radiocarbon decay	Isotope radioactive decay	0–50 ky
	K/Ar decay		10 ky–3 Gy
	U disequilibrium		0–400 ky
	Rb/Sr decay		60 My–4.5 Gy
	U/Pb decay		1 My–4.5 Gy
	Fission tracks	U fission tracks counting	500 My–1 Gy
	Thermoluminescence	Accumulation of electron traps in crystals	0–500 ky
	Optically stimulated luminescence		0–500 ky
	Electron paramagnetic resonance		1 y–1 My

For most of human history, i.e. the last few million years (Fig. 2.24), thanks to the joint effort of many disciplines we now have a reasonable idea of the evolution of land masses, sea level, and overall temperature changes. The extent of the ice caps in the cold periods and the feedback between temperature, ocean streams, atmosphere composition, and climate variations are still much debated (Raymo *et al.* 2006, Raymo and Huybers 2008), though there is a recent surge of paleoclimatic studies focusing on an understanding of climate dynamics and feedback mechanisms, in the hope of establishing reliable predictions for global climate changes (Vrba *et al.* 1996, Petit *et al.* 1999, Saltzman 2001). The overall curve of temperature variation for the last few hundred thousand years is fairly well known thanks to the $\delta^{18}O$ values extracted from benthonic foraminifera and δD in deep ice cores (Fig. 2.25; Box 2.v). This may be considered as a starting reference frame for general considerations on past climate and landscape

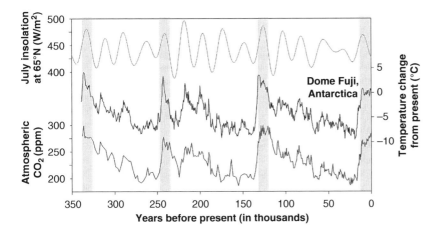

Fig. 2.25. Temperature variation in the last 350 ky as deduced from the hydrogen and oxygen isotopes in Antarctica ice cores. The data agree with the atmospheric CO_2 content trapped in the ice, with the calculated curves of solar irradiation, and with the independent curves derived from the measurement of oxygen isotopes in benthonic foraminifera. (Berger and Loutre 1991 and Kawamura et al. 2007; image courtesy of NOAA, available at: http://www.ncdc.noaa.gov/paleo/abrupt/data2.html)

of archaeological sites. The last glacial maximum is generally located between 25 and 15 ky before the present day, with two cold peaks at about 23–21 ky and 17–15 ky. After 14.5 ky a rapid deglaciation started, with most areas reaching a climate as warm and moist as today's climate at about 13.5 ky. Between 12.8 and 11.5 ky there was a sudden onset of a cool dry period (Younger Dryas), whose end marked the beginning of Holocene (Stage 1). A few more oscillations are observed before today's climate was established: the periods 9.0–8.2 ky and 8.0–4.5 ky were warmer and moister and at about 8.2 ky and 2.6 ky there were two cool phases.

Cross-linking between different information sources is a powerful and somehow unpredictable tool for novel discoveries and, above all, for setting new multidisciplinary perspectives on cultural heritage problems. To be effective, these tools commonly require extensive experimental databases, advanced data treatment and statistical tools, and a little ingenuity. A few examples are due.

The first application that comes to mind involves the use of detailed maps of stable isotope data (Box 2.v) of rainfall and water sources that are available for many geographical areas: under the assumption that living organisms are in isotopic equilibrium with the environment that they live, the isotope data may be used to trace population time dynamics, as has been recently done for a number of human remains of different ages, from the Neolithic communities (Bickle and Hofmann 2007) to the Iceman mummy in the Alps (Müller et al. 2003), to the Anglo-Saxon population in Yorkshire (Budd et al. 2001).

Chemical stratigraphy (i.e. the compositional evolution in time) may also reveal past features that are unexpected from conventional analysis. For example, the measured presence of transition metals in detailed soil stratigraphy and lacustrine sediments may be used as a proxy for past metallurgical activities, a method borrowed by environmental chemistry and geochemistry (Hong et al. 1996, Monna et al. 2004). At a different level, the interpretation of the complex evolution of glassmaking technology in Bronze Age Italy was achieved through fine tuning and detailed cross-linking of ages, typological seriation, chemical composition, and the geographical distribution of faience and glass beads (Artioli and Angelini 2009). In the latter case the chemical evolution of the glass has come to be known in such detail that sometimes it is possible to spot misidentified and unrecognized materials, or out of context glass beads. If confirmed, another surprising example is the very recently developed possibility of dating false artwork based on the presence of nuclear bomb-produced isotopes (^{137}Cs, ^{90}Sr). The technique has been developed as a result of the pressure from the burgeoning class of billionaire collectors in Russia who have recently fuelled demand and boosted prices for modern Russian artworks. The method, reportedly developed by Elena Basner, former curator of 20th century art at the Russian Museum in St Petersburg, seems simply to be based on the detection of isotopes in the painting materials (pigments, binders) that are not present in nature, since ^{137}Cs and ^{90}Sr were the results of the fallout from nuclear tests in the atmosphere carried out between 1945 and 1963. Any original painting made before World War II would show no sign of these isotopes, which are invariably dispersed and present in any material used for painting after the nuclear age.

One of the problems related to the absolute dating of archaeological samples is the **time resolution** required. For most of the techniques listed in Table 2.12 and described later (Sections 2.5.3 and 2.5.4) the errors resulting from the physical measurement make the uncertainty in the date much larger than the stringent requirements posed by art historians and archaeologist, who in many cases would like to have dates that are reliable to within 10 years or less. To date, in the best cases only dendrochronology yields dates reliable to within 1 year, and it is unlikely that most of the other available dating methods will reach such an accuracy in the near future.

Art historians sometimes need even higher resolution to decode undated documents and objects within the short time span of the artist's life. A very successful example (Giuntini et al. 1995) of high resolution indirect dating by chemical methods (PIXE in this specific case) is the story of Galileo's inks. By careful external beam PIXE analysis of the inks' chemical composition on Galileo's dated manuscripts, it has been possible to develop a very accurate map of the compositional variation in time, due to the trace transition elements present in different ink batches. Using this reference "ink stratigraphy" it is now possible to date most undated manuscripts with a resolution of a few months, and even date the notes and corrections made by Galileo on the margins at later times (Lucarelli and Mandò 1996).

2.5.2 Chemical time scales: the assessment of long term behaviour of cultural artefacts [B. Lavédrine]

Cultural properties are unique and irreplaceable, conveying all kind of significances and messages. Conservation professionals have the responsibility to preserve and promote that legacy, which unfortunately is continuously disappearing. Museum collections, in particular, suffered badly from the lack of understanding of ageing behaviour that we had on materials and on the techniques used for the manufacturing, the conservation and the restoration of artefacts that are now in critical condition. There is of course no need to reach the extreme of excluding the most fragile objects from museum collections, provided that problems are faced at the start and one is not caught by urgency and the management of intractable problems. A major task of conservation scientists is to better assess the ability of materials and treatments to withstand the test of time. To achieve this they adopt procedures developed by industry to test materials and processes, often adapting them to the specific requirements of the cultural heritage. Indeed, industrial objects and cultural artefacts by their nature and destination require different approaches and treatments. By remarking on the special status of cultural goods, whose degradation and ageing processes must be looked at from a different perspective with respect to currents objects used in daily life, the common ageing tests that are applied in conservation research laboratories are presented. They include thermal, photochemical and pollution ageing tests that are applied both for assessing and comparing materials stability and for predicting their life expectancy.

2.5.2.1 *Durability and changes*

Ageing is an inevitable fact for living beings: it brings development, but also a decline in which the final outcome is inevitable, even if scientific advances continually attempt to increase human life expectancy. For materials and artefacts the question might arise: as they are lifeless, will they inevitably age and disappear? And in such a case, how can we estimate the time they will last? In fact if, on the one hand time brings decay to materials, on the other hand time also enriches the object's significance and messages, then the assessment of the degradation and an acceptable limit may prove rather difficult to define. Such questions show the complexity of a field whose decisions are based on natural science and as well as human sciences.

The ageing of materials and artefacts

Any organized structure is subject to the laws of nature that lead towards its disorganization. This general finding concerning the impermanence of material things was formalized during the 19th century in the second law of thermodynamics: a closed system changes irreversibly towards a maximum state of disorder and in doing so it acquires a greater stability. The variable characterizing the disorder is called **entropy**, and any isolated system tends to its maximum entropy. Cultural goods are no exception, they face the test of the passing time; their destruction is inevitable and irreversible, the natural ageing of materials and objects being the consequence of the irreversibility of time. Since it cannot be avoided, ageing should be considered as an intrinsic dimension of the object, whether it has a positive impact (i.e. a valuable protecting patina) or a fatal result (active corrosion layers). Furthermore, the time layers physically stratified in past artefacts are often a substantial part of their added value, and actually what needs to be interpreted and preserved. Indeed, in Japan, some imperfections consequence of the natural impermanence of things are valued and respected, the so-called "Wabi-Sabi" [詫び寂び] (Juniper 2003). This inability to reverse the action of time has been a source of debate and reflection in regard to the restoration of cultural property, its purpose and limits. The issue continues to produce strong passions with opposing attitudes on what to do in terms of restoration of cultural artefacts. For Ruskin restoration was a fallacy: "restoration is as impossible as to raise the dead" (Ruskin 1849). Between Ruskin's views, which are respectful of minimal intervention preserving the authenticity of a work in its temporal dimension, and the more interventionist approach of Viollet Le Duc, who does not hesitate to reshape missing parts even when they never existed, there is a gradation of actions guided by scientific, aesthetic, or historical justifications. But the appreciation of the trace of passing time does not necessarily mean the desire of it, and it is exactly the role of the conservation scientist to propose solutions for avoiding it and, if not avoid, at least slow it down. To be effective, it is necessary to understand better the process that leads to deterioration. Changes are the result of a combination of many factors; one can distinguish the mechanical (manipulation, stress, repeated shocks), biological, and chemical factors. Mechanical and biological damage are most often accidental (vandalism, falls, vibration, floods…) and the effects are quickly noticed. The rates of chemical changes are generally slower and thus less obvious; they are maintained through the energy of light (photochemical damage), or ambient heat (thermal degradation).

It is reasonably possible to decrease the risk of accidents by a policy of prevention, though reducing the rate of chemical deterioration is more complex and may sometimes lead to limiting exposure or access to the artefact. It implies continuous and sometimes substantial action, which is based on the careful apprehension of two parameters. The first parameter is objective, it belongs to the field of natural sciences: the assessment and measurement of the ageing of materials and structures, i.e. the mechanisms involved and the rate of deterioration. The mechanisms of chemical degradation have been and still are a topic for fundamental research. Among the main alteration paths, one can distinguish the oxidation (radical reaction) from the hydrolysis one (ionic reaction). The other parameter is more subjective: it consists of defining what the degradation is and when it becomes relevant, thus introducing depreciation criteria for an object or some of its constituents. Appraisal of these two parameters is the first step to ensuring the preservation of cultural goods in a proactive way and it allows sustainable development of cultural heritage by applying a risk assessment strategy.

Alteration versus degradation

As soon as it is created an object begins a continuous process of ageing: at what point these natural changes become perceptible, and are called degradation, is a rather complex question, as it relates to the role of the object in its cultural context and in society. At what level of yellowing does a paint varnish or paper become unacceptable? The simplest approach is to consider that deterioration is present when alteration is noticeable, thus defining the **lowest observed adverse effect level** (LOAEL). But one's eyes are not always accurate enough to record or notice a change without a proper reference for comparison. And if the change is appreciated, when should we take this as degradation? The term degradation encompasses the notion of decline to an inferior state, such judgement being based on a set of criteria whose relevance is neither universal nor permanent. For objects that have a utilitarian function the concept of degradation or life expectancy is straightforward: it corresponds to the state or the length of time after which the objects no longer properly guarantee the functions for which they were conceived. We commonly say about a light bulb, a battery or a car it is "dead" at a certain point. Our common sense tells us when it's time to repaint shutters or renovate a facade because the level of degradation is no longer compatible with the protective function and aesthetics of these finishes. Approaching the degradation of cultural goods is somewhat more complex. It is recognized that an old object is marked by time and usage, though it is not always the degradation itself that may be a problem, but rather the speed at which it occurs. The museum visitor readily accepts that a two-centuries-old document has yellowed, but they will be surprised and may complain if this occurs on a print that is a few months old. Therefore, the change of the condition of an object over time does not necessarily mean the loss of value or functionality. Time shapes artefacts and with its marks it creates culturally significant signs, which are sources of values, like a scratch on the Stradivarius of Mstislav Leopol'dovič Rostropovič, which is a carefully kept track of a spur of Napoleon's boot. The graffiti on the walls of Pompeii, the cracks on a Flemish primitive painting, or a Renaissance majolica are esteemed and far more appreciated than identical cracks on the facade of a contemporary building. The emotional, historical, or aesthetic significances attached to cultural artefacts, just to name some of the values they embody, are neatly distinct from the scope of the rational considerations applied to objects of consumption.

The difficult task for conservation scientists is to identify the relevant damages that may occur over time and eventually link them to the time lapsed (i.e. determine the change rate), and to the specific environment or conditions of use. The conservation scientist also attempts to assess the stability or permanence of materials and treatments, though one should be cautious about using such a vocabulary. The terms "stability" as well as "permanence" are ambiguous, if one refers to their definition, "resistant to change". We know that nothing lasts for ever and so each professional category defines its own time scale. In his famous metaphor Bernard le Bovier de Fontenelle (1686) nicely reminds us of the subjective perception of time: "If the roses, which last only a day, wrote histories and left memoirs for one other to read...they would say: 'we have always seen the same gardener, he is the only one within rose-memory, he has always been the same as now, surely he does not die as we do, he even does not change'". For museum professionals and artists, human life was certainly the reference scale. The objects that survived several generations would be described as stable, as opposed to those that change drastically during the lifetime of their creator or owners. This would explain the commonly cited threshold of 100 years found in the literature and that R. Feller (1994) defined as the limit beyond which a material is regarded as having an "excellent stability". Nevertheless, this is not a standard and in the world of industry an excellent stability might be three or four decades (consider reinforced concrete), or even just ten years (how long should a car last?). To avoid such misunderstandings, conservation scientists have attempted to obtain more precise figures on the rates of alteration of materials and artefacts, and have contributed greatly to developing accelerated ageing tests adapted the cultural heritage field.

While it is easy to assess the behaviour of ephemeral materials by simple observation after a few years of use (natural ageing), for materials that have a longer lifetime it is necessary to set up experimental procedures known as "accelerated ageing tests." For the sake of thoroughness, some prefer the name "artificial ageing tests", as it seems quite difficult to simulate the effects of time. In fact natural ageing is a combination of factors (cycles of temperature, humidity, light, radiation, pollutants, stresses...), some of which can act synergistically, so that it is extremely difficult to anticipate their effects. Attempting to replicate these processes in a short time span can lead to erroneous results. Accelerated ageing is not a foolproof method but it is one of the few tools available to conservation scientists, even though it has shown its limitations, and for this reason it is a hot topic for much criticism and mistrust.

2.5.2.2 Artificial ageing methods

Ageing could be defined as all phenomena that consist of a slow and irreversible evolution of the structure and/or composition of a material under its own instability, interaction with the environment, stresses or a combination of several of these cases. It may be useful to distinguish physical ageing from chemical ageing. In the first case, there is no alteration in the chemical structure; in the second case, there is a change in the chemical structure of the constituents.

Accelerated ageing is applied to materials (or objects) in order to reproduce, in a controlled way and a relatively short time, their behaviour under normal conditions of use. In the field of cultural heritage, these tests can be made to predict the behaviour over time of new materials but also to judge the effectiveness of some conservation treatments and predict the influence of the environment (humidity, pollution, radiation...) on materials and artefacts. In industry, furniture is submitted to mechanical ageing tests by accelerating the frequency of use: an automatic mechanism opens a drawer several thousand times to determine the number of openings after which breakage or damage occurs. A chair will be subjected repeatedly to pressure by a heavy weight so as to identify weaknesses in its structure that could lead to a break over time. Such tests are meant to induce a fast ageing. For museum objects chemical ageing is often more relevant than mechanical ageing. In fact, stored and displayed items are rarely subjected to intense stresses. If this is the case, for example when objects are moved for exhibitions, the priority would be, above all, to reduce these constraints (even though mechanical properties can be used to measure the chemical ageing of a material, for instance through the loss of elasticity or tensile strength). Weathering is also rarely applied to simulate indoor museums conditions. Conservation scientists separate the factors of aggression in order to understand their respective effects better and discriminate the risks associated with each of them. In order to do so, they individually apply thermal, photochemical (light, radiation), or biological (microorganisms) ageing, and

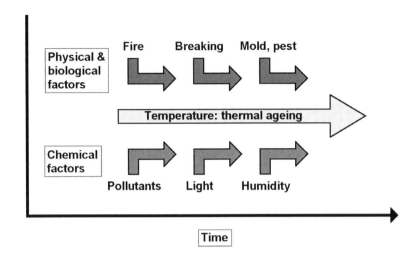

Fig. 2.26. Factors influencing the ageing of an object. Natural ageing of matter is a result of the combination of an intrinsic continuous thermal ageing with other constraints, applied permanently, shortly or regularly. These could be physical, mechanical, biological or chemical factors. The uncertainty of their respective importance and the difficulty in accelerating their effects makes an accelerated ageing test sometimes unreliable.

chemical tests (NO_x, SO_2, ozone, etc.) (Fig. 2.26). Thus, when one intends to expose coloured artefacts to intense light, the light fading and photochemical behaviour properties are first addressed,. In other cases, scientist will focus on resistance to pollutants, or ageing in the dark (thermal stability). If several of these aspects are relevant, several tests will be carried out.

Thermal ageing

Influence of temperature on the ageing of materials

By comparing objects (books, photographs, plastics artefacts...) stored in tropical climates, to similar artefacts kept in temperate latitudes, we notice that the latter are often in better condition. The heat (and humidity) seems to accelerate changes in many organic materials. Science provides an explanation to this phenomenon. Even if solid matter presents a static appearance at the macroscopic scale, at the atomic scale atoms and molecules are subjected to an agitation, called thermal agitation, which increases with temperature (Fig. 2.27). When the temperature drops thermal agitation decreases, then stops almost completely at absolute zero (i.e. zero kelvin, 0K = −273.15 °C), apart from a small residual vibration called the zero point energy, which arises from the fundamental properties of quantum oscillators. The disorder due to thermal agitation is of course far more important in gases and liquids than in solids. It is the source of collisions between atoms. The chemical changes are the results of these collisions, when the chemical species collide under favourable orientations and the energy is high enough. The energy necessary to "activate" the transformation is called the **activation energy** (Ea), and is expressed in joules per mole (J/mol) (Table 2.13): the lower the activation energy needed for a chemical reaction, the faster are the changes. At room temperature, the collisions having the required level of energy occur less frequently than at a higher temperature. Thus by increasing the temperature, the molecular agitation increases, the collisions with higher energy are statistically more numerous,

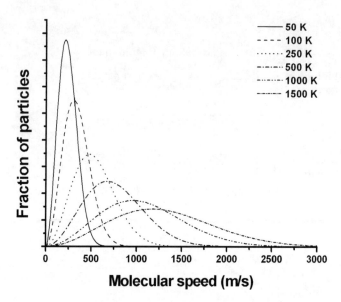

Fig. 2.27. Maxwell–Boltzmann distribution of the kinetic energy of molecules according temperature. (See for example Barry *et al.* 2000, p. 1056–1079.)

Table 2.13. Observed activation energy values for the thermal ageing of a number of materials (from Michalski 2002)

Material	Activation energy
Colour degradation	80–117 kJ/mol
Cellulose acetate hydrolysis	84–92 kJ/mol
Video tape binder deterioration (PUR binder)	68–83 kJ/mol
Yellowing of dammar resin	63–70 kJ/mol

and the transformation of matter is more rapid. By lowering the temperature, thermal agitation is slowed down and the deterioration is slower. This principle is applied for the long term conservation of unstable artefacts that are stored in cold storage to slow down their degradation process: as we do when we store food in a home refrigerator. Cold storage is routinely used for movies, films and colour photographs, though in special cases it is also used to preserve unique items, such as the Eneolithic Iceman mummy, which was exceptionally preserved in ice for more than five thousand years in an Alpine glacier, and must now forcefully be preserved at a low temperature (http://www.archaeologiemuseum.it/f01_ice_uk.html). Conversely, by exposing materials to higher temperatures they age faster.

Accelerated ageing tests

The procedure for thermal ageing consists of hanging materials in a climatic chamber (an oven with a controlled temperature and humidity) set at a temperature that is higher than the room temperature. It is chosen to be high enough to cause significant changes in a short time (a few days to a few weeks) but low enough not to induce changes that will never happen under normal conditions of use. Two problems may indeed occur if the test temperature is too high. First it may change the physical state of the material (a polymer for instance), which ages in conditions that are different from room temperature. Such a situation can lead to a wrong prediction. For instance, in a study of laser print stability, at the tested temperature the toner binder softened and the inks sunk into the paper fibres. As a result, the inks had a much better resistance to abrasion that they would have had under natural ageing. Migration of plasticizers from the polymers may also occur at high temperature. But the most common risk is to promote reactions that never occur at room temperature. Indeed nature offers several reaction paths and a higher temperature may provide enough energy to induce one mechanism over another. The didactic example is the chicken's egg, assuming that we want to find the conservation properties of an egg at room temperature by applying an accelerated ageing test. If the egg is incubated at 100 °C for a few minutes, one get a hard-boiled egg, such an experiment certainly does not tell us how to predict what will happen to an egg at room temperature. This is without doubt one of the major causes of failure of predictions based on tests at elevated temperatures. We must therefore choose the best compromise between a temperature that is as close as possible to room temperature to avoid unexpected phenomena and a relatively high temperature so that the test does not last too long. The range of tested temperatures commonly spans from 50 °C to over 100 °C, depending on the materials and on the time that can be devoted to experimentation. For example, the established paper standards Tappi plans for 105 °C (dry heat) or 90 °C and 25% RH, i.e. relative humidity (moist heat), while the ISO proposes 80 °C and 65% RH.

From a practical point of view, a series of similar samples is hung in a test chamber at a given relative humidity and temperature. Samples are regularly removed to assess intervening changes. This is done through monitoring specific properties: mechanical (elasticity, flexibility, tensile strength, viscosity...), chemical (acidity, level of oxidation, depolymerization...), colour (fading, yellowing....). The results are plotted on a graph of time versus changes (%). The chemical alterations generally precede physical changes, and therefore chemical properties commonly characterize the stages of ageing much more finely. For a polymer, the shift of the macromolecular weight distribution occurs before a tensile strength change is noticeable. Indeed ageing can have effects that can be measured either from a chemical point of view (molecular state and distribution), a physical point of view (strength), or an aesthetic point of view (colour, texture, shape, etc.): it is necessary to determine which of these criteria are important. Paper discolouration is less critical if it involves only the aesthetic changes of enclosures, whereas it may be crucial to signal and assess for a drawing's supports. Therefore it is necessary to select properties that are most relevant to accurately characterize the ageing as well as their relevance to the object's deterioration.

The comparison of the behaviour of materials at high temperature illustrated in Fig. 2.28 allows one to draw conclusions about their respective stability at room temperature. It may be satisfying when it is possible to make a choice between various materials in order to select the more efficient. This comparative approach does not prevent the risk of errors. If A is

144 *Overview of the analytical techniques*

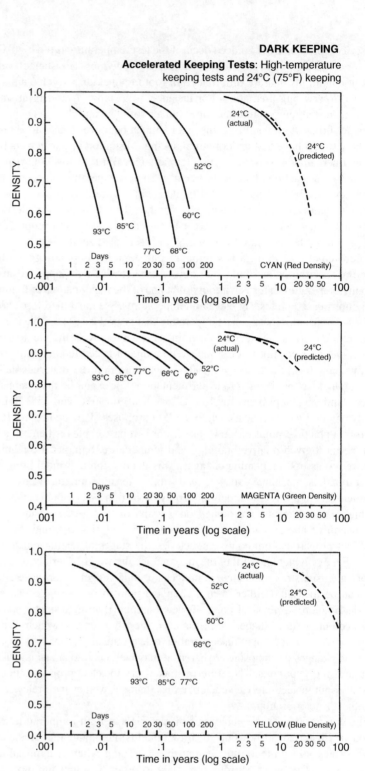

Fig. 2.28. Thermal ageing of colour photographs: fading of Yellow, Magenta and Cyan Dyes in a Kodak EKTACOLOR 74RC paper. The density loss of the three dyes over time is obtained by artificial ageing at 93, 85, 77, 68, 60, 52 °C and room temperature. Accelerated ageing tests show that the Cyan dye is the first to fade during dark storage, then the yellow dye, the magenta being the most stable. The predicted fading curve at room temperature is extrapolated by using Arrhenius relationship. (Source : Evaluating Dye Stability of Kodak Colour products. Costumer Technical Service. CIS n° 50-1, Rochester: Eastman Kodak Company, January 1981)

more stable than B in a test at 80 °C, it is not proven that the same is true at 20 °C; the causes and explanation of this phenomenon will be given below. In other words, the winner of a race in the tropics may not be the favourite in a temperate climate. The main disadvantage of an ageing test at a single temperature is the impossibility of obtaining an absolute time scale of the changes: i.e. the number of years of natural ageing that correspond to a few weeks of ageing at 80 °C? To solve this problem we have to delve into the measurement of kinetic rates.

The prediction of life expectancy

Answering the simple question "How long will it last?" requires a specific approach that in the past led to the introduction of the concept of **life expectancy** (LE). This expression was borrowed from the statistical measure of the average life span of a specified population. It is true that the use of the word "life" might seem inappropriate with non-living matter, but this terminology is inherited from the field of science and industry where it is very often used for matter, i.e. the "shelf life" of a product or "life time and half life" for the decay of radioactive atoms. In the same manner the term "life expectancy" has been applied to cultural heritage. For archival documents, "the glossary of terms pertaining to stability" (*ANSI/NAPM IT9.13-1996*) defines life expectancy as "a length of time for which information can be retrieved without significant loss when properly stored under extended term storage conditions". A life expectancy of 100 years (note that LE = 100) means that for 100 years the material will fulfil its initial functions properly. If the function of a utilitarian object might be easy to determine, it becomes more complex for a museum object (Section 2.5.2.1). The LE quantifies the persistence over the years of a property of the material or the object, Thus, contrary to what the words might suggest, once the time corresponding to the life expectancy is reached, it does not mean that the object is completely destroyed but simply that one of its properties has reached a threshold. It might be a mechanical or chemical property: tensile strength, acidity, colour, etc., and it corresponds, possibly, to noticeable damages.

The assessment of life expectancy is based on a law introduced in the late 19th century by Svante Arrhenius to describe the kinetics of all reactions that are activated by temperature. The eponymous equation shows how the kinetic rate constants vary with temperature. If the mechanism of the reaction does not change with temperature, the measurement of the isothermal deterioration rate of a material at a few high temperatures allows extrapolation at other temperatures. In practice, accelerated ageing at four or six different temperatures (for example: 60, 70, 80, 90 °C) is carried out. The length of time taken to reach the threshold of change (for example, a colour fading of 10%, 20% or 30%, Lavédrine 1996) at each temperature is then marked on a graph (i.e. inverse of the temperature in kelvin versus logarithm of time). The points are on a straight line that can be extended to lower temperatures. It is thus possible to determine the time needed to reach the same level of damage at any other temperature (Fig. 2.29). The slope of the line allows an estimate of the activation energy of the reaction. In much the same way, performing the test under different humidity conditions gives an appraisal of the role of humidity on ageing.

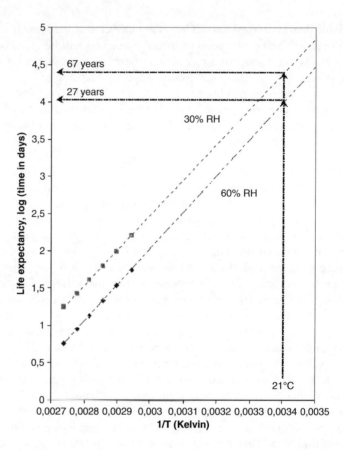

Fig. 2.29. Arrhenius plot and extrapolation of the life expectancy of a cellulose triacetate film base. Result of accelerated ageing at elevated temperature (92, 87, 82, 77, 72, 67 °C) at 30% and 60% relative humidity allow the plotting of the Arrhenius line and a prediction of the life expectancy (time to reach the vinegar syndrome decay) at any other temperature.

Some scientists prefer to use alternative extrapolations, such as the Eyring–Polanyi equation, rather that the Arrhenius one, though the experimental practice remains the same. Whatever the theory adopted and care with which the tests are conducted, these methods have their limitations. Indeed, they are based on a number of simplistic assumptions (reaction kinetics of the first order, validity of the law, no interactions with physical phenomena...).

There are cases where Arrhenius' law is not followed, a so-called "non-Arrhenius" behaviour (see for example Bilz and Grattan 1996). Furthermore, as pointed out earlier, higher temperatures can induce different mechanisms: the data obtained at these temperatures will affect the slope of the straight line and so induce inaccurate extrapolations. Finally, uncertainties related to the experiment may results in a margin of error so wide that any extrapolation is unjustified. For example, the life expectancy of a motion picture film may be 80 years, with a confidence range from 10 years to 5000 years! Nevertheless, this laborious approach has advantages: it allows one to not only estimate the duration in years but also to verify that the relative stability of materials is the same at high temperatures and at room temperature. When the Arrhenius straight lines of two materials crosses between the test temperature and storage temperature, the order of comparative stability is reversed before and after the point of intersection. Such a phenomenon goes unnoticed during a test at a single temperature. This methodology has also been standardized to estimate the stability of colour photographs, compact discs (CDR, DVD,...; Nugent 1992) and so on. Comparison with natural ageing indicates a good correlation, even though differences have been noticed. These quantitative approaches to the durability of materials and artefacts are the subject of much criticism, but the results are a valuable aid to the knowledge of materials and to collections management. The most relevant studies on this subject are those on paper with Sebera's isopermanence curves, on cinematographic films, colour films and on climate (*Climate Notebook*, http://www.climatenotebook.org/) by the Image Permanence Institute (Rochester, NY).

Application to collection management: the influence of thermo-hygrometric change and the index of permanence

Assuming that we can reliably estimate the effects of a constant temperature or humidity, predicting the effect of their changes or cycling is another difficulty. It is nevertheless very important to determine what fluctuation may be allowed in a collection without causing further deterioration. There are heated debates between conservation scientists on this topic (Camuffo 1998, 2004). A material subject to 50% RH, 20 °C will deteriorate slowly if it is kept in conditions under cycling ranging from 40 to 60% RH and 15 °C to 25 °C? Have seasonal fluctuations an impact on preservation compared with conditions stabilized at 50% RH, 21°C? Indeed the heat of summer accelerates the speed of deterioration but the cool temperatures in winter slows it down. How can one assess such impacts? From a theoretical point of view, it is easy to calculate the permanence index (p) from the kinetic equations (using Arrhenius or Eyring theory and simple kinetics) if we know the activation energy. The permanence index is the ratio between the reaction rate under reference conditions (let's says 50% RH, 21°C) and the rate of deterioration under other conditions. If p is greater than one, preservation is improved, if p is less than one, artefacts will deteriorate faster than the reference condition. For instance, $p = 2$ means that the artefact will last twice as long as in the reference environmental condition, conversely, $p = 0.5$ means that the degradation process is twice as fast. The permanence index is calculated as follows:

$$p = \frac{k_0}{k_x}$$

The rate constant of the reaction is expressed by:

$$k = A \cdot exp\left[\frac{-Ea}{RT}\right] \cdot (HR)$$

The permanence index is then:

$$p = exp\left[\frac{Ea}{R} \cdot \left(\frac{1}{T_x} - \frac{1}{T_0}\right)\right] \cdot \frac{(HR)_0}{(HR)_x}$$

where E_a is activation energy (J/mol), A is the frequency factor (constant), R is the gas constant (8.314 J/K·mol), T_0 is the reference temperature (K), $(HR)_0$ is the reference RH, T_x is the operation temperature (K), $(HR)_x$ is the operation RH.

For some materials, such as cellulose, the "humidity" factor may interact to the power 1.4; however even if such a value slightly amplifies the role of moisture, it does not substantially change its impact on the life expectancy compared with the impact of the temperature, which remains the predominant parameter. If one considers that the relative humidity fluctuates from 20 to 80% RH, the ratio varies in the range 0.25–4.0. Thus, the index will in the worst case be permanently divided by 4 and, in the best case, it is multiplied by 4. This is not as significant as the impact of a slight temperature change. For the sake of convenience, D. Sebera uses a graphical representation where each curve represents an isopermanence curve. When climatic conditions are changing (i.e., increasing humidity and temperature decreases, or vice versa), as long as one stays on the same line, the life expectancy of the subject matter remains the same. When one moves off the line, another line is found with a higher or lower permanence index. It is possible to calculate the index of permanence in fluctuating conditions, daily or seasonal climate changes. The calculation of the index of permanence is then made using an average speed of alteration. It can be assessed using the formula:

$$\frac{1}{p} = \sum_i \alpha_i \frac{1}{p_i}$$

where p is the average permanence index, α_i is the fraction of time for which collections are submitted to the stable condition (i), and p_i is the permanence index at the temperature (T_i) and at the relative humidity $(HR)_i$.

Application to collection management: the acetate film base as reference material

A similar approach was introduced by the Image Permanence Institute (IPI: http://www.imagepermanenceinstitute.org/). This work originates in the study of the deterioration of cinematographic films on a cellulose acetate support. The polymer hydrolyses over time giving out acetic acid, the "vinegar syndrome", so-called by film archivists. The total loss of the film occurs some time after the release of the acetic acid. Life expectancy is based on an acidity level at which the degradation start to increase exponentially. After performing multiple accelerated ageing tests on acetate base films, IPI was able to produce, by applying the Arrhenius relationship, a set of consistent data giving the life expectancy of a film, whatever the conditions of temperature and humidity . IPI then proposed using this data to reflect the impact of climate on any type of organic-made collection. In other words, if the life expectancy of acetate base films increases under given climatic conditions, this would be true for any other collection consisting of organic materials. The figures in years should not be regarded as representing the life expectancy of a specific material in the collection but as a reference indication about preservation. It is called the **preservation index** (PI) when thermo-hygrometric conditions are stable and the **time weighted preservation index** (TWPI), a sort of average of preservation indices, when environmental conditions fluctuate. This tool is somehow more specific for decision makers since it gives a scale in years, rather than a rate of acceleration. By dividing the preservation index by that at reference climate conditions, one gets a hint of the Sebera permanence index (Sebera 1994).

Once again, the benefit of this approach is not so much in getting absolute values but in determining the impact of thermo-hygrometric conditions on the collection to make decisions on the environment of the collection. Nevertheless these tools have some limitations. Assuming that the kinetic models adopted to establish these data are verified, only chemical ageing has been taken into consideration and physical and biological ageing has been ignored. With high relative humidity, there are of course biological risks that can lead to the development of micro-organisms, but also physical changes. Dry air induces shrinking, brittleness, and cracking in some organic materials. Frequent and rapid cycling in temperature and/or humidity can lead to adsorption and desorption of moisture, thus the impact on the chemical degradation is still not clearly identified and is not covered by this approach. For instance, it is known that in papers and textiles, peroxides are formed at the dry–wet interface that is at the origin of the development of brown lines (the so-called "tidelines").

Photochemical ageing

Influence of light on the ageing of materials

Electromagnetic radiation (γ-rays, X-rays, UV, light, etc., Section 2.1, Fig. 2.2) interacts with matter and the consequences of the interaction are to be specifically considered. The effect caused by radiation is called photochemical damage because the energy that brings about the deterioration comes from photons and not by thermal agitation. For museum objects, light, with or without ultraviolet radiation, is the main factor in photochemical degradation. While many collections such as those in libraries and archives are usually kept in the dark, others are permanently exposed to natural or artificial lighting. Thus carpets, tapestries, curtains, costumes, prints, drawings, photographs, paintings, etc. have often suffered significant fading: a comparison between the part of an object exposed to light and another that was in the shade, reveals the deleterious action of light, which adds and sometimes cumulates in thermal ageing. This finding is not new but it seems that it has been somewhat forgotten in the 20th century where overexposure has led to the rapid deterioration of many artefacts. In the dark, these reactions do not occur.

The light can be imagined as a stream of photons, whose energy depends on the wavelength. Blue light (short wavelength) is made up of photons with higher energy than green or red light (long wavelength). The simple relationship between the energy of the photons and the wavelength of the associated wave has been treated in Section 2.1.

The more energy the radiation has, the more it can affect matter, provided that absorption is present (Grotthus–Draper's Law, Section 2.1.1). In fact, the photochemical action is not just a question of energy; matter can only absorb some well-defined quantity of energy (quantum theory). For example, a silver photographic emulsion (black and white photographic paper) is not sensitive to red light, even if we increase the intensity (in this case only the number of photons increases, not the energy of each photon). On the other hand if it is exposed to a blue light or ultraviolet radiation, even at low intensity, a photochemical reaction will take place. A molecule absorbs one photon and then passes to an "excited" unstable state whose life expectancy is only a fraction of a second (about a nano-second). Several possibilities exist: this energy is either transferred to another molecule, or is transformed into heat (dissipation, relaxation), is re-emitted in the form of light (fluorescence, phosphorescence) or, and most critical for the object, it will provoke some chemical changes (photochemical reaction). Thus, not all photons have a chemical effect: the ratio of the number of molecules damaged to the number of photons absorbed is called the quantum yield. The quantum yield remains fairly low: for rubber (latex) it is 0.0004, for cellulose acetate it is 0.0002.

One can easily understand that the coloured materials might be vulnerable to lighting because they are coloured and are obviously absorbing some visible wavelengths. The energy provided by these absorbed photons can contribute to the destruction of the dye. It would be wrong to conclude that materials that do not absorb light (colourless materials) are not sensitive to light. First, natural daylight contains some ultraviolet radiation, very energetic radiation absorbed by some colourless organic compounds and leading to rapid deterioration. Then, the phenomena of radiation–matter interactions are complex. The degradation occurs through an intermediate compound: light energy is absorbed primarily by a chemical species in the surroundings, then it transfers the energy or reacts with the material itself. This is called a photo-initiated reaction. This can be a constituent present in very small amounts as an additive or impurity, but can also be a gas such as oxygen, water vapour or a pollutant, etc. It has been clearly demonstrated that some metallic impurities and pigments can accelerate the degradation. For instance, titanium oxide catalyses polyethylene degradation. The photochemical degradation occurs as a result of various mechanisms, such as photo-oxidation, photo-reduction or photolysis. Photo-oxidation is the most common (Fig. 2.30). Thanks to the energy from the light, the material reacts, for instance, with the surrounding oxygen. In all these reactions specific chemical species play a role: radicals (such as hydroperoxyl radicals, etc.)

Light ageing tests

The tests designed to assess the photochemical stability of a material or an object are based on the reciprocity principle of light action: short exposure to strong light has an action equivalent to long exposure under dim lighting, if the total dose is identical (as well as the light characteristic). In other words, as the total energy received by the object is the same—if it is applied within a few weeks or several years, continuously or discontinuously—the damaging consequences will be similar.

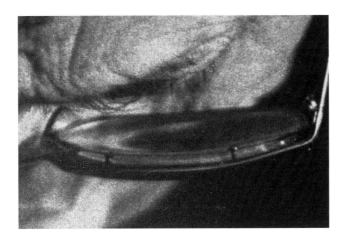

Fig. 2.30. Microscopic orange spots on silver photographic print. In the resin coated photographic paper, the so-called "RC paper", a subbing layer is made on polyethylene pigmented with TiO_2. During exposure to natural light, the pigment promotes the formation of radicals that deteriorate the polymer and oxidize the silver image causing the formation of microscopic orange spots. (Courtesy of B. Lavédrine, CRCC)
Original colour picture reproduced as plate 6.

There are two different methods for performing ageing tests. The first method, comparative, consists of exposing all the samples at the same time to strong lighting and comparing the effect of light on the samples after a certain length of time. The degradation can be measured regularly or only once at the end of the exposure. The first sample to deteriorate will be considered to be the least stable and, conversely, the least damaged will be considered to be the most stable for natural exhibition conditions. This approach, the so-called "window test", has often been applied in a simple way. Samples are simultaneously exposed to natural light behind a window for weeks or months until significant deterioration occurs. The samples' irradiation is not monitored and such a test allows only a comparative evaluation. The use of climatic ovens with controls over lighting, temperature and humidity, ensures a test's reproducibility, and the calculation of life expectancy. To perform the calculation it is necessary to measure the energy received by the samples. The units used for energy are usually the joule per square metre but such units do not convey the sensory impact of the light. Therefore, for convenience in the field of conservation science, lux·hours are used instead. It is then easier to establish a correlation with the exposure conditions in museums and recommendations rules for the exhibitions of artworks.

Energy (joules/m² or W·s/m²) = *Irradiance × Exposure time*

Light dosis (lux·hours) = *Illuminance × Exposure time*

To speed the ageing up by a factor of one hundred, one needs to multiply the illuminance by a hundred: 10 days of exposure at 10 000 lux, correspond to 1000 days of exposure at 100 lux. If a textile discolours by 30% after being exposed for 10 days at 10 000 lux (total dose = 2 400 000 lux·hours), it is likely that the same change occurs after about 3 years (1000 days) in the museum, considering 240 lux illuminance, 10 hours a day. Such a simple principle allows one to extrapolate to how many years it will take to observe the same damage under normal lighting conditions. To perform the accelerated ageing, it is important to use the same type of lighting as that used to exhibit the artefact.

If this is not the case, the effects might be different. Thus to simulate daylight exposure, scientist select lighting with a similar spectral distribution, such as a xenon source (arc lamp). It is also important to pay special attention to the content of the UV radiation, sometimes reducing or removing them using suitable filters in order to get a similarity with the lighting in normal conditions. A photochemical ageing test can be performed in combination with a particular environment: anoxia (oxygen depletion), low or high humidity, etc. The goal remains to determine the benefits or risks for the artefacts.

As for the thermal ageing test to determine ageing temperatures, the question is how far can we reasonably go in increasing the lighting intensity. The reciprocity principle seems respected in most cases, but there are a few exceptions (reciprocity failure), caused by phenomena of a physical or chemical nature. First, light heats the sample surface and might dry it. To reduce these effects it is preferable not to exceed nine times the solar irradiance and to use climatic ovens where humidity and temperature of the sample are kept at a certain level. Oxygen depletion was also cited as a source of disparity between natural and artificial ageing, especially if the materials are massive. Indeed when the luminous flux is intense, the photo-oxidation reaction is accelerated and oxygen in the substrate is quickly consumed. Thus the chemical environment (water, oxygen...) within the substrate can influence the test results and lead to erroneous data if it is different from the conditions of use. If the oxygen concentration is insufficient due to a low substrate permeability caused by a drying up of the substrate, the material appears to be more stable than in the natural conditions of use. Other reactive compounds such as ozone can be present only in particular conditions, thus creating a disparity between the natural environment exposure and the tests. If formed during the tests, the ozone action may increase the photochemical action, if absent it could not reproduce the natural ageing.

In 2002, inkjet inks were marketed as supposedly having an excellent light ageing ability. Accelerated ageing tests were conducted using the traditional protocol and these new inks showed excellent performance from the point of view of stability to light. A few months after their introduction on the market, they were withdrawn because consumers complained about fast light fading. Ozone was then identified as the predominant cause of alteration during light exposure, so that nowadays inks for inkjet printers are also evaluated for their resistance to ozone. Ozone is not the only oxidizing gas that can produce alterations when exposed to artefacts. In outdoor conditions nitrous oxides are equally damaging, though ozone occurs more frequently as it is quickly generated by UV radiation. Finally, under real exposure conditions, light cycles alternate with dark cycles. Some materials that have been damaged by light have the property of recovering, at least partially, when stored in the dark. Therefore continuous or alternate exposure to light may not produce the same result. This phenomenon has been observed with the Prussian blue pigment, which recovers after light bleaching. In contrast, some chemical reaction induced by light can continue in the dark well after exposure. Many of these phenomena are rather difficult to assess, so that any observed differences between accelerated and natural ageing may be quite difficult to interpret.

Assessment of light stability for museum objects

The ageing tests described above are destructive and are not applied to museum objects but to replicas or dummies with no value. Based on the results obtained, it is sometimes possible to extrapolate to museums object. However, the chemical composition of an old artefact is not always completely known and, furthermore, the age and the specific history of the object may induce a different ageing behaviour with respect to modern replicas even when they are made with very similar compositions. The Carnegie Mellon Research Institute (USA) has introduced a micro-fading test (see also Box 2.q) to assess in situ the light sensitivity of coloured artefacts. For this purpose a strong light beam is focused thanks to an optical fibre on to a micro-spot (less than 1 mm^2) while another optical fibre, connected to a spectrophotocolourimeter monitors colour variations on the spot as a function of time. The test is stopped when the colour difference is equal to 5 (delta E, CIE E = L*a* b*; Box 3.e). Such a low variation over a small area is imperceptible to the naked eye. The length of time required for such a colour change is recorded. The experiment is replicated on the blue wool standard samples. The stability of the measured area of the museum artefact is then given in reference to the blue wool sample, which exhibited about the same speed of fading.

Blue Wools Lightfastness Standards consist of eight samples of wool dyed with blue colorant whose light stability increases with number, blue wool standard n.1 being the least stable and blue wool standard n.8 being the most light stable (ASTM D5398-97 2003: *Standard practice for visual evaluation of lightfastness of art materials by the user*; and ISO 105-B08 1995). This standard was originally introduced to meet the needs of the textile industry. It has been adopted by the conservation world and used as a tool to define the scale of light sensitivity of different materials, and also as a dosimeter to evaluate the lighting conditions in historic houses and museums. Figure 2.31 shows the experimental values of a microfading test performed on three of the blue wool reference materials.

Museum objects are categorized into three families with increasing stability (Feller and Johnston-Feller 1978). The first category corresponds to the most light sensitive artefacts that have a similar light fastness to BWS1–BWS3; the second category includes artefacts with a light fastness of BWS4–BWS6, and the last category corresponds to more light stable objects having a light fastness of BWS7–BW8. For these three categories, annual exposure limits (in lux·hour) has been set in museums' policies. This is the reason why it is very convenient to connect the light fastness of an object to a blue wool standard sample.

As a conclusion to this part, we have to bear in mind that the microfading test is limited to the monitoring of a colour change. If the object becomes fragile or cracks, or has a change in gloss during light ageing, it will not be assessed. This method is also an accelerated ageing test using an extremely high illuminance. It suffers of the same disadvantages and limitations as the more conventional light ageing test described above (reciprocity failure, etc.).

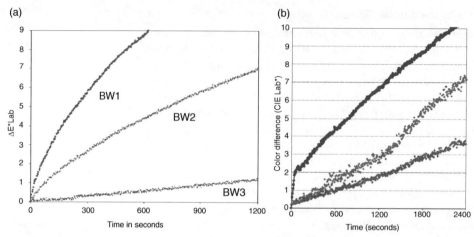

Fig. 2.31. (a) Microfading tests on standard samples No. 1, 2 and 3 of blue wool. (b) Examples of chemical changes: dye fading on an early colour photograph (Autochrome). The microfading test on dyed starch grains shows that the blue grains (blue curve) are fading faster than the red (red curve) and green grains (green curve).
Original colour pictures reproduced as plate 7.

Chemical aggressions

Influence of pollutants on the ageing of materials

In their environment, artefacts are in contact with volatile compounds that can play a role in their degradation process. Oxygen and moisture are natural constituents of air and we have already evaluated and discussed their importance during ageing. They are taken into account when implementing thermal and photochemical accelerated ageing. Ageing tests can be performed in anoxic conditions to evaluate the benefit of an oxygen-free environment for storage or exhibition. For common pollutants, specific tests have to be conducted. While it is recommended that the artefacts are protected by keeping them in a clean environment, it remains necessary to determinate their susceptibility in order to design appropriate conservation strategies. Chemical aggression has various origins, whether indoors or outdoors. Internal pollution sources include the construction materials and materials in the vicinity of the object. These are volatile organic compounds, organic acids, aldehydes, ketones (...), which contribute to the same extent as urban pollutants, to damage museums collections. The ageing tests have shown that papers deteriorate faster in a stack compared with separated sheets because of the accumulation of volatile organic compounds. Thus, pollution ageing test have often been applied to leather bindings, parchments, graphics documents and photographs, metal objects (...) to evaluate the benefits of a preventative treatment (paper deacidification, protective coating, barrier envelope, etc.) or the susceptibility of the material itself.

Pollution ageing tests

Artefacts or materials are put in a climatic chamber regulated at room temperature (Fig. 2.32), with concentrations ranging from a few ppm to several tens

Fig. 2.32. View of pollution chambers at the Centre de Recherche sur la Conservation des Collections (CRCC). Pollution tests for paper documents. (Courtesy of B. Lavédrine, CRCC)

of ppm of one or more gaseous pollutants such as sulphur dioxide, nitrogen oxides, ozone, hydrogen sulphide, hydrogen peroxide, organic acids and so on. After exposure, the effects on materials are evaluated and compared. The test allows one to determine those that deteriorate faster and those that are more resistant to the pollutants' aggression. Another approach is to assess the load ($\zeta = $ *Pollutant concentration* $\times$ *Exposure time*) that produces a certain level of alteration, and then to apply the reciprocity principle. The life expectancy defined previously can be calculated by a rule of thumb. For example, if a fading of 10% is measured on an object after 5 days to 10 ppm of nitrogen oxides, one concludes that in an urban atmosphere of 0.05 ppm, such changes occur after 1000 days ($\zeta = 1000 \times 0.05 = 10 \times 5 = 50$ ppm·day). Again one may argue that this does not reflect the real conditions of ageing, especially if the synergic effect of other gases in the atmosphere such as other pollutants or moisture can speed up or catalyse the chemical reaction in a compound. Sometimes under dry conditions some pollutants might be less damaging.

Weathering tests

The industry has developed weathering tests to evaluate and compare the durability of materials exposed to outdoor conditions. These tests are very drastic and combine visible and UV radiation, heat, water, humidity, pollutants, condensation, rain, salt, fog, freezing, etc., in order to compare the effects on the materials tested. Although developed for industrial products such as cars and construction materials, weathering may also find applications in evaluating materials used for historical monuments and objects exposed outdoors (sculptures, decorative elements). Weathering is performed either outdoors in places with adequate sunshine (south of France, Florida, Arizona...) and more or less rainy, or in weathering chambers by submitting samples to cycles of light and UV, heat and/or humidity (Fig. 2.33). These tests are described in ASTM (D1014, G90, G26, D3359, D2244, D523) or ISO 11507, among

Fig. 2.33. Weathering tests at the LRMH. Weathering of stone samples on the roof of Rouen cathedral, France. (Courtesy of LRMH)

others... Weathering provides interesting data to compare materials' and objects' behaviour, but the "acceleration factor" in relation to natural ageing is difficult to estimate. Interesting long term weathering experiments have been performed to experimentally test the degradation rate of archaeological materials in outdoor earthworks (Bell *et al*. 1996, Caple 2001). Such data may be profitably used to evaluate the missing portion of an archaeological record, i.e. most of the organic-based and easily degradable materials that are long gone before the excavation.

Conclusion

It is common to hear that the objects produced in the past were of a much better quality, and more stable than those that were produced recently. Artists were indeed not very likely to produce instable artwork but one should oppose the theory that the effect of passing time has left us with only the more robust objects. Examples show that some ancient materials may also pose serious problems, such as iron-gall inks that were used for more than five hundred years that caused the destruction of the paper substrate, and some dyes used in the past have poor light fastness. It is true, nevertheless, that for a long time craftsmen were able to capitalize on and pass on their experience. Among the materials that nature offered them, they selected the most suitable for their uses. The durability of objects depends not only on the raw material but on the way of producing the object. Practices were codified, craftsmen and artists followed specific rules associating materials, techniques and gestures. Nevertheless, the ability of an object to resist the arrow of the time depends greatly on the life expectancy of the materials themselves. This empirical knowledge based on the transmission of traditional knowledge dissipated with the industrial revolution in the 19th century and with the endless introduction of new materials and products whose long term behaviour had not yet been proven. Artists and craftsmen have gradually abandoned the rules and traditional practices to follow their inspiration without restriction and without worrying about the long-term consequences of their choices. At the same time, museums, libraries, and archives started to collect all kinds of evidence of human activity: among them many were unstable, or not really made to last long. The many problems that have arisen in cultural heritage, especially with the introduction of new synthetic polymers, led industrial and conservation scientists to develop a series of tests to assess long term behaviour better. Some tests have been very useful for choosing suitable materials or treatments or in the decision-making process to determine the most suitable storage or exhibition conditions.

However, despite the quality of research carried out, we always face the difficulty of replicating the laboratory scale results for all processes that take place during natural ageing and interact with each other to modify the appearance and structure of cultural properties. Without taking extreme attitudes, either totally rejecting the validity of these tests or going on blind trust, we must admit that, so far, we have no better alternative. Users should be careful about the significance of the tests and the ways of using these experimental results. Only time will give us the answer and if one wants to get a greater benefit from natural ageing, it is necessary to document the physical and chemical condition of artefacts well and monitor their changes better over time. Colour measurements are standardized and have routinely been used over the last few decades, providing a unique and non-destructive tool to monitor colour changes of artefacts in collections. To be useful in the future such measurements must be comparable to the measurements made with future devices. The development of standardized procedures and references is a must. The same requirements are applied to multispectral imaging, gloss measurement (an important criterion which has not yet been taken into account) and to chemical characterization. Finally, finding deterioration markers, based on chemical or physical properties that characterize and inform us on the state of deterioration of a material or an artefact is a very promising approach that will contribute to better access and to understanding the ageing and deterioration of cultural artefacts.

2.5.3 Relative methods: fundamentals

As mentioned in Section 2.5.1, relative dating methods produce information on the relative age of an object or event under investigation. They do not tell us when the object was made or when the event occurred, but they allow us to set the objects and the events in temporal order (time sequence). All relative methods (Table 2.12) are essentially based on the concept of stratigraphy, derived from geology.

The basics of **stratigraphy** tells us that in normal sedimentary sequences younger layers are deposited on top of older layers. This is the law of superposition, formulated in 1669 by the Danish naturalist and theologian Nicholas Steno. However, every geologist also knows that most natural sequences are discontinuous, faulty, defective, incomplete, and sometimes even reversed by geological tectonics (i.e. older layers are found at the top of the sequence).

Archaeological layers are hardly completely reversed, though in many cases they are certainly disturbed, removed, or contaminated by subsequent human and natural activities (Harris 1979, Renfrew and Bahn 2008). The tools and concepts developed by geological stratigraphy to recognize and interpret disturbed layer sequences therefore may turn out to be very useful to field archaeologists. The key of stratigraphy is in the development of a **master layer sequence** (the analogue to the geological "stratigraphic column"), painstakingly composed through the identification and description of each layer. In the case of sedimentary geology the layers are identified by lithological characteristics and by the fossils they contain. In the case of archaeological layering we may have layers defined by excavation operations, by soil micromorphological or mineralogical characteristics, by pollen species, by human artefacts, by organisms or biological markers and so on. Each one of these sequences may be compared with a master sequence of the same kind (if available) and then jointly interpreted. The classic and controversial example of the correlation between the Thera eruption in the Aegean sea and the tephra layers found in Greenland ice cores, or the frost tree rings records in California and Northern Ireland reminds us that the correlations may not necessarily be local (LaMarche and Hischboeck 1984, Hammer *et al.* 1987, Baillie and Munro 1988, Aitken 1990). Correlation between sequences is performed using marking layers (layers having well recognized features) or, when these are not available, on the sequence chain itself. To compare long sequences statistical methods and autocorrelation functions are available, such as those routinely used in dendrochronology.

Sometimes the layers are not evident at the macroscopic level (by colour, grain size, texture, etc.), and analytical tools may be used to discriminate hidden layering (fired clays by FTIR, specific mineral components by XRD, chemical tracers by XRF). Some of these analyses may also be performed in the field, thus providing further or complementary evidence of human activity.

The most diffused methods of relative dating are briefly reviewed. Archaeological seriation techniques are treated in detail in archaeological textbooks.

2.5.3.1 Palynology

Palynology is the discipline investigating the microscopic remains of plants in soils (palynomorphs), most of which are composed of pollen grains (Fig. 2.l.1) produced by flowering plants. They are homogeneously dispersed by the wind so that they are widely distributed and statistically well represented in the soil of vegetated areas. Pollen grains are very durable in most soil conditions, so that they are commonly abundant both in archaeological layers, in natural deposits such as soil sequences and peat-bogs cores. Pollen may derive from naturally occurring plants (natural pollen) or from cultivated plants (artificial pollen). Since agriculture and other human activities may severely affect the pollen record, the pollen stratigraphy from archaeological sites must always be carefully compared with the natural profiles in the region.

Relative dating is carried out by establishing the frequency profiles of the various plant species in the stratigraphy (West 1971). Each layer unit is individually analysed, and all the pollen grains are classified by plant species and counted. At each single time, the observed plant association yields a good indication of the vegetation pattern and the climate present in the area, and the vegetation evolution along the sequence can be correlated with climatic variation and/or human activity (i.e. deforestation, agriculture, use of territory). Detailed pollen sequences have been established for many areas in the world, especially for the post-Pleistocene period.

Figure 2.34 shows a simplified pollen record from an Alpine peat bog, as an example to show how a pollen diagram works for the reconstruction of the palaeoenvironmental history of the last 15 000 years and the biochronological dating of pollen events. On the left side of the figure, a vertical scale represents the depth (in cm from the field surface). A lithological log shows the stratigraphical succession of deposits, from minerogenic fine silt sediments, to gyttja (= fine organic, limnic mud) to peat (= in situ accumulation of macrophytic plants). The uncalibrated radiocarbon ages supporting the chronostratigraphic frame of the record are also shown. Only the most representative trees, shrubs, herbs and indicators of anthropogenic activities are presented here. The complete pollen record may be consulted in the original publication (Pini 2002). The selected pollen curves in colour (green, red and yellow) are expressed as a percentage of the total pollen contained in the samples studied, excluded aquatic and peat-forming plants. The percentage curves of trees, shrubs and herbs are exaggerated ×3 (in grey), anthropogenic indicators ×5 (in yellow). The exaggerated curve is optional and frequently omitted. The main events affecting the forest history occurred at a depth of 800 cm (reforestation of the site after the last ice age), 602 cm (abrupt expansion of *Picea-Abies* forests), 100 cm (expansion of *Castanea* and *Juglans* during the late Roman times and Middle Ages). Estimated ages for these points can be derived by an age–depth model built from calibrated ^{14}C ages. The stratigraphic intervals of homogeneous pollen content are distinguished by local pollen zones (LPZ), numbered from base to top.

The black curves show the concentration of pollen grains and of charcoal particles per cm^3 of sample studied. The concentration of charred particles clearly displays two peaks, the first one occurred in the early Late Glacial period, before 12 320 ± 60 ^{14}C y BP, while a second peak during the Middle Ages represents the record of anthropogenic fire close to the peat bog, largely matching the expansion of crop fields (compare with the *Cerealia* percentage).

The last two curves before the column LPZ display the abundance of plants related to the development of the sedimentation basin: *Pediastrum* (a planctonic algae) expanded as soon as the lake turned green after the last ice age, and disappeared at the time the lake filled and the onset of peat accumulation (Gramineae and *Sphagnum* expansion).

2.5.3.2 Obsidian and quartz hydration

Obsidian is a natural anhydrous silica glass produced by rhyolitic volcanism. It is generally black and it has been one of the preferred materials used for cutting tools since prehistory because of its hardness and conchoidal fracture, which contribute to producing flakes with extremely sharp edges. The use of obsidian hydration measurements for archaeological chronometry was first proposed by Friedman and Smith (1960), and it is still very much used especially in regions where radiocarbon dating and dendrochronology cannot be used. It is based

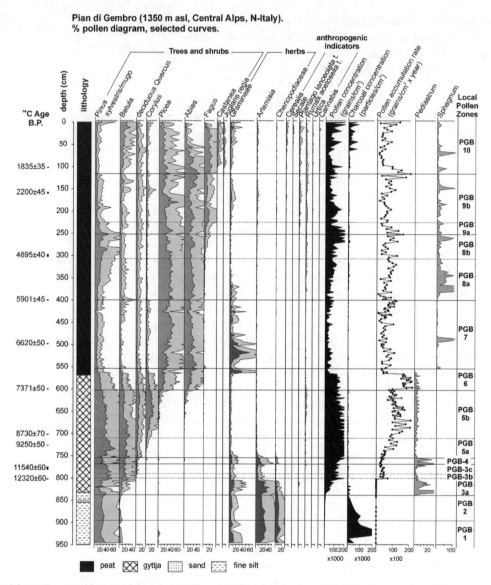

Fig. 2.34. Simplified pollen record from an Alpine peat bog. The palaeoclimatic and palaeoecologic significance of the pollen sequence is discussed in the text. (Courtesy of R. Pini, University of Milano Bicocca. Image available at: http://www.disat.unimib.it/Palinologia/Home-eng.htm)

on the principle that, when a freshly exposed surface of obsidian is exposed to the environment, water molecules present in the atmosphere or in the soil diffuse into the glass at a predictable rate. As the water diffuses into the glass it forms a micrometres-thick hydration layer (the so-called hydration rim), which can easily be recognized and measured on thin cross-sections of the sample surface using optical microscopy, by virtue of the different refractive index in

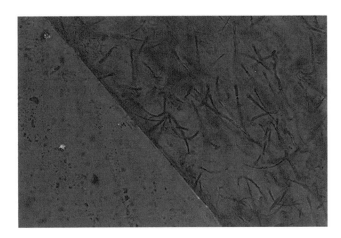

Fig. 2.35. Boundary layer separating the hydration rim from the anhydrous body at the surface of an obsidian tool. (Image available at: http://www.pacificsites.com/~hamilton/)

the hydrated layer from that of the anhydrous bulk. The boundary between the hydrated and anhydrous parts is usually sharp (Fig. 2.35) and the thickness of the hydration rim can provide estimates of the time since the flaking of the tool, if the rate of hydration is known or calculated. Alternative methods for measuring the penetration of water are SIMS (Box 2.t) and tritium exchange.

Of course the rate of hydration is dependent on a number of factors, including obsidian chemistry, the original water content of the nominally-anhydrous glass, and of course temperature and relative humidity. The multiplicity of the parameters affecting the hydration process is such that the method produces nearly-absolute dates only in exceptionally well-studied cases. Rogers (2008) reviews the accuracy and limitations of the method, which has proven to be a reliable routine tool for the relative dating of obsidian samples.

The concept of silica surface hydration has also been proposed to date quartz artefacts (Ericson 1982, Ericson *et al.* 2004). Ion-beam based NRRA and ERDA (Box 3.i) provide non-invasive techniques for the measurement of the hydrogen penetration into the material, and for the assessment of the thickness of the hydration layer. The ERDA analysis has recently been applied to the relative dating of the puzzling quartz skulls allegedly of Mesoamerican craftsmanship (Calligaro *et al.* 2009).

2.5.3.3 *Lichenometry*

The measurement of the average growth rate of lichens has been proposed as a method for dating exposed rock surfaces, and it was first applied by Roland Beschel to estimate the age of glacial moraines (Worsley 1990). It is based on the observation that the growth of each lichen species depends essentially on environmental factors and on the petrology of the rock substrate and, if these are constant, then the lichen growth rate of that species can be reliably calibrated. Of course each growth curve (Fig. 2.36) needs apposite local calibration with dated surfaces such as gravestones, buildings, landslides, etc., paying attention to environmental factors such as elevation, proximity to the sea, rock exposure, etc.

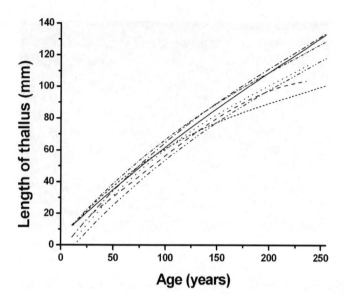

Fig. 2.36. Close agreement of seven lichenometric dating curves from southern Norway, suggesting a good overall correspondence between thallus size and glacial moraine age. (Adapted from Matthews 1994)

The method has been mostly applied to geomorphological surfaces that are hundreds of years old (Innes 1985), though in principle it could also be applied to much older surfaces, maybe up to 10 ky old. The precision and accuracy of the method greatly depend on the calibration and on the measurement techniques (http://mc2.vicnet.net.au/home/date/web/lich.html), nonetheless lichenometry could be profitably used to date surfaces that are hard to date with other techniques, such as abandoned quarries and, especially, rock art (Bednarik 1996). There is in fact great potential in the use of lichenometry in dating carved rock surfaces, especially considering that it may be combined with other techniques that are applicable to organic materials, such as aminoacid racemization and radiocarbon dating techniques. It is therefore unfortunate that many petroglyph surfaces are totally cleaned of lichens, on the presumption of their damaging and degradation effects. On the contrary, there is a rising feeling that lichen removal actually helps to accelerate the chemical attack of the rock surface, besides destroying invaluable time information (Childers 1994, Tratebas and Chapman 1996, Walderhaug and Walderhaug 1998).

2.5.3.4 Chemical changes in materials

Any continuous chemical change may be used to define the direction of the arrow of time and thus to distinguish older from younger materials. The more the rate of the reaction is known, is predictable, or calibrated, the more the method becomes quantitative and the estimated dates reliable. All chemical changes that are monotonically increasing or decreasing the concentration of a chemical species in a given compound may be used, though by far the most studied variations are related to phosphatic bone materials (Mays 1998).

Bone (Section 3.7.1) is an inorganic–organic composite material. The phosphate inorganic biomineralization is hydroxylapatite, $Ca_5(PO_4)_3(OH)$, which after burial undergoes the well-known OH ↔ F substitution generally called

bone fluorination. In time the hydroxylapatite–fluorapatite solid solution $Ca_5(PO_4)_3(OH,F)$ becomes progressively enriched in fluorine (Johnsson 1997), the F-intake in the phosphate depending, among other parameters, on the burial conditions, the fluorine concentration in the soil fluids and the temperature. Depending on the ionic concentration in the percolating solutions, other chemical changes may take place at the apatite cation structure sites, involving for example Ca ↔ U or Ca ↔ Sr substitutions. The organic, protein-based part in fresh bone is largely dominated by collagen, which decomposes during diagenesis so that the bone progressively loses the organic components, especially nitrogen. After burial therefore the bones become enriched in F, U, Sr and depleted in N. The relative amounts of these elements may therefore be used to build a chemical stratigraphy of loosely found objects or fragments if they have been exposed to the same burial environment. The analysis is sometimes called the fluorine–uranium–nitrogen test (FUN test) and it may be applied to bone, antler, ivory and teeth. Although largely superseded by modern techniques based on radioactive decay, a large number of analyses exist on fossil remains (Oakley 1980).

The method was used in 1949 to expose the Piltdown forgery (Feder 2008). By showing that the fluorine content of the jaw and skull of the alleged Piltdown man were radically different Kenneth Page Oakley, a professor of anthropology from Oxford University, chemically confirmed earlier conclusions based on bone morphology that the 'missing link' fossil was actually a hoax made of a fossil ape jaw and a modern human skull.

The fluorine uptake method has also been applied to archaeological flints, by measuring the depth profile of F in the material (Walter *et al.* 1992).

2.5.3.5 *Aminoacid racemization method*

All but one (glycine) of the aminoacids forming the proteins of the living organisms are chiral, that is to say they can assume two molecular structures that are mirror images of each other, called the laevo (L) or left form and the dextro (D) or right form, exactly as are our human hands. For unknown reasons, attributed by many to the extraterrestrial origin of life, the greatest majority of organisms proteins are not formed by the thermodynamically stable statistical mixture of the two aminoacid forms (i.e. D/L = 1, called a **racemic mixture**), but they are composed only of the left handed (L) forms. The L-only molecular mixture (D/L = 0) is metastably maintained by the molecular organization of life until the organism dies, at which point thermodynamics takes over and slowly reverts to the racemic mixture (**racemization process**). After death therefore the D/L ratio of each amino acid increases with time from 0 to 1. Racemization is a chemical reaction and, like all other chemical reactions, occurs faster at higher temperature. Furthermore the rate of conversion depends on the molecular structure, so that different aminoacids have different rates of L ↔ D conversion. Racemic mixtures of aminoacids have been found in old organisms that have reached equilibrium, such as old fossils that are millions of years old.

Amino acid geochronology is best suited as a relative-dating tool, or as a calibrated-dating method in conjunction with other dating techniques. It is applicable to a wide range of fossil organisms (molluscs, ostracodes, foraminifera, bone, egg shells, and teeth) and, depending on the burial and diagenetic conditions, it is also applicable to a wide range of time scales (decades to millions of years). It is particularly useful for organisms older than the range of radiocarbon dating (about 50 ky, Table 2.12), for which few alternative geochronological tools are available (Rutter and Blackwell 1995, Wehmiller and Miller 2000). Despite the fact that the conversion rate is highly dependent on the environment and therefore the technique needs cross-validation and calibration to be quantitative, it has been successfully applied in dating a number of interesting cases, including the age of the sediments at Pakefield, UK, which is the earliest record of human occupation north of the Alps (Parfitt et al. 2005).

2.5.3.6 Geomagnetic polarity reversals and archaeomagnetism

Archaeomagnetic and paleomagnetic dating techniques rely on the fact that the Earth's magnetic field varies over time. The Earth's magnetic field is produced by the dynamo effect of electrically charged particles moving by convection within the Earth's molten outer core, and by conduction within the Earth's solid inner core (Anderson 1989, Jacobs 1995). The inner–outer core and the core–mantle systems are very dynamic environments that make the magnetic field produced by the rotating Earth very variable and unstable in time (Coe et al. 1995). Two main kinds of movements are observed in the Earth's magnetic dipole: the first is the 180° change in orientation of the dipole (**polarity reversal**), occurring rapidly (within a few hundred or thousands of years) between two periods with opposite polarity that are stable for hundreds of thousands of years (Fig. 2.24). Our period of normal polarity abruptly started about 780 ky ago and is called the Bruhnes period; it follows the Matuyama period of reverse polarity. Paleomagnetism studies the polarity inversions of the past, and yields magnetostratigraphic columns that may be used for relative dating and for correlating different layer sequences. Many of the hominid-containing sediments in East Africa have been put into sequence using geomagnetic polarity reversals, and have then been attached to an absolute scale by K/Ar and Ar/Ar dating of a few volcanic ash layers (Aitken et al. 1993, Walter and Aronson 1982, Walter 1994).

However even within a single polarity period the magnetic field is unstable, because it changes intensity and continuously shifts in direction with respect to the geographic pole (**magnetic polar wander**), by as much as 15 km/year (Fig. 2.37). At present the magnetic North Pole is inclined by about 11° to the geographic pole, which marks the Earth's axis of rotation. Many types of rocks and sediments contain microscopic mineral particles containing iron, mostly in the form of magnetite (Fe_3O_4), hematite (Fe_2O_3) or other iron oxides and hydroxides. When the particles are crystallized, as in the case of lava cooling, or are deposited by sedimentary processes, their magnetic field is oriented parallel to the Earth's field, so that a so-called **remnant magnetization** is frozen in the rock. The measurement of this relict magnetic field along igneous or sedimentary stratigraphic sequences allows the construction of the charts of polar wandering in time. Databanks of this type have been created by

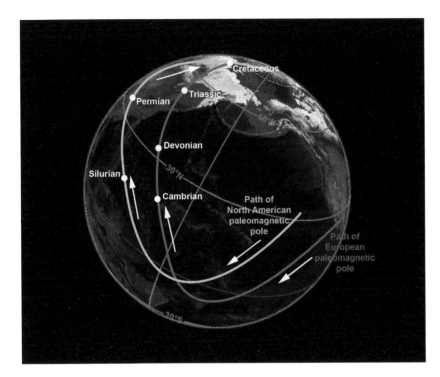

Fig 2.37. Apparent wandering of the North Pole measured from the data collected in North America and Europe. The discrepancy is reconciled if the continents are modelled following the original position before the opening of the Atlantic Ocean. (Data source: from A. Cox and R.R. Doell, Review of Paleomagnetism, GSA Bulletin, Vol 71, Fig. 33, p. 758)

geologists and geophysicists who are interested in the movement of the planetary poles (http://www.ngdc.noaa.gov/geomag). These charts of magnetic polar changes in time can effectively be used as master records to track the moment recorded in the material during its formation. This technique is called **archaeomagnetism** (Eighmy and Sternberg 1990, Sternberg 2008), and when based on adequate polar wandering charts it may provide reliable absolute date estimations.

The application of archaeomagnetic dating techniques requires (1) a thorough understanding of the remnant magnetization process and of the crystal phases involved (2) the careful recording of the absolute orientation of the samples with respect to the present magnetic field, and (3) the availability of polar wandering charts for the period of interest.

Two kinds of materials are routinely dated by archaeomagnetic techniques. The first and more common materials are fired clays present in bricks, furnaces, or burned soils (Bellomo 1993, Gose 2000). The method relies on the fact that many materials are paramagnetic (i.e. the electron spins are randomly oriented so that there is no net magnetic moment) at high temperature, and they become ferromagnetic (i.e. the electronic spins align in the same direction producing a net magnetic moment) at low temperature, below the so-called Curie point. Each crystalline phase that is capable of permanent magnetism has a well-defined temperature at which it becomes ferromagnetic (the Curie point). If the transition takes place in an external field such as the Earth's magnetic field, then the dipole of the cooling particle becomes partly or totally oriented parallel to it. The most common magnetic phases in fired materials are the iron oxides magnetite (Fe_3O_4) and hematite (Fe_2O_3). Their respective Curie points are 578 °C and 675 °C (McElhinny and McFadden 2000), which are temperatures consistently reached in open pits or furnaces that are widely used for pyrotechnological processes, such as the production of ceramics, glass and metals. All natural clay minerals invariably contain Fe in some amount, and clays are one of the basic ingredients of the starting mixture for ceramics (Section 3.1). When clays and other Fe-containing phases (including iron oxides already present from the start) decompose at high temperature during firing, several iron oxides are formed depending on the oxygen fugacity in the furnace, and many of them become permanent magnets upon cooling by crossing the Curie point (thermo-remnant magnetization), thus recording the Earth's field orientation at the time of cooling. Each time the object or the furnace is re-heated, the remnant magnetization is reset, so that archaeomagnetic dating only provides information on the last time that the Curie point was crossed.

Another kind of material that has been tentatively dated by archaeomagnetism is the fine iron oxide particles present in red pigments, mainly ochre, applied to wall paintings and frescoes (Chiari and Lanza 1999). The method relies on the fact that part of the nanometric particles included in the fine-grained pigments are dispersed in the fluid media during the application and when the paint dries the particles settle with their magnetic component oriented parallel to the Earth's field. The paint's layer therefore records the direction of the Earth's magnetic field at the time of the painting. Of course the method cannot be applied to paintings that have been removed from their original location, or that have been retouched or restored at later times.

2.5.4 Absolute methods: fundamentals

The ability to attach an absolute chronological date to an object, a site or an event is fundamental to archaeology and to many other disciplines. Absolute dating methods measure a physical quantity that is proportional to the time elapsed from a specific starting point in the history of the material. This therefore allows one to go beyond the limitations imposed by historical methods, based on the simple use of objects with dates inscribed on them and written documents, both related to the chronologies and **calendars** established by ancient cultures. However, to date an object does not mean dating its context. Very often dated objects such as coins only indicate the ***terminus post quem*** (i.e. the deposit cannot be older than the object, though it could be younger), whereas sometimes the finding of an object in a well dated context determines the **terminus ante quem** (i.e. the object cannot be younger than the context, though it could be older). In extreme cases, the date on the object could be counterfeited and the absolute dating method might be used to expose fakes.

To be valuable, all historical chronological systems require careful and complete reconstruction, and they must be cross-linked with our present calendar. Much in the same way, all absolute dating methods based on modern scientific techniques must be calibrated to link the measured parameters with the time line. The earliest recognized method for absolute dating was dendrochronology, based on counting tree rings, which was developed early last century. Most of the more sophisticated methods developed in the last forty years are divided into two broad categories: those based directly or indirectly on radioactive decay processes, and those based on the accumulation of electron defects in solids. A brief account is given of the main absolute dating methods and recent developments.

2.5.4.1 *Dendrochronology*

Dendrochronology is the name given to the archaeological dating technique that uses the growth rings of longevous trees as a calendar. During their lifetime trees grow by adding each year wood cells and other organic material (mainly cellulose) as an outside layer surface known as a growth ring (Fig. 2.38). The thickness of the ring is directly dependent on the species (Fig. 2.39(a, b, c)), on the nutrient and moisture availability, and on the climatic conditions. Fall and winter layers are readily distinguishable from spring and summer layers because of their density, cell size and thickness. Furthermore trees in the same area add thin rings during dry years and thick rings during wet years, so that a tree ring stratigraphy can be developed by measuring the sequence of ring thickness over time. If we start from living trees and ancient wood fragments are available so that part of the sequence in both wood samples overlaps, then a precise and continuous sequence of tree rings can be derived, recording the living period of each tree with annual resolution. Each ring sequence contains a record of rainfall throughout the tree's life, expressed in density, trace element content, stable isotope composition, and intra-annual growth ring width (Hillam 1998). Exceptional climatic events are recorded as missing, unusually thin, or deformed rings (Fig. 2.40).

In order for this to be a reliable method for dating, several requirements must be satisfied. (1) The species studied must produce only one ring per growing season or year, and this should be clearly defined. This is commonly possible outside the tropics, where there are pronounced differences between seasons. (2) One dominant environmental factor should be the main cause of hindered or increased growth. (3) The dominant environmental factor should vary each year so we can see the changes clearly in every ring. (4) The environmental factor must affect a large geographic area so testing can be compared easily.

The sequence of tree rings is normally measurable on fresh and dried wood, but also on charcoal (Fig. 2.38(d)), so that the technique is applicable to most wood remains from natural or anthropogenic soils, and to the wooden components of buildings and sites (Baillie 1995). Recent applications include the dating of human activities by year or seasonal resolution (Eckstein 2007) and dating of timber used in prehistoric mines (Grabner *et al.* 2007). Dendrochronology is to date the only technique reaching annual time resolution and, along with modern radiocarbon techniques, is by far the most widespread dating method in temperate and arid lands. As with all techniques, caution must be exerted when interpreting the data, since although archaeologists can date the piece of wood and when it was felled, it may prove difficult to definitively determine the age of a building or structure that the wood is in. In fact the wood

Fig. 2.38. A cross-section of a paulownia tree (*Paulownia tomentosa*) about thirty years old, having well distinguishable tree rings in the sapwood between the pith (dark cm-sized internal spot attacked by parasites) and the external bark.

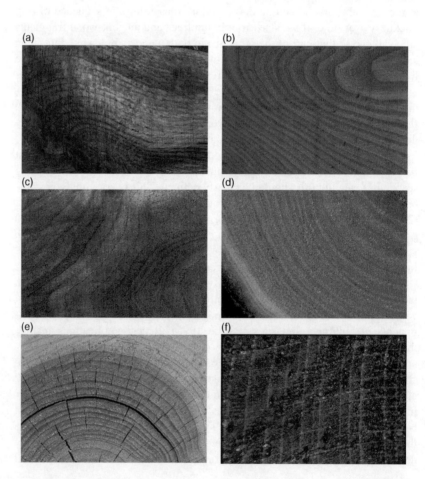

Fig. 2.39. Examples of tree rings from different species: (a) cherry tree (*Prunus cerasus*), (b) European ash (*Fraxinus excelsior*), (c) field elm (*Ulmus minor*), (d) princess tree (*Paulownia tomentosa*) and (e) cedar (*Cedrus libani* var. *atlantica*). Picture (f) shows that tree rings are also detectable in charcoal samples, such as this charcoal fragment of pinyon (*Pinus edulis*) collected from southwestern New Mexico. (Photo (f) courtesy of H.D. Grissino-Mayer, image available at: http://web.utk.edu/~grissino/gallery.htm)

could have been reused from an older structure, it may have been felled and left for many years before use, or it could have been used to replace a damaged piece of wood.

Tree-ring dating was one of the first absolute dating methods, and was discovered in the early decades of the 20th century by the astronomer Andrew Ellicott Douglass (the founder of the laboratory for tree-rings research (LTRR) at the University of Arizona, http://www.ltrr.arizona.edu/) and archaeologist Clark Wissler. In 1901, Douglass began investigating tree ring growth as an indicator of solar cycles. Douglass believed that solar flares affected climate, and hence the amount of growth a tree might gain in a given year. His research culminated in proving that the width of a tree ring varies with annual rainfall. Not only that, it varies regionally, such that all trees within a specific species and region will show the same relative growth during wet years and dry years. Using local pine trees, Douglass built up a 450-year record of tree ring variability. Clark Wissler, an anthropologist researching Native American groups in the Southwest, recognized the potential for such dating and brought Douglass subfossil wood from Puebloan ruins. Unfortunately, the wood from the Pueblos did not fit into Douglass's record, and over the next 12 years, they searched in vain for a connecting ring pattern, building a second prehistoric sequence of 585 years. In 1929, they found a charred log near Show Low, Arizona, that connected the two patterns, thus making it possible to assign a calendar date to archaeological sites in the American southwest covering over 1000 years. Dendrochronology has been extended in the American southwest to 322 BC by adding increasingly older archaeological samples to the record.

Over the last hundred years or so, tree ring sequences have been built up all over the world, with the longest to date consisting of a 12 ky sequence in central Europe completed on oak and pine trees by the Hohenheim Laboratory (Friedrik *et al.* 2004). There are fairly complete dendrochronological master sequences for several other areas in Europe (http://www.wsl.ch/dendro/dendrodb.html), the Aegean, and many are being developed for Asia and Northern Africa. The International Tree Ring Database maintained by the by the World Data Center for Paleoclimatology (http://www.ngdc.noaa.gov/paleo/treering.html) has contributions from 21 different countries.

Dendrochronological studies not only provide valuable dating techniques, but they are also extremely important sources of information about past climate, since the dynamics of tree growth is intrinsically related to environmental factors. Rainy or arid periods may be recognized in tree ring records, and the application to climatic reconstructions is essential both to interpret past human activity in a broader environmental context, and also to insert future climatic changes in perspective with the recent climatic history (Fritts 2001).

An important implication of dendrochronology is the possibility of establishing cross links with other dating techniques. For example there are attempts to relate tree ring dating and thermoluminescence (Dykeman *et al.* 2002). Even more fundamental is the contribution of dendrochronology to radiocarbon dating, because precisely dated tree rings supply the primary organic material to reconstruct the curve of original atmospheric ^{14}C for calibrated radiocarbon dates (Box 2.u).

Fig. 2.40. Details of the tree rings from a Siberian pine (Pinus sibirica) in Mongolia, which records the years AD 534–539 (left to right). The narrow, distorted rings for the years 536 and 537 indicate a drastic cooling in the northern hemisphere that froze sap in the cells during the growing season. Evidence for this abrupt climate change points to a massive eruption of the volcanic precursor of Krakatoa (D'Arrigo *et al.* 2001). (Sample by G. Jacoby, photo by D. Breger, image available at the Tree-Ring Laboratory at Lamont-Doherty Earth Observatory of the Columbia University website: http://www.ldeo.columbia.edu/news/2004/06_03_04.htm)

2.5.4.2 Methods based on radioactive decay processes

It is necessary to remind ourselves that the elements of the periodic table are indicated as $^A_Z E$, where E is the chemical symbol of the element, Z is the atomic number (i.e. $Z = p$, the number of protons) and A is the atomic mass (i.e. $A = Z + n = p + n$), which is the sum of the protons (p) and neutrons (n) in the nucleus. $^{14}_{6}C$, for example, is the important case of radiocarbon, which will be discussed in detail. When the same chemical element has two or more nuclides with the same atomic number but different masses, they are called **isotopes**. For example the three isotopes of carbon ($^{12}_{6}C$, $^{13}_{6}C$ and $^{14}_{6}C$) all have six protons but different numbers of neutrons in the nucleus. Isotopes of the same element have identical chemical behaviour, but different physical properties. Some isotopes have stable nuclides (stable isotopes), whereas others (radioactive isotopes) have metastable nuclei that spontaneously disintegrate and transform into nuclides of different elements: this process is called **radioactivity**.

Just as an atom can exist in any one of a number of excited electronic states (Section 2.1), so too the nucleus can have a set of quantized excited nuclear states. The behaviour of nuclei in transforming from excited to more stable states is somewhat similar to electronic transitions, but there are some important differences (Clayton 1968). First, the differences between energy levels is much greater in nuclear transitions, a notion that most people instinctively gather from the fact that the most powerful known energy sources, that is nuclear reactors including the Sun and nuclear bombs, are related to nuclear processes. Second, the lifetime that an unstable nucleus spends in an excited state has a much wider time range (from 10^{-14}s to 10^{11}y), whereas the lifetime of excited electronic states are usually in the nanosecond range (about 10^{-8}–10^{-9}s). Third, during decay the excited atoms emit photons, whereas the excited nuclei may emit photons or particles of non-zero rest mass. Nuclear decay can be considered to be a special type of nuclear reaction, which in general describes the way that nuclei change their state spontaneously or due to the interaction with other nuclei and particles (Table 2.14). Nuclear reactions must obey general physical laws, conservation of momentum, mass–energy, spin, etc. and conservation of nuclear particles.

The use of radiogenic isotopes and natural radioactive processes for dating is one of the great successes of physics, with important implications for a variety of disciplines, including geology, archaeology, and all archaeometric applications (Faure 1998, Dickin 2005, Garrison 2001).

The **radioactive decay** of nuclides from excited states takes place at a rate that follows a basic first-order law (Faure 1998, Wagner 1998):

$$\frac{dN}{dt} = -\lambda N$$

where N is the number of radioactive nuclides of a given species at the time t, and λ is the decay constant, which represents the probability that a nucleus disintegrate in the time interval dt. The minus sign simply indicates that N decreases in time by the radioactive decay process (Fig. 2.41). Interestingly,

2.5 Time and dating

Table 2.14. Principal types of nuclear reactions

Reaction	Process	Example
α-decay	Emission of $^{4}_{2}He$ nuclei (α-particles). The nucleus decreases both A and Z.	$^{238}_{92}U \rightarrow ^{234}_{90}Th + ^{4}_{2}He$
β-decay	Emission of an electron (e⁻) or a positron (e⁺). The nucleus changes Z but A is the same.	$^{14}_{6}C \rightarrow ^{14}_{7}N + e^{-}$ $^{12}_{7}N \rightarrow ^{12}_{6}C + e^{+}$
γ-decay	Emission of a high energy photon (γ-ray) during the transition from an excited state (*), commonly produced as the daughter of a previous α-decay, β-decay, or β-capture event, to the ground state. A and Z are unchanged.	$^{208}_{81}Tl^{*} \rightarrow ^{208}_{81}Tl + \gamma$
β-capture	Addition of an electron (e⁻) to the nucleus. It has the reverse effect of a β-decay. The nucleus changes Z but A is the same.	$^{40}_{19}K + e^{-} \rightarrow ^{40}_{18}Ar$
Spontaneous fission	The nucleus (parent) splits into two or more heavy nuclei (daughters). Neutrons are emitted in the process.	$^{238}_{92}U \rightarrow ^{142}_{56}Ba + ^{93}_{36}Kr + 3n$

the decay rate is dependent only on the energy state of the nuclide; it is independent of the history of the nucleus, and is essentially independent of external influences such as temperature, pressure, etc.

If N_0 is the number of parent radioactive nuclides present at time $t_0 = 0$, and $D = (N_0 - N)$ is the number of daughter nuclei (**radiogenic species** produced by the decay process) at time t, integration of the rate law indicates that

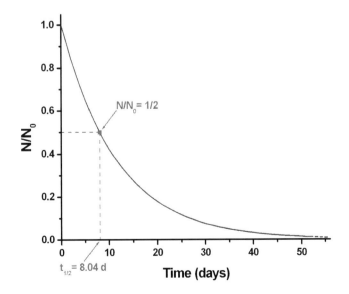

Fig. 2.41. Universal exponential curve for spontaneous radioactive decay. The time at which the number of decaying nuclides (N) reaches half the original value ($N = N_0/2$) is called the half-life. The time scale is gauged to the decay of $^{131}_{53}I$, with a half-life of 8.04 days.

$$N = N_0 e^{-\lambda t} \Rightarrow t = \frac{1}{\lambda} \cdot ln\left(\frac{N_0}{N}\right)$$

or, by substituting $N = (N_0 - D)$ and $N_0 = (N + D)$,

$$D = N_0(1 - e^{-\lambda t}) \Rightarrow t = \frac{1}{\lambda} \cdot ln\left(\frac{N_0}{N_0 - D}\right)$$

and

$$D = N(e^{\lambda t} - 1) \Rightarrow t = \frac{1}{\lambda} \cdot ln\left(\frac{D}{N} + 1\right)$$

These equations allow us to compute the variation of N/N_0, D/N_0, or D/N in time (Fig. 2.42). These curves tell us that if we know the decay constant (λ) and obtain any two of the values of N_0, N or D by experimental measure, then we can calculate the time t that has elapsed since the closure of the system.

If the daughter nuclei produced by the reaction are not stable but are radioactive themselves, a decay chain can be started in which every step proceeds with its own decay constant (λ_1, λ_2,...) until a stable end-member is reached and accumulated in the system. In time, if the half-life of the parent nuclide is much longer than that of the daughters (i.e. $\lambda_2 \gg \lambda_1$), the complex decay chain may reach an equilibrium (called the **secular equilibrium**) between the intermediate products, where all the activities of the radiogenic species are the same (i.e. $\lambda_1 N_1 = \lambda_2 N_2 = \lambda_3 N_3 = ...$). Under these conditions, the number of radiogenic daughters that have accumulated can be treated as though the initial parent species decays directly to the stable end-daughter (Faure 1998).

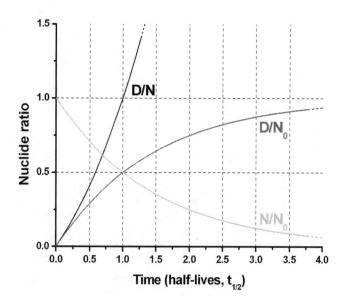

Fig. 2.42. Evolution of the N/N_0, D/N, D/N_0 ratios with time for a generic parent-daughter (N→D) decay, in terms of half-lives (redrawn from Wagner 1998).

The experimental values for the solution of the time equations can be obtained by conceptually different measurements, depending on the system. The most used techniques are:

A. (A) measurement of the effective radioactivity of the sample by α-, β-, or γ-spectrometry (i.e. Dn/dt ratio), which is proportional to the amount of radiogenic nuclides n present at time t;
B. (b) direct counting of the present isotopic abundances of the parent and daughter species (N, D) by mass spectrometry (Box 2.t). Commonly **isotopic ratios** are measured in place of absolute abundances.
C. (C) indirect measurement of the radiogenic and daughter nuclides by induced radioactivity (neutron activation) or particle tracks counting (fission track and α-recoil track dating).

The different measurement methods of course greatly vary in their practical use, especially with respect to the amount of sample required, the limit of detection of the radionuclides, and the resulting precision and accuracy. The application of the different methods based on radioactive decay in general is dictated by two main constrains: (1) the nature and chemistry of the sample, which determines the type of radionuclides that are incorporated in the material, and (2) the decay constant (λ) of the nuclear reaction, which determines the time interval that can be accessed (Table 2.15). As a rough guide, the datable time period is approximately five to eight half-lives.

The use of conventional thermo-ionization mass spectrometry (TIMS, Box 2.t) and particle- or photon-counting techniques in dating measurements severely limited in the past the quality of the data and the minimum amount of material required for the measurements. Many of these limitations have now been overcome by using modern accelerator mass spectrometry (AMS, Box 2.t), which remarkably improves the sensitivity, precision and accuracy of the most widely used radiometric techniques (K–Ar, U series and ^{14}C: Ludwig and Renne 2000), which have greatly improved in sensitivity, precision and accuracy, so that they may be applied to much smaller amount of sample and produce higher resolution data. Furthermore a number of novel techniques, especially those involving cosmogenic nuclides (^{3}H, ^{10}Be, ^{26}Al, ^{32}Si, ^{36}Cl, ^{39}Ar, ^{41}Ca, ^{81}Kr, Wagner 1998) have found a much wider application because of the improved detection limits.

The array and power of radiometric techniques now available for dating Quaternary materials and processes, including those related to human activity, is so vast that with a little ingenuity and effort we may date the whole history of materials from their natural occurrence, through geological processes, to their extraction, utilization, and dismissal by humans. Application examples include radiocarbon dating of organic materials entrapped in the rock varnish of petroglyphs and sediments, original plant remains entrapped in phytoliths, and both the organic and inorganic carbon in bones and eggshells. The application of novel radiometric techniques based on cosmogenic isotopes allows the dating of the different landscapes accompanying or being transformed by human activity, such as agricultural soils, geoglyphs, alluvium mixed with archaeological soils, moraines and glacial sediments. No doubts the powerful

Table 2.15. Summary of the most used systems of radioactive decay for dating purposes. Half-life values are from Faure (1998), Wagner (1998) and Magill *et al.* (2007)

Parent/Daughter system	Decay constant (y^{-1})	Half-life (y)	Materials that can be dated
$^{3}_{1}H \rightarrow ^{3}_{2}He$	5.64×10^{-2}	1.24×10	moraines, ocean water mixing
$^{32}_{14}Si \rightarrow ^{32}_{15}P$	4.95×10^{-3}	1.40×10^{2}	diatoms, groundwater, glacier ice
$^{39}_{18}Ar \rightarrow ^{39}_{19}K$	2.58×10^{-3}	2.69×10^{2}	groundwater, snow compaction, ocean water mixing
$^{14}_{6}C \rightarrow ^{14}_{7}N$	1.21×10^{-4}	5.73×10^{3}	organic materials, carbonates, organics in sediments and soils, mortar, wood and charcoal, bones, shells
$^{230}_{90}Th \rightarrow ^{226}_{88}Ra$	9.19×10^{-6}	7.54×10^{4}	rocks, ores, carbonates, stalagmites, diatoms, bones, teeth, corals
$^{41}_{20}Ca \rightarrow ^{41}_{19}K$	6.73×10^{-6}	1.03×10^{5}	bones, cave carbonates
$^{81}_{36}Kr \rightarrow ^{81}_{35}Br$	3.03×10^{-6}	2.29×10^{5}	groundwater, glacier ice
$^{234}_{92}U \rightarrow ^{230}_{90}Th$	2.83×10^{-6}	2.45×10^{5}	rocks, ores, carbonates, stalagmites, bones, teeth, corals
$^{36}_{17}Cl \rightarrow ^{36}_{18}Ar$	2.30×10^{-6}	3.01×10^{5}	moraines, impact rocks, groundwater, ice cores, evaporites
$^{26}_{13}Al \rightarrow ^{26}_{12}Mg$	9.37×10^{-7}	7.40×10^{5}	rock surfaces, alluvium, moraines
$^{10}_{4}Be \rightarrow ^{10}_{5}B$	4.59×10^{-7}	1.51×10^{6}	rock surfaces, loess, alluvium, moraines, soils, ice cores
$^{235}_{92}U \rightarrow ^{207}_{82}Pb$	9.85×10^{-10}	7.04×10^{8}	carbonates, stalagmites, bones, teeth, corals, shells
$^{40}_{19}K \rightarrow ^{40}_{20}Ca$	4.91×10^{-10}	1.41×10^{9}	rocks
$^{238}_{92}U \rightarrow ^{206}_{82}Pb$	1.55×10^{-10}	4.47×10^{9}	rocks, soils, Pb pigments
$^{40}_{19}K \rightarrow ^{40}_{18}Ar$	5.66×10^{-11}	1.28×10^{10}	volcanic glass, tephra, obsidian, basalts, tektites
$^{232}_{90}Th \rightarrow ^{208}_{80}Pb$	4.99×10^{-11}	1.39×10^{10}	rocks, ores
$^{176}_{71}Lu \rightarrow ^{176}_{72}Hf$	1.83×10^{-11}	3.78×10^{10}	rocks
$^{187}_{75}Re \rightarrow ^{187}_{76}Os$	1.59×10^{-11}	4.35×10^{10}	rocks
$^{87}_{37}Rb \rightarrow ^{87}_{38}Sr$	1.42×10^{-11}	4.75×10^{10}	rocks, volcanic materials
$^{147}_{62}Sm \rightarrow ^{143}_{60}Nd$	6.53×10^{-12}	1.06×10^{11}	rocks

mix of geology-based and archaeology-based techniques will provide a much more complete comprehension of the recent past. The three most important radiometric dating methods widely applied to materials related to archaeology and palaeoanthropology are briefly detailed: the K–Ar, U series and ^{14}C methods.

The potassium–argon method (K–Ar) was among the earliest quantitative methods applied to dating materials of paleoanthropological interest (Ludwig and Renne 2000). The decay constant of ^{40}K into ^{40}Ar (Table 2.15) is well suited to the time-scale of human evolution; furthermore potassium is an element that is invariably contained in minerals (mica, hornblende and feldspar, especially sanidine) and glass derived from volcanic activity, such as tephra, ashes, ejecta and tuffs. The K–Ar method is based on the precise MS measurement of both the parent (K) and the daughter (Ar) species in the sample. It can be used to date the volcanic layers deposited in the sedimentary sequences containing human and animal remains. The absolute dating of volcanic events above and below the fossil-containing layer effectively brackets the age of the remains with lower and upper age limits. Following the early applications to Olduvaian fossils (Evernden and Curtis 1965), K–Ar has become the most applied and reliable method (to about 1% precision or better) for dating volcanic rock from 100 ky to 10 My years old, though in principle it covers the whole age of the solar system. Recently a variant of the method (called the ^{40}Ar/^{39}Ar method) has been introduced, which measures both argon species that are thermally driven off a single crystal extracted from the sample (Deino *et al.* 1998, McDougall and Harrison 1999). The use of step-wise heating to release Ar from isolated crystals solves many of the issues related to the standard K—Ar method, including low temperature loss of radiogenic argon and the inheritance of ^{40}Ar from older grains. The technique is called the single crystal laser fusion (SCLF) ^{40}Ar/^{39}Ar method, and it is virtually the most reliable method to use to absolute date relative sedimentary sequences, including historical eruptions and the palaeomagnetic georeversals scale (Baksi *et al.* 1992).

The U series methods of dating (also termed the U decay series, U disequilibrium, or the uranium–thorium method) actually include all methods related to the radioactive decay of ^{235}U and ^{238}U into a series of daughter isotopes, radioactive themselves, whose end stable products are ^{207}Pb and ^{206}Pb respectively (Ivanovich and Harmon 1992, Ivanovich 1994). ^{232}Th also decays into ^{208}Pb (Table 2.15), and all three decay series involve a large number of short-lived daughter/parent intermediates such as ^{230}Th/^{234}U, ^{231}Pa/^{235}U and ^{234}U/^{238}U (Fig. 2.43), each of the intermediate processes representing an independent dating method. If the uranium–thorium decay system is closed, it reaches secular equilibrium in which all species have the same decay activity. If the system is perturbed, most commonly because of geochemical fractionation during geologic processes, such as weathering, alteration, transport or sedimentation, then it is possible to determine the time of the perturbation by measuring the distance to equilibrium (Wagner 1998, pp. 84–87). The U series method can be used in the time range from a few years to about 1 My, the time ranges being dependent on the decay constant of the specific daughter/parent couple that has

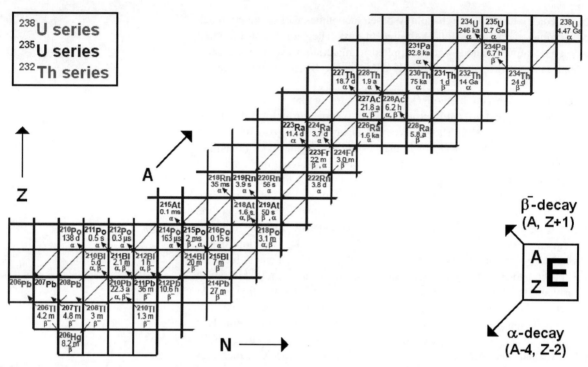

Fig. 2.43. Part of the chart of the nuclides showing the series of decays that occur as 238U, 235U and 232Th are transformed to 206Pb, 207Pb and 208Pb respectively. Arrows indicate the type of decay linking parent and daughter nuclides.

been adopted. The most commonly and successfully used daughter/parent is the ^{230}Th/^{234}U (decay constant 245 ky, Table 2.15), which is widely used to date marine and freshwater carbonates, bones, teeth, peat deposits, volcanic sediments and several other materials, with a focus on deep sea sediments, corals and terrestrial carbonates. The method relies on the fact that in subsurface and marine water uranium is highly soluble, whereas thorium is not. Carbonates precipitating from oxygenated waters, such as marine sediments, oolites, corals, travertines, tufa formations, cave concretions and speleothems therefore incorporate radiogenic ^{234}U as a substitution for Ca in the carbonate crystal structure, but negligible amounts of the radioactive daughter ^{230}Th. Fossil bones, teeth, and mollusc shells mainly incorporate uranium after deposition from groundwater, so that the method has been successfully used in palaeoanthropology (Schwarcz 1992, Cheng and Edwards 1997). It can be used to about 350 ky with a precision of 5% using α-spectrometry, and to about 550 ky with a precision of 1% using TIMS (Box 2.t). Limitations derive from possible ^{234}U/^{238}U ratio disequilibrium in the initial system (i.e. a poorly known starting ^{234}U/^{238}U ratio), presence of ^{230}Th in the original material and, above all, an opening of the system at some point in the history of the material. This could be caused by a number of processes, for example the aragonite–calcite transformation in corals and molluscs, so that initial events must be properly assessed.

The radiocarbon, or ^{14}C method, is obviously the privileged dating method for organic materials in archaeology, and it is safe to say that it has truly revolutionized prehistoric chronologies (Renfrew and Bahn 2008, pp. 141–149), though it has also been applied to inorganic carbon-containing materials, such as cave and soil carbonates. It was first developed in 1949 by Libby at the University of Chicago (Arnold and Libby 1949, Libby 1952, Taylor 2000, Currie 2004) and is based on the fact that radiogenic ^{14}C is produced in the upper atmosphere, mostly at an altitude of 12–15 km, by cosmic rays converting ^{14}N into radiocarbon. The average annual production is estimated to be about 7.5 kg, though it is highly dependent on time and latitude variations (Wagner 1998). The underlying assumption is that atmospheric radiocarbon is cosmogenically replenished, and it is therefore homogeneously distributed in natural reservoirs, such as breathable air and drinkable water, so that it is constantly transmitted to the biosphere through the biological processes included in the carbon cycle, mainly respiration, nutrition, and photosynthesis. The organisms are thus in equilibrium with the atmosphere during their lifetime, and when the continuous exchange is interrupted at death, the system starts to decrease the $^{14}C/^{13}C$ and $^{14}C/^{12}C$ ratios because of the radiocarbon decay reaction having a half-life of 5730 y (Table 2.15). The universal radioactive decay equation may then be used to calculate the age at the death of the organism, provided that the original (N_0) amount of radiocarbon is known and that the present amount (N) is measured (Box 2.u). Experimentally, the quantity of ^{14}C present in the sample can be measured by β-spectrometry to about 40 ky or, much more rapidly and precisely, by AMS (Box 2.t). Mass spectrometric analyses employ a much smaller amount of sample (Table 2.16), and in favourable cases dating can be extended to over 60 ky with a precision of 1% (Wagner 1998, Tuniz 2001). Application of the radiocarbon method to archaeological problems must necessarily take into account the problem of contamination, usually by younger material, and the fact that some time may elapse from the death of the organism until the burial in the sediment or the use by man. A typical example is the problem of use or re-use of old wood as fire fuel or as timber in constructions.

Table 2.16. Typical amount of material in mg required for ^{14}C dating by AMS, from Tuniz (2001). Amount of material required for measurement by β-spectrometry is commonly one thousand times more

Material	Quantity (mg)
Wood, charcoal	3–5
Beeswax	1–2
Pollen	1
Plants: grass, seeds, leaves, grains	5–10
Shells, eggshells, biomineralizations	10
Paper, textile fibres	5–10
Hair, skin	5–7
Bone	500
Teeth, tusk, ivory	500–700

Box 2.t MS Mass spectrometry

Mass spectrometry (MS) is a technique designed to separate charged atoms and molecules on the basis of their mass or, more precisely, on the basis of the mass-to-charge ratio (m/z, where m is the mass of the ion and $z = Q/e$ is the number of elementary charges (e) on the ion with total charge Q). Although this is mostly used for the quantification of atoms and molecules at trace levels, it is also the only technique that can identify and quantify individual isotopes of the same chemical element. It is therefore the essential technique to measure both radioactive (Section 2.5.4) and stable isotopes (Box 2.v).

In the simplest design, the spectrometer must include: (a) a device to positively or negatively ionize atoms and/or molecules, (b) an electromagnetic field analyser that physically separates species having different mass, and (c) a counter or ion collector. The principle of mass separation is very simple (Fig. 2.t.1): ions that are accelerated to have the same kinetic energy pass through a magnetic field deflecting their path as a function of their masses, i.e. light ions are deflected more. As an example, charged ions of the same element Z can be measured and separated into isotopic abundances (Section 2.5.4). If absolute concentrations of chemical elements are sought, the measured isotopic concentrations are normalized to the natural isotopic abundances of the elements (see for example http://ie.lbl.gov/education/isotopes.htm). Accurate analyses need careful calibration using reference standard materials with known concentrations, as in all chemical analyses. However, when relative abundances are sought, as in the case of stable isotopes or dating methods based on radioactive decay, the ratios between isotope concentrations are measured (Platzner 1997), so that high instrumental precision is of the outmost importance. Subtle instrumental effects causing biased measurements must be avoided or corrected, the most common being the **mass fractionation** that takes place during ionization, which caused depletion of lighter elements in time (Habfast 1998).

From the instrumental point of view, there are many different experimental configurations and technical choices, depending on the specific applications. The main differences are related to the geometry of the ion optics (Burgoyne and Hieftje 1996), and to the use of static or dynamic magnetic fields in the analyser, or even the use of two or more coupled mass analysers, as in **tandem mass spectrometry (MS/MS)**, where multiple fragmentation processes occur, so that the technique is mostly used to analyse large fragments of organic molecules (McLuckey *et al.* 1988). One class of mass spectrometers using combinations of spatially separated static electric or magnetic sectors as a mass analyser is **sector instruments**. A popular configuration of these sectors

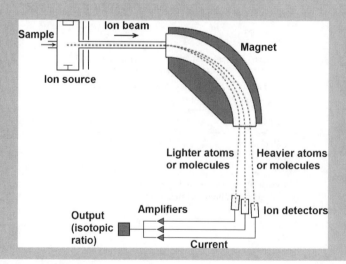

Fig. 2.t.1. Scheme of the deflection of light and heavy ions by a magnetic field used in mass spectrometric analysis. (Modified from the Science Education Resource Center at Carleton College website, original image available at: http://serc.carleton.edu/research_education/geochemsheets/techniques/gassourcemassspec.html)

has been the BEB (magnetic–electric–magnetic). Most modern sector instruments are double focusing instruments in that they focus the ion beams both in direction and velocity. **Accelerator mass spectrometers (AMS)** are sector instruments that incorporate a high energy ion accelerator, commonly an electrostatic tandem accelerator, and its beam transport systems among the components used for mass analysis (Tuniz 2001). AMS remarkably enhances the sensitivity of MS to the ultra-trace regime, making it the technique of choice for the detection of nuclides that cannot be analysed with conventional instruments, and for the measurement of extremely small quantities of sample.

Another class of instruments accelerate the ions to the same potential and then measure the time it takes to reach the detector: the lighter atoms are faster and reach the detector first. These instruments are called **time-of-flight** analysers (**ToF-MS**). **Quadrupole** mass analysers (**QMS**) use oscillating electrical fields to selectively stabilize or destabilize the paths of ions passing through a radio frequency quadrupole field. Only a single mass/charge ratio is passed through the system at any time, but changes to the potentials of the quadrupolar rods allow a wide range of m/z values to be swept rapidly, either continuously or in a succession of discrete hops. Fig. 2.t.2 shows a m/z scan obtained by a QMS instrument.

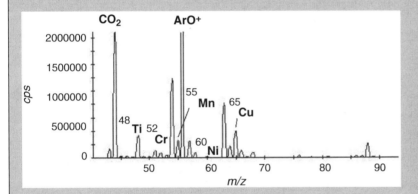

Fig. 2.t.2. Intensity vs. m/z ratio obtained by a mass spectrometer using a quadrupolar mass analyser. The sample is a lacustrine sediment; quantification of elemental concentration yields for example 52Cr = 29.6, 55Mn = 16.9, 60Ni = 1.98, 65Cu = 242 µg/l, showing the high sensitivity of the technique. (Courtesy of S. Recchia, University of Insubria)

A Fourier transform ion cyclotron resonance mass spectrometry (**FT-ICR-MS**) measures mass by detecting the image current produced by ions travelling in the closed orbit of a cyclotron magnetic field (Marshall *et al.* 1998). The ions are injected into a static electric/magnetic ion trap and detectors at fixed positions along the circuit measure the electrical signal of the passing ions, producing a periodic signal whose frequency depends on the m/z ratio. The time frequency is deconvoluted by Fourier Transform analysis of the signal. The technique has the advantage of high sensitivity, because each ion is counted repeatedly when circling.

The capabilities of mass spectrometry of resolving individual ionic or molecular masses may be enhanced by using it in combination with separation techniques, such as liquid- or gas-chromatography, where the ion source of the mass spectrometer is fed directly by the chromatographer stream. The techniques are called **liquid chromatography-mass spectrometry** (**LC-MS**, Box 3.o) and **gas chromatography-mass spectrometry** (**GC-MS**, Box 3.p). Another very interesting development of mass spectrometry is the combination with a primary ion microbeam (commonly made of O^-, O_2^+, Ar^+, Xe^+ or Cs^+), which hits the surface of the sample and causes atoms from the samples to be ejected as secondary ions that are analysed by a sector, quadrupole, or time-of-flight mass spectrometer. The technique is called **secondary ion mass spectrometry** (**SIMS**, Benninghoven *et al.* 1987).

(cont.)

Box 2.t (*Continued*)

It has recently become very popular for the surface analysis of cultural heritage materials (Spoto 2000), especially using time-of-flight instruments (**ToF-SIMS**, Rutten *et al.* 2006). The technique is ideal for chemically or isotopically mapping a surface at the microscale, and to characterize alteration, corrosion, and diffusion layers by depth profiling.

Several instrumental combinations of ion sources, mass analysers, and detectors have been optimized for specific applications, so that they have become recognized standard configurations for particular applications. In the analysis of large organic molecules, the **matrix-assisted laser desorption/ionization** (**MALDI**) soft source or the **electro-spray ionization** (**ESI**) are routinely employed, thus making the **MALDI-ToF** and the **ESI-MS** very popular for the analysis of biomolecules and macromolecules, such as proteins. At the other end of the source spectrum, a number of very efficient ionization sources such as the **inductively coupled plasma** (**ICP**), **thermal ionization** (**TI**) and **spark source** (**SS**), are the preferred choice for trace element and isotopic ratio analysis. Common instrumental configurations used in archaeometric applications include **thermal ionization mass spectrometry** (**TIMS**, Wintle 1996), which is mainly used in radiometric dating using the uranium series and is suitable for species with a low ionization energy such as Sr and Pb. Also included is **inductively coupled plasma-mass spectrometry** (**ICP-MS**, Holland and Tanner 2001), which is commonly used for the analysis of trace elements. The latter technique may be expanded to the micro-invasive analysis of surfaces using the **laser ablation** (**LA**) sampling technique, which employs a small laser beam to extract ions from the sample surface, generally producing small micro-holes that are visible only by optical or electron microscopy. The techniques (**LA-ICP-MS** and **LA-ICP-ToF-MS**) have been shown to be effective in the micro-invasive analysis of precious objects and in the micro-mapping of compositional and isotopic variations across the sample (Müller *et al.* 2003, Duwe and Neff 2007, Resano *et al.* 2009).

Finally, in the field of very precise measurement of isotopic ratios two problems are often encountered. The first is the time instability of the plasma source when using a single detector in **QMS-ICP-MS** that commonly limits the precision to around 1%, which is insufficient for most radiogenic isotope systems. The second is the mass fractionation effect due to ionization of the sample by **TIMS**, and the further inability to analyse a few systems (for example Hf-W and Lu-Hf, Table 2.15), due to the high ionization potential of the elements involved. To overcome these problems, a system based on plasma ionization and employing a **multiple detector** array was developed, to provide, at the same time, an efficient ionization source and simultaneous detection of several isotopes. This technique, defined **multi collector-inductively coupled plasma-mass spectrometry** (**MC-ICP-MS**) is the preferred choice for the measurement of precise isotopic ratios as it is independent of time fluctuation in the source (see for example White *et al.* 2000, Schoenberg *et al.* 2000, Kleinhanns *et al.* 2002). The instrument may of course be coupled to a laser ablation sampler (**LA-MC-ICP-MS**: Junk 2001).

For in-depth information

Boyd B., Bethem R. and Basic C. (2008) Trace quantitative analysis by mass spectrometry. John Wiley & Sons, New York.

Platzner I.T. (1997) Modern isotope ratio mass spectrometry. John Wiley & Sons, London.

Speakman R.J. and Neff H. (eds) (2005) Laser ablation ICP-MS in archaeological research. University of New Mexico Press, Albuquerque.

Tuniz C., Bird J.R., Fink D. and Herzog H.G. (1998) Accelerator mass spectrometry: ultrasensitive analysis for global science. LLC, CRC Press, Boca Raton.

Box 2.u ^{14}C calibration procedures

Radiocarbon dating is a radiometric dating method (Section 2.5.4) that uses ^{14}C to determine the age of carbon containing materials up to about 50–60 ky old. Though many laboratories still measure the activity (i.e. β decays/unit time) of radiocarbon present in the sample, most measurements nowadays rely on **accelerator mass spectrometry** (AMS: Box 2.t) techniques, which require a very small amount of sample (Table 2.16), and produces much more precise data. From the measured amount of ^{14}C present in the sample, the age is conventionally computed by the universal decay curve:

$$t = 1/\lambda \times \ln(N_0/N) = 8033 \times \ln(^{14}C_0/^{14}C)$$

where $^{14}C_0$ is the original amount of radiocarbon present in the organism at the time of death and is assumed to be in equilibrium with the atmospheric carbon. The **conventional age** follows the original Libby model (Arnold and Libby 1949), which was based on a measured half life of the $^{14}_{6}C \rightarrow ^{14}_{7}N$ reaction of 5568 y and on the assumption that the radiocarbon content in the atmosphere is constant. This convention is maintained by the measuring laboratories even if the correct half life has since been corrected to 5730 y (Table 2.15), and it has been long shown that there are long-term variations in the atmospheric ^{14}C/C ratio (De Vries 1958). It is now known that there are both long-term and short-term natural variations in atmospheric radiocarbon, in addition to those caused by anthropogenic activities, such as the 20th century burning of ^{14}C-deprived coal and oil after the industrial revolution, and the emission of high ^{14}C spikes during the nuclear bomb tests carried out between 1950 and 1964 (Wagner 1998). The long-term natural variations are thought to be caused mainly by the deflection of cosmic rays by the Earth's magnetic field, which varies in time, and by palaeoclimatic feedback between oceanic circulation and the atmosphere, which controls the release of ^{14}C poor carbon dioxide from deep ocean waters. Mid-term variations are caused by variation in solar activity (the so-called "Suess wiggles") and consequently in the flux of cosmic rays. Further, the conventional calculated age (in years BP, or before present) is always referred to year AD 1950, to avoid confusion between measurements performed at different times.

To convert conventional ages into proper **calendar dates**, calibration curves are used that are based on the direct and precise measurement of the starting ^{14}C/C ratio on dendrochronologically dated tree-ring sequences (Section 2.5.4). The most recent calibration curve (INTCAL04) based on terrestrial tree-rings goes back to about 12.4 calibrated ky BP, and it extends further by the use of the marine carbon record (corals and foraminifera) to cover the whole period to about 26 calibrated ky BP (Reimer et al. 2004). Figure 2.u.1 shows a portion of the curve for the calendar years 2050–4050 BC, clearly showing the observed oscillations in the atmospheric radiocarbon.

The INTCAL04 calibration curve supersedes the previous INTCAL98 curve; it has a much higher resolution and it is continuously updated and ratified by the international IntCal working group (Reimer et al. 2002). The IntCal04 or similar curves are also embodied in publicly available programs that may be used for calculating the calibrated age starting from the conventional age BP (CALIB, CalPal, OxCal). The conventional and calibrated dates are meant to be reported as

$$11\,000 \pm 100\ ^{14}C\ BP\ (95\%\ 12\,820 - 13\,100\ cal\ BP)(10\,973 \pm 117\ cal\ BC)$$

Where the first date is the conventional uncalibrated date based on the measurement, the reported range shows the age range calculated from the calibration curve that has 95% statistical probability. The last date is the corresponding calibrated calendar age. The relative graphical conversion is shown in Fig. 2.u.2, which also shows the discrepancy with the calibration performed using the previous IntCal98 curve.

(cont.)

Box 2.u (*Continued*)

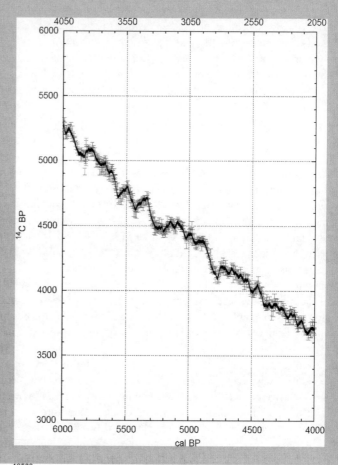

Fig. 2.u.1. The IntCal04 calibration curve for the calendar period 2050–4050 BC. (From Reimer *et al*. 2004)

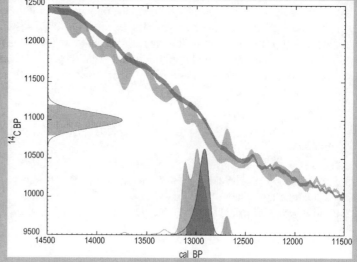

Fig. 2.u.2. Calibrated probability distributions for a hypothetical 14C age of 11 000 ± 100 14C BP calibrated with IntCal98 (light grey) and IntCal04 (dark grey) showing 95% cal age ranges. (From Reimer *et al*. 2004)

For in-depth information

CALIB Radiocarbon Calibration software: http://calib.qub.ac.uk/calib/

CalPal Cologne Radiocarbon Calibration and Paleoclimate Research Package http://www.calpal-online.de/

OxCal Oxford Radiocarbon Calibration: http://c14.arch.ox.ac.uk/embed.php?File=oxcal.html

Wagner G.A. (1998) Age determination of young rocks and artifacts. Physical and chemical clocks in Quaternary geology and archaeology. Springer, Berlin. p. 466.

Two other dating methods are indirectly related to radioactive decay: the **fission track** and **α-recoil track** methods (Wagner and Van den Haute 1992, Wagner 1998). They are based of the observation that each α-decay and nuclear fission reaction produces particles with high kinetic energy (He nuclei, neutrons, heavy nuclei), which leave a track of structure damage within the hosting crystal structure. When ^{238}U atoms located in crystal lattice sites undergo spontaneous fission into two heavy fragments, each one of them forms a latent track of broken chemical bonds and lattice defects with length of the order of 10–20 µm (Fig. 2.44). Much in the same way the α-particles emitted during uranium and thorium α-decay processes produce similar tracks, as well as a smaller track in the proximity of the emitting nucleus due to the recoil energy. The recoil α-particles tracks have lengths of the order of 0.01 µm. All the produced tracks accumulate in the crystal structure until a bleaching event occurs, usually some kind of thermal annealing that supplies enough energy to speed up atomic diffusion and reforming of the broken bonds. The latent tracks are commonly amplified by chemical etching and then counted under an optical microscope at high magnification, thus providing an estimation of the time passed since the formation of the material or the last annealing event. The track number is measured as track density/unit area, and it assumes that all tracks are counted and there has been no loss or fading during accumulation. Figure 2.45 shows well evident fission tracks in an apatite crystal. A problem with fission-track dating is that the rates of spontaneous fission are very slow, requiring the presence of a significant amount of uranium in a sample to produce useful numbers of tracks over time. Additionally, variations in uranium content within a sample can lead to large variations in fission track counts in different sections of the same sample (Wagner and Van den Haute 1992, Van den Haute and De Corte 1998). Because of such potential errors and because of the need to know the concentration of the parent nuclide, most experimental measurements of fission track dating use a form of internal calibration based on the comparison of spontaneous and induced fission track density against a standard of known age. The principle involved is no different from that used in many methods of analytical chemistry, where comparison to a standard eliminates some of the poorly controlled variables. The method relies on the irradiation of the sample and of the reference material, usually glass, by thermal and fast neutrons, which induce fission in ^{238}U and ^{235}Th respectively. The dose, cross-section, spontaneous and induced decay constant, and measured uranium isotope ratio are combined into a single time equation (Wagner 1998):

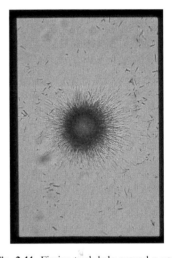

Fig. 2.44. Fission-track halo around a very small uranium-rich particle from stream sediment below a uranium mine. The particle is much smaller than the starburst pattern. Size of the image is 150 µm (Image available at the USGS website: energy.cr.usgs.gov/other/uranium)

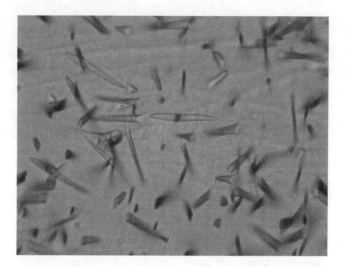

Fig. 2.45. Etched surface of an apatite crystal showing the fission tracks produced by uranium spontaneous decay. (Courtesy of A. Carter, UCL; image available at: http://www.es.ucl.ac.uk/research/thermochronology/methodology.html)

$$t = \frac{1}{\lambda_d} \cdot ln\left[1+\left(\frac{\rho_s}{\rho_i}\right)\cdot\left(\frac{\lambda_d}{\lambda_f}\right)\cdot\left(\phi\frac{^{238}U}{^{235}U}\sigma\right)\right]$$

where ρ_s and ρ_i are the spontaneous and induced fission track areal densities (cm^{-2}), ϕ is the thermal neutron flux (cm^{-2}), λ_d and λ_f are the constants of total decay and spontaneous fission of ^{238}U, and σ is the ^{235}U fission cross-section for thermal neutrons.

Suitable minerals that commonly incorporate sufficient amounts of fission elements and retain the tracks are quartz, apatite, mica, titanite, and zircon. Volcanic glass can also profitably be dated by fission track counting. The most common archaeometric applications are the dating of the thermal annealing events including heated obsidian and flint objects, fired ceramics, and burned stones and soil grains. This annealing temperature varies from mineral to mineral and the details of the temperature vs. time history may be complicated, though typical temperatures are 70–110 °C for apatite, 230–250 °C for zircon, and 300 °C for titanite. The fission-track dating of titanite grains from the fired layers in the Zhoukoudian caves, near Beijing, is a classic example of the application allowing the dating of the use of fire by the *Homo erectus pekinensis* at 462±45 and 306±56 ky (Guo *et al.* 1991).

Box 2.v Stable isotopes

Geology has long employed isotope methods to interpret and trace geological processes (Faure 1998, Hoefs 2008). Radiogenic isotope systems undergoing natural decay have been mostly used for dating (Section 2.5.4), whereas stable isotope systems (Table 2.v.1) have been largely applied to archaeological reconstruction (Pollard 1998), especially to define the origin and diffusion of materials and to the interpretation of palaeoecology, palaeodiets, palaeoclimates, and population dynamics. The possibility of using stable isotopes for the interpretation of natural or anthropogenic processes is based on the observation that, starting from a given isotopic abundance, during chemical reactions the isotopes of an element undergo mass fractionation because of the mass dependence of the reaction kinetics. As a consequence the extent of fractionation is strongly dependent on the temperature and on the relative mass difference between the isotopes of the reacting element. The underlying assumption of stable isotope studies is that the measured isotope ratios provide information on the processes leading to fractionation.

As far as the experimental aspects are concerned, the measurement of **isotopic ratios** must be performed using mass spectrometry (Box 2.t), and the relative abundances of the isotopes in the investigated sample are compared with those in a certified reference standard material (Table 2.v.2). The experimentally measured values are commonly reported in terms of delta notation (Coplen 1996, Slater *et al.* 2001, Werner and Brand 2001), which is the relative difference in parts per thousand between the sample isotope ratio and that of the international standard. For example, for carbon isotopes relative to the VPDB reference standard (Table 2.v.2), the measured isotope ratio in the sample is expressed as

$$\delta^{13}C_{VPDB} = \left(\frac{\left(\frac{^{13}C}{^{12}C}\right)_{sample} - \left(\frac{^{13}C}{^{12}C}\right)_{VPDB}}{\left(\frac{^{13}C}{^{12}C}\right)_{VPDB}} \right) \times 1000$$

The earliest applications of isotope chemistry to archaeological problems were mainly devoted to **provenance** issues, stemming from the pioneering use of lead isotopic ratios in pinpointing the origin and diffusion of metal and glass (Brill and Wampler 1967). Apart from ^{204}Pb, the other lead isotope are not constant in time, because they the end daughters of the U-Th decay series (Table 2.15) and their concentration varies as a product of spontaneous radioactivity. Though these series are mostly used for chronological purposes in geology (Section 2.5.4), the measured ratios of some of the long-lived isotopes are considered to be stable in the short time-span pertaining to archaeology, so that they may effectively be used as tracers reflecting the variation in age of geological ore deposits. The use of lead isotope ratios as tracers reflecting the original source therefore has been widely used for the provenance studies of metals (Section 3.5), though recently they have also been used to recognize the origin and the toxic role of lead in human tissues derived from metallurgical activity and pollution (Budd *et al.* 2004). Other stable isotopes used for provenancing inorganic materials are carbon, oxygen, and strontium. Typical applications include the discrimination of the Mediterranean marble quarries used in classic antiquity to produce marble blocks for statues and buildings (Herz 1992, Maniatis 2004).

Table 2.v.1. Stable isotope systems used for archaeological interpretation. Modified from Stos-Gale (1995) and Pollard (1998)

Stable isotope system	Naturally occurring isotopes*	Commonly measured isotopic ratios	Investigated materials	Application
Hydrogen	^{1}H, ^{2}H=D, ^{3}H=T	D/^{1}H	organic molecules, water	palaeoclimate, plant metabolism
Carbon	^{12}C, ^{13}C, ^{14}C	^{13}C/^{12}C	organic molecules, carbonates, soil	palaeodiet, habitat, plant metabolism, provenance
Nitrogen	^{14}N, ^{15}N	^{15}N/^{14}N	organic molecules, soil, groundwater	palaeodiet, palaeoclimate, pollution

(cont.)

Box 2.v *(Continued)*

Stable isotope system	Naturally occurring isotopes*	Commonly measured isotopic ratios	Investigated materials	Application
Oxygen	$^{16}O, ^{17}O, ^{18}O$	$^{18}O/^{16}O, ^{17}O/^{16}O$	organic molecules, biocarbonates, biophosphates, sediments	palaeoclimate, plant/animal metabolism, provenance
Sulphur	$^{32}S, ^{33}S, ^{34}S, ^{36}S$	$^{34}S/^{32}S$	organic molecules, sulphates, sediments	paleodiet, pollution
Copper	$^{63}Cu, ^{65}Cu$	$^{65}Cu/^{63}Cu$	copper metal, copper minerals	provenance, types of smelted minerals
Strontium	$^{84}Sr, ^{86}Sr, ^{87}Sr, ^{88}Sr$	$^{87}Sr/^{86}Sr$	bone, teeth, sediments, water	palaeodiet, provenance
Lead	$^{202}Pb, ^{204}Pb, ^{205}Pb, ^{206}Pb, ^{207}Pb, ^{208}Pb$	$^{208}Pb/^{206}Pb, ^{207}Pb/^{206}Pb, ^{206}Pb/^{204}Pb$	metals, minerals, soils biominerals, bones	provenance, pollution

* The short-lived radioisotopes, such as $^{11}C, ^{13}N, ^{15}C, ^{67}Cu, ^{85}Sr, ^{210}Pb$, etc. are not included.

Table 2.v.2. Some of the reference isotope ratio values certified for internationally recognized standards. Data from Slater *et al.* (2001), Werner and Brand (2001) and original certifications of the reference materials

Certified standard material	Label	Isotope ratio	Certified value
Vienna Standard Mean Ocean Water		$D/^1H$	$1.5575(8) \times 10^{-4}$
	VMOW	$^{18}O/^{16}O$	$2.00520(45) \times 10^{-3}$
		$^{17}O/^{16}O$	$3.799(8) \times 10^{-4}$
Standard Light Antarctic Precipitation		$D/^1H$	8.9089×10^{-5}
	SLAP	$^{18}O/^{16}O$	1.89391×10^{-3}
Vienna Pee Dee Belemnite		$^{13}C/^{12}C$	$1.11802(28) \times 10^{-2}$
	VPDB	$^{18}O/^{16}O$	$2.0672(21) \times 10^{-3}$
		$^{17}O/^{16}O$	3.860×10^{-4}
Atmospheric nitrogen	AIR-N_2	$^{15}N/^{14}N$	$3.6782(15) \times 10^{-3}$
Vienna Canyon Diablo Troilite meteorite	VCDT	$^{34}S/^{32}S$	$4.41509(117) \times 10^{-2}$
NIST Standard Reference Material 976	SRM-976	$^{63}Cu/^{65}Cu$	$2.2440(21)$
Institute for Reference materials and measurements IRMM-632	IRMM-632	$^{63}Cu/^{65}Cu$	$2.8921(92) \times 10^{-3}$
European Reference Material ERM-AE647	ERM-AE647 IRMM-647	$^{65}Cu/^{63}Cu$	$4.4560(74) \times 10^{-1}$
NIST Standard Reference Material 987	SRM-987	$^{87}Sr/^{86}Sr$	$7.1034(26) \times 10^{-1}$

Besides materials diffusion and provenance, the variation of stable isotope ratios during chemical processes has been fruitfully employed to investigate the past interaction between organisms, including humans, and the environment (Griffiths 1998). These kinds of studies, originated by the seminal paper of Urey (1947) on the thermodynamics of isotopes during chemical reactions, have found straightforward application in the investigations of past temperatures based on marine sediments, organisms, and ice cores. In fact, the recent temperature and climatic global variations have been mapped through the changes in hydrogen and oxygen isotope ratios (Figs 2.24 and 2.25). The changes of the isotopic composition of water within the water cycle due to fractionation during precipitation and evaporation provide a recognizable signature, relating such water to the different phases of the cycle (Gat 1996). As a result, meteoric waters are depleted in the heavy isotopic species of H and O relative to ocean waters, whereas waters in evaporative systems such as lakes, plants, and soil waters are relatively enriched. If the organisms are in equilibrium with the water system they live in, then the measurement of the hydrogen and oxygen isotope ratio provides important clues to their living habitat. Similarly, since the pioneering investigation of Bender (1968) and Smith and Epstein (1971) on carbon isotope pathways in plant photosynthesis, the fractionation of hydrogen, carbon, and nitrogen isotopes along the food chain solidly provide information on the palaeodiet of the organism, especially the so-called **trophic level**, that is the degree of carnivory with respect to dietary proteins (Hedges and Reynard 2007, Reynard and Hedges 2008).

There are of course a number of issues that must be taken into account when working on the dietary or environmental reconstruction from the fossil remains, though it is generally acknowledged that stable isotope results are less prone to biases due to contamination and diagenetic processes with respect to trace element data (Sandford 1993, Pollard 1998, Lee-Thorp 2008). From the point of view of materials science, a major issue is whether the material actually analysed is inorganic (i.e. bones and teeth) or organic (mostly bone collagen, but also hairs and nails). Stable isotope studies use carbon, nitrogen, and non-exchangeable hydrogen in collagen, carbon bound within the crystal lattice of bone mineral, oxygen in phosphate and carbonate, and strontium bound to phosphate. Depending on the nature of the material, the patterns of isotope fractionation and the subsequent interpretation may be different. For example, isotope data from mineral apatite and from the surviving collagen fraction of the same bone are known to show systematic differences due to their different biological and metabolic pathways (Krueger and Sullivan 1984, Lee-Thorp *et al.* 1989, Schwarcz 1991). Far from being a problem, the measurement of both signals may actually provide deeper insights into the protein versus carbohydrate fraction of the diet. Systematic mass spectrometric data resolved on the different molecular fractions (for example through combined chromatographic–mass spectrometric measurements; Box 2.t) of the analysed compounds may in principle offer unprecedented information on past human behaviour. In much the same way, spatially resolved isotopic data (for example through the use of laser ablation MS; Box 2.t) offer the possibility of following human migrations or dietary changes through lifetime. The deduction of regional mobility from isotope data is based on the assumption that different tissues record different periods of an individual's history. For example bone tissue is continuously reformed during metabolic turnover, thus reflecting the latter period of life of the individual, whereas dental tissue, especially enamel, is formed early in life and it remains essentially unaltered in subsequent periods. The changes in Sr and O isotopic ratios between teeth and bones, or between dentine and enamel are potential indicators of migrations during a lifetime, and the migration path could be followed by matching the measured stable isotope ratios to those of the environment (water, rocks, food). A number of successful studies of population dynamics in the literature (Schoeninger 1995, Müller *et al.* 2003, Bentley 2006) demonstrate the potential of the stable isotope techniques, though it is evident that proper understanding the human behaviour requires a wealth of information on environmental, nutritional, bio-behavioural, and general ecology. These are the kinds of information that stable isotope analysis can provide, assuming regional contextual studies and data bases are available.

For in-depth information

Griffiths H. (ed) (1998) Stable isotopes. Integration of biological, ecological and geochemical processes. βios Scientific Publishers, Oxford.

Hoefs J. (2008) Stable isotope geochemistry. 6th Edition. Springer, Berlin.

Sharp Z. (2007) Principles of stable isotope geochemistry. Prentice Hall, Upper Saddle River, NJ.

2.5.4.3 Methods based on electron trap accumulation

There are three dating methods based on **radiation dosimetry**, which is the measurement of the time-dependent accumulation of radiation-induced defects in crystal structures. The basic processes producing the electron defects in solids are conceptually the same for the three techniques, but they differ in the nature and energetics of the detected defects and in the measurement methods (Table 2.17). The methods are based on the slow and continuous accumulation of electron traps in solids produced by ionizing radiation (α- and β-particles, γ-photons) caused by natural radioactive nuclei embedded or surrounding the material, or cosmic radiation. The incoming radiation produces a number of physical effects on the atoms (Section 2.1.1) including ionization, scattering, photoelectric and pair-production processes. Energy is transferred to the material, ending up in heat and electron–hole pairs. The resulting free charges (i.e. free electrons and holes) diffuse through the lattice until they recombine or they are trapped by defects that have charge deficits (vacancies, interstitial species, Schottky–Frenkel pairs). Each defect with a trapped negative or positive charge is a potential **emitting centre** of secondary luminescence or a **resonance centre** process exploited by the radiation dosimetry methods. The energy and life time of the electrons traps will vary: some traps are sufficiently deep to store charges for hundreds of thousands of years, although for much longer irradiation times a saturation of the available defect centres is reached. From the theoretical point of view it is not yet clear whether the traps probed by the different techniques are the same or different ones, nonetheless each technique reads the **natural dose** (ND) accumulated in the material since the last bleaching event. The time is computed by:

$$t = \frac{ND}{NDR}$$

where *NDR* is the **natural dose rate**, i.e. the dose the material has absorbed per unit time, which can be calculated from the observed chemical composition (U, Th, K) and the known cosmic ray activity, or directly measured (Aitken 1985, Chen and

Table 2.17. Dating methods based on time-dependent defect accumulation in solids. See Boxes 2.w and 2.x for experimental details

Method	Physical process for the measurement of the defects	Applications
Thermoluminescence (TL)	Recombination of centres by thermal energy	ceramics, bricks, fire places, sediments, cave carbonates, shells, metal slags
Optically stimulated luminescence (OSL)	Recombination of centres by light stimulation energy	dunes, loess, colluvium, archaeological soils, tephra, ceramics, rock surfaces
Electron paramagnetic resonance (EPR) or electron spin resonance (ESR)	Resonance activation by microwave irradiation	cave deposits, travertine, shells, teeth, bones, corals, sediments, flints

Box 2.w TL and OSL luminescence

Thermoluminescence (TL) and **optically stimulated luminescence (OSL)** dating is based on the quantification of the accumulated radiation dose in materials. Natural crystalline materials contain imperfections: impurity ions, stress dislocations, and other phenomena that disturb the regularity of the atoms in the crystalline lattice. These chemical and physical imperfections are commonly treated as alternative levels in the energy potential (electron traps) that can attract and trap a free electron (Fig. 2.w.1). Any flux of ionizing radiation, such as cosmic radiation and radioactivity from decaying nuclides, may produce electron–hole pairs, exciting electrons from atoms in the crystal lattice into the conduction band where they can move freely. Most excited electrons will soon recombine with lattice ions, but some will be trapped, storing part of the energy of the radiation in the form of trapped electric charge (Chen and McKeever 1997). The depth of a trap is determined by the energy required to free an electron from it, and the storage time of trapped electrons depends on the trap depth. Some traps are sufficiently deep to store charge for hundreds of thousands of years, and this is the basis for charge accumulation in time.

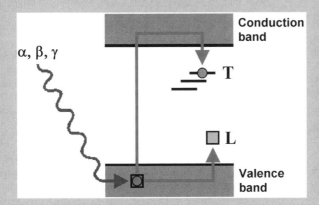

Fig. 2.w.1. Energy band model for electron traps in crystalline solids. The incoming ionizing radiation produces an electron–hole pair, promoting electrons from the valence band into the conduction band, where they can diffuse and be captured.

The long-lived traps are measured in radiation dosimetry as secondary luminescence emitted when energy is supplied to the system, either in the form of heat or as light stimulation. In both cases the recombination of the electron–hole pairs is induced, with the emission of luminescent radiation whose energy and intensity are proportional to the type and number of trapped electrons, respectively (Fig. 2.w.2). In TL, the material is heated during the measurement, and the weak light signal emitted is recorded by CCD elements or a photomultiplier tube as a function of temperature, as increasingly deep traps are progressively emptied at higher temperatures. The recorded curve is called the **natural glow curve** (Fig. 2.w.3), and it is the natural TL signal (**natural dose**, *ND*) accumulated in the sample since the last bleaching (by light) or annealing (by heat) event (Section 2.5.4). Depending on the instrumentation, it is desirable that the signal is spectrally resolved, in order to identify the type of the recombination centres (Fig. 2.w.4). Finally, to calculate the elapsed time by $t = ND/NDR$, the **natural dose rate** (*NDR*), i.e. the rate of trap production in the sample by the incoming radiation flux per unit time, needs to be evaluated by additive or regenerative techniques (Wagner 1998, Wintle 1998). Commonly, the measurement of the α radioactivity (i.e. the uranium and thorium content) and the β and γ emission (mostly by ^{40}K) of the sample are taken. Often the gamma radiation field at the position of the sample material on site is also measured, or it may be calculated from the alpha radioactivity and potassium content of the sample environment, and the cosmic ray dose is added in.

(cont.)

Box 2.w (*Continued*)

Fig. 2.w.2. The energy supplied by heat (TL) or laser radiation (OSL) makes the electron–hole recombination possible, with simultaneous emission of luminescent radiation, which is measured in radiation dosimetry.

Required conditions for the correct calculation of the age of the starting bleaching or annealing event are that (a) the event completely resets the electron traps, (b) the radiation flux producing the defects has been sufficiently constant in time, and (3) the relationship between signal and natural dose rate can be described by a known mathematical function that is possibly linear or quasi-linear. Furthermore, it is desirable that the accumulated traps are stable in time, though in many cases they have a mean lifetime determined by thermally-controlled kinetics. Optimal materials for TL dating are quartz and feldspar grains.

TL has been widely applied to ceramics and burnt clays since it can be used to date the last firing event, generally related to the production of the ceramic object or use of the kiln (Aitken 1985, Wagner 1998, Martini and Sibilia 2001). For archaeological ceramic sherds, TL can be often compared with dates obtained by radiocarbon on charcoal and vegetable remains in the same layer. The TL of ceramics is also commonly used to spot recent fakes and unlikely archaeological finds (Carriveau and Han 1976, Craddock and Bowman 1991, Craddock 2007). Burnt flints and stones are also suitable for dating by TL, especially for the time span that includes the Middle and Lower Paleolithic, which cannot be easily accesses by radiocarbon (Valladas 1992, Mercier *et al.* 1995). Other materials that are potentially datable by TL are metallurgical slags, using the quartz grains separated by the heterogeneous fayalite-loaded slags (Haustein *et al.* 2003), metallurgical kilns (Zacharias *et al.* 2006), and glass mosaic tesserae (Galli *et al.* 2004). Interestingly, if the date of the ceramic is known, TL can be used to retrospectively assess the absorbed dose during high emission events, such as nuclear contaminations and tests (Bailiff 1997).

Optically stimulated luminescence (OSL), also referred to as optoluminescence or **optical dating**, is conceptually very similar to TL, the main difference being the optical stimulation of the trapped electrons (Aitken 1998, Wagner 1998). The continuous or modulated source can be an ion laser, an arc lamp with suitable filters, a halogen lamp, or infrared and green diodes. Since the luminescence centres in OSL are erased by light, it is potentially very interesting to use the method to date the last exposure to daylight of the object's surface. The OSL traps are emptied within minutes of exposure so that, on the one hand, events involving even weak exposure to light can be detected, though on the other, the technique is much more sensitive to visible radiation and samples are to be taken in dark conditions. OSL offers several advantages with respect to TL: the initial optical bleaching is always complete, so that in principle the lack of residual signal allows very recent events to be dated (i.e. less than 100 y, Wagner 1998). Further, its sensitivity allows single aliquot and single grain methods to be optimized. Pre-heating

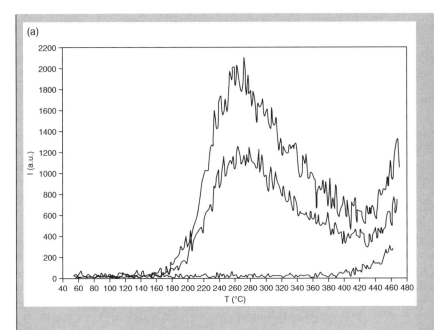

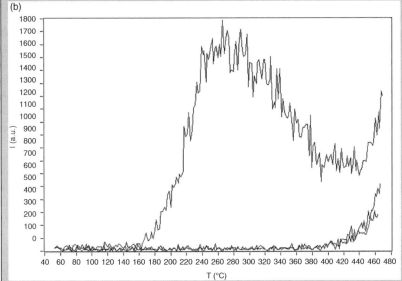

Fig. 2.w.3. TL signal for an ancient ceramic (a) and a modern fake (b). The measured TL curves for sample A are: background (bottom), natural dose (middle), and natural dose after 6 Gy beta irradiation (top). The total absorbed dose is 15.2±0.2 Gy, proving that it is original Etruscan ware, with an approximate age of 2700–2500 y. The same curves measured on sample B show that the natural glow curve is absent, the top curve is the natural dose after 15 Gy beta irradiation. The total absorbed dose is less than 0.5 Gy, so that the sample is straightforwardly recognized as a modern forgery. (Images and data courtesy of E. Sibilia, University of Milano Bicocca)

and regenerative-dose protocols have been proposed to remove unstable components, especially in quartz and feldspars (single-aliquot regenerative-dose protocol, or **SAR**, Wintle and Murray 2006).

Most materials that are suitable for TL, such as ceramics, can also be dated by OSL. In addition, OSL requires much less material, it is potentially more sensitive, and it can be extended to other type of materials such as quaternary sediments (Duller 1996, Wagner 1998, Fuchs and Lang 2001, Murray and Olley 2002), metallurgical slags (Gautier 2001), and even wasps' nests (Roberts *et al.* 1997).

(*cont.*)

Box 2.w (Continued)

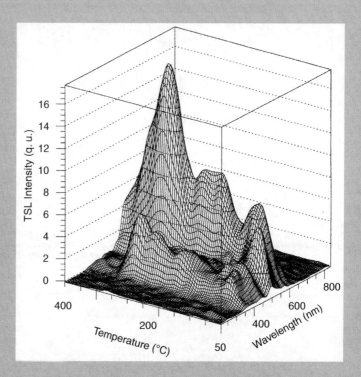

Fig. 2.w.4. Spectrally resolved TL signal as a function of temperature induced on a Roman ceramic sample by a 25 Gy beta irradiation. The identification of the emission wavelengths of a glow curve is essential to assess the nature of the recombination centres. (Images and data courtesy of E. Sibilia, University of Milano Bicocca)

For in-depth information

Aitken, M.J. (1985) Thermoluminescence dating. Academic Press, London.

Aitken, M.J. (1998) Introduction to optical dating. Oxford University Press, Oxford.

Pagonis V., Kitis G. and Furetta C. (2006) Numerical and practical exercises in thermoluminescence. Springer-Verlag, Berlin.

Wagner G.A. (1998) Age determination of young rocks and artefacts. Springer-Verlag, Berlin.

Box 2.x EPR Electron paramagnetic resonance spectroscopy [M. Brustolon]

A key to quickly grasping the principles of Electron Paramagnetic Resonance (EPR) spectroscopy, also known as Electron Spin Resonance (ESR) spectroscopy, can be its similarity with the better known Nuclear Magnetic Resonance spectroscopy (NMR), also used for medical imaging, for readers familiar with NMR. In both spectroscopies magnetic spin moments are oriented by an external magnetic field, and the spins acquire an extra energy due to the interaction of their magnetic moments with the magnetic field (Zeeman interaction). In NMR and EPR the magnetic moments of nuclei and electrons, respectively, are studied as they are oriented by the magnetic field. Each electron has two spin states, α and β, corresponding to two orientations of their spin moments (with quantum number $m_s = \pm 1/2$) (Fig. 2.x.1). We can appreciate the similarity with nuclear magnetic moments: in fact some nuclei, as the proton ^1H, have also two spin states α and β (with quantum number $m_H = \pm 1/2$), whereas for other nuclei the spin states can

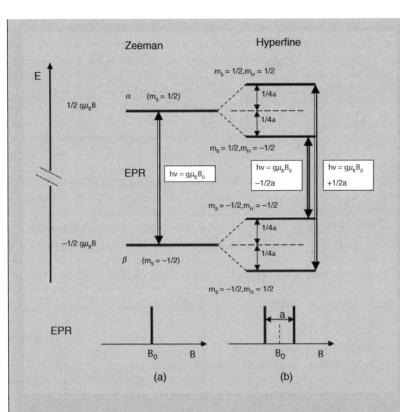

Fig. 2.x.1. A scheme of the energy levels and EPR transitions for an unpaired electron: (a) uncoupled, (b) coupled to a nuclear spin ½. In (a) the transition between the two levels corresponding to the two orientations of the electron spin in a magnetic field gives a single EPR line. From the pair of values of B_0 and ν the g *factor* can be obtained. In (b) two EPR transitions occur between pairs of spin states with the same nuclear magnetic moment ($\Delta m_s = \pm 1$, $\Delta m_I = 0$). The two EPR lines are separated by the *hyperfine splitting*, as the EPR spectra are obtained by varying the magnetic field, a is expressed generally in gauss or mTesla (1 mT = 10 gauss).

be more than two. For example, ^{14}N has three possible spin states, $m_I = +1, 0, -1$; ^{55}Mn has six possible spin states, $m_I = +5/2, +3/2, +1/2, -1/2, -3/2, -5/2$, etc.).

In general, electrons are present in matter in pairs, as chemical bonds are given by two electrons with opposite spins α and β, so each electron pair has a zero moment (i.e. it is in a *singlet* state). Therefore, pure samples of common chemical compounds are *diamagnetic*, they don't contain unpaired electrons, and they don't show any EPR signal.

Unpaired electrons are found only in particular **paramagnetic** chemical species: the most common and interesting here are *free radicals*, *paramagnetic ions* and *paramagnetic defects in solids*. In complex and composite materials paramagnetic species are very often present as traces and impurities. The extremely high sensitivity and resolution of EPR allows one to exploit these species present in traces as *spin probes*, providing information on their environment, i.e. on the bulk material. Moreover, the particular association of paramagnetic impurities in materials is a fingerprint revealing their origin.

For the two states α and β of an unpaired electron the energies in a magnetic field B are respectively $1/2 g\mu_B B$ and $-1/2 g\mu_B B$, in general therefore $E_{m_s} = m_s g\mu_B B$, where g is a small number called the g *factor*, and μ_B is a physical constant, the *Bohr magneton* ($9.27400949 \times 10^{-24}$ J·T^{-1}). Between the two states α and β there is an energy difference of $g\mu_B B$, and transitions between the two states can be driven by a microwave radiation of frequency ν. A photon with energy $E = h\nu$ can be absorbed by the spins when its energy matches the difference in energy of the two spin states ($h\nu = g\mu_B B$) and the spin system is then said to be 'on resonance'. The pairs of values of ν, B causing resonance correspond to a spin transition and they produce absorption lines in the measured spectra at specific g values (Fig. 2.x.2).

The pairs of values of ν and B used in commercial EPR spectrometers are such that the radiation frequency is in the microwave wave range, i.e. with a wavelength λ of the order of cm, and a frequency of some GHz (10^9 Hz); the

(*cont.*)

Box 2.x (Continued)

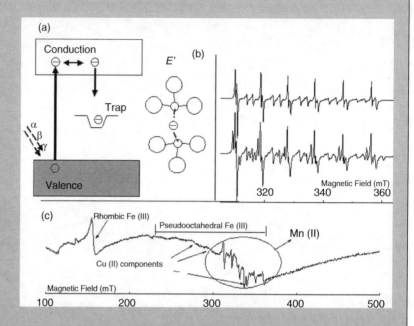

Fig. 2.x.2. (a) The mechanism of defect formation: radiation promotes electron transitions from the valence band (VB) to the conduction band (CB). A trap, energetically localized in the band gap, catches a CB electron. The figure on the right represents the so-called E' centre in SiO_2, a typical paramagnetic defect used for ESR dating. (b) Mn(II) EPR signals in white marbles from capitals stored in the Civic Museum of Padua. The upper spectrum is typical of calcite marble, the lower of dolomitic marble. The different profile is related to different quarries for the marble of the two capitals. (c) EPR spectrum of a paper fragment of the 16th century with indication of the various signals: the broad line ranging from 200 to 350 mT is due to pseudooctahedral Fe(III); the rhombic Fe(III) is evident in the sharp line at 157 mT; the Mn(II) sextet between 300 and 370 mT is circled and small Cu(II) components can also be seen.

magnetic field is around some thousands gauss (1 G = 0.1 mTesla). The spectrometers are named on the microwave "band", the most common is the X-band, corresponding to $v = 9.797$ GHz and $B = 3500$ gauss, for a 'free' electron (with $g = 2.0023$).

The *g factor* is one of the parameters identifying a paramagnetic species. A "free" electron has a pure spin moment, whereas the moment of an electron in a paramagnetic species has some contribution deriving from its orbital motion around the nuclei. The deviation Δg of the *g factor* from 2.0023 indicates how large this contribution is; it is increasing for heavier atoms. In organic radicals the orbital contribution is small, with a maximum value of $\Delta g \approx 10^{-2}$. Inorganic radicals often have g factors in the range 1.9–2.1, while for paramagnetic ions the g factor can be very different from 2 ($g = 0 \div 4$). It should be noted that for a paramagnetic species in a solid matrix the effect of the magnetic field on the energies of the spin states depends on the orientation of the species with respect to the magnetic field (anisotropy of the Zeeman interaction). From the mathematical point of view, this means replacing the *g factor* in the equations by a *g tensor*, characterized by three *principal values*. The shape of the EPR spectrum in the case of a disordered distribution of the paramagnetic species (as in any polycrystalline, glassy, dispersed or powdered material) is determined in part by these three values (g_x, g_y, g_z), which can therefore be obtained by the features of the spectrum.

In addition to the interaction with the external magnetic field, in general in a paramagnetic species there is the interaction of the electron magnetic moment with *nuclear* magnetic moments in the sample. This leads to the formation of a **hyperfine structure**, i.e. the breaking up of the single EPR line into multiple components. The nuclear magnetic moments influencing the electron magnetic energy are said to be *coupled* to the electron, and the coupling intensity is measured by a *hyperfine coupling constant a*. Consider, for example, the effect on the EPR spectrum of the coupling with a single H nucleus. We know that the nuclear spin can be α or β, corresponding to a nuclear spin quantum number $m_H = \pm 1/2$. The contribution to the energy of the electron magnetic moment will depend on the sign of the hyperfine coupling constant a and on the relative orientation of the two moments. The energy of an electron spin state will therefore be: $E_{m_s} = m_s g \mu_B B + m_s m_H a$. Each Zeeman energy level corresponding to one of the two values $m_s = 1/2$, is now split in two, thanks to the magnetic effect of the coupled proton spin in one of its two orientations.

The EPR transitions occur by reversing the electron spin, and leaving the nuclear spin *unchanged*. Therefore the selection rules are $\Delta m_s = \pm 1$, $\Delta m_I = 0$, and in the previous case two transitions are allowed, corresponding to $\Delta E = g\mu_B B \pm 1/2a$ (Fig. 2.x.1). As half of the protons are α and half are β, the two lines have the same intensity. If the coupled nucleus has more spin states, the hyperfine structure of the spectrum shows a number of lines that are equal to the number of nuclear spin states. For example, an electron coupled to a ^{14}N nucleus has an EPR spectrum with three lines, of equal intensity; if the coupling is with a ^{55}Mn the EPR lines are six, etc. For a coupling with several nuclei the pattern of the EPR spectrum can be obtained by imaging sequential separations of each line due to the next nucleus. When more nuclei are coupled to the electron with the same hyperfine coupling, some transitions occur at the same frequency and the effect is that the lines have different intensities. For the coupling with two equivalent protons the EPR spectrum is a triplet with intensities 1:2:1, for three equivalent protons 1:3:3:1, and so on. Also the hyperfine interaction is *anisotropic* for an oriented paramagnetic species in a solid, and therefore the hyperfine coupling constant should be substituted by a *hyperfine tensor*, characterized by three *principal values*. Again, the shape of the EPR spectrum in the case of a disordered distribution of the paramagnetic species is determined in part by these three values (A_x, A_y, A_z).

The main applications of EPR to cultural heritage materials to date are in the areas of (i) dating (ceramics, bones, teeth), (ii) characterization of materials by means of their peculiar content of paramagnetic ions and defects (white marbles, paper), and (iii) assessment of the degradation degree of the materials (paper).

1) Dating exploits the natural formation of defects in solids by environmental radioactivity. The paramagnetic defects are trapped electrons and electron vacancies (Fig. 2.x.2(a)). The analysed material is used as an EPR dosimeter, the total absorbed dose of radioactivity is obtained, and by knowing the annual dose the age of the material is estimated. The time span of the ESR dating depends on the material and the conditions, but it is generally in the range 10^3–10^6 y (see, for example, Grün 1993), with typical errors in the range 15–30%. Increasingly ESR datings are combined with $^{230}Th/U$ methods in order to validate the uranium-uptake assumptions (McDermott *et al.* 1993).

2) The most frequent paramagnetic ion to be detected is Mn^{2+} as this ion in natural materials substitutes in traces the more common Mg^{2+} and Ca^{2+}. Figure 2.x.2(b) shows the measured EPR spectra of Mn^{2+} in two samples of white marble from different quarries. A database of the Mn^{2+} spectra in marbles from the most used ancient quarries in the Mediterranean area is available (Attanasio 2003, Maniatis 2004). A comparison between the spectrum of the unknown sample and the database spectra, performed on the basis of several parameters extracted from the spectra and statistically processed for a linear discriminant analysis, allows one to assign the provenance of the marble.

3) Ancient paper displays EPR signal from Mn^{2+}, Cu^{2+}, Fe^{3+} species and radicals (Fig 2.x.2(c)). Fe^{3+} and Cu^{2+} ions are related to some extent to paper degradation, e.g. it has been observed that Cu^{2+} signals are generally found in ruined or degraded paper. On the other hand, Mn^{2+} may be a useful probe to mark the differences in paper production or origin.

For in-depth information

Brustolon M., Giamello E. (eds.) (2009) Electron Paramagnetic Resonance, A practitioner's toolkit. John Wiley & Sons, New York.

Ikeya M. (1993) New Applications of electron spin resonance. Dating dosimetry and microscopy. World Scientific, Singapore-New Jersey-London-Hong Kong.

Weil J.A., Bolton J.R. and Wertz J.E. (2001) Electron paramagnetic resonance: elementary theory and practical applications. Wiley-Interscience, New York. 2nd Edition (2007).

McKeever 1997, Aitken 1998, Wagner 1998, Wintle 1998, Bøtter-Jensen et al. 2003, Pagonis et al. 2006). The *ND* value is derived from the experimental measurement of the intensity of the TL, OSL or EPR signal (Boxes 2.w, 2.x).

TL and OSL cover a time interval of a few hundred years up to 1 My. They may therefore be applied beyond the limits of radiocarbon methods, and they may be used to date rather special events (Aitken 1989, Roberts 1997, Wintle 2008). In the interpretation of the signal, two concepts must be taken into account: (1) the traps continuously accumulate after the material is formed, due to environmental radioactivity and cosmic rays; and (2) heating processes and exposure to light represent bleaching events for both TL and OSL techniques.

As a consequence both TL and OSL are perfectly adequate to date the original formation of cave carbonates or volcanic tephra, whereas they may be used to date the last heating event of fired objects such as metallurgical slags, heated stones, and products of pyrotechnology such as ceramics, bricks, furnace linings, etc. Further, sediments and sands can be dated by OSL if the defects have been erased during particle transport by light bleaching, and provided that the samples were never exposed to light between burial and measurement (Lian and Roberts 2006). Extensive light exposure may also significantly attenuate the latent TL signal, together with other partially understood signal-modifying processes such as anomalous fading, especially in feldspars, supralinearity, and phototransfer TL (Chen and McKeever 1997, Wagner 1998, Wintle 2008).

By far the most important recent advances in luminescence dating are single-aliquot and high-resolution single-grain OSL measurements, which allow careful screening of perturbed and mixed signals, and selection of the thermally stable OSL signal. The combination of these methods, called the single-aliquot regenerative dose (SAR) protocol (Wintle and Murray 2006), has greatly enhanced the sensitivity of and the applicability of the technique, so has the high spatial resolution strategy (Greilich and Wagner 2006), which allows the dose rate of individual grains to be calculated. These improvements are leading the way to exciting and holistic investigations of recent sediments related to human activities, encompassing interpretation of past environments, climate, agricultural practices, and landscape use. Seminal examples are the reconstruction of the links between Holocene soil erosion and historical farming in southern Greece (Fuchs and Wagner 2005), and the comparison between human occupation and climate variations in East Africa (Scholz et al. 2007). It even looks possible to date the burial of granite stone surfaces in soil or walls after they were exposed to sunlight bleaching (Greilich et al. 2005).

2.5.5 Advantages and pitfalls

Following up on the early success of dating techniques, recent scientific and technological advances offer a truly impressive array of powerful tools for absolute dating. The same tools may be used by archaeologist to date artefacts, layers, sites, and by art historians for authenticity testing and historical sequencing and correlation of events.

Each method needs to be applied by the careful matching of the material, the available amount of sample, the required precision and accuracy, and possible comparison with reference time scales. Keeping in mind that each method has limitations due to its conceptual and/or physical foundations, the best possible strategy is to try and combine several methods within the investigation. Each individual method can hardly stand alone, and it may provide a faulty date for one reason or another. The list of the possible pitfalls is as long as that of the success stories. Radiocarbon dates are easily biased by sample contamination. Luminescence dating may suffer from anomalous fading. Dendrochronology needs complete master sequences and exclusion of re-use processes. Assumptions of a closed system or starting isotopic ratios for radioactive clocks may be incorrect. Finally, for physical reasons none of the absolute dating methods can be applied to metals.

Relative methods may be deceptive too, since site stratigraphies are invariably disturbed and often difficult to read, and even chronological markers may turn out in unusual and apparently inconsistent situations. The finding of Neolithic axes in secure Roman contexts is an instructive example of artefact recycling through purely religious and cultural actions (Adkins and Adkins 1985), which of course cannot be explained by natural processes.

The safest strategies to use to resolve the issues and limit mistakes are: (1) to anchor firmly all observations within the investigated context; (2) as much as possible to use a variety of techniques based on different physical measurements, that is cross-dating (or "never trust the uncritical acceptance of ages derived from a particular approach": Grün and Stringer 1991), and (3) to link the available measurements with the understanding of the processes that created the object (or layer, or site).

The amazing advantage of advanced techniques is that now we can attempt the absolute dating of several materials from the same context, or the dating of the same material with different techniques, for example applying pollen, radiocarbon, and aminoacid racemization methods to the organic fraction of soils. We should thus aim for a consistent and cross-validated reconstruction of the time sequence encompassing the object, the surrounding context, the landscape, and their global evolution.

Further reading

Amelinckx S., van Dyck D., van Landuyt J. and van Tendeloo G. (eds) (1997) Handbook of microscopy. Applications in materials science, solid-state physics and chemistry. Methods I. VCH, Weinheim.

Barry R.S., Rice S. and Ross J. (2000) Physical chemistry. Oxford University Press, New York.

Barry R.S., Rice S. and Ross J. (2001) Structure of matter: An introduction to quantum mechanics. Oxford University Press, New York.

Brandon D. and Kaplan W.D. (2008) Microstructural characterization of materials. 2nd Edition. John Wiley & Sons, Chichester.

Brothwell D.R. and Pollard A.M. (eds.) (2001) Handbook of archaeological sciences. John Wiley & Sons, Chichester.

Ciliberto E. and Spoto G. (eds) (2000) Modern analytical methods in art and archaeometry. John Wiley & Sons, New York.

Janssens K., Van Grieken R. (eds) (2004) Non destructive microanalysis of cultural heritage materials. Comprehensive Analytical Chemistry, Vol. XLII. Elsevier, Amsterdam.

Lindon J.C., Tranter G.E., Holmes J.L. (2000) Encyclopedia of spectroscopy and spectrometry. Academic Press, London.

Stuart B.H. (2007) Analytical techniques in materials conservation. John Wiley & Sons, New York.

Wagner G.A. (1998) Age determination of young rocks and artifacts. Physical and chemical clocks in Quaternary geology and archaeology. Springer, Berlin.

Materials and case studies: how to meet the needs

3

3.1	Structural materials I: lithics, rocks, stones, structural clay products, ceramics	209
3.2	Structural materials II: cements, mortars, and other binders	242
3.3	Pigments	266
3.4	Glass and faience	278
3.5	Metals	305
3.6	Gems	348
3.7	Organic materials	355
3.8	An example of complex composite materials and processes: Photography [D. Stulik and A. Kaplan]	419

The history of the use of materials by mankind for the fabrication of tools is a complex one, though it is fundamental to one's understanding of the development of technical abilities and the interaction and modification of the natural environment. The association of tools with human remains is commonly taken as the proof of their use and is therefore linked to the very same definition of the human species.

Intuitively, we assume that it all started with the use of raw natural materials, with little or no human change or manipulation. Then, as technical abilities advanced, the properties of the materials as found in nature were transformed and optimized by physical intervention, mainly through the use of mechanical forces (cutting, grinding, pounding) and pyrotechnology (fire). To what extent and how fast this happened is a matter of debate, the relationship between cultural and technical advances and their evolutionary, geographical, environmental, and social causes being the objective of active archaeological and archaeometric research. As an example, only recently has scientific attention been focusing on the problem of how to identify deliberate heat transformations of natural materials by man such as soil burning by fireplaces, as opposed to natural or uncontrolled events and operations.

Very early in prehistory the knowledge concerning the competent use of **composite materials** was also acquired, composite materials being defined as engineering materials that are made from two or more constituents that remain separate and distinct on a macroscopic level while forming a single component or tool. The composite has different properties from each of its original constituents. Technically, many natural materials such as bone and wood are actually composites at the microscopic level, and therefore the use of objects intentionally made of composite materials is referred to as the voluntary assemblage of different components with the aim of obtaining man-made materials with properties that are different from any one of the naturally available constituents. Bricks formed from clay and straw, ceramics formed from clay and temper, and even arrows formed from wood, flint, rope, and resin can all be considered to be composite tools. Iron and steel reinforced concrete is undoubtedly the most diffused modern composite material, and a rather complex example is exemplified by the skilled combination of support and light-sensitive chemicals used in photography. The latter case is treated in detail in Section 3.8.

Figure 3.1, developed by Ashby (1987), is one of the few attempts to assess the relative use of different materials by man as a function of time. The whole of prehistory is dominated by natural polymers (wood, skins, fibres, bones) and by ceramics, in a general engineering use of the term, which includes all kinds of natural rocks and minerals, pottery and, later on, even glasses and binders. Flint and obsidian of course had a major role in early times.

The discovery and use of metals had great consequences in prehistoric societies (Young et al. 1999, Diamond 1999, Scarre 2005), so that metals rapidly became the preferred technological material, a source of welfare and power. Metals, mainly copper alloys and iron, were the dominant material component of tools and weapons for the largest part of human history until steel lead the way for the modern industrial revolution, which has subsequently witnessed the development of alloy steel and superalloys. The domination of metals in the world of materials was such that, in the words of Ashby, 'materials science' was for a long time synonymous with 'metallurgy'. The scene started to change with the very recent introduction of light-weight resistant materials in place of steel: high-strength polymers, modern ceramics, and structural and reinforced composites. The knowledge of materials nowadays is such that the future will be dominated by specifically-designed materials, conceived and prepared to

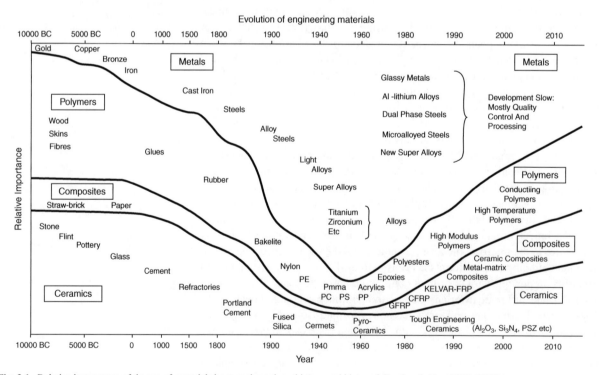

Fig. 3.1. Relative importance of the use of materials by man through prehistory and history following Ashby (1987, 2005).

meet the needs of technology-based societies. Understanding the role of empiricism, rational modelling, and the use of mathematical and computational tools in modern materials development (Ashby 2001) also greatly helps to rationalize the conceptual development of past technologies.

The scientific investigation of materials in archaeology is by far the archaeometric application with the longest history. Chemical analyses of artefacts, namely the chemical composition of archaeological coins and glasses, were performed by Klaproth in Germany as far back as late 1700, Davy in the UK analysed pigments from Pompeii in the early 1800s and, at about the same time, Faraday studied different archaeological materials, including metals, glasses, organics, and glazed potteries (see Caley 1949, 1951, Craddock in: Bowman 1991, and Pollard in: Pollard *et al.* 2007).

The aim of the analyses is to obtain as much information as possible on the nature, structure, and status of the studied artefacts, so as to answer the questions related to the life cycle of the objects. The basic questions involved are: how, when, and where was the object produced? (Tite: in Brothwell and Pollard 2001), though of course physico-chemical analysis of the materials yields only part of the information and it is to be combined to archaeological and excavation data, ethnographic information, and results from experimental archaeology for the full reconstruction of the object's history. Further, despite all possible technical analyses and experimental inquiries, there will always be substantial intangible parts of the materials and artefacts that are related to human nature (Bouchenaki 2004, Kirshenblatt-Gimblett 2004) and that can hardly be quantified on the basis of scientific inquiry alone. The tangible and intangible components are inevitably and inextricably related (Bouchenaki 2003), and it can only be hoped that scientific methods investigating the tangible materials culture will contribute to our understanding and appreciation of the intangible part, mysterious and immaterial as it may be.

Starting from the point of view of materials, it is hard not to fall into the trap of focusing just on the technical properties of the constituents. Especially because as materials scientists we are so used to select, transform, synthesize and characterize materials based on their mechanical, chemical and/or physical properties (Askeland 1988), that most or all other values are neglected. However, we just need to consider, for example, that the market values of gold and diamonds are far overestimated with respect to their actual abundance and specific physical properties, in order to realize that much more should be sought than the properties of the matter alone. Nonetheless the starting point for every investigation is bound to be the characterization of the material: we must always have a clear comprehension of the nature and properties of the investigated objects in order to understand why and how they work, how they came into use in the past, and how we may approach their conservation. Then we might proceed into the interpretation of the more intangible properties.

Based on the assumption that different classes of materials have entered human history at different times and places depending on the complex interaction patterns between the availability of natural resources, technological abilities, and cultural needs (Burenhult 1993–1994, Diamond 1999, Scarre 2005), we'll try to summarize our present knowledge on (1) the physical nature of each materials class, in order to understand its properties and the techniques best used to characterize it, (2) the available evidence of their use in prehistory and history, and (3) the technology needed to produce and use the material profitably and efficiently.

From the point of view of materials science, substances are generally divided into classes depending on their properties (Askeland 1988, Ashby *et al.* 1995). The properties are broadly defined as **mechanical**, based on how the material responds to external forces, and **physico-chemical**, which includes all properties related to the atomic, structural, and microstructural nature of the material. There are at least four levels of description of the real structures: (1) the electronic configuration at the sub-atomic scale, (2) the elemental composition at the atomic scale, (3) the crystallographic structure at the tens or hundreds of nm scale, and (4) the textural description encompassing the shape, dimension, and orientation of the crystalline domains. These level broadly correspond to three of the scales (nanoscopic, microscopic, and mesoscopic scales) illustrated in Fig. 2.4, with the nanoscopic scale encompassing both the electronic and atomic levels, though the electronic level involving chemical bonds should be actually treated at the sub-nanoscale. Figure 3.2 attempts to clarify how each level is actually the magnification of the one below. Even minimal changes in each of the structural levels may result in profound changes in one or more of the observed macroscopic properties of the bulk material. Table 3.1

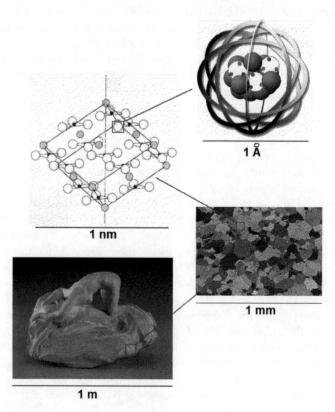

Fig. 3.2. Different levels of description of matter.

Table 3.1. Typical measurable macroscopic properties of materials (modified from Askeland 1988)

Mechanical properties		Physico-chemical properties	
Creep	creep rate fracturing stress–rupture effects	Chemical	reactivity alteration corrosion
Ductility	elongation volume reduction	Density	
Fatigue	fatigue life endurance limit	Electrical	conductivity dielectric insulation ferroelectricity piezoelectricity
Hardness	scratch resistance wear rates	Magnetic	ferrimagnetism ferromagnetism
Shock	absorbed energy toughness impact resistance	Optical	colour absorption photoconduction reflection, refraction transmission
Strength	elastic modulus tensile strength yield strength	Thermal	heat capacity thermal conductivity thermal expansion

lists some of the mechanical and physical properties that are commonly taken into account in materials evaluation.

Engineers commonly subdivide materials depending on their mechanical and physical properties (Fig. 3.3, see also Askeland 1988), though this may not be an absolute and generally accepted classification. A petrologist would be horrified by the insertion of concrete and rocks into the same category, and any organic chemist would be embarrassed by the classification of polymers of entirely different nature into a single group. Nonetheless these classifications have the very practical function of indicating which material will perform well in a specific application. When shaping an object, manufacturing a tool, or constructing a building we must make sure that the constituent materials meet the requirements and respond accordingly. Knowledge of materials science is rapidly developing, and special materials are continually produced and selected to optimize or change their natural properties. If not on the market, at least in scientific laboratories nowadays we may encounter materials that can hardly be classified by using traditional material classes. Examples of this are organic conductors, metal glasses, carbon nanotubes and nanosheets, silicates that expand upon compression, and oxides that shrink upon heating. The basics of superconducting and nano-materials with spectacular properties seem to be common knowledge and they are driving fast-developing processing technologies for societal and industrial needs. This is not to divert attention from the main task of understanding the materials and the technologies used in the past. Rather, the message to convey is that the fine properties of materials are extremely variable and complex, and that a great deal of ingenuity is needed to understand and control them, the final outcome being dependent both on

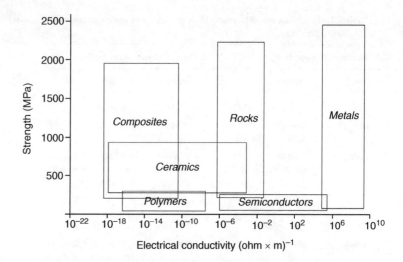

Fig. 3.3. General classification of material classes based on their mechanical and electrical properties.

the structural nature of the compound and on the process used to produce it. We may thus talk about complex **structure-processing-properties** relationships (the *materials paradigm*) both for modern and ancient materials. Such relationships are often the key to understanding the past use and production of materials.

3.0.1 Pyrotechnology

Although the Greek philosopher Plato and the Latin poet Lucretius already had some perception of the atomistic nature of matter, the submicron knowledge of matter before the 20th century was very limited and indirect. In the absence of clear-cut structure–properties relationships we therefore assume that the selection and usage of materials in the past was largely dominated by cultural transmission and empirical testing. Within this conceptual framework, it is probably safe to state that the development of fire-based processes (**pyrotechnology**) was the key factor enabling the modification of existing materials into advanced substances with controlled properties to be used for specific functions, and not previously found in nature. Fire is at the basis of all major early developments in technology: ceramics, metallurgy, glass and binders (Rehder 2000). To understand past technologies it is therefore important to appreciate how combustion in furnaces works, and how we can assess the efficiency and the outcome of ancient pyrotechnologies. In principle our knowledge of high temperature physico-chemical transformation in materials and the details of furnace technology should guide our interpretation and reconstruction of past processes, be they firing clays for ceramics, roasting limestones for lime, reducing metals from ore minerals, or producing glass and faience from silica sand. Even before the mass production of fired materials (pottery, metals, glass), small scale pyrotechnology was used to heat-harden flint and obsidian points, and to decompose limestones and gypsum to produce plaster.

Primary sources of ancient pyrotechnologies before the industrial revolution are very scarce. We mostly rely on G. Plinius Secundus *Naturalis Historia* of AD 77 (Pliny the Elder, Natural History), Vannuccio Biringuccio *Pirotechnica* (1540), which apparently originated the etymology, and Georgius Agricola *De re Metallica* (1556). They are also the main sources of information concerning ancient metallurgy. Recent fundamental works reconstructing the history of furnace technology are Wertime and Wertime (1982), Craddock and Hughes (1985), and Rehder (2000).

Three aspects of pyrotechnological processes need to be briefly outlined here: (1) the thermodynamics of combustion and oxidation reactions, (2) the importance of fuel, and (3) the design of the furnaces.

The thermodynamics of combustion is fundamental to the understanding of why carbon-based materials such as wood and coal are used as fuels, and what the minimum temperature is that is needed to be reached to process different compounds. The information is enclosed in the well-known Ellingham diagrams (Ellingham 1944), which plot the standard free energy of a reaction as a function of temperature. Originally the values were plotted for oxidation and sulphidation reactions for a series of metals, relevant to the extraction of metals from their ores. However these reactions can be used in the general case involving the reaction of an oxidizing gaseous phase (normally air) with condensed phases that are reduced in the presence of carbon (the oxidized phase). By using the diagrams (Figs 3.4 and 3.5) the standard free energy change for any included reaction can be found at any temperature, so that we may compute the equilibrium composition of the system: above each line the oxide is stable, below each line the metal is stable. Alternatively, from the diagram we may also calculate the partial pressures of oxygen needed to oxidize an element or reduce the metal oxide, at any given temperature. For example the point of intersection between the line of each metal–metal oxide and that of carbon–carbon monoxide marks the minimum conditions (i.e. temperature and oxygen pressure) for reduction of the oxide into the metal by carbon monoxide (i.e. combustion).

As for fuel, the burning of biomasses was the only source of manageable heat in antiquity. The oxidation reaction of carbon:

$$C + O_2 = CO_2$$

to produce carbon dioxide is strongly exothermic, and it is the major source of heat during combustion, provided there is a sufficient supply of oxygen, sometimes implemented as a forced stream by means of air conducts, tuyeres, and bellows. If the air supply is insufficient, then the concomitant reactions

$$2C + O_2 = 2CO, \quad C + CO_2 = 2CO$$

become prevalent under conditions of high carbon/oxygen ratio, also because the product CO_2 streams over hot coal yielding CO, which is an endothermic

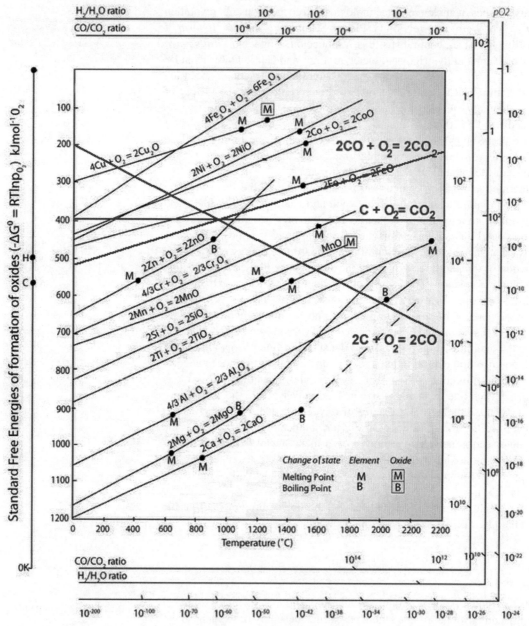

Fig. 3.4. Ellingham-type diagram showing the standard free energy of formation of several metal oxides (Ellingham 1944). At any given temperature the vertical difference between the ΔG values of two lines gives the value used in redox reactions. Inserted in the diagram are also the lines for the combustion reactions involving carbon, and the partial oxygen pressures (pO_2) needed to reduce the metal oxide or oxidize the element (scale at the right). Oxygen pressure data are from Richardson and Jeffes (1948).

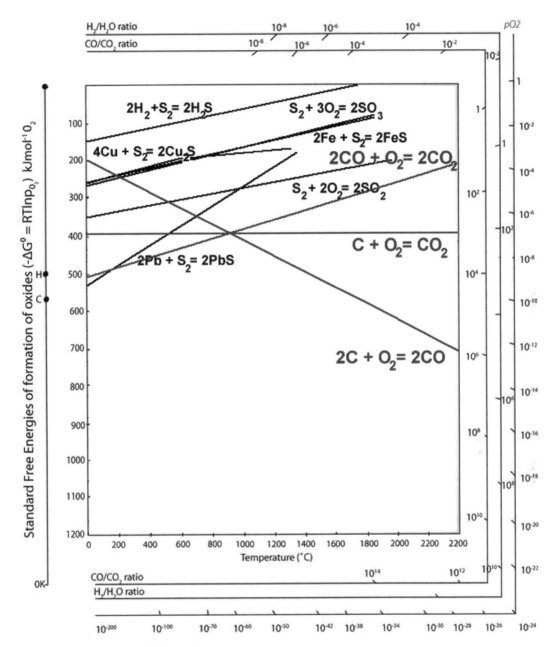

Fig. 3.5. Ellingham-type diagram for sulphide reactions.

reaction and therefore contributes to lower the furnace temperature. The amount of fuel, the oxygen pressure (pO_2) derived from the air supply, and the way that they are mixed in the furnace are crucial parameters that must be carefully controlled to produce the necessary high temperature in the whole volume of the fired charge (Rehder 1986, Rehder 2000). If the combustion reaction is adequately powered with a congruent air stream, it has been experimentally proven that temperatures in excess of 1000 °C can easily be reached with either crucible, pit, or furnace techniques (Livingstone Smith 1991, Gosselain 1992). With regard to the fuel, it has been argued that the mass development of pyrotechnology in antiquity was the prime cause for deforestation, inducing drastic changes in the landscape and in the use of territory (Wertime 1983, Redman 1999).

A large number of ancient furnaces of different ages and geographical areas have been excavated and reconstructed (Craddock and Hughes 1985, Tylecote 1987, Tylecote 1992, Rice and Kingery 1997). For most of the firing technologies used in the Iron Age or later, for which ample archaeological evidence and some direct description exists, the reconstructed process and operation sequences seem quite reasonable and satisfactory. On the other hand, when dealing with prehistoric and protohistoric pyrotechnologies, the archaeological evidence is very fragmentary, and there is of course no direct description, written account, or detailed depiction, besides a few general images found in Egyptian hieroglyphs or Greek vases (Craddock and Hughes 1985). The field evidence of pyrotechnology can be fragmentary, puzzling, and hard to recognize (McDonnell 2001), so that the interpretation of the firing processes in most cases is therefore still very tentative. Sometimes the reconstruction is plausible and seems to work in practice, at least based on modern attempts to reproduce the operations (see for example: Tylecote and Merkel in Craddock and Hughes 1985, pp. 3–20; Rothenberg *et al.* 1979; Pryce *et al.* 2007). In other cases the archaeological evidence is apparently contradictory and hard to replicate, such as in the case of Late Bronze Age copper smelting in the Eastern Alps, where the presence of large furnaces (Fig. 3.6) coexisting with different kinds of smelting slags still pose problems in the reconstruction of the operational sequences.

Fig. 3.6. One of the Late Bronze Age metallurgical furnaces at Acqua Fredda, Passo del Redebus, Trentino, Italy. Details of the site can be found in Cierny *et al.* (2004). (Courtesy of Archivio della Soprintendenza per i Beni Archivistici ed Archeologici della Provincia Autonoma di Trento; photo R. Perini)

3.1 Structural materials I: lithics, rocks, stones, structural clay products, ceramics

For most of human history and prehistory the materials used for buildings and architectural structures for human activities were derived from cheap, widely occurring natural materials. Even today, for the mass production of structures such as roads, houses, and social edifices the materials of choice are, for the vast majority, selected as a compromise between cost, engineering properties, aesthetics and social values. Which one of these components was a priority in determining past choices is frequently a matter of discussion. Scientific analysis may easily assess the engineering and physical properties of the material at different scales (Ashby *et al.* 1995, Gibson *et al.* 1995), so that we may have an idea of the technical part of the selection criteria. We may also expand the assessment to the abundance, distribution, and accessibility of the source materials, in order to insert the alleged source location into proper geographical and geological contexts. The reconstruction of the extraction and transport processes greatly helps in inserting the past use of the structural materials into an overall anthropological, social and economic frame, thus attempting to evaluate the cost of the technical selection. However, concerning the other more intangible components including aesthetic, anthropological, and social values, the assessment of culture-related selection criteria may only be performed in conjunction with other disciplines. This is more difficult the more we adventure into prehistory and lack any direct or written report. To return to the example cited in Section 1.2 concerning Frank Geary's widespread use of titanium in modern architecture, there are apparently no sufficient technological and economic justifications to support the choice, and future archaeometrists might have substantial difficulties in understanding and explaining the practical selection of Ti as an architectural metal were it not through a complete assessment of the present trends in architectural philosophy and of the overall aesthetic and cultural values imposed by society and derived from the media.

It is thus evident that the information derived from scientific methods concerning structural materials addresses only a few basic questions. (1) What are the properties of the materials? (2) Are the properties functional to its usage? (3) What are the technicalities of their provisions and use? (4) Is the material representing the most economic and practical choice at the time, to our knowledge?

This information must then be combined with a great deal of additional knowledge towards the complete understanding of the relative importance of cost, properties and cultural values in the past technical choices. Finally, the analytical data obtainable on past structural materials are fundamental in the diagnostic stages for conservation, and in the design of appropriate intervention for the stabilization of the components and a slowing down of the alteration processes.

This section focuses on widely occurring geological materials that were and are used in their pristine state, or shaped and fired to produce large elements used for constructions and tools. The large variety of geological materials (Rapp 2002) and their properties are treated in detail in geosciences courses. They are so important in the history of mankind that possibly every archaeologist or conservation scientist ought to have at least a course on geomaterials in his/her curriculum (Herz and Garrison 1998, Garrison 2003). Here we will use an extremely simplified scheme, treating materials from the point of view of the way they are used; that is **solid rocks** of different kind that are shaped into usable fragments, tools or blocks, and unconsolidated **clay-based materials** that can be plastically shaped with water and then hardened by fire (Table 3.2).

In the case of solid, consolidated, or naturally cemented materials (stones, rocks), the action of humans is simply that of selecting the right substance and shaping it to their needs. Size and shape depend on the application, and may vary from microartefacts (Dunnell and Stein 1989) such as small chips of obsidian and flint used for prehistoric blades and arrow points (Fig. 3.7), to very large stoneblocks used for buildings and statues, such as the impressive Preseli dolerite bluestones of Stonehenge (Williams-Thorpe et al. 2006), the Kachiqhat red granite and Rumiqolqa andesites used in the Inca walls of Ollantaytambo and Cuzco (Protzen 1985, Protzen and Nair 1997), the massive basalt blocks of Sardinian nuragi (Fig. 3.8), or the striking moai made out of the Rano Raraku tuffs by the Rapanui people on Easter Island (Baker 1993).

In the case of unconsolidated geologic materials, the role of clay minerals is fundamental, because their specific mineral properties make them the ideal material to be plastically shaped when mixed with water. The final shape is maintained through high temperature reactions and mineral transformations. The other clay-free unconsolidated natural materials such as sand and gravel are historically important mostly for binders and concrete, which are treated separately (Section 3.2), or as minor component (temper) of ceramics.

3.1.1 Lithics, rocks, stones

The systematic listing of rocks and minerals used in the past is not of interest here. The excellent volume by Rapp (2002) should be used for the overall understanding of the variety of geomaterials used in ancient times, and their detailed physico-chemical and structural properties can be found in all textbook on mineralogy and petrology (for example, Deer et al. 1996, Blatt et al. 2005). To keep up with the philosophy of the volume, only a few materials will be mentioned (Table 3.3) whereas the concepts that will be outlined are (1) how do we investigate geological materials, (2) what kind of information may we derive from the scientific analyses, and (3) is there a relationship between the properties of the employed materials and their use?

Rocks are assemblages of one or more mineral phases, and they are commonly investigated by the tools developed within mineralogy, petrology, and

Table 3.2. Natural materials used to produce structural elements and tools

Class	Materials		
	Geological materials		**Examples**
Rock-based materials (**shaped**)	Sedimentary rocks	Terrigenous clastic	sandstone, greywacke, siltstone
		Carbonatic	limestones, bioclastic carbonates, travertine
		Evaporites	gypsum–alabaster
		Other	flint, chert, ironstone, coal
	Metamorphic rocks	Regional metamorphism	schist, shale, amphibolite, eclogite, serpentine, quartzite
		Contact metamorphism	marble
	Igneous and volcanic rocks	Intrusive	granite, gabbro, diorite, syenite
		Extrusive	basalt, andesite, trachyte, rhyolite
		Volcanic glass	obsidian, pumice
		Pyroclastic	tuff, pozzolan
	Fired products		**Use**
Clay-based materials (**shaped and fired**)	Structural clay products	Bricks	facing wall (common) paving
		Tiles	structural floor wall (glazed) flue roofing
		Pipes	sewer and drain chemical resistant conduit drain tiles
		Architectural terracotta	plain glazed
	Ceramics	Earthenware	pottery, faience, terracotta, majolica
		Stoneware Porcelain Gres	china

geochemistry (Kempe and Harvey 1983, Herz and Garrison 1998, Garrison 2003). They can be fairly simple (monomineralic, i.e. composed of one type of mineral), such as limestones and marbles composed of calcium carbonate ($CaCO_3$) present as mineral calcite, or they can be extremely complex, composed of tens of mineral phases (polymineralic), especially if their geological history has brought them through a number of stages at different

Fig. 3.7. Prehistoric arrow points made of different silica-based materials: (a) chert (from Mali), (b) quartz (from Castel-Grande, Bellinzona, Switzerland; photo A. Carpi, courtesy of Archivio fotografico Ufficio dei beni culturali – Bellinzona), and (c) obsidian (from the Giara di Gesturi, Sardinia, Italy).

Fig. 3.8. Bronze Age nuraghe made with large basaltic blocks. Santu Antine, Sardinia, Italy.

pressures and temperatures (Tables 3.2 and 3.3). To a mineralogist and a petrologist, the identification of the mineral components of a rock directly suggests the type of rock, where it may come from, its geological history and its basic properties. Bulk or point chemical analyses, and a number of spectroscopic and diffraction techniques may be performed to refine the interpretation.

Table 3.3. Some types of minerals and rocks used in the past for objects of different dimensions and shapes

Object type		Dimensions	Required properties	Materials used
Ornaments, vessels, carved decorations		10^{-3}–10^0 m	Fine-grained polycrystalline materials with good workability. Transparent single crystals with nice colour and appearance. Worked by chipping, carving, polishing, drilling, etc.	SiO_2-based materials (chalcedony, carnelian, amethyst) carbonate-based materials (calcite, travertine, marble, malachite, turquoise) gypsum-based (alabaster, selenite) feldspar (amazonite) pyroxene (jade) amphibole (nephrite) phyllosilicate-based (serpentinite, steatite, soapstone) gems and precious stones (lapislazuli, emeralds, aquamarine, ruby)
Lithics	blades, knives, points, choppers	10^{-2}–10^0 m	Glassy, microcrystalline or crystalline materials with conchoidal fracture and high hardness. Mostly worked by flaking and chipping.	SiO_2-based materials (obsidian, chert, jasper, agate, quartz, quartzite, radiolarite, silicified fossil wood)
	polished axes, hammerstones	10^{-2}–10^0 m	Fine- to medium-grained polycrystalline materials, mostly worked by chipping, pounding, and polishing. Toughness and hardness are required.	sedimentary rocks (chert, sandstone) metamorphic rocks (quartzite, greenstone, gneiss, eclogite) igneous rocks (basalt)
Ground stones, mills, querns, weights		10^{-1}–10^0 m	Medium- to coarse-grained polycrystalline materials. Shaped by flaking and pounding, then abraded and polished by use.	sedimentary rocks (sandstones, limestones) igneous rocks (trachyte, quartz-rhyolite)
Refractory stones, casting moulds		10^{-1}–10^0 m	Good workability and heat resistance.	fine-grained sandstones metamorphic rocks (amphibolite, serpentine)
Architectural components		10^0–10^1 m	Coarse-grained large blocks with good workability. Colour and texture are essential.	igneous rocks (granite, sienite, diorite, basalt, trachyte) sedimentary rocks (sandstones, limestones. marbles) metamorphic rocks (gneiss, breccia)

In general, investigation of geomaterials starts with the preparation of a cm-sized 50-micrometres thin section of the specimen, which is suitable in sequence for optical microscopy, scanning electron microscopy, and electron probe micro-analysis. This series of analyses is generally sufficient to identify the mineralogy, the petrology, and the major and minor element geochemistry of the rock (Table 2.10). For specific problems, we may proceed to investigate the trace elements (by XRF, NAA (Box 3.a), AAS, OES, or MS), especially to discriminate between similar rocks from different outcrops, or even the isotopic signatures (by MS) to check for the rock's history and age. IR and Raman spectroscopies have also been shown to yield useful textural and trace element patterns in stone tools, including cherts (Long *et al.* 2001) and greenstone axes (Smith 2006), though they are clearly not the most obvious tools for multimineralic rock identification. Beside optical microscopy, X-ray diffraction may be the required and most straightforward technique to identify the specific mineral phases present in complex multiphasic and multi-elemental samples.

All these data are useful to identify the possible source of the material, to reconstruct its origin and diffusion, and ultimately to start the insertion of the specimen in a geographical and cultural landscape (Shackley 2008). Possibly we would like to know whether local or imported material was used. Was it worked before or after travelling? By foreign or local people? Was it reused in any way? The main objectives of the analyses of geomaterials employed by man therefore are:

- identification of the nature of the material, including authentication;
- investigation of its provenance;
- insertion of the extraction and working processes into an appropriate environment and socio-economic system;
- interpretation of the trade and diffusion routes;
- understanding of the properties–selection–use relationships;
- definition of the past and present alteration/degradation processes;
- evaluation of restoration/conservation strategies and materials.

Box 3.a NAA-PGAA-NRCA Neutron-based analysis

Neutron activation analysis (NAA) is one of the earliest techniques for chemical analysis to be applied systematically for archaeometric applications (Speakman and Glascock 2007). NAA as an analytical technique was developed within the research programmes with neutron sources following the Second World War. The start of its application to archaeology is precisely dated to 1954, when the fruitful collaboration between J. R. Oppenheimer of Princeton and R. W. Dodson of Brookhaven National Laboratory started (Harbottle and Holmes 2007). Large archaeometric research programmes, mainly in the area of the chemical characterization and provenancing of pottery, obsidian, chert, glass and metals, proliferated in the following twenty years, profiting by the numerous large and small reactor sources available, and of the advantages provided by the technique, namely (a) the ease of preparation of the bulk samples, and the possibility of sample-free non-invasive analysis on the object themselves, (b) the high analytical precision, (c) the large number of detected elements (Fig. 3.a.1), (d) availability of calibration materials and inter-laboratory protocols, and (e) the lack of matrix interference effects (Hughes *et al.* 1991,

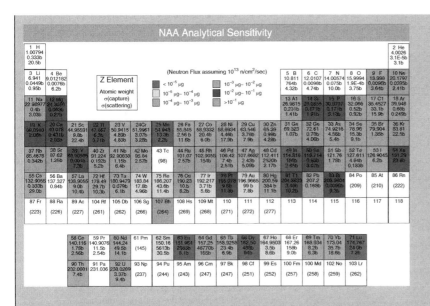

Fig. 3.a.1. Periodic table of the elements evidencing the isotopes that can be analysed by NAA, with the associated detection limits. (10^{-1} μg/g = 0.1 ppm)
Original colour picture reproduced as plate 48.

Glascock and Neff 2003). In fact accuracy and precision of NAA are so good (Fig. 3.a.1) that it is the technique mostly used to certify elemental concentration in standard reference materials.

The experimental procedure involves the irradiation of the sample by a neutron beam, the irradiation time depending on the available neutron flux. The reference calibrating materials are inserted in the irradiation area together with the sample, so that the absorbed dose is the same. The nuclei of the elements in the sample and in the standard materials capture a neutron and turn into metastable radioactive isotopes, each species decaying with a characteristic half life and emitting an energy quantum in the range of γ-rays (nuclear processes, Table 2.1). When thermal neutrons are used as incident particles, the excited nucleus decays following the universal decay rate and the gamma re-emission process is delayed (Section 2.5.4.2). The probability of capture depends on the activation cross-section specific for each isotope. After irradiation the delayed γ-rays emitted from the radioactive sample are measured at specific times: NAA is effectively γ-ray spectroscopy performed by high resolution solid state detectors (Li-drifted or high purity germanium or silicon detectors), the energy of the gamma peak identifies a specific element, whereas the area of the peak is related to its concentration. The number of starting nuclei of an element can be calculated from its capture cross-section, the decay constant, the length and flux of irradiation, and the length of measurement (Glascock and Neff 2003). The neutron autoradiography techniques (Box 2.m) that are widely used for the analysis of paintings are basically just a two-dimensional extension of neutron activation analysis.

If the incident beam is composed of fast energetic neutrons or charged particles (i.e. protons) then γ-rays are quickly re-emitted by the excited nuclei during irradiation. The measurement of these γ-rays is called **prompt gamma activation analysis** (**PGAA**). The technique allows access to a different set of chemical elements and sensitivities (Fig. 3.a.2), and can it also be profitably used in combination with neutron diffraction (Box 2.h) to perform simultaneous non-invasive chemical and phase investigation of cultural heritage samples (Kasztovszky *et al.* 2007). A recently applied variation of the technique implies the measurement of prompt γ-rays emitted at isotope-specific energies, where neutrons are captured at resonance peaks usually detected by time-of-flight systems. The technique is called **neutron resonance capture analysis** (**NRCA**) and it promises to be a fully non-destructive method to determine the bulk elemental composition of any sample. NRCA is applicable to almost all stable isotopes, it does not require any sample preparation, and it results in a negligible residual sample radioactivity (Postma *et al.* 2004, 2007).

(*cont.*)

Box 3.a (*Continued*)

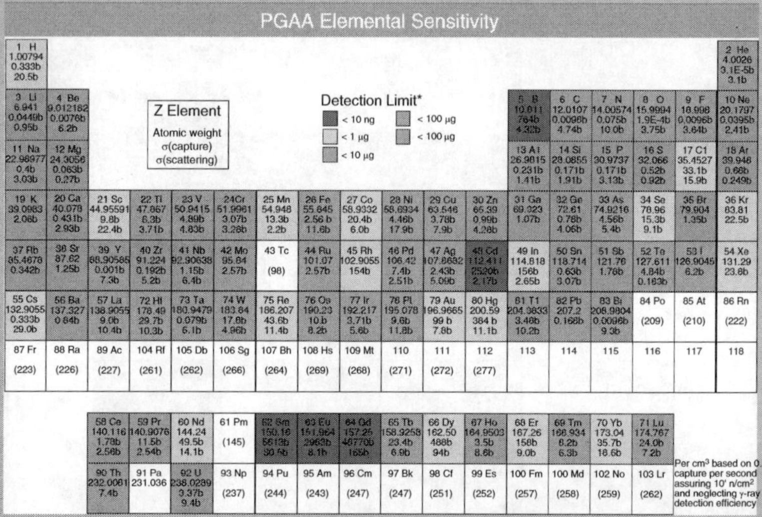

Fig. 3.a.2. Periodic table of the elements evidencing the isotopes that can be analysed by PGAA, with the associated detection limits. (1 μg/g = 1 ppm)
Original colour picture reproduced as plate 49.

NAA has been a reference technique for extended characterization programmes of a variety of archaeological materials, including pottery, glass, metals and geological materials (Harbottle 1990). With regard to ceramics in particular, most of the research reactors in the period 1960–1980 were involved in the systematic chemical analysis of pottery groups for provenance and authentication purposes. Many of the extensive analytical databases of ceramics developed in that period are now being assembled and made available to researchers through the MURR (http://archaeometry.missouri.edu) and Bonn (http://www.iskp.un-bonn.de/gruppen/mommsen/top.html) programmes. They include the data generated at MURR (Glascock *et al.* 2007), Manchester (Newton *et al.* 2007), Brookhaven National Laboratory (Harbottle and Holmes 2007) and Lawrence Berkeley National Laboratory (Perlman and Asaro 1969, Asaro and Adan-Bayewitz 2007) on ancient ceramics of different periods and geographic areas. Countless important contributions have been produced through these and other projects of systematic analyses of pottery sherds and geomaterials. Noteworthy are the studies of Near East pottery contributing to our understanding of archaeology in Biblical times, including interpretation of Eastern Mediterranean trades and the social environment of the Qumran scrolls (Brooks *et al.* 1975, Yellin and Maeir 2007, Gunneweg and Balla 2003, Balla and Gunneweg 2007).

For in-depth information

Hughes M.J., Cowell M.R. and Hook D.R. (eds.) (1991) Neutron activation and plasma emission spectrometric analysis in archaeology. Techniques and applications. British Museum Occasional Papers 82. British Museum Press, London.

Molnár G.L. (2004) Handbook of prompt gamma activation analysis with neutron beams. Kluwer Academic Publishers, Amsterdam.

Parry S.J. (1991) Activation spectrometry in chemical analysis. Chemical Analysis Series Vol. 119. John Wiley & Sons, New York.

Speakman R.J. and Glascock M.D. (eds) (2007) Acknowledging fifty years of neutron activation analysis in archaeology. Archaeometry, Special Issue **49**. pp. 179–420.

Expert geo-scientists who are knowledgeable about the outcrop's distribution in a region and the mineral properties of the outcropping rocks can distinguish their provenance straightforwardly, and sometimes also identify the particular period in which that specific rock was quarried and used. An excellent example is the Bugini and Folli (2008) systematics of the architectural stones of Milan. The authors' encyclopaedic knowledge of the Central Alps geology and petrology ensures the direct identification of almost any stone used in Milan for building purposes since Roman times. Jackson and Marra (2006) produced a full geological description of each stone and a list of extant ancient Roman buildings that used that stone. Similar systematic studies have been applied, for example, to the classical marbles used in Imperial Rome (Gnoli 1971, De Nuccio and Ungaro 2002), to the Greek classical marbles (Maniatis 2004, Lazzarini 2006), to the coloured marbles and jaspers used in the Baroque decorations of Sicilian churches and palaces (Montana and Gagliardo Briuccia 1998), and to the building stones of ancient Egypt (Klemm and Klemm 2001). These investigations are especially profitable for areas where geological maps and information are available, so that geological and historical information can laboriously be assembled. However, a general knowledge of all types of rocks in a vast area is hard to come by. The first step is to identify and sample all possible sources, subsequently the correct identification is made by means of scientific analyses. With regard to the first step, all kinds of information can be useful to identify existing rock outcrops: geological maps, regional and historical lithoteques, University collections, stone dealers, and sometimes a great deal of field work and some luck. As a personal case, the determination of the rock source of a Stone Age hand-axe (Fig. 3.9) found on the shore of an ancient lake in the Libyan Sahara was just the result of a fortuitous visit to one of the locally known deposits of silicified wood (Fig. 3.10).

With regard to the second step, the task of locating the rock source is relatively easy when dealing with local materials. The location of the source quarry is straightforward, for example, in the case of the Cusa quarries, Sicily (Fig 3.11) where in Greek times (5th and 6th centuries BC) the local limestone formations were directly carved on site to the final round shaped blocks used for the construction of the nearby Selinunte temples (Peschlow-Bindokat 1990). Another case is the well-known Aswan red granite quarry, Egypt (Fig. 3.12) where abundant evidence of ancient activities and several unfinished monuments are still present, including sarcophagi and a huge obelisk (Engelbach 1923, Ward-Perkins 1972). On the other hand, the search for provenance may be difficult and unpromising if the material was imported from outside sources. We know that materials travelled a long way even in prehistoric times, and after the Bronze Age routine trade routes were well established (Scarre and Healy 1993). Transport models and provenance assumptions ought to be thoroughly tested by making the best use of all information. As for other materials, trace elements, isotopes, defects and unusual textures of the rocks may be profitably used for identification of sources. For example, concerning the classical white marble trade in the Mediterranean, where several well known quarries

Fig. 3.9. Stone Age (Palaeolithic) hand-axe obtained from extremely hard microcrystalline silicified wood from a Libyan Sahara source (Fig. 3.10). The chopper was found in the desertic Gara Ouda area (Fezzan, Libya), which was a lake basin between the 7th and the 5th millennium BC. By the 2nd millennium BC the area was completely dried up (Cremaschi 1998, Cremaschi 2003).

Fig. 3.10 (a, b) A Libyan source of silicified fossil wood used as a source for tools in prehistoric times.

are known in Italy, Greece, and Turkey, the reliable identification needs a complete cross-validation of optical microscopy data, O and C stable isotope ratios, trace element patterns, luminescence and electron paramagnetic effects (EPR) and textural data, such as a maximum grain size analysis (Herz and Waelkens 1988, Lazzarini 2004, Maniatis 2004). Provenance studies have of course been performed on many other types of rocks, including Mediterranean millstones and grindstones (Peacock 1980, Williams-Thorpe and Thorpe 1993, Renzulli *et al.* 2002), greenstone polished axes (D'Amico and Starnini 2006), Roman granite columns (Williams-Thorpe 2008), and green Thessalian ophicalcites used both in Roman and Bizanthine times for sarcophagi and decorative architectural parts (Melfos 2008).

A vast number of examples testify to the ingenuity and hard work necessary to exhaustively map the possible sources and trace the provenance of rocks. Obsidian is the best natural glass available for producing extremely sharp blades and tools, so it has been a valued material since prehistoric times. As a consequence obsidian is one of the preferred materials of study (Shackley 1998, 2008), first because the sources are limited in number and geologically well located, and secondly because each obsidian source has a distinct geochemical character derived from the volcanic system, sometimes even the single obsidian flow (sub-source) can be discriminated (Eerkens and Rosenthal 2004). Furthermore it has a conspicuous appearance and obsidian tools are hard to be missed during excavations. The production and diffusion of obsidian

Fig. 3.11. Cusa, Sicily. (a, b) Large rounded limestone blocks cut from the quarry layers and ready to be carried to the Selinunte temple site (c), some 8 km away.

tools have been investigated all over the world, for example in Europe and the Mediterranean (Williams-Thorpe 1995, Kilikoglou *et al.* 1997, Tykot 2004), in the Rio Grande region of the US (Church 2000), North-West Alaska (Clark 1995), Mesoamerica (Cobean 1991, Glascock 1994, 2002), Andean Peru (Brooks *et al.* 1997, Glascock *et al.* 2007), and the Pacific (Weisler and Clgue 1998, Ambrose *et al.* 2009, Carter *et al.* 2009). It is to be noted that chemical analytical techniques measuring a large number of tracing elements have been employed to characterize obsidian, including XRF, AAS, OES, PIXE (Box 3.i) and NAA (Box 3.a).

The analysis of small-scale lithics artefacts is a discipline of its own because of the importance of such tools in the prehistory of mankind (Church 1994, Kardulias and Yerkes 2003, Odell 2004, Andrefsky 2005). In addition to tracing the source provenance, their macroscopic dimensions, shape and surfaces are widely studied from the points of view of (a) statistics, to understand typological classifications and populations, (b) fracture mechanics, to understand how they were made, and (c) surface microwear, to interpret their use (Andrefsky 2005). Despite the common usage of the

Fig. 3.12. Unfinished obelisk carved into the red granite quarry at Aswan, Egypt (Engelbach 1923).

terms, there is no totally clear cut distinction between lithic and groundstone artefacts (Table 3.3). The first are supposed to be prepared by flaking and chipping, in order to have sharp cutting edges, though carefully polished prehistoric axes are supposed to be more efficient in wood cutting (Coghlan 1943), and rounded stone hammers are better tools for mining and stone dimensioning operations than sharp tools (Fig. 3.13). The ground stones are supposed to be shaped by abrading, polishing, and milling, especially through continuous use (Fig. 3.14), though this does not reflect the fact that early shaping from nodules or boulders is performed by flaking, pecking, and pounding. Nonetheless the traditional usage of the terms has been retained in the broad tool classification of Table 3.3. Ground stones artefacts have long received attention because they are an essential part of domestic tool kits and production technologies, such as food and ore processing. The studies have advanced considerably in the last years because of the application of novel analytical technologies, focusing mainly on the determination of function and use-wear (Adams 2002, Rowan and Ebeling 2008). A further expanding area yielding novel and exciting results is the analysis of organic residues connected with stone tools. Although the early

3.1 *Structural materials I* 221

Fig. 3.13. Early Copper Age stone hammers presumably used for the extraction of Cu minerals from Ligurian Eneolithic mines. Note the subrounded irregular shape.

Fig. 3.14. Ground stones presumably used for the treatment of cereals and other food. (a, b) Pastoral period, Libyan Sahara, (c) Nuragic period, Sardinia.

data obtained by such innovative techniques as haemoglobin identification (Loy 1983) have been severely challenged on grounds of analytical protocol, later results on blood (Loy and Hardy 1992, Loy and Dixon 1998, Smith and Wilson 2001) and starch (Piperno *et al.* 2004) show that extraction of reliable information is indeed possible, casting a whole new vision on past activities.

Table 3.3 shows that there is a reasonably broad correlation between the kind of geomaterial used and its specific properties for the making of tools, though this relationship can only be used as a general guideline for the interpretation of the technological choices. Effortless re-use of ready-made blocks and existing parts from previous building in many ways superseded the priorities determined by the structural properties of the material itself (Fig. 3.15(c)). The rapid or slow dismantling and destruction of buildings is part of a process intrinsic to cultural succession, and re-use of construction stones is just one aspect of the exploitation of locally available resources. Examples can be found from prehistoric times to the present. Just to name a few: Béatrice Caseau (2004) fascinatingly tells the story of rural temples in late antiquity, a time were contrasting religious beliefs were determinant for the fate of many classical monuments; David Petts (2002) convincingly revises the alleged use of prehistoric stones and slabs in Middle Age Western Britain; D. Peacock (1997) tells us how Charlemagne requested black Roman columns from the Mercia King Offa to be used in contemporary constructions.

In addition to mechanical and physical properties, sometimes the use of a specific material involves parameters such as availability, location, and even indirect properties that are hard to imagine from measurable parameters. This is exemplified by the curious case of the porous andesitic rocks from Assos in the Troad region known as "*lapis sarcophagus*" (Lazzarini 1994). These rocks have a high porosity, and were commonly employed as coffin stones and exported around the Mediterranean in Roman times because, as reported by Plinius (*Naturalis Historia*, XXXVI, 131), they seemed to favour the fast decomposition of bodies. It is reported that soldiers especially required the use of these specific stones for their sarcophagi. As a consequence of this tradition, in North-Eastern Italy the Romans using local trachytic rocks carefully selected the quarries supplying the more porous rocks for burials, whereas the more dense rocks were used for road paving. The selective application of different trachytic rocks was continuous and systematic (Capedri *et al.* 2002, Capedri and Venturelli 2003).

The stones selected for high temperature applications, such as in metallurgical casting moulds or in cooking slabs and pots ("pietra ollare" in the Italian terminology), often reflect their thermal history, being for the most part metamorphic rocks that recrystallized at relatively high temperatures and variable pressures, often known as greenstones (amphibolites, serpentinites, chlorite-schists, eclogites). Having previously experienced high temperature equilibria, they are more resistant than other rocks to thermal gradients, though inevitably rapid thermal shocks will cause fracturing by differential expansion and contraction. Depending on their thermal properties some stone moulds can hardly be used more than once, whereas greenstone moulds may often be reused for several casting operations. In ancient times the mould blocks were often reused and recycled several times by simply erasing or reshaping the carved grooves.

The stones used for architectural decorations were often rather soft polycrystalline rocks that can be easily modelled, drilled, and polished. Depending on the importance of the construction, local or imported stones were selected, compromising availability, aesthetic properties (colour, texture, grain size) and cost. The location of the stone block in different parts of the building often dictates the minimum uniaxial compression resistance to be required, so that finely carved decorations with little mechanical resistance are generally located in structurally unimportant parts of the architectural complex. Mechanical and working properties are for the most part at variance, so that a compromise must be reached in the selection. It is straightforwardly realized why fine-grained carbonates (marble, limestones) (Figs 3.15(a) and 3.15(b)) and sandstones are and were universally selected for

Fig. 3.15. Romanesque marble decorations of the Pieve di S. Siro, Cemmo, Capo di Ponte, Italy. (a) The lunette of the main entrance, (b) particular of a side column of the door, with bas-relief figures of fantastic animals. The Romanesque church is dated to the 11th–12th century AD, though the re-employ of previous romanic stones (c) and the presence of thousand of rock petroglyphs (Fig 3.b.1) in the surrounding Valcamonica area testify to previous human presence and activities.

Fig. 3.16. Romanesque decorations in the church of S. Pietro di Zuri, Ghilarza, Oristano, Italy: (a) the decorated lintel and friezes in the front portal, (b) detail of the capital decorations, (c) corner frieze in the façade showing a line of dancing figures. All the building blocks and the carved decorations are made of the local red trachyte of the Bidonì quarry, which is still used for artistic stone carving. Interestingly, the 12th century S. Pietro church was saved in 1926 from the waters of the nearby artificial lake by a stone-by-stone rescue operation, in much the same way that the more famous Abu Simbel temple in Egypt was saved from the Lake Nasser filling.

Fig. 3.17. Portrait of Eleonor of Aragon by Francesco Laurana, dated to about AD 1421, in the Galleria Regionale della Sicilia, Palazzo Abbatellis, Palermo. It is one of the finest examples of Renaissance sculpture.

the fine decorations of churches and palaces, especially in Romanesque and Gothic architectures, though any kind of fine-grained rocks were actually employed (Fig. 3.16). Accordingly, the marble quarries with high carbonate purity and homogeneous texture, such as Carrara (Italy), Paros, Naxos, Thasos, and Mount Pentelic (Greece), Aphrodisias, Afyon, and Marmara (Turkey) were the prime choices to extract the blocks for fine statues and temples in Roman, Greek and later times (Figs 3.17 and 3.18). However, it is sometimes amazing to observe how in the past rocks that would be normally considered unsuitable for fine carving, such as coarse grained dolerites, granites and breccias, have been patiently modelled into outstanding artworks. Egyptian statuary, vessels, and objects are excellent examples of past working abilities (Fig. 3.19).

With regard to architectural conservation (Jokilehto 2002), the field is vast and challenging, given the variety of material and conditions encountered around the world (Kumar and Kumar 1999). It is shocking to witness the rate of stone decay and the devastating alterations to monuments, especially outdoor statuary. The general perception is that acid rain is the main culprit, but the real story is much more complex: the mechanisms and rates of alteration depending on chemical composition, the geologic nature of the stone, exposure, architectural detailing, and maintenance. Stone conservation has become a research discipline in its own right (Price 1996, Baer and Snethlage 1996, Henry 2006), encompassing the measurement and understanding of decay mechanisms (air pollution, salts, biodeterioration, etc.), and the long-term assessment of available treatments (cleaning, desalination, consolidation, surface coating, etc.).

Several issues related to stone degradation and conservation have stirred lively debates, for example concerning the nature and role of the calcium oxalate patinae found on calcareous rock surfaces (Alessandrini 1989, Realini and Toniolo 1996). The issue is definitely not yet resolved, with publications appearing every year favouring each one of the theories. The patinae are commonly a mixture of whewellite (calcium oxalate monohydrate, $CaC_2O_4 \cdot H_2O$) and weddellite (calcium oxalate dihydrate, $CaC_2O_4 \cdot 2H_2O$), mostly mixed with calcite ($CaCO_3$) and gypsum ($CaSO_4 \cdot 2H_2O$). They produce a mm-thick yellowish-brown layer on the surface of the carbonate, sometimes acting as protective layers, and sometimes detaching as thick films causing extensive surface damages. The nature of the discussion resides essentially in the formation mechanism, especially on the surfaces of old monuments: the formation of the oxalates has been attributed to bioformation (by fungi and lichens), to pollution and particulate deposition, and finally to the chemical reaction between the carbonate surface and organic substances applied for protective treatments. Despite the extensive literature and the numerous experiments there is no general consensus on the nature of the process, though most authors probably agree that many of the old patinae must be related to ancient treatments of the surfaces with organic substances. Another issue for debate is whether the patinas are still protecting the surfaces or contributing to their deterioration (Alvarez de Buergo and Fort Gonzalez 2003). In fact the proposal has been made actually to induce the formation of the oxalate patina through the use of ammonium oxalate to form a shallow protecting layer on calcareous surfaces and wall paintings (Matteini et al. 1994). However, the positive side of the discussion is that it stimulated a very active research, so that oxalate patina are a good example of surface coatings characterized by an impressive array of microanalytical tools, such as ultra-thin sections and micro-XRD (Appolonia et al. 1996), optical microscopy, SEM-EDS, LIBS, FTIR, and GS-MS (Rampazzi et al. 2004). Further, the oxalate layers promoted the development of advanced conservation treatments, such as laser cleaning (Siano et al. 1997, Bromblet et al. 2003), a technique that was also proposed for the treatment of sulphated marble surfaces and black crusts (Maravelaki-Kalaitzaki et al. 1999, Potgieter-Vermaak et al. 2005). Interestingly, the gypsum-loaded sulphated crusts on carbonates may also be cleaned through the action of sulphur-reducing bacteria (Gauri et al. 1992, Ranalli et al. 1997, Cappitelli et al. 2007). The special case of the conservation of remote outdoor rock art and petroglyphs is especially challenging (Box 3.b, see also Section 2.5.3.3).

Fig. 3.18. The marble statue of a young man (the "Giovanetto di Mozia" or "Efebo di Mozia") in the Whithaker Museum, Mozia, Marsala, Sicily. Of outstanding quality, the statue is a curious example of Greek statuary found in a Punic context. It was excavated in 1979 in the Punic colony of Mothya (G. Falsone in: Bonacasa et al. 1988, pp. 9–28) and on the basis of stable isotope composition (Box 2.v) it is thought to be made of Anatolian marble, like the methopae of the nearby temples of Selinunte (R. Alaimo and M. Carapezza in: Bonacasa et al. 1988, pp. 29–37).

Fig. 3.19. Theriomorphic vessel in the shape of a bird made of limestone breccia, Late Predynastic–early 1st Dynasty ca. 3100 BC. British Museum-BM35306. (Courtesy of J. Bodsworth, image available at: http://www.egyptarchive.co.uk/html/british_museum_index.html)

Box 3.b Rock art characterization and preservation

To most people the term *rock art* is evocative and mysterious. It refers to depictions and symbols present on rock surfaces of all continents and possibly dating as far back as 40 ky ago (Chippindale and Taçon 1998, Anati 2000, Clottes 2002). The millions of rock paintings and engravings distributed all over the world is a tremendously important heritage testifying to the vision and perception of the life and cosmos of traditional people now long vanished. Unfortunately, most of these art works are located in remote and unprotected environments, constantly challenged by time, nature and man.

The paintings and engravings that we know today are but a small part of those produced in the past. The causes of deterioration are countless. In the first place the exposure to natural elements (sun, rain, snow, temperature changes, etc.) for thousand of years severely damages or completely erases most of the unsheltered works. All degradation processes acting on the rock surfaces obviously do the job: water runoff, surface flacking, microfracturing, salt efflorescence, freeze and thaw cycles, etc. Biodegradation is also a major factor (Ciferri *et al.* 2000, St Clair and Seaward 2004), including large scale deteriorations due to plant roots, birds' nests, animal abrasions and human graffiti. In recent times human activities are the most frequent accidental or intentional causes for damages: pollution, acid rain, urban development, submersion by dam construction, quarrying, and even pure vandalism. Disputable "cultural tourism" also frequently jeopardizes art works in many different ways, from retouching of the drawings to make them visible, to careless copy-casting, to chemical contamination (Clottes and Agnew in: Clottes 2002, Kröpelin 2002).

Preserving the rock cultural heritage is a daunting task. Conservation resources are limited, and they are to be shared between characterization, interpretation, conservation, monitoring, documentation, education and other activities. Priorities are often unclear. To date, only about fifteen areas with rock art have been inserted in UNESCO's World Heritage List (http://whc.unesco.org/), and countless sites are unprotected from all possible points of view.

Cultural heritage management and strategic policies are of course needed to improve the situation, and they may differ from place to place, depending on the nature and distribution of the art works, however isolated they are, their relevance for the local culture and history, and many other parameters. Within the process of valorization and preservation of rock art, scientific research can certainly contribute through (1) the physical characterization of rock art works (survey, documentation, dating), and (2) the definition of the potential hazards, the understanding of the degradation mechanisms, and proposals for preservation strategies. Concerning the physical characterization, the information to be sought are (a) the nature of the substrate, (b) the technique of carving and engraving (if the art has been produced by mechanical removal of material from the surface, Fig. 3.b.1), (c) the technique of painting and the nature of the pigments (if the art was painted on the natural or prepared substrate, Fig. 3.b.2), (d) the sequence of the drawing and possibly the dating.

Assigning absolute dates to rock art is a difficult task. For petroglyphs carved in the black desert varnish (Fig 3.b.3) the dating of the varnish layer may provide a terminus post quem (that is the carving cannot be older than the engraved varnish). If the varnish is deposited later and it covers the engravings or the surface of an artefact, then it may provide a terminus ante quem (that is the carving cannot be younger than the varnish). Therefore the dating of the varnish layers can be used to bracket the age of the rock art or the lifetime of the coated artefact (Zerboni 2008).

Attempts to derive ages of the varnish have been made by electron microscopy (Harrington and Whitney 1987), by AMS radiocarbon dating (Dorn *et al.* 1989, Beck *et al.* 1998), and by cation-ratio methods (Loendorf 1991, Harry 1995) among others (Watchman 2000, Sowers 2000). Despite a certain disagreement on the reliability of the absolute dates, it is quite clear that desert varnishes may nonetheless provide useful relative timing and complementary information on past climate and environment (Dorn 1994, Cremaschi 1996, Zerboni 2008). In other cases, where carbon-based pigments are present in the drawing or inorganic pigments have been applied using

3.1 Structural materials I 227

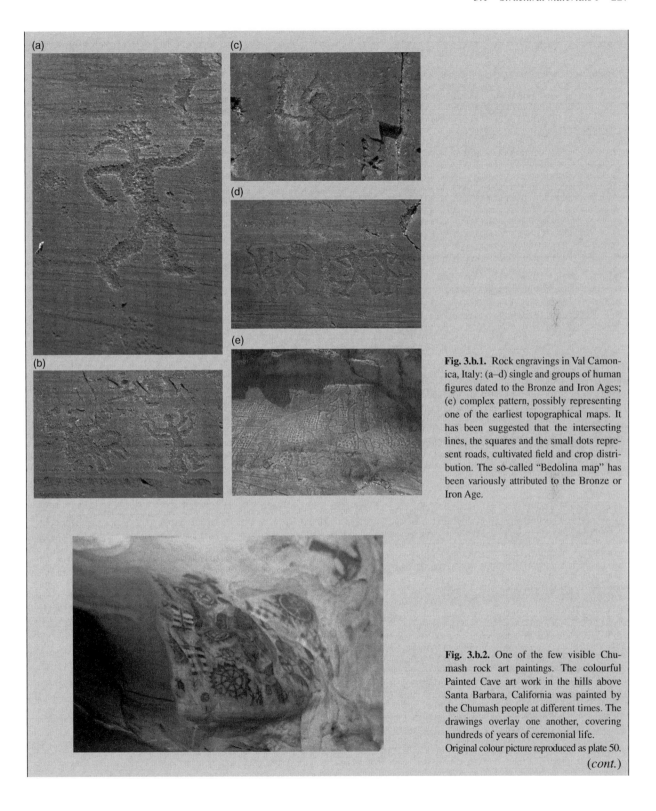

Fig. 3.b.1. Rock engravings in Val Camonica, Italy: (a–d) single and groups of human figures dated to the Bronze and Iron Ages; (e) complex pattern, possibly representing one of the earliest topographical maps. It has been suggested that the intersecting lines, the squares and the small dots represent roads, cultivated field and crop distribution. The so-called "Bedolina map" has been variously attributed to the Bronze or Iron Age.

Fig. 3.b.2. One of the few visible Chumash rock art paintings. The colourful Painted Cave art work in the hills above Santa Barbara, California was painted by the Chumash people at different times. The drawings overlay one another, covering hundreds of years of ceremonial life.
Original colour picture reproduced as plate 50.

(*cont.*)

Box 3.b (*Continued*)

Fig. 3.b.3. Rock engravings at Butler Wash, San Juan River, Utah. The figures were made by the San Juan Basketmaker people by removing the black desert varnish made of manganese and iron oxides to expose the bright orange sandstone substrate. Original colour picture reproduced as plate 51.

organic binders, then radiocarbon dating is possible (Loy *et al.* 1990, Lorblanchet *et al.* 1990, Valladas 2003). Radiocarbon dating has also been applied to the oxalate layers covering the pictures (Watchman 1991).

With regard to conservation, investigations focus on the attempt to arrest or slow down the degradation process that is inevitably present on the exposed rock surface. As mentioned in Section 2.5.3.3, it is believed that in many cases cleaning the soil and vegetation cover from the rock art surface may even cause acceleration of the degradation processes. Cave engravings and paintings are less subject to erosion and abrasion, though they are often covered with post-depositional carbonate, phosphate or oxalate layers (Russ *et al.* 1999) that in extreme conditions may totally obscure and embody the art work. Cleaning and applying a protective layer is generally not a viable solution. In the long run all surface consolidants prove to be ineffective or even deleterious, since the modification of the physico-chemical conditions regularly alter the local equilibrium reached in thousand of years, thus triggering novel and often unpredictable side alteration processes. Further, every removal of the original material and application of modern components likely precludes future investigations of the surface layers. When feasible, little intervention with continuous monitoring and control may well be the best solution, especially if it is inserted in a framework of cultural education and growing awareness.

One further contribution of science to the virtual preservation of rock art (Box 2.q) is helping in the documentation and in the reproduction of art works. The Lascaux caves in France are leading the way, as an example of an outstanding total reproduction of the site, where the tens of thousands of annual visitors are actually looking at the exact replica of the cave (Lascaux II), thus leaving the original one undamaged. The reproduced drawings sometimes can even be closer to the original art work whenever scientific analyses (i.e. high resolution photography, UV and IR spectroscopy, reflectography, etc.) allow us to gain or recover lost information. Finally, the virtual part of the reproduction can be easily accessed in the best possible conditions, so that it serves educational and cultural purposes (http://www.culture.gouv.fr/culture/arcnat/lascaux/en/index.html).

Several national and international projects are actively engaged in the documentation and study of rock art around the world: the World Archives of Rock Art (WARA, http://rockart.iworm.co.uk/waraindex.htm) started by

> the Centro Camuno di Studi Preistorici, Italy and now co-operating with the Swedish RockCare project (Swedish Archive for Rock Art Research, http://www.rockartscandinavia.se/uk/news/swedish_archive_english.htm) based at the Tanums HällristningsMuseum; the Rock Art Archive at the University of California, Los Angeles based at the Cotsen Institute of Archaeology (http://www.sscnet.ucla.edu/ioa/rockart/); the Rock Art Research Institute based at the University of Witwatersrand, South Africa (http://web.wits.ac.za/academic/science/geography/rockart/); The Australian Rock Art Research Association based in Victoria, Australia (AURANET, http://mc2.vicnet.net.au/home/aura/web/index.html); and many others.
>
> **For in-depth information**
>
> Clottes J. (2002) World rock art. The Getty Conservation Institute, Los Angeles.
>
> Updated list of Rock Art organizations through the world: http://www.rock-art.com/orgs.htm

3.1.2 Structural clay products, ceramics

Modelling wet clays into different shapes was a very early human activity, possibly developed independently several times in prehistory. In a number of European Palaeolithic caves the ancient designs are also traced into wet clays, other than painted on the walls. The French site of Tuc d'Audoubert preserves two moulded but unfired bison shapes (MacCurdy 1912). The earliest evidence of terracotta figurines hardened by fire is the remarkable statuettes of the Gravettian site of Dolní Věstonice, Czech Republic, including the famous Venus (Vandiver *et al.* 1989). At this site well-modelled shapes of animals and humans were expressly inserted in the fire when still wet, possibly to produce an explosive tribute to the gods (Soffer *et al.* 1993). Thermoluminescence dating of these samples produced dates close to 30 ky BC (Zimmerman and Huxtable 1971). Another very early figurine dated to about 17 ky BC has been found in a rescue operation at Maina, Russia (Vandiver and Vasil'ev 2002). Although very fragmentary, this evidence shows that the Late Palaeolithic humans recognized that wet clay was plastic, it could be shaped, and it retained its form when dried or fired. The basic properties of clay materials were thus already known.

The earliest evidences of the use of fired pottery made of earth materials for everyday tools are in China (Yuchanyan and Miaoyan sites, dated to 14.5–13.8 ky BC; Zhao and Wu 2000), Japan (Incipient Jomon period, dated to about 14.5–13.5 ky BC; Habu 2004), Russian Far East (Amur River Basin, dated to 14.1–12.8 ky BC; Kuzmin 2006) and Siberia (Ust-Karenga 12 site, dated to 12.5–10.5 ky BC; Kuzmin and Vetrov 2007). Given the chronology uncertainties, it appears that fired ceramics appeared almost simultaneously in several places in Eastern Asia at about 14000 y BC.

The appearance of pottery vessels that were used to contain and store food is generally associated in Europe and the Near and Middle East to the Neolithic changes, involving the adoption of food production vs collection, and living in permanent settlements vs. temporary camps. Although there is no evidence of strict causal relationship between the rise of agriculture, the settlement into villages, and the making of pottery, there is a neat observed prevalence of ceramic production and use in sedentary societies as opposed to nomadic ones (Rice 1987, p. 9; Kuzmin 2003). Furthermore, there is ample evidence that pottery firing in the Middle East stemmed from a long period of technological innovation involving incipient craft specialization and economic complexity, which includes the widespread production and use of lime and gypsum plasters (Section 3.2) during the Pre-Pottery Neolithic, phase B, dated to approximately 8.7–7.0 ky BC (Kingery et al. 1988).

It is more difficult to trace the origin of the use of unfired or sun-baked clay products, because they leave scarce and poorly recognizable archaeological records. We may assume that wet clay linings applied to straw huts, baskets, and ground holes to make them impermeable started well before the widespread used of the fired products, so the origin was probably the use of sun-dried bricks for construction (**adobe**) and the use of mixing burnt lime or gypsum with water to create pastes that harden upon drying (**plaster**, Section 3.2). In its most simple form, without the use of pyrotechnology, a wall can be composed of sun-dried bricks bound by a moist layer of mud that, on drying, makes the wall a solid mass of dry clay. The tradition of using earthen architecture based on a mixture of clay, water, and straw is still largely present in many rural areas of the world, the splendid Al Mihdar Mosque of Tarim, Yemen and the Mosque of Djenne, Mali being two splendid examples of earthen architectural achievements (Fig. 3.20; see Avrami et al. 2008 for a review). The technological basis of modern earth constructions is pretty much the same as the one that we infer from the prehistoric record (Oates 1990).

Irregularly shaped sun-dried bricks are reported to have been used in the Pre-Pottery Jericho, Israel, at least from the 9th millennium BC (Bar-Yosef 1986) and in Nemrik, Iraq at least from the 8th millennium BC (Kozlowski and Kempisty 1990). Rectangular mould-based bricks were introduced in Mesopotamia at least from the 4th or 3rd millennium BC (Tell Braq, Syria: Oates 1990; Halaf, Syria: Davidson and McKerrell 1976), rapidly becoming decorated or inscribed. Fired bricks bearing inscriptions have been found in most Middle Eastern excavations, including Babylonia (Oppenheim 1965), Nimrud (Oates 1960), and Nippur (McCown 1952). The Sumerian, Babylonian, Assyrian, and Hittite civilizations widely used sun-dried clay tablets (Fig. 3.21) for cuneiform inscriptions, which was the written system for most of the languages of the Mesopotamian region (Walker 1987, Reade 1996). Large archives of tablets are available starting from about the 4th millennium BC (more than 130 000 in the British Museum alone) and they present a number of conservation problems, because they are fragile and salt-loaded (Organ 1961). Only a very small number have been fired, mainly because of accidental and uncontrolled fires in the buildings in which they were stored in antiquity. Paradoxically, high temperature helps their preservation, so that modern electrical heating up to 740 °C seems to be one of the best conservation treatments to stabilize these materials (Thickett et al. 2002): the clay tablets are now undergoing the systematic high temperature processing that they never had in their lifetime. Alternatively, a large project has been launched to preserve the tablets in virtual format, using the most modern techniques of 3D surface laser scanning (Box 2.q) and producing

3.1 *Structural materials I* 231

Fig. 3.20. (a) Earthen architecture (adobe) in the medina of Ghat, Libya. (b) Even in arid climates that have extremely low precipitation regimes, the poorly bound clay–water–straw mixture needs continuous management to survive.

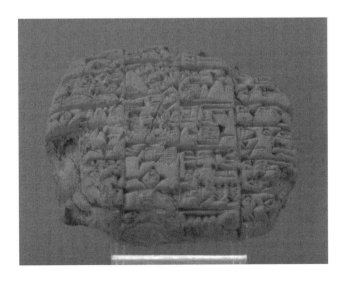

Fig. 3.21. Clay tablet, ca. 2400 BC, found in Telloh (ancient Girsu). Letter sent by the high-priest Lu' to the king of Lagash (maybe Urukagina), informing him of his son's death in combat. (Image available at: http://en.wikipedia.org/wiki/File:Letter_Luenna_Louvre_AO4238.jpg)

"immortal" virtual rendering of the inscriptions (The Cuneiform Digital Library Initiative: http://cdli.ucla.edu/index.html).

Table 3.2 purposely groups structural clay products such as bricks, tiles and pipes together with pottery because they share common basic constituents, processing and overall structural properties.

An excellent introduction to the properties and use of structural clay products is the text by Brownell (1976). In fact, having the same composition, the physical and mechanical properties of both classes of fired products critically depend on (a) the relative quantity of the starting components, and (b) their thermal history. This chapter introduces a number of concepts related to the production and study of clay-based products, so that their importance for the interpretation of archaeological and technical problems may be assessed.

3.1.2.1 *Chemical and mineralogical composition*

The chemical composition of a substance indicates the relative proportions of the chemical elements forming it. For oxides and silica-based materials such as clays the composition is commonly expressed as oxide wt%, i.e. the relative proportions of the oxides within the material, because all elements are chemically bonded to oxygen. However, this common convention of expressing chemical proportions does not yield any information on the structural relationship between the chemical components that may be arranged in different spatial configurations (or mineral phases, or crystalline arrangements). If no component is lost or acquired during high temperature processing, then the chemical composition is virtually independent of the thermal history of the material, whereas the mineral phases present, each one having a well-defined thermodynamic stability field, is crucially dependent of the time–temperature path during processing. Therefore the chemical and the mineralogical compositions of the ceramic material are both complementary and equally important in determining its properties. Both chemical and mineralogical analyses are thus necessary to understand the history and properties of the material being investigated.

The representative chemical compositions of most structural ceramic products are in the range (wt%): SiO_2 50–70, Al_2O_3 10–20, CaO 0–15, Na_2O+K_2O 0–3, MgO 0–5, Fe_2O_3 0–10, TiO_2 0–2. Besides oxygen, three elements (i.e. Si, Al, Ca) make up most of the ceramics, so that a ternary diagram showing the relative amount of these oxides is sufficient to give a fairly complete idea of the main composition of the material. The main classes of clay products can be described using such a diagram (Fig. 3.22). The high Ca materials, close to the lower left vertex, are the calcium aluminates and silicates that are the main constituents of cements. The high Al materials, close to the lower right vertex, are the basic alumina-rich refractory industrial materials. The upper vertex is pure silica (SiO_2), which in most materials derives from mineral quartz and is the main component of glass (Section 3.4); the closer the compositional field of the material is to this vertex, the higher the proportion of starting quartz in the mixture.

With regard to the original mineralogical components, which are of course the source of the constituting elements as well, we may essentially describe

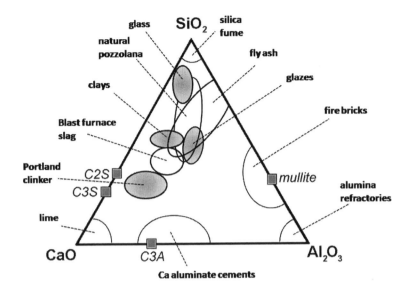

Fig. 3.22. The approximate location of typical products (cements, lime, glass, glazes, refractories) and the raw materials used in the formulation of clinkers (silica fumes, pozzolans, blast furnace slags) in the SiO_2-CaO-Al_2O_3 phase diagram.

any ceramic mixing with three major mineral sources: clays, feldspar, and quartz (Fig. 3.23), representing the plastic, the fluxing, and the temper components, respectively. The relative proportions and compositions of the three sources determine the physical properties of the ceramics, since each source has a specific chemical and structural nature.

Clay minerals are the major source of the *plastic component*, allowing the object to be shaped in the desired form with the addition of water. Clays are very fine grained (less than 2 μm) aluminosilicate minerals, whose lamellar structure allows strong surface interaction between particles, and plastic to colloidal behaviour in the hydrated state (Velde 1992, Bailey 1988, Velde and Druc 1999). In general, the main distinction of the source clays is based on the type of clay mineral (kaolinite, smectite, illite, halloysite, montmorillonite,

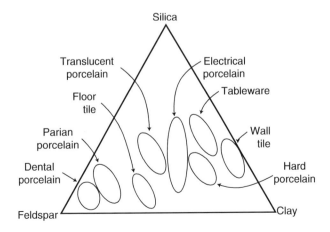

Fig. 3.23. The location of typical clay products in the silica–clay–feldspar phase diagram (redrawn from Askeland 1988).

etc.), on the amount of carbonate (Ca-rich, Ca-poor), and on the presence of impurities (mainly Fe-oxides).

Natural alkali feldspars frequently act as a fluxant during the firing process. The role of the *fluxing agent* is to lower the melting point of many mineral components, thus allowing the object to be hardened by sintering and partial vitrification.

Finally, a certain amount of structural *non-plastic materials* (the so-called **temper,** or filler) is added to the starting mixture. The temper modifies the mechanical properties of the clays, allowing the water to evaporate more smoothly and preventing shrinkage and cracking during drying and firing (Kilikoglou *et al.* 1995). Quartz, mostly in the form of sand grains or crushed quartzite, is frequently used as a temper, but also other minerals (carbonates, pyroxenes, feldspars, etc.) found in sands, or even fragments and sherds of previously fired pottery (**grog**, or chamotte) are used. In ancient ceramics organic materials were also used (Skibo *et al.* 1989), besides crushed shells, crushed rocks, crushed pottery, sand, grass, and straw. Experiments have shown the effect of different temper materials on the mechanical resistance of the pot (Bronitsky and Hamer 1986). Figures 3.24–3.27 show thin sections of archaeological ceramics with different types of texture and temper material taken by optical microscopy in transmission mode.

3.1.2.2 *The physical-chemistry of the firing process*

The stabilization of pottery and its long-lasting properties derive from the pyrotechnological process of heating to high temperature (Section 3.0.1). During firing the rigidity and strength of the ceramic body increase. This is mainly due to the removal of interparticle water molecules (drying) and to reduction of the pore size between particles (sintering). Figures 3.28 and 3.29 show the effect of volume and density changes due to the release of water and sintering. Other volume changes are caused by the structural decomposition of the mineral phases during heating (i.e. dehydroxylation of the clay minerals, loss

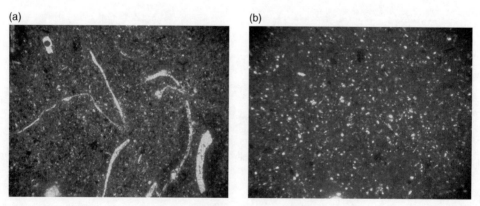

Fig. 3.24. Thin section of a Late Bronze Age pottery sherd from the Frattesina terramara (Rovigo, Italy): (a) plane polarized light (PPL), (b) crossed-polarized light (CPL). Horizontal width of the image is 4 mm. The matrix is glassy with finely dispersed crystalline phases; the temper is formed by crushed shells.
Original colour pictures reproduced as plate 15.

3.1 *Structural materials I* 235

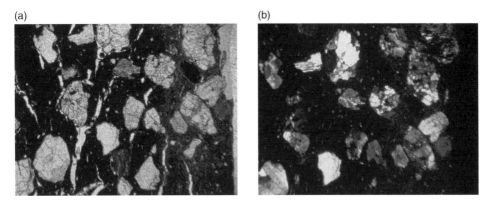

Fig. 3.25. Thin section of a Late Bronze Age pottery sherd from the Frattesina terramara (Rovigo, Italy): (a) PPL, (b) CPL. Horizontal width of the image is 1 mm. The matrix is glassy with scarce crystalline phases, the temper is formed by large sand grains, mostly composed of recrystallized mosaic quartz.
Original colour pictures reproduced as plate 16.

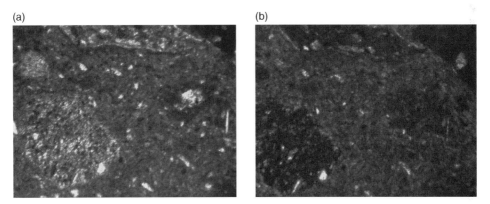

Fig. 3.26. Thin section of a Late Bronze Age pottery sherd from the Frattesina terramara (Rovigo, Italy): (a) PPL, (b) CPL. Horizontal width of the image is 4 mm. The matrix is glassy with abundant crystalline phases, the temper is formed by large grains of crushed pottery (grog).
Original colour pictures reproduced as plate 17.

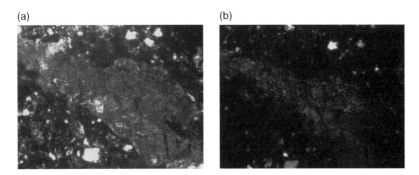

Fig. 3.27. Thin section of a Late Bronze Age pottery sherd from the Frattesina terramara (Rovigo, Italy): (a) PPL, (b) CPL. Horizontal width of the image is 1 mm. A large fracture in the ceramic body has been filled by secondary Fe-phosphates of the vivianite group, possibly of organic origin (Maritan *et al.* 2009).
Original colour pictures reproduced as plate 18.

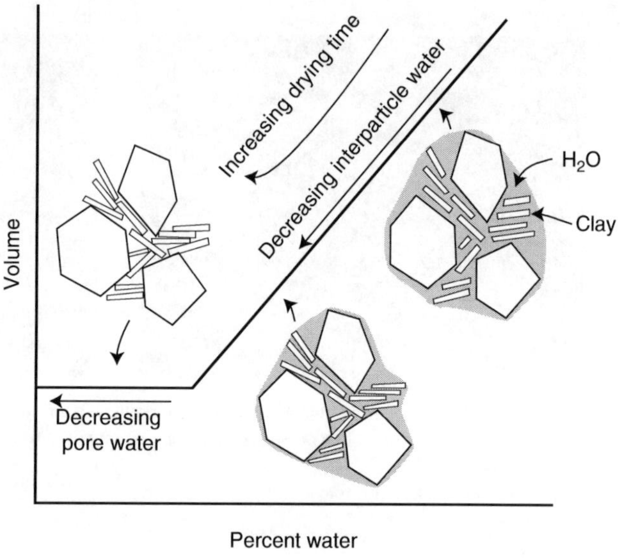

Fig. 3.28. The change in volume of a ceramic body as the water is removed during drying. Dimensional changes stop after the interparticle water is gone (redrawn from Askeland 1988).

Fig. 3.29. During sintering, diffusion produces bridges between the particles and eventually causes the pores to be filled in. If the melting temperature is reached, then glass is formed and this provides an additional source of bonding (redrawn from Askeland 1988).

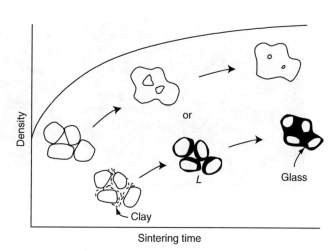

of CO_2 from the carbonates, loss of SO_2 from the sulphates, etc., see Table 3.4) and to mineral reactions and phase transformations (i.e. the alpha–beta transition in quartz, the formation of high temperature phases such as gehlenite, anorthite, etc., Table 3.4).

Drying may occur at ambient conditions, i.e. before firing, or during the initial stages of firing. It causes large dimensional changes and shrinkage of the ceramic body. The temperature and humidity should be carefully controlled during drying to minimize stresses, distortion and cracking. Interparticle water or surface-adsorbed water is lost at room temperature or during firing below 100–150 °C, whereas the structurally bound water and hydroxyls are lost at temperatures in the range 400–700 °C, depending on the initial minerals and composition.

Table 3.4. Scheme of the physico-chemical reactions involving the mineral phases during firing

Approx. temperature range (°C)	Loss of water molecules	Decomposition reactions	Chemical reactions	Phase transformations
$T < 100$	loss of unbound and adsorbed H_2O			
100–200	loss of weakly bound structural H_2O (hydrous salts: gypsum; some zeolitic water)			
200–400	loss of zeolitic and interlayer H_2O (clays)			
400–600		loss of SO_2 (sulphates)	oxidation of organic compounds	573 °C alpha-beta quartz
600–800	loss of OH (clays, micas)	loss of CO_2 (carbonates)	MgO and CaO are produced by clay and carbonate decomposition	
800			oxidation of sulphides liquid phase may start to form	
800–1000			oxidation of ferrous oxides	MgO and CaO react to produce HT phases (spinels, gehlenite, diopside)
1000–1200				formation of mullite, anorthite formation of cristobalite

Sintering causes additional shrinkage of the object as the pore size is reduced. The microstructural features of the final sintered ceramic, in terms of grain size, pore size, pore shape and amount of glass formed at high temperature is what determines the macroscopic properties of the pottery (porosity, mechanical strength, etc.). The microstructure is controlled by an appropriate choice of sintering temperature and time, the initial particle size, and the fluxes. During sintering, ions first diffuse along grain boundaries and surfaces to the points of contact between particles, providing connection of the individual grains. More grain boundary diffusion reduces the pore size and increases the density, while the pore shape becomes more rounded. When the pores become so small that they are no more in contact, then further shrinkage is possible only through volume diffusion in the solid and grain growth occurs. When sufficient fluxes are present, then vitrification or melting occurs. The liquid phase helps eliminate the remaining porosity and changes to the glass after cooling.

3.1.2.3 *Physical properties and classification*

Depending on the type, chemical composition, and amount of starting material and on the time–temperature path during the firing process, the final ceramic product may have very different macroscopic appearance and physical properties. The most important characteristics determining the final use and performance of the material are porosity and mechanical resistance. In general we may say that higher firing temperatures and/or finer initial particle sizes produce more vitrification, less porosity, and higher density. The higher density of the ceramics ensures better mechanical strength, higher impermeability, and lower thermal insulation qualities (Table 3.5).

Properties such as surface appearance, smoothness, colour, lustre and other aesthetical parameters are very important for modern materials and for fine objects such as majolicas and porcelains, much less so for the archaeological everyday objects such as earthenware and stoneware. Earthenware are porous clay bodies fired at relatively low temperatures; little vitrification occurs, the porosity is very high and interconnected, and the object does not retain liquids for a long time. To avoid leaking, the ceramic surface must be covered with an impermeable glaze: the glaze-covered earthenware is called *majolica*. Stoneware must be fired at higher temperatures, producing a higher degree of vitrification and less porosity. The material can be used to produce drainage and sewer pipes.

Depending on the grain size, ceramic materials can be grouped into (a) **coarse ceramics** (i.e. most of the structural clay products such as heavy clay wares: bricks, pipes; and low temperature earthenware, i.e. porous and easily-scratched pottery, usually very red-coloured from Fe-oxides), and (b) **fine ceramics** (i.e. stoneware that is mildly porous and partially vitrified pottery, porcelain that is highly vitrified materials and, eventually, high temperature refractory bodies).

Other kinds of materials used in the ceramic industry are really mixtures of crystalline and glassy components (Section 3.4.1): *enamels* and *glazes* are

Table 3.5. Relationship between porosity, strength and firing temperature for typical ceramic materials

Ceramic product	Porosity	Mechanical strength (kg/m^2)	Approx. firing temperature (°C)
Earthenware (terracotta, majolica)	10–25%	90–130	800–1100
Stoneware	5–18%	100–180	1150–1300
Fire-clay	9–13%	500–1500	1230–1250
China, porcelain	0.5–3%	5000–15000	1250–1450
Porcelain-gres	<0.5%	10 000–20 000	1300–1450

silica-based materials treated at high temperature, commonly composed of a glass matrix and several crystalline phases, they are mostly used for surface finishing and impermeabilization; *frits* are intermediate highly heterogeneous glass materials produced by melting and air-quenching, they are commonly re-ground and used as base materials to produce glazes and glass.

Slips and *engobes* are particular types of clay-rich coatings generally applied directly on the surface of the ceramic body or in-between the body and the true-glaze layer. Their application is meant to enhance surface adhesion, hide defects, and improve colour-saturation effects from the pigments embedded in the glaze. Red slips are generally made out of fine Fe-rich clays, whereas white slips are made of kaolinite-rich clays. Slips and engobes alone improve the aesthetic quality of the object, but they do not affect the permeability.

3.1.2.4 *Characterization methods and interpretation*

The investigations of archaeological ceramics tend to answer primarily the three basic questions of archaeometry: how, when, andwhere was the material produced? Once this starting information is safely obtained, more far reaching inferences may be deduced (Kingery 1981).

How?

The firing processes can be reconstructed from the characteristics of the final product. For example, if the surface is vitrified or glazed, temperatures above 900 °C must have been achieved. High temperature minerals in the ceramic matrix (gehlenite, cristobalite, mullite, anorthite) are also indicative of the maximum temperature of firing. The original temperature can also be assessed by re-heating the ceramic fragments under controlled conditions, and by monitoring the triggering of mineralogical or microstructural changes (Kingery and Frierman 1974).

The use of combined mineralogical and petrographical tools is essential in the characterization of ceramic bodies. A common sequence of analysis includes (a) the mineralogical and petrographical analysis by optical microscopy in a thin section, (b) the mineral phase identification by X-ray powder diffraction, (c) chemical analysis (by XRF, NAA, AAS, OES, etc.), (d) chemical mapping (by EPMA, SEM-EDS, etc.) and (e) the use of spectroscopic tools to evaluate the valence state of oxidizing cations (mainly Fe by Mössbauer spectroscopy).

The optical microscopical analysis on a thin section of the materials allows one to directly evaluate the main phases present in the temper, the approximate filler to matrix ratio, and the overall texture of the ceramic body. The microtextural and petrographic analysis of the material is the starting point for all subsequent analyses (Kingery 1987, Middleton and Freestone 1991, Philpotts and Wilson 1994). The next step is to identify the mineral phases present as temper, as newly formed phases in the matrix, or as reaction phases at the matrix–grain boundary. The knowledge of the phases and of their crystal chemistry is essential to interpret the firing process on the basis of the mineral reactions and of the known phase diagrams (Heimann 1989, Riccardi *et al.* 1999). Most of the temper phases can be readily identified in thin section, but the newly formed

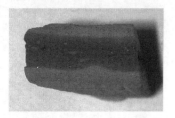

Fig. 3.30. Section of an Iron Age pot sherd of Etruscan–Roman type from the Hospital excavation, Este, Italy, showing the typical black core and red rim of pottery fired at low temperature or for insufficient time. (Courtesy of L. Maritan, University of Padova) Original colour picture reproduced as plate 19.

phases are commonly dispersed in the glassy matrix or at the matrix grain boundary and are invisible by optical microscopy. X-ray diffraction or electron beam analysis (SEM-EDS, EPMA) are required to analyse in detail the mineralogy and the crystal chemistry of such phases. Recent advances using Raman and micro-Raman spectroscopy (Box 2.e) successfully allow identification of crystalline and amorphous phases both in the body and in the glaze of the ceramics (Colomban 2004, Colomban *et al.* 2006). Compositional zoning and phase textural relationship are important to assess the firing technology, as they provide information on high temperature mineral reactions, and on chemical intracrystalline and intercrystalline diffusion during the process. The use of equilibrium phase diagrams must be made with caution, because the firing rarely attains equilibrium conditions, and the transformation processes at high temperature are dominated by kinetics.

The redox conditions in the kiln atmosphere may be qualitatively inferred by the pot colour, i.e. complete oxidation produces a uniform light or red colour, mainly due to the presence of trivalent Fe in haematite or other phases, whereas partial reduction due to carbon causes a dark grey or black colour. It is frequently observed that ancient and modern ceramics show a black core and a red surface layer (Fig. 3.30). This indicates that the temperature was too low to oxidize the clay body fully, or that the internal part was kept in reducing conditions by excess organic matter, or else the duration of the firing was insufficient to complete the oxidation (Brownell 1957). This may frequently happen in an open kiln.

The oxidation and structural state of the Fe cations in the ceramic may be quantified through the use of Mössbauer spectroscopy, which can yield information on both the maximum temperature and the kiln atmosphere (Murad and Wagner 1998; Wagner *et al.* 1998; Molera *et al.* 1998). In general, we may state that in Ca-poor ceramics (i.e. those produced with carbonate-poor clays) the colour-producing hematite is present at low temperatures, whereas if the material is fired at sufficiently high temperatures (i.e. above 1000 °C), then mullite is formed, which incorporates trivalent iron cations in the structure, producing a light creamy colour. If a small amount of CaO is formed in ceramics produced from carbonate-bearing clays, then anorthite is formed in place of mullite (Kreimeyer 1987). Since anorthite hardly incorporates Fe in the lattice, hematite is formed as a free phase and red-colouring pigment. If the starting material is very Ca-rich (i.e. carbonate-rich clays, marls) then abundant CaO is formed and a number of Ca silicates are formed: fassaitic pyroxene, anorthite, melilite (gehlenite, ferri-gehlenite), spinels, wollastonite. Most of these phases, besides anorthite, may contain Fe in different amounts, so that free Fe-oxides are not formed and the ceramic body has a colour (whitish, pink, brown, grey) that depends of the newly formed minerals and impurities present (graphite, minor Ti- or Mn-oxides, etc.).

Finally, more specific analyses are necessary to investigate the fine details of the surface layers (enamels, glazes, and pigments), especially in recent objects produced by advanced technologies. The methods and techniques of analyses are rather similar to those used in glass studies, and they aim to understand the composition and making of the surface coatings, including the colouring agents. Trace element chemical analysis (SRIXE, NAA, XPS, AAS, SIMS, etc.) provide information on the chemical composition of the surface layers, grazing incidence X-ray diffraction (GIXRD) and transmission electron microscopy (TEM) allow distinction of the diffraction signal of the surface from the bulk, and surface spectroscopies (EELS, XPS, SEXAFS, etc; See Boxes 3.f, 3.g, 3.h) provide an indication of the local electron state and coordination of the colouring atoms. Microbeam techniques are particularly useful in the study of unusual ceramics coatings, such as the "black gloss" of classical age, or the "lustre" decoration of Late Middle Age and Renaissance pottery (Perez-Arantegui *et al.* 2001).

When?

The dating of ceramics is most frequently performed on the basis of the changes in macroscopic properties derived from visual examination (typology), such as shape, style, colour, decoration, and overall composition (presence of slip, glaze, etc.). Relative chronology based on shape and style have been established by archaeologists for most cultures (Rice 1987, Sinopoli 1991). The assumption is that characteristic ceramic types are linked to a particular source, thus indicating the provenance of the object, and that they evolve in time, thus providing a chronological scale. As an example, many prehistoric cultures in Northern Italy and elsewhere are named after specific features of the pottery they produced (i.e. cultura dei "vasi a bocca quadrata", cultura del "bicchiere o vaso campaniforme" [bell beaker culture]).

The absolute chronology of pottery can be determined by indirect dating of the materials associated in the layer (radiocarbon dating of organics in the soil, K/Ar dating of volcanic layers, etc.) or by direct dating of the pot sherd by luminescence techniques (Box 2.w), fission tracks counting (Garrison *et al.* 1978), electron spin resonance (Maurer 1980), and even the smart use of rehydroxylation kinetics (Wilson *et al.* 2009). By far the most widely used method for ceramics dating is thermoluminescence, since firing resets the termoluminescent clock to zero.

Where?

The bulk chemical and mineralogical analysis of the ceramics provides specific information on where the object was manufactured. The same techniques may also be used to prove authenticity of archaeological materials. The mineralogy and petrology of the mineral components, from the clays but more likely from the temper, may help identify the sources of raw materials, and thus associate the object with a specific site of production. When the temper is composed of sands or natural rocks that have undergone mechanical fragmentation, then the identification of the mineral components and of their crystal chemistry is straightforward in most cases, and this may help to identify the source of the material by comparison with the mineralogy and the chemistry of the outcropping rocks. On the other hand, when the temper is composed by transformed materials such as ceramics, slags, glasses, or organics, then the assessment of the provenance is much more difficult. The issue is further complicated by the fact that several materials may be mixed and transformed by the potter. Mineralogical and petrographical data are essentially obtained by optical microscopy (coupled with computer-assisted image analysis), X-ray diffraction, and SEM-EDS.

The bulk chemistry of the ceramics, especially as far as the trace elements are concerned, may also be used to distinguish ceramic groups and to locate their origin by comparison with samples from known production sites (the so-called reference group). Patterns and groups can be based on concentrations of major and minor elements, on ratios of selected chemical species, or on more complex statistical treatment of the chemical database (cluster analysis, multivariate analysis, neural networks). The major element concentrations are generally measured by XRF, although a number of other methods can be used. Minor and trace elements may be measured by AAS, NAA, ICP-MS, etc. Very successful systematic studies have been performed on ceramics of all ages and cultures in order to support typological classifications with chemical parameters (Box 3.a).

3.2 Structural materials II: cements, mortars, and other binders

Cements based on Portland-type clinkers, mortars (pastes and plasters prepared with fine aggregates), and other binders form an important class of construction material: they are all supplied as powders and when mixed with water they form a fluid mass (**paste**) that can be shaped, moulded, added to other components, or attached to the surface of other materials. The paste then hardens spontaneously at normal environmental conditions.

Binding materials are used in buildings with the aim of (a) making structural elements for constructions, (b) increasing the resistance of the construction by linking the structural and architectural elements, (c) increasing waterproof and protecting masonry surfaces from environmental degradation, and (d) preparing substrates for artwork and decorative purposes. Excluding last century's binders and adhesives based on polymeric compounds, the binders used in antiquity are based on carbonates (calcite, dolomite), sulphates (gypsum) and alumino-silicates (cements). Table 3.6 gives an overall classification of ancient and modern binders based on their chemical nature and reaction processes.

3.2.1 Lime-based materials

The first binders used by mankind seem to be the limestone and gypsum based plasters widely used in the Near and Middle East in the 7th and 8th millennia BC (Frierman 1971, Gourdin and Kingery 1975, Kingery et al. 1988). The technological basis of plaster material is very simple: the reactive compound (either quicklime or plaster of Paris, Table 3.6) is obtained by burning limestone or gypsum at the appropriate temperature, and then the heated block is ground to a fine powder and slaked with water to form a slurry or a paste, depending on the solid/water ratio. In Roman times long aging of the slaked slurry in excess water was a priority with respect to the modern attitude of grinding and mixing the paste at the time of application.

Table 3.6. Main classes of binding compounds produced by pyrotechnology

Starting reactive material	Production process	Material–water mixture	Final product	Mineral phases in the hardened aged material
Lime plaster (quicklime)	Calcinations of limestone	Slaked lime (lime putty)	Lime plaster	Calcite
		Slaked lime + fine aggregate	Lime mortar	Calcite + aggregate
		Slaked lime + fine aggregate + pozzolan	Hydraulic mortar (Roman opus caementitium)	Calcite, zeolites, C-S-H + aggregate
	Calcination of dolomite	Slaked magnesia-lime	Dolomitic or magnesian plaster	Calcite, brucite, periclase
Gypsum plaster (plaster of Paris)	Calcination of gypsum	Bassanite (± anhydrite)	Gypsum plaster	Gypsum
		Bassanite + fine aggregate	Gypsum mortar	Gypsum + aggregate
Portland clinker	Calcinations of limestones + clay	Portland cement paste	Portland cement	Portlandite, C-S-H, calcite
		Portland cement paste + fine aggregate	Portland cement mortar	Portlandite, C-S-H, calcite + aggregate
		Portland cement paste + fine and coarse aggregate	Concrete	Portlandite, C-S-H, calcite + aggregate
		Cement paste + fine aggregate + pozzolan	Pozzolanic Portland cement mortar	Portlandite, C-S-H, calcite, Ca-aluminosilicates

The reactions involved in the production and use of lime are:

$CaCO_3$(calcite) + heat $\rightarrow$ CaO(lime) + CO_2(carbon dioxide)   [production of quicklime]

CaO(lime) + H_2O(water) $\rightarrow$ $Ca(OH)_2$(Portlandite) + heat   [quick hydration]

$Ca(OH)_2$ + CO_2 $\rightarrow$ $CaCO_3$ + H_2O [long term carbonation]

The temperature needed to produce CaO must be above 898 °C, though the decomposition reaction of the carbonate can also proceed at slightly lower temperatures (780–800 °C) in reducing conditions. Operational temperatures of lime-kilns are in the range 920–1000 °C in order to speed up the decarbonation reaction. Excessive temperatures are avoided because they produce unreactive "dead-burned" lime. The production of lime therefore is a very energy-consuming process and it requires a substantial amount of biomass fuel. It has been estimated that the ratio of fuel biomass/quicklime is in the range 2–5 (Kingery et al. 1992, Hauptmann and Yalcin 2001), so that about 4–8 tons of wood would be required to produce the quicklime necessary for one house. Considering the diffusion of lime plaster resulting from the excavations in the Levant, the production of lime could have caused a serious impact on the environment (Rollefson and Köhler-Rollefson 1992, Redman 1999). Similar claims raised for the Mayan Lowlands have recently been challenged (Wernecke 2008).

Traditionally the burning of the carbonates (limestones, dolomites, travertine, marbles, but also shells and corals) is performed in lime-kilns, which are massive furnaces sometimes several metres high, charged from above with decimetre- to metre-sized blocks of limestone, and then fired for days by adding wood or charcoal in the combustion chamber at the base (Oates 1998, Williams 2004). Several ancient lime-kilns have been excavated from Roman (Dix 1982, Coulson and Wilkie 1986), to Late Classic Maya (Abrams and Freter 1996), to more recent times (Williams 2004). An accurate description of lime burning operations in Roman times is supplied by Marcus Porcius Cato (Cato the Elder: *On Agriculture*, XXXVIII).

The fired blocks are then ground to obtain the fine powdered quicklime that, however, is rather unstable in normal humidity conditions and tends to hydrate to portlandite (calcium hydroxide, $Ca(OH)_2$). If the CaO powder is mixed with an exact (i.e. stoichiometric) amount of water (lime/water = 75.7/24.3 = 3.12 by weight) the product is fine dry powder and the process is called *dry hydration* because there is just the right amount of water to produce portlandite. If the CaO powder is mixed with excess water then a smooth paste is obtained in a slurry form, and the process is referred to as *lime slaking*. The portlandite paste (*slaked lime* or *lime putty*) can then be used as a binder and an architectural component (filler, adhesive, cracks sealer, floor consolidant, surface smoother, etc.) or as a raw material for modelling objects, vessels and even artwork. After the application the paste slowly dehydrates and reacts with atmospheric CO_2 producing a hard material composed of microcrystalline calcite. The kinetics of carbonation is slow (Van Balen 2005) so that in recent samples the reaction is not complete and remaining crystals of portlandite may be observed. If an inert component (usually fine quartz sand) is added to the slaked lime then a lime mortar is produced (Table 3.6). The quality of the binder depends on a variety of parameters, including the composition, porosity and impurity content of the fired limestone, the maximum temperature and the time–temperature path of the firing, and the conditions of slaking. The starting limestone should have a non-carbonate mineral content (usually silicates and clays) of less than 5–10 wt%, and the carbonate should be pure calcium. If the carbonate contains magnesium, deriving from the presence of magnesian calcite ($Ca_{1-x}Mg_xCO_3$ with $x < 0.1$) or dolomite ($CaMg(CO_3)_2$), then the material is a magnesian- or dolomitic-lime. The periclase (MgO) produced together with lime during the firing has a much slower rehydration kinetics with respect to CaO, so that in the magnesian putty both periclase and brucite ($Mg(OH)_2$) are present with portlandite (Bläuer-Böhm and Jägers 1997).

Since during carbonation and hardening the microstructure of the slaked lime paste changes significantly (Arandigoyen *et al.* 2006), the identification of lime-derived calcite from an unconsolidated archaeological layer of ground calcite is essentially based on the carbonate particles' dimensions (in the range 0.1–2.0 μm), its texture, and its mechanical properties (Kingery *et al.* 1988, Karkanas 2007). Very careful FTIR work may also be used to discriminate lime derived-calcite from natural calcite (Chu *et al.* 2008) (Figs 3.31(a)–(d)).

The earliest well-characterized example of quicklime production is the Hayonim Cave in Israel (Kingery *et al.* 1988, Goldberg and Bar-Yosef 1998), dated to the Natufian period at about 10.4–10.0 ky BC. The numerous other reports of lime-plaster in the Pre-Pottery Neolithic sites of the Levant (8.7–7.0 ky BC) have been carefully reviewed and characterized by Kingery and co-workers (1988), demonstrating beyond any doubt that there was a widespread use of lime-based plaster in the Near East coastal area from Palestine to Anatolia. Especially striking are the plastered faces from 'Ain Ghazal in Jordan, originally modelled on human skulls (Griffin *et al.* 1998), and the exceptional sculptured head found in Jericho (Garstang 1936; Fig. 10 of Kingery *et al.* 1988). The coeval plasters used in the Middle East from Syria

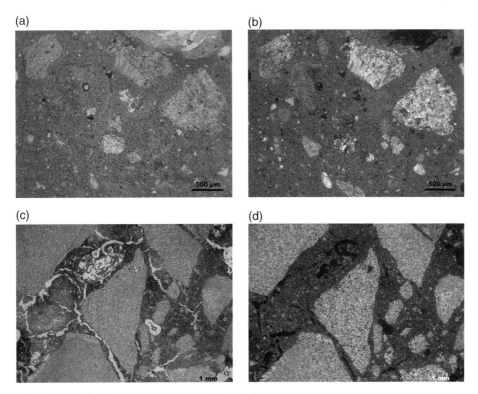

Fig. 3.31. Middle Age lime mortar from the AD 1150–1250 fortress of Sacuidic, Carnia, Friuli, Italy (Gelichi *et al.* 2008). The optical microscopy images [(a, c) plane polarized, (b, d) cross-polarized] clearly show the presence of abundant mm-sized clasts of limestone, the microcrystalline matrix formed out of the carbonated lime plaster, and a few carbonated lime lumps (Figs 3.31a, 3.31b: centre and upper left) that are typical features for the identification of lime mortars.
Original colour pictures reproduced as plate 20.

to Iraq (Kingery *et al.* 1988) as well as most of the early Egyptian plasters (Lucas 1968, Klemm and Klemm 1991) seem to be gypsum-based plasters, though there is evidence of the use of lime plaster in Egypt as an adhesive for hafting flint blades starting from at least the Eighteenth Dynasty (1550–1292 BC) (Endlicher and Tillmann 1997). The use of lime plaster apparently diffused through Greece and Crete to the Etruscans and then to the Romans (Blezard 1998, Pecchioni *et al.* 2008). They all systematically used lime-based compounds both for brickwork and for surface and wall coating. Interestingly, the production and use of lime plasters and mortars through history has gone through a number of evolutionary steps, the most important being mixing reactive material in addition to the quartz sand. In early Roman times there was the widespread belief that aging of the slaked lime was crucial to the quality of the plaster. Vitruvius (*De Architectura*, Book VII) had realized the importance of properly mixing the lime with a hoe to ensure good homogeneity and above all slaking for a sufficient time in pits. Similarly, Plinius (*Naturalis Historia*, Book V) claimed that lime putty needed to mature for at least three years to produce a good binder. Long homogeneous mixing and aging definitely helped in producing a slaked compound with a very fine grain size and little porosity, and this is one of the "secrets" of the excellent quality of Roman mortars, which are still astonishingly solid and hard. However, Greek and Etruscans already mixed some volcanic ash or finely ground pottery (*cocciopesto*) with the mortar, resulting in materials with higher strength and durability. The volcanic tuff of Santorini (the so-called Santorini earth) was a reputed material for the preparation of water resistant mortar, and silica-lime reactions were also activated, probably by chance, in one of the early instances of pyrotechnology at Aşıklı Höyük, Turkey where lime and volcanic ash particles were mixed in the plaster (Hauptmann and Yalcin 2001). The possibility exists that the silica phytoliths and other alkali minerals derived from plant ashes indeed could have induced pozzolanic-type reactions in other cases of primitive plaster technology.

The Romans, during the period between 300 BC and AD 200, crucially improved this technology by using slaked lime in a mixture with high alkali volcanic ash (*pozzolana*), first from the banks of the river Tevere, and then from the volcanic sands found near Naples, at Pozzuoli; hence the name **pozzolan**. Recent studies of the evolution of Roman mortars from Republican through Imperial times reveal neat changes in the composition and use of mortar components: detailed investigation of the ash outcrops within and around ancient Rome (especially the Pozzolane Rosse ignimbrite; Jackson *et al.* 2007) show that specific zeolite-rich tuffs with highly reactive properties were carefully selected especially in later Imperial age to produce exceptionally hard and durable mortars (Figs 3.32(a)–(d)). These important studies not only confirm the early chronology of Roman mortars identified on macroscopic observations (Van Deman 1912a,b), but also confirm the incredibly detailed description of the materials that Vitruvius indicated as ideal for the preparation of quality mortars (black and red sands, or *harenae fossiciae*: *De Architectura* 2.4.1). The experience and skill of Roman builders eventually led to what is universally known as **pozzolanic mortars** (Massazza 1998),

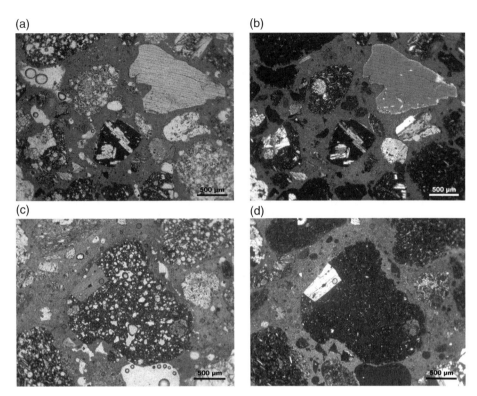

Fig. 3.32. Typical Roman pozzolanic lime mortar of Julian–Claudian Age (48 BC–AD 41) from a wall plaster of the Temple of Venus, Pompeii, Italy. Optical microscopy images [(a, c) plane polarized, (b, d) cross-polarized]. The cryptocrystalline calcite matrix contains abundant millimetre- and submillimetre-sized inclusions of feldspars, pyroxenes, biotite and volcanic particles. The so-called pozzolanic material is represented by the well visible composite volcanic particles containing abundant glass fragments and zeolites, mostly rounded analcimized leucite. (Courtesy of R. Piovesan and M. Secco, University of Padova)
Original colour pictures reproduced as plate 21.

where the improved properties in the hardened state are based on chemical reactions between the slaked lime and the amorphous aluminosilicates of volcanic origin, mainly zeolites or high alkali glass. In his fundamental work *De Architectura*, which was considered a handbook for Roman builders, Vitruvius described this material as one that hardens both in air and underwater, opening the way to the so-called **hydraulic mortars** (i.e. mortars able to harden under water, *De Architectura* 2.6). The physico-chemical and engineering characteristics of mortars used for the construction of Roman harbours are presently investigated in the frame of the ROMACONS project (Brandon *et al.* 2005, Oleson *et al.* 2006; http://web.uvic.ca/~jpoleson/Harbour Concrete/HarbourConcrete.html). Not only the outstanding properties and durability are confirmed, but surprisingly the tuffs used in the mortars of the harbor of King Herod in Caesarea seem to have been transported all the way from the Bay of Naples. The Romans are known to have used all the known occurrences of zeolitic tuffs, including Thera and the volcanic tuffs present in the German Eifel area known as Trass. When volcanic sand was not available, finely ground pottery and ceramics was used to induce hydraulicity to the material, a technique that originated in Minoan Crete (Moropoulou *et al.* 2000).

It is to be remarked that the world cement (*Opus caementicium*) in ancient Roman times referred to the concrete masonry of monuments composed of centimetre sized brick and tuff fragments (*caementa*), which are bonded by hydraulic mortars with alkali-rich, calcium–alumino–silicate volcanic ash sands. Only in recent times has the meaning changed to refer to modern clinker-based materials. The Romans also developed the concept of lightweight concrete by casting jars into wall arches as well as extensively using pumice aggregates, which were obtained by crushing porous volcanic rocks. The arches of the Colosseum and the Pantheon dome are reported to be made of such materials. The art of producing and using excellent mortars was slowly lost after the fall of the Roman Empire. Lime-based mortars continued to be used through the Middle Ages, though in many cases the mortars were of rather low quality, made of partially burned lime, and unaged and poorly slaked putty. The careless preparation would make them mostly porous and degradable. Many of the Saxon, Norman and Longobardic materials are of this kind. The standardized production of the Roman Empire was followed by very local productions, mostly having very low technological content, and it was only in specific and prestigious construction sites that high-quality binders were produced, as in the case of the Byzantine mosaics of Ravenna, and the Leaning Tower of Pisa (Franzini *et al.* 2000).

Outside the Mediterranean world, no evidence is reported on the early use of plaster or mortar in the South American Inca world, where the massive constructions of polygonal stones were assembled by smoothing the edges to make close contacts. However in Central America the use of lime plaster is reported in Preclassical and Classical Central Mexico and Classical Maya period (Hyman 1970, Maglioni *et al.* 1995, Abrams and Feter 1996, Hansen *et al.* 1997, Houston 1998, Barba *et al.* 2009). Interestingly, there is ample evidence that in Mesoamerica the lime plaster was prepared by mixing different kinds of organics extracted from a number of local plants such as cactus (Littman 1960, Laws 1962). The type and extent of the effect of the organic molecules on the putty is still debated; on one hand there are claims that the organics chemically enhance the pozzolanic reactions by destabilizing portlandite and releasing Ca ions, on the other hand the physical effect of packing is emphasized, with the claim that the organics provide for a more homogeneous microstructure, smaller portlandite crystals, and ultimately a higher strength development. The debate is open.

In China, lime mortars were extensively used in ancient architecture, with examples reported in the Qiantan river dam, Dutifulness Monument, Sticky Rice Bridge and others, mainly completed before the Ming (AD 1368–1644) and Qing dynasties (AD 1644–1911) (Huang 2003, Yu and Chen 2004, Zeng *et al.* 2008). According to written sources (Song and KaiWu 1982) the lime plaster production followed a traditional standard formula mixing lime and sticky rice solution up to 15–20 wt%, and the plaster today is reported to be in very good conditions after a few centuries. Sticky rice is a type of rice grown in Southeast and East Asia and it is mainly composed of amylopectin. Again, it seems that the organic influences the calcite dimensions during carbonation (Zeng et al. 2008).

A significant change from traditional lime mortars was made by John Smeaton in England in 1756, when he was involved in the reconstruction of the Eddystone lighthouse (Blezard 1998). He was driven by the need to develop a masonry construction that would be durable in a marine environment and composed of a binding lime mortar that would not dissolve in seawater. Among several attempts, he departed from the Vitruvian recommendations of using pure white limestone, and by using clay-rich carbonates of marly composition he obtained better hydraulic properties than lime. This class of materials may be defined as hydraulic limes and can be considered intermediate between slaked lime and modern Portland cement. Eventually the material that Smeaton selected for the lighthouse was a mortar prepared with equal proportions of local argillaceous limestones (blue Liassic limestones) and Italian pozzolana from Civitavecchia. The hydraulic properties of the mortar are derived from the aluminosilicates formed during the firing. Slightly improved mixtures of this kind were in use until the introduction of Portland cement (Section 3.2.2).

With regard to the characterization of mortars and plasters, they may roughly be considered as artificial geological material and studied in much the same way that rocks are analysed. Preliminary analyses frequently rely on optical microscopy (OM), X-ray diffraction (XRD) and scanning electron microscopy (SEM-EDS) for the identification and evaluation of the mineral phases present. Petrographic analysis of the texture (i.e. lime lumps) is crucial to recognize the phases formed by carbonation reactions of the slaked lime (Reedy 1994, Leslie and Hughes 2002, Hughes et al. 2001, Karkanas 2007). The binder fraction is often also analysed by FTIR spectroscopy, Raman analysis (Edwards et al. 2006), grain size analysis and thermal analysis (Bakolas et al. 1998, Ellis 2000). Advanced electron microscopy is also used for the in situ characterization of the Portlandite–carbonate reactions and porosity (ESEM: Radonjic et al. 2001).

A possible analytical protocol for mortars has been proposed by Crisci et al. (2004) encompassing:

A. Preliminary observations and measurements of mechanical resistance;
B. Measurement of specific density in dry conditions and hydrostatic force of the water-saturated sample
C. Measurement of the capillary water absorption coefficient, water imbibitions capacity, permeability to water vapour
D. Thin section analysis of phases, microtexture and reaction layers (optical microscopy, SEM/EDS)
E. Measurements on powder sample (XRF, calcimetry, thermal analysis, XRD);
F. Dissolution in HCl and analysis of non-carbonate fraction;
G. Ultrasound disaggregation, separation and measurement of the fraction below 63 μm (the binder fraction)
H. ^{14}C dating of the carbon fragments or dating of the carbonate binder.

3.2.1.1 Dating of mortars

The dating of mortars and cements is a strategic (Hale *et al.* 2003, Rech 2004) though difficult area of research to which substantial efforts have been devoted in the last thirty years or so (Folk and Valastro 1976, Van Strydonck *et al.* 1992, Ambers 1987, Zouridakis *et al.* 1987, Heinemeier *et al.* 1997). Since the binders are ever-evolving systems, sometimes incorporating older materials from the start (inert, fillers) and often dissolving and recrystallizing during their lifetime, it is hard to safely analyse by radiocarbon methods (Box 2.u) pristine fragments (charcoal, carbonate) that are incorporated or produced at the time of their emplacement. The analytical attempt is to remove all possible contaminants such as older carbonates (i.e. fossils: van Strydonck *et al.* 1986) or younger carbonates that crystallized after the mortar formation (Lindroos *et al.* 2007, Chu *et al.* 2008). Recent results do indeed demonstrate that reliable dating is possible if adequate selection and treatment of the sample is used (Frumkin *et al.* 2003).

3.2.2 Gypsum-based materials

The technological basis of gypsum plaster is very similar to that of lime plaster. Gypsum blocks are heated in the kiln to produce the reactive bassanite (calcium sulphate hemihydrate), or a mixture of bassanite and anhydrite (anhydrous calcium sulphate) if the temperature is too high or the firing time is too long. Anhydrite is not commonly desired because it is much less reactive than bassanite. An advantage over lime is that the dehydration of gypsum to bassanite takes place at relatively low temperature (nominally at 128 °C), generally in the range 100–160 °C, depending on the water pressure (P_{H_2O}), and therefore the production of the plaster of Paris requires much less energy and biomass fuel than does lime plaster.

However, the gypsum–bassanite–anhydrate transitions are fairly complicated from the kinetic point of view. Depending on the pressure and on the thermodynamic path, different forms of bassanite can be formed (α-, β-, γ-bassanite), which differ in thermodynamic and kinetic properties because of structural ordering, microstructural features such crystal size and morphology, and defect density. The α-form is produced by dehydration of gypsum in conditions of high P_{H_2O}: it is commonly fairly crystalline, it requires less water to rehydrate, thus producing a dense plaster with good mechanical properties. The β-form is produced by dehydration at low P_{H_2O} or in vacuum, it is nanocrystalline and has a large surface area, thus requiring much more water to rehydrate into gypsum, and the plaster has consequently more volume shrinkage. The γ-bassanite is generally produced by slow hydration of anhydrite. Anhydrite also has several forms depending on the temperature of bassanite dehydration (α-anhydrite: 160–200 °C, β-anhydrite: 250–300 °C, γ-anhydrite: 300–600 °C): the solubility of anhydrite decreases with the temperature of formation. Above 900 °C the sulphate starts decomposing into CaO and SO_2.

The powderized bassanite (plaster of Paris) is very hygroscopic and needs to be stored in a dry place. The plaster is used by mixing bassanite and water, causing the dissolution of bassanite (and eventually the associated anhydrite) and the precipitation of gypsum. The growth of the gypsum crystal creates the interlocked crystal grid that makes the hardened plaster (Adams *et al.* 1992). The hardening process may be slowed down by adding salts that increase the solubility of gypsum, or it may be accelerated using a mixture of gypsum seeds and potassium sulphate, favouring nucleation and precipitation. The presence of additives, the relative proportions of bassanite and anhydrite forms in the mixture, their crystal size and morphology, and the temperature of dehydration are the parameters affecting the reactivity and the final type of plaster, i.e. its ability to hydrate fast or slow, its crystallinity and appearance, its workability.

In the case of calcium sulphate the reactions are:

$$CaSO_4 \cdot 2H_2O \text{ (gypsum)} + \text{heat} \to CaSO_4 \cdot 0.5H_2O \text{(bassanite)} + 1.5H_2O$$
[production of plaster of Paris]

$$CaSO_4 \cdot 0.5H_2O + \text{heat} \to CaSO_4 \text{(anhydrite)} + 0.5H_2O$$
[production of anhydrite]

$$CaSO_4 \cdot 0.5H_2O + 1.5H_2O \to CaSO_4 \cdot 2H_2O \quad \text{[quick hydration]}$$

$$CaSO_4 + 2H_2O \to CaSO_4 \cdot 2H_2O \quad \text{[slow hydration]}$$

As mentioned in Section 3.2.1, most of the plaster used in Pre-Pottery Neolithic Middle East far from the coast (Kingery *et al.* 1988) and Egypt was gypsum-based, though is has been shown that lime plaster was the rule in the Near East, and sporadically used in Egypt too. Since Egypt has abundant and easily accessible limestone sources (Klemm and Klemm 2001), the reason generally assumed for the prevalent use of gypsum is the scarcity of fuel (Lucas 1968). Interestingly, gypsum plaster in ancient Egyptian architecture was used, for example between the large blocks of the pyramids, more as a lubricant to slide the blocks into place rather than as a binder to hold them together.

Just as update information on the recent pyramid controversy, it should be said that the recent claims that the stones used for the pyramids were actually made of artificial limestone–geopolymer mixtures (Davidovits 1987) rather than carved natural stones have been systematically disproved on the basis of the mineral phases present, the texture, and the lack of features expected to be found in artificially-produced mortars, such as the reacted lime lumps (Ingram *et al.* 1993, Harrell and Penrod 1993, Campbell 2007): The pyramids and the Sphinx are made of natural limestone blocks.

3.2.3 Clinker-based materials

In the first half of the 19th century the search for optimal hydraulic binders was actively pursued in several countries (Blezard 1998, Bentur 2002). In Britain, John Smeaton attempted alternative formulations of hydraulic mortars (Section 3.2.1) and J. Parker introduced and patented a so-called "Roman cement" (Patent by James Parker, of Northfleet, Kent, No. 2120 (1796) Cement or tarras to be used in aquatic and other buildings and stucco work), which was made by calcinations of nodules of argillaceous limestones (known as septariae) and produced a quick-setting cement. In France, Louis Vicat's experimentations lead to the preparation of hydraulic lime by calcination of a mixture of high-grade quicklime (produced by the chalk of the Upper Cretaceous carbonatic formation of the Paris Basin) and clay (L. J. Vicat, Mortier et ciment calcaires, Paris, 1828). His formulation, called the "twice-kilned" process, met with considerable success and lead his son Joseph Vicat to establish the well known Vicat Cement company. This is considered by many to be the predecessor of Portland cement.

A large number of patents were issued around the same time to establishing plants in Southern Britain, including the London area. The most famous one is the one related to the three-stage process of Joseph Aspdin (1824), who described his product as **Portland cement**, because at that time Portland limestone had a reputation among builders for quality and durability, and he wanted to capture the similitude between his cement and Britain's favoured quarried stone. Portland cement was marketed as an improvement in the mode of producing artificial stone. In one of the several plants established by Aspdin or his son, the temperatures were running high enough to produce partial or complete vitrification and crystallize alite, as shown by the retrospective analysis of the type of clinker material from Aspdin's kiln (Blezard 1981). The **clinker** is the reactive product formed by cooling from the high temperature processing (1450 °C) of the limestone and clay mixture within the kiln. The temperature is such that partial fusion occurs and the reacted molten material lumps into nodules that are partly crystalline and partly vitrified, the whole material being highly reactive with water. The basic four-phase composition of the clinker is reported in Table 3.7, though the real situation is far more complicated by the existence of several polymorphs for each phase, and by compositional deviation from stoichiometry. The basic chemical composition of Portland cement is shown in Fig. 3.22, where it is evident that the composition of the two major phases (alite and belite) lies on the CaO–SiO_2 join as a result of the lime–silica reaction.

As a curiosity, there appears to be one place in the world (Maqarin, Jordan) where this reaction occurred naturally, as the result of high temperature/low

Table 3.7. Main phases of ordinary Portland clinkers (Taylor 1997)

Phase	Composition	Cement notation	wt %
Alite	$Ca_3SiO_5 = 3CaO \cdot SiO_2 =$ tricalcium silicate	C3S	50–70
Belite	$Ca_2SiO_4 = 2CaO \cdot SiO_2 =$ dicalcium silicate	C2S	15–30
Aluminate	$Ca_3Al_2O_6 = 3CaO \cdot Al_2O_3 =$ tricalcium aluminate	C3A	5–10
Ferrite	$Ca_2AlFeO_5 =$ tetracalcium aluminoferrite	C4AF	5–15

Box 3.c On-site investigation of masonry [M. R. Valluzzi]

Masonry is a composite and non-homogenous material, characterized by irregularities in morphology and composition, which influence its mechanical behaviour, which is often critical under medium–high severe actions. Among existing structures, historical masonry constructions deserve a specific approach for their preservation, which should take into account the peculiarities of the geometry, structural details and material conditions, as well as the identification of weaknesses and vulnerability, both at global and local level. Moreover, historical constructions are subjected to a never-ending process of modification and transformation, due to the current interaction with environmental and anthropic actions, and also to previous damage and deterioration, lack in maintenance, past interventions and reuse in time, which add new materials and volumes to the original architectural and structural layout (Binda *et al.* 1999). Historical constructions belong to various typologies (churches, towers, common buildings, palaces, castles, fortified walls, bridges), often combined as rows or more complex aggregates in villages and city centres. Furthermore, the multiplicity of their structural components (walls, columns and pillars, arches, vaults and domes, floors and roofs) and the large variability of constituent materials (brick/stone, earth, various type of mortars, timber, iron) and arrangements (various textures, multi-leaf sections, lacks in connections, etc.), make particularly complex the univocal classifications and the standardization of methods for diagnosis, assessment and intervention (Modena 1997).

In such a context, after the preliminary steps of knowledge on the history of the building, the recognition of the current conditions by using suitable procedures of in situ investigation should be applied. They can lead to the identification of the specific problems to solve and, consequently, to the suggestion of the most appropriate intervention strategies. This approach should guarantee the respect of the criteria for the architectural heritage preservation, as stated by the ICOMOS Charters of Venice (1964, http://www.icomos.org/venice_charter.html) and Victoria Falls (2003, http://www.international.icomos.org/charters/structures_e.htm), aimed at the conservation of authenticity, and thus mainly based on the concepts of minimum intervention, "reversibility" (meant as removability or replaceability), compatibility and durability (Valluzzi 2007).

The diagnostic phase should consider as a priority the minimization of the potential loss of material by selecting the least invasive experimental procedures. On the other hand the choice of test positions, their number and distribution should be optimized, in order to achieve significant results that are to be implemented in models at various scales. Therefore, a minimum investigation programme should be proposed, based mostly on the proper combination of the **minor destructive tests** (MDT) and **non-destructive tests** (NDT) approaches, in order to validate the results by the crosschecking of different methods. In fact it is unfortunately the case that the lower the obtrusiveness of the technique, the less the reliability of the results. Cross-checking can thus reduce the approximation in the interpretation of the data, especially when qualitative outcomes are provided. For the characterization of chemical, physical and mechanical properties of single and compound materials and elements in masonry structures, several MDT and NDT are available.

On-site investigation commonly starts with a visual inspection aimed at the recognition of the respect of the "rule of art" (Giuffrè *et al.* 1999), e.g. for a masonry wall: proper dimensions and shape of resisting elements, texture (misalignment of the head joints) and presence of diatons (or connecting elements), quality and sufficient quantity of mortar, presence of horizontal courses or wedges, etc. These aspects, together with the presence of improvements or constructive traditional measures for a good synergy between components (ties, pilasters, buttresses), can guarantee a good global behaviour or, if absent, point out an intrinsic vulnerability of the building (weakness for out-of plane actions, lack of strength of materials, etc.). Then, some qualitative and quantitative experimental procedures can be applied.

In the area of application of NDT on existing masonry structures, **sonic pulse velocity** represents a valid experimental procedure, generally aimed at: (a) evaluating the current consistency by identifying the presence of possible anomalies and defects in the thickness (cracks, voids, other inhomogeneities); (b) providing a base of quantitative comparison

(cont.)

Box 3.c (*Continued*)

between diverse portions of masonry in the same conditions (e.g. before a consolidation intervention); or (c) of the same portion in different conditions (e.g. before and after injection). It is based on the transmission of a sonic wave produced by an instrumented hammer and received from sensors kept in adherence with the surface of the masonry (RILEM Recommendations, TC 127-MS, 1996). The higher is the measured velocity, the more is the compactness of the wall. Sonic wave transmission is particularly suitable for such composite materials as masonry, in comparison with ultrasonic tests, which are more reliable for homogeneous (e.g. stone components). Direct (i.e., through the thickness) or indirect (i.e., superficial) modes of testing are possible, as well as combinations of measures elaborated through tomography (Fig. 3.c.1). Despite the reliability of direct comparison between velocities e.g. before/after intervention, the basic character of the results remains quite qualitative. Despite the recent diffusion of masonry as a well-known subject of research, it is still very difficult to propose an absolute catalogue of masonry typologies where sonic velocities can be straightforwardly related to the conditions of walls (consistency, damage), thus reducing confidence in the reliability of the method (Berra *et al.* 1992, Forde *et al.* 1985, Riva *et al.* 1997). Nevertheless, besides the advantages listed above, the method turns out to be very helpful in a more complex diagnosis programme, for the preliminary check of the conditions of walls, in order to identify the areas where the presence of weaknesses are more probable, therefore limiting the possible subsequent application of localized MD methods such as coring or flat-jacks only to specific points of the structure, thus minimizing damage and optimizing costs (http://www.onsiteformasonry.bam.de).

In the field of MDT, **flat-jack tests** have great potential for the identification of local mechanical parameters of

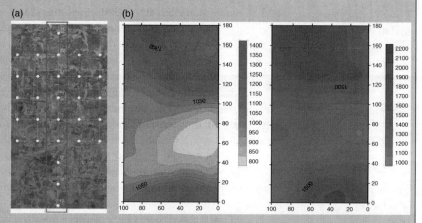

Fig 3.c.1. Sonic test grid (a) and results of tomography in a wall section before (b, left) and after (b, right) grout injections: the increase in velocity denotes the good filling of voids in the thickness. (Courtesy of M. R. Valluzzi, University of Padova)
Original colour pictures reproduced as plate 52.

masonry under compression, i.e. the stress condition (single flat-jack, SFJ), or the deformability and the stress–strain behaviour up to the elastic limit (double flat-jacks, DFJ) (Rossi 1982, RILEM Paper LUM-D2 1991, RILEM Paper LUM-D3 1991, ASTM C-1196 1991, ASTM C-1197 1991). SFJ tests are based on the stress relaxation caused by a cut executed in a horizontal slot in which the thin flat-jack is inserted and subsequently inflated until the original distance between the two edges is restored. DFJ are composed of two parallel flat-jacks inserted at about 50–60 cm among them; such a configuration is able to simulate a uni-axial compression test, by keeping the area around the test fairly undisturbed (Ronca *et al.* 1997). Cyclic loads are repeated and progressively increased to identify the elastic modulus of the masonry. It is important to remind ourselves that the minor obtrusiveness is related solely to the cutting of joints, thus the test should not induce further damage in the masonry (Binda *et al.* 2007). Actually, although some limits persists in the application and interpretation of the results, SFJ and DFJ tests are recognized as a reliable method that can provide the most significant mechanical parameters from a masonry, actually considered as composite material, thus avoiding both destructive tests on-site and laboratory tests on full-scale samples taken from the wall. Rectangular or semi-circular flat-jacks are usually employed in masonry walls; FJ dimensions for walls are commonly about $350 \times 250 \times 4\,mm^3$, but smaller (for head joints in arches) or larger (in the case or non-homogeneous patterns, e.g. irregular stone masonry) elements can be produced for specific cases.

Research is increasingly focused on the possibility of proposing in situ ND procedures to identify quantitative parameters in masonry (e.g. a drilling energy test operated on mortar to determine the masonry strength), whose reliability still need further correlation studies with destructive tests before possible standardization and use. Nevertheless, qualitative procedures are particularly suitable in the diagnosis phase on historical masonry, and novel methods are proposed. Among NDT it is worth mentioning: (1) the Ground Penetrating Radar system (GPR, Box 2.p), which allow one to identify the thickness of the wall and the presence of inclusions or anomalies, moisture (even if the presence of humidity can falsify the outcomes if not properly considered) by elaborating the electromagnetic impulses detected by an antenna system moved on the surface of investigation (a precious surface should be protected with cardboard or plastic sheets) (Binda *et al.* 1994, Valle *et al.* 1998); and (2) the Infrared Thermography (IRT, Box 2.p), which allows one to detect superficial (up to about 10–15 cm) discontinuities, moisture, delaminations and inhomogeneities, by providing (with flash lights or a heater) a transient temperature gradient on the surface (Maldague 1993, 2001). This method is particularly suitable for delicate surfaces as it does not require any contact with the original material. Moreover, in the light of the results obtained by ND procedures, when sampling of a material is possible or allowable, limited cores optimized in number and position, can be useful for direct inspection (visual or with boreholes or endoscopies) and to provide material for characterization in laboratory. When possible, for visual internal inspection of walls, simple local disassembly can be suggested, with the opportunity of placing back the resisting elements (bricks or stone), with new suitable mortars.

It is worth to remind that, in the preservation approach to historic constructions, the experimental in-situ procedures can be properly used before and after intervention, in order to evaluate the effectiveness of the techniques, especially in case of inclusion or diffusion of consolidation materials (e.g. for grout injections, pointing or re-pointing of mortar joints) (Schuller *et al.* 1994). In many cases, qualitative procedures, e.g. sonic tests for injections, can provide quantitative useful evaluations by comparison of the conditions before and after the intervention, thus exploiting all the potential of the investigation method itself (Fig. 3.c.2).

In order to provide durability, materials employed in consolidation have to respect as much as possible the fundamental requirement of full compatibility (chemical, physical and mechanical) with the original ones, rather than looking for high-strength solutions. In this connection, it is proved that the use of high-performance consolidation material (e.g. cement grouts in masonry) may be useless, or even dangerous (chemical reactivity inducing salt crystallization or secondary ettringite-thaumasite), as the particular features of the original walls will not be able to exploit the whole increase in strength expected by the use of modern or innovative materials (Valluzzi *et al.* 2007). As a final step in the

(a)
(b)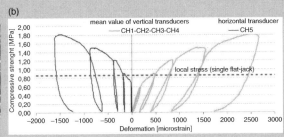

Fig. 3.c.2. Double flat-jack test set-up and superimposition with results of SFJ, denoting the margin between the current state of stress and the strength at the elastic limit of the masonry. (Courtesy of M. R. Valluzzi, University of Padova)
Original colour pictures reproduced as plate 53.

(*cont.*)

> **Box 3.c** (*Continued*)
>
> preservation process, the upgraded conditions of the construction should then be properly monitored in time, in order to protect the Cultural Heritage from further deterioration process.
>
> **For in-depth information**
>
> Binda L. and De Vekey R.C. (eds) (2003) On-site control and non-destructive evaluation of masonry structures. International RILEM workshop. Mantova, Italy, 13–14 November 2001. Published by RILEM Publications.
>
> McCann D.M. and Forde M.C. (2001) Review of NDT methods in the assessment of concrete and masonry structures. Vol. 34. Pp. 71–84.
>
> ONSITEFORMASONRY, Project n. EVK4-2001-0091, Contract n. EVK4-CT-2001-00060, December 2004, http://www.onsiteformasonry.bam.de, deliverable D11.1 (2004) "Technical guidelines for an appropriate use of the suggested equipment."

pressure metamorphism of marls and limestones. The Maqarin site, located along the Yarmouk river along the Jordan–Syrian border, comprises Cretaceous–Tertiary carbonate rocks overlain by Quaternary basalts, soils, and alluvium. The river has exposed by erosion a Bituminous Marl Formation overlain by the Chalky Limestone Formation. For some reason during the recent geological past (estimated at 600 ky ago) a spontaneous combustion was triggered in the Bituminous Marls, possibly caused by the exothermal oxidation of pyrite igniting the kerogen, and producing the unique formation of thermally metamorphosed "cement zones" which are very close analogues of industrial cements. The subsequent interaction with groundwater caused the formation of hyperalkaline waters and C–S–H phases in the rock fractures, which are much studied to understand the long-term leaching and cation mobility in the only known analogue case of a radioactive-waste repository. The investigation project is called The Maqarin Natural Analogue Project (Khoury *et al.* 1992, Alexander and Smellie 1998, Alexander and Blaser 2002; http://www.natural-analogues.com/maqarin.htm).

Following Blezard (1998), the historical evolution of clinkers after the illustrated pioneering experiments moved from *proto*-Portland materials (i.e. Aspdin's original patent), which shows little interaction between CaO and SiO_2 because of the limited temperature in the kiln, to *meso*-Portland materials, which are very heterogeneous materials showing some silica–lime interaction and containing mostly belite and some alite, and finally to *normal*-Portland cements as we know them today. The main characteristics of modern Portland clinkers are actually derived from the use of rotary kilns in place of the traditional shaft kilns, a technical development that also allows continuous production in place of the batch process. The carefully-controlled initial formulation of calcareous and argillaceous components, together with the use of the rotary kiln that determines a long permanence at high temperature and the continuous mixing, favours (a) the minimization of the amount of unreacted lime, (b) the maximization of the alite/belite ratio, (c) the appropriate crystal size of the mineral components, commonly in the range 10–40 µm. Modern clinkers contain about 60–65 wt% of C3S, and less than 2 wt% of unreacted free lime.

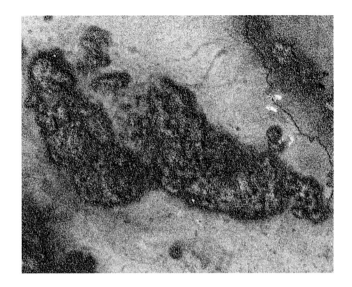

Plate 1. High resolution false-colour image taken from the IKONOS satellite of the Guatemalan lowlands [Fig 1.7].

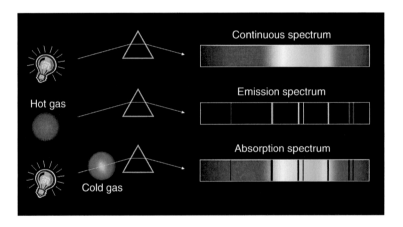

Plate 2. Mechanisms of production of the continuous spectrum, the emission spectrum and the absorption spectrum [Fig. 2.5].

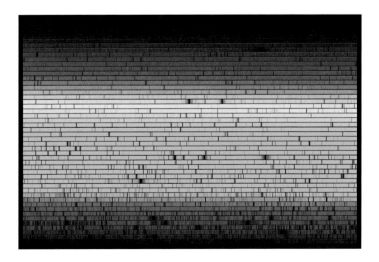

Plate 3. Dark absorption lines in the continuum visible emission spectrum of the Sun's atmosphere [Fig. 2.6].

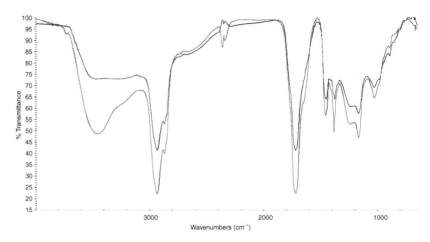

Plate 4. FTIR spectra of Baltic amber [Fig. 2.18].

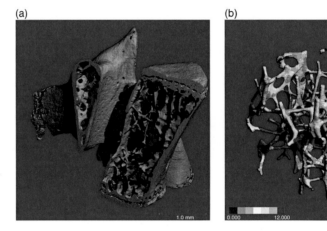

Plate 5. The amount and distribution of the inorganic part (apatite) in bone may be imaged and quantified using radiographic and tomographic techniques [Fig. 2.19].

Plate 6. Microscopic orange spots on silver photographic print [Fig. 2.30].

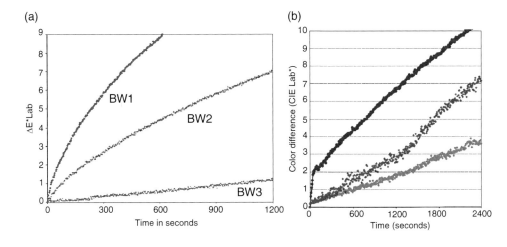

Plate 7. (a) Microfading tests on standard samples No. 1, 2 and 3 of blue wool. (b) Examples of chemical changes: dye fading on an early colour photograph (Autochrome) [Fig. 2.31].

Plate 8. Optical emission spectra of hydrogen (H, Z = 1) and iron (Fe, Z = 26), showing the characteristics emission lines [Fig. 2.c.2].

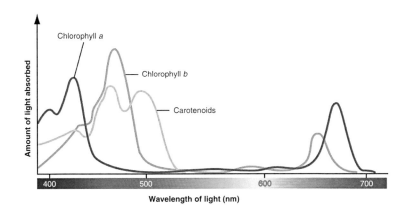

Plate 9. Vis absorption spectrum of light from molecular compounds contained in plant leaves [Fig. 2.f.1].

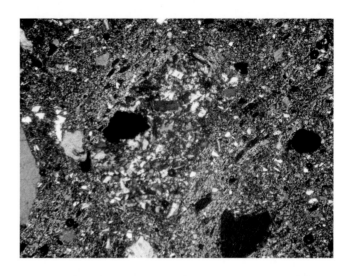

Plate 10. Cross-polarized transmitted-light OM image of a ceramic shard under the microscope (horizontal field of view is 4 mm) [Fig. 2.k.2].

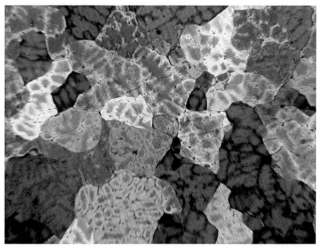

Plate 11. Cross-polarized reflected-light OM image of a thermally recrystallized copper sample showing clear evidence of the original dendrites formed during casting [Fig. 2.k.3].

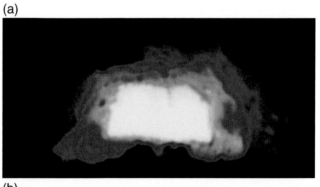

Plate 12. Heavily altered and corroded protohistoric copper ingots [Fig. 2.n.4].

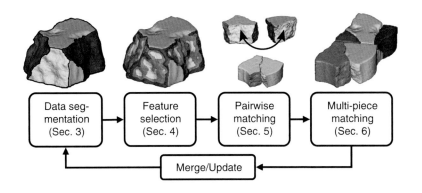

Plate 13. The generic process of object reconstruction. It consists of four steps [Fig. 2.q.3].

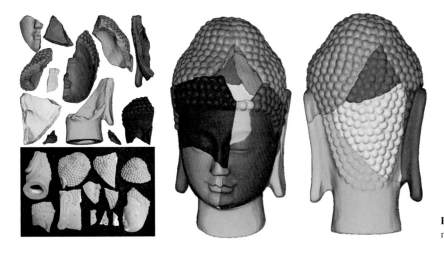

Plate 14. Reassembling a fractured head model [Fig. 2.q.4].

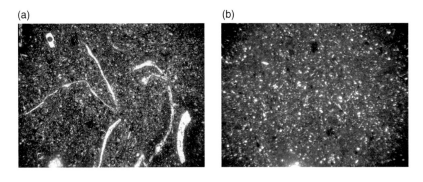

Plate 15. Thin section of a Late Bronze Age pottery sherd from the Frattesina terramara (Rovigo, Italy) [Fig. 3.24].

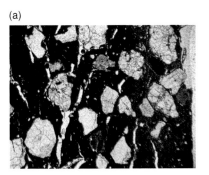

Plate 16. Thin section of a Late Bronze Age pottery sherd from the Frattesina terramara (Rovigo, Italy) [Fig. 3.25].

Plate 17. Thin section of a Late Bronze Age pottery sherd from the Frattesina terramara (Rovigo, Italy) [Fig. 3.26].

Plate 18. Thin section of a Late Bronze Age pottery sherd from the Frattesina terramara (Rovigo, Italy) [Fig. 3.27].

Plate 19. Section of an Iron Age pot sherd of Etruscan–Roman type from the Hospital excavation, Este, Italy [Fig. 3.30].

Plate 20. Middle Age lime mortar from the AD1150–1250 fortress of Sacuidic, Carnia, Friuli, Italy [Fig. 3.31].

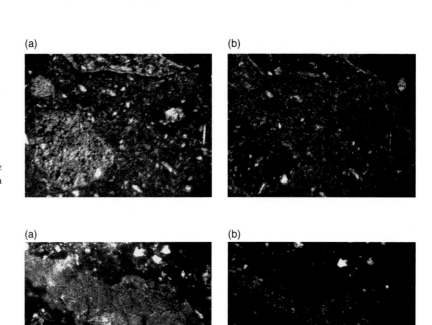

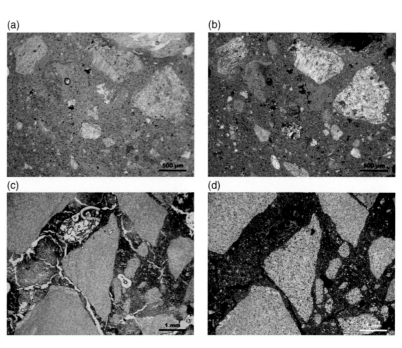

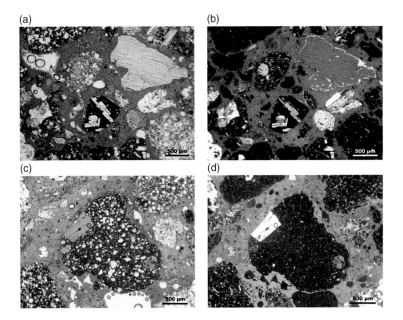

Plate 21. Typical Roman pozzolanic lime mortar of Julian–Claudian Age (48 BC–AD41) from a wall plaster of the Temple of Venus, Pompeii, Italy [Fig. 3.32].

Plate 22. Figurine head from Jaina Island, Campeche, Mexico, still bearing traces of Maya Blue pigment [Fig. 3.34].

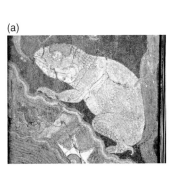

Plate 23. Enlarged figures in the wall painting of Templo Rojo of Cacaxtla, Tlaxcala, Mexico: (a) toad (photo M. Zabé), (b) Quetzal, sacred bird of the Maya [Fig. 3.36].

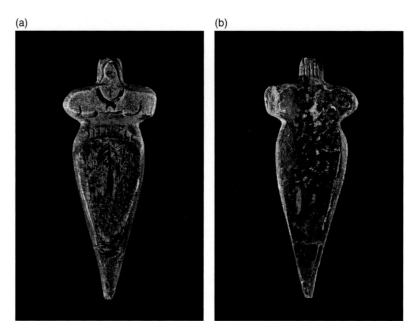

Plate 24. The so-called "Venere del Gaban". A small Neolithic female figurine carved out of antler and painted on both sides with red ochre pigment [Fig. 3.37].

Plate 25. Moldavite: a tektite. Tektites are natural glasses ejected at large distances from meteoritic impacts [Fig. 3.39].

Plate 26. Libyan desert glass: a light coloured impact glass found in the Western Desert between Egypt and Libya [Fig. 3.40].

Plate 27. (a) Tiffany Glass and Decorating Company: vase, ca. 1900, blown favrile glass, (b) vase, ca. 1903, favrile glass, (c) vase, ca. 1903, favrile glass [Fig. 3.42].

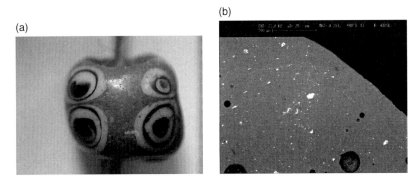

Plate 30. (a) Eyed glass bead from Sardinia, (b) SEM backscattered electron image showing a very homogeneous glass matrix with few crystalline and metal inclusions [Fig. 3.45].

Plate 28. Face-shaped pendants were spread throughout the entire Mediterranean Sea region by the Phoenicians from the 6th century BC [Fig. 3.43].

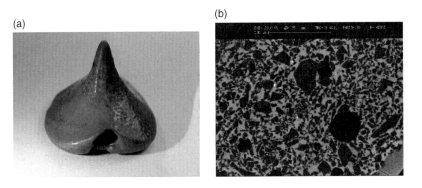

Plate 31. Middle Bronze Age conical buttons made of glassy faience are peculiar objects and materials produced in Northern–Central Italy [Fig. 3.46].

Plate 29. Modern art exploiting the versatile properties of glass [Fig. 3.44].

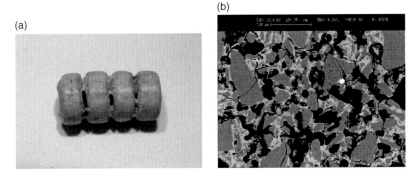

Plate 32. Example of a typical Early Bronze Age faience bead from Lavagnone, Italy [Fig. 3.47].

Plate 33. The famous polychrome-glass tilapia fish produced at Amarna, in Middle Egypt (Shortland 2009) and conserved at the British Museum in London [Fig. 3.49].

Plate 34. Blue balsam container in the shape of a dove, from Groppello Cairoli (first decades of the 1st century AD) [Fig. 3.50].

Plate 35. Portrait of a young man. Glazed terracotta by the workshop of Luca della Robbia ca. 1445 [Fig. 3.51].

Plate 37. Blue glass turning red: example of surface copper reduction ($Cu^{+2} \rightarrow Cu^0$) on a blue glass due to firing under reducing conditions [Fig. 3.53].

Plate 38. Red glass turning blue: example of oxidation of copper ($Cu^{+1} \rightarrow Cu^{+2}$) in a cuprite-coloured red glass ingot from Qantir-Piramesses, Egypt; Late Bronze Age [Fig. 3.54].

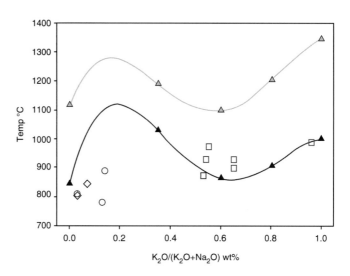

Plate 36. Experimental values of the melting temperature (yellow curve) and the working temperature (red curve) for a series of synthetic glasses with varying $K_2O/(K_2O + Na_2O)$ composition (Polla 2006) [Fig. 3.52].

Plate 39. Well crystallized transparent minerals from Afghanistan pegmatites: (a) pale blue aquamarine, (b) polychrome tourmaline and (c) pink tourmaline [Fig. 3.70].

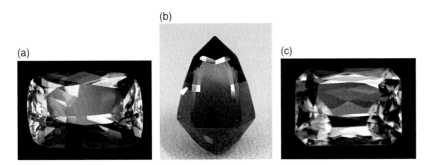

Plate 40. Cut gems from minerals similar to those shown in the previous picture: (a) aquamarine, (b) polychrome tourmaline, (c) morganite [Fig. 3.71].

Plate 41. Fragments of blue-stained bones from the excavation of a Terramara site (Montale, Italy) [Fig. 3.74].

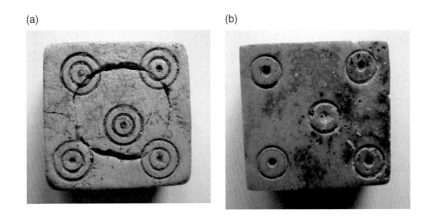

Plate 42. Etruscan dice pieces from Tarquinia: (a) hollow bone piece, unstained; (b) stained ivory [Fig. 3.75].

Plate 43. Example of *foxing*. Archive of Icpal Chemistry Laboratory [Fig. 3.82].

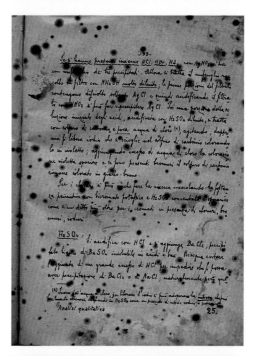

Plate 44. Wall painting, Mogao Grottoes, Cave 85 [Fig. 3.91].

Plate 45. A finished coupon, simulating possible stratigrafies of *Gardenia augusta* pigment (PI) and inorganic pigments. 1 and 11 [Fig. 3.92].

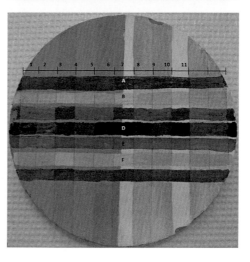

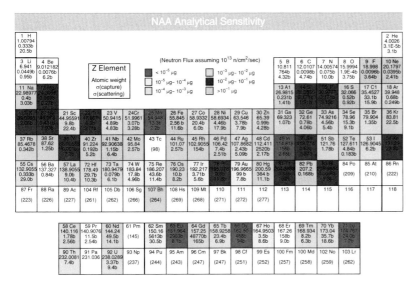

Plate 46. Archaeological textile, cotton, Guane, Northern Andes, Colombia, museum *Casa de Bolivar,* Bucaramanga, Colombia [Fig. 3.93].

Plate 48. Periodic table of the elements evidencing the isotopes that can be analysed by NAA, with the associated detection limits [Fig. 3.a.1].

Plate 47. *Clavus* in multicoloured tapestry weave on green tabby ground [Fig. 3.94].

Plate 49. Periodic table of the elements evidencing the isotopes that can be analysed by PGAA, with the associated detection limits [Fig. 3.a.2].

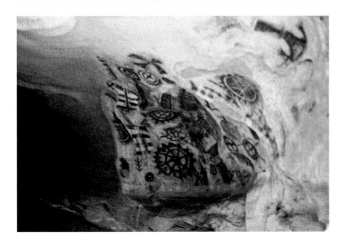

Plate 50. One of the few visible Chumash rock art paintings [Fig. 3.b.2].

Plate 51. Rock engravings at Butler Wash, San Juan River, Utah [Fig. 3.b.3].

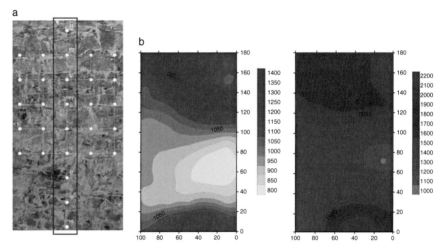

Plate 52. Sonic test grid (a) and results of tomography in a wall section before (b, left) and after (b, right) grout injections [Fig. 3.c.1].

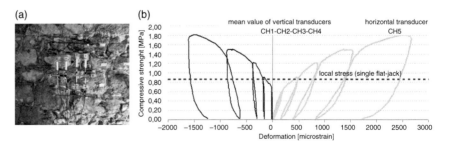

Plate 53. Double flat-jack test set-up and superimposition with results of SFJ, denoting the margin between the current state of stress and the strength at the elastic limit of the masonry [Fig. 3.c.2].

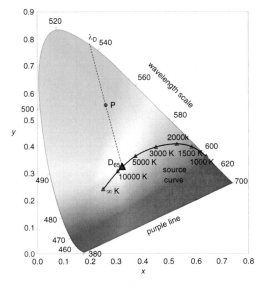

Plate 54. CIE chromaticity diagram. It is the most versatile diagram to express colour components [Fig. 3.e.1].

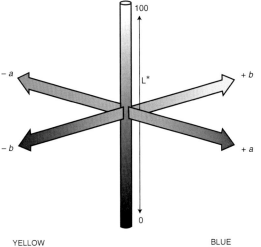

Plate 55. The CIELAB coordinate systems, separating the L* brightness axis from the two antagonistic chromatic systems: red-green (a^*) and yellow-blue (b^*) [Fig. 3.e.2].

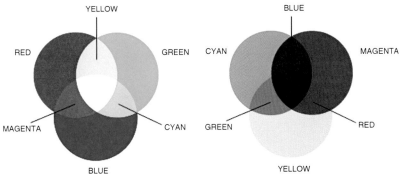

Plate 56. The (a) red-green-blue (RGB), and (b) cyan-magenta-yellow (CMY) colour codes [Fig. 3.e.3].

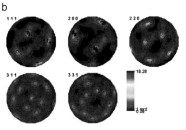

Plate 57. Chalcolithic copper axe from Castelrotto/Kastelruth, Bolzano, Italy (Museo Archeologico dell'Adige) [Fig. 3.l.5].

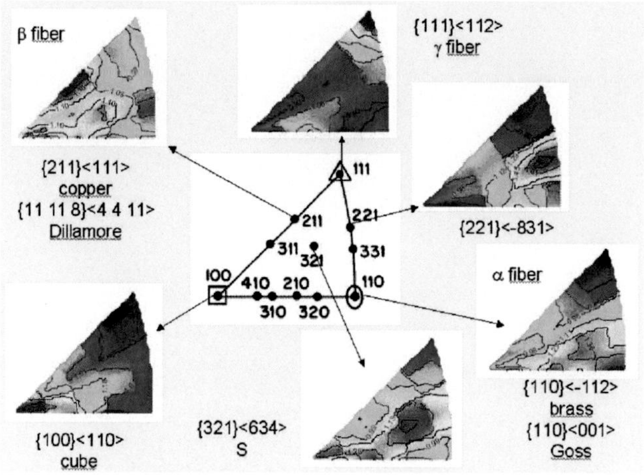

Plate 58. Inverse pole figures showing some of the experimentally observed textures in chalcolithic copper axes [Fig. 3.l.6].

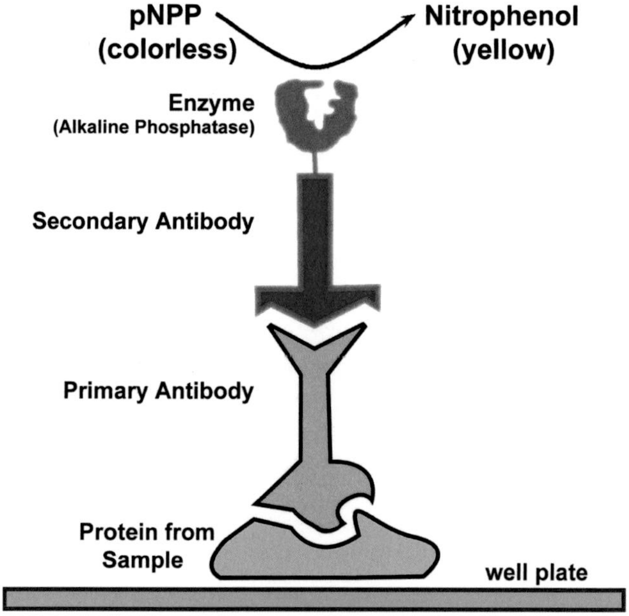

Plate 59. The protein binds to the well plate and the primary antibody binds to the protein. The secondary antibody binds to the primary antibody and pNPP turns yellow [Fig. 3.q.1].

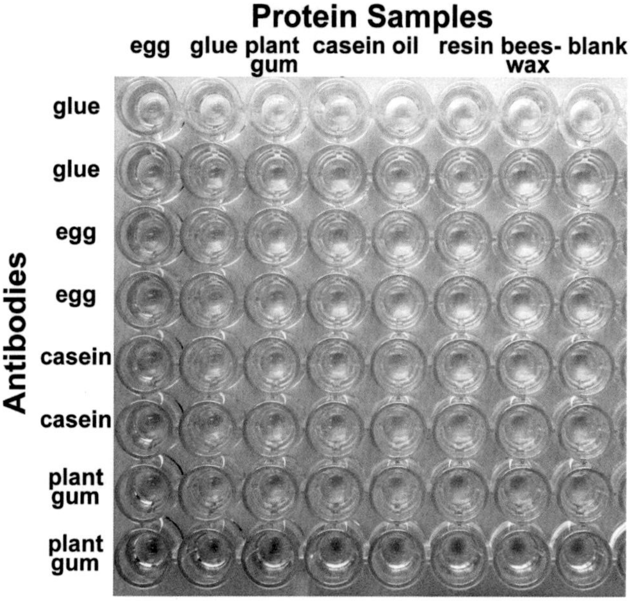

Plate 60. Photograph of ELISA well plate results from the cross-reaction experiment [Fig. 3.q.2].

The standard reaction properties of modern Portland clinker are essentially due to a high alite/belite ratio and to careful control of the grinding of the clinker into a fine powder.

The very reactive Portland cement powder is mixed with water to produce a final hardened material through a series of complex reactions, the so-called **hydration process**, involving dissolution of the crystal phases, surface reactions, gel formation, precipitation of new phases and textural changes (Taylor 1997, Gartner et al. 2002). The different crystal phases present in the clinker have very different reactivities, C3A having the highest and C2S the lowest. In fact C3A is so exothermally reactive that it can cause unwanted rapid setting of the paste (flash set), with consequent loss of workability. Therefore calcium sulphate (commonly gypsum) is usually added to the clinker in the amount of about 4–8 wt% as a set-controlling agent. The finely powdered mixture of clinker and gypsum is marketed as standard **Portland cement** (or OPC, ordinary Portland cement), which is then mixed with water in the ratio water/cement > 0.38, which is the least amount of water necessary for complete hydration of the cement phases. When an aggregate material is added to the paste we have **mortar** (fine aggregate, such as sand with grain size below 1 mm) or **concrete** (fine and coarse aggregate, such as coarse sands or gravel). They both can be considered to be cement composites, with the aggregate having the role of reducing the formation of fractures during the shrinkage that accompanies the hydration and hardening; for the same reason temper is added to pottery clays before firing (Section 3.1.2.1). The aggregate may be composed of any type of loose or ground rock, and it is assumed not to react with the cement paste (hence it is inappropriately called the *inert* component), though this is not necessarily the case.

Upon mixing with water the clinker phases react at different times to produce first a series of intermediate phases (the so-called AFt and AFm phases) such as ettringite ($Ca_6Al_2(SO_4)_3(OH)_{12} \cdot 26H_2O$) or monosulphoaluminate ($Ca_4Al_2(SO_4)O_6 \cdot 12H_2O$), and then eventually the reaction progresses to the final hydrous phases: portlandite and amorphous calcium silicate hydrate (C–S–H), which are the phases forming the interlocked gridwork of the material and producing high mechanical resistance. The completion of the hydration process may take days, months, or even years, depending on the crystal size, the defectivity and polymorphism of the phases, the porosity of the paste, environmental conditions, etc. With time portlandite will eventually convert into calcite through a carbonation reaction with atmospheric carbon dioxide, so that abundant calcite is found in old cements. Figure 3.33 shows environmental scanning electron microscope (ESEM) evidence of the growth of ettringite and C–S–H at different times on the surface of the clinker grains.

Strength development is related to the degree of hydration and to the speed of the process. In general, the slower the hydration kinetics, the higher the developed compressive strength. This is sometimes observed in the difference between modern clinkers and those produced in the early part of the last century. Modern cement is required to develop high resistance to compression (over 40 MPa) within a few days from the emplacement, whereas early 1900 cements developed high strengths (in the range 20–30 MPa) in much longer

Fig. 3.33. (a) Prismatic crystals of ettringite growing amidst C-S-H felt during the hydration of a Portland cement, and (b) well-developed platy crystals of Portlandite. Images obtained by ESEM. (Courtesy of D. Salvioni, Mapei S.p.a.)

times, because of the higher belite content. This has some consequences for durability (Box 3.d). High compression resistance combined with tensional resistance are of course obtained by **steel reinforced concrete** (SRC), which is a composite where the concrete is poured and solidified around a metal skeleton. The French–Swiss architect Le Corbusier was one of the first to understand and exploit the properties of SRC in modern architecture.

Nowadays only about 35% of the cements industrially produced for the global market are OPC, and most of the materials are special composite formulations (Odler 2000, Chatterjee 2002), where different components are added to enhance specific physico-chemical or mechanical properties (Table 3.8). The main added constituents are (a) latent hydraulic components (granulated blast furnace slags, metakaolin, class C fly ash) that have self-cementing properties that need to be activated by OPC, (b) pozzolanic aluminosilicate components (class F fly ash, silica fume, pozzolan) that have no self-cementing properties and need to be activated by portlandite, and (c) non-reactive or poorly reactive components that modify the grindability of the clinker or the rheology of the paste.

Furthermore, the rheological and working properties of the cement pastes are now invariably controlled with the use of chemical admixtures (water-soluble organic polymers) that allow good fluidity of the paste with lower water/cement ratio. The net result is also a lower porosity of the paste and a higher mechanical strength: high performance concrete and high strength concrete are made this way, they can reach compression strengths of over 150 MPa. Organic and inorganic additives are used to control specific properties of the mortars and cements, such as air entrainment (for a better resistance to freeze and thaw cycles), acceleration or retardation of the setting (to control the time of progressive increase in mechanical resistance), plasticity and viscosity (plasticizing or superplasticizing effects to control the workability of the paste) and even resistance to chemical attack. Modern additives are designed for specific actions, though the change of the paste properties resulting from the addition of external compounds has long been known, as briefly mentioned

Table 3.8. Main types of common and special cements

Cement type	OPC (wt%)	Composition	Properties
Portland cement (OPC) [CEM I*]	95–100	Mixed with minor constituents	Common cements
Special cements [CEM II*]	65–94	Mixed with blast furnace slags, silica fume, natural or calcined pozzolans, fly ash, burnt shale, limestone	Blended cements with enhanced durability
Blast furnace cement [CEM III*]	5–64	Mixed with blast furnace slags	
Pozzolanic cement [CEM IV*]	45–89	Mixed with pozzolan, silica fume, fly ash	
Composite cement [CEM V*]	20–64	Mixed with blast furnace slag, pozzolan, fly ash	
High belite cements	95–100	Belite, alite obtained by fast quenching of clinker	Slow steady increase in strength, low heat production
Calcium aluminate cement (CAC) or high alumina cement (HAC)	0	Made by calcinations of limestone and bauxite	Rapid increase in strength, superior sulphate resistance, resistance to high temperature
Fibre cement	95–100	Mixed with reinforcing fibres (polymeric, steel)	Crack control, reinforcement, reduction of plastic shrinkage, shotcrete applications
Cement–polymer systems	variable	OPC or HAC mixed with water-soluble polymers	Macro-defect free (MDF) materials, high compressive and flexural strength
Ultra-rapid hardening cements	85–90	$C_{11}A_7CaF_2 > 20$ wt% 8–12 wt% anhydrite	Very rapid development of strength, jet cement applications
Expansive cements	variable	Contain $3CA \cdot CaSO_4$ (kleinite), lime, or high-C3A	The expansive component is used to compensate OPC shrinkage

* Classification (UNI EN 197-1, 2007)

Box 3.d Degradation and conservation of binders. A tale of pores and water

Like all materials, binders, mortars and cements undergo degradation processes, alteration and structural modifications, as they tend to equilibrate with the thermodynamics of the surrounding (Section 2.5.2). The amount and kinetics of the degradation processes depend (a) on the nature of the binder, and (b) on the environmental conditions acting upon it. In most cases the degradation processes of chemical nature are slow and continuous, though of course catastrophic events and mismanagement can severely and abruptly damage the artefact (Figs 3.d.1 and 3.d.2). As a first approximation, most of the alteration and degradation processes are caused or mediated

(*cont.*)

Box 3.d (*Continued*)

Fig. 3.d.1. Internal stucco decoration of the civic theatre of Schio, Vicenza, Italy. The Art Nouveau establishment was built in 1907–1909, and was abandoned for several years after 1968. It suffered severe damages caused by roof collapse and subsequent weathering. The theatre is now undergoing complete restoration. The stucco decorations were made with a gypsum-based plaster strengthened by fine limestone powder.

by water, therefore the prime parameter affecting the speed of degradation is the **porosity** of the material, i.e. its capacity to absorb and diffuse water in the bulk (Hall and Hoff 2002). The amount, dimension, distribution and connectivity of the pores in the material have a profound effect on his long term chemical and mechanical behaviour. The *measurement of porosity* is therefore one of the first steps in understanding the active processes and assess the degree of alteration. Porosity (P) is defined as $P\% = 100 \cdot (V_p / V_a)$, i.e. the percentage volume of pores in the examined volume of material. Since porosity is a multi-scale property (i.e. nano-pores often coexist with micro-, meso- and macro-pores) one analytical technique is usually insufficient to measure porosity: each method can be applied to measure pores in specific dimensional ranges. Macro- and meso-porosity (with pore diameters in the range $4 \times 10^{-2} – 4 \times 10^2 \, \mu m$) is usually measured by mercury intrusion porosimetry (or mercury picnometry) (Cook and Hover 1999), though it is known that mercury-based methods only accesses part of the pores and severely underestimates the pore volume and distribution (Diamond 2000). The porosity measured by water absorption simply measures the difference between the dry sample and the water-loaded one, or else exploits Archimedes' principle of volume displacement (Lowell *et al.* 2004). The test is often performed at different water partial pressures to distinguish accessibility by vapour and liquid water. Micro- and nano-porosity (down to the $10^{-4} \, \mu m$) are measured by water vapour or gas (nitrogen or helium) absorption porosimetry coupled with the BET (Brunauer *et al.* 1938) or BJH (Barrett *et al.* 1951) modelling of the data. The results may be substantially different depending on the interpretive model (Robens *et al.* 2002, Collet *et al.* 2008). 2D and 3D image analysis is also widely used to assess porosity at different scales, depending on the nature and resolution of the measured images (AFM maps; SEM-BSE images: Wong *et al.* 2006; micro-tomographic reconstructions: Gallucci *et al.* 2007, Promentilla *et al.*

Fig. 3.d.2. Close up of the gilded decorations of the civic theatre of Schio, Vicenza, Italy. Besides abrasions and scratches, there is a diffuse surface weathering due to water dripping from the collapsed roof.

2008; OM images; Photographic images; etc.). Fractal models are sometimes used to describe the pore distribution and surface areas (Niklasson 1993, Collet *et al.* 2008).

In the binder, the amount and distribution of the pores is controlled by the nature of the paste and the aggregate system. Key factors affecting porosity are: (a) the water/binder ratio in the paste, (b) the paste/aggregate ratio in the composite mortar or concrete, (c) the nature of the aggregate, (d) the use of additives as dispersants and surfactants, (d) the environmental conditions of setting and hardening. A typical operation causing an excess porosity and micro-fracturing in the binder from the start is the use of excess water in the paste to increase workability. Besides bleeding (segregation of water from the paste causing highly inhomogeneous materials, aggregate lumps and surface layering), excess water also causes the formation of unwanted capillary porosity and shrinkage-related meso- and macro-fracturing during hardening. The presence and distribution of pores then critically determines the fate and duration of the binder, since most of the degradation phenomena are related to water absorption and transport, including dissolution of structural components and binder phases, dissolution and recrystallization of soluble salts, formation of secondary phases, progression of the carbonation reactions, corrosion of the steel reinforcements, and many other.

The humidity of masonry and cement artefacts may have several causes (Mora *et al.* 1977): (1) water dripping and infiltration from exposed external walls, defective pipes or damaged roofing; (2) rise by capillarity from humid soil; (3) condensation on cold surfaces; (4) humidity gradients due to heterogeneous hygroscopic materials; and (5) transport by humid air from soil. At saturation ($RH > 20\%$), the relative humidity (RH) of structural components (porous rocks, bricks) is the same as that of the binder. At lower RH the humidity proportion between bricks and mortar may vary considerably depending on the nature of the materials and their positions in the architectural structure. The measurement of humidity and the interpretation of the transport mechanisms could be rather difficult, because different water sources ((1)–(5) above) may be active at different times. The on-site measurements must therefore be repeatedly carried out for extensive periods, in order to assess variability, amplitude, and frequency of the humidity changes and their relationship to temperature gradients.

In fact, in addition to the intrinsic characteristics of the material, it is the interaction between the artefact and the environment that determines the kinetics of chemical changes (Section 2.5.2; Camuffo 2004). The basic concept is that any material or composite undergoing thermodynamic shocks, including the manufacturing of the object, the variation in the location, the change of its use, the conservation treatment, etc. slowly reaches thermodynamic equilibrium with the surroundings following a universal kinetic decay curve (Fig. 3.d.3) whose rate is described by Boltzmann statistics (Section 2.5.2); this transformation is what we call **degradation** process. The shape of

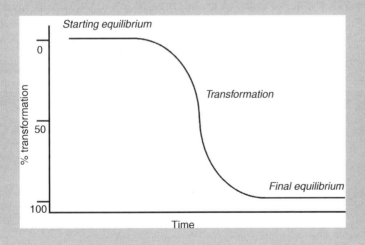

Fig. 3.d.3. Universal kinetic decay curve whose rate is described by Boltzmann statistics (Section 2.5.2). Independently of the time scale, the curve is followed by all simple chemical transformations, including degradation and alteration processes.

(*cont.*)

Box 3.d (*Continued*)

the curve is largely independent of the time scale of the process, which can vary from seconds (i.e. destruction of gypsum plaster submerged into water) to millennia (i.e. slow weathering of igneous stones such as granite) (Viles 2001). If the intervention (i.e. our measurement) is far from the initial slope and close to the equilibrium plateau, then the situation attained over time is the most stable, and probably the best possible condition for the preservation of the material. In such a case any intervention, including conservation treatments, is bound to negatively perturb the lifetime of the material. If the measurements indicate that strong chemical/physical gradients are acting on the artefact, then the material is quickly undergoing chemical changes and the process is following the steep part of the kinetic curve. Intervention is therefore necessary in order to slow down the degradation process, i.e. to flatten the kinetic curve. The direction of the intervention (i.e. resetting the initial equilibrium or stabilizing the new one) however depends on the selected conservation philosophy.

In the case of masonry and binders, humidity and temperature fluctuations are natural components of the environment. Day/night and seasonal climatic changes induce thermal gradients, expansion shocks, and numerous cycles of evaporation/condensation through the **dew point**, inducing water formation and transport through the structures. We may distinguish between weakly physiosorbed (**absorbed**) water molecules and the more tightly chemisorbed (**adsorbed**) water molecules. In all cases the changes in the water content of the material is going to activate chemical reactions, mobilization of components and mechanical damages. The most common ones are related to the soluble salts already dissolved in the water, such as in marine spray (i.e. sulphates, chlorides), or dissolved in the material by the presence of water: the most mobile salts in cements and mortars are the sulphates. Specifically, sulphate-loaded water percolating through cement clinker deleteriously produces the formation of secondary ettringite and/or thaumasite (Section 3.2.3; Collepardi 1999, Gaze and Crammond 2000). Salt recrystallization within the pores and micro-fractures of the binder is one of the most pernicious alteration mechanisms of building materials. The crystallization pressure exerted by the salt growing in confined spaces (Scherer 2004, Steiger 2005) is not dissimilar to that produced by the freeze and thaw cycles of water and ice: the volume of the crystal is always larger that the volume of the same amount of molecules in the liquid state, because of the ordered arrangement due to chemical bonding in the crystal lattice. **Salt weathering,** however, induces other damages related to differential thermal expansion, osmotic swelling of clays, hydration pressure and enhancement of wet/dry cycles (Goudie and Viles 1997, Charola 2000, Doehne 2002, Al-Naddaf 2009). The use of surfactants as agents to interfere with the salt crystallization behaviour is being tested (Rodriguez-Navarro *et al.* 2000, Selwitz and Doehne 2002).

Furthermore diffusion of chlorine into modern steel reinforced concrete (SRF, Section 3.2.3), mostly derived from marine water or rock salt dispersed in winter on the roads for defrosting, causes enormous damages to internal steel reinforcements because Cl is a very effective catalyser of steel oxidation. The most common result is the detachment of the whole concrete layer covering the steel backbone (Fig. 3.d.4). The progressive surface carbonation of the concrete is related to recrystallization of the binder matrix, opening of pores, and slow penetration of

(a) (b)

Fig. 3.d.4. Corrosion of steel reinforcement in concrete causing the detachment of the outer concrete layer and the collapse of the structure: (a) a reinforced concrete structure erected in the 1950s in the industrial area of Marghera, near Venice, Italy under severe attack by marine salted water; (b) close up of a concrete beam with evident damage from metal corrosion and concrete fragmentation.

the chlorine to the metal reinforcement. Conservation and restoration of 20th century OPC and SRF, materials that were originally thought to be very durable, is becoming an issue in most of industrialized countries. The number of cases of premature degrading and disintegration of OPC structures is endless.

Concerning historical gypsum-based mortars (Section 3.2.2), even low humidity may be a problem if combined with high temperature. In fact gypsum is a dihydrate phase, and at $T > 30\,°C$ and $RH < 30$–40% it may undergo slow dehydration. The process has been detected in the plaster of the Nefertari tomb in the King's Valley, Egypt (Plenderleith et al. 1970, Preusser 1991, McDonald 1996).

The chemical equilibrium of lime-based mortars is related both to absorbed water and to atmospheric carbon dioxide. The CO_2 dissolved in the water forms carbonic acid, which dissolves calcite and produces soluble calcium bicarbonate that migrates and re-precipitates elsewhere as Ca carbonate, in many cases producing a thin white veil on the surface. Such a carbonate layer frequently covers frescoes and cave paintings. The major problems involving wall paintings of all ages are: pigment alteration and fading, detachment of the painted layer and plaster support, surface corrosion, and salt precipitation. (Figs 3.d.5.–.7).

In the restoration of buildings, masonry and wall plasters the choice of the materials to be employed must be consistent with the conservation strategy. The conservation treatment must be durable and aesthetically acceptable, though above all it must be the least invasive as possible on the existing parts of the artefact, be they original or

Fig. 3.d.5. Damaged painted decoration in the box-wall of the civic theatre of Schio, Vicenza. The painted layer suffered from surface abrasion and severe detachment of the plaster support.

(a)

(b)

Fig. 3.d.6. The South façade of the church of S. Vigilio, Pinzolo, Trento, Italy. The long fresco below the roof is a remarkable *danse macabre* painted by Simone Baschenis in 1539. It is composed of over forty characters and numerous inscriptions in the vulgar language of the time. The picture (a) was taken in 1978, shortly before the artwork was severely vandalized. (b) Close up of the upper right corner of the painted frieze, showing that even before the vandalistic attack the artwork had been subject to weathering, partial destruction in the lower part due to architectural modifications, and detachment of the plaster support.

(*cont.*)

Box 3.d (*Continued*)

Fig. 3.d.7. Detail of exfoliation of a painted surface on the north wall of Cave 85, Mogao Grottoes (Agnew 1997; See also Fig. 2.17, Fig. 3.91). Exfoliation is characterized by the lifting and resulting loss of paint, ground and upper plaster layers. This phenomenon caused by cycles of deliquescence and recrystallization of soluble salts is particularly severe in areas previously treated in the 1970s with synthetic and impermeable fixatives. (Courtesy of N. Agnew, GCI, ©J Paul Getty Trust)

restored by previous interventions. Decisions upon the conservation strategy are inevitably quite difficult because of the large number of parameters involved, the uncertainties concerning the properties and the durability of available materials, and the hypersensitivity of the management and the public to the final impact of the project. Each conservation project is a unique and irreproducible event. The most basic guidelines for the restoration of cultural heritage buildings are the documents approved by two of the ICOMOS Charters (Venice 1964 and Victoria Falls 2003, see Box 2.c). They clearly state that "The conservation and restoration of monuments must have recourse to all sciences and techniques which can contribute to the study and safeguarding of the architectural heritage" (Venice Charter, Art. 2) and that "Where traditional techniques prove inadequate, the consolidation of a monument can be achieved by the use of any modern technique for conservation and construction, the efficacy of which has been shown by scientific data and proved by experience" (Venice Charter, Art. 10). The Venice Charter therefore amply supports the use of scientific analyses and techniques to back up conservation strategies. The Victoria Falls Charter goes into more details concerning **reversibility** (Art. 3.9: "Where possible, any measures adopted should be reversible so that they can be removed and replaced with more suitable measures when new knowledge is acquired"), **compatibility** and **long-term effects** (Art. 3.10: "The characteristics of materials used in restoration work (in particular new materials) and their compatibility with existing materials should be fully established. This must include long-term impacts, so that undesirable side-effects are avoided"). In the specific case of restoration binders, the Charter states that cements containing salts can only be used when there is no risk of damage to the masonry and particularly to its surfaces, and there may be a problem of leaching soluble salts from the mortar resulting in efflorescence on the surface of brickwork. The statement is an obvious reference to the salt recrystallization effects discussed above, and following the guideline no hydraulic mortar used in restoration and conservation treatments should contain gypsum-added cements, including OPC. This is the reason why most available conservation materials for masonry and wall plasters have formulations close to ancient Roman or pozzolanic mortars. A final note concerning compatibility. Given that the concept is obviously correct and desirable, it may be somewhat difficult to apply it consistently throughout the four basic stages of conservation (cleaning, stabilization, repair and restoration). The operational limits of the reversibility concept are often debatable in practice (Oddy and Carroll 1999), and they need to be continuously cross-checked and validated through analyses and experimentation.

> **For in-depth information**
>
> Aligizaki K.K. (2006) Pore structure of cement-based materials: testing, interpretation, and requirements. Taylor & Francis, London. p. 432.
>
> Bartos P.J.M., Groot C.J.W.P. and Hughes J.J. (eds) (2000) Historic mortars: characteristics and tests. Proc. RILEM Intern. Workshop. RILEM Publications, Cachan, France.
>
> Hall C. and Hoff W. (2002) Water transport in brick, stone and concrete. Spon Press, London–New York. p. 320.

for the ancient Mesoamerican and Chinese recipes for lime plasters (Section 3.2.1). Egyptians, Minoans, and Romans also used to modify the lime mortars with Arabic gum, animal glue, fig's milk, egg yolk and many other organic substances in an attempt to improve the mechanical or working properties (Sickels 1981, Arcolao and Dal Bò 2001). The practice was passed down to the Middle Ages and Renaissance, especially regarding quality mortars for decorations and applications exposed to weathering agents. A large variety of organic and proteic substances were tested, with mixed results: cereal dough, animal blood, fermented wine, beer, milk derivatives including cheese, bee's wax, lard, and many others. The modern range of plasticizing additives, mostly confined to polyvinylic alcohols, methacrylic polymers, naphthalene and ligno-sulphonates, and a few others (Chandra 2002) are certainly more effective, but much less interesting from the culinary point of view.

The techniques used to investigate cement-bases binders are somehow similar to those used for mortars. The analysis is finalized (1) to understand the original composition and formulation of the bulk and the paste, (2) the measurement of the physical characteristics of the paste and the aggregate, and (3) the identification of the active degradation processes and evaluation of compatible materials for conservation issues (Box 3.d). Cement and concrete being rather complex heterogeneous materials, almost all available analytical techniques may be profitably used to characterize the samples at different scales. The identification and quantification of the crystalline phases that are present is traditionally performed by optical microscopy (Campbell 1999), though advanced full-profile analysis of XRPD data is now preferred (De la Torre and Aranda 2003, Scrivener et al. 2004, Gualtieri et al. 2006). Physical measurements include the evaluation of surface area, heat of hydration, permeability and porosity (Box 3.d), electron microscopy is used to visualize and interpret the microstructures (Richardson 2002), and chemical methods, especially XRF, are generally used to evaluate the bulk chemistry. Advanced Raman (Garbev et al. 2007), XPS (Black et al. 2006) and solid-state NMR (Skibsted and Hall 2008) techniques have been recently introduced to investigate the subtleties of the cement hydration process. A very interesting application investigated clinker phases with simultaneous SEM microscopy and Raman spectroscopy using the so-called SEM-SCA analyser, though sample preparation seems to be experimentally demanding (Black and Brooker 2007).

3.3 Pigments

Colour is one of the most fascinating properties of matter, as testified by the very early use of pigments in prehistoric paintings, the importance of surface decoration in artwork and everyday objects, and the widespread use of colourants in ancient and modern cosmetics. The physical origin of colour is nested in the absorption and interference phenomena and it has been briefly treated in Section 2.6 and Box 2.f. The measurement of colour is described in Box 3.e. Organic dyes are treated in Section 3.7.5. It is worth remarking on the difference between pigments and dyes: pigments are finely divided insoluble materials that act as colouring agents when dispersed in a medium such as water as in watercolours, oil as in oil paintings, or a number of other organic media (Gettens and Stout 1942). Dyes, on the other hand, are soluble organic complexes that cannot be used alone, rather they are used as colourants of white or colourless particles, such as clays. The substance made by the dyed particles is called the lake.

The analytical and conservation issues related to inorganic pigments will be reviewed here, the most fundamental and practical being: (1) what is the physico-chemical nature of the material?, (2) how is the pigment included or bound in the paint layer?, (3) is it in the original state or is it undergoing alteration?, and (4) what are the strategies to stop deterioration and to stabilize it?

Pigment analysis is a fundamental aid to art history, authentication, and conservation management. Of course colour has so many aesthetical, philosophical, and anthropological implications that the materials science side is just a starting point in the broader colour science (see for example Technè no. 9–10 (1999) and no. 26 (2007)). The emphasis here will be confined to basic pigment investigation and methods.

Pigment identification and characterization

The analytical techniques for identification of crystal phases described in the volume are mostly used to answer the first two diagnostic questions. Inorganic pigments have a definite chemical composition and crystal structure, so that they may be identified by chemical analysis (SEM-EDS, EPMA, XRF, AAS, OES, MS, LIBS), by diffraction (XRD, ND), and by molecular spectroscopy (IR, RS).

XRD and Raman scattering (RS) are possibly the most used techniques for pigment identification, because of their sensitivity for crystallographic structure types, molecular environment, and structural polymorphs. Both techniques are so widely employed that several experimental databases of reference compounds are available (Bell *et al.* 1997, Hochleitner *et al.* 2003) for identification purposes. The analyses are generally performed on microsamples of pigment material analysed in the laboratory. Raman scattering has the obvious advantage that it can also be performed in situ in totally non-invasive mode (Smith *et al.* 2000, Smith 2006), though serious attempts are being made to make XRD portable too (Box 2.r). Non-invasive in situ investigation of pigments is of course a critical analytical issue, since any sampling or alteration of the sample during the analysis compromises the nature and appearance of the object. Furthermore, the paint layers are usually very thin, so microscopic or microbeam techniques are required to resolve the distribution of the pigment phases.

In many cases, small objects such as painted vases, illuminated manuscripts and fragments of frescoes can easily be analysed and imaged in 2D or 3D by existing facilities, especially employing versatile microbeams such as synchrotron beams (Box 2.a) or accelerated ions (Box 3.i). Chemical and mineralogical maps and depth profiles at synchrotron sources can be routinely obtained using these advanced sources (Janssens and Van Grieken 2004, Woll *et al.* 2005, Box 2.a) by simultaneous μ-XRD, μ-XAS and μ-XRF analyses down to the sub-micrometre spatial resolution (Box 2.a). However, architectural painted surfaces, large monuments and paintings, and frescoes must essentially be analysed in situ, and a great deal of effort is devoted to developing portable instrumentation, such as the mobile laboratory project of EU-ARTECH (Brunetti *et al.* 2007). Complementary information combining the confocal Raman signal with XRF chemical analysis can be obtained using the microbeam of the recently developed PRAXIS instrument (Van der Snickt *et al.* 2008). XRD and/or Raman signals are very important for the identification of the pigment, since chemical information alone is generally not sufficient for identification. Several distinct phases may have very similar chemical compositions, and they may be present at the same time, so that element-specific techniques can hardly distinguish them unambiguously. The presence of copper, for example, can indicate several possible compounds (Table 3.9) and additional information from diffraction and/or spectroscopy is required for complete characterization of the pigmenting phases.

In addition to being characteristic of a given crystal phase, the colour is also determined by the presence of crystal defects, crystallite and grain size, distortion of the atomic coordination, chemical substitutions, etc. One of the most extreme cases is the iron oxide mineral hematite (Fe_2O_3), which is black in large crystals, and bright red when ground to a fine powder. The colour is not only given by absorption of the trivalent iron (electron transition and charge transfer), but by a complex interaction between absorption and surface reflectance, so that the particle shape and morphology critically determine the scattered light from the hematite pigment (Cornell and Schwertmann 1996). A similar effect is found in glass and glaze that is coloured by metallic particles (Fig. 1.5), in which the size and composition of the metal nanoparticles (Figs 2.i.3 and 2.l.4) critically determines the colour properties of the diffused light.

Therefore sometimes phase identification is not sufficient to enable us to understand the nature of the colouring mechanism, and detailed characterization of the local (short range) molecular environment and atomic electron configuration (oxidation or valence state) is required. Advanced spectroscopic methods should be used for this sub-nanoscale investigation, such as XPS (Box 3.f), EELS (Box 3.g) or XAS (Box 3.h). For specific cations, especially iron, Mössbauer spectroscopy may also be used. If the size, shape and distribution of the colouring particles are to be studied, HR-TEM is a necessary technique.

Concerning the nature of the pigments, they can be natural or synthetic. Although the earliest investigated occurrences in Paleolithic caves indicate the use of a limited number of natural pigments (Table 3.9) such as red (iron oxides) and black (charcoal, manganese oxides) pigments (Chalmin *et al.*

Table 3.9. List of the main inorganic pigments used in antiquity and historical times (min = found as natural mineral; synth = synthetic)

Colour	Pigment name	Nature	Chromophore ion	Notes (alternative names)
White	barium white	min		$BaSO_4$ (baryte)
	bone white	min		$Ca_3(PO_4)_2$ (apatite, bone ash) made from bone calcination
	chalk	min		$CaCO_3$ (calcite, lime white)
	gypsum	min		$CaSO_4 \cdot 2H_2O$
	lithophone	synth		co-precipitated $ZnS+BaSO_4$, introduced in 1874
	titanium white	synth		TiO_2 (titanox) the main modern pigment, introduced in 1795
	white lead	synth		$Pb_3(CO_3)_2(OH)_2$ (hydrocerussite, flake white) synthesized in antiquity
	zinc white	synth		ZnO (zincite, Chinese white) introduced in 1782
Red	cinnabar	min	Hg^{+2}	α-HgS
	litharge	min	Pb^{+2}	PbO (tetragonal)
	realgar	min	As^{+2}	α-As_4S_4
	hematite	min	Fe^{+3}	Fe_2O_3 (iron oxide red)
	madder	synth	alizarin	($C_{14}H_8O_4$) (madder lake) natural dye from the roots of *Rubia tinctorum* in ancient times, synthetic from 1868
	ochre (red)	min	Fe^{+3}	hematite + clay + quartz mixtures (red earth, Indian red)
	red lead	min	Pb^{+2},Pb^{+4}	Pb_3O_4 (minium)
	vermilion	synth	Hg^{+2}	HgS – artificial cinnabar
Orange	Mars orange	synth	Fe^{+3}	artificial ochre
	chrome orange	synth	Pb^{+2},Cr^{+4}	$PBCrO_4 + Pb(OH)_2$, introduced in 1809
Yellow	barium and strontium yellow	synth	Cr^{+6}	$BaCrO_4$ and $SrCrO_4$ (lemon yellow), introduced in 1809
	cadmium yellow	synth	Cd^{+2}	CdS (mineral greenockite), synthetic from 1817
	chrome yellow	synth	Pb^{+2},Cr^{+4}	$PBCrO_4$ (mineral crocoite), synthetic from 1809
	cobalt yellow	synth	Co^{+2}	$K_3[Co(NO_2)_6] \cdot 1.5H_2O$ (aureolin), introduced in 1848
	gamboge	synth	gambocic acid	$C_{38}H_{44}O_8$ and $C_{29}H_{36}O_6$ – yellow gum resin extracted from plants of genus *Garcinia*, historically used in the Far East, imported in Europe after AD 1600
	Indian yellow	synth	euxanthic acid	$Mg(C_{19}H_{16}O_{11}) \cdot 5H_2O$ (purrée) Mg salt of euxanthic acid precipitated from cow's urine, used in Bengal 15th cent
	lead tin yellow	synth	Pb^{+2},Sn^{+4}	Pb_2SnO_4 – used between 13th and 18th centuries
	massicot	min	Pb^{+2}	PbO (orthorhombic)
	Naples yellow	synth	Pb^{+2},Sb^{+5}	$Pb_2Sb_2O_7$ – synthesized in antiquity
	ochre (yellow)	min	Fe^{+3}	FeO(OH) goethite + clay + quartz mixtures (raw sienna, limonite)

	orpiment	min	As^{+3}	As_2S_3
	saffron	synth	crocetin	$C_{20}H_{24}O_4$ – carotenoid dicarboxylic acid extracted from the stigmas of *Crocus sativus*. Widely used for illuminating manuscript
	zinc yellow	synth	Cr^{+6}	$ZnCrO_4$ – introduced in 1809
Green	atacamite	min	Cu^{+2}	$Cu_2Cl(OH)_3$
	chromium oxide	synth	Cr^{+3}	Cr_2O_3 – introduced in 1809. The anhydrous form is opaque, the hydrous variety ($Cr_2O_3 \cdot 2H_2O$) is transparent (emerald green, viridian, Guignet's green)
	chrysocolla	min	Cu^{+2}	$CuSiO_3 \cdot nH_2O$
	cobalt green	synth	Co^{+2}	Co-doped zinc oxide (Rinmann's green)
	green earth	min	Fe^{+2}, Fe^{+3}	Fe-aluminosilicates (celadonite, glauconite)
	malachite	min	Cu^{+2}	$Cu_2(CO_3)(OH)_2$
	verdigris	synth	Cu^{+2}	$Cu(C_2H_3O_2)_2 \cdot 2Cu(OH)_2$ (Cu-acetate, also hydrous)
	Egyptian green	synth	Cu^{+2}	mixture of green glass and $CaSiO_3$ (wollastonite), synthesized in antiquity
Blue	aerinite	min	Fe^{+2}	Complex Fe-aluminosilicate, used in the Pyrenees and Catalunya
	azurite	min	Cu^{+2}	$Cu_3(CO_3)_2(OH)_2$
	cerulean blue	synth	Co^{+2}	$CoO \cdot nSnO_2$
	cobalt blue	synth	Co^{+2}	Co-doped alumina (Thénard's blue)
	Egyptian blue	synth	Cu^{+2}	$CaCuSi_4O_{10}$ (cuprorivaite), synthesized in antiquity
	Han blue	synth	Cu^{+2}	$BaCuSi_2O_6$ (Han purple), synthesized in antiquity
	indigo	synth	indigotin	indigotin ($C_{16}H_{10}N_2O_2$) extracted from plants of genus *Indigofera*, such as *I. tinctoria*
	Maya blue	synth	indigotin	Palygorskite+indigo complex
	Prussian blue	synth	Fe^{+3}	$Fe_4[Fe(CN)_6]_3 \cdot 14{-}16H_2O$, introduced in 1704
	ultramarine	min	S_3^-	$Na_3CaAl_3Si_3O_{12}S_n$ (mineral lazurite, also called lapis, lapislazuli), synthetic from 1814
	smalt	synth	Co^{+2}	Co-doped silica glass
Violet	cobalt violet	synth	Co^{+2}	$Co_3(PO_4)_2$ and/or $Co_3(AsO_4)_2$ – introduced in 1859
	Tyrian purple	synth	indigotin	6,6′-dibromo-indigotin ($C_{16}H_{10}Br_2N_2O_2$) extracted from mollusks (*Murex brandaris*, *Purpura haemostoma*) in antiquity
Black	carbon black	min	C	graphite
	ivory/bone black	synth	C	Ivory or bone powders charred in close vessels
	lamp black	synth	C	Amorphous carbon produced by burning resins, oil, tar, pitch, mostly in lamps
	magnetite	min	Fe^{+2}, Fe^{+3}	Fe_3O_4

2003, Edwards 2005), there is evidence that the Egyptians already produced synthetic powders specifically for cosmetics (Walter *et al.* 1999, Tsoucaris *et al.* 2001). Subsequent civilizations developed sophisticated technologies to prepare especially appealing or chemically resistant pigmenting substances. The long-investigated case of the Maya blue stands as an example of how difficult the interpretation of ancient technologies might be. The start of the fascinating Maya blue story can be dated back to over seventy years ago, when

Merwin (1931) first analysed the blue pigment from the Maya wall paintings at Chichén Itza, Yucatàn. He already noted the uniqueness of the pigment, which was different from all other known blue colourants, and its apparent resistance to time compared with other paints. Gettens (Gettens and Stout 1942, Gettens 1962) raised interest in the material and confirmed its unusual resistance to acid treatment, establishing that the inorganic base was attapulgite, later confirmed to be palygorskite, and soon proposed to be an organo-clay complex (Shepard 1962). Since then, a huge amount of effort has been devoted to understanding the exact nature and properties of the pigment, which is still striking in wall paintings and figurines for its peculiar tinge and brilliant hue (Figs 3.34–3.36). I will not report all the experimental investigations published to date, which amount to over thirty or so, but rather I'll try to summarize the most recent achievement, also obtained using the most advanced techniques including HR-TEM, PIXE, synchrotron XRD, neutron diffraction, XANES, thermal analysis, Raman spectroscopy and many other methods. The readers are invited to browse the reference list of the most recent articles (Chiari et al. 2008, Sánchez del Río et al. 2008) to unravel the tortuous path of the investigation. My understanding of the story at this stage is that:

(a) Maya blue is indeed an organo-clay complex, where the clay is the troublesome palygorskite mineral, having at least two polymorphs (one monoclinic and one orthorhombic: Chisholm 1992), probably a good deal of disorder and stacking faults, and is possibly associated in some samples with sepiolite-type minerals.

(b) The organic part is undoubtedly the indigo molecule extracted from the *Indigo suffruticosa* plant (popularly known as *añil* or *xiuhquilitl*), though careful spectroscopic work has yet to determine whether the molecule is present in the complex as a monomer or a dimer.

(c) The ancient mode of preparation of the pigment out of the clay–plant mixture has been apparently sorted out, first by Constantino Reyes-Valerio (1993) and then replicated (Sánchez del Río et al. 2006).

(d) There seems to be only one Yucatan source of palygorskite for the Mayan pigment (the region around Uxmal: Sánchez del Río et al. 2009), whereas a similar pigment used by the Aztecs must have exploited a different source area, based upon mineralogy.

(e) The complexity of the mineral structure and of the organic–inorganic interaction has baffled researchers for years, though now there is at least a plausible model for the locking of the indigo molecules on the surface of the palygorskite fibres (Chiari et al. 2008). The details of the dynamics of the molecules and of the precise structural arrangement are still to come, and the story is progressing.

Existing collections of past and modern pigment samples stand as invaluable sources of reference materials. Two of the most extensive pigment collections are that at the Getty Conservation Institute, Los Angeles, patiently put together through a number of research projects, and the Forbes' Pigment Collection, assembled by the late Edward Waldo Forbes, former Director of the Fogg Art Museum at Harvard University. Currently, the latter is housed in

3.3 *Pigments* 271

Fig. 3.34. Figurine head from Jaina Island, Campeche, Mexico, still bearing traces of Maya blue pigment. It dates to the Late Classic (ca. AD 600–900) and it is stored at the Field Museum, Chicago, USA.
Original colour picture reproduced as plate 22.

Fig. 3.35. Wall painting from Templo Rojo of Cacaxtla, Tlaxcala, Mexico, showing ample use of the Maya blue pigment. (Image courtesy of G. Chiari, GCI)

the Straus Center for Conservation at Harvard University, while Forbes' private collection of pigments resides at the Institute for Fine Arts Conservation Center at New York University. It contains over 1000 colourants and pigments (Carriveau *et al.* 1984). Table 3.9 lists some of the most commonly used ancient inorganic pigments. As an introduction to colouring substances, the colour-of-art (http://www.artiscreation.com/Color_index_names.html) and pigments-in-paintings (http://www.webexhibits.org/pigments/) websites contain very useful information on most commercially-available pigments,

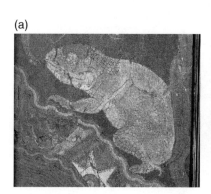

Fig. 3.36. Enlarged figures in the wall painting of Templo Rojo of Cacaxtla, Tlaxcala, Mexico: (a) toad (photo M. Zabé), (b) Quetzal, sacred bird of the Maya (image courtesy of G. Chiari, GCI).
Original colour pictures reproduced as plate 23.

their codes and historical names. The historical part can subsequently be completed by perusal of the excellent classic volume by Gettens and Stout (1942).

It is worth reminding ourselves that pigment analysis is a fascinating subject of research (Chiari and Scott 2004), and that just as every piece of artwork is unique, so are the materials used to produce it. Each detailed investigation is bound to carry surprises and deliver new challenges. Figure 3.37 shows both sides of a recently investigated figurine carved out of antler. The so-called Venus of Gaban (Pedrotti 2009), from the Alpine shelter where it was found, is dated as Neolithic and, very interestingly, the XRD analysis showed that the ochre pigment (reddish iron oxides) was applied over a layer of finely ground calcite. It appears to be one of the earliest known examples of a preparatory substrate made expressly to fill the bone rugosity and obtain a smooth surface for painting.

Moreover, would you expect to find a real butterfly used as a pigment in ancient paintings? Check the paper by Berthier *et al.* (2008) to find out how this was actually done, and maybe you'll be eager to know why butterfly wings are so coloured...

Pigment binder

The presence of organic binders in the pigment layer, such as proteins, lipids, or gums is commonly assessed using the invasive or micro-invasive techniques of organic chemistry (FTIR, GC-MS, HR-LC, ELISA), though an array of advanced non-invasive in situ techniques are being developed, such as fluo-

 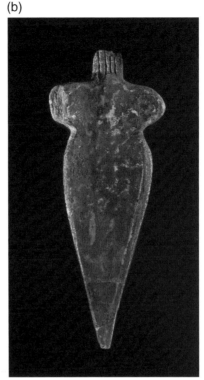

Fig. 3.37. The so-called "Venere del Gaban" (Pedrotti 2009). A small Neolithic female figurine carved out of antler and painted on both sides with red ochre pigment. Interestingly on the back (b) the ochre has been laid over preparatory layer of microcrystalline calcite. It was found in the Gaban repair, near Trento, Italy. (Courtesy of P. Chistè, University of Trento)
Original colour picture reproduced as plate 24

rescence life-time imaging (FLIm: Comelli *et al.* 2004) and fibre optics fluorescence spectroscopy (FOFS: Clementi *et al.* 2006). This is the subject of an ongoing international project coordinated by the Getty Conservation Institute (*Organic Materials in Wall Paintings*, Piquè 2005; http://www.getty.edu/conservation/science/omwp/index.html) and it is undergoing rapid instrumental and methodological developments. The combination of FTIR and X-ray microbeams at synchrotron sources might be a powerful tool for the complementary analysis of organic and inorganic components in the painting layers (Cotte *et al.* 2009).

Pigment degradation and conservation

The equilibrium or non-equilibrium state of the material requires of course the identification of pigment phases and the knowledge of their thermodynamic stability in the actual condition of the artwork. Since the paint layers are generally very thin, the complete interpretation of the pigment stratigraphy and the interpretation of the acting chemical processes require an accurate spatial mapping of the phase particles, and also of their chemical state. Very often the pigments involved contain elements with mixed-valence states, or cations that are prone to redox reactions. Since many of the fundamental light-absorption mechanisms producing colour depend on electron transfer, the intrinsic cause of colour is often related to the possibility of easy changes of the oxidation state of the involved cations (See also: "Photochemical ageing", Section 2.5.2.2). The mapping of the redox state of the elements (Section 2.1.3) is not a trivial experiment and can be performed only by the use of valence-sensitive spectroscopies (XAS, XPS, EELS: see Boxes 3.f, 3.g, 3.h) coupled with microbeam probes. Synchrotron radiation is virtually the only source that provides at the same time enough photon flux, microbeam collimation and energy tunability (Box 2.a). Furthermore, very often what is left in the pigmenting layer is not the original material, but the altered products. The search for the original pigment and the interpretation of the degradation process may indeed be a difficult task (Chiari and Scott 2004). The identification of the pigment, the interpretation of the active processes, and the careful characterization and imaging of the actual state of the painted layers are of great technical help in developing conservation strategies (Cather 1991, 2003).

Two excellent case studies are briefly reported. The first is the detailed synchrotron-based study of the degraded cadmium yellow (CdS) in the paintings by James Ensor (Van der Snickt et al. 2009), conclusively showing the presence of small globules of hydrous cadmium sulphate ($CdSO_4 \cdot H_2O$), possibly produced by local dissolution and reprecipitation processes induced by the wrong conservation microclimate. The second case concerns the progressive blackening of the red cinnabar paintings in the walls of Pompeii (Cotte et al. 2006). The assumed isochemical conversion of red cinnabar (HgS) into black metacinnabar is found to be inconsistent with the observations at the microscale, where the newly formed black layer coexists with the unaltered cinnabar. The formation of the dark layer is tentatively attributed to a pollution-induced process similar to that producing the typical black crusts on marbles and limestones (Section 3.1.1).

It should be noted that for many pigments the exact historical use is known, even the preferences of many painters for specific pigments or mixtures. The analysis of pigments may therefore greatly help in recognizing the original layers with respect to retouches or later repaintings. Authentication can also be supported by technical analysis (Ainsworth 2005). Incidentally, the short-lived ^{210}Pb isotope of lead (half-life = 22.3 y) formed by ^{222}Rn has been proposed as a marker of Pb-containing pigments: pigments that are older than 100–150 y should be radioactively dead, whereas modern fakes should contain younger active lead (Keisch et al. 1967).

Box 3.e The measurement of colour

Colour is an essential property of matter. The physical mechanisms of colour generation are complex (Nassau 1983, 1998), and the visible colour is the sum of several phenomena contributing to it. Spectral absorption by electron transfer is the most common mechanism (Section 2.1, Fig. 2.f.1), though surface diffusion, inter-particle interference, molecular motion, defect trapping and several other physical processes can contribute to the modulation of the observed visible spectrum. The human eye is sensitive to chromatic components (i.e. the wavelengths of the reflected light) and to brightness (i.e. the black/white perception).

Each stable pigment must have a well-defined chemical composition and structure in order to reproduce the colour under actual conditions, i.e. defined lighting source and observer's eye or detector. The pigment may vary the colour depending on the way it is prepared: particle size, particle shape, chemical impurities, structural defects, etc. significantly contribute to the absorption or scattering part of the spectrum in the visible range. By decreasing the grain size of a powder, for example, it is possible to decrease the colour saturation, because there is more white component due to the increased scattering surface of the particles and vice versa: if we wet the surface of the particles with water or another transparent coating, the colour saturation is increased because of the larger absorption related to multiple reflection. As an example of experimentally observed curves, the spectral signature of several non-white pigments proposed for cool surfaces (i.e. surfaces with both high thermal emittance and high solar and near-infrared reflectance) that can be used for energy-saving coloured roofs is reported in the LBNL database (http://coolcolors.lbl.gov/LBNL-Pigment-Database/database.html). Since electron and atomic mobility increase with temperature, it is difficult to maintain the colour in a material that must be processed at high temperature. The production of high temperature pigments is a notorious problem for industrial processes, for example ceramic tile coating. No organic pigment or dye will stand temperatures above 500 °C, so high temperature coatings must rely on inorganic substances.

The range of visible electromagnetic wavelengths is divided into colour regions (Table 3.e.1). The colour that is actually perceived by the human eye is the sum of the wavelengths reflected by the object's surface, and it is a complex function of the scattering and absorption coefficients. Mathematically this is usually treated by the use of the Kubelka-Munk theory (Nobbs 1985). The sum of all the visible waves is perceived as white, whereas the absence of emission due to absorption of all visible waves is perceived as black.

Table 3.e.1. The wavelength range of the colours of the visible light spectrum

Colour	Wavelength (nm)	Frequency (THz)
Red	700–635	430–480
Orange	635–590	480–510
Yellow	590–560	510–540
Green	560–490	540–610
Blue	490–450	610–670
Violet	450–400	670–750

The quantitative measurement of the intensity–energy relationship in any part of the electromagnetic spectrum is called **spectrophotometry** (Section 2.1). The specific case of measurements in the visible part of the spectrum (Table 3.e.1) is called **colorimetry** (Bacci 2004) because it is intimately related to our perception of colour. When a reflectance spectrum from an object's surface is recorded under standard operative conditions (Box 2.o), the curve can be used for the quantitative measurement of colour and for identification of the mechanism (and therefore the pigment) producing it.

(cont.)

Box 3.e (*Continued*)

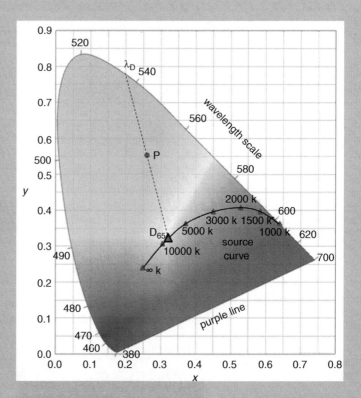

Fig. 3.e.1. CIE chromaticity diagram. It is the most versatile diagram to express colour components. The source curve is the locus of the possible source points, depending on the temperature; the standard D_{65} average daylight illumination is marked with the red triangle. P(x,y) are the chromaticity coordinates of any measured colour, and the λ_D value obtained by intersection of the D_{65}–P line on the upper boundary of the diagram is the dominant wavelength. Original colour picture reproduced as plate 54.

The reference guidelines for the measurement and representation of colour are provided by the series of documents approved in the last century by the International Commission on Illumination (Commission Internationale de l'Eclairage, CIE: http://www.cie.co.at/index_ie.html) and continuously updated. The CIE standard implies: (1) a reference illumination source, (2) well-defined measurement conditions, and (3) univocal calculation of colour coordinates on the chromaticity diagram (Fig. 3.e.1). The most commonly used lighting source is D_{65}, i.e. the average daylight illumination with a temperature of 6500K, though other illuminant curves are available, such as the A curve (tungsten-filament lamp with a colour temperature of 2854K) or the B curve (model of noon sunlight with a temperature of 4800K). The reference D_{65} curve can be downloaded from the CIE site (http://www.cie.co.at/publ/abst/datatables15_2004/sid65.txt). The tristimulus values used to define colour are defined as:

$$X = k\int \beta(\lambda)S(\lambda)\bar{x}(\lambda)d\lambda, \quad Y = k\int \beta(\lambda)S(\lambda)\bar{y}(\lambda)d\lambda, \quad Z = k\int \beta(\lambda)S(\lambda)\bar{z}(\lambda)d\lambda$$

where $\beta(\lambda)$ is the measured reflectance spectrum, $S(\lambda)$ is the spectra power of the source, and $\bar{x}, \bar{y}, \bar{z}$ are the colour matching functions taking into account the observer's eye response (http://www.cie.co.at/publ/abst/datatables15_2004/CIE_sel_colorimetric_tables.xls). From the measured data and the tristimulus equations above, it is possible to calculate the chromaticity coordinates (Fig. 3.e.1):

$$x = \frac{X}{X+Y+Z}, \quad y = \frac{Y}{X+Y+Z}, \quad z = 1-x-y$$

Every point P(x,y) in Fig. 3.e.1 corresponds to a unique colour. The dominant wavelength (λ_D) can be found by drawing a line from the source (D$_{65}$) through the point P to the wavelength-scale upper boundary. If the point lies in the region below the source line (i.e. the purple–red region), the intercept along the purple line is not a real wavelength, and the colour is characterized by the complementary wavelength ($\lambda_C = -\lambda_D$). The chromaticity diagram is important to compare the colour of different materials, for example industrial products or the colour compatibility of restoration materials. The colour changes during fading and accelerated tests (Section 2.5.2.2) are also quantified using the CIE diagram. In the so-called 1976 CIE L*a*b* Space (CIELAB) the colour difference (ΔE) between two samples or between the measurements of the same sample at different times (micro-fading tests, Box 2.q) is defined as:

$$\Delta E = [(\Delta a^*)^2 + (\Delta b^*)^2 + (\Delta L^*)^2]^{\frac{1}{2}}$$

with

$$a^* = 500\left[\left(\frac{X}{X_0}\right)^{\frac{1}{3}} - \left(\frac{Y}{Y_0}\right)^{\frac{1}{3}}\right], \quad b^* = 200\left[\left(\frac{Y}{Y_0}\right)^{\frac{1}{3}} - \left(\frac{Z}{Z_0}\right)^{\frac{1}{3}}\right], \quad L^* = 25\left[100\left(\frac{Y}{Y_0}\right)^{\frac{1}{3}}\right] - 16$$

where $X_0 = 95.017$, $Y_0 = 100$, $Z_0 = 108.842$ are the nominal values for white using D$_{65}$. Figure 3.e.2 clearly shows that the CIELAB system separates brightness (L* axis) from the two opponent chromatic red–green and

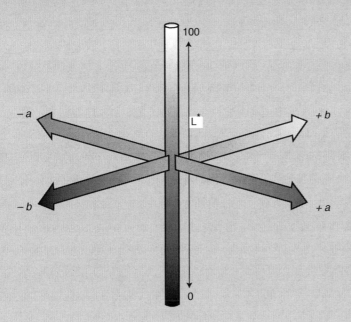

Fig. 3.e.2. The CIELAB coordinate systems, separating the L* brightness axis from the two antagonistic chromatic systems: red-green (a^*) and yellow-blue (b^*). Original colour picture reproduced as plate 55.

(*cont.*)

Box 3.e (Continued)

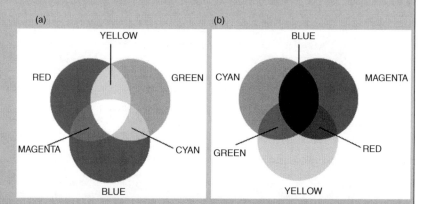

Fig. 3.e.3. The (a) red-green-blue (RGB) and (b) cyan-magenta-yellow (CMY) colour codes. Both models fall short of reproducing all the colours we can see, and can map just a small subset of the CIE chromaticity diagram. Furthermore, they differ to such an extent that there are many RGB colours that cannot be produced using CMY and, similarly, there are some CMY colours that cannot be produced using RGB.
Original colour picture reproduced as plate 56.

yellow–blue systems (a^* and b^* axes, respectively). These chromatic parameters seem to be directly related to the retinal colour stimuli of the eye, which during the nervous transmission to the brain are actually translated into distinctions between light and dark, red and green and blue and yellow. CIELAB has become very important for desktop colour. Like all CIE models, it is device-independent, unlike the RGB and CMY systems that are widely used for printers (Fig. 3.e.3), it is the basic colour model in Adobe PostScript© (level 2 and level 3), and it is used for colour management as the device-independent model of the ICC (International Color Consortium) device profiles.

Other more visual systems of colour evaluation based on charts are much diffused for practical and field use, the most common ones being based on the Munsell colour system (Munsell 1905, http://www.xrite.com/top_munsell.aspx). Although dating back to the end of the 19th century and devised more by intuition than exact science, it is still an internationally accepted, leading colour system (Nickerson 1976). Commercially available Munsell tables can be used to define the colour of a sample by visual comparison. In geology, sedimentology and archaeology, a subset of the Munsell charts are currently used to estimate the colour and properties of soils (Melville and Atkinson 1985). Of course the Munsell values can be converted to the CIE international system (McCamy 1992).

For in-depth information

Hunt R.W.G. (1998) Measuring colour. 3rd Revised Edition. Fountain Press Ltd, Kingston Upon Thames, UK.

Nassau K. (1983) The physics and chemistry of color. The fifteen causes of color. John Wiley & Sons, New York.

Nassau K. (ed.) (1998) Color for science, art, and technology. Elsevier, Amsterdam.

Wyszecki G. and Stiles W.S. (2000) Color science: concepts and methods, quantitative data and formulae. 2nd Edition. Wiley Series in Pure and Applied Optics. Wiley-Blackwell, New York.

3.4 Glass and faience

Glass is defined as a solid material that does not have long-range order in the atomic arrangement, as opposed to crystalline solids, which possesses ordered atomic configurations on a lattice (Giacovazzo 2002). Independent of composition, the term glass can be extended to all **amorphous** (i.e. non-crystalline) solids (Doremus 1994, Shelby 2005). In glass molecular groups (i.e. [SiO$_4$], [PO$_4$], [BO$_3$], etc.) and ions locally have a well-defined chemical and geometrical configuration, exactly as in the crystal structure, though we cannot define periodic lattice vectors and translational symmetry (Fig. 3.38) (Zachariasen 1932, Rao 2002). Because of the configurational entropy contribution, glass has a higher Gibbs free energy than a solid with the same composition. Glass therefore tends to transform into the crystalline form with time in order to lower its free energy; the process is called devitrification. Since for common glass compositions the ordering process requires the breaking of covalent bonds, the kinetics of transformation at ambient temperature is very slow, and the glass appears thermodynamically stable over the human lifecycle, exactly like diamonds that are stable at non-ambient pressures. Glass and diamonds are actually metastable solids.

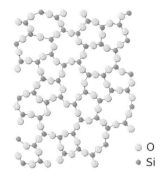

Fig. 3.38. Zachariasen (1932) model for the structure of silica glass. Each [SiO$_4$] tetrahedron complies with local polyhedral symmetry, though there is no sign of long-range lattice periodicity.

Glass can also be considered to be an undercooled liquid, i.e. a liquid that has solidified below melting temperature but did not have enough time to reach an ordered configuration because of the slow diffusion kinetics of the atoms. Glass therefore is produced by rapid cooling from the melt, compared with the crystallization time. There is no critical temperature for the transformation, rather there is a temperature region (defined as T_g or the **glass transition temperature**) where the thermodynamic variables (volume, enthalpy, and entropy) change continuously without showing sharp discontinuities. The temperature interval between the lower point showing the enthalpy of the equilibrium liquid and the upper point showing the enthalpy of the frozen solid is called the **glass transformation region**. It is to be noted that, though ancient glasses were always produced by high-temperature melting technology, nowadays there are alternative ways of producing glass materials at low temperature, such as vapour deposition, sol–gel processing of solutions, and neutron irradiation of crystalline solids.

The complete definition of glass is therefore that of an amorphous solid, completely lacking long-range periodicity, and exhibiting a region of glass transformation behaviour (Doremus 1994, Shelby 2005). From the point of view of composition, there are few oxides that can form a continuous network of oxygen-sharing polyhedra (Fig. 3.38) and at the same time exhibit glass behaviour. They are the oxides of 3- or 4-coordinated cations: B$_2$O$_3$, SiO$_2$, GeO$_2$, P$_2$O$_5$. The fact that only the cations that like to be surrounded by a small number of oxygens (triangular or tetrahedral coordination) form glass was initially rationalized by Zachariasen (1932), following the earlier ideas of Golschmidt and Pauling, by the so-called **random network theory**, which is still much used today to illustrate simply and understand the structure of glass (Rao 2002). The essential principles are that (a) the cations must be surrounded by 3 or 4 oxygens, (b) the coordination polyhedra must be connected only by vertices (no edge or face sharing), and (c) part of the oxygens must connect two cations, in order to form a continuous network. Further, the character of the cation–oxygen bonds must not be totally covalent (i.e. too rigid), nor totally ionic (i.e. no directionality). The cations forming good glass networks should have nearly 50% of covalent bond character. B, Si, Ge, and P are the cations that form oxides satisfying these conditions, and therefore they are natural **network formers**. Nowadays we can also form glass by rapid

Fig. 3.39. Moldavite: a tektite. Tektites are natural glasses ejected at large distances from meteoritic impacts. (Image available at: http://gemcompendium.blogspot.com/2008/08/moldavite-and-tektites.html)
Original colour picture reproduced as plate 25.

quenching or vapour deposition using the oxides of Bi, As, Sb, Te, Al, Ga and V. It is also possible to produce chalcogenide glasses of S, Se, and Te, and halide glasses of BeF_2 and ZrF_4. However these are all very recent innovations in materials science, and in nature and throughout history only silica-based glasses are found. For the most part, glass science is the mineralogy, chemistry, and physics of silica (Heaney et al. 1994).

Natural glasses are formed at high temperature by rapid quenching of a silica-rich melt. There are three main geologic processes to form glass: rapid cooling of a lava flow (obsidian, pumice), jet cooling of ejecta from a meteoritic impact (tektites, Fig. 3.39), and quenching of the molten layer at the site of the shock impact (impactite, desert glass, Fig. 3.40). All three types of natural glass have been used by humans one way or another. As examples, let's remind ourselves that obsidian was one of the best cutting tools in prehistory (Section 3.1.1), and polished fragments of the yellow impactite (desert glass, Giuli et al. 2003 and references therein) found in abundance in the Egyptian Western Desert has been shown to be used as gem in Tutankhamun's pectoral (De Michele 1998), in addition to being widely used for the manufacturing of small blades in prehistoric times (Fig. 3.41).

Fig. 3.40. Libyan desert glass: a light coloured impact glass found in the Western Desert between Egypt and Libya. This material has been used to carve the scarab present in Tutankhamun's pectoral (De Michele 1998). (Image available at: http://gemcompendium.blogspot.com/2008/08/moldavite-and-tektites.html)
Original colour picture reproduced as plate 26.

Since it is virtually impossible with old pyrotechnologies (Section 3.0.1) to reach the melting point of quartz (1610 °C) or of the high temperature polymorphs (trydimite 1703 °C, cristobalite 1723 °C), in order to produce silica glass with ancient technologies, the melting point needs to be substantially lowered. This was (and still is) achieved by adding elements with a very low electronegative character that form highly ionic bonds with oxygen, commonly alkali elements (Na, K) or Pb. These elements never act as network formers, but their strong interaction with oxygen modifies existing networks by compensating the anion charge and thus distorting the network configuration. They are called **network modifiers**. The oxides favouring glass formation are also called

Fig. 3.41. Prehistoric blades made of desert glass. (Image available at: http://gemcompendium.blogspot.com/2008/08/moldavite-and-tektites.html)

fluxants. Table 3.10 shows how the working temperature ranges are modified by the addition of network modifiers. The fact that monovalent cations (Na^+, K^+) can and are used to lower the melting point of silica has important consequences for the physical properties of glass, including the viscosity and the resistance to leaching. Silica glass that only contains alkali cations is easily degraded, and it must be stabilized by cations with a higher ionic strength (i.e. higher charge/radius ratio), most commonly alkaline earths (Mg^{+2}, Ca^{+2}).

Some other elements with intermediate electronegative character between the network formers and the modifiers are called **intermediates**. Intermediate elements such Al, Ti, Zr, never act as network formers, though they can substitute Si in the tetrahedral sites of existing networks. Finally, the only other intentional components in the formulation of ancient glass are the colourants, commonly oxides of the transition elements (V, Cr, Mn, Fe, Co, Ni, Cu) with partially filled 3d orbitals or rare earth elements (lanthanides, actinides) with partially filled f orbitals (Nassau 1983). Table 3.11 lists the major components used today and in the past for glass formulation.

The value of glass and glass technology to humans is immense. In modern times only plastic has had a similar impact on everyday life. In order to appreciate it let's try to imagine the modern world without glass: no transparent

Table 3.10. Approximate reference temperatures for glass working. Viscosity is expressed in SI unit system (Pa·s = pascal·second = 10 poise). The working range is the temperature interval between the working point and the softening point (modified from Holloway 1973, Shelby 2005)

Reference point	Viscosity (Pa·s)	Definition	Pure silica (°C)	Soda-lime silica glass (°C)	High-lead glass (°C)
Melting point	1–10	Solid–liquid transition	1610	1400	1100
Working point	10^3	Sufficiently soft for shaping	1560	1000	850
Softening point	$10^{6.6}$	Glass bar bends in a flame	1500	700	500
Annealing point	10^{12}–$10^{12.4}$	Internal stress removed, grain sintering	900	500	300

282 *Materials and case studies*

Table 3.11. Summary of the glass-forming properties and functions of commonly used elements and oxides

Function	Ancient glass	Modern glass	
		Main components	Special components
Network formers (glass former)	SiO_2	B_2O_3, SiO_2, P_2O_5	Al_2O_3, GeO_2, Bi_2O_3, As_2O_3, Sb_2O_3, TeO_2, V_2O_5
Network modifiers (fluxants)	Na_2O, K_2O, (PbO)	Na_2O, K_2O, PbO	Li_2O, Rb_2O, Cs_2O
Property modifiers (stabilizers)	MgO, CaO	MgO, CaO	Al_2O_3
Colourants	Cu, Fe, Co, Sb, Ag, Au, (Mn)	V, Cr, Mn, Fe, Co, Ni, Cu, Au, Ag, REE	decolourants
Fining agents (bubble removal)			NaCl, CaF_2, NaF, Na_3AlF_6

windows, no car windshields, no optical lenses, no mirrors, no electrical bulbs, no network cables, no chemical glassware, no cameras... and in the pre-industrial era the situation would be even worse: no bottles, no drinking vessels, no impermeable containers for long term storage, no spectacles, no cathedral stained windows... All this ought to make us cherish the early efforts in developing glass-making technologies, and properly prize the long intimate association between mankind and glass (Marshall 1990), from the

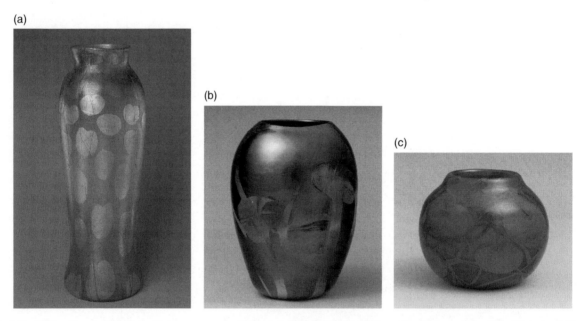

Fig. 3.42. Tiffany Glass and Decorating Company: vase, ca. 1900, blown favrile glass (The Metropolitan Museum of Art, Gift of Robert and Gladys Koch, 1999 (1999.412.1)), (b) vase, ca. 1903, favrile glass (The Metropolitan Museum of Art, Gift of Louis Comfort Tiffany Foundation, 1951 (51.121.8)), (c) vase, ca. 1903, favrile glass (The Metropolitan Museum of Art, Gift of Louis Comfort Tiffany Foundation, 1951 (51.121.18)). (The images were provided by The Metropolitan Museum of Art through the Images for Academic Publishing IAP initiative.) Original colour pictures reproduced as plate 27.

early use of natural obsidian to modern glass art. To come a full circle in time, the so-called *favrile iridescent glass*, a type of art glass patented in 1894 by Louis Comfort Tiffany, is a nice example of modern glass whose technology was stimulated by the desire to imitate old Roman and Syrian glass vessels, by Tiffany's own admission. The iridescent effects were obtained by mixing different colours of glass together while hot (Fig. 3.42). Glass as a versatile material obviously can satisfy the most diverse religious or artistic inspirations (Figs. 3.43, 3.44).

To define an investigative approach to the multifaceted materials science of glass in cultural heritage, we should try to answer a few basic questions. (1) What is the nature and composition of the vitreous materials used at different times? (2) What were the source materials and the manufacturing techniques? (3) What are the alteration and degradation processes, and the conservation strategies? (4) What are the techniques that should be used to answer these questions?

Fig. 3.43. Face-shaped pendants were spread throughout the entire Mediterranean Sea region by the Phoenicians from the 6th century BC The large eyes are remarkable, and lead one to believe that the beads were amulets that served to ward off the evil eye. 5th – 4th centuries BC; originally from the Easter Mediterranean Sea Region or Carthago. (Dutch National Museum of Antiquities, Rijksmuseum van Oudheden, Leide, Holland)
Original colour picture reproduced as plate 28.

3.4.1 The nature and composition of vitreous materials

Having defined the nature, structure, and composition of glass from the point of view of materials science, it is important to understand that common glass is easy to recognize: it is hard, transparent to light and fragile. It has irregular conchoidal fracture with sharp edges. It can be moulded to any shape at high temperature. However, historical and archaeological vitreous matter encompass a wide variety of very different and composite materials, variously defined as faience, frits, vitreous paste, enamels, glazes, and so on. Some terms have a precise technological meaning, others are often used inappropriately and are confusing. Much of the confusion stems from the fact that many of the vitreous artefacts have not been properly characterized by scientific means. Let's try to sort things out:

1) To make glass we start from natural crystalline materials, mostly quartz as a network former and alkali elements as network modifiers. If we complete the glass-forming process by high temperature melting and subsequent quenching, then we obtain a true **homogeneous glass**.

2) It is difficult to produce homogeneous glass, especially with primitive pyrotechnologies, and therefore the glass-making process is usually carried out in steps, by grinding the heated material, mixing it, and then reheating it at high temperature. The steps are repeated several times, each time obtaining a sintered, partially molten material, which is called **frit**. The term frit in the strict sense should be referred to materials obtained during the first stages of glass-making. Some pigmenting materials produced in ancient Egypt (Egyptian blue and Egyptian green, Table 3.9), are sometimes inappropriately called frits, because they are multicomponent materials produced in a similar way (Tite and Shortland 2008).

3) Some glass contains a certain amount of crystals (Artioli *et al.* 2008). This may be due to incomplete melting of the starting raw materials left unreacted during the glass production process, to crystalline phases deliberately or accidentally formed during manufacturing (i.e.

Fig. 3.44. Modern art exploiting the versatile properties of glass. (Cathy Lybarger, Aardvark Art Glass, http://aardvarkartglass.net/about_us.htm)
Original colour picture reproduced as plate 29.

Fig. 3.45. (a) Eyed glass bead from Sardinia, (b) SEM backscattered electron image showing a very homogeneous glass matrix with few crystalline and metal inclusions. (Courtesy of I. Angelini, University of Padova; sample from the Museo Nazionale Archeologico di Nuoro, courtesy of M. A. Fadda, Soprintendenza per i Beni Archeologici per le province di Sassari e Nuoro)
Original colour pictures reproduced as plate 30.

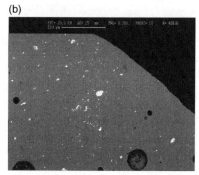

crystallization during slow cooling), or to crystalline phases added in the last stages of glass working to change the macroscopic properties of the final material (i.e. opacifiers). Depending on the relative proportion of glass matrix and crystalline components we may distinguish at least three classes of materials: (a) fairly homogeneous **glass** with some or no crystalline inclusions (Fig. 3.45), (b) materials with similar proportions of glass and crystals, sometimes defined as **glassy faiences** (Angelini *et al.* 2004) (Fig. 3.46), and (c) materials with a small layer of glass covering a bulk of sintered crystals (Fig. 3.47). These are called **faiences** (Tite and Shortland 2008). When the relative volume fractions of the glass matrix (V_G) and the crystalline components (V_C)

Fig 3.46. Middle Bronze Age conical buttons made of glassy faience are peculiar objects and materials produced in Northern–Central Italy: (a) bright blue conical button from Mercurago, Italy, (b) SEM backscattered electron image showing abundant glass matrix (light grey), and a nearly equal proportion of quartz grains (dark grey). (Courtesy of I. Angelini, University of Padova; sample courtesy of M. Venturino, Soprintendenza per i Beni Archeologici del Piemonte e del Museo Antichità Egizie)
Original colour pictures reproduced as plate 31.

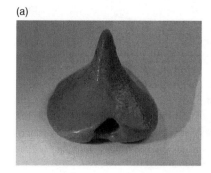
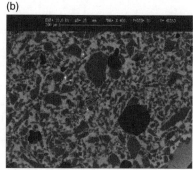

Fig 3.47. Example of a typical Early Bronze Age faience bead from Lavagnone, Italy (sector B, us 776, Early Bronze Age I C, ca. 1900–1800 AC; De Marinis 2002): (a) pale blue/green segmented bead, (b) SEM backscattered electron image showing scarce glass matrix (light grey), a number of quartz grains (dark grey), and a very porous texture. (Courtesy of I. Angelini, University of Padova; sample courtesy of R.C. De Marinis, University of Milano)
Original colour pictures reproduced as plate 32.

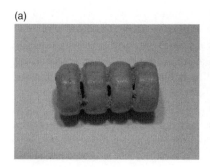
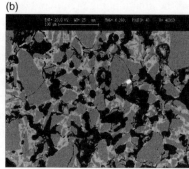

are quantified, usually by image analysis (Box 2.q), it is possible to propose tentative limits between the three materials classes (glass = $X_M = V_G/(V_G + V_C) > 0.75$; glassy faience = $0.40 < X_M < 0.60$; faience = $X_M < 0.25$; Artioli and Angelini 2009). It is important to remark that these three groups of materials have been produced with entirely different technologies and scopes.

4) When a thin layer of glass is applied to the surface of a different material, such as stone or structural ceramics (Table 3.2, see also Section 3.1.2.3), it is called a **glaze**. Glazing is performed to change the properties of the material, for example to make it impermeable to liquids, more resistant to scratch, or more aesthetically appealing. Glazes normally contain high content of Pb to get a melting point that is lower than that of the ceramic or glass substrate (Table 3.10). When the glaze contains a thin layer of metal particles (usually silver and/or copper), special chromatic effects are obtained and the glaze is called **lustre** (Figs 1.5, 2.i.3, 2.l.4).

5) If the glass is deliberately produced in coloured form, ground to a fine powder, applied to a metallic surface and then fused to the substrate by firing to about 750–850 °C, it is called **enamel**. Enamelling produces a smooth, durable and intensely coloured surface and is mostly used for decoration. It has also been applied to glass objects (enamelled glasses) and ceramics (enamelled pottery, see Section 3.1.2.3). Powdered blue glass was also used with organic binders to produce pigments for paintings (smalt) (Table 3.9).

6) **Mosaics** are made of very different and often heterogeneous materials. The Romans had a flourishing production of glass especially for mosaic tesserae, and in later Byzantine and Middle Age mosaic recycled Roman glass is often found. Thick-coloured enamel is often used to produce mosaic glass; the opacity is obtained by in situ crystallization of the opacifiers, especially antimonates (Lahlil et al. 2008). To obtain tesserae with a special gilded colour or metallic effects, gold and silver foils were at times laid on a glass substrate and then embodied in transparent glass by double firing in a kiln.

Ancient glass is generally classified into compositional groups, mostly based on the relative amount of fluxant (Na, K, Pb) and stabilizing elements (Mg, Ca). The compositional groups correspond to chronologically or geographically-distinct glass productions, and reflect the use of different components in the glass formulation. Sayre and Smith (1961) already distinguished several compositional groups, based on the chemical and archaeological evidence available at the time. Their observations are still largely valid (Henderson 2000, 2001), though subsequent analytical data allowed refinement of the distribution and evolution of early vitreous materials. Table 3.12 reports the recognized compositional groups of ancient glass. Modern ordinary soda–lime silica glass is very similar to standard Roman glass.

Concerning the history of vitreous materials, the earliest available evidences are glazed stones (quartz and steatite) produced in Egypt, the Near East, and the Indus Valley during the 5th millennium BC, (Tite and Bimson 1989, Barthélemy de Saizieu and Bouquillon 2001, Tite and Shortland 2008). Faience production

Table 3.12. Major compositional groups of archaeological and historic silica-based glass. Modified from Sayre and Smith (1961) and Henderson (2000, 2001). The significant chemical signatures distinctive of the group are in bold

Glass type	Chemical signature (oxide wt%)					Note	Alkali source
	Na$_2$O	K$_2$O	MgO	CaO	Other		
HMG	8–20	**0–3**	**2–10**	3–10		**High-Mg glass** – Near East (1500–800 BC)	plant ash
LMHK	0–8	**4–18**	**0–1**	0–4		**Low-Mg High-K glass** – Europe (Bronze Age)	plant ash
LMG (SLS)	13–20	**0–1**	**0–1**	5–10	0–1 Fe$_2$O$_3$ 0–2 MnO	**Low-Mg glass** – Near East (after 800 BC) and standard Roman-type soda-lime silica glass (AD 50–300)	natron salts
HIMT	16–20	0–1	1–2	5–9	**1–3** Fe$_2$O$_3$ **1–3** MnO **0–1** TiO$_2$	**High-Fe-Mn-Ti glass** – Mediterranean area and North-Western provinces (AD 300–500)	natron salts
LEG	10–15	0–1	0–1	8–12	**1–3** Al$_2$O$_3$	**Levantine-type** – near East, Mediterranean area, North-Western provinces (AD 500–800)	natron salts
HSB	15–21	0–1	0–1	4–6	**0–2** Sb$_2$O$_3$	**High-Sb glass** – Near East, Mediterranean area (AD 100–300)	natron salts
HKEG	0–14	**2–14**	0–5	6–20		**High-K Medieval glass** – Europe (stained window glass)	plant ash
HLEG	0–1	**3–10**	1–3	4–16	**20–65** PbO	**High-Pb Medieval glass** – Europe	plant ash
HMEIG	10–18	1–3	**3–7**	6–12	0–1 MnO	**High-Mg Early Islamic glass** (AD 840–1000)	natron/ plant-ash
HLIG	8–10	0–2	0–1	4–5	**30–40** PbO	**High-Pb Islamic glass** (AD 1000–1400)	natron/ plant ash
HLHB	2–7	0–4	0–1	0–3	**15–40** PbO **5–15** BaO	**High-Pb High-Ba glass** – China (Han Dynasty 206 BC – AD 221)	witherite?
HAG	2–12	**4–16**	1–2	2–6	**2–4** Al$_2$O$_3$	**High-Al glass** – India (1st millennium AD) – maybe several groups	plant ash
FDV	12–15	2–4	1–3	4–10		**Venetian "cristallo"** and **Dutch "façon de Venise"** (16th–17th century)	soda-rich ash

started towards the end of the 5th millennium BC in the Near East (Hamoukar and Tell Brak, Syria) and soon after in Egypt (Badarian graves) (Bimson and Freestone 1987, Caubet and Pierrat-Bonnefois 2005, Caubet 2007, Tite and Shortland 2008). The technological connection between glazing technology and copper metallurgy is intriguing (Hauptmann et al. 2001). Egyptian blue and green frits (Tite and Shortland 2008) are by all means pigmenting materials (Table 3.9): they are complex multicomponent materials made by fritting techniques similar to those used for glass, the final composition of Egyptian blue is mainly mineral cuprorivaite ($CaCuSi_4O_{10}$), Egyptian green is a copper rich glass with abundant wollastonite ($CaSiO_3$). They were produced by firing mixtures of quartz, lime, copper compounds, and alkali fluxants. They were used to produce sintered blue and green faience-like objects, or were simply ground and used as pigments with a binder, the latter use continuing at later times in Greece (Kakoulli 2009), in Etruria (Bordignon et al. 2007), and even in medieval Rome (Lazzarini 1982, Gaetani et al. 2004).

Fig. 3.48. An amazing Early Bronze Age parure composed of 2178 faience beads. Otomani Culture, Nižná Myšľa, Slovakia (Furmánek and Pieta 1985, Bátora 1995). The LMHK faience beads have been analysed by Angelini et al. (2006).

By the end of the 3rd millennium BC, faience production from Egypt and the Near East spread north- and west-wards to Minoan Crete, and then to Cyprus, Rhodes, and Mycenaean Greece (Mirtsou et al. 2001, Panagiotaki et al. 2004, Jackson and Wager 2008). By the first half of the 2nd millennium BC there is evidence of faience in the Northern Caucasus and Caspian region (Shortland et al. 2007), Poland, Slovakia (Fig. 3.48), Italy, Switzerland (Angelini et al. 2006, Tite and Shortland 2008) and up to France and Britain. The earliest glazed clay objects (pottery, bricks) are from the Mesopotamian area (Nuzi, Tell Atchana) and dated to 15th–14th century BC (Moorey 1994, Paynter 2009). The clay glazing technology seem to have derived from the glass-making industry, and it evolved to such splendour, producing amazing wonders like the glazed Babylon walls and the Ishtar gate, now at the Pergamon Museum in Berlin. Glassy faience (LMHK composition) seems to be a European material, essentially produced in Northern and Central Italy in the mid-2nd millennium BC to make the typical conical buttons (Fig. 3.46, Artioli and Angelini 2009). Texturally similar materials but with different composition are found later in Southern Italy, the Aegean and Egypt, including the Uluburun shipwreck.

Glass beads in Egypt (Qau, Badari) and Mesopotamia (Megiddo, Ajjûl) are reported starting from the end of the 3rd to the beginning of the 2nd millennium BC (Lilyquist 1993, Lilyquist and Brill 1993, Moorey 1994), then after about 1500 BC an advanced production of coloured glass took place in Amarna (Nicholson and Shaw 2000, Shortland 2000) yielding amazing pieces such as the famous polychrome tilapia fish (Fig. 3.49, Shortland 2009). Detailed information on Egyptian glass-making techniques in Ramesside times stemmed from the investigation of the evidence excavated at Qantir-Piramesses, in the Eastern Nile Delta (Rehren et al. 2001, Rehren and Pusch 2005). Mainly because of the crucibles found in Amarna, and the matching glass ingots found in the Uluburun shipwreck off the Turkey coast (Nicholson et al. 1997), there is an emerging picture of a very dynamic glass industry production and trade in the Eastern Mediterranean in the Late Bronze Age (Jackson 2005). Reflections of these flourishing exchanges are found in the European glass composition of the Middle and Recent Bronze Age, which neatly corresponds to the Aegean and Near Eastern HMG glass signature

Fig. 3.49. The famous polychrome-glass tilapia fish produced at Amarna, in Middle Egypt (Shortland 2009) and conserved at the British Museum in London (Inv. EA 55193). It is one of the most well known ancient glass objects. It is dated to the Reign of Akhenaten, 1353–1336 BC.
Original colour picture reproduced as plate 33.

Fig. 3.50. Blue balsam container in the shape of a dove, from Groppello Cairoli (first decades of the 1st century AD). Origin: Necropolis, Zone B, Sector I, cremation Tomb V, site Marone, Groppello Cairoli, Pavia, Italia. Musei Civici del Castello Visconteo, Pavia. (inv. St. 7373).
Original colour picture reproduced as plate 34.

(Artioli and Angelini 2009). A sharp change is evident in the European record starting from the Final Bronze Age, where glass with a distinct LMHK signature appears, probably reflecting the local production associated with the site of Frattesina, Italy, which is still the only known Bronze Age glass-producing site in Europe (Henderson 1988, Towle *et al.* 2001, Angelini *et al.* 2004).

Slightly after the Bronze Age–Iron Age transition there are further compositional and technological changes, the most important being the introduction of mineral sources (natron, see Section 3.4.2) for the alkali elements. This innovation induces a gradual shift towards Mg-free and K-free compositions (Table 3.12), and the standardization of the Roman-type soda-lime silica glass (SLS) composition that will continue almost unchanged through the Roman Empire until the beginning of the Middle Ages. Very interesting artefacts, such as large vessels with carved motives and peculiar **mosaic glasses**, are produced in the Hellenistic world after the death of Alexander the Great, with the establishment of important production centres in the Syro–Palestinian coastal area, especially Sidon (or Saïda), and then in Alexandria, Egypt. Antimony as a **decolourant** (Section 3.4.2) was introduced in the 7th century BC and apparently discontinued after the 4th century AD. The similar use of manganese became widespread in the 1st century AD. The other major innovation introduced during the 1st century BC was **glass blowing** (Israeli 1991), a technique that allowed glass vessels to be made cheaply and rapidly (Fig. 3.50). In the beginning the blowing relied only on free handling, after this the technique of mould blowing started in the Syrian–Palestinian area. In Roman times glass became a commodity and it was produced industrially with ample recycling in several centres along the Mediterranean. The case of the Roman ship *Julia Felix*, discovered off Grado in Italy in 1987 (*Julia Felix* 2000), yielded plenty of evidence fragments of different composition and provenance of glass that were ready to be recycled in the large nearby production centre of Aquileia (Silvestri *et al.* 2008). A few compositional groups have been identified in the Late Antiquity period (4th–8th century AD); they are slight variants with respect to standard SLS Roman glass: a high iron–manganese–titanium group (HIMT, Freestone 1994), and a Levantine group (Freestone *et al.* 2000), with lower soda and higher lime content than previous Roman glass. Coeval with the early Roman production there is a remarkable production of coloured glass in Asia. The two most-studied groups of materials are the high lead and high barium oxides glasses produced during the Han Dynasty (206 BC – AD 221) in China (HLHB: Jiazhi and Xianqiu pp. 21–26, in Bhardwaj 1987, Brill and Martin 1991), and the vast bead production starting from the 2nd century BC in India, especially at Arikamedu, and in the larger Indo-Pacific region, where at least one high-aluminum glass group has been recognized (HAG: Brill pp. 1–28, Lal pp. 44–56, in Bhardwaj 1987), though there are few analyses and the overall picture will undoubtedly be richer in the future.

At the onset of the Middle Ages there is a comeback of high-Mg glass in the Islamic world (HMEIG, Table 3.12), possibly caused by a scarcity in the supply of natron (Henderson 2002) as well as a gradual increase of potassium in the European glass, witnessing the use of different plant ashes as sources of alkali. Medieval European production resulted in a High Medieval use of high-K glass (HKEG, Table 3.12; also called forest glass or fern glass) especially for the glass windows of churches and cathedrals, with positive advantages for their durability and conservation (Section 3.4.3). A range of post-Medieval glass compositions has been characterized (Henderson 1998).

All the 2nd millennium BC glazes on pottery in the Near East are of HMG composition (Paynter 2009), reflecting the coeval glass technology, even if the soda-rich glaze is known to shrink during cooling and produce cracking (or crazing) due to misfit to the pot surface. The earliest Pb-based glazes had been introduced in China by about the same time (Wood 1999, 2009), and the glazed stoneware remained the main production for over two thousand years, slowly being replaced by porcelain after the 7th century AD. During the Han Dynasty (corresponding to later republican to imperial Roman periods) the composition of stoneware glazes reached a lead content of up to 53–60 PbO wt%, which is not far from the eutectic composition (i.e. 70 PbO wt%, 30 SiO_2 wt%) of comparable materials introduced in the Roman world and later in the Byzantine, Islamic and then in Hispano–Moresque productions (Tite *et al.* 1998, Molera *et al.* 2009). The use of compositions close to the eutectic leads to lower melting temperatures (700–900 °C), better glaze-body fit, ease of application and firing, and also a better appearance. It is interesting to observe that virtually the same transparent high-Pb glaze composition, opacified by tin oxide, was used by Renaissance masters such as the Della Robbia family for the finishing of their terracotta masterpieces (Fig. 3.51), and that it was the fine-tuned application of glaze, pigments, and opacifiers that made their art so unique. Recent analyses of the Della Robbia glazes by non-invasive PIXE shows a sum of PbO + SnO_2 oxides in the range 20–50 wt% (Zucchiatti *et al.* 2002, 2006).

Fig. 3.51. Portrait of a young man. Glazed terracotta by the workshop of Luca della Robbia ca. 1445. (Museo Civico Gaetano Filangeri, Napoli, Italy)
Original colour picture reproduced as plate 35.

3.4.2 Source materials and manufacturing techniques

The source material for glass must encompass: (a) a source of silica, (b) a source of network modifier (alkali or lead), (c) stabilizers and, eventually, (d) pigmenting elements as colourants.

The source of silica can be silica sand, crushed quartzite or biogenic sediments (diatomaceous earths). Pure silica sands are very hard to come by, since most fluvial or marine sands are very impure mixtures of quartz with other minerals (calcite, feldspars, micas, amphiboles, etc.) or with organic and bio-inorganic matter (foraminifera, algae, shells, etc.). Plinius (*Naturalis Historia*, book XXXVI, pp. 193–194) specifically mentions two localities allegedly used in Roman times: the mouth of the river Belus in Israel and the shore near the mouth of the river Volturno in Italy. The Belus site is traditionally reported as the location of the accidental discovery of glass, by the fortuitous fusion of sand and natron. The exploitation of these historical deposits has been well debated, especially concerning the eventual treatments to be used to purify the sand (Turner 1956, Henderson 1985, Silvestri *et al.* 2006). In fact it is normal to find high levels of non-quartz minerals in sedimentary deposits, most of them (micas, clays, feldspars) contributing to the Al_2O_3 content of the final product, as well as containing a sufficient amount of Fe to give the typical green colour of bottle glass. Al_2O_3 contents above 1–2 wt% and FeO contents above about 0.5–1.0 wt% can be used to indicate sand or mixed silica sources (Tite and Shortland 2008, see especially Table 3.1, pp. 38). It is to be noted that some desert sands are of extremely pure quartz, and they are still used as source of silica in modern industry. Beach sand also contains carbonatic fragments (shells, foraminifera), which may substantially contribute to the lime content. However, impurities may also be diffused into the glass from reaction with the crucibles or the clay containers used for fritting, therefore care should be exerted in interpreting that low level of impurities are derived from the raw silica source. Contamination from the substrate reaction is especially important for glazes that are applied in thin layers at high temperature.

Quartzite dikes or sorted fluvial pebbles may constitute a substantially cleaner source of quartz, and were likely the main source of silica in antiquity. It is reported that the reputed "cristallo" (high-quality glass) fabricated in Venice from the 15th century was made out of clear quartz pebbles from the Ticino river (Swiss–Italian Alps) and soda-ash imported from Syria. The glass was so transparent and valued that it soon started to be locally imitated ("façon de Venise" glass) in several places including London (Mortimer 1995) and Antwerp (De Raedt et al. 2001). The use of biogenic diatomites has been reported as an unusual source of very reactive silica in Slovak prehistoric faience (Angelini et al. 2006). Both sands and quartz fragments were duly ground to fine powders to increase the chemical reactivity.

The source of alkali elements can be inorganic (minerals, salts) or organic-derived (plant ashes). In both cases the final ingredients that are mixed with silica to make the frit must be oxides that are very soluble in the SiO_2 matrix. There are three major ancient sources of alkali: (a) natural evaporite minerals (i.e. natron, basically a mixture of sodium carbonates, sulphates, and chlorides; Shortland 2004), (b) soda-rich plant ashes, generally derived from salt-tolerant halophytic plants growing in coastal areas and desert regions (i.e. *Salsola soda, Salsola vermiculata, Anabasis sp., Halopeplis sp.*, etc.; Barkoudah and Henderson 2006), and (c) ashes of different plants, which usually contain similar proportions of sodium and potassium and are therefore called mixed alkali sources (Jackson et al. 2005). This also includes ashes of trees and ferns, which often produce much higher potassium contents.

Despite the regional variation of the plant composition, the three sources can be distinguished mainly by the Na_2O/K_2O ratio, which is of the order 50–100 for natron, in the range 2–10 for soda-rich plants, and below 1.5 for mixed alkali ashes. Natron and soda-rich plant ashes are the sources for the groups of ancient glass that show negligible potassium content: HMG, LMG, SLS, and HMEIG (Table 3.12). The compositional change between HMG and LMG glasses and faiences observed early in the 1st millennium BC is though to be due to the Near East introduction of natron in place of high-soda plant ash, though the soda-plant/natron transition is still largely unexplained from the technological point of view (Shortland et al. 2006, Tite and Shortland 2008). The use of a mineral- or plant-derived alkali in the glass composition is signalled by the magnesium level. Also the distinct composition of the ancient LMHK European glasses and faiences with respect to the Eastern materials is thought to be due to the use of mixed alkali plant ashes (Henderson 1988, 2001; Tite and Shortland 2008), the low content of divalent cations (Mg, Ca) being derived from hot-water treatment of the ash and re-precipitation of monovalent cations hydroxides. It is to be remarked that alkali ratios may vary during the glass-making process, due to different solubilities of the salts, diffusion, or external contamination (Rehren 2008).

Since glass with a high content of monovalent cations is highly prone to dissolution and alkali leaching, it needs to be stabilized by divalent alkali earth cations. Lime has always been the most common stabilizer. It is generally assumed that the low lime level in prehistoric faience is derived from the calcium impurities present in the sand or the plant ashes (Tite and Shortland 2008, Pp. 43). However early HMG and LMHK glasses already have a substantial amount of lime as stabilizer, and SLS glass was definitely stabilized by the deliberate addition of lime as shell fragments, as noted by Plinius (*Naturalis Historia*, book XXXVI, p. 66). Shell carbonate or limestone is one of the most common sources of lime for plaster and it is certainly a materials technology that was available very early in prehistory (Section 3.2.1).

Some of the furnace types used to produce glass are reviewed by Henderson (2001). Here, we would like to remark that glass production is a very energy consuming process, since plant material is necessary both for the production of ashes for the alkali, and for the heat necessary for the glass-making process (fritting, melting, boiling, mixing, blowing, etc.). Because of the high viscosity of glass materials and the critical dependence of viscosity on temperature (Table 3.10), it is very easy to see that small temperature gradients produce a rapid departure thermodynamic equilibrium, incomplete reactions, and compositional inhomogeneities (Messiga and Riccardi 2001). Glass-making is therefore a highly technological process. The fact that many of the ancient glass compositions are close to the cotectic lines of the liquidus surfaces of the soda–lime–silica diagram indicate, on one hand, the direct dependence of the partial melting and final glass composition on the temperature path and, on the other hand, the excellent mastery of the process by ancient glassmakers in producing high quality glass (Rehren 2000, Shugar and Rehren 2002). One related point deserves attention. The melting and viscosity properties do not vary linearly in the compositional join between soda–silica glass and potassium–silica glass. In fact both the melting temperature curve and the working temperature curve have two minima (Fig. 3.52), one at the maximum Na_2O content (i.e. $K_2O/(K_2O + Na_2O) = 0$) and the other at about $K_2O/(K_2O + Na_2O) = 0.60$. It is quite impressive to observe that the measured working temperatures for Near East ancient glass compositions (HMG, LMG) lie close to the first minimum, and the experimental working temperature values measured for European ancient glass compositions (LMHK) coincide with the second minimum (Polla 2006, Artioli and Angelini 2009). The two major groups of ancient glasses are simply the two most viable solutions to obtaining silica glasses that are workable at temperatures as low as 850 °C. For thermodynamic, economic, and historical reasons the high-soda solution (SLS) was the one adopted in the Roman world and discontinuously carried on to our times.

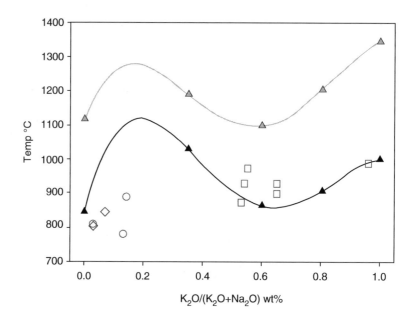

Fig. 3.52. Experimental values of the melting temperature (yellow curve) and the working temperature (red curve) for a series of synthetic glasses with varying $K_2O/(K_2O+Na_2O)$ composition (Polla 2006). It is observed that both curves, together with the curve of the sintering temperatures (not plotted) do not vary linearly with composition, but rather display minima at 0.0 and 0.6. Accordingly, the experimental values of working temperatures measured on archaeological samples nicely fall close to the reference curve: high-soda HMG and LMG glasses (blue circles) fit with the first minimum, whereas high-potassium LMHK glasses from Frattesina (red squares) fit with the second minimum (Artioli and Angelini 2009). The experimental values for two SLS Roman glasses from Trento (purple diamonds) are also shown for comparison. Original colour picture reproduced as plate 36.

With regard to the aesthetic qualities of glass, the two major properties that need to be controlled during production are colour and transparency. The glass colour is the outcome of all the chemical and mineralogical impurities contained in the raw materials (Weyl 1951, Bamford 1977, Henderson 2000, Doremus 1994). In fact glass can be pigmented by a very small amount of chemical additions: Fe, Co and other transition metals may colour glass at a few ppm concentration, and for glassmakers it is much more complicated to make colourless transparent glass than coloured glass. Even today shining transparent high-quality glasses are much more expensive than green or brown low-quality bottle glass. Incidentally, take care when recycling and keep the two glass colours distinct or you will degrade the whole mixed lot!

Transparent glass needs to be coloured by carefully dissolving the chromophore ions in the silica matrix, so that they can selectively absorb some light bands and produce colour (Section 2.1.1, Box 3.e), but still avoid particle scattering and diffusion. The proof is high-lead glass: lead is very absorbing, but if the ions are all isolated in the silica matrix there is no metallic bond, no absorption, and the glass is simply more brilliant because of the increased average density (i.e. higher refractive index). A short list of the most used chromophore ions responsible for colouring transparent glasses is given in Table 3.13. Note that the colour effect of a chemical element depends both on its oxidation state and chemical environment. Most of the ancient faience and glasses were coloured by divalent copper ions, hence the universally known turquoise-green colour of the early artefacts. The relative proportion between Cu^{+2} (blue), Cu^{+1} (colourless) and Fe^{+3} (yellow) is what determines the faience shades from turquoise, to green, to pale brown, to yellowish. The tinge could be further varied by the presence of Mn and As, and the typical opaque turquoise colour was

Table 3.13. Some chromophore ions used to colour glasses (Weyl 1951, Bamford 1977)

Element	Cation	Colour in tetrahedral coordination (network former)	Colour in octahedral coordination (network modifier)
Chromium	Cr^{+3}	–	green
	Cr^{+6}	yellow	–
Copper	Cu^{+1}	–	colourless to red, brown fluorescence
	Cu^{+2}	yellowish-brown	blue
Cobalt	Co^{+2}	blue	pink
Nickel	Ni^{+2}	purple	yellow
Manganese	Mn^{+2}	colourless to pale yellow, green fluorescence	weak orange, red fluorescence
	Mn^{+3}	purple	–
Iron	Fe^{+2}	–	green, blue, IR absorption
	Fe^{+3}	brown	yellow to pink
Uranium	U^{+6}	yellowish-orange	weak yellow, green fluorescence
Vanadium	V^{+3}	–	green
	V^{+4}	–	blue
	V^{+5}	colourless to yellow	–

obtained by adding calcium antimonate opacifier to the Cu^{+2}-coloured faience (Shortland 2002). Deep blue glass was obtained by the use of cobalt, apparently extracted from the cobalt alums present in the Western Egyptian Oases (Shortland et al. 2006). The sources of cobalt for glass produced in other areas are unknown. The colour palette of the early glass and faience was practically limited to all possible combinations of a few pigmenting ions (Cu, Co, Mn) in their different oxidation states with a few available opacifiers (lead antimonate yellow and calcium antimonate white) (See Table 1 of Shortland 2002). Antimony and manganese were also intentionally used in antiquity to counteract colouration due to iron in glasses that were intended to be clear and colourless (Sayre 1963). Reduction of both cations successfully enhance iron oxidation, hence decolourizing the glass.

A very unusual deep blue glass coloured by Fe^{+2} in distorted octahedral coordination was produced in Baroque Sicily by the fortuitous introduction of flint-containing limestones in lime-kilns. The combined effect of the highly reducing environment in the combustion chamber and the presence of alkali elements from the plant fuel yielded blocks of high quality blue glass that were widely used in Palermo for church decorations, as scrupulously recorded by Wolfgang Goethe in his *Italienische Reise* (Artioli et al. 2009).

Red glasses were produced from blue glass by different reduction mechanisms of the Cu^{+2} ions mainly through the formation of cuprite (Cu_2^+O) or metallic copper (Cu^0) particles (Angelini et al. 2004, Barber et al. 2009). Figures 3.53 and 3.54 show two very interesting cases of archaeological glasses where the reduction and the oxidation processes of copper have been frozen for posterity.

294 *Materials and case studies*

Fig. 3.53. Blue glass turning red: example of surface copper reduction ($Cu^{+2} \rightarrow Cu^0$) on a blue glass due to firing under reducing conditions. Raw glass block from Frattesina, Rovigo, Italy. Final Bronze Age (Angelini *et al.* 2004).
Original colour picture reproduced as plate 37.

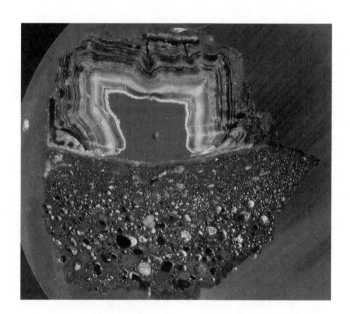

Fig. 3.54. Red glass turning blue: example of oxidation of copper ($Cu^{+1} \rightarrow Cu^{+2}$) in a cuprite-coloured red glass ingot from Qantir-Piramesses, Egypt; Late Bronze Age. The surface layer turned green-blue by alteration (Rehren *et al.* 2001, Rehren and Pusch 2005). (Photo B. Schoer, Project Qantir - Pi-Ramesses, courtesy of Th. Rehren)
Original colour picture reproduced as plate 38.

Any particle that is larger than the wavelength of light and is included in the glass produces light diffusion, reduces the amount of light passing through the material, and makes the glass opaque. This has interesting consequences. First, to make an opaque glass we have to induce crystallization within the glass (by slow cooling, crystal addition, etc.): the opacity of the glass depends on the type, density, number, morphology and distribution of the particles. Opaque glass in ancient times was produced by inducing crystallization or by the direct addition of high-density crystalline materials (opacifiers), typically antimony- or tin-based opacifiers (lead stannate yellow Pb_2SnO_4 and $PbSnO_3$, tin dioxide SnO_2, lead antimonate yellow $Pb_2Sb_2O_7$, calcium antimonate white $Ca_2Sb_2O_7$ and $CaSb_2O_6$; Shortland 2002, Tite et al. 2008). Opacifiers are of course fundamental to cover the substrate colour in glazes and enamels. Opaque white materials in protohistoric glass were also obtained by mixing quartz crystals to the almost-finished glass, and when Ca-rich silicates such wollastonite are observed the addition of calcite or lime is suspected (Artioli et al. 2008). Opacifiers used in later times include lead arsenate, calcium phosphate, sodium phosphate, calcium fluoride and other crystalline phases (Henderson 2001).

By playing with the particle type-size-distribution parameters one can make very interesting pigmenting effects, and this is exactly what ancient glassmakers were doing. The most striking examples are obtained by the use of metal nanoparticles (Liz-Marzán 2004, Mirguet et al. 2008), whose crystallization in the glass is obtained by complex temperature-oxidation paths. Generally a heating stage in oxidizing conditions is required to dissolve the metal in the silica matrix, and then a heat-reducing step is used to induce the metal reduction and the formation of nanoparticles. The process is quite the same as that used today to produce metallic nanoparticles in advanced composite silicate glasses used for optical devices. Metal micro- or nano-particles have been used in the past to obtain astonishing effects in glasses, for example to turn blue Cu-coloured glass into red glass (Fig. 3.53, Final Bronze Age LMHK glass from Frattesina: Angelini et al. 2004), to make dichroic vessels that change in colour from reflected to transmitted light (the famous Lycurgus cup of the 4th century AD that is conserved at the British Museum; Barber and Freestone 1990), and to produce wonderful iridescence effects in glazed lustre pottery (Hispano-Moresque, Islamic and Renaissance lustre pottery; Pérez-Arantegui et al. 2001, 2004, Fredrickx et al. 2004, Mirguet et al. 2008). In the lustre glazes it was possible to obtain all colour of the spectrum from yellow to red, gold, and brown (Fig. 1.5) by fine-tuning the Au–Cu composition, size, and distribution of the particles (Figures 2.i.3, 2.I.4).

3.4.3 Glass alteration and degradation processes

The assessment and quantification of changes in glass materials contribute to allow us to (a) potentially date the artefact, if the rate of change is known, (b) understand the present condition of the artefact and predict its future evolution for conservation purposes, (c) understand how alteration affects the physical properties and the artistic, archaeological and historical interpretation of the object and its manufacturing technology (Freestone 2001). The alteration of glass commonly operates by two principal mechanisms:

1) Ion exchange between environmental water (in the form of hydronium ions H_3O^+) and alkali cations in the glass. This process is especially active in neutral or acidic conditions (pH < 7). This process produces a hydrous layer at the surface of the glass, and it is at the root of hydration dating methods (Section 2.5.3.2). The reaction can be written as

$$H_3O^+ + O_3Si-O^-Na^+ \rightarrow O_3Si-OH + H_2O + Na^+$$

and it implies Na ions diffusing out of the glass and H_2O molecules diffusing into the material. The hydroxyl groups are covalently-bonded to silicon in the tetrahedral network, whereas the water molecules are hydrogen-bonded in the nano-pores. The smaller radius of hydrogen compared to that of sodium (or potassium) and the change from ionic to covalent bonding to oxygen leads to a net contraction of the glass network, more pore space, and increased water percolation.

2) hydrolysis of the silica glassforming network. The process involves breaking of the Si–O bonds and slow dissolution of the glass network. The reaction is of the kind

$$O_3Si-O-SiO_3 + OH^- \rightarrow O_3Si-OH + O^--SiO_3$$

and it especially active under alkaline conditions (pH > 8). The higher the pH the faster is the reaction kinetics. Substantial quantities of stabilizing divalent ions (Ca, Mg) are necessary to slow down the network dissolution. These processes have long been studied as they have implications for the storage and durability of nuclear waste storage glasses. The glass composition (including the pigmenting metal oxides), the silica saturation in the environment, the pH, the humidity, are all parameters influencing the speed of the alteration reactions (Adams 1984, Heimann 1986, Newton and Davison 1989).

The two alteration mechanisms are in competition. If dissolution is prevalent, then the surface is progressively etched away and the surface of the glass appears to be heavily corroded. This condition may be prevalent in atmospheric conditions, where the alkali diffused from the glass are concentrated as salts at the surface, and the alkaline conditions favour aggressive network dissolution. If ion exchange is prevalent over dissolution, we'll have an alkali-poor surface hydration layer. This is normally the case for archaeological glasses, which are buried in soils with neutral or slightly acidic conditions, and the diffused alkali are leached out by the percolating water. In homogeneous glass the hydration layer appears as an iridescent skin with opal-like optical effects (Section 3.6). The altered layer may vary from a thin layer depleted of alkali ions (Fig. 3.55), to a millimetre-thick crust of hydrous opalescent glass (Fig. 2.15), to the total disappearance of glass (Fig. 3.56). The hydrous layer contains up to 20–25 wt% of water as H_2O molecules and OH groups bound to silicon atoms. When the glass is brought to

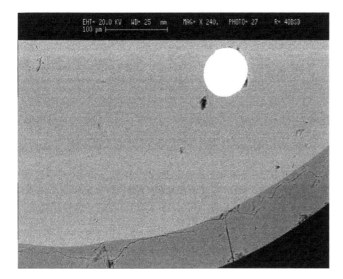

Fig. 3.55. Bronze Age homogeneous LMHK glass showing a thin K-poor alteration layer at the surface. The layer is about 70 μm thick. Backscatterd electrons SEM image. (Courtesy of I. Angelini, University of Padova)

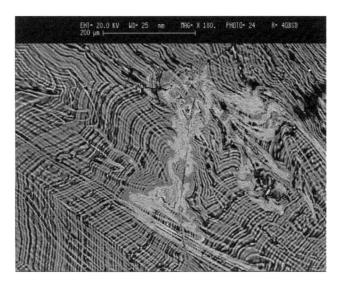

Fig. 3.56. Example of a totally altered Bronze Age HMG glass. There is no pristine glass left in the sample; the backscattered electron SEM image shows only the typical banding in the hydrated silica layer. (Courtesy of I. Angelini, University of Padova)

a low humidity environment (removal from ground, rescue from water, etc.) it rapidly dehydrates, normally causing flaking of the layer off the surface and mechanical degradation of the artefact. Measurement of the hydration layer is performed by destructive microscopy techniques (OM, SEM, TEM), or by non-invasive hydrogen profiling techniques (SIMS, Box 2.t; NRRA, ERDA, Box 3.i), exactly as in the case for hydration dating. Attempts to use the number of bands in the hydration layer to date the artefact, using a dendrochronology-like method, have failed because of the complex kinetics of the process, which makes the thickness of the bands unrelated to regular time cycles (Newton 1966). The thickness of the bands is of the order of hundreds of nanometres, thus producing the light interference effects observed as opaline iridescence (Fig. 3.57). The effect is aesthetically rather pleasant, and the layer is commonly not removed during the conservation treatment. In

Fig. 3.57. Roman glass vessels, showing the typical opalescence effects due to light interference with the surface hydration layer: (a, b) sprinkler flasks, 3rd–4th centuries AD, height 13.5 and 9.4 cm, respectively (inv. n. 71.AF.81, 71.AF.82); (c) amphora, 3rd century AD, height 25.5 cm (inv. n. 78.AF.35). (Courtesy of the J. Paul Getty Museum, Villa Collection, Malibu, California)

fact, after the 19th century AD there were several attempts to reproduce such rainbow-coloured surfaces artificially, the most notable examples being Art Nouveau artefacts (Fig. 3.42).

Glasses that have different compositions greatly differ in alteration mechanisms and rates. In general, plant-ash glasses (HMG, LMHK, HKEG, HLEG, HLIG: Table 3.12) are more subject to degradation and tend to develop a thick hydration layer, though this of course depends on environmental variables and their variable lime-content (Freestone 2001). High lead glass also seems to be susceptible to alteration and/or corrosion.

The high-K Medieval stained glass windows have been the subject of careful investigation, because of their alarmingly altered conditions, especially on the outer surfaces (Schreiner 2004). Pollution seems to contribute substantially to the acceleration of the process, which causes the development of weathering crusts at the surface, an evident decrease in the panel's transparency, and sometimes intense discolouration (Schreiner 1991). The alteration crusts are composed of hydrated silica with re-crystallised alkali salts (gypsum $Ca_2SO_4 \cdot 2H_2O$, syngenite $K_2Ca(SO_4)_2 \cdot H_2O$), the sulphate resulting from atmospheric SO_2. Surface analytical techniques are crucial to characterize the materials and the processes prior to conservation and intervention (Schreiner 2004).

3.4.4 Analytical techniques for glass studies

Having briefly discussed the variety, properties and composition of vitreous materials, it is clear that proper characterization must be commensurate with the complexity of the material and the problems involved.

If the glass is homogeneous, then bulk techniques may be used, either for chemical analysis (XRF, AAS, OES, NAA, MS) or for structural characterization (wide-angle XRD or ND, total scattering). Information on the local environment of specific ions can be performed by XAS, XPS, and Mössbauer spectroscopies. Surface techniques are seldom useful, because of the alteration layer invariably present at the surface of the glass: the very same hydration processes that are useful for dating silica-based artefacts (Section 2.5.3.2) are actually deleterious from the point of view of degradation processes (Fig. 2.15). Very frequently the surface chemical analysis of ancient glass materials only shows silica, because the alkali components that are so useful for characterizing the material (Table 3.12) are totally leached out. The preferred choice is microsampling of a tiny amount of material, commonly of the order of a few hundred micrometres, in order to carry out the analyses on pristine glass (Bronk and Freestone 2001).

If the sample is heterogeneous, optical and electron microscopies (OM, SEM, TEM) must be used to visualize the textural features, the alteration layers, and eventually to quantify the individual amorphous and crystalline phases by image analysis. Point analyses and chemical maps, which are especially useful in characterizing reaction surfaces and submicron inclusions such as metal nanoparticles, may be obtained by EPMA, PIXE, SIMS, SEM-EDS, TEM-EDS.

Finally, the complexity of glass colour must be investigated by advanced spectroscopies that probe the electron states of the chromophore ions: UV-Vis, FT-IR, XPS, EELS, XANES spectroscopies are invaluable aids in decoding subtle charge transfer effects and localized electron states. Stable (O, Sr) and radiogenic (Pb, Nd) isotope studies by mass spectrometry are used to study the nature and provenance of the raw materials used for glass production (Freestone 2006, Degryse *et al.* 2009, Henderson *et al.* 2009).

> **Box 3.f XPS X-ray photoelectron spectroscopy**
>
> X-ray photoelectron spectroscopy (**XPS**), also known as electron spectroscopy for chemical analysis (**ESCA**) is a quantitative spectroscopic technique that measures the elemental composition, though it is much more used to characterize the chemical state (valence) and the electronic state (electronic levels) of specific elements in a material. XPS spectra are obtained by irradiating a material with a beam of X-rays, and by measuring the number and kinetic energy of the electrons escaping from the top surface of the material being analysed (1–10 nm). XPS requires ultra high vacuum measuring conditions. Since many materials are easily contaminated by adsorbed molecules and ions, the material is often treated just before analysis or during the measurement (by ion beam etching, laser ablation, UV irradiation, ion beam implantation, or other means) in order to clean off some of the surface contamination, or even to check the material's reaction to bombardment, deposition, or treatment. It is a typical surface technique, thus it is the technique of choice to characterize semiconductors, ion implanted materials, vapour-deposited layers, surface contaminations, or corrosion products.
>
> XPS detects all elements with $Z > 3$; it cannot detect H and He. Detection limits for most of the elements are in the 10^2–10^3 range. Detections limits in the ppm range are possible under special conditions in the laboratory or using synchrotron beams.
>
> The experiment is conceptually very similar to the one performed over a hundred years ago to investigate the photoelectric effect by Hertz and Hallwachs: a photon source (gas-discharge lamp, X-ray tube, synchtroton source) hits the solid surface and the emitted electrons are analysed in terms of their direction, kinetic energy and momentum. The polarization of the incident beam is a useful property in angle-resolved experiments.
>
> (*cont.*)

Box 3.f (*Continued*)

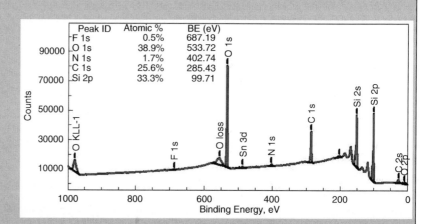

Fig. 3.f.1. Example of a wide energy XPS scan. All electron transitions of the elements present in the surface of the sample are observed as peaks corresponding to the electron binding energies.

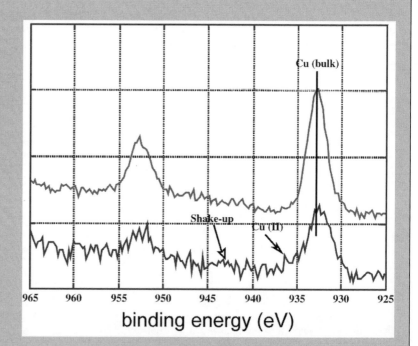

Fig. 3.f.2. Example of the detailed XPS peaks measured for copper present in a prehistoric glass. The measurements indicate that the metal is dissolved in the glass as Cu^{+1}, i.e. in a cuprite-like configuration. (Modified from Angelini *et al.* 2004)

Figure 3.f.1 is an example of a wide energy XPS scan, showing the allowed emission peaks of all elements present in the surface of the sample. The technique is used to detect the present elements and characterize their oxidation state. The detailed study of the electron peaks around 930–950 eV in a copper-coloured prehistoric blue glass (Fig. 3.f.2, Angelini *et al.* 2004) indicate for example that the metal in the sample is present as Cu^{+1}, most likely in a cuprite-like structural arrangement.

In archaeometry (Hubin and Terryn 2004), XPS has been mostly applied to the characterization of surfaces, especially metals, glass and glazed pottery. Metal surfaces are mostly analysed to understand corrosion processes (Ingo *et al.* 2004), evidence of ancient patinas (Garbassi and Mello 1984, Gillies and Urch 1985) and to distinguish alteration products from treatments applied to the object in its lifetime (Paparazzo and Moretto 1999, Ingo *et al.* 2004). The products of the accelerated corrosion of metals has also been analysed with similar methods (Casaletto *et al.* 2006). The depth profiling is commonly performed by carving the surface with an ion beam.

The composition and the structure of special surface treatments in ceramics have also been characterized by XPS spectroscopy. Examples are the investigation of the black gloss on Attic vases, showing the well known presence of iron oxides having different valence states (Fe^{+3}, Fe^{+2}, Fe^0), which are finely dispersed in a vitreous Al–K- and Fe-enriched silicate where also Fe–C and C–C carbon atoms are present (Ingo *et al.* 2000), and the characterization of the metallic nano-particles present in the glazed layer of renaissance lustre ceramics (Pérez-Arantegui *et al.* 2004).

For in-depth information

Briggs D. (1977) Handbook of X-ray and ultraviolet photoelectron spectroscopy. Heyden, London.

Ellis A.M., Fehér M. and Wright T.G. (2005) Electronic and photoelectron spectroscopy: fundamentals and case studies. Cambridge University Press, Cambridge.

Grant J.T. and Briggs D. (ed.) (2003) Surface analysis by auger and X-ray photoelectron spectroscopy. IM Publications, Chichester, UK.

Hüfner S. (2003) Photoelectron spectroscopy: principles and applications. 3rd Edition. Springer-Verlag, Berlin–New York.

Box 3.g AES – EELS Electron spectroscopies

The Auger electron spectroscopy (**AES**, beware of the similar acronym concerning atomic emission spectroscopy, Box 2.c) and the electron energy loss spectroscopy (**EELS**) are both techniques measuring the energy of electrons re-emitted from surfaces after bombardment by an electron beam. The electrons are however produced by different processes in the two cases, and therefore the two measurements yield complementary information on the electron structure of the atoms in the material.

The Auger effect, as it has come to be called even if it was first observed by Lise Meitner, involves energetic electrons emitted from an excited atom after a series of internal relaxation events (Fig. 3.g.1). Auger electron energies for the different atoms are known and tabulated (Jenkins and Chung 1970, Larkins 1977; see also the X-Ray Data Booklet: http://xdb.lbl.gov/Section1/Sec_1-4.html). In EELS, some of the primary electrons of the incident beam will undergo inelastic scattering, which means that they lose energy and change their trajectory. The measured energy difference between the incident electrons and the anelastically scattered electrons can by interpreted in terms of the interaction mechanism involving phonon excitations, inter- and intra-band transitions, plasmon excitations, inner shell ionizations and Čerenkov radiation. The inner-shell ionizations are particularly useful for detecting the elemental components of a material, because the measured energy loss corresponds to the

(*cont.*)

Box 3.g (*Continued*)

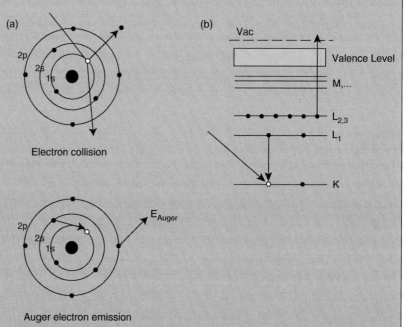

Fig. 3.g.1. Scheme of the generating process of Auger electrons: (a) atomic orbital diagram, (b) band diagram.

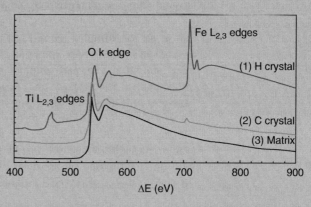

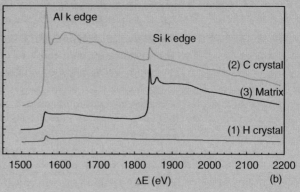

Fig. 3.g.2. EELS spectra of ceramics slips of terra sigillata Roman–Gaul pottery. The electron microbeam analysis reveals critical differences between coexisting matrix and crystals. (Modified from Sciau *et al.* 2006)

amount of energy needed to remove an inner-shell electron from a specific atom, these values are also known and listed in available databases (http://www.cemes.fr/~eelsdb/).

AES and EELS are intrinsically surface techniques, they operate under vacuum conditions, they need a very small amount of material and, in principle, may yield information both on surface composition, and on the element oxidation state and environment. Interestingly, by using raster techniques it is possible to obtain chemical images of the surface composition and oxidation state.

Though there are specific electron microscopes designed for Auger spectroscopy (scanning Auger microscopes, **SAM**), in many cases electron detectors are applied to existing scanning and transmission electron microscopes as additional capabilities. Experimental limitations derive from surface charge effects and energy losses, especially plasmon bands. The AES application of analysis to cultural heritage materials mainly focused on the characterization of metal surfaces (Paparazzo 2006, Northover *et al.* 2008).

EELS applications are more commonly associated with TEM measurements of micro- and nano-samples, in order to combine crystallographic phase determination with chemical analysis. Examples concerning the detailed microstructural and microchemical characterization of archaeological pottery are available (Fig. 3.g.2; Sciau *et al.* 2006, Mirguet *et al.* 2009). The technique has also been applied to modern analogues of Renaissance lustre, in order to understand the production technology (Fredrickx *et al.* 2004).

For in-depth information:

Brydson R. (2001) Electron energy loss spectroscopy. Garland/BIOS Scientific Publishers.

Hedberg C.L. (ed.) (1995) Handbook of Auger electron spectroscopy. Physical Electronics, Eden Prairie, MN.

Grant J.T. (2003) Surface analysis by Auger and X-ray photoelectron spectroscopy. IM Publications, Chichester, UK.

Box 3.h XAS X-ray absorption spectroscopy

X-ray absorption spectroscopy (**XAS**) is a powerful technique for determining the local geometric and the electronic structure of matter. The measurement needs an energy scan of the incident beam in an extended region around the absorption edge therefore, although in principle XAS can be performed using the bremsstrahlung of a laboratory X-ray tube, the available photon intensity is very weak, and practically the only intense and tuneable X-ray sources are synchrotrons (Box 2.a).

XAS data are obtained by scanning the photon energy using a crystal monochromator about a range where core electrons can be excited. Alternatively, a polychromatic beam can be focused on a linear detector in **energy-dispersive XAS**. The hard X-ray range (4–100 keV) encompasses most of the K, L and M edges (Fig. 2.b.1) of the chemical elements with $20 < Z < 86$, the K edges of light elements ($5 < Z < 20$) must be accessed in the soft X-ray region (0.1–5 keV). Figure 3.h.1 shows some absorption edges for light elements (C, N, O) in the 0.3–0.5 keV region, and for heavier elements (Ir, Pt) in the 11–14 keV region. A table of all electron binding energies is given in the famous X-Ray Booklet that is available at the LBNL web site (http://xdb.lbl.gov/).

A typical XAS experiment simply measures the incident (I_0) and transmitted (I_T) photon flux through the sample as a function of energy. Commonly the $\ln(I_0/I_T)$ vs. E absorption curve is reported (Fig. 3.h.2). The analysis of the XAS data takes into account three distinct regions: the modelling of the dominant edge feature, together with the pre-edge (few hundred eV below the edge) region is called the X-ray absorption near-edge structure (**XANES**) or

(*cont.*)

Box 3.h (*Continued*)

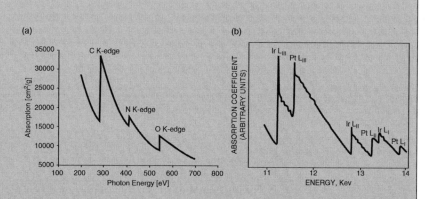

Fig. 3.h.1. (a) K-absorption edges of some light elements (C, N, O) in the 0.3–0.5 keV region and (b) L- and M- absorption edges of heavier elements (Ir, Pt) in the 11–14 keV region.

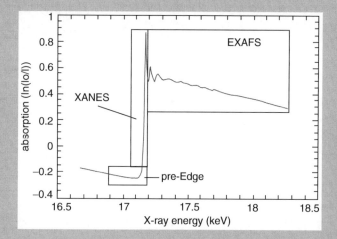

Fig. 3.h.2. Measured XAS spectra showing the absorption coefficient ($\ln(I_0/I_T)$) as a function of energy. The data are commonly modelled and interpreted in three distinct regions: the pre-edge, XANES and EXAFS regions.

near-edge X-ray absorption fine structure (**NEXAFS**). This region is extremely sensitive to coordination geometry and oxidation state. The change in charge may result in the energy shift of the edge up to few hundred eV. The edge position is thus a good indicator of the average valence state of the probed element.

The interpretation of the modulated region extending from the edge to higher energies and containing the positive and negative interference effects is called the X-ray absorption fine structure (**XAFS**) or extended XAFS (**EXAFS**) region. It corresponds to the transition to the continuum resulting in the ejection of a photoelectron that can backscatter off neighbouring atoms, probing the local atomic environment and yielding information on the local chemistry and geometry, i.e. the type and distance of near and next-near atoms (Rehr and Albers 2000). Information on the vibrational dynamics of atoms can also be extracted from high-resolution EXAFS data because the interference between the emitted and backscattered photoelectron waves is sensitive to the correlated and uncorrelated motions of the scattering atoms.

The samples can be in the gas-phase, solution, or solid condensed matter. The resulting local information around the absorbing atom provides average coordination, independent of the short- or long-range order of the material. Furthermore the technique is chemically highly selective, and can prove the local arrangement even in ultra-dilute concentrations.

There are a few variations of the techniques, which specialize in probing special samples: **SEXAFS** (surface-EXAFS) is XAS applied to surface phenomena, and the probed volume under glancing or grazing geometry is the thin surface layer of the specimen. Fluorescence EXAFS is a technique measuring the fluorescence yield in place of the transmitted signal.

XAS applications in archaeometry have been mostly focused on glass and glazes because of the advantage of probing the local geometry of specific elements in disordered materials. It should be remembered that diffraction and absorption spectroscopies are very complementary techniques, and furthermore they can be performed in combined or simultaneous mode at synchrotron beamlines, because in most case they exploit the same energy range. Probing at the same time the short-range (local geometry: by EXAFS) and the long-range (crystal volume: by XRD) structure of a compound is an extremely powerful characterization combination in materials science.

Examples of XAS applications on ancient materials concern the specific behaviour of network-modifier or colourant elements in glass and glassy materials (Section 3.4), such as the role of Ca ions in high-lead soda-lime glazes (Veiga and Figuereido 2008), the role of Pb and Sb ions in lead-rich tin-opacified yellow glazes (Figueiredo *et al.* 2005) and decorated tiles (Figueiredo *et al.* 2006), and the role of Zn on Portuguese tile glazes (Veiga and Figueiredo 2008). The XAS technique has also been applied to the Fe cations in the Maya blue pigment (Section 3.3), with uncertain results (Polette *et al.* 2002).

A very interesting investigation attempted to resolve the issue of the blackening of the copper resinate in 15th century Italian easel paintings (Cartechini *et al.* 2008) using a combination of XAS, UV-Vis and FTIR techniques. The discolouration of the copper resinate seems to be related to local modification of the copper coordination structure as evidenced by the observation of an increase of the Cu–Cu and Cu–C distances in the EXAFS spectra.

For in-depth information

Koningsberger D.C. and Prins R. (eds) (1988) X-ray absorption: principles, applications, techniques of EXAFS, SEXAFS, and XANES. John Wiley & Sons, New York.

Rehr J.J. (1994) X-ray absorption in bulk and surfaces. World Scientific, Singapore.

Stöhr J. (1992) NEXAFS Spectroscopy. Springer, Heidelberg.

Teo B.K. (1986) EXAFS: basic principles and data analysis. Inorganic Chemistry Concepts, Vol. 9. Springer-Verlag, Berlin-New York.

3.5 Metals

There is no doubt that metals have been essential materials in the history of mankind. From the engineering point of view (Fig. 3.1) they had a slow start but then they became the prevalent material and dominated much of the applications in materials science, especially after steel became mass produced at the onset of the industrial revolution. The interpretation of the social impact of metal technology (metallurgy) is a topic that has gone through several important cultural and philosophical phases. The emphasis of discussion throughout most of the last century was on technological determinism: metal developments and metal control in prehistory and history have much to do with technological skills, societal organization, and ultimately economical and political power. Starting from Thomsen's early definition of the "Three Age" system (i.e. Stone, Bronze, and Iron Age) Killick (2001) cogently summarizes the shift of thinking from the era of unrestrained speculation, when leading views were not much supported by empirical and archaeological evidence, to an era were speculation on past metallurgy started to be based on ethnographical, archaeological, and technical grounds. Following Killick, I tend to agree that Theodore A. Wertime and Cyril Stanley Smith were fundamental to the evidenced paradigm's shift (Wertime 1964). Papers challenging the standard views (Rowlands 1971) but especially large field projects exploiting expertise from very different scientific areas such as archaeology, metallurgy, mineralogy, chemistry, etc. were the foundations of modern archaeometallurgy. The benchmarks of recent history to be recall are the reference volumes of Ronald Tylecote (1976 repr. 1992, 1987) and the extensive Timna project directed by Beno Rothenberg, which lead the way to truly interdisciplinary field collaboration (Rothenberg et al. 1979). The Faynan project started soon after (Hauptmann 2007). Many of present day active archaeometallurgists stem from this cultural scene.

What is then modern archaeometallurgy? The discipline deals with all aspects of metal production, diffusion and usage and extends no less than throughout the history of mankind (Rehren and Pernicka 2008). Present day academic specialization is often in contrast with the broad expertise needed to interpret past metallurgical evidences. Wide-area surveys, field excavation, ancient pyrotechnology, ceramics, geology and mineral deposits, high temperature mineralogy and petrology, phase transformations, metal science, chemistry, ethnography, anthropology, history are among the encyclopaedic knowledge required to interpret pre-industrial metallurgy. Only large interacting groups can cope with such a vast array of requirements, and this is a major problem that is largely unresolved. As an example, when browsing the literature I came across a recent map of the principal ore deposits of Europe and the Mediterranean region (Fig. 9.1 of Kassianidou and Knapp 2005) that has little to do with actual geology. The marked copper deposits are clearly derived from the archaeological literature and among several geological aberrations, the whole Eastern Alps and Northern Balkans seem to be deprived of copper, whereas in reality the Austrian Alps (Schwaz, Mitterberg), South Tyrol (Val Venosta, Chiusa, Calceranica, Val dei Mocheni) and Serbia (Majdanpek) were fundamental both to the early diffusion of chalcolitic copper in Central Europe (Höppner et al. 2005) and to the extensive production of the Late Bronze Age (Weisgerber and Goldenberg 2004). The overall problem of the scarce interaction between archaeologists and scientists, repeatedly remarked on in this volume, is particularly acute in the archaeometallurgical field, whose complexity encompasses geomineralogical information, mining techniques, geochemical and isotopic tracers, smelting pyrotechnologies, environmental pollution, alloying, metal working, patination and surface coating, diffusion and trade patterns, use and wear assessment, intangible values, archaeological context, typology-chemistry relationships, degradation and corrosion, conservation treatments and many other issues (Fig. 1 of Rehren and Pernicka 2008).

In order to simplify the description of practical issues in the investigation of ancient metallurgy, we will limit the following treatment to (1) the materials science aspects of the *metals* investigation, (2) the problems related to *ores*, mines, and metal extraction, and finally (3) technical aspects related to the characterization and diffusion of the *objects*.

Fig. 3.58. Metallography of a thermally annealed copper sample, showing evident slip systems produced by a mild mechanical working subsequent to the thermal treatment. (Courtesy of I. Angelini, University of Padova)

3.5.1 Metal science

Metals are elements, alloys and compounds characterized by atoms that readily lose electrons to form positive ions. The ions in the crystal lattice are surrounded and kept together by delocalized electrons, forming interatomic **metallic bonds** and are free to move in the conduction band. The electron mobility is at the root of the chemico-physical properties of metals: high electrical conductivity, high thermal conductivity, high density, light absorption, and with a lustrous surface appearance (lustre) (Verhoeven 1975, Reed-Hill and Abbaschian 1991). The other major property is that metals under stress can deform without cleaving, instead they form internal dislocations by a relative shift of the lattice planes (slip planes and bands), and interplanar bonds are reformed. Metals therefore can deform with a ductile behaviour, absorbing shocks and dissipating the strain energy through defect formation (Fig. 3.58). The observation of the presence of slip systems is an important feature allowing a reconstruction of the mechanical history of the metal (Box 3.l).

The first evident feature when approaching ancient metallurgy is the limited number of metals and alloys that were used in the past, compared with the plethora of modern metallic and intermetallic compounds. Table 3.14 lists most of the metals and alloys used before the 18th century, when chemists discovered and studied many of the elements of the periodic table. Most of the metals that were unknown or unconsciously used in the ancient world, which are now familiar commodities in everyday life, were identified and isolated after the 18th century AD. These include Ti, Al, Mn, V, Cr, Co, which entered mass production in the 20th century.

The other surprising feature is that, excluding iron, all the metals used in antiquity have a rather low average concentration in the Earth's crust (Wedepohl 1995). There is an apparent contradiction in the fact that the earliest exploited resources (Cu, Pb, Au, Ag) are among the least abundant common metals in rocks, whereas some of the most abundant metals (Al, Ti, Mn,

Table 3.14. List of metal elements and alloys used in the past, with the approximate time of widespread use (modified from Killick 2001)

Compound	Elemental composition	Earliest reported widespread use	Place
Copper	Cu	7000 BC (native) 5000 BC (smelted)	Near East
Lead	Pb	6th millennium BC	Balkans, Near East
Gold	Au	5th millennium BC	Balkans
Antimony/copper alloy (Sb-bronze)	Sb/Cu	5th millennium BC	Near East
Arsenic/copper alloy (As-bronze)	As/Cu	5th millennium BC	Balkans, Near East
Silver	Ag	4th millennium BC	Balkans, Near East
Tin	Sn	4th millennium BC	Near East
Tin/copper alloy (Sn-bronze)	Sn/Cu	4th millennium BC	Near East, Iran
Zinc/copper alloy (brass)	Zn/Cu	3rd millennium BC	Lesbos
Zinc/tin/copper alloy (gunmetal)	Cu/Sn/Zn	3rd millennium BC	Iraq
Nickel/copper alloy	Ni/Cu	2nd millennium BC	Near East
Iron	Fe	2nd millennium BC	Near East
Tin–lead (pewter)	Sn/Pb	ca 1500 BC	Egypt
Low carbon-bloom	C/Fe	ca 1300–1500 BC	Near East
Mercury	Hg	1st millennium BC	China
Zinc/tin/copper alloy (gunmetal)	Cu/Sn/Zn	ca 500 BC	India
Wootz steel	C/Fe	ca 300 BC	India
Sulphur/copper (niello)	S/Cu	1st century BC	Roman world
Zinc/copper alloy (brass)	Zn/Cu	1st century BC	Roman world
Noric steel	C/Fe	ca 50 BC	Austria
Platinum	Pt	ca AD 100	South America
Sulphur/silver (niello)	S/Ag	4th century AD	Roman world
Gold/copper alloy (tumbaga)	Au/Cu	ca 500 AD	Central America
Damascus steel	C/Fe	900–1100 AD	Near East
Zinc	Zn	10th century AD	India
Sulphur/silver/copper/lead (niello)	S/Ag/Cu/Pb	11th century AD	Byzantine/Islamic
Gold/copper (shakudo)	Au/Cu	12th century AD	Japan
Bismuth/copper alloy	Bi/Cu	ca AD 1500	Peru

Cr) were extracted and exploited only in very recent times. The normally accepted explanation, introduced by Charles (1985), makes use of the Ellingham diagrams (Figs 3.4 and 3.5), clearly showing that only some of the least abundant oxides (i.e. the ones with lines above Zn) can be easily reduced to metals using primitive pyrotechnologies, assumed to reach about 1200–1300 °C in early furnaces or crucibles. All the metals with lines below Cr need much higher temperatures to be reduced. It is thus found that the historical sequence of metal use corresponds well with the ease of dissociation of the metal–oxygen bond. The widespread introduction of a metal is directly related to the technical ability to reach a sufficiently high temperature in a reducing atmosphere.

One other very important aspect of metallurgy is that the physical and mechanical properties of a metal are critically determined by the **microcrystalline texture** of the material, i.e. the dimension, shape, and distribution of the crystals in the bulk (Askeland 1988, Brandon and Kaplan 2008). In turn, the texture is a direct consequence of the thermal and mechanical history of the metal, i.e. the temperature and pressure gradients that the metal underwent during metallurgical processing. I guess that everybody is familiar with the fact that if a copper wire is repeatedly bent in opposite directions, it will always break at the bending point. This is a direct consequence of the rearrangement of the crystallites shape and dimensions during bending and of the accumulation of slip defects in the grains, which positively affect the tensile strength of the metal in the mechanically worked part. Analogously, cast and thermally annealed metals that are virtually devoid of defects are soft, tenacious, and easily deformable. On the other hand, mechanical working of the metal induces a volume reduction, which is incorporated in the material as increased grain surface (i.e. smaller grain size) and a higher density of defects. The worked metal is hard, fragile, and hardly deformable.

By fine tuning the thermal cycles and the direction and strength of the applied forces, it is possible to obtain very different textures in the metal, and therefore to control the mechanical properties of the material. This is a major scope of physical and engineering metallurgy (Reed-Hill and Abbaschian 1991). Some simple effects induced by temperature and/or pressure on the texture of a one-component metal (Cu) are illustrated in Fig. 3.1.2. Metal textures are experimentally measured and quantified by metallographic techniques (Box 3.1). Engineers commonly process metals and then measure the resulting texture, in order to understand and control the physical properties. In a sort of inverse sequence, archaeometallurgists commonly measure the texture in order to reconstruct the original manufacturing process.

One further parameter is extremely important for one to determine the physical properties of the metal: composition. **Alloying** is based on the production of metals with two or more components. In the strict sense, alloys should form a **solid solution**, i.e. the two alloying elements should substitute each other in the crystal lattice of the metal. In this case the metal is formed by one crystal phase, with mixed composition. In a more general sense, the metals formed by partially soluble components or even immiscible components can also be considered to be alloys, they are formed by two or more crystalline phases with pure or intermediate compositions between the end-members. More complex cases exist in three or more components systems. The reader unfamiliar with phase diagrams should peruse the excellent metal-oriented textbook by Porter *et al.* (2009) or the great manual by Ehlers (1972), which helped to train generations of geologists and petrologists.

The Cu–Pb system is a typical immiscible system (Fig. 3.59), i.e. during cooling each component crystallizes at the proper crystallization temperature (i.e. the same as the melting temperature) forming crystals of pure Cu and

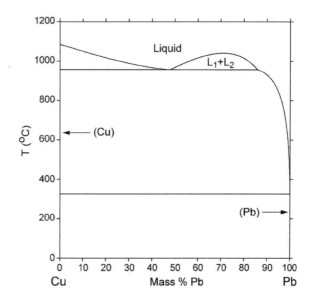

Fig. 3.59. Binary phase diagram of the Cu–Pb compositional join.

Pb composition. The two crystals coexist separately in the solid and, since the crystallization temperature of Pb (327 °C) is much lower than that of Cu (1084 °C), the lead crystals are invariably interstitial between the copper crystals.

The Cu–Sn system on the other hand has limited solubility of Sn in the copper α-phase (Fig. 3.60) up to about 15 Sn% at about 600 °C. The Cu–Sn phase diagram, which is relatively more complex than the Cu–Pb one, also shows that between 500 and 800 °C the β- and γ-phase are formed, both incorporating a substantial amount of Sn in the structure, and thus depending on the starting liquid composition and the cooling path, a variety of textural features are formed in tin–bronzes. Figure 3.61 shows the metallographic image of a Sn–bronze, with abundant β-phase segregation amidst the grid of α-phase dendrites. It is this interlocking network of the two phases that increases the mechanical strength of Sn–bronze with respect to pure copper.

By varying the initial composition and the time–temperature cooling path from the melt it is possible to produce a large variety of metal alloys with a range of physico-chemical properties. Common impurities in ancient copper are As, Ni, Bi, Sb, Ni, Co, Ag and Pb. Some of them enter the copper structure as solid substitutions, others form immiscible droplets that are commonly found as inclusions in the metal (Fig. 2.16).

A few alloys of historical importance are worth reminding ourselves of: **tumbaga** is an alloy with different proportions of copper and gold, typical of pre-Columbian Central America, which was commonly etched on the surface to remove the copper layer and enhance gilding (Scott 1995); **niello** is a general term defining several black-coloured mixtures of sulphides of changing composition through history (Table 3.14), from the 1st century BC Roman

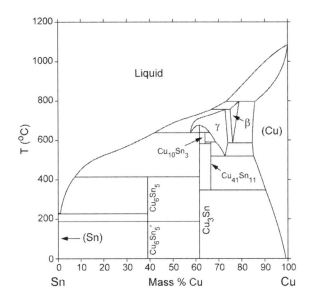

Fig. 3.60. Binary phase diagram of the Sn–Cu compositional join.

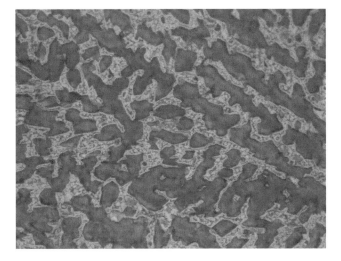

Fig. 3.61. Metallography of a tin bronze showing dendritic crystals of the α-phase (reddish) and δ-phase segregations (grey). (Courtesy of I. Angelini, University of Padova)

copper sulphide, to the 4th century AD silver sulphides. It became a ternary sulphide mixture in the 11th century AD through the interaction of European, Byzantine, and Islamic cultures (Northover and La Niece 2009). Niello was and is still commonly applied in the Far Eastern Asia as an inlay to decorate metalware such as silver table ware, and gold and silver jewellery; **shakudo** is a gold–copper alloy used in Japan for its beautiful dark blue-purple patina, and is mostly employed to decorate the Japanese swords guards and accessories (Oguchi 1983), it is part of the more general group of alloys called *irogane* (O'Dubhghaill and Jones 2009); Chinese **wu tong** is another similar black silver-containing patination applied on copper sheets (Wayman and Craddock

1993); and finally **Corinthian bronze** (also "*hsmn-km*" in Egyptian, or "*black copper*") is a Cu-based patina containing Au, Ag, As, or Fe and chemically treated in various way to assume shades from purple and blue to black. It has been recognized on various ancient objects from Egypt (Giumlia-Mair 1996) and the Roman world (Craddock and Giumlia-Mair 1993), and it has been proposed that it is at the origin of shakudo and wu tong.

The phase diagram of Fe–C (Fig. 3.62), which is the basis for ancient and modern steel, is even more complex as it contains a number of phases formed at different eutectic points using different temperature paths: *ferrite* (α-Fe, up to 0.035 wt% carbon, low temperature BCC structure), *austenite* (γ-Fe, up to 1.7 wt% carbon, FCC structure), δ-*ferrite* (δ-Fe, up to 0.08 wt% carbon, high temperature BCC structure), *cementite* (Fe_3C, contains 6.67 wt% carbon, intermetallic compound). A number of other compounds with specific microtextures are actually a mixture of phases: *pearlite* is a fine lamellar microstructure of ferrite and cementite, *bainite* is a non-lamellar mixture of ferrite and cementite obtained by large undercooling of austenite, *martensite* is a phase due to the diffusion-less transformation of austenite into a body-centred tetragonal phase.

Of course the variety of phases and microstructures in steel means that (1) each steel is a unique material, especially if formed with uncontrolled or partially controlled primitive processes, and (2) each steel has very different physi-

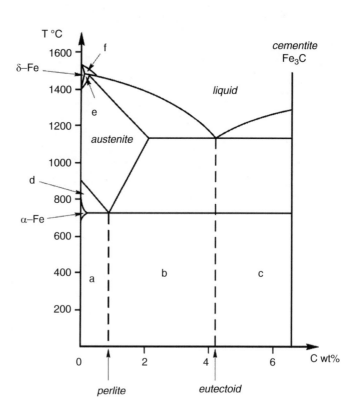

Fig. 3.62. The Fe–Fe_3C phase diagram for iron and steel. Phase formed are: **a** = ferrite + pearlite, **b** = pearlite + eutectoid, **c** = eutectoid + cementite, **d** = ferrite + austenite, **e** = δ-ferrite+ austenite, **f** = δ-ferrite + liquid. At the eutectoid austenite and cementite are formed.

cal and mechanical properties in terms of brittleness, hardness, tensile strength, etc. (Askeland 1988). Modern stainless steel has a much higher resistance to corrosion due to its chromium content (Cr > 12 wt%).

It must be clear to anyone that non-ferrous elements and alloys can all be produced and worked through a melting stage, or a vapour stage for the very volatile elements such as Zn. This is because the melting point can easily be reached by ancient pyrotechnologies. This is not the case for iron, whose melting point at 1540 °C precludes easy reaching of the liquid stage. Thus there are schematically two conceptually different technologies: for non-ferrous metals the gangue and the slag are more viscous than the matte and the metal, so the running phase is the metallic one; for iron (that is before the production of cast iron, see below) the metal remains solid and the fluid phase to be squeezed out by **forging** is the slag (Tylecote 1987).

Box 3.i PIXE Proton induced X-ray emission and ion beam analysis

Proton induced X-ray emission or particle induced X-ray emission (**PIXE**) is a technique used in the determination of the elemental composition of a material or sample. When a material is exposed to a beam of charged particles (protons, deuterons, α-particles, heavy ions) accelerated to energies in the range 200 keV–4 MeV, then atomic interactions occur, emitting photons in the X-ray range (characteristic fluorescence emission, Box 2.b), charged particles and neutrons. The technique was first proposed by Sven Johansson of Lund University (Johansson 1989, Johansson *et al* 1995). The optimal energy of the beam for analytical purposes is around 1–2 MeV, though higher energies may be necessary for the excitation of the K-lines of heavier elements (Denker and Opitz-Coutureau 2004).

From the instrumental point of view the technique is conceptually simple, since there is an ion beam hitting the surface of the sample and a solid-state X-ray detector at an angle analysing the emitted secondary fluorescence signal (Dran *et al*. 2000, Calligaro *et al*. 2004, Demortier 2004, 2005). However, the production of the ion beam requires an accelerator, commonly a Van de Graaff one, and few laboratories have dedicated machines as such: the AGLAE accelerator at the C2RMF, Louvre, France (Dran *et al*. 2004, Salomon *et al*. 2008) which is part of the Eu-Artech network (http://www.eu-artech.org/) is one of the few sources totally dedicated to the analysis of cultural heritage. The LARN laboratory at Namur (http://www.fundp.ac.be/en/sci/physics/larn/page_view/presentation.html) and the LABEC laboratory in Florence (Mandò 2007) (http://labec.fi.infn.it/index.html) are two other PIXE sources shared with other scientific and industrial applications. Other sources are available at nuclear centres. Most of the experimental ion beamlines have a so-called external micro-beam (Ioannidou *et al*. 2000), which allows the measurement of large samples in air, generally mounted on movable solid stages equipped with micro-translation motors. These recent extensions of PIXE, coupled with the use of micro-focused beams (**micro-PIXE**) and raster techniques allow chemical micro-mapping of any sample. The minimum dimensions of the beam on existing facilities are of the order of 1 μm, though it could be improved (Jamieson 2001).

The protons' interaction with the sample also emits *γ-rays* from nuclear processes and *protons* that are backscattered from the sample after elastic collisions. Detection of the particle-induced gamma-ray emission (**PIGE**) is a technique that can be used to detect some light elements that are hardly measurable with PIXE. Detection of the backscattered protons (Rutherford backscattering, **RBS**) yields information on the sample thickness and composition. The bulk sample properties allow for the correction of X-ray photon loss within the sample. PIXE, PIGE and RBS can be performed using the same ion beam and different detectors (Fig. 3.i.1).

(*cont.*)

Box 3.i (*Continued*)

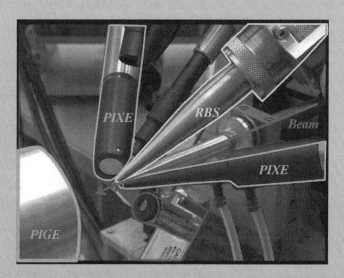

Fig. 3.i.1. Detection set-up for combined simultaneous PIXE-PIGE-RBS measurements at the LABEC laboratory in Florence. PIXE: standard set-up with two X-ray detectors at ~135° to the beam. RBS: Si PIN photodiode detector at ~140° to the beam, mounted in a vacuum case to minimize energy straggling along the path of scattered protons. The detector case, sealed with an aluminized 100-nm Si_3N_4 window, is evacuated to ~ 0.5 mbar. PIGE: HPGe γ-ray detector at ~ 0° to the beam, very close to the target (~ 4sr solid angle).

The advantages of the PIXE analysis stem from its non-destructiveness and its high sensitivity to the trace elements contained in the material. The main difficulty when applying PIXE may be in the interpretation of data. The results reveal all the elements present in the analysed volume but not their chemical state (Fig. 3.i.2). Moreover, in the case of heterogeneous materials such as stratified painting layers, metals, or glass the signal is an average of the textured sample, and PIXE results are to be considered bulk analyses. Some kind of depth profiling can be performed using ions accelerated at different energies, in which the particles probe volumes at different depths depending on their energy (Neelmeijer *et al.* 1996). The limitation is the escape depth of the secondary X-ray fluorescence signal.

PIXE is a powerful and non-destructive elemental analysis technique now used routinely in a number of different scientific areas, especially geology and environmental sciences. In archaeometry and conservation it finds fundamental application to surface and medium-depth analysis of metals, pigments, manuscripts, decorated ceramics, corrosion and alteration layers, patinas, etc. The penetration of the beam depends (a) on the energy of the beam, and (b) the composition of the sample. However, even if the ion beam penetrates 15–25 μm within the sample, the emitted X-rays escape only from the upper 1 μm or so of the material (Demortier 2004).

A large part of the PIXE publications in the literature is concerned with metal analyses (for example: Demortier and Ruvalcaba-Sil 1996, Narayan *et al.* 1996, Neff and Dillmann 2001, Demortier 2004, Ynsa *et al.* 2008). Interestingly, the measured pattern of trace elements in aluminium has been recently used to distinguish the electrochemically produced aluminium introduced in 1890 from the earlier metal obtained by reduction of aluminium chloride (Bourgarit and Plateau 2005).

Trace element analysis has been successfully used to determine the nature and provenance of semi-precious stones and gemstones, such as jade (Chen *et al.* 2004, Ruvalcaba-Sil *et al.* 2008) and many others, including emerald, ruby, and garnet (Calligaro 2005). The story of the ruby eyes of the small alabaster statuette of the Goddess Ishtar excavated in ancient Babylon and stored at the Louvre is especially fascinating (Calligaro *et al.* 1998): the red ruby cabochons inlaid in the eyes and navel were proved to be from Burma, on the basis of the measured trace element patterns. Trace element discrimination actually can be applied virtually to all kind of materials, and for many different purposes. Two examples based on PIXE analysis are the distinction of pozzolanic material in historical mortars (Sonck-Koota *et al.* 2008) and the authentication of Qing Dynasty porcelains (Huansheng *et al.* 1996, Cheng *et al.*

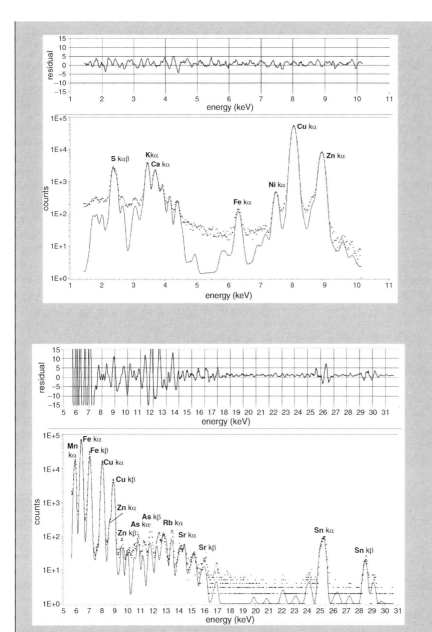

Fig. 3.i.2. PIXE raw data and peak fitting: (a) low energy range spectrum of a Cu reference material, (b) high energy range spectrum of a prehistoric copper smelting slag sample. The data were collected at the AGLAE laboratory.

2002). Finally, PIXE chemical mapping is an invaluable tool for the analysis of pigmented and inked surfaces, such as paintings and manuscripts (Olsson *et al.* 2001, Remazeilles *et al.* 2001; Figs 2.9 and 2.10).

The specific application of ion beams to **depth profiling** deserves to be outlined. As introduced in Section 2.1.3, the measurement of the chemical profiles from the surface of the material towards the bulk yields a wealth of information concerning alteration, diffusion processes, usage of the object, and relative dating. The diffusion profile at depth is commonly sought for many elements: F, Sr, U, in bones (Section 2.5.3.4, Box 3.m), Na, K

(*cont.*)

Box 3.i (*Continued*)

alkali dissolution and alteration in glass (Section 3.4), metal patina, Cl, and oxides in metals (Section 3.5, Box 3.j), sulphate and chloride salts in mortars and building materials (Section 3.1, Box 3.d). The special case of light elements, especially hydrogen, is instrumentally demanding because they cannot be detected by many common spectroscopic techniques. However hydrogen is a very important element, not only for relative dating of silica-based materials such as obsidians, flints and quartz (Section 2.5.3.2), but because it is related to the chemical alteration of materials in humid or aqueous environments (Scott 2002, Schreiner 1991). Ion beam techniques are commonly employed to measure the diffusion profiles of hydrogen and other light elements following two different kinds of beam–material interactions: nuclear reaction analysis (**NRA**, also nuclear resonance reaction analysis, **NRRA**) and energy recoil detection (**ERD**, also energy recoil detection analysis, **ERDA**). The NRRA technique exploits resonance nuclear reactions specific to each element. For example, for hydrogen the $^1H(^{15}N,\alpha\gamma)^{12}C$ reaction is commonly used having a resonance at 6.40 MeV (Becker *et al.* 1995), where ^{15}N ions are accelerated to resonance energies and targeted on the sample surface, and the emitted γ-rays of 4.44 MeV are counted at the detector. Since the resonance energy is very isolated in the element cross-section, the reaction only occurs when the energy of the beam is exactly at resonance. By progressively increasing the energy of the beam, at each energy step a particular depth in the sample is probed, and the γ-rays counted are proportional to the number of hydrogen atoms in the layer. The measured γ-counts vs. energy curve therefore corresponds to a hydrogen vs. depth profile (Fig. 3.i.3). The technique has been used to assess the deterioration of medieval glass artefacts (Schreiner *et al.* 1988, Schreiner 1991) and the elemental profiles through the patina of copper alloys (Ioannidou *et al.* 2000). The ERDA technique (Serruys *et al.* 1996) analyses the energy of the recoiled hydrogen atoms scattered by ions that are heavier than protons, usually He ions. The energy of the hydrogen atoms scattered in the forward direction carries information on the depth of the event, so that the energy spectrum yields a concentration depth profile. The technique has been recently used to assess the hydration layer in quartz artefacts, namely the rock crystal skulls thought to be manufactured by pre-Columbian Mesoamerican cultures (Calligaro *et al.* 2009). There exist about ten of such movie-inspiring objects, which appeared on the market in the 19th century, none of them found in archaeological contexts. On the basis of stylistic considerations (Walsh 2008), careful analysis of the working marks (Sax *et al.* 2008), and now also ERDA and microtopographic

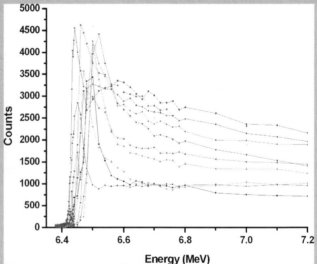

Fig. 3.i.3. Hydrogen profile measured by NRRA on calcium trisilicate, showing the progressive hydration in time. The energy of 7.0 MeV corresponds approximately to a depth of 250 nm. The data were collected at the Dynamitron Tandem accelerator of the Ruhr Universität, Bochum, Germany.

investigations (Calligaro *et al.* 2009) all of the analysed quartz skulls have been proven to be not older than the age of their appearance. Only a small anthropomorphic quartz head conserved at the Musée du quai Branly in Paris has a deep hydration layer testifying to its older origin, and possibly representing a genuine pre-Columbian artefact.

For in-depth information

Calligaro T., Dran J.-C. and Salomon J. (2004) Ion beam microanalysis. In: Janssens K. and Van Grieken R. (eds) (2004) Non-destructive microanalysis of cultural heritage materials. Comprehensive analytical chemistry series. Vol. XLII. Elsevier, Amsterdam. pp. 227–276.

Dran J.-C., Calligaro T. and Salomon J. (2000) Particle-induced X-ray emission. In: Ciliberto E., Spoto G. (eds) Modern analytical methods in art and archaeometry. John Wiley & Sons, New York. pp. 135–166.

Johansson S.A.E., Campbell J.L. and Malmqvist K.G. (1995) Particle-induced X-ray emission spectrometry (PIXE) Chemical Analysis: A Series of Monographs on Analytical Chemistry and Its Applications. Wiley-Interscience. John Wiley & Sons, New York.

Box 3.j Metal corrosion

Corrosion, as opposed to other forms of alteration and degradation, involves an electrochemical reaction usually between a metal and the water-loaded surrounding. Following the electrochemical convention, reactions are distinguished in *anodic*, involving oxidation reaction, and *cathodic*, involving reduction reactions. Metal ions at the anode undergo oxidation, thus building up a layer of solid oxide products, or else dissolving in the water solution. Metal oxidation lowers the free energy of the system and keeps the reaction going; this is why it is so difficult to stop the corrosion once it has started. The most common example of corrosion is rusting, which transforms iron and steel into non-metallic corrosion products (Figs. 2.m.3, 2.n.5). Corrosion reactions, common in all open environments, have enormous economic consequences because of their pernicious effects on most metallic industrial products used in transportation (rails, cars,...), buildings (steel reinforced concrete, Fig. 3.d.4b), communication and electrical wires, infrastructures (bridges, metal pipes,...), etc. In the field of cultural heritage the analysis of the corrosion products can (a) yield information on the original nature and structure of the artefacts, (b) indicate whether the corrosion products are natural or fakes made to artificially age the object, (c) indicate the best strategy for conservation. The relative stability conditions of the different compounds in the system at a given pressure and temperature are commonly plotted in the **stability diagram** (or Pourbaix diagrams: Pourbaix 1974; Fig. 3.j.1) as a function of the pH and the electrode potential (Uhlig and Revie 1985, McNeil and Selwyn 2001). The example shown in Fig. 3.j.1(a) is the very simple diagram for the copper–water system, with the concentration of dissolved copper at 10^{-6} molar and reduced sulphur at 10^{-2} molar, though the diagram can be plotted with any desired concentration. The marked upper part of the hatched parallelogram indicates common near-surface water conditions (McNeil and Little 1992). Where corrosion is dissolving the copper metal, ionic species in solutions are stable (i.e. $HCuO_2$ (aq)). Where the metal is stable, there is no corrosion; where the oxide is stable, there is commonly a layer of cuprite (Cu_2O) forming at the surface and acting as a protective barrier (the so-called **passive layer**) and slowing down the process. However the layer needs to be in close contact with the metal, without intervening pores or fracture where fluids can move. If the oxide layer is non-adherent, corrosion continues and the film is non-protective.

Figure 3.j.1(b) shows the stability field with no reduced sulphur, but containing chlorine in solution, to model seawater (McNeil and Little 1992). Chlorine is especially important in iron, copper and bronze corrosion also in freshwater conditions, because Cl ions tend to diffuse following the reaction potential and concentrate at the anodic

(*cont.*)

Box 3.j (Continued)

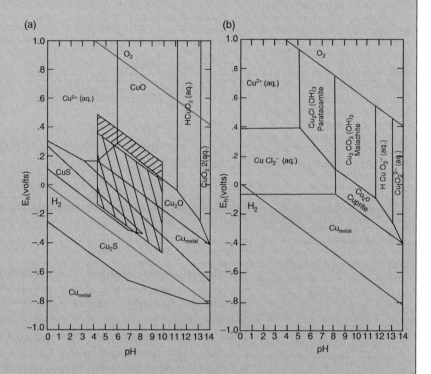

Fig. 3.j.1. Stability diagram (or Pourbaix diagram) of the water–copper system (a) with reduced sulphur and (b) with dissolved chlorine. The fields define the stability conditions of the different phases. Where ions are defined (i.e. $Cu^{+2}(aq)$) conditions for ion dissolution are present, and corrosion is acting (Modified from McNeil and Little 1992). The parallelogram in (a) marks the conditions of surface water, excluding extreme cases such as acidic bogs and mines. The upper hatched area indicates conditions of stagnant pond waters normally found in superficial environments.

region, greatly enhancing the localized corrosion (Scott 2002). At the interface between copper/bronze metal and the cuprite alteration, microanalytical techniques invariably allow observation of a thin layer of chloride compounds, principally nantokite (CuCl: Fig. 3.j.2) and other copper oxy-chlorides such as atacamite, clinoatacamite, paratacamite and botallakite (Scott 2000, 2002). The same process is one of the main causes of concrete degradation, especially when chlorides are used in winter as ice melters (Box 3.d). It is extremely important that chlorides are removed during the cleaning of an archaeological artefact, or else the corrosion process will soon restart after the conservation treatment, as moisture and oxygen reactivate what is called "bronze disease" or "active corrosion".

pH conditions present in natural environments can vary from extremely acidic (pH < 5, peat bogs, sulphide mines, etc.) to very basic (pH > 9, water in contact with limestones). The electrode potential conditions (E_h) are more related to the oxygen concentration: subsurface environments and water in contact with the atmosphere have higher E_h (i.e. oxidizing conditions), whereas deep and/or clay-rich sediments, which preclude water circulation, easily form anoxic conditions with little or no oxygen activity and much lower E_h (i.e. reducing conditions) (Garrels and Christ 1965). The electrode potentials in nature are limited by the stability of water: above the O_2 line water will dissociate producing free oxygen, below the H_2 line it will produce free hydrogen. The latter conditions can actually occur because of the local microbial activity.

Micro-organism activity is very important, especially when sulphate-reducing bacteria are at work in an S-containing environment (Fig. 3.j.1(a)). Copper sulphides are frequently present in copper or bronze objects as inclusions (Fig. 2.16), due to poorly refined copper smelting or a reaction between sulphur and other metal ions. Owing to bacterial activity a thin biofilm of reduced sulphides and sulphur is formed both in a seawater environment and in soil burial (McNeil et al. 1993). If the product sulphides are soluble (i.e. chalcocite Cu_2S) the corrosion will progress because the layer is unstable, whereas if the sulphides are stable (i.e. djurleite $Cu_{31}S_{16}$) the film acts as a protective passive layer. Copper sulphides are of course invariably present in the various formulations of niello decorations (Section 3.5.1) and sometimes the formation of sulphides replacing the copper metal are actually responsible for the conservation of the artefacts, such as in the case of the 2400 years old copper nails of the *Ma' Mikhael* shipwreck (Shalev *et al.* 1999).

Oxidation of surface or bulk sulphides (such as pyrite Fe_2S formed in anoxic conditions, especially in wood) after removal of the artefacts from the oxygen-poor burial conditions can create disequilibrium and activate severe chemical damage ("pyrite disease"). The conservation of sulphide-loaded artefacts in oxidizing and partially humid conditions such as those present in indoor museum environments is not an easy task, and one of the major headaches of conservators. Reducing humidity slows the reactions, but does not solve the problem. Oxidation of pyrite produce iron sulphate and sulphuric acid, with a net increase in volume (physical cracking) and acidity (increased chemical attack). The process is deleterious in archaeological objects, especially wooden objects (Jespersen 1989), and systematically destroys pyrite and marcasite specimens in mineralogical museum collections (Rimstidt and Vaughan 2003, Jerz and Rimstidt 2004).

The stability diagrams can be used to interpret the stability of the phases formed during corrosion under different conditions. In the presence of carbonate anions copper and bronze usually form malachite and/or azurite, whereas iron forms siderite. The presence of phosphates in the layer can induce the formation of copper (libethenite) or iron phosphates (vivianite).

(a) 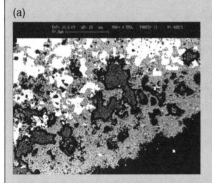 (b)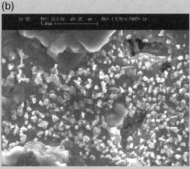

Fig. 3.j.2. (a) Nantokite islands (dark grey) at the interface between metal (white: upper left) and cuprite (pale grey: lower right). Chloride anions enhance the corrosion processes at the anodic region because of anion diffusion. (b) Close up of the nantokite (CuCl) crystals in the reaction layer. (SEM images courtesy of I. Angelini, University of Padova)

A few comments are warranted towards the proper characterization of metal surfaces (Giumlia-Mair 2005). The archaeometrical problems involved are complex and various, including the assessment of the corrosion processes, identification of original patinas, reconnaissance of re-deposited phases, identification of fake alteration layers, and so on. In principle many of these problems are related. What it really needs to be assessed is:

the *original structure and composition* of the metal, including microstructural features inherited by the manufacturing process, such as the common phenomenon of apparent superficial Sn-enrichment (Tylecote 1985), or the presence of original patinas applied for aesthetic or protective purposes (Section 3.5.1), such as the Japanese shakudo and the Chinese wu tong, also used in the ancient Western world for patination (La Niece and Craddock 1993);

the *evolution of the surface in time*, including metal corrosion, selective dissolution, and phase precipitation during burial. The understanding of the intervening processes yields as a by-product the identification of fake alteration or corrosion patinas, and the validation of ancient ones (McNeil and Selwyn 2001).

the *present state* of the metal, including an understanding of the active processes to design cleaning, stabilization, and conservation procedures.

(*cont.*)

Box 3.j (*Continued*)

Sometimes the identification of external contamination or diffusion phenomena related to corrosion is straightforward. In other cases the identification of the actual processes is much more difficult. For example, the common enrichment of Sn at the surface of bronze can be produced by inverse segregation during casting, by preferential dissolution of copper with respect to tin during corrosion, or by explicit superficial tinning of the object (Tylecote 1985). Chemical analysis is sometimes not sufficient to resolve the issue and careful analysis of the micro-textural features (such as the interstitial position of the crystal, chemical coring, morphology of the re-precipitated phases, etc.) are needed. Sometimes surface embrittlement of silver alloys due to discontinuous reprecipitation is a safe sign of antiquity. The presence of the modern ^{36}Cl isotope produced by recent nuclear explosion is an unequivocal sign of modernity, though there is always the possibility of recent absorption or modern patination of an ancient object. Crystal morphology due to dealloying by long water exposure is another sign of antiquity... the list of scientific tricks is almost as long as the tricks used to produce forgeries!

For in-depth information

Doménech-Carbó A. and Doménech-Carbó M.T. (2009) Electrochemical methods in archaeometry. Conservation and Restoration. Monographs in Electrochemistry. Springer-Verlag, Berlin–New York.

Scott D.A. (2002) Copper and bronze in art. Corrosion, colorants, conservation. The Getty Conservation Institute, Los Angeles.

Uhlig H.H. and Revie R.W. (1985) Corrosion and corrosion products. John Wiley & Sons, New York.

3.5.2 Ore, mines, smelting

The links between ore deposits, mining activities, and metallurgical sites for smelting and metal working are fascinating and mysterious. For many, they carry a sense of intrusion into the secrets of the Earth, and an alchemic sense of creation every time shining metal is born out of dark and mostly uninteresting minerals. Primo Levi's tale on "Lead" is particularly intriguing in this respect (in: *Il Sistema Periodico*, 1975; Engl. Translation: *The Periodic System*; named in 2006 by the Royal Institution of Great Britain the **best science book ever**). Yet mining and ore treatment activities in the past were far from being a catching gothic adventure and were most likely even more hard and inhuman than today. We just need to take a brief look at the present day ant-like frantic activity in the search for gold in the Brazilian Serra Pelada, the child exploitation for the extraction of coltan (i.e. columbite–tantalite minerals) in the Congo, or the painful journey of Indonesian sulphur workers up the slopes and into the helly vapours of the Ijen volcano complex, East Java, to get insights of what could have been the mining and smelting activities in the past. Images as such (Fig. 3.63–3.65) show the human side that is largely untold by the ancient records (Plinius, Agricola, Biringuccio, etc.), which are mostly confined to the more technical and elevated part of the story.

Getting back to Earth, besides the well-known cited historical accounts there are several pieces of archaeological evidence left from ancient metallurgical activities: mines, roasting and smelting sites, technical equipment, smelting slags, furnaces and crucibles, raw metal. They yield a great deal of information

Fig. 3.63. Panning for precious stones at Ilakaka, Madagascar.

Fig. 3.64. Deep pits hand-dug by tourmaline searchers at Mahaiza, Madagascar.

on the organization of the work and on the technical processes, though they commonly share the same problem: they are almost never found in the same place, so that we have plenty of small fragments of the puzzle, but no complete image helping to put them together. We have mines without slags, slags without furnaces, furnaces with no tuyeres or metal, raw metal with no ores, and so on. Building the whole picture means: (a) locating the mineral deposits

Fig. 3.65. Hand carved gallery for search for tourmalines and beryls at Ambohitralanana, Madagascar. (Photo courtesy of A. Guastoni, University of Padova).

exploited in antiquity, (b) linking the ores to the processing sites and the smelting slags, (c) understanding the different stages of ore treatment (beneficiation, separation, fragmentation, roasting), (d) reconstructing the pyrotechnological operations (furnace smelting, slagging, blooming, crucible smelting or melting, refining, etc.), and (e) inserting the whole chain into a socio-economical and cultural background.

In the most general case it is virtually impossible to complete all aspects of the investigation for an individual site or area, simply because of time, cost, and record biases. Only a few extensive long-term projects come close to delineating all the major features of the metallurgical chain in a geographical area, from the mine survey to the production and diffusion of metal. These are the cases of the Arabah (Timna and Faynan) region in the Near East where, over thirty years, long projects have investigated and detailed out all available evidence at different levels (Rothenberg and Merkel 1995, Rothenberg 1999, Weisgerber 2003, Hauptmann 2007), reaching a sound and complete interpretation of the regional archaeometallurgy. Such projects should be used as examples for planning and designing wide ranging, and far reaching, investigations.

Irrespective of the geographical and geological settings, the general simplistic story circulating on the prehistory of metals goes that native metals were exploited first, because they do not need any refined pyrotechnology for metal extraction, and that the early metal tools imitated the lithic artifacts that were fashionable at the time. Now, in many places there indeed is some similitude between early copper axes and coeval or older stone axes, but for the most part the story is not supported by archaeological or geological evidence. In many places native metal objects are unknown until the sudden appearance of very advanced metallurgy exploiting bronze alloys at their best. Scandinavia is totally lacking a Copper Age, and this seems also the case for China, despite several much-debated claims (Muhly 1988, Killick 2001). Furthermore, in several areas the early metal objects do not at all imitate the lithic tools in use. In the Alpine area and Central Europe, for example, a substantial fraction of the early copper tools are small awls (Pearce 2000, Höppner *et al*. 2005, Table 2 of Pedrotti 2001), which may have had better functional properties, especially for flint flaking, than tools made with other materials, such as antler (Pearce 2000). In many other places, analogously to ceramics (Section 3.1.2) and glass (Section 3.4), the earliest metal objects are not functional tools but rather aesthetically-oriented display goods (Eluère and Mohen 1991, Chernykh 1992), so that the functionality paradigm can hardly be applied. We should remind ourselves that copper and bronze, even when heavily worked, make very efficient tools but do not have hardness Mohs values above 4–5, whereas the obsidian and flint range is 5–7 (Table 3.17). Unlikely modern hard steel, the advantageous properties of prehistoric metals were therefore not related to hardness, but rather to tenacity, ductility, and the possibility of easily reshaping and thermally releasing the strain on the object after use (Artioli 2007). Once a lithic artefact is broken, the only possibility left is to re-use smaller and smaller fragments, whereas when a metal is deformed or broken, it could virtually be re-cast and reused forever. The real change in metal properties came much later with the slow advent of carburized iron (Stech Wheeler *et al*. 1979).

The popular story proceeds by claiming that upon depletion of the native metal resources at the surface, extractive metallurgy developed in order to sequentially exploit the ore stratification of the mineral deposits: first the altered part (gossan), mainly composed of oxidic ores (oxides, carbonates, sulphates, arsenates, etc.), and then progressively deeper and unaltered sulphidic layers (sulphides, arsenites, etc.). See Table 3.15 for a general list of the minerals that are mostly used for metal extraction. Unfortunately not even this story stands up to archaeological and geological evidence. Many of the Alpine, Irish, and Welsh copper deposits do not have an altered layer at all, and only sulphides or very small quantities of oxidized minerals are available. This is the case for many copper deposits known to have been exploited in Chalcolithic or Early Bronze Age such as Saint Véran (France: Bourgarit *et al*. 2008; Fig. 3.66), Brixlegg-Mariahilfbergl (Austria: Höppner *et al*. 2005), Cabrières (France: Ambert and Vaquer 2005), Libiola and Monte Loreto (Italy: Maggi and Pearce 1998; Fig. 3.67). In fact the Chalcolitic co-smelting of polysulphides and oxides (Rostoker *et al*. 1989) is a very attractive proposal that seems to account well for the available evidence of early crucible and primitive reactor smelting in many sites from Spain, across the Alps, to the Balkans (Bourgarit 2007). The recent results based on the replication of Chalcolithic crucible smelting seem to indicate that solid reactions between oxides and sulphides are more important than the forced oxygen flux to control the redox conditions (Burger 2008, Burger *et al*. 2010). The main function of the forced air is to keep the combustion going. Careful control of mineral composition of the charge may thus be the key factor determining successful or unsuccessful small-scale smelting.

Smelting reconstructions are one technical field of investigation where theoretical and very practical issues converge to design a plausible (read "working") story. Thermodynamics meets kinetics. The armchair approach (Killick 2001) and the experimental approach really must come together in order to squeeze copper out of sulphides...malachite is not a problem, as metal copper can be reduced out of carbonates and oxides (malachite, azurite, cuprite,

Table 3.15. List of principal minerals used for metal extraction in the past

Metal	Native	Oxidic ore	Sulphidic ore
Ag	silver Ag	Ag-rich argentojarosite $AgFe_3(OH)_6(SO_4)_2$	achantite–argentite Ag_2S proustite Ag_3AsS_3 pyrargyrite Ag_3SbS_3 Ag-rich galena (Pb,Ag)S
Au	gold Au		aurostibite $AuSb_2$ gold-amalgam (Au,Ag)Hg sylvanite $(Au,Ag)_2Te_4$
As	arsenic As	ludlockite $(Fe,Pb)As_2O_6$ scorodite $FeAsO_4 \cdot 2H_2O$ conichalcite $CaCu(AsO_4)OH$ austinite $CaZn(AsO_4)OH$ erythrite $Co_3(AsO_4)_2 \cdot 8H_2O$ beudantite $PbFe_3(AsO_4)(SO_4)(OH)_6$	skutterudite $(Co,Ni)As_{3-x}$ realgar AsS orpiment As_2S_3 arsenopyrite FeAsS gersdorffite NiAsS cobaltite CoAsS
Cu	copper Cu	cuprite Cu_2O tenorite CuO delafossite $CuFeO_2$ malachite $Cu_2(CO_3)(OH)_2$ brochantite $Cu_4(SO_4)(OH)_6$ azurite $Cu_3(CO_3)_2(OH)_2$ olivenite $Cu_2(AsO_4)OH$ chrysocolla $(Cu,Al)_2H_2Si_2O_5(OH)_4 \cdot n(H_2O)$ chalcanthite $CuSO_4 \cdot 5(H_2O)$ atacamite $Cu_2(OH)_3Cl$	chalcopyrite $CuFeS_2$ chalcocite Cu_2S digenite Cu_9S_5 covellite CuS bornite Cu_5FeS_4 cubanite $CuFeSnS_4$ tennantite $(Cu,Fe)_{12}As_4S_{13}$ tetrahedrite $(Cu,Fe)_{12}Sb_4S_{13}$ bournonite $CuPbSbS_3$ enargite Cu_3AsS_4
Fe		goethite FeO(OH) lepidocrocite FeO(OH) magnetite Fe_3O_4 hematite Fe_2O_3 siderite $FeCO_3$ ankerite $Ca(Fe,Mg,Mn)(CO_3)_2$ jarosite $KFe_3(OH)_6(SO_4)_2$	pyrrhotite $Fe_{1-x}S_x$ pyrite-marcasite FeS_2 arsenopyrite FeAsS
Hg		montroydite HgO	cinnabar HgS
Pb		litharge-massicot PbO cerussite $PbCO_3$ anglesite $PbSO_4$ phosgenite $Pb_2(CO_3)Cl_2$ laurionite Pb(OH)Cl	galena PbS boulangerite $Pb_5Sb_4S_{11}$ semseyite $Pb_9Sb_8S_{21}$ bournonite $CuPbSbS_3$
Sb		cervantite $Sb_3Sb_5O_4$ kermesite Sb_2S_2O senarmontite Sb_2O_3 stibiconite $Sb_3Sb_{25}O_6(OH)$ valentinite Sb_2O_3	stibnite Sb_2S_3 tetrahedrite $(Cu,Fe)_{12}Sb_4S_{13}$ bournonite $CuPbSbS_3$ boulangerite $Pb_5Sb_4S_{11}$ semseyite $Pb_9Sb_8S_{21}$ pyrargyrite Ag_3SbS_3 ullmannite NiSbS
Sn		cassiterite SnO_2	stannite Cu_2FeSnS_4
Zn		zincite (Zn,Mn)O smithsonite $ZnCO_3$ emimorphite $Zn_4Si_2O_7(OH)_2 \cdot (H_2O)$ hydrozincite $Zn_5(OH_3,CO_3)_2$	sphalerite-wurtzite (Zn,Fe)S

Fig. 3.66. One of the bornite veins exploited in the Early Bronze Age at Saint Véran, Queyras, French Western Alps.

Fig. 3.67. One of the Late Neolithic–Early Chalcolithic galleries at Monte Loreto, Liguria, Italy.

brochantite) in one step at relatively low temperatures under reducing conditions. Despite several claims, however, it is impossible to get copper metal from sulphides alone (chalcopyrite) in one slagging step. The first product is always Cu-enriched sulphidic matte (black copper), that must be subsequently refined. The reducing operation is instead possible using a mixed chalcopyrite–malachite charge. Further details on copper smelting processes are described through the investigations of slags (Box 3.k).

The major problems involved in the investigation of metal extraction from ore minerals are the following.

1) Can we trace the provenance of the metal to the ores?
2) Did extractive metallurgy diffuse from a single centre or rather was it an independent invention in several places?
3) How do we recognize and investigate smelting sites? What information can we extract from scientific analyses in order to interpret ancient smelting processes?

3.5.2.1 *Provenance*

The *vexata quaestio* of metal origin and diffusion has lead to a number of fierce discussions and several headaches over the last twenty years or so, with a climax being reached in the 1990s with a number of papers arguing over the validity of Pb-isotopes methods for provenancing metals and other materials. I will not replicate the discussions, but rather direct the reader to one fairly balanced account of the method itself (Gale and Stos-Gale 2000), and to a few papers summarizing the pros and cons of the whole debated issues (Tite 1996, Pollard 2009, Gale 2009). To the Pb-isotope aficionados these papers and the long list of references contained therein should be sufficient to obtain the essence of the discussions.

What I care to put straight in this context is that the so-called Pb-isotope method is not at stake. I described in Section 2.5.4.2 how the radioactive decay process is used to date rocks and other materials. The Pb isotopes (^{204}Pb,^{206}Pb, ^{207}Pb and ^{208}Pb) are the final products of the U, Th decay series, so that the measurement of the lead isotope ratios measured on the metal provide an indication of the age of the ore, provided that there is no isotopic fractionation during smelting. In geology the ^{206}Pb/^{204}Pb, ^{207}Pb/^{204}Pb and ^{208}Pb/^{204}Pb are most commonly used, whereas in archaeology the ^{206}Pb/^{204}Pb, ^{207}Pb/^{206}Pb and ^{208}Pb/^{206}Pb are more frequent.

This we take for granted: all experimental evidence seem to confirm that the reduction and refinement stages of the metal do not change the isotopic signal, therefore the Pb ratios measured on the metal and the slags are the same as those of the smelted minerals, within experimental error. The reader will note that I stated "the age of the ore", not the geographical provenance. In fact this is exactly the information that the Pb isotope method is providing: the age of formation of the ore minerals used to extract the metal. What is the problem then? The problem, as always, is related to the way that the method is used, and precisely (1) to the fact that there are commonly many ore bodies formed at the same geological age, and the Pb-isotopes alone cannot distinguish between them. The ill-defined and overlapping clouds of experimental points relative to the reference ore deposits are a warning to this point. Further (2) there are many problems related to the ore sampling. Schematically, an ore deposit is composed of primary sulphides (formed at the starting age of the deposit), remobilized sulphides (recrystallized during subsequent metamorphic events), secondary high temperature minerals (formed during late-stage hydrothermal events), secondary low temperature minerals (formed at much more recent times during low-temperature surface alteration) and sometimes even native metal (formed in supergenic conditions). All these generations of metal-carrier minerals have a distinct age. What is then the reference age of the deposit? It is a stream of points along the general isochron line, starting from the isotope ratios corresponding to the first event, and following the curve up to the last minerogenetic event. That's why there is hardly a narrowly defined isotopic field for an ore deposit, but rather a scattered cloud of experimental points. That's why careful sampling of the deposit is crucial to get a reliable reference signal. The worst case is represented by modern reference sampling of the primary sulphides (these are often the only minerals left in over-exploited mines that have been repeatedly worked at different times), when the metal of the artefact was extracted in ancient times by long gone carbonates and native copper. Interpretation of the evolution of the isotopic signal of the deposit during its geologic history is clearly required. Specific to copper deposits, many Pb-isotope data in the literature are of course collected on Pb-rich minerals of the deposit (Table 3.15), but there is no guarantee that the Pb minerals were formed at the same time as the Cu-minerals, from which the metal was extracted. Again, detailed minerogenetic interpretation is needed, and extraction of the Pb isotopic signal from Cu-rich minerals alone is advised. Finally, (3) ancient objects are often alloyed (Sn-bronze, As-bronze, brass) and possibly made with recycled metal. In such cases the signal of the original copper ore may be totally erased by the mixing of different components, depending on their Pb content and extent of alloying/mixing. This may be a major problem for the interpretation of the Pb-isotope signal.

There are many other points that should be touched upon, and indeed many of these are thoroughly discussed in the cited papers. I hope, however, that the message is clear: the Pb isotope method works, but it needs to be used with due competence and caution.

To move on from Pb isotopes, in addition to age there are of course many other geochemical links between the metal, the slags, and the ores. It has long been realized that trace elements following the main metal during extraction could be used to (a) define composition-based groups of artefacts, (b) determine the geologic nature of the ore from which the metal was smelted, (c) distinguish similar artefacts made with different ores and possibly originating from different areas, and (d) maybe date the metal objects based on the compositional evolution in time.

Many of these concepts are widely used regionally to develop a time stratigraphy and space distribution of metal compositions. The readers are referred to the excellent reviews of Pernicka (1999, 2004), describing in detail the use of chemical tracers in the provenancing of the artefacts and the reconstruction of the smelting processes. Here I will recall that normally a dozen or so chemical elements are experimentally measured for copper and bronzes: Cr, Mn, Fe, Co, Ni, Zn, As, Ag, Sn, Sb, Pb and Bi. These elements are commonly selected because of their behaviour during sulphide smelting, and because they are indicative of mineral associations in the starting ores, namely the presence of specific polysulphides in the charge. Interestingly, the archaeological literature is pervaded with the term fahlore (or fahlerz), which irrespective of its geological origin (from German "pale ore", strictly any grey-coloured ore mineral mainly composed by tetrahedrite-tennantite minerals) is now casually and often inappropriately used to indicate any mineral assemblage supplying minor or trace elements to the smelted copper, regardless of the fact that tetrahedrite and tennantite invariably contain As and Sb, sometimes Ag and Zn, very rarely Bi, Cd, Hg, or Te (Ixer and Pattrick 2003), but the sources of Cr, Co, Ni, and Pb are likely to be very different minerals. The elemental data obtained on the metal objects are systematically compared with the elemental abundances measured on representative specimens of the ores, the concentrations of the different elements spanning several orders of magnitude. The concentration patterns are taken as indicative of the source (Ixer 1999, Pernicka 1999), assuming that the chalcophile elements (i.e. those that have affinity with Cu and follow it during smelting) follow copper during the pyrotechnological smelting processes and there is no fractionation during the copper metal reduction (Pernicka 2004 and references therein). Depending on the local geology and the nature of the investigated metal a number of other elements can be analysed, provided they are analytically detectable in the metal and geologically significant.

The trace element strategy as provenance tracers can of course be extended to all metals. Gold has been widely investigated (Guerra 2004, Guerra and Calligaro 2004).

Despite the fact that the results of large analytical programs of ancient metal objects, such as the one launched by the Würtembergisches Landesmuseum in Stuttgart that analysed over 20 000 prehistoric metal objects from all over Europe (Junghans *et al.* 1968–1974), have met with controversial acceptance, there is plenty of information to be tapped from the developed database. The main limitation of the existing database is that it contains poorly checked and sometimes obsolete archaeological information, so that the chemical data are sometimes hard to compare with archaeological models. The consistency of the large database is also a critical point (Section 2.4.1). Nonetheless, it is clear that the data resulting from the chemical characterization of the objects contain valuable information on the original ores, though they may suffer from possible elemental fractionation during smelting (for example As loss in the furnace atmosphere, Zn loss to the slag, etc.), contamination from alloying and metal recycling, and a difficult comparison with mineralogically heterogeneous deposits.

The best solution of course is to use both the Pb-isotope methods and the chemical tracer methods, in order to complement the information and get more reliable indications of the deposit (age + mineralogical assemblage). Several examples of successful combined approaches are available concerning copper or iron artefacts (Schwab *et al.* 2006, Degryse *et al.* 2007).

By extending this logic, we may think of adding further chemical and isotopic parameters in order to constrain the characteristics of the original mineral assemblage even further. Two interesting extensions have been proposed. The first is to use more chemical tracers, even some elements present at trace or ultra-trace level (Section 2.3). Recent works show that a whole range of elements at the ppm level or below are detectable in metals, and they carry a very distinctive minerogenetic signal either from the ore minerals, or derived from the host rocks (Artioli *et al.* 2008). Elements such as Ga, Ge, In, Te that are rarely analysed in metals do indeed carry valuable information for specific mines. The whole suite of rare earth elements (REE), widely used as tracers of petrological processes, add valuable information concerning the host rocks of the ores. The other recent proposal is to use different isotopic ratios as indicative of the source and/or the type of mineral assemblage. Tin isotopes (^{116}Sn, ^{117}Sn, ^{118}Sn, ^{120}Sn, ^{122}Sn: Begeman *et al.* 1999), copper isotopes (^{63}Cu, ^{65}Cu: Gale *et al.* 1999, Klein *et al.* 2004, Colpani *et al.* 2007), osmium isotopes (^{186}Os, ^{187}Os, ^{188}Os: Junk and Pernicka 2003). The fractionation of copper isotopes in particular is known to be related to the temperature of the ore-forming environment (Zhu *et al.* 2000, Larson *et al.* 2003) and it is a very promising method of distinguishing the mineralogical assemblage use in the smelting charge. In each deposit there is in fact a neat enrichment in the lighter isotope in the sequence bornite < chalcopyrite < malachite, cuprite < native copper, so that the δ^{65}Cu is a good indicator of whether primary sulphides or secondary minerals were smelted.

Advanced statistical analysis of the largest possible number of chemical and isotopic tracers (Table 3.16) seems to be highly discriminant for the ore mineral districts, and the signal can be followed through the slagging and smelting process to slags and reduced metal. The method has been applied as a test case to the Agordo area, Italian Alps (Artioli *et al.* 2008).

Table 3.16. Main chemical and isotopic tracers used to link metals and ores

Tracer	Nature of the information	Application
Pb isotope ratios	age of the ore deposit	all metals
Cu isotope ratio	temperature of formation of the Cu-minerals	copper
Sr isotope ratio	nature of the ore	iron
Sn isotope ratios	coexisting Sn-oxides/Sn-sulphides fractionation temperature of formation of the Sn-minerals	tin
Os isotope ratios	type of deposit (placer or vein deposits)	gold
minor elements (Cr, Mn, Fe, Co, Ni, Zn, As, Ag, Sn, Sb, Pb, Bi)	ore mineral assemblage in the charge	all metals
REE	host rocks of the deposit	copper
trace elements (Sc, V, Ga, Ge, Se, Y, Zr, Nb, Ru, Rh, Pd, Cd, In, Te, W, Re, Os, Ir, Pt, Au, Hg, Tl)	genetic processes of the deposit and interaction with host rocks	copper

3.5.2.2 Diffusion vs multiple invention

The firm conviction that extractive copper metallurgy started in the Near East before 5000 BC and spread from there to all other areas of the Old World dominated the archaeological and archaeometallurgical thinking for the good part of the last century (Killick 2001).

We now have evidence of the first use of copper in Anatolia in the late 9th millennium BC (Çayönü Tepesi), and at least other three copper-using sites dated to the 8th millennium BC (Yalçin 2000). In addition to early copper metal circulation, whose nature is hardly distinguishable between primitive co-smelting and native metal, a recent summary of smelting activities from Late Neolithic to Late Chalcolithic (approximately the 2nd half of 5th millennium BC to the 2nd half of the 3rd millennium BC, Table 1 of Bourgarit 2007, and references therein) show that early smelting of oxidic minerals took place Arslan Tepe-Anatolia, Timna-Arabah, Los Millares, Almizaraque and Cabezo Juré-Spain. Smelting of mixed sulphidic–oxidic ores has been recognized at Murgul and Norsun Tepe-Anatolia, Abu Matar and Shiqmim-Israel, Faynan-Arabah, Dolnoslav-Bulgaria, Brixlegg-Tyrol, and Cabrières-France. Smelting of sulphides has been recognized at Milland, South-Tyrol, and Gaban, Trentino, Al Claus-France, and La Ceñuela-Spain.

There is therefore ample evidence of diffused metal extraction activity starting from the end of the 5th millennium BC in several areas, the main centres being in the Eastern Alps, Balkans, Anatolia, and the Near East. Southern France and Spain seem to enter the game slightly later, at the beginning of the 3rd millennium BC, and Britain shows the first smelting evidence in the last part of the 3rd millennium BC. With respect to Renfrew's map of early copper metallurgy (Renfrew and Bahn 2008, p. 346), the recent results tend to narrow the gap between the "isolated centres" of copper production. Whether this means a rapidly spreading technology, or else a diffuse tendency throughout most of the prehistoric world to attempt copper extraction by any means, remains a matter of debate and a stimulating topic for future investigations. I tend to favour the latter hypothesis, based on the following points: (1) all possible kinds of ores, from purely oxidic to purely sulphidic, were being exploited in different places with rather different and generally inefficient smelting techniques, (2) the early available evidence in each site is rather scarce and is mostly confined to crucibles and a few very heterogeneous slags (Box 3.k), and (3) the technological developments follow similar paths but at rather different times. For example when furnaces for large scale operations were being built in Faynan (Bronze Age II, ca.3100–3000 BC, the earliest known: Hauptmann 2007), many of the other sites were still using small-scale ore smelting in crucibles, including all Anatolian sites. However at about the same time small furnaces were apparently in use at Cabezo Juré-Spain and La Capitelle du Broum, Cabrières-France (Bourgarit 2007). The sites of Milland and Gudon, South Tyrol, Italy, radiocarbon dated to few hundred years later (2800–2700 cal. BC), allowed the recovery of tens of kg of slags and, although no furnace was found, the slags were certainly not the product of small-scale crucible smelting.

The panorama therefore offers a variety of developing technological experiences, each one coping with different geographical environments, specific geological resources, and technical and cultural traditions. The pyrotechnological shift from primitive reactor smelting to proper furnace smelting was common to all of them.

During the 2nd millennium BC independent development of copper smelting processes occurred in the New World (Killick 2001) and by the early 1st millennium BC in Niger, West Africa (Killick et al. 1988). In Asia, a well developed tin–bronze metallurgy is present in China by the early 2nd millennium BC (Erlitou-Erligang cultures), and in Thailand by the mid-2nd millennium BC (Khao Wong Prachan Valley, Pigott and Ciarla 2007). The origin of this advanced bronze metallurgy is under intense debate, though influences from the Northern Steppe borderlands are reasonable hypotheses (Linduff 2004). Besides a few contested finds in Central China (Jianjun and Yanxiang 2003), it is to be noted that a diffused brass (Cu–Zn) and gunmetal (Cu–Sn–Zn) metallurgy dedicated to small ornaments, pins, and a few tools was present in South-Western Asia (Iraq, Georgia, Turkmenistan, Uzbekistan, Azerbaijan) between the late 3rd millennium BC and the early 2nd millennium BC (Thornton 2007).

Finally, iron was known and had a very limited use for most of the 2nd millennium BC, though for some unknown reason the processing of carburized iron started to be mastered and the products more appreciated than bronze during the last centuries of the millennium. How this happened remains an open question. Iron had a very slow start, but when solid-state absorption of carbon at high temperature was under control, then it became apparent than the material was harder than bronze and more competitive. The reason for limited early success is the intrinsic difficulty of the process: Tylecote (1987) indicates that 6 hours carburization at 900 °C into the forger's hearth makes carbon travel about 2 mm into the iron. No wonder that early smiths frequently failed to produce it properly. Examples of surface carburized iron tools are present in Cyprus, in Egypt, and in the Near East, though for the most part the iron archaeological record is very limited because of corrosion processes. The earliest iron-pervaded culture in Western Europe is the Hallstatt one, though most burial weapons are in pretty bad condition.

During the last part of the 1st millennium BC the high temperature carburization and quenching process was optimized in different ways in Asia (wootz steel, later diffusing in the Middle East as Damascus steel) and in Europe (various types of carburized iron, the most famous being the Noric steel or "Ferrum Noricum", produced in the Austrian part of the Roman Empire). Cast iron (i.e. iron produced by eutectic melting) started in China during the Han Dynasty (206 BC–AD 220), and the few examples available in Roman Europe are thought to be imported from the Far East (Tylecote 1987). The first real production of cast iron by the bloomery process in Europe was in Sweden in the 13th century AD, apparently deriving the expertise from the Far East through Mongolian contacts. The earliest documented blast furnace was established at Ferriere, Genova, Italy in 1464.

3.5.2.3 *The investigation of metallurgical sites*

Metallurgical sites are frequently removed from the mines. There are just a few cases where slags and furnaces are in the immediate vicinity of the mines: for example, Cabrières in France, where a small "metallurgical village" is located close to a widespread extractive area (Ambert and Vaquer 2005) and Ross Island, Southern Ireland (O'Brien 1996). The surviving physical evidence of early smelting however is so ephemeral and inconsistent that in most cases it is overlooked or misinterpreted. A number of mines are known that were exploited in ancient times, but there is no corresponding mineral processing site or evidence of smelting. The Libiola and Monte Loreto mines in Liguria, Italy (Maggi and Pearce 1998) and the Ai Bunar mines in Bulgaria (Cernych 1978) are examples of heavily worked mines with no signs of metallurgical activity. The Early Bronze Age mine of Saint Véran, Queyras, France (Rostan and Rossi 2002, Bourgarit et al. 2008) is an example of a mine that is estimated to have produced several tons of metallic copper per year, but only a few kg of slags have been found in the vicinity to date. If the hypothesis of missing or overlooked archaeological records is plausible for the Late Neolithic and Early Chalcolithic mines, because of its physical inconsistency, this is of course not possible for Bronze Age mines, where mass copper smelting produced tons of very weathering-resistant fayalite-based slags.

In fact at many Bronze Age sites, the problems are quite the reverse: there are huge heaps and tons of copper slags, but no mine in sight, and often no mine anywhere near. A couple of clarifying examples: the site of Pletz von Mozze, near Luserna, Western Alps, is an Italian Final Bronze Age site with meter-thick layers of fayalitic copper slags, a few evidences of fire activity, but the closest copper mines (Calceranica) are more than 11 km away and 1300 m down the valley. Many similar coeval situations are found in the Trentino and South Tyrol area: why did the smelters undertake the effort of bringing the minerals up the valley and carry out smelting activity on this remote plateau? One other example is Chrysokamino, Crete (Betancourt 2006, Muhly 2007). Smelting activity took place in the Early Minoan III period (ca 2300–2100 BC), but there are no copper deposits on the Island, or on the nearby isles: why did the metalworkers bring the ore to that remote promontory, 30 m above the coastline? The generally adopted explanation is that fuel was the driving requirement. So much fuel was necessary that it was easier to bring the ore to the fuel than the reverse. This is certainly the case of the slags in the Golfo di Baratti, Tuscany, Italy where the Etruscan necropolis of Populonia and the copper slag layer on the beach (Cartocci et al. 2007) was completely covered by a 7–10 m thick layer of iron slags produced in later Roman times. The Etruscan tombs were actually discovered when the iron slags were removed and reused for industrial purposes in the first decades of last century. The copper minerals first, and the iron minerals later were transported from nearby Elba Island, incredibly rich in minerals, but at that time already depleted of forests. This is apparently not the case in Cyprus, where the well-studied Apliki, Skouriotissa and Mavrovouni mines, producing virtually all the copper oxhide ingots circulating in the Mediterranean in the Aegean Late Bronze Age (ca 1600–1100 BC: Muhly 1977, Stos 2009) still show heaps of ancient slags of different ages (Fig. 3.68).

The location of the smelting sites and their relationship with the mining areas are thus the first problems to be faced in the investigation of the sites. Leaving the excavation to the archaeologists of course, next in line comes the interpretation

Fig. 3.68. (a) Heaps of ancient copper smelting slags in proximity to the famous Skouriotissa mines in Cyprus. (b) Detail of a large poured block of very fluid fayalitic slags.

of the metallurgical activity, which might well start during the excavation, by checking the layers and the finds by a materials science approach. Small fragments of fused gangue are hardly distinguishable from heated residuals of common furnaces and need to be carefully looked at. Early very heterogeneous and partially reacted copper slags (Box 3.k) might be very hard to discriminate from partially fused soil. Hammer scales from iron metallurgy can not always be separated from the soil using a pocket magnet... Craddock (1999) reviews a few of the paradigms and protocols that can be a useful background for improving field work and the interpretation of archaeological evidence.

I will just remind some of the recurring questions concerning the interpretation of archaeological smelting sites: where is the technical evidence? In fact in most cases there is little or no evidence of furnaces, bellows, tuyeres, or even crucibles. It can be very puzzling. There are places with meter-thick layers of slags and not one single broken tuyere... Anyone carrying out experimental reconstructions (Timberlake 2007) knows how painful it is to prepare tens of ceramic tools (crucibles, tuyeres, etc.) and have them lost at each firing. Were the early smelters so much technically better than we are? Probably yes, but yet the almost total absence of broken ceramics at many smelting sites is daunting.

In other cases the ceramic is there, but what was it used for? The shape and temperature marks do not at all fit our reconstruction model. Particularly intriguing cases are the perforated ceramic cylinders excavated in the Thailand Khao Wong Prachan Valley (Pigott *et al.* 1997, Pryce 2009), and the coupled crucible-furnace of the French Late Bronze Age IIb–IIIa (the so-called "four-creuset": Mohen and Walter 1994). See Bayley and Rehren (2007) for a rational classification and interpretation of technical crucibles. Careful mineralogical and petrological analysis of firing ceramics may yield information on the temperature gradients, location and extent of the heating source, nature of the charge and type of processes carried out (melting, refining, alloying, etc...).

One further point is worth remarking on: smelting and metallurgical activities do not go unnoticed, even in modern times. Large industrial metallurgical plants established after the industrial revolution have invariably caused a substantial stirring of the job market from the social point of view, together with extensive environmental disasters from the ecological point of view. While the ancient social effects are unfortunately intangible for the most part, the pollution effects can be observed and measured with targeted scientific analyses of soils, lake sediments, and ice cores. Several studies have been carried out on ancient smelting sites and their environments in order to quantify and model the extent of the activities (Hong *et al.* 1996, Monna *et al.* 2004). The understanding of the palaeoecological effects of the ancient metallurgical activities may lead to a more complete understanding of the social, economic, and anthropological sides linked to metal production. Furthermore the metal-polluted layers in the sediment sequences can provide information on past metallurgical activities that can not be detected by other evidence (Mighall 2003, Mighall *et al.* 2007).

Box 3.k Smelting slags: the key to ancient metals extraction

Slagging is an essential step of metal production and/or refinement, and the investigation of slags may reveal a great deal of information on the metallurgical processes. Slags are heterogeneous materials (Fig. 3.k.1) produced by a partially or totally molten state; most ancient slags were derived from:

1. copper smelting,
2. iron smelting and smithing,
3. lead smelting and cupellation.

The investigation of slags is essentially carried out by mineralogical–petrological methods (Bachmann 1982) and it allows identification of: the smelted or refined metal type, the process employed (crucible, furnace), the type of charge (oxidic, sulphidic), the maximum temperature and redox conditions of the process, the cooling speed (slow cooling, rapid cooling, quenching). This information is combined to reconstruct the ancient metallurgical processes.

Fig. 3.k.1. Example of cross sections of heterogeneous Chalcolithic slags produced during copper smelting from sulphides. (a) Milland, Brixen, South Tyrol, Italy. (ca. 2600–2700 BC), (b) Riparo Gaban, Trentino, Italy (ca. 2600 BC). Large unreacted grains of the charge minerals are visible, mostly quartz.

Cu-slags

Copper can be extracted from oxidic or sulphidic ores (Table 3.15). If the copper is reduced from pure oxidic minerals such as malachite or cuprite (Hauptmann 2007), very little slags are produced, mostly deriving from the used flux, such as iron oxides and silica, and composed of fayalite:

$$2FeO + SiO_2 \rightarrow Fe_2SiO_4$$

If no flux is needed, such as in the Cu reduction from pure carbonates, then slags are almost inexistent. If pure flux is used the slags have a fayalitic compostion, and if gangue minerals (carbonates, silicates) are present a series of silicates are formed with various composition (melilite, pyroxene, monticellite, etc.). Typical examples are the Timna and Faynan slags, derived from oxidic (paratacamite, malachite, cuprite) and silicate (chrysocolla, dioptase) ores (Bachmann 1980, Hauptmann 2007), and those from Afunfun and Ikawaten, Niger, derived from mixed oxidic-silicate ore with some native copper (Killick et al. 1988).

In copper smelting from chalcopyrite ($CuFeS_2$), which is the most common sulphidic copper ore, the reduction of Cu^{+2} to metal (Fig. 3.4) requires (a) desulphuration by roasting, and (b) iron to be stabilized in the Fe^{+2} oxidation state by the addition of silica and formation of fayalite (Fe_2SiO_4) following the reactions:

$$CuFeS_2 + 3O_2 \rightarrow FeO + CuO + 2SO_2 \text{ (dead roasting or total roasting)}$$
$$2CuFeS_2 + 3O2 \rightarrow 2FeO + 2CuS + 2SO_2 \text{ (partial roasting)}$$
$$2FeO + SiO_2 \rightarrow Fe_2SiO_4 \text{ (slagging)}$$
$$CuO + CO \rightarrow Cu + CO_2 \text{ (reduction)}$$
$$CuS + O_2 \rightarrow Cu + SO_2 \text{ (reduction)}$$

Copper smelting slags therefore essentially have a fayalitic composition (Fig. 3.k.2), with the olivines having very different morphologies depending on their thermal path and cooling history (Donaldson 1976). Copper slags may vary from very heterogeneous materials such as the one shown in Fig. 3.k.1, containing abundant unreacted charge and quartz grains together with the reaction products (mainly olivine, but also pyroxenes, magnetite, copper droplets, etc.; Fig. 3.k.5), to very homogeneous materials such as the typical "plattenschlake" of the Alpine Late Bronze Age (Fig. 3.k.3), which are almost completely composed of pure fayalite. In both cases the molten part is essentially fayalitic, sometimes with different generations of olivines testifying to a complex temperature/time path. Hopper and skeletal olivines (Figs 3.k.2(b) and 3.k.4(c)) have experienced slow cooling rates (CR <50 °C/h), chain olivines (Figs 3.k.2(a) and 3.k.4(b)) have CR in the range 50–200 °C/h, and feathered or branching olivines (Fig. 3.k.4(a)) may have experienced very fast cooling in excess of 500 °C/h (Donaldson 1976). The latter olivine morphologies are invariably found in the flat Alpine slags (Fig. 3.k.3), which were rapidly cooled out of the furnace (Anguilano et al. 2002, Mette 2003).

(a) (b)

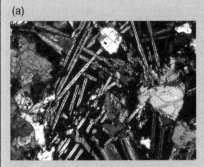 

Fig. 3.k.2. Microphotograph of olivine textures in Chalcolithic slags from Brixen: (a) chain olivines crystallizing out of the quartz grains, (b) skeletal olivines. Optical microscopy, polarized light. (Courtesy of F. Gallo, University of Padova)

(cont.)

Box 3.k (*Continued*)

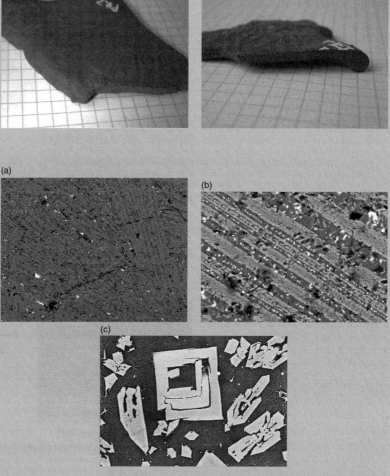

Fig. 3.k.3. Example of a flat slag ("plattenschlake") of the Alpine Final Bronze Age. (Courtesy of L. Anguilano, ETCBrunel)

Fig. 3.k.4. SEM-BSE images of fayalite morphologies in Final Bronze Age: (a) spinifex-type feathered olivines, (b) chain olvines, and (c) skeletal olivines. The morphology is directly related to the speed of cooling (Donaldson 1976).

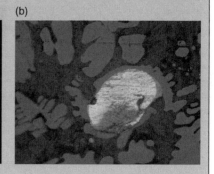

Fig. 3.k.5. Sulphide and metal inclusion in smelting slags: (a) partially transformed residual chalcopyrite, (b) copper metal droplet. (Courtesy of L. Anguilano, ETCBrunel)

Fe-slags

Slags are produced during **iron smelting** as molten slags (tapped slags), partially fused residuals of the charge in the furnace (cinder), or reaction between smelting products and the furnace linings (bear) (Morton and Wingrove 1969). Cinders and slags have a rather similar composition made of 60–70 wt% iron oxides, 20–30 wt% silica, and some impurities, but very different textures: cinders are porous and lumpy, sometimes with fragments of partially reacted charge or charcoal, whereas tapped slags show the smooth and black appearance of free-flowing melt. The slags produced during the first part of **iron smithing** and forging by high temperature squeezing of the slag out of the bloom, are compositionally similar to the smelting slags. They are very difficult to distinguish. The ones produced during the later stages of smithing and forging are usually smaller (**hammer scales**) and more enriched in iron oxides. The phase diagram $FeO–SiO_2$ (Fig. 3.k.6) has the eutectic point close to the fayalite composition. Accordingly, iron smelting slags are essentially composed of wüstite and fayalite, or wüstite and glass of fayalitic composition (Fig. 3.k.7).

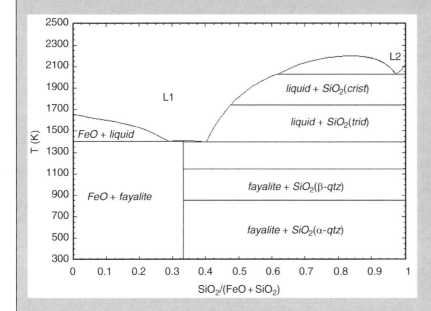

Fig. 3.k.6. Phase diagram of $FeO–SiO_2$. Most iron slags fall into the left part eutectict, crystallizing wüstite and fayalite or glass with fayalitic composition.

Pb-slags

Complete slagging of lead ores requires temperatures over 1000 °C, though the Pb metal has a low melting point (328 °C), so that a substantial amount of lead can be extracted at temperatures in the range 500–1000 °C. The main lead ore, galena (Table 3.15) is commonly roasted to the oxide (PbO) before reduction to metal. The slags derived from Pb smelting are often the results of inefficient extraction, and contain a considerable amount of metallic lead and lead oxide, other than a number of alteration phases, including cerussite ($PbCO_3$), anglesite ($PbSO_4$) and phosgenite ($Pb_2(CO_3)Cl_2$). To extract the silver contained in Ag-rich galenas, the extracted Ag-containing lead metal is oxidized forming litharge (PbO) and reduced metallic silver. The step is called **cupellation**.

The investigation of slags usually encompasses bulk chemical analysis (XRF, ICP-OES, AAS, PIXE, MS) to get the mean composition of the material, preliminarily understand the metallurgical process, and evaluate the efficiency of the extraction (i.e. how much metal is left in the slags). The powdered material is then

(cont.)

Box 3.k (*Continued*)

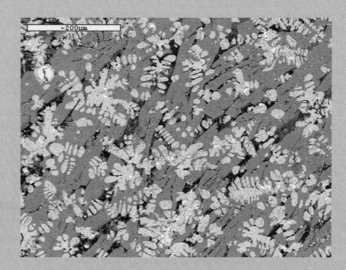

Fig. 3.k.7. Iron slag showing wüstite (FeO) and interstitial glass of fayalitic composition (Fe$_2$SiO$_4$).

analysed by XRD to identify the product crystalline phases, which are highly indicative of the smelting reaction processes. Small unreacted charge grains are hardly detectable by XRD, so that thin sections or thick sections of the materials are then carefully investigated by optical and electron microscopies, in order to visualize textural relationship between phases, crystal morphologies, reaction boundaries, residual charge grains (Fig. 3.k.5a) and inclusions of reaction products (Fig. 3.k.5b). Sometimes EPMA or PIXE are used to obtain point analyses on the single crystals or microinclusions. Spectroscopic techniques (Mössbauer, XPS, XAS) may be used to evaluate the oxidation state of specific elements (especially iron) and assess the environmental conditions during smelting.

Dating of residual quartz grains is also possible by TL and OSL (Box 2.w), since the smelting process efficiently anneals the electron traps and resets the dating clock (Gautier 2001, Haustein *et al.* 2003, Hauptmann and Wagner 2007). Other macroscopic measurements may also be useful, such as specific density, porosity, or remnant magnetism. It has been observed that there is a systematic difference between the slags produced by copper smelting processes at Late Neolithic or Chalcolitic times (Bourgarit 2007) often due to co-smelting of oxidic and sulphidic ores and rather inefficient processes carried out in crucibles, and those produced at later times, mostly in smelting furnaces with large scale operations. Among the systematic differences observed are: macroscopic dimensions of the slags, heterogeneity and incompleteness of the reactions (immature slags), specific density, absence of fluxing, and low reducing conditions (Bourgarit 2007, Artioli *et al.* 2007). The measured density of many Chalcolitic slags is in the range 2.6–3.0 g/cm^3 (less than the calculated value of 3.5 g/cm^3 for an ideal slag composed of 50 wt% quartz (density 2.6 g/cm^3) and 50 wt% fayalite (density 4.4 g/cm^3)), compared with the Late Bronze Age plattenschlake measuring 3.8–4.4 g/cm^3, which compare well to the value for pure fayalite.

The whole set of measured information is necessary to interpret the nature of the smelting process from slag evidence. A possible identification scheme is presented below, which takes into account the oxidic/sulphidic nature of the charge (O/S) and the presence of delafossite as indicator of crucible smelting (Fig. 3.k.8; Bachmann 1982, Bourgarit 2007, Burger 2008, Burger *et al.* 2010).

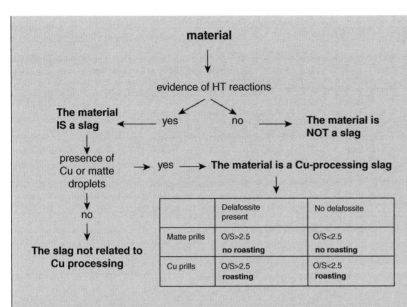

Fig. 3.k.8. Proposal for a classification scheme of copper smelting slags. Slags derived from Cu refining are normally connected to ceramics and more oxidized.

For in-depth information

Bachmann H.-G. (1982) The identification of slags from archaeological sites. Occasional Publication No. 6. The Institute of Archaeology, London.

3.5.3 Characterization of metal objects

Analysis of metal objects is performed to: (1) analyse the chemical composition, (2) evaluate the microstructure and relate it to physical properties, (3) measure the physical properties (hardness, porosity, conductivity, etc.), and (4) assess its internal and surface state for conservation purposes.

3.5.3.1 Chemical composition

The major problem involving the analysis of metals is that they tend to oxidize rapidly (Box 3.j), so that metal surfaces are always covered by a layer of alteration products, incorporating in various amounts elements present in the bulk of the sample and foreign species. Therefore most surface analytical techniques require the surface to be expressly polished, in order to bring the unaltered material to the surface. This is a major problem with portable analytical instruments (Box 2.r), which inevitably (a) excite the signal of the surface alteration together (hopefully) with that of the bulk, and (b) get a decreased signal from the sample because of the surface attenuation.

The only truly non-invasive method that can penetrate the bulk dense metals is NAA (Box 3.a). In principle any object can be irradiated with the neutron beam, and the induced radioactivity on the sample depends on the composition, i.e. on the half-life of the metastable radiogenic nuclides generated in the sample. Copper and bronze commonly fall back to normal background radioactivity within a few days. However, if the sample contains silver or other elements producing long-lived radionuclides the sample may be activated for months or years, depending on the concentration of the element and the irradiation dose.

All of the other analytical techniques based on the excitation of the fluorescence signal (XRF, PIXE, SEM-EDS, EPMA) suffer from the problems described above: either we polish the sample, and therefore the analysis is correct but the technique is invasive, or we operate directly on the untreated surface and therefore the technique is non-invasive, but the results are at best only qualitative. The thickness of the alteration layer plays a major role in determining the quality of the analytical results. The intermediate solution that is probably the best protocol at the moment is to operate in microsampling mode (Section 2.2). Any microbeam technique will therefore operate correctly by probing the sample through the small hole produced by a laser or ion microbeam (LIBS, LA-MS, SIMS), or by analyzing a microsample detached from the object and measured under standard conditions, either by microbeam or by bulk techniques (ICP-OES, AAS, ICP-MS).

Before deciding the analytical strategy (where to sample and how to measure), it is advisable to carry out a diagnostic survey by 2D and/or 3D imaging techniques (Section 2.1.3, Box 2.n). Careful imaging of the sample is required for conservation purposes, to identify and select the part to clean, the part to stabilize, or the part to remove.

The final problem to take into account is that many chemical components are immiscible in metals (Section 3.5.1), so that they segregate and produce a very heterogeneous material (Fig. 2.16). If the elemental analysis is carried out by dissolution of the sample, the result will correctly reflect the mean composition of the sample; however, if microbeams are used, extreme caution should be used in order to obtain a correct volume-averaged composition, either by analyzing a statistical number of points, or by randomly moving the beam over an area sufficiently large to contain representative inclusions and segregations.

The isotopic composition of the sample must be obtained by mass spectrometry (Box 2.t).

3.5.3.2 Physical properties and microstructure

The physical properties of the metal are critically dependent on its microstructural features (Section 3.5.1). The physico-chemical properties that can be easily measured in non-invasive mode are density, conductivity, and ferro- and ferri-magnetic properties; many of these can be indicative of the nature of the sample (oxide or sulphides inclusions, etc.) or its alteration state. Most of the mechanical properties (Table 3.1) can not be measured directly in non-invasive mode, and generally only the measurement of hardness through Vickers micro-indentation or similar methods is performed (Mott 1957, Buckle

1959). Figure 3.69 shows the image of a microindentation test performed on a copper tool produced by casting for reference purposes, the obtained test values (HV = 50–85) are representative of the values for cast copper in the unworked conditions. Alloying by Sn or As (Lechtman 1996) or mechanical working by substantial volume reduction both produce a sensible increase in hardness, so that the HV values may be doubled. Table 3.17 compares the microhardness test values with the reference Mohs scale. The values measured on the reference samples compare well with the HV range measured on unworked and worked Chalcolithic and Early Bronze Age alloyed axes (Kienlin *et al.* 2006).

The physical measurements can be related to the microstructural features measured on the polished and etched sample by RL-OM metallography, or obtained in non-destructive mode by CTA (Box 3.l). The physical and textural informations are the key to interpreting the manufacturing history of the artefacts and their time evolution. It is a remarkable experimental advancement that metal textures and microstructures can not only be visualized, but also quantified and compared. One word of warning on a particularly complex problem: natural native copper has textures that are very similar to those obtained by artificial thermal recrystallization after mild working. Careful compositional checks are required to sort out the nature of the sample.

As with many other materials, the technical and analytical characterization of metals is just the start of the investigation and, despite its absolute necessity, the risk of looking at past processes with modern over-technological eyes is always present. Intangible values are obviously to be taken into account, especially for such versatile, shining and inspiring materials as metals. The unravelling of the relationship between the described measurable properties and the practical consequences of metal technology on society is exactly one of the fundamental tasks of modern archaeometallurgy.

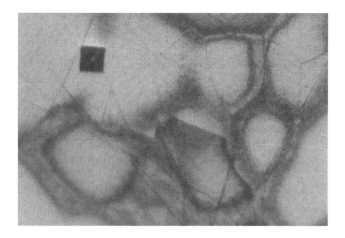

Fig. 3.69. Reflected light optical micrograph of the Vickers microindentation test performed on a cast copper reference. The small square is the mark left by the diamond tip after the measurement.

Table 3.17. Correspondence of the Mohs hardness scale for minerals and the Vickert-type microhardness test (HV_{HM} = Hodge and McKay 1934, HV_T = Taylor 1949, HV_{BCW} = Broz et al. 2006). Measured HV values on metals are from Lechtman (1996), Kienlin et al. (2006), Williams and Edge (2007), and our own measurements

Mohs scale	HV_{HM}	HV_T	HV_{BCW}	Measured HV values on artefacts
1 – talc	1	47	41	
2 – selenite (gypsum)	11	60	177	
				reference cast copper (50–80)
				reference cast As_{10}–bronze (85)
				unworked Sn–bronze axes (100–130)
3 – calcite	129	128	432	
				work-hardened copper 50% reduction (120)
				wrought iron (120–170)
4 – fluorite	143	188	580	
				modern stainless steel (140–180)
				modern bronze AZhMts 10–3–1.5 (170–200)
				work-hardened As_{10}–bronze 50% reduction (220)
				work-hardened Sn–bronze axes (200–280)
				wootz crucible steel (300–550)
5 – apatite	517	659	1574	
				quenched steel 0.2–0.4 C–content (500–700)
6 – orthoclase	975	714	1991	
7 – quartz	2700	1181	3536	
8 – topaz	3420	1648	5101	
9 – corundum (sapphire)	5300	2402	5681	
10 – diamond				

3.5.3.3 Surface analysis

In addition to the characterization of the bulk, for conservation purposes it is especially important to assess the processes that are currently or potentially active at the surface (Box 3.j). Chemical analysis (SRIXE, NAA, XPS, AAS, SIMS, etc.) provides information on the mean chemical composition of the surface layers, whether they are the original patinas, protective layers, corrosion products, or the original metal. Grazing incidence X-ray diffraction (GIXRD) and all surface-sensitive techniques (SEM, XPS, EELS, SEXAFS, DRIFT, PIXE, etc.) are used to characterize the surface or for depth profiling. Special materials such as ancient metal-coated embroideries (Migliori et al. 2008, Karatzani and Rehren 2009) or surfaces that have been expressly acid-treated for special aesthetic effects such as tumbaga jewels are especially intriguing (Gillies and Urch 1985).

Box 3.l Metallography and crystallographic texture analysis

Metallography based on reflected light optical microscopy (RL-OM, Box 2.k) is a well-developed standard technique for the analysis of archaeological metals (Scott 1991), and it has been used to understand the manufacturing technology for almost a century (Mathewson 1915). Recently **electron backscatter diffraction** (EBSD, Box 2.i) has been introduced, although to date it shows some limitations in the interpretations of the diffraction patterns of multiphasic materials. Furthermore, both RL-OM and EBSD must necessarily be performed on etched or polished metal surfaces, so that invasive microsampling or cleaning of the object surface is required. In the past invasive sampling for metallographic and/or chemical analysis was unavoidable (Fig. 2.14), whereas nowadays several non-invasive or microinvasive technical and methodological analytical developments are available and help to preserve the integrity of cultural heritage artefacts (Section 2.2). Substantial efforts therefore have recently been made in order to transfer and optimize the techniques commonly used in **crystallographic texture analysis** (CTA) of modern materials to the analysis of ancient metal objects, to provide a totally noninvasive metallographic technique for the cultural heritage. The advancements were made possible by a combination of factors: The recent theoretical developments in the field of CTA, the rapid improvement of source and instrumentation at large scale facilities (Box 2.a), and the close interaction between materials scientists, modern metallurgists, software developers and archaeologists. The outcome is a powerful non-destructive tool for the analysis of ancient metals. Here, we will briefly review the concepts and advantages of the three available techniques for texture analysis: metallography, EBSD, and CTA.

Fig. 3.l.1. Example of a reflected light optical metallography. The copper sample shows small grains perfectly recrystallized at high temperature after mechanical working. The thermal annealing was not sufficient to delete all the strain systems formed during working. Evidence of the direction of working is provided by the swarm of iso-oriented sulphidic inclusions, all flattened and elongated along two directions weakly inclined from the horizontal. (Courtesy of I. Angelini, University of Padova)

Metallography

Reflected light optical metallography is the most widespread technique yielding a variety of information on the nature and the thermomechanical history of the investigated metal object. It is inexpensive, rapid and is routinely used in modern and ancient metallurgical analysis. The parameters obtained from the analysis of the optical images are (Scott 1991): identification of the metal phases, visual estimation of the shape and size of the metal crystals, visual estimation of the presence and distribution of inclusions, alteration phases, and deformation patterns, such as slip banding, or mechanical twinning (Fig. 3.l.1). The main disadvantages are the invasiveness of the measurement, and the fact that only a small two-dimensional section of the object is analysed, so that the results lack statistical

(*cont.*)

Box 3.1 (*Continued*)

significance and do not capture the three-dimensional nature of the material. Furthermore the characterization of the exact composition of the metal phases requires complementary point or bulk chemical analysis for the complete interpretation of the artefact.

Figure 3.1.2 shows RL-OM images taken at different stages of metal processing of a cast object that underwent several cycles of thermal annealing and cold and hot working. It must be pointed out that, once mechanical forces are applied to the object and strain is propagated through slip systems, it is virtually impossible to erase the developed tex-

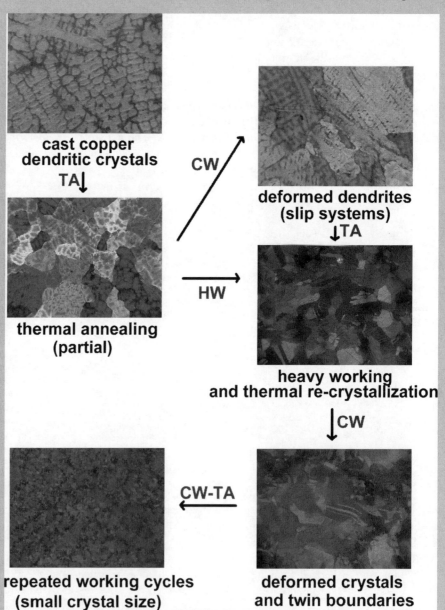

Fig. 3.1.2. Reflected light optical metallography: diagram of the microtextural evolution of copper and copper-alloys from casting (upper left), through several stages of thermal annealing (TA), cold mechanical working (CW), and hot working (HW). Working cycles induce strain systems (slip systems) and thermal annealing induces recrystallization and the formation of twin boundaries (Scott 1991).

ture without re-melting the object. Purely random crystallite orientation is possible only through casting and thermal annealing alone. The reader is referred to Scott (1991) and Ottaway and Wang (2004) for the detailed interpretation of the metallographic features.

EBSD

Electron backscatter diffraction analyses the diffraction signal induced by the electron beam (SEM, TEM, Box 2.i) on a small surface volume of sample (Schwartz *et al.* 2000). The analysed volume is of the order of the dimensions of the electron beam (5–10 nm) to a depth of 50–100 nm (Fig. 2.3). The Kikuchi lines present in the recorded images (Figs. 2.i.2) correspond to the crystallographic planes in the crystal structure of the probed metal grain. If the structure is known, the orientation of the crystal with respect to the beam can be calculated. EBSD investigates a raster sequence of points on a polished surface of the sample and determines the orientation of the crystallite at each point. Texture mapping by EBSD produces maps of iso-oriented points corresponding to the crystallite domains (Fig. 3.l.3). Laboratory single crystal X-ray texture diffraction, which probes the surface of the material, produces similar maps. Iso-orientation maps can be used to compute statistical diagrams of the orientation density of the crystallites.

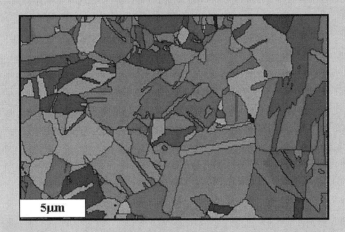

Fig. 3.l.3. An experimental crystal orientation map measured on thermally re-crystallized copper using EBSD. The grains are colour-coded by the crystal direction normal to the measured surface plane.

CTA

Crystallographic texture analysis is a technique based on the measurement and interpretation of the diffraction signal of each individual crystallite in the probed volume of material (Boxes 2.g, 2.h). Probing metals therefore requires a penetrating incident beam, usually provided by neutrons or high energy synchrotron radiation, with one or more detectors measuring the diffraction intensities in reciprocal space. Each crystallite within the metal, defined as a region of the material that diffracts coherently, has a specific shape, size and orientation in space derived from the crystallization history, the orientation being described as the orientation of the crystallographic axes with respect to the Cartesian coordinate system describing the object. As the intensity of the diffracted beam in any specific crystallographic direction depends on the volume of the diffracting crystallite, the measurement of the distribution of the diffracted intensities along the Debye cones (Giacovazzo 2002) allows direct evaluation of the volume fraction of the crystallites producing the diffraction effects. In fact the size, shape, and orientation of the crystallites determine the textural properties of a solid, and they critically influence many of the physical properties of the material. Since the early days of X-ray crystallography therefore the crystallographic analysis of materials has been used

(*cont.*)

Box 3.1 (*Continued*)

to measure the structural and microstructural properties of crystalline solids (Warren 1969), and in particular the analysis of crystallite texture using diffraction methods has developed with specific reference to metals (Hatherly and Hutchison 1979, Kocks *et al.* 1998, Bunge 2000), because of their technological and industrial importance, and also because of the highly symmetric cubic structures of most of them, which simplifies the calculation of the orientation distribution function (ODF), which describes in 3D how many crystallites in the probed volume of the sample possess a given orientation. Only very recently in fact has the analysis of complex multiphase materials composed of low symmetry crystal structures become possible, thanks to the measurement of high resolution diffraction patterns at large facilities, to the improved ODF calculation algorithms, and to the development of the full pattern methods of analysis. It is nowadays possible to investigate the texture of rather complex materials such as polymineralic rocks and industrial components (Wenk 2002, Wenk and Van Houtte 2004). In metallurgy, the CTA is generally performed after metal processing, in order to interpret how specific temperature–pressure paths influence the microstructural properties of the metal. In archaeometallurgy, the inverse operation is performed. The textural properties of the archaeological object are measured in order to reconstruct its thermal and working history. With respect to RL-OM and EBSD metallography, the texture information extracted by CTA is intrinsically more significant both from the point of view of the crystallite statistics, because of the large number of crystallites contributing to diffraction, and from the point of view of the three-dimensional sampling, because the signal analysed is from the whole volume sampled by the incident beam within the object.

The experiments are totally non-invasive, insofar as the whole object can be probed by the beam without any preparation (Figs 2.22 and 3.1.4). The diffraction data are collected and analysed by full profile analysis, the intensity differences being modelled by texture coefficients. The orientations of the crystallites in real space are visualized through the **pole figures**, which represent the density of crystal poles as a continuous function of the polar and azimuthal angles. They are two-dimensional projections of the three-dimensional ODF in the sample space. Also, an inverse pole figure can be defined in crystal coordinate system as the probability that a sample direction (conventionally coded RD, TD, or ND) is parallel to an arbitrary crystal direction [*hkl*]. The inverse pole figures may be thought of as the two-dimensional projections of the orientation density in the crystal space. In practice they are very useful for describing the orientation density of crystallographic poles along a specific direction of the object, e.g. for revealing fibre textures.

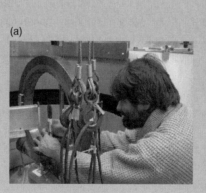

Fig. 3.1.4. (a, b) The Iceman axe mounted directly on the goniometer of the D20 beamline at the ILL neutron source, Grenoble (Artioli *et al.* 2003). The CTA analysis performed by penetrating neutron beams is totally non-invasive.

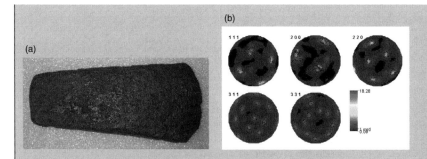

Fig. 3.l.5. (a) Chalcolithic copper axe from Castelrotto/Kastelruth, Bolzano, Italy (Museo Archeologico dell'Adige). (b) Pole figures for the Castelrotto axe measured at ILL (Artioli *et al.* 2003). The pattern indicates that the axe is almost a single crystal, with an extreme iso-orientation of the copper crystallites (Artioli 2007).
Original Colour pictures reproduced as plate 57.

Neutron CTA has been recently employed to decode the metal textures and the metallurgical processes in Chalcolithic copper axes (Artioli *et al.* 2003, Artioli 2007). Figure 3.l.5 shows the experimental pole figure for five crystallographic directions (111, 200, 220, 311, 331) relative to a measured prehistoric copper axe. The peaked density of the poles indicates a nearly complete iso-orientation of the crystallites, probably due to extreme slow cooling in the mould during casting.

Interpretation of the measured textures for over twenty chalcolithic axes (Artioli 2007) indicate that:

(a) most Copper Age axes show clear features related to cold working and subsequent thermal annealing. In a few cases the re-crystallization is complete, though in most objects there is only a partial re-crystallization of the copper metal into the soft state,
(b) the intensity and distribution of the textural features related to mechanical working, if compared with standards produced by progressive reduction in thickness (Kockelmann *et al.* 2006), seem to indicate a reduction in thickness of the axes in the range 5–10 %. Combined with the fact that most axes have been partially annealed and softened, the interpretation is that the mechanical working in the Copper Age was not used to harden the metal, but rather to slightly reshape the axe after casting, probably to mask surface casting defects and porosity, or as a consequence of deformation during use. Figure 3.l.6 shows the different types of inverse pole figures actually observed in the prehistoric copper axes,

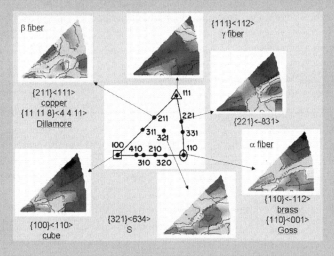

Fig. 3.l.6. Inverse pole figures showing some of the experimentally observed textures in chalcolithic copper axes (Artioli 2007).
Original colour picture reproduced as plate 58.

(*cont.*)

> **Box 3.1** (*Continued*)
>
> (c) two of the axes, namely the Similaun and the Kollman axes, do not show textural patterns above background. Since it is virtually impossible to erase the orientation texture of the crystallites once it has been produced, it is deduced that these two axes never underwent mechanical working or thermal recrystallization, and they both are in the as-cast state, although they were certainly used, as shown by the crystallite size and distribution near the blade area resulting from high-energy X-ray diffraction (Artioli *et al.* 2003),
>
> (d) three of the axes (including the Castelrotto axe shown in Fig. 3.1.5) show highly iso-oriented crystallites, with orientation of the 111 poles parallel or slightly tilted with respect to the main plane of the flat axes. This is interpreted as produced by very slow cooling during casting.
>
> The data are in good agreement with textural analysis of chalcolitic axes performed by conventional RL-OM investigations (Budd 1991, Kienlin *et al.* 2006).
>
> **For in-depth information**
>
> Kocks U.F., Tomé C.N. and Wenk H.R. (1998) Texture and anisotropy. Cambridge University Press, Cambridge.
>
> Schwartz A.J., Kumar M. and Adams B.L. (2000) Electron Backscatter Diffraction in Materials Science. Springer Verlag, Heidelberg.
>
> Scott D.A. (1991) Metallography and microstructure of ancient and historic metals. The Getty Conservation Institute, Los Angeles.
>
> Wenk H.R. and Van Houtte P. (2004) Texture and anisotropy. Rep. Prog. Phys. **67**, 1367–1428.

3.6 Gems

Humans have been fascinated by brightly coloured stones since prehistory. Gemstones became associated with folklore and superstition very early (Jangl and Jangl 1989), without any relationship to their physico-chemical properties or an appreciation of the geological processes that formed them. Even today, crystal therapy believers associate gemstone type and colours with non-existent properties causing health, sickness, luck, or misfortune. The association of stones with the zodiac signs, the tribes of Israel, the Twelve Apostles, the days of the week, or the habit of assigning birthstones derive from traditional beliefs of this kind originated in medieval times or even earlier (Rapp 2002, O'Donoghue 2006). In the Middle Age some gems were thought to reflect the morality and health of the bearer, and that their colour would vanish following misbehaviour or sickness of the owner (Kunz 1971). Gemstones as cultural and social indicators rapidly came to be associated with status, power, role, or profession of social groups or individuals. Among Aztecs, for example, gold was highly valued for its brilliance and warm colour, and because "it leads reaches on earth" (Sahagùn 1950–1982, Book 11, p. 234), however it was not as highly esteemed as the precious green and blue stones, which were deliberately selected to signal enduring otherworld qualities, both in high-ranking persons and in representations (Evans 2004, McEwan *et al.* 2006). Turquoise was a hallmark of lordly divine status across Mesoamerica, and in central Mexico and in Yucatán turquoise diadems were conferred on nobles of the highest rank (Harbottle and Weigand 1992).

Because of their value and their limited availability, gemstones started very early to be transported and traded. Since they are very recognizable among other goods in the archaeological record, they often serve as tracers of exchange and trade patterns both locally and over long-distances. The conceptual methods used to trace the provenance of gems are just the same as those described for lithics and rocks: a good geological knowledge of the distribution of the mineral deposits is required, together with a complete reference database of well-characterized materials from all of them. Compared with other kinds of geological materials, gemstone investigations have the advantage that many source localities are well known and their material has been carefully analysed, because of their importance for the modern gem market. The physical and chemical parameters used to identify the correct origin are trace and ultra-trace elemental patterns, isotopic ratios, defect and inclusions. All analytical techniques for trace element and isotope analysis are used (Table 2.10), in addition to sensitive spectroscopies that may detect defects, substitutions, and local electronic states that are hardly detectable at the chemical level (UV–Vis, Raman, FTIR, XPS, Mössbauer, XAS, etc.).

A number of successful examples are available concerning the reconstruction of old trade patterns. Surprisingly, they often show that gemstones in ancient times travelled much further that previously thought. The emerald trade routes from Gallo–Roman times to the 18th century AD, reconstructed on the basis of stable oxygen isotopes (Box 2.v), clearly point to the ancient use of deposits supposedly discovered in the 20th century (Giuliani *et al.* 2000). Lapislazuli, widely used as a blue pigment (Table 3.9), is diffused in much of the Old World, though the only known source in antiquity was Afghanistan (Herrmann 1968). Rubies also travelled all the way from Burma to ancient Babylon (Calligaro *et al.* 1998), as evidenced by PIXE analysis (Box 3.i).

Though any well-crystallized and transparent mineral species (Fig. 3.70) may be cut and transformed into enchanting gems (Fig. 3.71), traditionally some minerals are more valued or requested than others because of their colour, their brightness derived by particular internal refraction, or simply because of their traditional appeal. Very seldom is the value connected with rarity or availability. Table 3.18 lists some of the most used gem and precious stones in antiquity. The identification of the mineral, besides provenancing, is quite important for authentication. Commercial gem analysis often relies on the measurement of the light refractive index (by total-reflectometry at the critical angle), infra-red analysis, and sometimes diffraction. Because of their value, sampling is out of question and non-invasive techniques must necessarily be used, such as all the techniques analysing the surface of the sample (DRIFT, surface-XRF, PIXE). One of the advantages of cut gems is that the surface is always well polished and unaltered, so that surface techniques for elemental analysis can generally be applied with confidence to identify the mineral species. However, simple chemical techniques may not be sufficient to distinguish natural varieties and synthetic fakes of the same composition. Furthermore, many natural gems (including most of those on the market) display enhanced colour due to

Fig. 3.70. Well crystallized transparent minerals from Afghanistan pegmatites: (a) pale blue aquamarine, (b) polychrome tourmaline and (c) pink tourmaline.
Original colour pictures reproduced as plate 39.

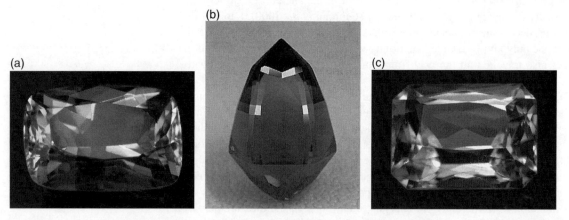

Fig. 3.71. Cut gems from minerals similar to those shown in the previous picture: (a) aquamarine, (b) polychrome tourmaline, (c) morganite.
Original colour pictures reproduced as plate 40.

thermal treatment or artificial irradiation (Nassau 1994). These supplementary treatments are much harder to spot and recognize, though with some effort detailed analysis of microcracking produced by heat and characterization of the spectral signature of electron–hole defects produced by irradiation may do the job.

Altered and fake gemstones are known from archaeological records (Nassau 1984, Rapp 2002). The Romans burned light-coloured chalcedony to oxidize

Table 3.18. List of the most common mineral species and varieties used as gems in antiquity. Valid mineral species (unbracketed) are distinguished from varieties (in brackets), whose unofficial names entered the gemological jargon

Mineral	Composition	Colourless	Red–pink orange	Yellow	Green	Blue–purple	Black
Apatite	$Ca_5(PO_4)_3(OH,F)$	apatite	apatite	apatite	apatite	apatite	
Azurite	$Cu_3(CO_3)_2(OH)_2$					azurite	
Beryl	$Be_3Al_2Si_6O_{18}$	(goshenite)	(morganite)	(heliodor)	(emerald)	(aquamarine)	
Calcite	$CaCO_3$	(spar)	(Co,Mn)–calcite	calcite			
Cassiterite	SnO_2						cassiterite
Cordierite	$Mg_2Al_4Si_5O_{18}$					(iolite)	
Corundum	Al_2O_3		(ruby)	(sapphire)	(sapphire)	(sapphire)	
Crysoberyl	$BeAl_2O_4$		(alexandrite)	crysoberyl	(alexandrite)	(alexandrite)	
Diamond	C	diamond	diamond	diamond	diamond	diamond	carbonado
Diopside	$CaMgSi_2O_6$		diopside	diopside	diopside		
Dioptase	$CuSiO_3·H_2O$				dioptase		
Enstatite	$MgSiO_3$	enstatite		enstatite	enstatite		
Epidote	$Ca_2XAl_2(SiO_4)(Si_2O_7)O(OH)$		piemontite (X=Mn)	epidote	(pistacite) (X=Fe), (clinozoisite) (X=Al)	zoisite, (tanzanite)	allanite, (ortite)
Fluorite	CaF_2	fluorite	fluorite	fluorite	fluorite (demantoid)	fluorite	(melanite)
Garnet	$(X)_3(Y)_2(SiO_4)_3$	pyrope (X=Mg, Y=Al), grossular (X=Ca, Y=Al)	almandine (X=Fe, Y=Al), (hessonite), grossular (X=Ca, Y=Al), pyrope (X=Mg, Y=Al), spessartine (X=Mn, Y=Al)	andradite (X=Ca, Y=Fe) grossular (X=Ca, Y=Al)	andradite (X=Ca, Y=Fe) grossular (X=Ca, Y=Al) uvarovite (X=Ca, Y=Cr)		
Hematite	Fe_2O_3						hematite
Jadeite	$NaAlSi_2O_6$				(jade)		
Kyanite	Al_2SiO_5					kyanite	

(*Cont.*)

Table 3.18. Continued

Mineral	Composition	Colourless	Red–pink orange	Yellow	Variety Green	Blue–purple	Black
Lazulite	$(Mg,Fe)Al_2(OH)_2(PO_4)_2$					lazulite	
Lazurite	$Na_3CaAl_3Si_3O_{12}S$					lazurite, (lapislazuli)	
Malachite	$Cu_2(CO_3)(OH)_2$				malachite		
Olivine	$(Mg,Fe)_2SiO_4$			(peridot), (chrysolite)	(peridot), (chrysolite)		
Quartz	SiO_2	(rock crystal)	(rose-quartz)	(citrine)	(chrysoprase) (prasiolite) (aventurine)	(amethyst)	(smoky-quartz)
Rodochrosite	$MnCO_3$		rodochrosite				
Rutile	TiO_2		rutile	rutile			
Scheelite	$CaWO_4$		scheelite	scheelite			
Spinel	$MgAl_2O_4$		spinel		spinel	spinel	
Spodumene	$LiAlSi_2O_6$		(kunzite)		(hiddenite)		
Titanite	$CaTiOSiO_4$			titanite	titanite		
Topaz	$Al_2SiO_4(F,OH)_2$	topaz (acroite)	topaz (rubellite)	topaz elbaite (X=Li,Al)		topaz elbaite (X=Li,Al)	
Tourmaline	$(Na,K,Ca)(X)_3Al_6(BO_3)_3 Si_6O_{18}(OH)_4$		elbaite (X=Li,Al)		(indicolite), elbaite (X=Li,Al)		schorlite (X=Fe), dravite (X=Mg)
Turquoise	$CuAl_6(OH_2PO_4)_4·4H_2O$					turquoise	
Zircon	$ZrSiO_4$	zircon	(hyacinth)	zircon	zircon	zircon	

iron and obtain the much requested darker variety. Pliny describes how Romans skillfully used to imitate different gemstones with glass, imparting the wanted colour in many ways. Despite the fact that relatively few ancient gems have been thoroughly analysed, the high quality technology of the Lycurgus cup (Section 3.4) should remind us not to underestimate the technical abilities reached in antiquity. In the Middle Age small fragments of precious stones were glued to imitations and partly concealed by carefully designed mountings, so as to make only the real gem visible. These fakes were known as "doublets" (Bauer 1968, Rapp 2002).

In addition to the listed minerals, which have geological origin, well-defined composition, and may produce gem-quality crystals, there are other materials that have been used as precious stones, though they have different nature and/or formation processes.

An important group of materials are composed of microcrystalline, cryptocrystalline, or amorphous silica. The chemical composition is the same as quartz, but the texture and microstructure are entirely different. The micro- and crypto-crystalline varieties are called chalcedony, carnelian, jasper, agate, onyx, and chert. Other names used in the past are sard, sardonyx, plasma, heliotrope, bloodstone, banded agate, prase, hornstone. Many of these names do not have specific physical or genetic significance, rather they are just derived from the colour, the use, or the place of origin, like sard (i.e. from Sardinia). In antiquity microcrystalline quartz (especially carnelian, jasper, and agate) was extensively used for seal stones and bead decorations. Opal is partially hydrous amorphous silica (Heaney *et al.* 1994), whose intense iridescence is caused by the interparticle interference of light. Opal colour can vary from whitish to pale blue, pink, bright red (fire opal), to almost black.

Carbonate-based biomineralizations (pearls, nacre, shells, coral) have also been used as precious stones at various times. Shell ornaments are well known from prehistoric sites, and red-coloured *Spondylus* seem to have been a much sought material, perhaps for its connection with blood (Seferiades 1995). There is established evidence of the use of pearls since Ptolemaic Egypt, though they were possibly used as ornaments before this, together with nacre (or mother-of-pearl). Polished nacre has been used throughout history for surface decorations and inlays.

I leave technical descriptions of gems and minerals to specialist textbooks (Hurlbut and Kammerling 1991, Read 2005, O'Donoghue 2006, Schumann 2006). Being a mineralogist and crystallographer from the start, it is tempting to transform this short chapter on gems into a major part of the book because I like minerals, though not necessarily in the cut-and-polished form, but above all because they are beautiful and inspiring. Human kind has always been fascinated by light, by colours, and of course by the materials that make light and colour meet and stimulate imagination: crystals. Beyond aesthetics, beyond art, the stones sometimes reach some unfathomable level within us. A materials science volume is surely not the place to probe the anthropological, psychological or religious meanings of gems, but there is a reason why gems independently of their rarity and abundance are so valued, so sought after, so linked to the act of offering. There is an interesting passage in the decadent novel *À rebours* by Joris-Karl Huysmans (1884), where he probes the relationship between gems, colours, and perception. In Chapter 4, the one before his famous description of Gustave Moreau's Salomé, he describes the young Des Esseintes' thoughts on gems:

> ...Le choix des pierres l'arrêta; le diamant est devenu singulièrement commun depuis que tous les commerçants en portent au petit doigt; les émeraudes et les rubis de l'Orient sont moins avilis, lancent de rutilantes flammes, mais ils rappellent par trop ces yeux verts et rouges de certains omnibus qui arborent des fanaux de ces deux couleurs, le long des tempes; quant aux topazes brûlées ou crues, ce sont des pierres à bon marché, chères à la petite bourgeoisie qui veut serrer des écrins dans une armoire à glace; d'un autre côté, bien que l'Église ait conservé à l'améthyste un caractère sacerdotal, tout à la fois onctueux et grave, cette pierre s'est, elle aussi, galvaudée aux oreilles sanguines et aux mains tubuleuses des bouchères qui veulent, pour un prix modique, se parer de vrais et pesants bijoux; seul, parmi ces pierres, le saphir a gardé des feux inviolés par la sottise industrielle et pécuniaire. Ses étincelles grésillant sur une eau limpide et froide, ont, en quelque sorte, garanti de toute souillure sa noblesse discrète et hautaine. Malheureusement, aux lumières, ses flammes fraîches ne crépitent plus; l'eau bleue rentre en elle-même, semble s'endormir pour ne se réveiller, en pétillant, qu'au point du jour.... Décidément aucune de ces pierreries ne contentait des Esseintes; elles étaient d'ailleurs trop civilisées et trop connues. Il fit ruisseler entre ses doigts des minéraux plus surprenants et plus bizarres, finit par trier une série de pierres réelles et factices dont le mélange devait produire une harmonie fascinatrice et déconcertante.

But it is even more interesting what Huysmans stated twenty years later, in the preface to his own work:

> In *La Cathédrale* I reworked the chapter on precious stones, but from the point of view of their symbolism. I gave new life to the lifeless gems of *À rebours*. Of course I do not deny that a beautiful emerald may be admired for the flashes which glitter in the fire of her green waters, but, if one is not versed in her symbolic language, is the stone not an unknown quantity, a stranger with whom one cannot converse and who herself remains silent because people do not understand her speech? But she is more, and better, than that.
>
> Without believing, with Estienne de Claves, an old writer of the sixteenth century, that precious stones, like human beings, propagate their species by means of seed scattered in the womb of the earth, one can perfectly well agree that they are minerals that signify, substances that speak; that they are, in a word, symbols.... The chapter in *À rebours* is, therefore, only superficial, the simple gem settings. It is not what it ought to be, a display of gems from the

world beyond reality. It is composed of matched jewellery that is quite well described, quite well displayed in a showcase; but that is all, and it is not enough.

I can not but agree that the mere description and analytical characterization of gems is far from being enough…a rather unusual and breathtaking novel makes the jump over the twilight zone, and actually imagines humans and other living organisms as the products of a crystals' dreams:

> They dream in flesh and sap, wood and bone and blood. And sometimes their dreams aren't finished, and so I have a cat with two legs, a hairless squirrel, and Gogol, who should be a man, but has no arms, no sweat glands, no brain. (Theodore Sturgeon, *The Dreaming Jewels*, 1950).

The boundary between inorganic and organic, lifeless and living is not a matter of simple definition, as Jacques Monod suggested a while ago (Jacques Monod, *Le hazard et la nécessité*, 1970).

3.7 Organic materials

Organic chemistry deals with carbon compounds. Since carbon together with half a dozen other elements forms over a million compounds, and they are involved in all aspects of life processes, it is not surprising that there is enormous interest in their chemistry. Most of science is focused on elucidating the subtleties of the relationship between the molecular structures and their properties, including food for nutrition, pharmaceuticals for health, oil and gas for energy, and the very essence of life, proteins and nucleic acids. Many observe that humankind has entered into the century of biology, much as physics dominated the scientific scene in the twentieth century (Pollard 2001). The enormous recent advances in the analysis of archaeological organic residues and remains and in the analysis and use of organics in conservation protocols surely indicates that this might well be the case.

Materials science instinctively is associated with hard materials, because of the traditional classes of compounds that are used to make long-lasting tools (Ashby 2001), and archaeology also traditionally assumes that most biological materials hardly survive in the past record except under unusual climatic conditions. It is not surprising therefore that most analytical techniques adopted in archaeology and conservation science have been focused and optimized for inorganic materials. Now the perception of materials has changed. Organic materials and organic–inorganic composites are in the spotlight. Natural evolution of course has known it since the dawn of life: biomineralization has always been very popular among all kind of organisms from bacteria to vertebrates (Mann 2001, Dove *et al.* 2003). Biomineralization processes involve the selective extraction of chemical elements from the environment, and their incorporation into functional structures under strict biological control. They are extremely important because they produce the most durable parts of organisms, the ones that are mostly found in the archaeological, palaeontological and forensic record: bones, teeth, shells, phytoliths, etc.

In order to keep in touch with the present trends and developments in the field of organics analysis, very popular topics in archaeology nowadays are: the analysis of organic residues (mostly lipids and proteins) in pottery and soils (Evershed et al. 2001, Evershed 2008, Pollard et al. 2007), the extraction and characterization of organic fluids (blood, saliva) from lithic artefacts and other tools (Section 3.1.1), the relationship between microbial activity and biodeterioration processes (Grupe 2001), and especially the "Jurassic park-reminiscent" ancient DNA studies (Renfrew 1998, Brown 2001, Willerslev and Cooper 2005). The fascinating field of bone analysis is treated in some detail in Section 3.7.1.

Despite the fact that technically these are all very challenging and demanding investigations (Pääbo et al. 2004, Gilbert et al. 2005), analytical access to the organic part of the archaeological record adds totally new dimensions to our view of the past: *colours* of cosmetics (Dayagi–Mendels 1993, Walter et al. 1999, Tsoucaris and Lipkowski 2003, Evershed et al. 2004), *smells* of flowers, perfumes and essences (Pybus and Sell 1999, Morris 1999, Manniche and Forman 1999, Joichi et al. 2005), *tastes* of foods and beverages (Alcock 2006, Barnard and Eerkens 2007), *notes* of past sounds (Zhang et al. 1999, 2004) and possibly some plausible virtual flesh around the excavated old bones. The organics that are part of everyday life also make the past really alive!

Conservation science is also much involved with organics, not only as primary materials for restoration and consolidation (resins, glues, etc.), but also in understanding the role of organics in ancient structural binders (Section 3.2.3), as binding materials in pigments (Section 3.3), as natural dyes (Section 3.7.5), and of course in the effort of preserving organic artefacts of very different natures (mummies, paper, textiles, etc.).

From the point of view of materials science, it is important (1) to understand the nature and properties of the organic materials involved, (2) to define their original state and the processes acting upon them, and (3) to assess their present conservation state, hazardous issues, and preservation protocols. Among all possible organic materials, a few have been selected to illustrate analytical issues and problems of interpretation.

3.7.1 Bones and ivory

Apatite is a complex calcium phosphate that is the basic mineral component of a number of composite biomineralizations: mammal bones and teeth (including enamel and dentin), ivory, brachiopod's shells, and several pathological human and animal mineralizations (calculi, stones). Structurally, apatite is composed of $[PO_4]$ tetrahedra and monovalent anions (OH^-, F^-, Cl^-) linked by the Ca polyhedra (Fig. 3.72). Several compositionally different mineral varieties are possible because of the structural substitution of the phosphate group by [CO$_3$] groups, and by extensive substitutions of the monovalent anions. Table 3.19 lists the most common phosphate-based mineral species sharing the apatite-type structure, other minerals with the same structure are based on [AsO$_4$], [SiO$_4$], [SO$_4$], or [VO$_4$] groups (Pan and Fleet 2002).

Though the basic constituent of mammalian bone and teeth is carbonate-hydroxylapatite (or dahllite) (McConnell 1952, 1960), the table shows that extensive crystallochemical substitutions are possible: this is the cause of the chemical changes occurring during diagenesis and allowing relative dating of the bones by chemical analysis (Section 2.5.3.4).

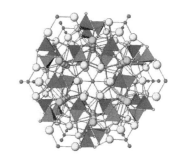

Fig. 3.72. The structure of apatite is composed of (PO$_4$)$^{3+}$ groups (pink tetrahedra), monovalent anions like F$^-$ and (OH)$^-$ (purple spheres) in "tunnels" parallel to [0001], and Ca^{2+} cations in two coordinations: seven- and nine-fold.

3.7.1.1 Bone material: major components and hierarchical organization
[F. Berna]

Bone, teeth and ivory belong to a family of biomaterials characterized by a common building block: the mineralized collagen fibril. Weiner and Wagner (1998) reviewed the material and mechanical properties of the mineralized collagen fibril and its arrangement in different biomaterials (e.g. bone, dentin, cementum, mineralized tendon). The mineralized collagen fibril is composed of three major components that are intimately associated in an ordered structure. These are:

(1) the fibrous protein collagen;
(2) dahllite (Table 3.19)—some authors suggest that bone dahllite actually lacks OH ions (Rey *et al.* 1995, Pasteris *et al.* 2004). Others suggest very low concentration of OH within a very complex stoichiometric composition (Aoba 2004); and
(3) water

Table 3.19. The most common chemical end-members and solid solutions of the apatite group minerals

Mineral	Composition	Notes
Hydroxylapatite	Ca$_5$(PO$_4$)$_3$OH	
Fluorapatite	Ca$_5$(PO$_4$)$_3$F	
Chlorapatite	Ca$_5$(PO$_4$)$_3$Cl	
Dahllite	Ca$_5$(PO$_4$,CO$_3$)$_3$OH	Carbonate-hydroxylapatite (also called collophane). Basic constituents of human bones and teeth
Francolite	Ca$_5$(PO$_4$,CO$_3$)$_3$F	Carbonate-fluorapatite (also called collophane, dehrnite, or lewistonite)
Alforsite	Ba$_5$(PO$_4$)$_3$Cl	
Pyromorphite	Pb$_5$(PO$_4$)$_3$Cl	
Strontium-apatite	Sr$_5$(PO$_4$)$_3$OH	
Fermorite	Ca$_5$(PO$_4$,AsO$_4$)$_3$OH	
Belovite	(Sr,Na,La,Ce)$_5$(PO$_4$)$_3$OH	

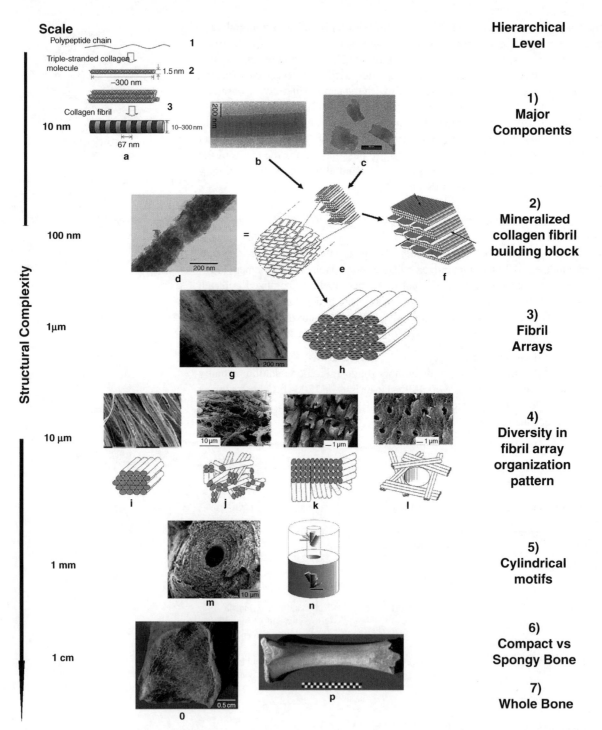

Fig. 3.73. The seven hierarchical levels of organization of bone materials according to Weiner and Wagner (1998). Level 1: major components (a, b, c): (a) schematic illustration of Type 1 collagen fibril organization (1: Polypeptide chain; 2: Triple-stranded collagen molecule; 3: Collagen fibril); (b) part of an unmineralized and unstained collagen fibril from a turkey tendon observed in vitreous ice in a TEM (modified from Weiner

Collagen constitutes the main frame of a three-dimensional matrix into which, and in some cases onto which, the dahllite crystals forms.

The proportions of the three components in the mineralized collagen fibril can vary considerably between different tissues, and the manner in which the mineralized collagen fibrils are organized into higher-order structures. In fact, this is the basis for differentiating between bone, dentin, cementum and mineralized tendon. Weiner and Wagner (1998) recognize several levels of hierarchical organization in bone (Fig. 3.73). These are (from the low to the high): Level 1: Major components (type I collagen fibrils, dahllite crystals and water); Level 2: Mineralized collagen fibril; Level 3: Fibril arrays (fibres); Level 4: Fibril arrays patterns; Level 5: Cylindrical motifs (Osteon); Level 6: Bone Type (spongy versus compact bone); Level 7: Whole bone (Anatomical). While dahllite is the only mineral phase in bone, type I collagen is one of the many proteins present. In fact there are two hundred or more so-called non-collagenous proteins (NCPs), the most abundant of which is osteocalcin (Smith *et al.* 2005). These proteins though generally comprise less than 10% of the total protein content. Lipids may also be present in fairly large amounts in some bones (Weiner and Wagner 1998). Fatty acids are major constituents of tissues and bone marrow that adhere to and/or fill the bones (Compston 2002).

3.7.1.2 *Mineralized collagen fibril* [F. Berna]

According to Weiner and Wagner (1998) mineralized collagen can be considered as a composite material in which collagen forms the fibrous framework and the dahllite plate-shaped crystals (mainly formed inside the fibrils) reinforce the composite. Type I collagen is characterized by its fibrous nature, with the fibrils made up of three polypeptide chains of about 1000 aminoacids (mainly glycine, proline, and hydroxyproline). The polypeptides chains are coiled together in a triple helix forming a cylindrically shaped molecule, with an average diameter of about 1.5 nm, and lengths of 300 nm. Owing to space of about 35 nm intercalated longitudinally between the triple-helical molecules

---

and Wagner 1998); (c) TEM image of isolated bone dahllite crystals (modified from Berna *et al.* 2004). Level 2: Building block: mineralized collagen fibril (d, f): (d) TEM image of a mineralized collagen fibril from a turkey tendon (from Weiner and Wagner 1998); (e) and (f) model showing the proposed location of the platy crystals in the channels (modified from Weiner and Wagner 1998). Level 3: Mineralized Collagen Fibril Arrays (g, h): (g) TEM image of a thin section of mineralized turkey tendon; (h) Schematic illustration (not drawn to scale) showing an arrangement of mineralized collagen fibrils aligned both with respect to the crystal layers and the fibril axes (modified from Weiner and Wagner 1998). Level 4: Diversity in fibril array organizational patterns (i–l): (i) SEM image of a mineralized turkey tendon (scale: 0.1 mm) and schematic illustration showing the fibril bundle; (j) woven fibre structure: SEM image of the outer layer of a 19-week old human foetus femur and a schematic illustration showing fibril bundles with varying sized diameters arranged in different orientations; (k) plywood-like structure present in lamellar bone: SEM micrograph of fracture surface of a baboon tibia and a schematic illustration showing that the fibrils in each layer are rotated relative to their neighbours (depicted by the change in direction of the ellipsoid cross-section); (l) Radial fibril arrays: SEM image of human dentin fractured roughly parallel to the pulp cavity surface. The tubules (holes) are surrounded by collagen fibrils that are all more or less in one plane. Schematic illustration of the fibril bundles arranged in a plane perpendicular to the tubule long axis. Within the plane they have no obvious preferred orientation. Level 5: Cylindrical motif (m, n): (m) SEM image of a single osteon from human bone. Note the concentric lamellae radiating from the haversian canal (at centre) and the intercalated lacunae; (n) schematic illustration (not drawn to scale) of a cylindrical osteon-like structure with the five sub-lamellar model superimposed. The orientation of the sub-layers is reversed on opposite sides of the cylinder, thus balancing the asymmetry of the lamellar structure. Level 6: compact and spongy bone: (o) Light micrograph of a fractured section through a fossilized (about 5500 years old) human femur (from Weiner and Wagner 1998). Level 7: Whole bones: (p) bovine tibia (scale: 10 cm).

these are staggered by 68 nm. This organization of the fibril results in the internal structure not having radial symmetry but differing in all three orthogonal directions. Fibrils associate with each other to form arrays of aligned fibrils to make up a larger structure called the fibre. Importantly the packing of fibrils in a fibre may vary from one tissue to another and influence the mechanical properties. The crystals of bone are plate-shaped, with average lengths and widths of 50 and 25 nm. Small-angle X-ray scattering analyses (SAXS) show that the plate thickness varies from just 1.5 nm for a mineralized tendon up to about 4.0 nm for some mature bone types (Wess *et al.* 2001). These crystals are therefore extremely small—in fact they are probably among the smallest biologically formed crystals known. Interestingly, these crystals are plate-shaped, even though dahllite has hexagonal crystal symmetry: the crystal morphology is in fact totally controlled by the biological environment of growth (Meldrum and Cölfen 2008). One proposed explanation is that they grow via an octacalcium phosphate (OCP) transition phase. In fact, OCP crystals are plate-shaped and have a structure that is very similar to apatite, except for the presence of a hydrated layer. More recently Mahmid *et al.* (2008) studied the continuously growing fin bony rays of the Tuebingen long-fin zebrafish as a model for bone mineralization and demonstrated the presence of an abundant amorphous calcium phosphate phase in the newly formed fin bones. They therefore proposed that amorphous calcium phosphate may be the precursor phase that later transforms into the mature crystalline dahllite mineral (Mahmid *et al.* 2008). Nevertheless the very thin plate shape confers to the bone dahllite crystals a high specific surface area–volume ratio that makes these crystals thermodynamically unstable (Berna *et al.* 2004). Water is the third major component of the bone family of materials. The water is located within the fibril, in the gaps, and between triple-helical molecules. It is also present between fibrils and between fibres. It is interesting to note that in the different bone family tissues collagen content is constantly ca.30% by weight while the content of water varies in inverse proportion to the content of dahllite. The importance of water for the mechanical functioning of bone cannot be underestimated since mechanical measurements of dry bone are different from those of moist bone (Weiner and Wagner 1998).

3.7.1.3 Ivory

Ivory is formed from the bulk dentine of mammals' teeth and tusks. Although true ivory should be referred exclusively to material from modern elephants (*Loxodonta africana*, *Elephas maximus*), and ancient mammoth (*Elephas primigenius*) and mastodonts (*Mammut sp.*), many archaeological and recent art works are made with hippopotamus (*Hippopotamus amphibius*), walrus (*Odobenus rosmarus*), sperm whale (*Physeter macrocephalus*), and narwhal (*Monodon monoceros*) materials. Tusks of wart hogs and pigs, and reindeer, elk, and deer antler have also been used to manufacture tools and to carve ornaments. The chemical structure of ivory from the teeth and tusks of mammals is the same regardless of the species of origin (Raubenheimer 1999), and it is a dense composite of carbonate-apatite and collagen. Antler has basically the same composition, but a very different texture, as it always contains a substantial part of spongy bone in the inner part, so that the dense fraction that can be used to carve solid tools and ornaments is much smaller.

Identification of the material is important for archaeometric reasons, to help in the characterization of archaeological contexts, the status or economic importance of a tool set or funerary goods, or to interpret trade and exchange routes. However now the Convention on International Trade in Endangered Species (CITES, http://www.cites.org/) is fortunately enforced, so elephant products are banned from the market in the attempt to halt the flourishing illegal trade in ivory that has contributed so much to the decrease of elephant numbers in the African continent. Therefore there is a very practical need to identify the original species of the traded objects.

Identification of large ivory pieces is generally performed straightforwardly by experts, because the characteristic patterns of tusk and teeth construction represent typical fingerprints for the attribution. A few classical texts are available for the macroscopic identification of structural patterns of ivory from different species (Penniman 1952, MacGregor 1985, Krzyszkowska 1990, Espinoza and Mann 1992). The identification of elephant from mammoth ivory is especially based on the angle between structural lines, the so-called Schreger lines (Espinoza and Mann 1992, Ábelová 2008). However, there are objective difficulties in the identification of small fragments or highly carved artefacts. To help in such cases several optimized spectroscopic techniques have been developed, mostly based on FT-Raman (Edwards *et al.* 1997, 2006, Shimoyama *et al.* 2003), FT-IR (Shimoyama *et al.* 2004) and XRF spectroscopies (Kautenburger *et al.* 2004). How much these techniques can be applied to degraded archaeological materials remains to be demonstrated (Edwards *et al.* 2005, Long *et al.* 2008). Interestingly, DNA techniques seem to be promising techniques for the identification of ivory species (Lee *et al.* 2009). Dating can be performed by radiocarbon and ESR.

Ivory material has been used since prehistoric times for tools and art carving. Almost every civilization shows the use, exchange, and diffusion of ivory artefacts, from the use of mammoth tusks in palaeolithic Siberia (Khlopatchev G., Scheer A., and Vasilev S. A. In: Cavarretta *et al.* 2001) to the ample diffusion and trade of hippopotamus and elephant ivory in Egyptian and Mycenaean Mediterranean (Krzyszkowska 1993, Krzyszkowska and Morkot 2000). In the Late-Bronze Age Uluburun shipwreck off the coast of Turkey a number of ivory items were recovered including elephant tusks, a dozen hippopotamus teeth, many decorative objects, and a trumpet carved from a hippopotamus incisor into the shape of a ram's horn. In the Classical world ivory use was so widespread that the demand probably triggered the extinction of the Syrian and North African elephant populations. In Asia, Southeast kingdoms included tusks of the Indian elephant in their annual tribute caravans to China, where ivory has long been valued for art and tools. As early as the first century BC ivory artefacts moved along the Northern Silk Road for consumption by western nations. In the West, before the advent of plastics, ivory was used to make billiard balls, piano keys, and a number of decorative and useful items, including artificial teeth.

Finally, it is worth mentioning a special kind of Miocene Age fossil ivory of mastodon found in Southern France that has absorbed metal ions

Fig. 3.74. Fragments of blue-stained bones from the excavation of a Terramara site (Montale, Italy). The sample is from a Recent Bronze Age layer dated to the second half of the 14th century BC. Before the analysis, the object was initially identified as vitreous material. (Photo by I. Angelini, University of Padova; sample courtesy of A. Cardarelli, University of Roma)
Original colour picture reproduced as plate 41.

(especially Mn) from the environment during diagenesis. This material is called **odontolite** (also bone turquoise, or fossil turquoise), and it has a bright blue colour thought in the past to be caused by vivianite crystallization within the ivory microfractures. It has been recently shown by an impressive array of experimental techniques (including XRD, XAS, PIXE/PIGE, TEM-EDS, UV-Vis reflectance spectroscopy, laser-induced luminescence spectroscopy, and others), that (1) two types of blue fossil apatite exist, odontolite and stained bones; and (2) odontolite owes its turquoise colour to a process of deliberate heating in air above 600 °C (Reiche et al. 2000, 2001, 2002). "True odontolite" therefore is mastodon fossil ivory that has absorbed Mn ions from the soil and produces the turquoise-coloured material by heating: the turquoise tinge derives from the tetrahedral $(MnO_4)^{-3}$ oxocomplex of Mn^{+5} substituting for $(PO_4)^{-3}$ in the hydroxyapatite structure. This material is so brilliant that it was extensively used in the Middle Age to decorate metal artwork and religious objects, such as reliquary crosses. There are 17th and 18th century known reports of French naturalists producing turquoise by fire processes (Reiche et al. 2000). Possibly, similar material had been used much before as a substitute for turquoise-beads in Neolithic Syria (Taniguchi et al. 2002), though the deliberate process is hard to demonstrate.

Stained bones can also show various shades of blue, green, or brown. They are very frequent in archaeological excavations, though they may often be misinterpreted as a different material when they are oddly shaped or carved (Fig. 3.74). Bone colour may have a different origin, such as deposition of secondary copper minerals from copper or bronze metallic objects present in the same layer, or metal ions (mostly Mn and Fe) absorbed during diagenesis and redeposited as oxides and hydroxides in the pores of the material. Accidental or intentional heating may cause changes in the mineral components and further complicate the issue (Yubao et al. 1993, Reiche and Chalmin 2008, Chadefaux et al. 2009). Examples of coloured apatites are presented in Fig. 3.75. The

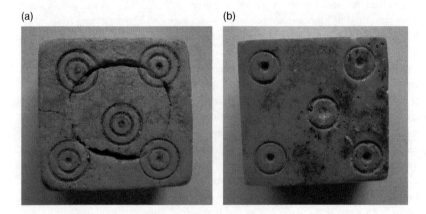

Fig. 3.75. Etruscan dice pieces from Tarquinia: (a) hollow bone piece, unstained; (b) stained ivory. (Photo courtesy of V. Nociti; samples from the National Museum of Tarquinia, Italy)
Original colour pictures reproduced as plate 42.

images show the corresponding face of two dice found in Etruscan Tarquinia: the white one is bone, with the hollow internal part completed by an inset; the green one is ivory, presumably stained on purpose for aesthetical reasons. The reader is referred to a recent dissertation (Nociti 2007) that presents extremely interesting data on an extensive set of ancient dice from Southern Etruria. Not only the studied samples represent a fairly complete dataset of archaeological dice, showing the shape, structure, nature, sign style and technique, and iconographical evolution of these objects, but the investigation shows that a number of different materials such as limestone and tuffaceous stones were in use to make dice pieces, though bone and ivory were by far the preferred materials. The choice of the material may have a connection with the origin of the game, which is a transformation of the popular game of throwing bones, perhaps for divination. In fact dice were probably originally made from the ankle bones (specifically the talus or "astragalus") of hoofed animals (such as oxen), colloquially known as "knucklebones", and still used in many places such as Mongolia and Africa for children's' amusement and fortune-telling. Accordingly, the earliest archaeological pieces are not cubes, but rather rectangular- or square-based prisms, very similar to the astragalus bone.

However, there is an even more interesting story stemming out of the dice investigation, which is an excellent example of important side information derived from scientific analysis. The set of dice was analysed for the statistical distribution of the numbers (one to six) on the faces. Without listing all the mathematical details (the reader is referred to the original work by Nociti 2007) it is here sufficient to remind that there are 15 different ways of distributing six numbers on the faces of the cube, assuming that the exchange of opposite faces is invariant (reminding the crystallographic point symmetry of the cube, which has 48 symmetry elements, the number is obtained by 6!/48=720/48=15). It turned out that over a set of over ninety samples, only two of the possible configurations are present in the studied set from Etruria, and precisely those two that satisfy the conditions (a) sum of opposite faces = 7, and (b) difference of opposite faces = −1. In both cases, that is in all investigated dice, the number 3 is always opposite to number 4. This can be put in direct relationship with the famous two Etruscan dice conserved at the Bibliothèque Nationale de Paris (Fig. 3.76(a)), which carry letters of the Etruscan alphabet in place of numbers. The letters (*thu, zal, ci, sa, mach, huth*) clearly correspond to the first six Etruscan numerals. The interpretation of four of them is quite agreed by scholars: *thu*=1, *zal*=2, *ci*=3, *mach*=5; however there have been fierce discussions about the allocation of the other two letters (*sa, huth*) to the numbers 4 and 6 (Agostiniani 1995, Pittau 1995). The experimentally observed configurations of numbers leave no space for interpretation: opposite to number 3 (*ci*) there must be number 4 (*sa*) (Fig. 3.76(b)). This is a great example of a straightforward archaeometrical solution of a puzzling linguistic problem (Nociti 2007).

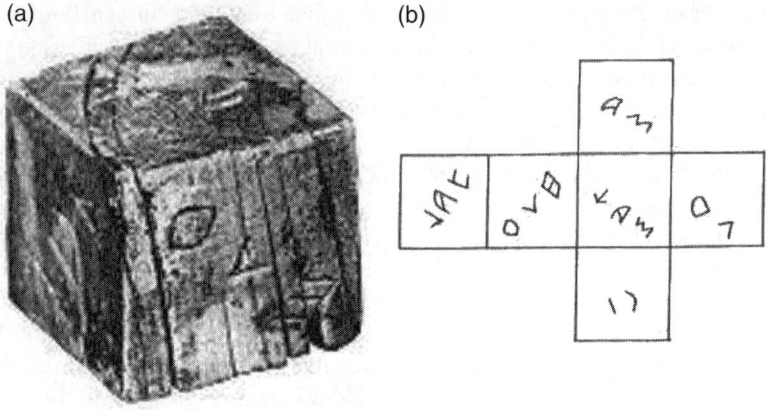

Fig. 3.76. (a) One of the two existing examples of an Etruscan dice carrying letters of the Etruscan alphabet in place of numbers. Close comparison of the letter sequence (b) with the observed distribution of numbers permits unambiguous interpretation of the numerical meaning of the letters.

Box 3.m Bone alteration and diagenesis [F. Berna]

Collagen and dahllite dehydration and transformation start soon after death with the ending of the bone regeneration processes (osteoclastogenesis). Upon deposition into the ground surface and subsurface the bone components alter and undergo several diagenetic processes (Hedges 2002). It is important to note that transformation and diagenesis of the bone material components are fairly rapid and are driven by the environmental conditions of deposition/conservation (Weiner and Bar-Yosef 1990, Hedges 2002, Berna *et al.* 2004).

Organic components

Collagen and other non-collagenous proteins (NCP) such as osteocalcin rapidly diminish and degrade by simple oxidation processes and/or intense biological activity (Trueman *et al.* 2004, Smith *et al.* 2005). Organic components of bone and tissues that seem to survive even in extreme environmental conditions such as those found in bogs and/or desert caves (e.g. Dead Sea Scroll parchment) undergo transformations and/or contaminations (e.g. Van der Plicht *et al.* 2004, Weiner *et al.* 1980). A very important discovery by Salamon *et al.* (2005) shows that relatively well-preserved ancient DNA and collagen are occluded within clusters of intergrown bone crystals that are resistant to disaggregation even by a strong oxidant such as 2.5% sodium hypochlorite (NaOCl). These findings not only allow the optimization of pristine DNA extraction but importantly show that bone is indeed a very complex material with portions having very specific physico-chemical properties. Amongst the NCP Smith *et al.* (2005) demonstrated that the protein osteocalcin can survive in bone in the archaeological record, and postulated that it has the potential to survive over geological time periods. In favourable conditions aminoacids may even be preserved for several millions of years such as those extracted from *T. rex* fossil bone (Schweitzer *et al.* 2007). Important transformations to the organic components of bone occur upon heating. In fact the structure of the collagen fibril is altered once the bone is heated to temperatures as low as 60 °C (Koon *et al.* 2003). At the temperature of 200 °C the molecule of collagen degrades and at 300 °C it pyrolizes into a charred phase that contain aromatic structures (Figs 3.m.1 and 3.m.2). Other organic by-products forming due to exposure to heat of bones are charred connective tissues and fat-derived condensed char (Berna *et al.*, in preparation). To date their formation and degradation mechanisms are only postulated. Charred materials, being rich in aromatic bonds and therefore toxic to micro-organisms, are resistant to atmospheric oxidation and to biological degradation and are expected to survive for several thousands of years. Never the less if deposited in alkaline environments (e.g. wood ash) charred materials, even charcoal with a highly ordered graphite-like structure, is subject to strong chemical oxidation and will degrade (Rebollo *et al.* 2008).

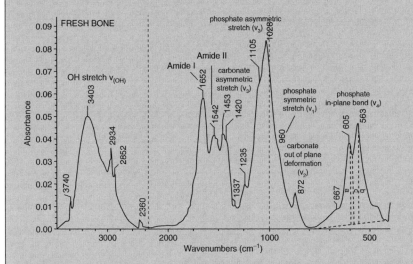

Fig. 3.m.1. Typical FT-IR spectrum of fresh bone with assigned IR bands. Notice the characteristic absorption of dahllite: PO_4 absorptions doublet at 565–603 (v_4) and 1030 cm^{-1} (v_3); carbonate absorptions at 875 cm^{-1} (v_2), and split at 1420–1450 cm^{-1} (v_3); collagen absorption (Amide I and II absorptions at 1655 and 1540 cm^{-1} and the CH_2 stretching at 2852 and 2934 cm^{-1}). The Infrared Splitting Factor (IRSF) is defined as (a + b)/c (Weiner and Bar Yosef 1990).

Inorganic components

As far as the alteration and diagenesis of the mineral phase is concerned, numerous studies showed that the dahllite crystals of bone and dentin undergo processes of recrystallization and uptake of carbonates, and other important elements such as F, Cl, U, Sr, K and Rare Earth Elements (REE) that results in an alteration of the original crystal structure and chemical composition (Hedges 2002, Kuhn *et al.* 2008, Millard and Hedges, 1996, Trueman *et al.* 2004). Berna *et al.* (2004) demonstrated that biogenic bone dahllite crystals will dissolve and re-precipitate into a larger and more stable phase of dahllite. In fact bone dahllite crystals due to their plate-shape and small dimensions [40 × 20 × (2 – 4) nm^3] are thermodynamically more unstable with respect to the geogenic (or synthetic) dahllite and dissolve in conditions where geogenic dahllite will form. This mechanism drives the recrystallization process described for the quasi-totality of archaeological bone in which bone dahllite plate-shaped crystals had transformed into rod-like apatite crystals (Hedges 2002, Berna *et al.* 2004). Significantly Trueman *et al.* (2004) showed that the size of dahllite crystals of bones deposited on the soil surface for a few tens of years increases significantly (especially along the long c-axis) at the expenses of the degrading collagen. Very importantly Trueman *et al.* (2004) showed that the increase in crystal size is proportionally reflected by increasing infrared splitting factors (IRSF). Trueman *et al.* (2004) also showed that carbonate and other elements present in the soil solution such as Ba, Na, and REE are absorbed into the mineral phase by capillary action. The uptake of fluorine into the lattice promotes the formation of francolite [$Ca_5(PO_4,CO_3)_3F$], the bone mineral composing dinosaurs fossil bone (Hulbert *et al.* 1996). The presence of F substituting OH can be detected by the increasingly stronger FT-IR absorption at the 603 cm^{-1} with respect to the 567 cm^{-1} band in the v_4 PO_4 (Geiger and Weiner 1993). There is less agreement about the effects of dahllite recrystallization and diagenesis on the isotopic composition (Lee-Thorp 2008). It has been observed that many of the standard indicators of diagenesis do not correlate with one another (Hedges 2002) and, furthermore, crystal changes measurement do not correlate with stable isotope alteration (Trueman *et al.* 2008).

(*cont.*)

Box 3.m (Continued)

Fig. 3.m.2. FT-IR spectra of unheated and heated (2 hours) bovine bone: (a) unheated bone (IRSF=2.6); (b) heated at 100 °C (IRSF=2.6); (c) heated at 200 °C. Note the relative reduction of the intensity of the Amide II band (IRSF=3.0); (d) heated at 300 °C. Note the disappearance of both Amide I and II absorptions substituted by a large absorption at 1620 cm^{-1} with shoulder at ~1550 cm^{-1} (IRSF=3.2); (e) bone heated at 550 °C. Note the drastic reduction of the absorption of the carbonates (at 875 and 1415–1450 cm^{-1}) and the organic components at 1600–1500 cm^{-1}. Also note the presence of weak but distinct O-H absorption at 632 and 3570 cm^{-1} of hydroxylapatite (IRSF=3.8). (f) Bone heated at 800 °C: note the absence of carbonate absorptions (at 872 and 1420–1450 cm^{-1}) and organic components (at 1500–1700 and 2852–2934 cm^{-1}), the general sharpening of the IR pattern, the shift of the v_3 PO$_4$ band to higher wavenumbers (1047 cm^{-1}), the strong PO$_4$ absorptions at 982 cm^{-1}, and the strong O-H absorption at 632 and 3570 cm^{-1} (IRSF=4.2).

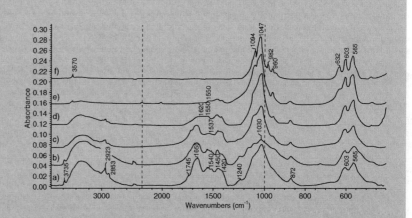

Irreversible transformations occur in bone mineral upon exposure to heat (Munro *et al.* 2007). The crystallinity of bone dahllite increases with increasing temperature as indicated by increasing values of crystallinity index such as the IRSF (Figure 3.m.2) or the XRD crystallinity index (Munro *et al.* 2007). At temperature above 700 °C the carbonates are calcined and the dahllite transforms into hydroxylapatite (Figure 3.m.2). Calcined bone is therefore likely to preserve better than un-calcined bone due the lower solubility product of high temperature hydroxylapatite with respect to the one of bone dahllite (Berna *et al.* 2004).

Experimental measurement of the state of preservation

Numerous analytical techniques can be used to characterize the state of preservation of bone and dentin. Fourier Transform Infrared spectroscopy (and micro-spectroscopy) is probably the most useful technique. It in fact gives important information on the water content and the composition of both the mineral and the organic phases. The assignments of infrared absorptions of bone and dentin have been widely investigated by numerous scholars with inter-calibrated studies (e.g. Rey *et al.* 1989, Bohic *et al.* 2000, Magne *et al.* 2001, Trueman *et al.* 2004, Kuhn *et al.* 2008). The quantity and quality of collagen, and the substitutions of carbonate, fluorine, hydroxyl and hydrogen phosphate can be estimated by identify specific IR bands, their shifts and/or by measuring specific IR absorbance ratios. Specifically the amount of protein contained in bone and dentin can be estimated by measuring the ratio between the Amide I band (v_1) at ca. 1650 cm^{-1} and the v_2 PO$_4^{-3}$ at ca.1030 cm^{-1} (e.g. Trueman *et al.* 2004). Similarly the amount of carbonate contained in the dahllite can be estimated by measuring the ratio between the v_3 CO$_3^{-2}$ absorbance at ca. 1415 cm^{-1} and v_3 PO$_4^{-3}$ at ca.1030 cm^{-1} (Wright and Schwarcz 1996, Ou-Yang *et al.* 2001). Very importantly the relative nature of the carbonate environment in the dahllite lattice can be identified by IR spectroscopy upon deconvoluting the spectra. In fact Rey *et al.* (1989) showed that bands in the v_2 CO_3^{-2} domain have three specific components. The major component at 871 cm^{-1} is due to carbonate ions located in PO_4^{-3} sites (type B carbonate); a second band at 878 cm^{-1} is exclusively assigned to carbonate ions substituting for OH ions in the apatitic structure (type A carbonate); and a third band at 866 cm^{-1} corresponds to a labile carbonate environment. The intensity ratio of type A to type B carbonate (I_{878}/I_{871}) is used to characterize different bone samples while the 866 cm^{-1} carbonate band seems to vary randomly (Rey *et al.* 1989). Similarly the deconvolution of v_3 and v_4 PO_4 domains is used to identify the presence of HPO_4 substituting for the PO_4 groups (Rey *et al.* 1990, 1991). Finally the presence of F substituting OH can be detected by the increasingly stronger FT-IR band at the 603 cm^{-1} with respect to the 567 cm^{-1} band in the v_4 PO_4 domain (Geiger and Weiner 1993). Apart from the change in chemical composition high importance for the assessment of the preservation of bone and dentin is given to the changes to the original crystal structure of dahllite. The crystallinity of bone and dentin mineral has been widely investigated by FTIR spectroscopy and XRD (Weiner and Bar-Yosef 1990). The infrared crystallinity index also known as the infrared splitting factor (IRSF) is probably the most used (Weiner and Bar-Yosef 1990, Wright and Schwarcz 1996, Munro *et al.* 2007). In fact due to its ease of acquisition procedure and its higher sensitivity to a smaller increase of crystallinity there exists a large database of bone showing that pristine bone IRSF have a typical value of between 2.6 and 2.8 (Berna *et al.* 2004).

Further analysis of protein material can be performed after extraction by dissolving the mineral phase either in diluted HCl (Shahack-Gross *et al.* 1997) or better by EDTA treatment (Salamon *et al.* 2005). On the other hand bone crystals can be separated and isolated by treating the bone material with NaOCl (Weiner and Price 1986).

For in-depth information

Dove P.M., De Yoreo J.J. and Weiner S. (eds) (2003) Biomineralization. Rev. Mineral. Geochem., Vol. 54. Mineralogical Society of America-Geochemical Society, Washington, DC. p. 381.

Hedges R.E.M. (2002) Bone diagenesis: an overview of processes. Archaeometry **44**, 319–328.

Mays S. (1998) The archaeology of human bones. Routledge, London.

3.7.2 Amber and resins [I. Angelini]

Resins, together with natural and synthetic derivates have been variously employed by humans since prehistory. In the last fifty years synthetic polymers have mostly replaced natural materials in a wide range of applications, though resinous materials still maintain scientific and economic interest. Resins and amber started being studied in the 18th century by chemists, geologists, palaeontologists, botanists and archaeologists, each one measuring different properties. By perusal of the literature, the first difficulty encountered is the range of terms used to characterize the resinous materials; the same word may be used by various authors with different meanings, and terms may change meaning over the course of the time. Furthermore, erroneous or improper usage of specific technical terms is frequent. As an aid to understand the literature, Table 3.20 lists the fundamentals terms with current definitions, secondary, obsolete and improper use.

Table 3.20. Main technical terms used for resinous materials with current definitions and notes on the secondary and improper use (Larson 1978, Poinar 1992, Vavra 1993, Mills and White 1999, Anderson and Crelling 1995, Pollard and Heron 1996, Beck 1996, Rice 2006, King 2006)

Term	Definition	Other use and notes
Resins	Unmodified exudation of a living plant, only volatile component may be lost for natural evaporation. Chemically it is a complex mixture of water-insoluble organic compounds, primarily terpenoid and/or phenolic compounds.	
Fossil resins	Transformed plant-derived resins that go through a fossilization process; the fossilization essentially consisting of a polymerization of the low molecular weigh compound and a contemporary lost of the volatile fraction.	
Amber	Now generally used to define any fossil resins.	Originally limited to describe Baltic Amber, specifically succinite.
Baltic amber	Amber present in the Baltic Region.	Often referred only to the principal type of Baltic amber "succinite"; the terms are used in some cases as synonymous.
Amberine	Yellowish-green chalcedony from Death Valley (California).	
Ambrite	Fossil resin that occurs on the North Island of New Zealand.	Not to be confused with the Kauri Copal.
Amberlite	Commercial name for a synthetic polymer used in analytical chemistry.	
Ambroid	Large masses of material formed by artificial compression at 160 °C and high pressure of small amber pieces. The first process was invented in 1880.	Also referred to as "Pressed Amber"; in some cases also named "Reconstructed Amber"
Reconstructed Amber	Material composed of approximately 90% synthetic compound and 10% resins.	Also named "Synthetic Amber"
Ambergris	It is not a resin. It is a waxy substance formed by the oxidation and hardening of a viscous liquid present in the hindgut of sperm whale.	Known to Arabs as "Anbar", also initially called "amber" in the West.
Ambrein	A cholesterol derivative, it is the principal components of ambergris.	
Resinite	Now generally used as synonymous for "Fossil resin" and "Amber". Actually derived from coal maceral term defined by International Committee for Coal and Geology and by American Society for Testing and Materials convention; some authors suggest that it has to be used only to describe material that can be petrographically identified.	In some cases "Amber" refers to macroscopic materials, and "Resinites" to microscopic materials.
Copal	Hard semifossilized resins, derived from the oxidation and polymerization of resins. It can be distinguished by amber for its physical proprieties. Sometimes referred also as "true copal".	Sometimes it is wrongly used for some types of incense, because the name derives from the Aztec word copalli (incense). Commercially used for resins from the Araucariaceae (Manila and Kauri Copals) from the Caesalpinioideae (Brazilian and African copals), or from other leguminous plants (Zanzibar, Congo, South American and Columbian Copal).
Succinite	Mineralogical name of the more important Baltic amber, characterized by the presence of a very high amount of succinic acid (3–8 wt%).	"Succinite" and "Baltic amber" are often used synonymously. In the old studies "succinites" is the term used to describe all ambers containing succinic acid.

Retinite	Term used to describe all the amber that does not contain succinic acid.
Retinellite	An Oligocene fossil resin from Great Britain.
Tar	Anthropogenic material made by heating resin, resinous wood or bark; it is the initial pyrolysato. Tar and pitch are generally referred to as pyroligeneous substances.
Pitch	Pitch is obtained from further heating of tar to remove the remaining volatile compounds.
Oleo-resins	Resins with a relatively high fluidity containing a high amount of volatile terpenes. Oleo-resins derive from many Pinaceae and Dipterocarpaceae genera.
Rosin = Colophony	The non-volatile fraction (mainly diterpenoids components) that remain after the distillation of resin from *Pinus* species. The distilled fraction, composed of a mixture of monoterpenes and small amounts of sequiterpenes, is called "Oil of Turpinene". Colophony is also found on Baltic shores, perhaps derived from the activity of ancient varnish company.
Balsam	Generally it is used for relatively soft and initially malleable resins. The term "Balsam" or "true Balsam" is used only for four resins: Canadian balsam, Malaysian balsam, Tolu and Perù balsam.
Elemi	Scented and initially semisolid resins, mainly derivate from different species of the Burseraceae. In some cases the term "Elemi" is restricted to Manila elemi from *Canarium luzonicum* (Philippines).
Incense	High scented resins; the commercial incenses (frankincense, myrrh and Mexican incense) derived from various genera of Burseraceae. Defined also as "gum-resins": a combination of components found in resins and polysaccharidic gum.
Mastic	Triterpenoid resins produced by trees of Mediterranea *Pistacia lentiscus*, of the Anacardiaceae family. Sometimes it also refers to resins of different *Pistacia* species; even a resin from India (*Bursera Gummifera*) has been described in some case with the same name. Mastic has been used also for resins from Peruvian *Schinus molle* (called "American Mastic"), and the herb *Thymus mastichina*. It is rarely used for cements, made from lime plaster or crushed brick and resins, tar or bitumen.
Dammars	Triterpenoid resin originating from trees of the sub-family Dipterocarpoideae (family Dipterocarpaceae). Some confusion about this term arises from the fact that "Dammar" is the Malay word for all resins. Sometimes it is used for resins from the Burseraceae.
Sandarac	Diterpenoid resins of *Tetraclinus articulata*, Cupressaceae (North Africa). "Australian Sandarc" also exists, which derives from the genus *Callitris*.

3.7.2.1 Resins

A number of plants spontaneously produce a water-insoluble exude that may be generated in variable amounts in response to specific problematic conditions. The exude generally called resin is sticky, smelling and waterproof; other exudes such as latex, gum and kino have different chemical and physical characteristics. The resins have various functionalities: as a defense against herbivorous or pathogens; they may help in the repair of broken parts or damage created by fire (Fig. 3.77); they act as a waterproofing agent controlling water evaporation and desiccation of the leaves; they also cover and defend seeds from environmental injury (Langenheim 1995, 2003, Pollard and Heron 1996).

Resin-producing trees are present in several families of both orders Conifers and Angiosperms. In the order Coniferales the families producing significant quantities of resin are: Pinaceae (genera *Pinus, Abies, Picea, Larix, Pseudolarix, Pseudotsuga, Cedrus,* in different amounts); Araucariaceae (genera *Agathis,* and in a minor quantity *Araucaria*); Taxodiaceae (genera *Sequoia, Matasequoia, Sequoidendron,* and *Taxodium,* which produce resin only as a consequences of trauma in the wood, Fig. 3.77); Podocarpaceae and Cupressaceae (in particular *Cupressus, Juniperus, Callitris* and *Tetraclinus* genera). Among the Angiosperms (flowering plants) the resin producing tree belongs to numerous families and species; those producing resins that are important in archaeology or cultural heritage are: Leguminosae, subfamily Caesalpinioideae (Detarieae tribe, genera *Copaifera, Daniellia, Guibourtia*); Dipterocarpaceae,

Fig. 3.77. Resin exude by a Giant Sequoia (*Sequoiadendron giganteum*, family Cupressaceae) as a reaction to a fire injury.

subfamily Dipterocarpoideae (genera *Shorea, Hopea, Vatica, Anisoptera* and *Dipterocarpus*); Burseraceae (genera *Protium, Dacryodes, Bursera,* and *Canarium*); Anacardiaceae (genus *Pistacia*); Hamamelidaceae (genus *Liquidambar*); Styraceae (genus *Styrax*); Umbelliferae (genera *Opopanax* and *Ferula*); Cistaceae (genus *Cistus*). (Langenheim 1995, 2003, Serpico and White 2000; and references cited therein).

Resin is generally produced in canals or pockets of secretory tissue (Langenheim 1995), and may be found in different parts of the trunk. The resin block is slowly transformed by the fossilization process into amber, often preserving the shape. Resin accumulations may be present in the inter-bark and the under-bark areas, and they may fill cracks or voids in the wood. The amber forms massive blocks, mainly showing irregular shapes, in some cases with elongated forms, and sometimes preserving the imprint of the bark. The exude resin creates drops, stalactite and stalactite-like ambers, in the last two cases with typical shell-like structures (Kosmowska-Ceranowicz 1984a, 2006).

Chemically resins are composed of a complex mixture of terpenoid and/or phenolic compounds. The terpenoid resins are the major class of resins, in which the terpenoid may be accompanied by volatile phenolics, acetogenins, fat and traces of aminoacids (Langenheim 1995, 2003). Terpene are complex organic molecules that may be formally described and classified into functions of the number of isoprene (C_5H_8) units; if terpene contains functional groups (such as hydroxyl, carbonyl, carboxylic groups), and therefore atoms other than C and H, the compounds are called terpenoids. Monoterpenoids (C_{10}) and sesquiterpenoids (C_{15}) are liquid volatile compounds that are responsible for the initial fluidity of the resins. Diterpenoids (C_{20}) and triterpenoids (C_{30}) and polymers formed from them are the structural compounds that constitute the terpenoid resins. The volatile fraction of resins is composed therefore of mono-, sesqui-, and some diterpenoids, whereas the non-volatile fraction is constituted primarily of unsaturated carboxylic diterpene acids, or triterpene acids. Alcohols, aldehydes and esters may also be present (Langenheim 1969). Di- and tri-terpenoids are not known to be present in the same resin, with the exceptions of the Umbrelliferae family (Serpico and White 2000). A detailed survey of the chemical characterization of natural resins is reported in Mills and White (1999) and in Langenheim (2003). The main type of resins used in antiquity, some of which still having commercial importance, are summarized in Table 3.20.

The chemical nature of a resin is strongly related to the botanical source; within the same specie the chemistry of constitutive resin (i.e. resin synthesized in the plant prior to damage) may differ from induced resin (i.e. resin produced following an injury) (Langenheim 1995). On the other hand the chemical and structural composition of fossil resins is strongly related to those of the ancestor resin, so that the analytical study of modern resins is the basis for attempting to identify the paleobotanical sources of fossil resins. Several Conifers families (Pinaceae, Araucariaceae, Taxodiaceae, Cupressaceae and Podocarpaceae) and Angiosperm families (Leguminosae, Burseraceae, Dipterocarpaceae, Hamamelidaceaea, Anacardiaceae and Styraceae) have been investigated as possible botanical source of amber.

In Cupressaceae and Pinaceae the diterpenoids pattern is different, therefore the analyses of the molecular composition, generally carried out by GC-MS, Py-GC-MS, Py-MS (see last paragraph, and Boxes 2.t, 3.o), are able to distinguish between the resins of the two families. Determination of the exact genera is not always possible, and actually only a few can be positively identified based on analytical methods. On the other hand, in the triterpenoid and other non-terpenoid resins, the different genera often show typical composition patterns; for example the resins of *Pistacia*, storax, myrrh, frankincense have been recognized in several archaeological samples by GC-MS, TLC, and IR analyses. The heating process necessary for the production of tar and pitch also creates typical molecular compounds such as retene that can be used to identify the original material (Serpico and White 2000, and references therein).

The chemical and physical characteristics of resins have been exploited since prehistory for pharmaceuticals, disinfectants, smell, waterproofing, adhesives and sealants applications. In the objects found with Ötzi, the Alpine Iceman (Boxes 2.v, 3.*l*; Figs. 2.22, 2.m.2), there is clear evidence of resinous substances used to fix the copper axe to the wooden handle, and to attach the flint arrow tips. The analyses show that the resinous material is a tar derived from Betulaceae bark (Sauter *et al*. 1992, Spindler 1993).

Mastic and Chios turpentine have been identified by GC, MS, GC-MS and FTIR analyses in Syro–Palestinian amphorae from the Late Bronze Age shipwreck of Uluburun (Turkey), testifying to the widespread use and trade of resins in the Mediterranean area since protohistory (Mills and White 1989, Stern *et al*. 2008). It has been demonstrated that in Egypt there was a wide use of resin, pitch and tar. In the Middle Kingdom jars from different sites have been shown to contain the remains of Cedar pitch, *Pistacia* resin and myrrh. *Pistacia* resin, *Pistacia* pitch and a mixture of *Pistacia* pitch and coniferous pitch were used during the New Kingdom to produce varnish for funeral equipment. In the mummification process the treatment of the body and the wrappings require the use of coniferous products of Pinaceae (pine, cedar or fir), beeswax and bitumen in different combinations and ratio according to the age (Serpico and White 2000). An important use of resins is related to the waterproofing of ships; in antiquity, since Etruscan and Roman times, the use of tarry materials and pitch derived from Pinaceae wood in ships' paints has been clearly identify by GC-MS, FTIR and XRD analyses (Robinson *et al*. 1987, Colombini *et al*. 2003).

3.7.2.2 Amber and copal

During ageing resins change their chemical and physical properties by oxidation, polymerization processes and loss of volatile components. The transformations occurring over time in the geological layers encompass diagenesis and fossilization. Fossil resins are organic amorphous materials that are much harder than the other resinous compounds (Table 3.20), however they may easily be subject to degradation by physical, biological and chemical process, especially oxidation (Section 2.5.2.2). Exposure to high temperatures, sunlight, air, rain and the action of bacteria and fungi are responsible for resin deterioration preventing resin from fossilizing into amber. The fossilization process only occurs in specific conditions: when the resin is resistant to the action of external agents, and when it is buried in anoxic soils or in water. The environmental diagenetic conditions, especially temperature, pressure and pH, strongly affect the rate of the fossilization process, and also the final physical and chemical characteristics of the fossil resin. Therefore it is virtually impossible to define a set of parameters that unequivocally identify the maturity of the resins. Age is not sufficient to distinguish resin from copal, or from amber. The fossilization process that converts fresh resin to fossil resin is a continuous process in which the transformation of the material proceeds at different rates depending on environmental conditions.

Amber is a fossil resin, therefore the chemical composition and structure is a function of the composition of the original resin and the paleobotanical source, other than the diagenetic process. For example it has been recently proposed, by GC-MS, Py-GC-MS and IR analysis, that rumanite (Romanian amber) is actually derived from the same botanical source as succinite (Baltic amber), though it has suffered partial thermal degradation due to metamorphic events (Stout *et al.* 2000). Amber is no longer classified as an organic mineral, but rather as a mineraloid and is generally defined as an organic-derived gemstone like pearl, coral and ivory. The name *amber* is derived from the Arabic name *anbar*, which was originally used to describe *ambergris*, an organic wax compound produced by the sperm whale (Table 3.20) and, through time, it has been variously named. The Greeks used the name *elektron*, whereas Romans use the term *succinum,* meaning "sap stone". Pliny was the first to correctly identify plant resins as the "source" of amber. Rice (2006, pp. 178–180) nicely tell the story of amber terminology through the ages.

Copal is a hard semifossilized resin (Poinar 1992), also derived from the oxidation and polymerization of resins, like amber. Owing to the origin of the name and to its widespread commercial use, the word is often used with different meanings in literature (see Table 3.20). The more correct usage of the term copal is to refer to a semifossilized resin, with hardness and a melting point lower that that of amber, and exhibiting a better solubility in organic solvents and a high fraction of volatile components (Table 3.21).

Both amber and copal show a wide range of colours and opacity. Baltic amber may show such a high amount of small bubble inclusions as to appear completely white (bone amber). The gases entrapped in amber have been studied by MS in an attempt to investigate the atmospheric composition of the past (Berner and Landis 1987, 1988). However doubts have been raised to the complete sealing by the amber matrix (Hopfenberg *et al.* 1988).

Mineral inclusions are also present in several amber types. In Baltic amber inclusion of iron sulphides (Rice 2006) and small amount of quartz (Merkevičius1 *et al.* 2007) have been detected, the presence of pyrite and/or marcasite being related to the reducing diagenetic environment.

The accepted theory for the origin of the extended fields of Baltic amber is based on the assumption that the North European region had a temperate to subtropical climate during the Eocene, so that the area was covered by a wide resin-producing forest. The forest extended over the whole pre-Fennoscandinavian continent, from Island and Britain to Ukraina, reaching as far south as parts of France and Germany; the Baltic area was above sea level. It has been estimated that the forest may have lasted about 5 million year (from Upper

Table 3.21. Main physical properties of amber and copal that may be used for distinguishing the two materials (data from Poinar 1992, Rice 2006). Many other microscopic parameters such as colour, opacity, refractive index and age can be very similar

Characteristics	Amber	Copal
Melting Point in °C	200–380	< 150 (generally in the range 120–130)
Hardness (Mohs scale)	2–3	1–2
Refractive index	1.5–1.6	1.5–1.6
Specific gravity	1.04–1.10	1.03–1.08
Solubility in acetone (drop on the surface)	Not soluble	Partially soluble (sticky surface)
Colour	From white to light yellow – to deep brownish. Also green and blue	Typically from yellow to brown
Opacity	From completely opaque to totally transparent	From transparent to translucent
Under UV light	Bluish and yellow colour. Occasionally green, orange and white	Faint light sheen
Reaction to fire	Dark smoke, steady flame	White smoke, sputtering flame
Smell from hot point	Acrid, burnt-resinous	Sweet resinous
"Indicative" age range	Carboniferous–Pleistocene	From Pliocene (the oldest) to 100 years old

Eocene to Lower Oligocene), thus explaining the huge amount of resin produced, the production and the distribution of the plants being controlled by climate changes. During the Oligocene a marine transgression covered the area, and the marine sediments created the appropriate conditions for the preservation of the resins. Subsequent coastal changes and the action of a large palaeo-river called Eridanus started the erosion of the resin-bearing sediments and their transport to the sea. Sediments of silt and clay sealed the resins in an anoxic swampland environment where the fossilization process took place. These sediments are called "Blue Earths", they constitute the secondary deposition of the resin, and they are the major source of Baltic amber. Major mines are in the Sambia Peninsula, near Kaliningrad, Russia. The marine streams moved the resin from the palaeo-delta of the Eridanus causing dispersion of the amber over a wide area towards south-west. In the Pleistocene, during the Ice Age, glaciers, moraines and glacial water eroded the Blue Earths, moving ambers, mud and stones to new depositions in Poland, Bielorussia, Lithuania, Latvia, Germany, up to the coast of England. This sequence of events reasonably accounts for the distribution and age of the Baltic amber deposits (Kosmowska-Ceranowicz 1984b, 1996, 1997, Rice 2006).

Amber nomenclature may also generate considerable confusion, since in the past various terms have been used, originating from the mineralogical literature. Sometimes they refer to the name of the discoverer or the seller, at other times they refer to historical terms, to the geographic origin, and also to fantasy names. Often the name of a new amber typology came into use before the complete chemical and physical characterization of the material. Many of the traditional amber typologies are now known to be the same material, thanks to modern analytical techniques, though they are still sporadically used. A complete review of the mineralogical amber names is a task exceeding the purpose of this volume, so that only the principal amber types and a few peculiar fossil resins are reported here, with the relative deposits' diffusion area and age, when known (Table 3.22). Starting in 1992 a novel rational classification method has been proposed based on the chemical and stereochemical structure of the fossil resin. It is not yet universally adopted, though it is now well diffused in the specialized literature and is encountering increased recognition (Anderson *et al.* 1992, Anderson and Botto 1993, Anderson 1994, Anderson and Crelling 1995). According to the latter classification, fossil ambers are divided into five classes as a function of the polymer or terpenoid main structure; class I (formed by fossil resin derived from labdanoid diterpene polymers) is split into three sub-classes depending on the succinic acid content and on the stereochemistry. This classification is extremely useful for the clear description of the amber structure. Unfortunately not all the molecular amber compositions have been studied or completely interpreted, therefore there are still some limitations in its applicability. The introduction of new names to describe recently discovered fossil resins is strongly discouraged: the material should be simply referred to as amber, or fossil resin, followed by the locality of provenance.

Amber may have very different ages: from Carboniferous to Pleistocene (Langenheim 1969, Kosmowska-Ceranowicz 1984b, Rice 2006). Amber can not be directly dated by radiocarbon, as amber is outside the detectable range of ^{14}C (Section 2.5.4.2), though radiocarbon has been used to detect forgeries and to establish the relative age of copal and amber (Burleigh and Whalley 1983, Grimaldi *et al.* 1994, Anderson 1996). The age is normally derived from the stratigraphic study of the amber deposits, although this may be complicated by the geologic history of amber, which rarely lies in its primary depositional setting, and is often redeposited by sea, fluvial water, or glaciers into a secondary, or even tertiary deposition, such as in the case of the Baltic amber discussed above. Amber deposits are numerous in Europe, North America, Japan and more scarce in Asia. Further, they seem to be concentrated in the Northern hemisphere, whereas in the Southern hemisphere there are mainly copal deposits. Maps of amber and copal deposit distributions as a function of age may be found in Poinar (1992), Grimaldi (1996) and Rice (2006), with detailed descriptions of the possible paleobotanical sources and characteristics of the geological deposit.

The oldest fossil resins reported in literature are from the Palaeozoic: the Carboniferous Period of Northumberland (UK); and the Permian Period from Cekarda River, Ural Mountains (Russia) though only small and poorly characterized fragments are available (See Fig. 2.23 for the geological timescale). Mesozoic amber deposits, especially from the Cretaceous period, are a little more numerous but deliver scarce quantity of material, with small nodules commonly affected by oxidation and erosion. Triassic ambers are known from Arizona (Petrified Forest Formation, Chinle Group); Bavaria, Germany (Mt. Leitnermoons); two sites in Austria (Tab. 3.22); and from the Dolomite Alps, Italy (Vavra 1993, Gianolla *et al.* 1998, Roghi *et al.* 2006). The Dolomitic

Table 3.22. List of the more common and a few peculiar amber names, with the relative distribution area and age. For several amber types different authors report contrasting information. The present compilation assembled data from: Ghiurca and Vavra 1990, Poinar 1992, Vavra 1993, Stout at al. 1995, Grimaldi 1996, Valaczkai and Ghiurca 1997, Kosmowska-Ceranowicz 1997, 1999, 2006; De Franceschi and De Ploëg 2003, Rice 2006

Name	Place/material	Geographic area	Age	Notes
Kochenit	Kokental	Tyrol, Austria	Triassic	
Copaline	Satzberg area	Austria	Upper Cretaceous –Triassic. Also some Palaeogene deposit	Wrongly used also for Eocenic amber from Gablitz, Vienna.
Chemawinite (or cedarite)	Cedar Lake	Manitoba, Canada	Cretaceous	"Grassy Lake amber": similar to cedarite, near Medicine Hat, Alberta.
Jelinite (or kansansite)	Smoky Hill River	Ellsworth County, Kansas	Cretaceous	Analyses show similarity with cedarite. They are considered of the same group.
Walchowite (or muchite or neudorfite)	Walchow, Moravia	Czech Republic	Upper Cretaceous	Muchite and neudorfite, primarily described as different ambers from the Czech Republic, are actually the same as walchowite.
Succinite	Poland, Russia, Germany, Ukraine, Lithuania, Latvia, Estonia, Denmark, Sweden, Great Britain and Holland	"Baltic Region"	In blue earth: Upper Eocene-Lower Oligocene. Redeposit in Late Tertiary	
Gedanite	Pale yellow –whitish, matt, whethered, brittle amber	Germany, France, Poland, North Siberia (Khatanga district)	In blue earth with succinite. Cretaceous in France and in Siberia	Also called "Brittle amber".
Gedano-Succinite	Friable, light color amber	Germany, Poland, Ukraine, Sambia Peninsula	In blue earth with succinite	Also called "Friable amber".
Stantienite	Opaque dark color, brittle amber	Germany, Poland, Ukraine, Sambia Peninsula	In blue earth with succinite	Also called "black resin".
Beckerite	Opaque dark color	Germany, Poland, Ukraine, Sambia Peninsula	In blue earth with succinite	Also called "brown resin".
Glessite (or sheibeite)	Hard amber, light brown color, generally opaque	Germany, Poland, Sambia Peninsula	In blue earth with succinite. In quaternary sediments from Lausitz	Today it is no longer possible to find glessite on the Sambia coast. The amber from Lusitz was also named sheibeite, but it is the same that glessite.
Goitschite	Opaque, dirty white-light beige amber	Germany, Poland, Ukraine	Oligocene	Sub-variety of succinite. Also called "white amber". Problematic identification, often confused with white succinite.

Name	Description	Location	Age	Notes
Siegburgite	Grey amber, soft but hard to cleave. Siegburg, near Bonn; also in Bitterfeld	Germany	Oligocene	The only known amber that probably derived from *Liquidambar*, Hamamelidaceae family (Angiosperm). Squankum, a fossil resin near New Jersey (USA) is comparable. Also similar amber in Montana.
Krantzite	Soft fossil resin	Germany (Saxony) and Baltic Kaliningrade area.	Eocene	Characterized by high sulphur (4-6%).
Oxykrantzite	Kranzite from Alexander mine, Nienburg, that show chemical differences from the other kranzite	Germany	Eocene	
Colophony	Young amber	Baltic region	Holocene	
Ajkaite	Pale yellow, brittle fossil resin from Ajka	Hungary	Upper Cretaceous	
Romanite	Buzau, Ploesci, Colți, Sibiciu and Mlajetu.	Romania	Oligocene	
Almaschite	Piatra Neamț, Neamț district	Romania	–	This is actually jet, not a fossil resin. Jet is a dark mineral substance produced by transformation of wood or coal under extreme pressures.
Schraufite	Vama, Suceava district; Carpathian area (Bucovina)	Romania. Also from Poland	Oligocene	
Telegdite	Săsciori, Alba district	Romania	Cretaceous	
Muntenite	Olănești, Vîlcea district	Romania	Eocene	
Delatynite	From Delyatyn and Myshyn, Carpathian area	Ukraine, Romania		
Simetite	In limestone layers in the area from Petralia and Enna, in the Simeto river delta. On the coast between Catania and Siracusa, in the southwest coast.	Sicily, Italy	Tertiary: Oligocene-Pliocene	
Plaffeiite	Höllibach near Plasselb, not far from Plaffeien (Fribourg)	Swisserland	Paleocene	
Birmit or Burmite		Myanmar (Burma)	Miocene (secondary deposition)	

amber is particular interesting because of its inclusions of fossil bacteria, fungi, algae and protozoans (Schmidt at al. 2006). Jurassic ambers are reported only from Chimkent, Kazakistan. Numerous Cretaceous amber deposits are present in North America from Alaska to New Mexico, the Cedar Lake, Manitoba, and the Kansas and Wyoming deposits are listed in Tab. 3.22. Cretaceous ambers are also present in the Middle East (Jordan, Israel, Lebanon) and Japan, where younger Oligocenic ambers are also reported. Cretaceous fossil resins are known from Austria, Spain (Peñalver at al. 2007) and Romania (Table 3.22, Ghiurca and Vavra 1990, Valaczkai and Ghiurca 1997). In France several amber deposits are reported, they are dated from the Carboniferous to the Tertiary Era (De Franceschi and De Ploëg 2003, Nel *et al*. 2004, Giuliano *et al*. 2006).

Tertiary amber deposits are very numerous in Europe, among them Baltic amber is by far the most well known fossil resin. In the same deposit different types of amber may be present, as in the Bitterfeld mines, Germany, and in the Sambian deposits, Russia. Succinite is the better known Baltic amber because of its wide and nice variety of colour (Fig. 3.78), its good workability and its abundance. Succinite represents about 97–98% of Baltic amber, though a small amount of different varieties are present in the Baltic deposit (Table 3.22, Stout *et al*. 1995, Kosmowska-Ceranowicz 1997, 1999, 2006; Rice 2006). Whether the differences are related to different botanical sources, as in the case of glessite and stantienite (Yamamoto *et al*. 2006), or to various diagenetic histories as in the case of gedanite and gedano-succinite (Stout *et al*. 1995), is still a matter of debate. Baltic amber types display rather different chemical characteristics, especially in the quantity of succinic acid: high (succinite: 3–8%), low (gedano-succinite: 1.1–1.7 %, down to trace amounts in several samples) and no succinic acid (gedanite, ajkaite).

Several European countries (France, Romania, Italy) have amber deposits of different ages. In Italy Triassic ambers are located in the Dolomites, Eastern Alps; Early Eocene ambers in the Lessini Mountains, near Verona (Trevisan

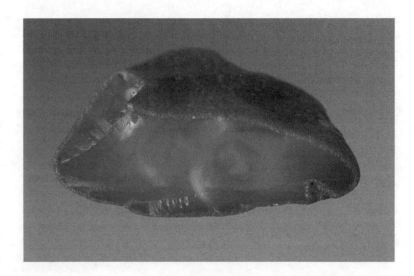

Fig. 3.78. Fragment of a nodule of Baltic succinite. The nodule section has a brilliant yellow-orange colour and a mm-thick oxidation layer at the surface.

Fig. 3.79. Nodule of amber from Italian Apennines, Prignano sul Secchia, Modena, still enclosed in the original sandstone layer. The dark red colour is typical of ambers from this area (Sample courtesy of Iames Tirabassi, Musei Civici di Reggio Emilia).

et al. 2005); simetite is found in Sicily (Tab. 3.22, Skalski and Veggiani 1990, Beck and Hartnett 1993); and numerous small deposit are located in the Northern and Central Apennine chain from Reggio Emilia (Fig. 3.79) to Foligno (Dalrio 1980, Skalski and Veggiani 1990, Angelini and Bellintani 2005).

The vast younger deposits of amber are commercially important. In Central America there are valuable Late Oligocene–Early Miocene ambers in Chiapas, Mexico, and Middle Oligocene–Early Miocene ambers in the Dominican Republic. Both are thought to be produced by *Hymenaea* species of the Leguminosae family. Some of the Dominican ambers are particularly valued for their nice colour varying from yellow–reddish to shining blue, and for their well preserved fossil inclusions. The blue tonality is due to a delayed fluorescence that generally disappears after few years; recent investigations by time resolved visible spectroscopy (Box 2.f) suggest that it may be due to aromatic compounds, in particular to perylene, which originated during the anaerobic diagenesis of the resin (Bellani *et al.* 2005).

Fu Schun Chinese ambers of Eocene age, Paleogenic Bornean amber, and Miocenic birmite, a fossil resin from Myanmar (Burma), are the other ambers that are worth listing here, because of the abundant quantity of material recovered and especially for their relevance in the productions of Asiatic amber artefacts.

The study of the fossils included in amber is an important field of scientific research. The primary aim is to investigate extinct organisms, though in some cases the identification of plants and animals permit the dating of the amber itself, and allows one to reconstruct the palaeo-environment. A large variety of living animals and plants are found: insects, bacteria, fungi, plant remains including pollen, spiders, worms, snails and even frogs and lizards. Fossils are badly preserved in Baltic amber, in several cases little or no organic tissue is left. Dominican, Burmese, Mexican and Borneo ambers on the other hand provide excellent specimens. After the entrapping of the insect the bacteria and the enzymes present in the organism continue their action on the organic material, sometimes completely damaging or dissolving the tissues. In rare cases, the organisms may be pyritized, due to the local sulphur content (Martìn-Gonzales *et al.* 2009). As a sequel to Jurassic Park, many attempts have been made to recover and study ancient DNA from organisms included in amber. As explained, it is not straightforward to extract well-preserved tissue and obtain it in sufficient amounts for DNA reconstruction. Cano *et al.* (1992) first published a notice of a DNA fragment extraction from a bee in Dominican amber, and later from a termite, though repetitions of the experiments produced contrasting results. Contamination of the sample and extreme fragmentation of the sequence are yet unresolved problems (Austin *et al.* 2007, Rice 2006).

The study of plant fossils in amber, together with the molecular analyses, is important for the identification of the paleobotanic source. A statistical study of pollen and plant remains indicate the species that were present in the original amber-producing forest, since even with modern analytical techniques for molecular amber chemistry the botanical origins of most amber types are still unknown. The best-studied case is that of succinite. The research indicated an ancestral plant probably belonging to the family of Araucariaceae (possibly *Agathis*), or Pinaceae, because succinite commonly contains a large number of remains of the conifer genera. On one hand, the wood fragments inclusions seem to be a "pine-like fossil wood" (Langenheim 1995), and succinic acid can also be present in Pinaceae resins. The plant ancestor thus has been named *Pinus succinifera*, meaning that it can be an extinct variety of pines or spruce even if doubts arise from the Pinaceae impossibility of massive resin production, and to the fact that succinite has a completely different resin molecular structure based on labdanoid polymer. *Agathis* on the other hand has been proposed as a possible source because the chemistry of its resin is very similar to that of succinite, and because it can produce a massive amount of resin, though no Agathis remains are present in succinite amber and its resin does not contain succinic acid (Langenheim 1995 and 2003, Rice 2006). *Peseudolarix* has also been proposed as a possible genus source because, even if it has a different stereochemical configuration, its resin shows a labdanoid diterpene structure (Anderson and La Page 1995). A recent proposal (Wolfe *et al.* 2009) indicated Sciadopityaceae as the family of the paleobotanic succinite source. They considered succinic acid as a diagenetic product crystallized from succinates produced in the fermentation processes that took place in the sedimentary environment. In fact the chemical structure of the resin of the only living plant of the family, *Sciadopitys verticillata*, seems to be close to that of succinite, as measured by FTIR and multivariate clustering analyses.

Amber is appreciated and used by humans since the Palaeolithic Age. Simple amber nodules have been found in different Middle Palaeolithic sites in Europe; it is believed that they may have been used as incense. Amber artefacts appear in Europe from the Late Palaeolithic, mainly concentrated in Northern Germany

(Meindorf, Uretep and Weitsche), and in Ukraine (Meinz, Dobranichevka and Mezhirach). The first anthropomorphic figure in amber was found in Dobranichevka (Burdukiewicz 1993). During the Neolithic Age the use of amber diffused through the whole Baltic area and towards Central Europe; coeval amber workshops are found in Latvia. In the Copper and Bronze Ages amber reached the Mediterranean countries. Amber objects have been found in sites from Norway and Finland to Italy and Greece, testifying to the existence of long-range trade. The detailed study of the amber routes used since the Bronze Age and the changes occurring in the different periods has been a hotly debated topic by archaeologists since the end of the 18th century (see for example: Capellini 1872, De Navarro 1925, Negroni Catacchio and Guerreschi 1970, Bouzek 1993). The discussion on the amber trade actually prompted the early scientific investigation of the succinite origin, when in 1875–1890 Otto Helm tried to identify and distinguish Baltic amber from the other types by succinic acid distillation. The work was based on the wrong hypothesis that succinic acid is only characteristic of Baltic amber, though it lead the way for all subsequent archaeometric investigation. (See the early literature cited in: Beck *et al.* 1966, 1967.)

Ancient Egyptian amber objects are scarce, mainly there are amulets with aspects similar to those of amber, but the real nature of the material has never been checked. In Northern Africa there are no deposits of amber, and the Lebanese amber deposits were not exploited in antiquity. Moreover fresh and subfossil resins present in Sudan were possibly known. Unfortunately there are very few analyses on Egyptian amber materials; a scarab of the British Museum, analysed by GC-MS, has show an Agathis-like structure, but the origin of the material is not unambiguously determined (Serpico and White 2000). Several analyses have been performed on amber from Palestine. The two Late Bronze Age objects analysed turned out to be one Baltic amber and, in the other case, a non-Baltic amber of undetermined origin. All other analysed beads are of Roman–Byzantine age and they are all succinites (Todd 1993).

3.7.2.3 *Analytical methods for the study of resins and amber*

Since ambers are all composed of a complex organic mixture (often with not too much dissimilarity among different amber types), elemental chemical analysis is not helpful for the characterization and distinction of the different amber sources. The chemical composition of amber fall in the range: C 74–83%, H 9.5–11.2%, N 0.004–0.07%, S 0.02–1.74%, O and other elements 4.7–14.6% (Beck 1986, Ragazzi *et al.* 2003, Bogdasarov 2007). Even if some specific trends in the chemical composition of ambers from different sources have sometimes been observed, these are not sufficient to support the unambiguous identification of the material. Moreover the elemental composition of amber is affected by the non-systematic presence of mineral and organic inclusions, so that even the application of modern and more accurate techniques such as OES, PIXE and NAA do not offer real benefits to the provenance research.

The major problem in the analyses of ambers is their insolubility. A total dissolution without complete deterioration of the components is impossible. Therefore, if the technique requires a solubilization of the sample (AAS, OES, MS), only the soluble fraction of the amber will be investigated. The variability

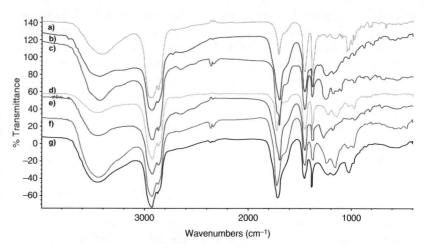

Fig. 3.80. FTIR spectra of: (a) amber from Monte Malo, Lessini, Italy; (b) Dolomitic amber, from Cortina d'Ampezzo, Italy; (c) amber from Apennine, Casal S. Pietro, Bologna, Italy; (d) amber from Apennine, Prignano sul Secchia, Modena, Italy; (e) Plaffeïite, Swizerland; (f) Rumaenite, Colți, Romania; (g) Walchovite, Moravia, Czech Republic. y-axis is Transmittance%. The spectra are shifted from the original baseline for comparison.

in the chemical structure of ambers and resins with different origins is immediately evident in a comparison of the measured infrared spectra (Fig. 3.80). The IR spectroscopy emphasizes the different functional molecular groups and it is therefore particularly useful in the characterization and identification of organic compounds (Box 2.e). Moreover it is not necessary to dissolve the samples, as the finely ground powder may be directly distributed on the sample holder, or mixed with KBr and pressed into a small pellet. IR may also work in non-invasive mode or with the use of a minimal amount of sample, using the ATR and DRIFT techniques (Fig. 2.18, Box 2.e).

It has long been known that succinite shows a very characteristic IR spectrum (Fig 2.18; Beck *et al.* 1964, 1965). Succinite exhibits a strong absorption peak at about $1160\,cm^{-1}$ preceded by a broad absorption band, called the "Baltic shoulder", and other two important peaks are recorded at about $990-1000\,cm^{-1}$ and $885-890\,cm^{-1}$. Only some North American ambers have similar spectra (Langenheim 1969). Because the composition is a fairly complex molecular mixture, not all amber types have a typical and well characterized (i.e. "unique") IR absorption spectrum such as succinite, and only a few amber typologies may be unambiguously identified (Beck and Hartnett 1993, Beck 1986, Stout *et al.* 1995, Ghiurca and Vavra 1990, Valaczkai and Ghiurca 1997, Angelini e Bellintani 2005, Giuliano *et al.* 2006). Furthermore, not all the amber deposits have been well characterized by IR spectroscopy, and of course a reference database of the variability of the sources is fundamental to provenancing. Identification problems include the fact that ambers from very different deposits may show similar spectra, and that some amber deposits from geographically close areas show unexpectedly different spectroscopic signals (Angelini and Bellintani 2005). In the absence of a complete deposit characterization and discrimination, it is easy to recognize whether an amber is Baltic succinite, though in the case of a negative response, it is much harder to assign its origin. Because most of the identified ancient amber is succinite, many scholars seem to think that virtually all prehistoric and protohistoric amber is Baltic. This is not correct. Although the vast majority of the artefacts are indeed succinite, it has been proven that in several specific archaeological occurrences local ambers have been employed (Beck 1971, Angelini and Bellintani 2006, Peñalver et al. 2007).

IR analysis is also an easy and fast method for the identifying fakes (Kosmowska-Ceranowicz 2003).

Even if the theoretical principles of Raman spectroscopy are rather different from those of IR spectroscopy (Box 2.e) the two techniques are quite similar in practice. Raman also requires a small amount of material and dissolution of the sample is not necessary. The Raman signal is strongly related to the skeletal polymeric structure forming the resin/amber. Therefore FT-Raman analysis cannot be used to distinguish fossil resins with different origins, though it has been applied to the investigation of copal/amber maturation (Brody et al. 2000, Winkler 2001) and also to the analysis of insects included in amber (Edwards et al. 2007).

Solid state ^{13}C-NMR analysis (Box 3.n) is particularly useful in the investigation of the polymer molecular structure of resin and amber, and therefore it is widely used for the identification of the botanical origin of the material (for example: Langenheim 1969, Lambert et al. 1988, Lambert and Johnson 1996, Martinez-Richa et al. 2000, Poinar et al. 2004). Because of the solubility problems described above, NMR analysis of amber would not be possible without the application of solid-state NMR. A relatively high amount of sample is necessary for the analyses, compared with IR or Raman spectrometry, so that the technique is not systematically used for archaeometric investigations.

MS spectrometry (Box 2.t) has been limitedly applied in the study of amber because of the strong fragmentation of the molecules during ionization, which results in unreliable identification of the molecular compositions of the resins/amber. This problem may be essentially avoided by using a field ionization spectrometer, but the instrument is expensive and therefore has limited application (Vavra 1993). Novel MS techniques have recently been applied to the study of amber in order to expand the potential in the field (Tonidandel et al. 2008). Isotopic studies of amber are scarce, and are generally limited to the measurement of the stable isotopes of light elements such as: H, C, S and O. The studies aim to obtain paleobotanical, paleoclimatological and paleoenviromental information (Nissenbaum and Yakir 1995, Gaigalas and Halas 2009).

GC (Box 3.o, 3.p) is able to separate the individual components of the samples and to identify the single components separated by the retention time. The best result in the identification of the molecular component is obtained by coupling the two techniques: GC-MS. In this way it is possible to have a mass spectrum for each separated component. The limit of the technique is the necessity of dissolving the sample in a solution, so only the volatile fraction of the fossil resins can be analysed. The only possibility for the characterization of the whole amber sample is to use pyrolysis (Py), which may be separately coupled to the MS or GC or to both, Py-GC-MS (Box 3.o, 3.p). Examples of the use of these techniques are reported in the majority of the cited work concerning the study of botanical sources of amber and the resin characterization.

They have also been applied to archaeometric investigation with successful results (Boon *et al.* 1993), though to date they have limited application because of the cost of the analyses, because there are no public reference databases and therefore lengthy standardization procedures are necessary to interpret the spectra and, finally, because the amount of material is of the order of 1–2 mg, which is an order of magnitude higher than the amount needed need for DRIFT (0.1–0.2 mg, Angelini and Bellintani 2005, 2006).

As in most cases when analyzing complex materials, the best way to characterize amber is the combined use of several techniques. The particularly difficult question of the identification of the amber paleobotanic source is approached using numerous combined techniques, especially ^{13}C–NMR, FTIR and GC-MS (Langenheim 1969, Stout *et al.* 1995, Clifford *et al.* 1997). In most archaeometric studies it is not possible to proceed to invasive- or multiple-sampling of the object so the selected technique should be the least invasive as possible, and produce data that can be compared with an extended reliable database of similar samples; this might well be the most complex issue.

3.7.3 Paper [M. Bicchieri]

Analysis and interventions related to conservation of cultural heritage are concerned with items composed of a variety of materials. These have undergone spontaneous and often non-reversible interactions, occurring over an extended period and often in unknown conditions (Section 2.5.2). Moreover, the above-mentioned "items" are unique masterpieces. In the specific case of library, archival or graphic heritage, a large amount of materials is involved in each and every book, scroll, drawing or artwork. Among the materials, paper occupies a predominant position. Paper is a substrate typically made of fibres of different origins that are held together by hydrogen bonds, sized (either with vegetal, animal, artificial or synthetic glues), filled (by means of minerals and salts), sometimes bleached or brightened and—if produced industrially—mechanically and chemically treated. The substrate is then written, illuminated, decorated and then bound. Each of these steps involves new materials and new reciprocal interactions between the substances concerned. The final object undergoes chemical and biological degradation as a consequence of being stored, used and displayed. Restoration is required when such degradation induces an unstable state, which means further interactions with new materials.

3.7.3.1 *Paper degradation*

Over the years, paper undergoes a series of alterations caused by both external and internal factors, summarized in Table 3.23 leading to cellulose (Fig. 3.81) degradation.

From the chemical point of view complete degradation leads to carbon dioxide and water. Partial degradation can also be dangerous and leads to a paper that is no longer usable. An understanding of the different chemical processes

Table 3.23. External and internal degradation factors for cellulose

External factors			
Environment	Use	Disasters	Conservation treatments
Temperature and relative humidity variations, light, ionizing radiations, dust, chemical and biological pollution	mechanical damages, improper use	fire, flood, earthquakes	wrong products or materials, wrong conservation procedures or methods

Internal factors		
Intrinsic instability	Impurities in paper	Writing media
evolution of pulping and paper-making processes, adhesives sizes and coating materials, fillers	non-cellulosic materials, oxidizing compound, metals, acidic salts	metallo-organic inks, metallic pigments, acidic media, printing oils

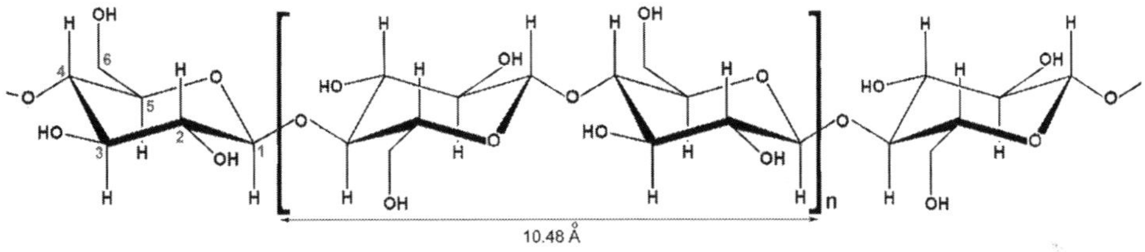

Fig. 3.81. Cellulose molecular structure.

is indispensable to prevent further damage and to choose the appropriate conservation method. Acids, strong alkalis and many oxidizing agents all cause partial degradation following different mechanisms leading to the loss of paper resistance (Porck and Teygeler 2001).

Hydrolysis occurs both in acid and in alkaline environments, attacking the β-glycosidic bond and thus causes depolymerization of the cellulose chain. Alkaline hydrolysis takes place only with strong alkalis at high temperatures and plays a very important role only if the cellulose contains oxidized groups. In this case β-alkoxyelimination takes place, leading to cellulose chain depolymerization. Following this mechanism, alkaline hydrolysis occurs even with diluted alkalis and it can become the main degradation process, especially if cellulose is submitted to strong deacidification.

Metal cations also play an important role in cellulose degradation, acting either as a catalyst in the homolytic scission of the cellulose peroxide (which is formed by a free radical mechanism), or with a donor–acceptor Lewis mechanism involving either the semiacetalic oxygen on the anhydroglucose unit or the β-glucosidic oxygen. In the first case cleavage of a glucopyranose unit occurs; in the second depolymerization of the chain immediately takes place.

The reactions of most oxidizing agents are non-specific and the study of the oxycelluloses is therefore very problematic. The result of the action of oxidants on the cellulose can lead to the glycosidic bond breaking, as with acids and strong alkalis, to the formation of carbonylic or carboxylic groups on the cellulose ring or to the formation of carbon–carbon double bonds on the anhydroglucose unit. In some cases even carbon–carbon triple bonds have been found as a consequence of strong oxidation. Oxidative degradation takes place both in acidic and in alkaline mediums. The only primary hydroxyl group in the cellulose anhydroglucose unit (C6 in Fig. 3.81) can generate an aldehydic group, which can be further oxidized to carboxyl; the two secondary hydroxyls (C2 and C3 in Fig. 3.81) can be oxidized to ketons and, if a cleavage of the C2–C3 bond occurs, two aldehydic groups can be obtained, with further possible oxidation to carboxyl.

Oxidation is a sly process, leading to the loss of paper strength, to its yellowing (if conjugated double bonds are formed in the cellulose chain or between two adjacent chains), or to the increase of acidity of the paper (if carboxyl groups are formed). Hydrolysis breaks the chain, and the paper becomes more fragile. It is to be stressed that almost all writing substances that are used may damage the paper, either producing an acid action (hydrolysis) or inducing chemical deterioration (oxidation), due to the presence of metals, such as iron and copper. Paper is also susceptible to biological attacks (Klemm *et al.* 2002). Cellulose chains aggregates *via* intermolecular hydrogen bonds forming high crystalline and hydrophobic micro-fibrils. These have good strength and are insoluble in most solvents. In native cellulose crystalline and amorphous regions coexist and the ratio between the mass of ordered regions and the total mass of the cellulose, is called the "degree of crystallinity".

The susceptibility of cellulose to chemical reagents or to enzyme action is a function of its state of aggregation: amorphous regions imply the presence of "free volumes" that are more available for external chemical or biological attacks. Cellulose microbiological degradation is brought about by bacteria, fungi and protozoa. The enzymatic attack is difficult, due to the insolubility of cellulose; nevertheless, enzymes called *cellulases* are able to induce a chiefly hydrolytic degradation mechanism. Several properties of the substrate influence the kinetics of such hydrolysis: degree of crystallinity, degree of polymerization, distribution of the molecular weights, accessibility of a surface for the enzymes and the microstructure of the cellulose surface.

A particular case of degradation, the appearance of spots and stains on paper, has been puzzling conservators and scientists since the 1930s (Fig. 3.82). Such a phenomenon is known under the name of *foxing* (Choi 2007). The term is applied to stains of reddish-brown, brown, or yellowish colour, generally of small dimensions, with sharp or irregular edges, most of which, if excited with UV light, show fluorescence. Fluorescence is also often present in areas in which the alteration is not yet visible under natural light, but that is probably in formation. Classification attempts have been tried, over the years, based on the merely aesthetical aspects (shape and colour) but it was only with a scientific approach and structural analysis that some improvements were made in the understanding

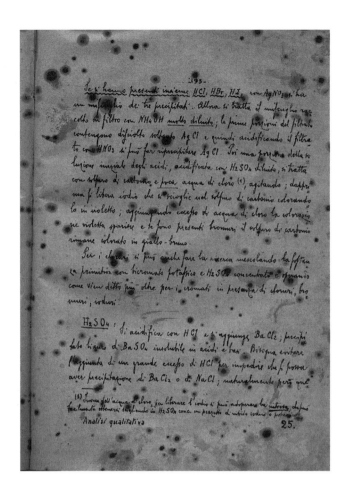

Fig. 3.82. Example of *foxing*. Archive of Icpal Chemistry Laboratory.
Original colour picture reproduced as plate 43.

of the phenomenon. Even though biological strain or metal traces have been reported in correspondence to foxing, the only structural factor that is actually in common amongst stains is a deep oxidation of the cellulose fibres in the spots (Bicchieri *et al.* 2002). Suggestions have been made that a correlation can be found between fibre deterioration and morphology for foxing stains, which mainly occur in the amorphous regions of the cellulose (Buzio *et al.* 2004).

The approach to library, archival or graphic heritage consists of different lines of research. They include a knowledge of all used materials and reciprocal interactions, characterization of their behaviour with respect to environmental conditions, study and simulation of degradation causes and kinetics. Many degradation patterns can usually be detected in a destructive way. In order to apply the obtained results to real conservation problems, the features recognized as significant must be mapped into markers that are detectable in a non-destructive way.

Methodologically, research activities can be regarded as two big families: the first group comprehends all the investigations starting from the materials in the original items, the second one includes the projects on materials to be used as restoration products. In both cases, analyses are first carried out on chemically modified (by exposure to acidic vapours, contaminating substances, oxidants, ionizing radiation and bio-deteriorating agents) model samples so as to simulate original materials. On these samples, as prepared and subjected to standard (UNI/EN/ISO and ASTM) accelerated aging, destructive methods can be applied and can be adapted and optimized for the identification and quantification of functional characterizing groups. Besides providing an insight into the material structure, model samples are indispensable in verifying how much information can be obtained, on the same specimens, in a non-destructive way. Because acidity and oxidation play a very important role in paper conservation and deacidification is a widely used method to remove acidity on aged papers, it is necessary to own or set-up methods to measure the amount of oxidized (mainly carbonyls) and/or oxidized-acidic functions (mainly carboxyls). Deacidification, in fact, should not be applied as a conservation treatment if a large amount of carbonyl groups is present in the paper. Aldehyde or ketone groups indeed promote an alkali catalyzed ß-alkoxy elimination mechanism, leading to cellulose degradation (Santucci *et al.* 2001). Deacidification is unable to act on carbonylic functions that must be removed using specific reduction treatments, before applying deacidification treatments.

3.7.3.2 *Paper analysis: destructive*
pH

The pH measures how much the paper alters the hydrogen–hydroxyl equilibrium of pure water as a consequence of the acidic functions present in the cellulosic support. Its evaluation is important because acidity affects the permanence of the paper. pH measurements can be carried out in three different standardized methods by hot or cold extraction or by surface measurements. The **hot extraction method** (Technical Association of the Pulp and Paper Industry: TAPPI T435) measures the pH of an aqueous extract of paper, after boiling an appropriate amount of paper in water for one hour. This method can be applied on writing supports, but additives (fillers, coating, sizes) and the effects of possible heat-induced hydrolysis can affect the measure, modifying the results. The **cold extraction method** (TAPPI T509) measures the pH of a cold aqueous extract of paper (one hour extraction). The method avoids the effect of the heat-induced hydrolysis and the results of the measurements are only empirically correlated with the end-user's needs and paper requirements. Some fillers or sizes that are soluble in water can modify the results. Both hot and extraction methods are destructive (changing the respective amount of water and paper, the two tests can be regarded as micro-destructive) and not particularly suitable for original documents. The **pH surface measure** (TAPPI T529) is a non-destructive method, even though spots—sometime irreversible—can be formed on the paper surface by the action of the water. One drop of water is applied to the surface of the sample and, after a short time, the excess is removed and the pH is measured by means of a flat electrode. Values obtained by applying this method to sized materials measure only the pH of the surface of these materials and should not be regarded as "true" pH values of paper. pH determinations by different methods are not comparable, due to the different factors affecting the measurements.

Carboxylic groups measurements

The exchangeable carboxylic group's content of a paper is a measure of the free oxidized acidic groups responsible for local or diffused acidification of the cellulosic support. It can be spectrophotometrically measured at 620 nm (American Society for Testing and Materials : ASTM D1926-89: Standard test method for carboxyl content of cellulose, 2006) on the basis of the reaction of cationic exchange between the carboxylic protons of the cellulose and the methylene blue ions. Results obtained give more precise information in respect to pH measurements on the deacidification needs of the support. The typical content of carboxylic groups in a paper in good condition is about 0.5–0.7 m.mol/100 g of paper. It is a destructive method, not suitable for original documents.

Carbonylic groups measurements

Direct measure in UV–Visible of carbonylic groups is impossible, due to the low molar absorbivity ε of C=O and to the insolubility of the cellulose. The determination of the amount of carbonyl can be performed *via* a redox reaction between 2,3,5-triphenyl-2H-tetrazolium chloride (TTC) and the reducing aldehydic functions present in the cellulose chain. After alkaline hydrolysis with potassium hydroxide and the subsequent breaking of the C2–C3 bond, all carbonylic groups of the cellulose become aldehydic. These react with TTC, reducing it to deep red triphenylformazan, which is insoluble in water, but is soluble in ethyl alcohol. The ethanolic solution is sized to a known volume and its absorbance at 480 nm is measured and compared with a calibration line obtained from water–alcohol solutions of glucose. The amount of formazan formed is directly proportional to the amount of carbonyl groups present in the paper. The typical content of carbonylic groups in a paper in good condition is about 0.4 m.mol/100 g of paper (Bicchieri *et al.* 1999). The method is destructive and the quantitative measure is unsuitable for original documents. If small original fragments are available, the method can be used in a qualitative way.

Degree of polymerization

The Degree of Polymerization is the measure of the number of monomers contained in a polymeric chain. For paper supports, the DP value is obtained by measuring the intrinsic viscosity of cellulose dissolved in specific solvents (ASTM D1795 – 96(2007)e1: Standard Test Method for Intrinsic Viscosity of Cellulose). The measure is micro-destructive and can be applied on original documents only if small fragments can be destroyed. Papers that are in a good state of conservation have a DP of about 800–1200. At values lower than about 400 the paper becomes brittle, fragile and the original document needs to be reinforced.

Fibre analysis

Small fragments of paper should be dispersed in water, applied to a slide for microscopy and then dyed with appropriate stains that provide contrast and aid

Table 3.24. Chemical data on books in different conservation states

Owner/Copy/Treatment	Chemical parameters		
Icpal Chemical Laboratory 19th c. book	Carbonylic groups [mmol/100 g paper]	Carboxylic groups [mmol/100 g paper]	pH [±0.02]
Copy for readers	6.02 ± 0.40	9.37 ± 0.08	4.95
Copy kept in safe	0.46 ± 0.01	7.83 ± 0.08	5.58
Restored copy for readers	0.18 ± 0.03	4.90 ± 0.10	8.21
State Archive, Rome 17th c. book	Carbonylic groups [mmol/100 g paper]	Carboxylic groups [mmol/100 g paper]	pH [±0.02]
Before restoring treatment	11.0 ± 0.2	18.3 ± 0.5	2.86
After restoring treatment	5.0 ± 0.3	4.7 ± 0.5	7.8

in the identification of pulp and paper fibres by light microscopy and even in the identification of the pulping process applied in the papermaking (ASTM D1030 – 95(2007): Standard Test Method for Fiber Analysis of Paper and Paperboard). The analysis is micro-destructive but gives an important indication on the state of the paper conservation (dimension, integrity, percentage fibrils) and on the amount of lignified fibres in the raw material. It should be remembered that lignin is responsible for extensive oxidative degradation of paper. Moreover cellulose is stable in alkaline ambient (around pH = 8.5) but the maximum stability of lignin is around pH = 5.5. This means that the action to follow for restoration is strongly dependent on the kind of fibres present in the sample and should be tuned accordingly.

Comments on destructive tests

Even though some authors would still refer to the degree of polymerization (DP) as the key parameter in order to identify and describe the "age" of a paper, this view is not totally correct. Original specimens spent most of their lifetime in unknown and uncontrolled conditions, experiencing a superimposition of spontaneous and irreversible interactions. It is not uncommon that texts, printed or written together and differently preserved, show completely contrasting chemical parameters, including DP as shown in Table 3.24, where the chemical parameters are reported for the same book that has been preserved in two different conditions and before and after conservation treatment and for an ancient archival document, before and after restoration. In both cases a simultaneous deacidification and reduction was carried out in a non-aqueous solvent (Bicchieri et al. 2001). It is evident from experimental data that no single parameter can be an absolute descriptor of "age".

It is definitely more appropriate to indicate DP and the other chemical parameters as indicators of the degradation processes suffered by the analysed specimen, and to use these parameters to choose the better intervention method. Destructive tests can not always be applied to original documents and it is of a great importance to find non-destructive techniques that are more appropriate to detecting the conservation state of original specimens (Łojewska et al. 2007).

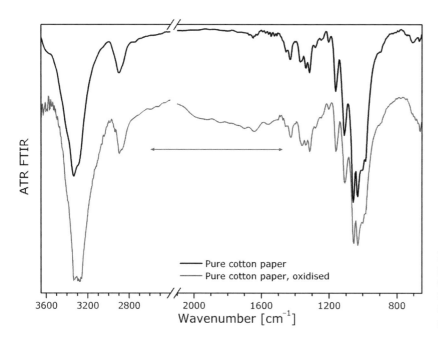

Fig. 3.83. ATR/FTIR spectrum of a well-preserved paper (black line) and of a degraded one (grey line). The arrow indicates the spectral region in which modifications of the spectrum show the effect of a strong oxidation of the support.

3.7.3.3 *Paper analysis: non-destructive*

FTIR and ATR-FTIR

Infrared spectroscopy is a widely used technique for the characterization of organic materials (Box 2.e). In paper application, modification of the cellulose spectrum in the 1550–2500 cm^{-1} region can be observed if oxidized functions are formed by aging or interactions with degradation agents (Fig. 3.83). Normal IR from pellets is micro-destructive and Attenuated Total Reflection Fourier-Transform InfraRed (ATR-FTIR) spectroscopy is a reliable method to obtain IR spectra of the surface of materials in a non-destructive way. The infrared radiation passes through an infrared transmitting crystal with a high refractive index, allowing the radiation to reflect within the ATR element several times, increasing the instrumental response. In the specific case of paper, where the refractive index varies slightly ($n_{paper} \approx 1.55$), depending on the fillers, the penetration depth achieved with ATR crystals is typically of a few micrometres. In this way, the technique allows one to distinguish the many layers constituting a written text, i.e. the fibre conservation state, the adhesives, inks, fillers.

Raman

An exact knowledge of the actual degradation occurring in the paper is of paramount importance, because the conservation treatments are very specific. Raman spectroscopy (Box 2.e) is an attractive technique for non-destructive analysis of paper since, in contrast to IR, the –C=O and –C=C stretching peaks are not superimposed on the H_2O bending peak and, therefore, the functional groups produced during oxidation can be detected. It is known that, in a polar-

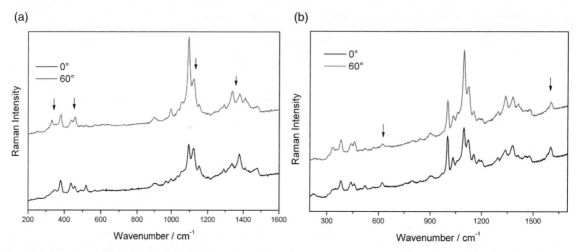

Fig. 3.84. Raman spectra of cellulose, after rotation of the fibres: (a) unoxidized paper—the arrows indicate the polarized bands that show modifications as a consequence of the rotation; (b) oxidized papers—the arrows show the bands that remain unaltered after rotation (marker of the oxidative process). Excitation $\lambda = 514.5$ nm.

ized configuration the intensity of some peaks has a strong dependence on the orientation of the fibres with respect to the polarization of the exciting beam. These effects are observed on several bands of the cellulose (Fig. 3.84(a)), but the typical bands of oxidized groups do not change their intensity with fibre orientation. In particular all the spectra collected from oxycelluloses show that the peak centred around 1577 cm^{-1} is always present and with the same intensity as the fibres are rotated, making this band a good "marker" of the oxidation process (Bicchieri et al. 2006; Fig. 3.84(b)). Moreover, the cellulose spectrum

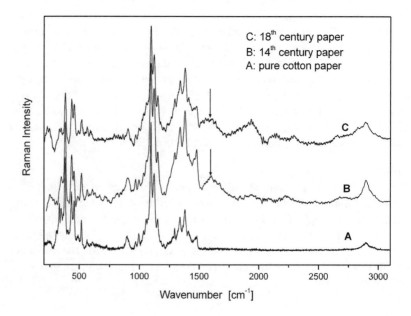

Fig. 3.85. Raman spectra of three different papers: well-preserved, pure cotton paper (A), (B) 14th century original documents and (C) 18th century original documents. The arrows indicate the 1577 cm^{-1} band, marker of the oxidative degradation. The other bands in the 1600–2800 cm^{-1} region, more intense for the 18th century paper, evidence strong oxidation of the support (carbonyl, carboxyls, carbon-carbon triple bonds). Excitation $\lambda = 785$ nm.

does not present Raman signals in the range 1500 – 2800 cm^{-1}. All the modifications occurring in that region indicate degradation processes (Fig. 3.85). Raman technique can be also very useful in the detection of sizes, adhesives, fillers, pigments ink and dyes, allowing for a complete characterization of the original artefact.

AFM

Atomic Force Microscopy (AFM) provides topographies with a higher resolution than those achieved with SEM. Such topographies are acquired as real three-dimensional datasets allowing statistical evaluation on the sample surface properties (roughness, symmetries...). Since all AFMs operate on very small samples (a few mm^2), this technique is generally regarded as micro-destructive. Completely non-destructive devices are however now available, thus widening the range of possible uses in cultural heritage analysis. AFM has been one of

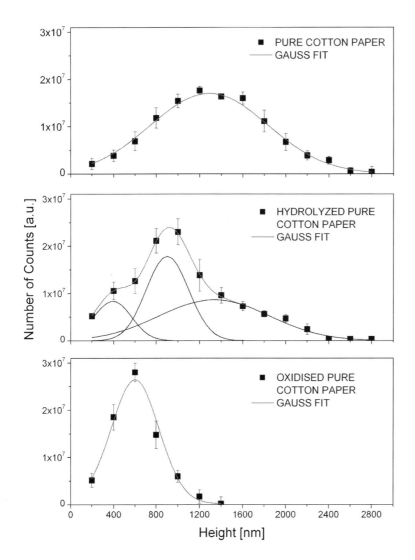

Fig. 3.86. Pure cotton paper surface height distribution. From top to bottom: unaged, hydrolysed and oxidized paper.

the leading techniques in the characterization of cellulose single fibres and crystalline surfaces, and has been applied to pulps and their main components, allowing for the identification of a fibres structure and surface properties. The technique has recently proved to be effective in assessing a fibres condition and in distinguishing the ongoing specific degradation pattern (oxidative or hydrolytical, Fig. 3.86).

Severe degradation of pure cellulose paper affects the surface morphology and is recognizable by AFM analysis. A dominant oxidation (chemical or biological) causes a surface height distribution that is well distinguishable from that induced by a mainly hydrolytical degradation (Coluzza et al. 2008). Oxidative processes give rise to the formation of a single peak at lower dimensions than that observed in a well-preserved paper. After hydrolysis the fragmentation is an expected effect of the observable degradation of the fibre cuticle, which releases fibrils and microfibrils, causing first a random spread of the surface distribution, and eventually an increase of the lower dimensions.

X-ray fluorescence

The emission of characteristic secondary X-rays from a material is widely used for elemental analysis (Box. 2.b). In conservation science applied to library, archival or graphic artworks the elemental composition is useful to detect if impurities are present in the cellulosic support, to characterize inks and pigments in order to obtain information complementary to those obtained with the molecular techniques (Fig. 3.87).

SEM-EDS

SEM measurement (Box 2.l) can be performed at different magnifications, giving information on the morphology of the scanned areas. In addition, a qualitative and quantitative chemical characterization of the inorganic constituents of the samples can be obtained by electronic dispersion spectroscopy (EDS),

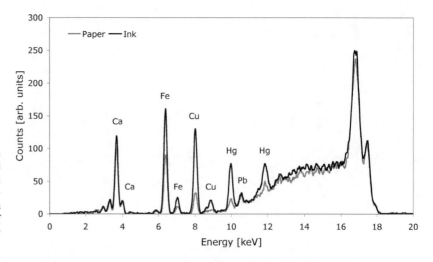

Fig. 3.87. XRF spectrum of an original 12th century document. Raman analysis showed the use of iron-gall ink. The presence of Cu and Hg lines in the XRF spectrum—especially in the ink—suggests the addition of a copper salt to obtain a more bluish colour of the ink and the use of a mercury salt (probably $HgCl_2$) to avoid microbiological attacks.

which allows for an X-ray scanning of the area focused in SEM images. In this way a compositional map of the surface can be obtained. The setup of some SEM instruments allows one to analyse samples without prior preparation by means of metallization. The differences in chemical composition, and therefore in average atomic number of the organic (fibres, fungi, spores, organic adhesives) and inorganic materials allow for well-contrasted observation of filling materials and ion distribution on paper surfaces (Pinzari *et al.* 2006). A limiting parameter in SEM analysis is the dimension of the specimen chamber: all samples must be of an appropriate size to fit in it and they should usually be mounted rigidly on a specimen holder (stub). For these reason SEM technique should be regarded as micro-destructive.

Comments on non-destructive tests

The problems presented by cultural heritage objects are multifaceted. Conclusions obtained from observations with the naked eye are as faulty as those drawn from limited diagnostics. No single technique can be claimed to be the decisive one and multidisciplinary analyses are required for a scientific and correct diagnoses (Missori *et al.* 2006). AFM microscopy has a potential application as a non-destructive technique for the evaluation of pure cellulose paper degradation: imaging of standard samples—after and during accelerated ageing—show a surface deterioration due to the superimposition of oxidative and hydrolytic processes that can be verified destructively, by means of quantitative chemical characterizations, and non-destructively, by means of spectrophotometric analyses. Moreover, the combined use of SEM-EDS and AFM in analysing foxing stains allows for the recognition and characterization of morphologies associated with the chemical or biological origin of the stains. Studies of paper surface with SEM give interesting information on the morphology of the fibres and their conservation state, on the presence of biological attacks and, if coupled with EDS, on the chemical composition of impurities or fillers in the support. This information can also be obtained by means of XRF, but usually on a larger scale. Molecular spectroscopic techniques, such as Raman and IR, are complementary: some vibrations can be more easily detected in Raman, others with IR. This gives information on the specific degradation process occurring in the paper. The measure or the discovery of chemical processes only provides degradation evidence allowing one to choose the correct conservation treatment.

3.7.4 Fibres and textiles

Compared with other materials, archaeometric investigations of textiles are surprisingly scarce. King (1978) a few decades ago stated that "little has been written about the analysis of textiles, fibres, and dyes as it specifically relates to archaeology". The situation has improved, especially since the publication of the dedicated volume, *Archaeological Textiles* (O'Connor *et al.* 1990). Archaeological textile studies are now recognized as an important source of information on past technologies and anthropological costumes, telling stories about cultural transfers, paleoeconomic and gender studies, and aesthetical traditions (Good 2001).

The main archaeometric problems in fibre analysis are (a) the identification of the type and strain of the fibre, (b) the evaluation of its antiquity, (c) the interpretation of the origin and provenance, and (d) the interpretation of past treatments, including dyeing, thread structuring and textile production techniques, where textiles are defined as interlaced threads produced using a loon. Conservation science of course is involved in the thorough assessment of the degradation of the fibres, much as described for paper in Section 3.7.3, and in the development of conservation strategies involving cleaning, stabilization, repair, and restoration.

Natural fibres of archaeological interest may have plant, animal, or mineral origin (Table 3.25). The conditions for the preservation of organic materials are either extreme aridity, freezing, an acidic microenevironment, such as that produced by metal objects, or nitrogen-rich bogs in which strongly anoxic (i.e. oxygen-poor) conditions develop. Each environment has a different set of physico-chemical parameters and, while preserving organic matter, also poses conservation problems. For example flax and other cellulose-based materials preserve better in alkaline conditions (pH > 9–10), whereas animal protein fibres such as wool preserve better in slightly acidic environments (pH $\cong$ 5–6).

Fibre and textile remains can be found in many forms (Good 2001). Direct evidence includes (a) actual intact objects, (b) pristine or degraded isolated threads, filaments, or fibres, (c) chemically degraded pseudomorphs. Indirect evidence includes (d) deliberate or accidental impressions in clay and plaster, and (e) designs and patterns found in ancient representations, art, or other media. Figure 3.88 shows a thread of wool preserved by secondary mineral encrustation over a Final Bronze Age object of the Chiusa di Pesio hoard (Motella De Carlo 2009). Figure 3.89 is an electron microscopy image of isolated fibres from the same sample.

There is direct evidence that fibre manipulation abilities were known in the Upper Palaeolithic on the Eurasian continent (Leroi-Gourhan 1982, Nadel *et al.*

Table 3.25. Nature of the most common fibres used in the past

Nature	Main component	Fibres used in antiquity	Modern fibres
Vegetable	cellulose	flax (linen) (*Linum usitatissimum*)	hemp, jute, ramie, sisal
	cellulose	cotton (*Gossypium hirsutum*)	
Animal	proteins	wool, hair	
	proteins	silk	
	collagen	sinew (tendon)	
	proteins	catgut (intestines)	
Mineral	Mg-aluminosilicate	crysotile asbestos	fibreglass, carbon fibres

3.7 *Organic materials* 397

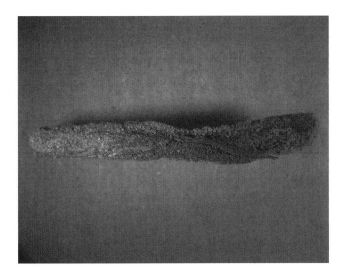

Fig. 3.88. Wool filament covered by secondary copper minerals from the bronze object to which it was attached. The Chiusa di Pesio hoard, including over 300 bronze objects, is dated to the last part of the Final Bronze Age. It is one of the oldest examples of wool fibres in Italy (Motella de Carlo 2009).

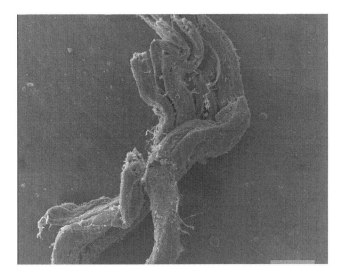

Fig. 3.89. Electron microscopy image of isolated wood fibres extracted from the sample shown in Fig. 3.88. (Motella de Carlo 2009).

1994, Adovasio *et al*. 1996) and moved to the Americas together with the first inhabitants. From the Mesolithic period in Europe (10th–8th millennia BC) we have evidence of the common use of cordage, nets, bags and baskets, which are well preserved in waterlogged sites, such as Noyen, France (Mordant and Mordant 1992) and Friesack, Germany (Gramsch 1992). By Neolithic times cordage basketry and nets are found in numerous dry sites in the Near East, such as Nahal Hemar and Jericho (Israel), Catal Hüyük (Turkey), Jarmo (Iraqui Kurdistan), and many others (Nadel *et al*. 1994, and references therein). The most used plant fibres in antiquity were cotton and flax, the latter made with the wild progenitor of the modern species (*Linum bienne*). The famous textiles from Catal Hüyük (Turkey) dated to 6 ky BC and originally thought to be woolen, proved by careful analysis to be bast (Ryder 1965).

With regard to animal fibres, wool and hair of different animals have long been used for threading and felt making. The difficult conservation of wool partly justifies the nearly non-existent archaeological record during the pre-Neolithic and Neolithic periods, the latter being a time when the development of animal breeding (sheep, goats) was well under way, and the use of wool fibre is only hinted at by the modification of loom weights. Mammoth hair could reach over 50 cm and it is quite unthinkable that it was not used for threading and weaving. The earliest direct evidences of wool are in Bronze Age Majkop culture, North Caucasus (Shishlina et al. 2002, 2003), Tell el-Amarna, Egypt (Ryder 1972), several sites of the Danish Bronze Age (Ryder 1983), and the Final Bronze Age hoard of Chiusa di Pesio, Italy (Motella De Carlo 2009). Numerous occurrences are known from Iron Age sites, such as Scythian Crimea and Siberia (Ryder 1973, 1990), the salt mines of Hallstatt, Austria (Ryder 1990), and the Etruscan tombs at Verucchio, Italy (Stauffer 2002). Sheep breeding and extensive wool use during and after the Iron Age are well demonstrated, and the careful investigation of fleece type and use has stimulated interesting considerations on the early domestication and fleece evolution in domestic animals (Ryder 1964, 1987), other than the relationship between textile production and socio-economical developments (Gleba 2008). Of course wool and hair of many animals other than sheep have been used. Examples are the extensive use of alpaca and llama wool in pre-Columbian Peru, rabbit hair in Mesoamerica, dog, mountain goat, and buffalo hair fibres in North America, and camel in Central Asia. Occasionally unusual animals have been used, such as in the case of the 2500 years-old wool found in the Celtic burial mound of the Hochdorf prince, near Stuttgart, Germany (Körber-Grohne 1988), reported to be made of fine badger's hairs.

With regard to analytical techniques, optical and electronic microscopy (SEM) at high magnification are the most common tools for the identification of fibres. However depending on the fibre integrity and degree of degradation it may be very hard to reach a positive identification of the species. Vegetable fibres are generally harder to identify than animal fibres, the exception being cotton that has a specific twisting, ribbon-like appearance (Emery 1980). If the fibres are woven into thread or cordage, the direction and degree of spin, the fibre content, and the presence of other components are measurable parameters that provide information on the manufacturing technology.

In the case of badly altered fibres more sophisticated analytical techniques are needed. The nature of the fibres can be identified by chromatography-mass spectrometry (LC-MS, HPCL, GCY) to inspect molecular fragments, infrared spectroscopy (FTIR) to recognize functional groups, chemical microanalysis (XRF, PIXE, MS) to discern trace-element patterns and, eventually, DNA analysis to confirm their genetic origin (Drooker and Webster 2000, Jakes et al. 2007). Very promising investigations using extremely small amounts of material have been performed using the small-angle scattering of X-rays (SAXS). The technique has been known for a long time (Giacovazzo 2002),

3.7 *Organic materials* 399

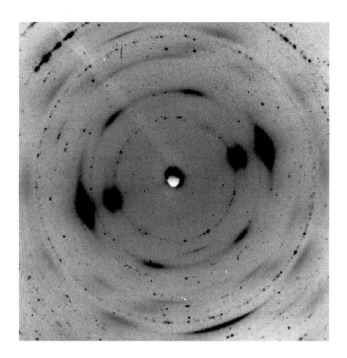

Fig. 3.90. X-ray microdiffraction diagrams of 2000 year old single textile fibres from the caves of Qumran at the Dead Sea. The large reflections constitute a cellulose fibre diagram (flax fibres); the sharp spots on diffraction rings are from crystalline soil particles sticking to the fibres (Müller *et al.* 2000, 2006, 2007). (Image courtesy of M. Müller, GKSS)

though the new microbeam developments at synchrotron sources (Box 2.a) make it possible to perform identification of the plant or animal that was used starting from a single fibre (Müller *et al.* 2000). Figure 3.90 shows the small-angle diffraction patterns obtained on a cellulose fibre of flax. The powerful technique using synchrotron microbeams has opened up a whole new chapter of textile fibre analysis, allowing precise identification even on fragmented or poorly preserved material, which would be impossible to characterize by microscopic methods. In fact the technique may well be used to quantitatively assess the decay of ancient fibres (Hermes *et al.* 2006).

The technique has been successfully applied to the identification of wool and linen from the scroll caves at Qumran and from the Cave of Letters at Nahal Hever (Müller *et al.* 2003, 2004, 2006).

3.7.5 Natural organic dyes and pigments in art [J. Wouters]

A natural colouring matter is an organic material, prepared from a plant or an animal that may confer colour to an object or part of an object, as a dye to textiles or as an organic pigment in a painting. In a dyeing process, the dye is chemically bound to an organic substrate (the yarn), which may have been preliminary mordanted, whereas the preparation of an organic pigment requires the precipitation of the dye on an inorganic base. Such a pigment may then be ground and suspended in a binding medium to make paint. In most cases both mordant and inorganic base are alum.

Natural dye and pigment analysis aims at identifying the biological source(s) used in the production and creation processes of objects of art and culture. This knowledge may contribute to the revealing of technology and its changes over time, to suggest trade, to authenticate, to conserve and to confront information found in historical treatises. However, several factors related to biology, geography, phytochemistry, production technology and analysis hamper the proper identification of these biological sources. The most important of these factors are as follows. A large number of sources has been available for use, especially when taking into account geographically limited local practices; one single source has present several products that contribute to colour and such product populations are often quite similar when different sources are observed; moreover, these products have often closely related chemical structures. Largely unknown factors pertaining to source processing include the timing of harvesting and treatments that follow harvesting in the short term, such as drying, extracting, fermenting, transport and storage. Dyeing procedures, pigment precipitations and paint preparations and applications may be carried out in many different ways. Further complications in identifying biological sources must be expected from ageing processes, which are often influenced by accompanying materials. Last but not least dye and pigment compositional changes may be inherently related to the analytical protocol applied, and interpretations of analytical results are often made troublesome by the scarce availability of adequate reference products, dyes and pigments and by samples of limited size made available for analysis.

This contribution aims at discussing the most important features to take into account when applying or developing natural organic dye and pigment analytical protocols, as well as illustrating the present potential of natural organic dye and pigment analysis, with special emphasis on high performance liquid chromatography (HPLC: Box 3.o).

Important features to be considered for the application or development of protocols

Three groups of features are considered, dealing with compositional, analytical and practical issues, respectively. They are represented in Table 3.26.

Compositional features refer to the wide variety of chemical classes of the major dyestuffs and to the fact that, in most cases, the dye as such is a minor portion in what is called the "sample". A yarn is an organic material itself (proteinaceous or cellulosic), present in more than 99% of the mass of the textile sample to be analysed for dyes. In such a case one will have to look for high selectivity for the detection of the dye in the presence of a high amount of matrix. The same is true when considering the organic dye part of a glaze layer of paint, accompanied by its own binding medium (protein, gum, oil, resin) and eventual other stratigraphic layers with their own pigments and binding mediums. Inorganic bases used for mordanting (dyes) or precipitation (pigment) are unlikely to pose analytical problems, but will influence the recovery procedure of the dye from its matrix.

Table 3.26. Potential parameters to be considered for the development of protocols for organic dye analysis

Feature		Relevance in organic dye/pigment analysis
Compositional features	presence of organic materials	yes
	presence of organic contaminants	yes
	presence of inorganic materials	yes
	chemical nature of compounds	quinoid (benzo-, naphta-, anthra-), flavonoid (flavones, flavonols, isoflavones, chalcones, aurones, anthocyanins, homoisoflavonoids), indigoid, carotenoid, alkaloid, tannin...
Analytical features	balance of detail of information to level of destructiveness	yes
	sample preparation procedure	mostly needed
	sample derivatization procedure	to be avoided
	preliminary chemical separation of components	yes
	spatial resolution, mapping and statistics	interesting
	quantification of individual components	only internal
	need for structural information	sometimes
	detection of alteration products	yes
	trace level detection of components	sometimes
	sensitivity	(sub)-microgram
Practical features	need for reference products	yes
	need for reference objects	yes
	need for data libraries	yes
	universal applicability	preferable
	tandem "hyphenated" techniques	interesting
	invasiveness/destructiveness	low

Analytical features refer to sampling, sample preparation and analysis. Sampling may be avoided when non-invasive analysis is applied, but that is no guarantee of non-destructiveness (Wouters 2008). Non-invasive approaches such as those featured by dedicated configurations of infrared-, fluorescence-, reflectance- and Raman spectroscopies are increasingly successful, but still suffer from the absence of sufficient analytical resolution needed to diagnose an analytical output as a function of the complex compositional features discussed earlier. Such analytical resolution is generated by HPLC and high-sensitivity diagnostic power is added through the coupling of photodiode array (PDA) and mass spectrometry (MS) detectors.

Samples should be prepared in such a way that the dye can be analysed with a minimal risk of loss or alteration of essential products or for external contamination. So liquid–liquid extraction prior to thin-layer chromatography or the boiling of a dyed yarn or a paint chip in an acidic methanolic solution preliminary to HPLC, may pose problems in this respect. However, these treatments are indispensable for avoiding tailed patterns or low recoveries, respectively. When such a procedure must be incorporated during sample preparation it should be carried out in a reproducible manner on references and unknowns for relevant comparisons.

Further **analytical options** such as spatial resolution, (chemical) mapping and statistical treatment of results are interesting, but are seldomly needed in textile dye research. They may gain in importance when dealing with glazes in multi-layered paint.

Quantitative dye analysis may be needed when studying the dyeing behaviour of a dye in the laboratory, involving possible economical aspects. So the determination of the amount of easily extractable dye in Mexican cochineal (up to 20% by weight of the body; Wouters and Verhecken 1989) could help in explaining the rapidly changing dyeing technology in pre-Columbian Peru (Wouters and Rosario-Chirinos 1992). However, in nearly all cases relative quantitation will be the most important feature to be considered. It has indeed been shown that the calculation of the relative ratio of minor components may represent a key diagnostic feature for the identification of scale insect red dyes (Wouters and Verhecken 1989).

The complexity of the biology and chemistry of natural dyes implies the need to have access to appropriate reference products. Such reference products may range from crude plant or animal material, through the prepared dye or pigment, the finished dyed yarn or paint, to pure chemical substances that are known to be components contributing to colour and that are commercially available or have been purified from sources or have been synthesized. Since many important dye components are not commercially available, their synthesis remains a high priority issue.

Issues such as the availability and exchangeability of data libraries, the universal applicability of an analytical technique and the established complementarity of several techniques are closely related practical features that, when appropriately dealt with, may lead to better analytical results on smaller amounts of samples, and hence to lower invasiveness/destructiveness.

3.7.5.1 HPLC-PDA as a protocol for the analysis of natural dyes

The usefulness of an analytical protocol should be evaluated according to its diagnostic power, to destructiveness/ invasiveness, and to representativity and reproducibility of the analytical result. HPLC-PDA scores high in all of these criteria but one, destructiveness/ invasiveness. Despite this shortcoming, HPLC-PDA has become the standard technique for the analysis of natural organic dyes and pigments in a majority of institutes involved in cultural heritage research, thus creating the additional benefit of allowing interlab comparisons.

The analytical procedure outlined below is valid for a wide range of natural dye components such as quinoids, flavonoids, indigoids and tannins. Carotenoids and other molecules with repetitive conjugated double bonds or large aromatic structures (such as is often the case with synthetic organic dyes, for instance) do not run smoothly in that system. Another disadvantage is that the rather harsh sample preparation procedure causes transformations and losses. However this may be partly compensated for by running references following exactly the same sample preparation protocol.

When organic dye and pigment analysis is a part of a more complete analytical approach in textile and paint studies, then the fact that the sample is irretrievable must be taken into account. Hence, HPLC-PDA of natural organic dyes and pigments should come late if not at the end of a multidisciplinary study.

A basic HPLC-PDA analytical protocol

To a sample of dyed yarn or pigment, 250 µl of water/methanol/37% hydrochloric acid (1/1/2, v/v/v) is added. The mixture is heated for 10 minutes at 105 °C in open Pyrex tubes in a heating block. After cooling, any particulate matter is removed by filtering through a porous polyethylene frit (Vac-Elut, Analytichem, USA). The clear filtrate is dried in an evacuated desiccator over NaOH pellets. The dry residues are taken up in 50 to 100 µl methanol/water (1/1, v/v) and 20 µl of this solution is injected in the chromatograph (the injected volume may be 5 µl or less if a mass spectrometer is used). The HPLC equipment consists of a high-pressure pump that is programmable for flow-rate, time periods and composition of the eluting solvents (Model M615, Waters, USA); a column with renewable cartridges of 4.0 × 125 mm Lichrosorb-RP18, 5 µm (Merck, Overijse, Belgium); a photodiode array detector (model 996, Waters, USA); a system for data storage, manipulation and retrieval (Millennium, Waters, USA). Three solvents are used: (A) water, (B) methanol, and (C) 5% (w/v) phosphoric acid in water. The elution program is 60A/30B/10C: 3 minutes; linear gradient to 10A/80B/10C: 26 minutes; flow rate: 1.2 ml/minute, creating a system back-pressure of 18 to 24 Mpa. The temperature in the chromatography laboratory is stabilized at between 20 °C and 22 °C. Any dye component is identified according to two criteria: the retention time and the UV–VIS spectrum.

The exact amount of coloured yarn needed to perform a significant analysis depends on the amount of colour on the fibre, the nature of the colourants and the fibre. In general, lengths from 2 to 5 mm of a wool yarn and 5 to 10 mm of silk thread must be available. The amount of pigment needed is even more difficult to estimate, but can be as low as a couple of sub-mm particles.

3.7.5.2 *Aspects of natural organic dye and pigment analysis in practice*

The four examples given in this chapter to describe approaches, objectives and the potential of natural organic dye and pigment analysis, are summarized from earlier work. The reader is invited to consult the respective references for more detailed information.

Fig. 3.91. Wall painting, Mogao Grottoes, Cave 85. (Courtesy of GCI, © J Paul Getty Trust)
Original colour picture reproduced as plate 44.

The Asian Organic Colourants project of the Getty Conservation Institute: a multidisciplinary approach (Grzywacz et al. 2008)

Rather than showing analytical results, this example describes the high input of resources and networks that are necessary to prepare reference materials that may be used for the development of sustainable sample preparation and analysis protocols, before precious historical samples should be analysed. The Getty Conservation Institute (GCI) has worked with the Dunhuang Academy at the Mogao Grottoes, Gansu Province, China since the 1990s (see Fig 3.91). GCI's wall-painting conservation team reported organic colourants based on UV photomicrographs of paint cross-sections. Identification of those colourants is a challenge because only a limited knowledge is available on the use of dyes and lakes in that geographical and historical (5th to 15th centuries CE) context (Yamauchi *et al.* 2007). From a screening of over 400 relevant papers, 116 biological sources for the production of Asian organic dyes and pigments were identified at the genus and species level. A shortlist of 24 best-hits was established by counting citations from independent publications. Most best-hit plants and resins were purchased at Chinese pharmacies in Dunhuang and Beijing, China and in the Los Angeles area. Items that were not available at

3.7 Organic materials   405

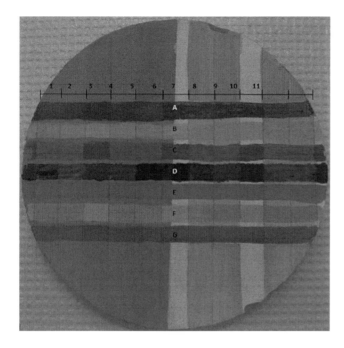

Fig. 3.92. A finished coupon, simulating possible stratigrafies of *Gardenia augusta* pigment (Pl) and inorganic pigments. 1 and 11: Pl/gum over intermediate protective layer; 2 and 10: Pl/gum; 3: plaster render; 4 and 8: Pl/animal glue over intermediate protective layer; 5 and 7: Pl/animal glue; 9: chalk/talc ground over render; A: azurite; B: orpiment; C: red earth; D: black; E: red lead; F: atacamite; G: vermilion (Courtesy J. Wouters, GCI; © J Paul Getty Trust)
Original colour picture reproduced as plate 45.

pharmacies were requested from botanical gardens in the USA and China. Some sources were purchased at the chemical market, Old Delhi, India, others at the garden of the Association Couleur Garance, Lauris, France. Reference bulk samples were prepared as dyed wool and silk and as pigments. Wall-painting mock-ups, consisting of local clay, a ground layer and seven stripes of paint made with the most frequently encountered inorganic pigments, were prepared at the Dunhuang Academy and shipped to Los Angeles. Additional organic paint layers were applied at the GCI laboratory, in such a way that all possible stratigraphic combinations were available (see Fig. 3.92). The latter allows for the development of analytical strategies to identify organic pigments in the Mogao Grottoes' paintings in the presence of potentially interfering materials, and by applying non-invasive as well as medium- and highly-destructive analytical techniques. Such a multifaceted approach is believed to lead to an analytical protocol that involves the best balance between the level of invasiveness/destructiveness and the detail of information.

A special approach to satisfy conservator's needs (Wouters, unpublished report)

In a study and conservation project of a 15th-century wood panel painting, representing The Marriage of the Holy Virgin, painted by an anonymous painter, the conservator had encountered at the highly damaged painted back of the panel a red overpaint, probably consisting of an organic pigment, overlaying an original red layer in which an organic pigment was also suspected. The conservator wanted to discuss an analytical approach that would confirm the suppositions and would identify the organic pigment in both the original paint and the overpaint. The conservator agreed to remove and collect overpaint until the transition zone with the original paint was encountered; then to proceed in the same way through the transit layer and finally collect material from the original paint. The three collected fractions were analysed separately. The overpaint almost exclusively consisted of alizarin, with traces of purpurin and kermesic acid; the original paint was composed of mainly kermesic acid, with traces of alizarin and purpurin; the transition layer contained kermesic acid, alizarin and purpurin in the ratio 26/67/7 as calculated from integration values registered at 255 nm. This result clearly showed that both layers were indeed composed of a red organic pigment, but that the original layer was a kermes lake (*Kermes vermilio*), whereas the overpaint was a madder lake (*Rubia tinctorum*).

Presence of the *Arrabidaea chica* red dye in precolumbian textiles produced in Colombian Andean cultures (Devia et al. 2002a, 2002b)

In the course of a project, aiming at investigating the use of natural organic dyes by Andean tribes in pre-Columbian Colombia, several red samples (see Fig 3.93) generated similar chromatographic profiles and UV–Vis spectra of three main components that suggested the presence of anthocyanidins, but failed to sufficiently match with the chromatographic profile and spectral characteristics generated by a reference cotton dyed with the indigenous *Arrabidaea chica* plant.

Three colourant peaks were found in the chromatograms from the historic samples. One peak was identified as the anthocyanidin carajurin. The spectrum from a second peak seemed to suggest another anthocyanidin, but of unknown structure. Both were present in the *Arrabidaea chica* reference dyed standard. A third peak could also not be identified and was spectrally different in historic samples and the *Arrabidaea chica* reference. Systematic research of the *Arrabidaea chica* dye was carried out, involving the study of traditional Colombian Andean dye recipes, a local plant search, purification and identification of (new) dye components, accelerated aging studies on pure component dyed models and, finally, an application of the new data to those obtained from the historic materials.

This work resulted in the structure elucidation of the two unknown components and in the identification of the photochemical breakdown patterns of the three anthocyanidins. The three principal colouring components of *Arrabidaea chica* dyed cotton were established to be carajurin, arrabidin and 3'-hydroxyarrabidin. Accelerated aging of cotton yarns dyed with each of the three purified components indicated that photochemical decay generated molecules that co-eluted with actual dyestuff components in the up until then used HPLC elution programme. An adaptation of the chromatographic conditions resulted in the complete separation of the three anthocyanidins and of their photodegradation components, so that the third, unassigned peak in historic reds could be identified as 3'-hydroxyarrabidin (see Fig. 3.93). All formerly cited red historic samples could now be identified as having been dyed with *Arrabidaea*

Fig. 3.93. Archaeological textile, cotton, Guane, Northern Andes, Colombia, museum *Casa de Bolivar,* Bucaramanga, Colombia; dyed with *Arrabidea chica*. (Courtesy of M. Cardale de Schrimpf) Original colour picture reproduced as plate 46.

chica. This example illustrates that scrutiny is required when comparing dyestuff compositions in historic samples and modern references.

Rubiaceae in Roman and Coptic Egyptian textiles (Wouters *et al*. 2008)

Of 97 samples taken from Roman Egyptian archaeological textiles dated 1st century CE, 88 turned out to contain components that are characteristic for madder plants (Rubiaceae) (alizarin and/or purpurin), in red, purple and black colours (see Fig. 3.94). The average ratio of alizarin/purpurin calculated from these samples using integration values recorded at 255 nm was around 30/70. This ratio was in-between the 50/50 found in samples taken from 16th to 17th centuries Bruges tapestries (Wouters 1987) and the 10/90 ratio resulting from calculations made on a large series of reference dyeings produced with wild madder or *Rubia peregrina* (Wouters 2001). The latter result was calculated from 35 samples and implied a standard deviation of 10% for alizarin, thus suggesting a relative amount of alizarin to be expected in *Rubia peregrina* dyed yarns ranging from 0 to 20%. These observations were used to recalculate the "madder" samples, but this time taking into account colour and the 0–20% alizarin interval. It appeared that the 0–20% alizarin range was specifically associated with purple colours, whereas higher alizarin ratios were found in reds and blacks. This amazing result led us to recalculate the composition of samples from the Coptic Egyptian era, carried out in a former study, in which madder had been detected (Wouters 1993). Again, the lower alizarin ratios were clearly associated with the purple colour.

It was tempting to ascribe these observations on the madder compositions in purples to the use of a specific plant species, for instance *Rubia peregrina*. However, also other factors should be taken into account such as a different extraction or dyeing procedure. Any influence of the indigo co-dye in the purple on the composition of the red component could be ruled out by noting that in blacks, where the relative amount of indigo was much higher as compared to that in the purples, the madder composition was the same as that in the reds, where no indigo was present. Following these observations on relationships between colour and specific madder compositions, applicable over an 800 years long period in the same geographical framework, it could be suggested that the use of madder to produce purple, in combination with an indigoid dye, was based on a technology for using the red dye, quite different from that used for reds and blacks, and possibly involving the use of another dye plant.

Fig. 3.94. *Clavus* in multicoloured tapestry weave on green tabby ground; purple weft: 94% madder components, 6% indigotin; pink weft: 92% madder components, 8% indigotin; greyish-purple weft: 85% madder components, 15% indigotin; green ground: 13% madder components, 87% indigotin; all figures calculated from integrations at 255 nm. (Courtesy of D. Cardon, UMR5648, CNRS, Lyon, France)
Original colour picture reproduced as plate 47.

Conclusion

Identification of organic materials in general and of natural organic dyes and pigments in particular remain among the most challenging facets of the analysis of art. Analysis of these materials in context may contribute to a better understanding of art, its creation process, conservation history and long-term preservation. Today, the most performant approach in terms of the information–destructiveness balance consists of a high-performance liquid chromatograph, coupled with a hyphenated photodiode array and mass spectrometric detectors. Interesting developments are under way in dedicated Raman (Leona and Lombardi 2007) and microfluorimetry (Claro *et al*. 2008) spectrometries.

Box 3.n NMR Nuclear magnetic resonance [L. Pel]

Nuclear Magnetic Resonance (NMR) is a very versatile technique that has found many applications both as an analytic technique for the structure of molecules as well as for imaging, then often referred to as Magnetic Resonance Imaging (MRI). Here we will only give the basics of NMR; for a more in-depth discussion of the various possibilities the reader is referred to the references given. NMR makes use of the magnetic properties of certain nuclei (Abragram 1961). Nuclei having an odd number of protons, an odd number of neutrons or both have an inherent spin (angular momentum) and therefore a magnetic moment. Nuclei having an even number of protons and an even number of neutron have no nuclear magnetic moment and thus cannot be observed using NMR techniques.

Normally the magnetic moments are randomly oriented. When a static magnetic field is applied the magnetic dipole moments tend to align parallel or anti-parallel to the field, giving an energy difference. Resonant absorption will occur when radio frequency radiation of the correct frequency to match this energy difference is applied (see Fig. 3.n.1). This will occur when the frequency equal to:

$$f_L = \frac{\Delta E}{h} = \frac{\gamma}{2\pi} B_o$$

This resonant absorption can be detected by NMR. The most important perturbation of the NMR frequency for applications is the "shielding" effect of the surrounding electrons. In general, this electronic shielding reduces the magnetic field at the nucleus. As a result the energy gap is reduced and the frequency required to achieve resonance is also reduced. This shift of the NMR frequency due to the chemical environment is called the chemical shift, and it explains why NMR is a direct probe of chemical structures. If the nucleus is more shielded, then it will be shifted upfield (lower chemical shift) and if it is more deshielded, then it will be shifted downfield (higher chemical shift). Hence protons attached to carbons have a different frequency from those attached to oxygen. Depending on their chemical environment the precise resonance frequency will be shifted away by a few parts per million, hence the name chemical shift. Therefore NMR spectroscopy can be used as an analytical technique to measure the chemical composition of liquid and solids.

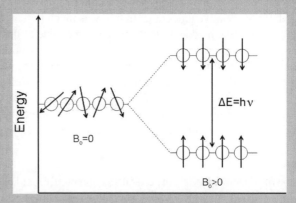

Fig. 3.n.1. A schematic representation of the energy of a spin 1/2 nucleus in a magnetic field.

In order to perform spectroscopy one needs a very homogeneous magnetic field and hence small samples are preferred. In addition, a high field is of advantage as the sensitivity is a function of the magnetic field. For this reason superconducting magnets are often used. The NMR sensitivity of certain elements is given relative to ^{1}H sensitivity, since this is the most sensitive nucleus for NMR. In Table 3.n.1 an overview is given for various nuclei.

3.7 Organic materials

Table 3.n.1. The NMR parameters for various nuclei

Nucleus	Spin	Natural abundance	Sensitivity		$\gamma/2\pi$ (MHz/T)
			Absolute	Relative	
^{1}H	1/2	99.98	1.0	1.0	42.58
^{1}D	1	0.015	9.65×10^{-3}	1.45×10^{-6}	6.53
^{7}Li	3/2	92.58	0.29	0.27	16.55
^{13}C	1/2	1.11	1.59×10^{-2}	1.76×10^{-4}	10.71
^{15}N	1	99.63	1.01×10^{-3}	1.01×10^{-3}	4.32
^{19}F	1/2	100	0.83	0.83	40.05
^{23}Na	3/2	100	9.25×10^{-3}	9.25×10^{-3}	11.26
^{27}Al	3/2	100	0.21	0.21	11.03
^{29}Si	1/2	4.7	7.84×10^{-3}	3.69×10^{-4}	8.46
^{35}Cl	3/2	75.53	4.7×10^{-3}	3.55×10^{-3}	4.17

Often a quantum mechanical approach is necessary to describe some features of NMR spectroscopy, but in most cases it is easier to explain the NMR technique in terms of the semi-classical model. In this case the nuclei are seen as small magnetic moments. If a static magnetic field B_o is applied an isolated magnetic moment will experience a torque. Because the magnetic moment and the angular momentum are coupled, as a result the moment will start to precess around the main magnetic field with the Larmor frequency, f_l. In an NMR experiment there will be an ensemble of moments and the sample will have a net magnetization. When a Radio Frequency (RF) field with the Larmor frequency is applied, the motion of the magnetization will be changed. This changing of the motion by an RF pulse is pivotal to NMR. By using different variations of pulses, also called pulse sequences, one can obtain information concerning the nucleus (Fukushima and Roeder 1981). After the distortion, the magnetization will return to its equilibrium by certain relaxation mechanisms.

A well known experiment is the spin–echo, in which first the magnetization is flipped over 90° followed by 180° after a certain time. The magnetization will go back to its equilibrium by two relaxation mechanism, the so-called transverse and longitudinal relaxation. Assuming that both these mechanisms give rise to a simple exponential relaxation and that spin lattice relaxation is much slower than the spin–spin relaxation, the magnitude of the NMR spin–echo signal is given by:

$$S \sim \rho G \left[1 - \exp(-TR/T_1)\right] \exp(-TE/T_2)$$

In this expression ρ is the density of the nuclei, G the sensitivity with respect to hydrogen, T_1 the spin–lattice or longitudinal relaxation time, TR the repetition time of the spin–echo experiments, T_2 the spin–spin or transverse relaxation time, and TE the so-called spin–echo time. Serious complications occur if the materials under investigation contain paramagnetic ions, like for example Fe, as is the case for many materials in cultural heritage. These ions will disturb the magnetic field and the shift in frequency will no longer be due to the chemical composition but will rather be due to the magnetic impurities. In these cases NMR spectroscopy can no longer be performed.

The analysis of the relaxation can be used to study the fluids in a porous media (Blümich and Kuhn 1992, Vlaardingerbroek and den Boer 1999). For pure water $T_{1b}=T_{2b}$ and in the order of 2–3 s. The NMR relaxation of the fluids in porous media is however largely controlled by the relaxation of the hydrogen at the pore walls. The molecules will diffuse and collide with the pore walls where they will relax by much faster relaxation mechanisms. The source for these relaxation mechanisms is the presence of magnetic impurities near the pore wall. This decay due to the pore interface does not depend on the pore shape but only on the surface-to-volume ratio, i.e.;

(cont.)

Box 3.n (Continued)

$$\frac{1}{T_{1,2}} = \frac{1}{T_{1,2b}} + \rho_{1,2}\frac{S}{V} \approx \rho_{1,2}\frac{S}{V}$$

where $\rho_{1,2}$ is the surface relaxivity for the T_1, or respectively for the T_2 relaxation. The surface relaxivity coefficient is characteristic for the magnetic interaction at the pore surface for a given material. Hence the relaxation is proportional to V/S, which is the structural length of a pore. So water in a large pore will relax slowly, resulting in a longer relaxation time in compared with water in small pores, which will relax much faster. Therefore the relaxation of water in porous materials can provide information on the pore water distribution. By an inverse transformation of the relaxation distribution one can obtain the pore size distribution using NMR. Hence by measuring the relaxation one can study the changing pore size distribution during a chemical reaction like the hydration of cements or the pore water distribution during drying of a porous material.

For materials with very high amounts of magnetic impurities T_1 will reflect the pore size distribution whereas the T_2 relaxation no longer reflects the pore size distribution but rather the internal magnetic gradients. In general, for a material under investigation the content of magnetic impurities is not known. Hence one has to be careful in interpreting the T_2 relaxation distribution as reflecting the pore size distribution. Often one has to perform additional measurements to determine the pore size distribution, like for example using mercury intrusion porosimetry.

In order to obtain a spin–echo one has to apply a magnetic gradient. As a result of this gradient the resonance frequency will be a function of the position and by changing the frequency one can obtain, for example, the moisture distribution without moving the sample. By applying the gradients in different directions one can obtain 2D and 3D distributions as with medical imaging. For materials with high amounts of magnetic impurities, magnetic field gradients of about 0.2 T/m (20 G/cm) are needed in order to achieve a spatial resolution in the order of 1 mm. As a consequence, in order to measure moisture profiles with NMR one cannot use standard NMR/MRI equipment and usually specially adapted equipment is used.

For cultural heritage it is important to differentiate between invasive and non-invasive NMR. With non-invasive NMR it is possible to measure contactless the moisture content and pore-distribution in materials. Whereas with invasive NMR one can also, e.g. do chemical analysis of deterioration mechanisms or structure analysis of paintings.

For in-depth information

Callaghan P.T. (1991) Principles of Nuclear Magnetic Resonance Microscopy. Clarendon, Oxford.

Hornak J.P. (1997) The Basics of NMR, a hypertext book on nuclear magnetic resonance spectroscopy. www.cis.rit.edu/htbooks/nmr

Slichter C.P. (1990) Principles of magnetic resonance. Springer, New York.

Box 3.o HPLC High performance liquid chromatography [J. Wouters]

The term chromatography is derived from the Greek words chroma (χρώμα), colour and graphein (γραφειν), to write. Invented and named by Mikhail Tswett in 1901, following a separation experiment of carotenoids and chlorophylls on calcium carbonate with petroleum ether/ethanol as an eluent, it has now become a general term for a prominent group of laboratory separation techniques.

A chromatography experiment aims to separate mixtures into individual chemical components by introducing the mixture into a system in which a solid stationary phase, normally sitting in a column, is in continuous contact with a mobile phase that is forced through that column. According to the affinity of a component for either the mobile or the stationary phase, it will leave the separation system sooner or later, respectively. When two different components display different affinities, they will pass through the separation system at different speeds and will be separated. Separated components may be collected.

In **liquid chromatography**, the mobile phase is a liquid. In high-performance liquid chromatography, the quality of all elements of the whole separation system is such that very efficient separation (high resolution) of complex mixtures can be achieved in a short amount of time (often not more than a few minutes). The great power of HPLC resides in the versatility of its constituents, which are schematically represented in Fig 3.o.1. Ideally, all essential elements of a HPLC system are set up, managed and run with one chromatography software package (S).

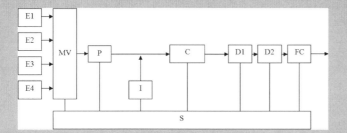

Fig. 3.o.1. Representation of a basic HPLC configuration; E1, E2, E3, E4: eluents; MV: mixing valve; P: pump; I: injector; C: column; D1, D2: detectors; FC: fraction collector; S: software. Arrows indicate the liquid flow; lines indicate programme data flow.

Most high-level commercial instruments provide containers for four eluents (E1–E4), which may be pure solvents such as acetonitrile or mixtures such as an aqueous buffer. The eluents can be mixed in any volumetric ratio in an electromagnetic mixing valve (MV) and that ratio can be varied over time. Elution is called isocratic when the eluent composition does not change over time. Alternatively gradients may be generated representing increased elution power as a function of time. The solvents used for HPLC separations must be of the highest quality grades (a particular HPLC grade is usually commercially available for the most popular solvents). Typical flow rates are between 0.1 and 1.5 ml per minute and depend on column dimensions and type of detector. Flow rates are normally not changed within a running chromatographic process. The composed eluent is pumped (P) to the column inlet, while providing the addition (injection) of the sample in that eluent line. Injection of sample (I) may be done manually or automatically. Typical injection volumes for analytical separations are between 5 and 20 µl, but may be of the order of 1000 µl for preparative purposes. The total volume of eluent sitting between the inlets of the electromagnetic mixing valve and the pump outlet is called the internal volume of the instrument. This volume should be as low as possible, especially for the generation of reproducible smooth gradients with high resolving power.

All elements from eluent containers to the pump inlet belong to the low pressure part of the instrument; all elements from the pump outlet to the column outlet are in the high pressure zone, although a fall of pressure will occur through the column. The highest pressures generated in an average HPLC system are of the order of magnitude of 200 times the atmospheric pressure (or 200 000 hPa or 2800 psi). All solvents must be thoroughly degassed before pumping because residual gas, solubilized at high pressure might generate gas bubbles when pressure drops; such bubbles would influence the quality of separation and detection.

The column (C) represents the real heart of the HPLC separation system. It normally consists of a heavy-walled stainless steel cylinder with internal diameters ranging from 1.0 to 4.6 mm and lengths between 20 and 250 mm. The cylinder is filled with the stationary phase, the chemical nature (functional groups) of which can be chosen as a function of the separation envisaged. The quality of separation may be influenced further by the column filling procedure quality (ascertained by a test chromatogram that should come with every column), particle size and size distribution, flow rate, temperature, secondary chromatographic effects and column ageing. By far the most

(*cont.*)

Box 3.o (*Continued*)

popular stationary phase is the so-called reversed-phase, consisting of a hydrophobic tail (mostly a linear C-18 moiety) grafted on a silica core material. Such a stationary phase will retard hydrophobic substances most and those will normally be eluted with a gradient of increasing hydrophobic strength (from 90/10 to 10/90 water/methanol (v/v) for instance). The expected lifetime of an average reversed-phase column is a couple of hundred injections but of course it also depends on the elution conditions, taking into account the alkali-lability of the silica and the acid-lability of the silica-C-18 bond. Hundreds of prepacked columns and column cartridges are commercially available. It is important to realise that stationary phases may display differences in chromatographic behaviour even when their naming would suggest a similarity. Such differences may then be due to particle parameters such as form, size and size distribution, porosity, carbon load and end-capping of residual silanol groups.

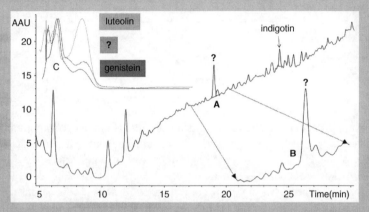

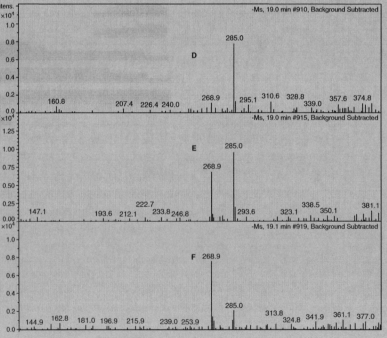

Fig. 3.o.2. HPLC chromatogram, registered at 255 nm, arbitrary absorption units (AAU) against time, full profile (A) and expansion (B); C: UV–Vis spectra of unknown peak (marked with ? in the picture), luteolin and genistein; mass spectra at the head (D), apex (E) and tail (F) of the unknown. (Courtesy of C. M. Grzywacz, GCI; © J. Paul Getty Trust)

A major aspect of versatility of HPLC chromatographic systems is the possible coupling to the separation system of any detector (D1, D2) that can handle fluids. Most popular are photodiode array detectors (PDA), fluorimeters and mass spectrometers (MS), but also conductometers (high-performance ion chromatography), evaporative light scatterers (ELSD) and even nuclear magnetic resonance (NMR) and infrared (IR) detection systems have been reported. In some cases the parallel or serial coupling of detectors to the separation system may be envisaged. The most popular of such is serial PDA-MS detection.

From the nature of the detectors it may already be concluded that HPLC will be applied mostly for the separation and identification of organic materials in objects of art, such as dyes, amino acids, proteins, carbohydrates, oils–fats–waxes, tannins, acids. When considering applications of HPLC for analysing art, it must be realised that the technique is destructive in all of its aspects; it requires the removal of a sample from the object and the sample is destroyed in the analytical process, but separated components may be recovered with a fraction collector (FC) for further study. Alternatively, analytical parameters such as sensitivity, reproducibility, resolution, quantification and identification potential are of very high quality and make HPLC a technique with a high ratio of detail and quality of information to the degree of destructiveness.

A typical example of synergism between liquid chromatographic separation, PDA and MS detection is given in Fig. 3.o.2. The chromatogram displays the analysis of a pinkish silk yarn sample, taken from a 17th century upholstery, recorded at 255 nm. The only components contributing to the colour were found to be indigotin and an unknown yellow. At the given retention time, the unknown yellow should correspond to luteolin, but the UV–Vis spectrum did not match. Scanning through the unknown symmetrical peak for M^- masses, revealed the presence of luteolin ($M^- = 285$) in the head of the peak, genistein ($M^- = 269$) in the tail, and both at the apex. This result pointed to the presence of the yellow dye of *Genista tinctoria* (dyer's broom), together with an indigoid dye.

For in-depth information

Lough W.J. and Wainer I.W. (eds) (1995) High performance liquid chromatography: fundamental principles and practice. Blackie Academic and Professionals, London. p. 276.

Meyer V. (2004) Practical high-performance liquid chromatography. 4th Edition. John Wiley and Sons, New York. p. 357.

Niessen W.M.A. (2006) Liquid chromatography—mass spectrometry. 3rd Edition. CRC Press, Boca Raton. p. 608.

Box 3.p GC/MS Principles of gas chromatography/mass spectrometry [M. Schilling]

Gas chromatography/mass spectrometry (GC/MS) is a hyphenated analytical technique that combines the resolving power of GC to separate complex mixtures of organic compounds (Box 3.o), with the capability of MS to identify unknown organic components and measure their concentration (Box 2.u).

In all chromatographic techniques, organic compounds *partition* themselves between a stationary phase and a mobile phase that flows through the chromatographic instrument. Compounds with higher affinity for the mobile phase will emerge (or *elute*) from the chromatograph before compounds with higher affinity for the stationary phase. The time at which each compound elutes, known as the *retention time*, is sensed by a detector. By this process, a chromatographic system is capable of physically separating organic mixtures into individual components.

In the case of gas chromatography (GC), *carrier gas* (the mobile phase, often helium or hydrogen) flows through a long capillary tube called a *GC column*. The inner wall of the capillary is coated with a thin film of a highly viscous organic liquid (the stationary phase). Samples are introduced into the GC column through a

(cont.)

Box 3.p (*Continued*)

sample inlet; typically, inlets are heated in order to vaporize the sample, although others permit injection of the sample directly into the GC column. Carrier gas transports vaporized sample components through the column, where partitioning and separation of the compounds occurs. The column is housed inside an oven that controls the column temperature, because elevating the column temperature greatly reduces the retention times of most compounds (Fig. 3.p.1).

The outlet end of the GC column is coupled directly to a mass selective detector (MSD), which is a high-vacuum device with three main components. An *ion source* uses a heated electrical filament to produce ionized molecules and fragment ions from gaseous sample molecules. Next, a *mass analyser* uses oscillating electrical fields to separate the sample ions and measure their mass-to-charge ratio (*m/z*); most MSDs reply on either quadrupole or ion trap technologies. A *detector* measures the abundance of each separated ion by recording either the charge or current when an ion passes or strikes the detector surface. A GC/MSD chromatogram is a display of the total ion current versus time, and the individual organic compounds after separation by the GC appear as a series of peaks. Finally, the intensity versus *m/z* data for each GC peak is stored in the form of a *mass spectrum*. Mass spectral databases, such as the NIST library, can be consulted to identify unknown compounds from their mass spectra (McLafferty 1980).

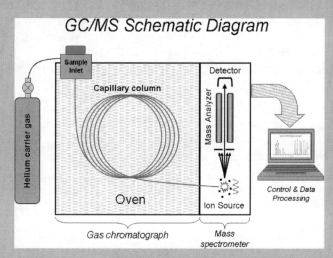

Fig. 3.p.1. Schematic diagram of a GC/MS instrument configuration.

Chromatographic results for a sample typically consist of a list of peaks characterized by retention time and peak area. By comparing the list of retention times with the results for a set of reference standards, one can determine if a sample contains a particular component; this is *qualitative analysis*. *Quantitative analysis* is required when identification is either based on the amount of a certain component, or the ratio of the amounts present in the sample. Quantitative analysis requires a set of calibration standards to calculate the detector response factors for each component in the sample mixture.

Sample preparation is perhaps the most important part of any GC procedure because samples must be in the volatile state in order to be analysable. Thus, only a limited number of organic compounds are amenable to GC analysis (roughly 10% of all known organic compounds). Methods of sample preparation depend on the nature of the organic components of interest. Gases and liquids often require no pre-treatment, whereas solids are first dissolved in volatile solvents. Large, non-volatile molecules may be broken into smaller fragments using chemical treatment, as in the depolymerization of proteins and plant gums by acid hydrolysis. Another option for high molecular weight materials (such as polymers) or intractable substances (such as aged tree resins) is to use a pyrolyser to degrade the material thermally prior to GC/MS analysis (as in Py-GC/MS).

Ideally, if sample molecules exit the GC inlet in a narrow band and do not chemically react with the inlet or column during the analysis, the resulting peaks in the GC chromatogram will be narrow and symmetrical. However, GC results for certain classes of compounds may be less than ideal, giving rise to broad, severely tailing peaks. Problematic materials include compounds that are especially prone to hydrogen bonding, such as alcohols, fatty acids, and amines. To improve the chromatographic results, these compounds may be converted to lower-boiling, thermally-stable species in a process known as *derivatization*. The type of derivative formed depends on the chemical structure of the components present in the sample, and is controlled by specific chemical reactions. For example, carboxylic acids are commonly converted to esters for GC analysis, whereas alcohols and amines are often acylated. Derivatization with silylating reagents to produce silyl ester and silyl ether derivatives is also very popular.

Knowledge of the advantages and disadvantages of GC/MSD is beneficial, especially when developing a new analytical method. For instance, a major advantage of capillary GC/MSD is the capability of resolving extremely complex mixtures rapidly due to the high efficiency of the columns. Nonetheless, when selecting the GC column, it is helpful to know beforehand what classes of chemical compounds the sample might contain so that the optimum resolution can be achieved and the type of derivative can be chosen.

Next, depending on how many ions are analysed at any given time, an MSD can function as a *universal* or a *selective* detector, which gives it certain flexibility. For example, an MSD set up to *scan* a mass range from 10 to 600 amu will be capable of detecting nearly all organic compounds; in the process, it will produce a mass spectrum for each GC peak that can be searched using a computerized spectral library. Therefore, SCAN mode is obviously beneficial to identify unknown compounds in a sample. However, to analyse trace levels of known compounds, it is best to set up the MSD to perform *selected ion monitoring* (SIM), in which only a few ions that are specific to the analyte(s) of interest are measured during the GC analysis. By doing this, detection limits in the picogramme range are achievable because the detector does not spend any time looking for ions that are not present in the analyte(s). It should be noted that, because of the extreme sensitivity of GC/MSD in SIM mode, it is imperative to minimize sources of contamination during sample preparation and testing.

For in-depth information

Barry E.F. and Grob R.L. (2004) Modern practice of gas chromatography. Wiley-Interscience, New York.

Rood D. (1991) A practical guide to the care, maintenance, and troubleshooting of capillary gas chromatographic systems. In: Bertsch W. (ed.) Chromatographic Methods series. Hüthig Buch Verlag, Heidelberg.

Box 3.q The use of antibodies for molecular recognition [J. Mazurek]

The precise identification of protein and polysaccharide containing binding media and adhesives is a frequently encountered challenge in art conservation science laboratories. The use of immunological techniques based on antibodies produced by mammals in response to invading organisms may be a valid answer for the identification of binding media in extremely small samples. Antibodies, or immunoglobulins, are proteins and are present in the serum and tissue fluids of all mammals (Crowther 2001). All antibodies are produced by animal immune systems in response to the presence of a foreign body (an antigen): usually virus or bacteria, but also foreign proteins, and organic molecules. The antibodies produced in response to a foreign protein can be artificially induced, rendering

(cont.)

Box 3.q (*Continued*)

them commercially available. Antibodies are used in the enzyme-linked immunosorbent assay (ELISA), and it is a commonly used medical diagnostic test. For example, it is used to detect the AIDS virus as well as a hormone in home pregnancy tests. ELISA, in particular, is an advantageous technique since it does not require expensive instrumentation, can distinguish between source or species of protein, and it is reliable and extremely sensitive with samples containing complex protein mixtures as well as with aged sample material.

There are thousands of different "primary antibodies", and countless can be utilized in ELISA for the analysis of proteins in works of art. Primary antibodies are proteins that bind to a specific epitope, or region, of another protein and tag it for recognition. In most ELISA procedures "secondary antibodies" are needed to identify and bind to the primary antibody. Five classes of secondary antibodies exist: IgG, IgA, IgM, IgD, and IgE. When selecting a secondary antibody, it must be from the same animal that produced the primary antibody. Secondary antibodies are modified by conjugation with "tags" that aid in detection, i.e. enzymes, fluorophores, gold nanoparticles.

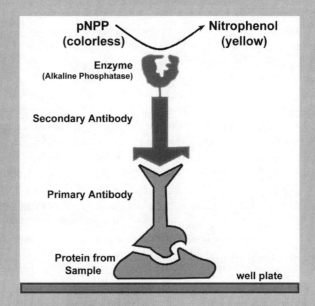

Fig. 3.q.1. The protein binds to the well plate and the primary antibody binds to the protein. The secondary antibody binds to the primary antibody and pNPP turns yellow. (Courtesy of A. Heginbotham) Original colour picture reproduced as plate 59.

Conservation scientists have long seen the value in using antibodies for the specific and simultaneous detection of four most common protein or polysaccharide types: collagen from glue, casein from milk, ovalbumin from egg white, and arabinogalactan from plant gums. Immunochemical methods of analysis offer several advantages over traditional techniques, though such investigations of artists' materials have been somewhat limited to date. Immunofluorescent techniques were first proposed to detect proteins in works of art several decades ago (Jones 1962), and the application of fluorescent antibodies were investigated (Johnson and Packard 1971). Antibodies have been used to successfully detect egg, animal glue, blood, plant gum and casein (Ramirez Barat and De la Vioa 2001, Hodkins and Hedges 1999, Cattaneo *et al.* 1995, Scott *et al.* 1996, Yates *et al.* 1996). Recently, ovalbumin was detected in the polychromy of a 17th century French cabinet (Heginbotham *et al.* 2006) and a detailed description of an ELISA technique was presented.

Optimal ELISA conditions vary depending on the specific antibodies employed and the nature of the sample material to be analysed; a procedure was developed to provide good results with four commercially available antibodies used simultaneously under the same conditions (Mazurek *et al.* 2008). For samples that contain unknown binding media, a test for four classes of proteins constitutes a good first approach. These are mammal collagen (present in glues and distemper paints), ovalbumin (present in egg white), arabinogalactan protein (AGP, detectable in many plant gum media) and casein (present in milk, and commonly found in paints and plasters). All four proteins can be tested on one sample. Extraction of the sample proteins into solution is carried out in a urea solution over a period of two to three days. The ELISA test plate is a standard 96 well microplate that has twelve columns (1–12) and eight rows (A–H) (Maxisorp Nunc-Immuno™ 96 MicroWell™). Normally, each sample is assigned to one column and is tested twice for each antibody (to minimize the false positives due to contamination or procedural error). The sample is run twice for each antibody. With this arrangement it is possible to analyse up to ten samples in the same plate. If there is sufficient sample material, the sample can be analysed in a series of dilutions, repeating each antibody test in several wells. In all cases, a positive result occurs when a well turns yellow at the end of the test. Blank wells (with no sample added) and positive controls (with appropriate antigen added) are always run alongside the samples in order to ensure that the procedure functioned properly. Once the assay is complete, each well is tested for positive reaction by measuring the optical density (OD i.e., absorbance) of the final solution with a spectrophotometer. The measured OD is compared with an experimentally derived baseline value to determine if a positive result has been obtained (procedure described below).

An ELISA procedure for detecting binding media in paint: Place 10 to 500 μg of each sample to be analysed into 2-ml micro-centrifuge tubes. Add 100 μl of elution buffer to each sample. Elution Buffer: 5 ml of 1M tris (hydroxymethyl)aminomethane hydrochloride (tris-HCl), 1 ml of 0.5M ethylenediaminetetraacetic acid (EDTA), 180 g urea, 25 ml of 20% sodium dodecyl sulphate (10 g in 50 ml deionized water) and deionized H_2O final volume 500 ml. EB pH adjusted to 7.4 using NaOH. The solution may be stored at room temperature. Prepare positive controls: Add 10 to 500 μg each of egg white, animal glue (typically bovine), cow's milk (casein) and gum Arabic into separate micro-centrifuge tubes. Add 20 μl elution buffer to each positive control and also to a sterile "blank" tube, containing no antigen. Allow samples, standards and blank to elute by standing for 2–3 days at room temperature. Next, add 160 μl of 100 mM sodium bicarbonate to each tube. 100 mM sodium bicarbonate solution: 0.42 g of $NaHCO_3$ brought to a total of 50 ml with deionized H_2O; the solution may be stored at room temperature. Mix well and let sit for 10 minutes; this allows any particulate matter to settle to the bottom of the tube. Prepare the ELISA plate for the samples by adding 60 μl of bicarbonate buffer to each of the wells in a 96-well polystyrene ELISA plates. ELISA plates are typically labelled with eight rows (A–H) and 12 columns (1–12). The plate should be labelled as follows: rows A and B – glue; rows C and D – egg; rows E and F – caesin; rows G and H – plant gum. Dilutions are recommended to verify the results of the assay. Add 20 μl of eluent from the first sample to each of the wells in column 1. Repeat this procedure for each subsequent sample by adding eluent to the successive columns of the plates, reserving columns 11 and 12 for blanks and positive controls. Next, add 20 μl of blank eluent solution to all the wells in column 11. After the blanks, add the positive controls as follows: add 20 μl eluent from the animal glue to wells 12A and 12B, 20 μl egg eluent to wells 12C and 12D, 20 μl casein eluent to wells 12E and 12F, and 20 μl gum arabic eluent to wells 12G and 12H. Cover the plates with Parafilm and incubate at 4 °C for 24 hours. After incubation, allow the plates to come to room temperature and empty the contents of each well with a multi-channel pipette. Be careful at this step, as it is possible to cross-contaminate nearby wells. If necessary, dry the pipette tips on clean absorbent paper between rows. Rinse the wells by adding 300 μl of phosphate buffered saline (PBS) to each well and allow to stand for 2 minutes. All PBS solutions used in this procedure were diluted 1:10 with deionized water from standard commercial 10× concentrate. PBS solutions may be stored at room temperature. Empty the wells into a waste receptacle by inverting the plate with a brisk shake. Repeat this rinsing operation twice. It is important to rinse thoroughly, so that all unbound sample material, including particulate residue, is removed. Tap the plate on a paper towel between rinsing to remove all of the PBS from of the wells. Add 300 μl Sea Block Buffer (Pierce #37527) diluted 1:10 v/v in PBS to all the wells. Allow

(cont.)

Box 3.q (*Continued*)

the trays to stand for 60 minutes at room temperature. Empty the wells into an appropriate receptacle with a flick of the wrist and pat the inverted plate dry on a paper towel. Next, add 80 μl of the appropriate diluted primary antibody to each of the wells in the rows described below. Dilutions of all the antibodies are prepared using 1:10 Sea Block Buffer solution mentioned above. In plate 1, add 1:400 dilution of #AB19811 (collagen) to rows A and B; add 1:800 #AB1225 (ovalbumin) to rows C and D; add 1:800 # RCAS-10A (casein) to rows E and F; and add 1:50 # JIM13 (general plant gum) to rows G and H. Allow the antibodies to bind for 2 hours at room temperature. Empty the plates with a flick of the wrist and rinse all wells three times with 300 μl of PBS as above. Next, add 80 μl of diluted secondary antibody to each row of wells as follows. In plate 1, add 1:400 #AB6742 (goat IgG) to rows A and B; add 1:500 #AP132A (rabbit IgG) to rows C through F; add antibody 1:100 KPL#05-16-03 (rat IgM) to rows G and H. Allow the antibodies to bind for 2 hours at room temperature. Empty the plates and rinse all wells three times as before, then add 80 μl of p-nitrophenyl phosphate pNPP solution. Wait until the controls are fully developed while making sure that the blanks remain clear (up to one hour). If desired, quench the reaction by adding 80 μl 0.75M NaOH to each of the wells. Measure the absorbance at 405 nm of the solution in each well using a spectrophotometer or automated plate reader such as Finstruments model #341 96-well microplate spectrophotometer (MTX Lab Systems, Inc., Virginia USA). A minimum OD_{405} value of 0.3 for both wells was routinely used as the threshold for all ELISA positive tests. In the case of strong responses, the results can be read qualitatively by eye.

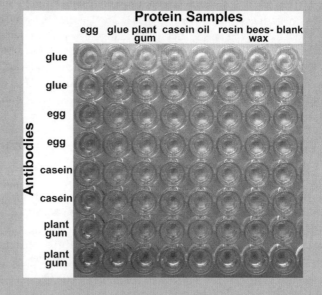

Fig. 3.q.2. Photograph of ELISA well plate results from the cross-reaction experiment. In column 1 (labelled egg) a solution containing a whole egg in elution buffer was added to all of the wells in the column. In column 2 (labelled glue) a solution of animal glue was added to all of the wells, and this was repeated for the rest of the columns. All of the wells in rows A–H were treated with one of the four primary antibodies for animal collagen: egg, casein or plant gum, listed in the far left column. Each primary antibody was run twice (two rows), and a yellow colour is a positive result.
Original colour picture reproduced as plate 60.

For in-depth information

Crowther J.R. (2001) The ELISA guidebook. Humana Press, New Jersey.

Mazurek J., Heginbotham A., Schilling M. and Chiari G. (2008) Antibody assay to characterize binding media in paint. ICOM Committee for Conservation. Vol 2. pp. 678–685.

3.8 An example of complex composite materials and processes: Photography [D. Stulik and A. Kaplan]

The so-called "classical" or "chemical" photography, as we once knew it, has been almost completely replaced by digital photography. This transition has a profound effect not only on the photographic industry and profession but also on our everyday life and on the way we create, manipulate, share, and store photographic images. These profound changes of technology will also have a major effect on future developments in photographic research and photographic conservation. Photo conservators in the future will have to deal not only with all the photographic artefacts and documents created during the era of chemical photography stored in museum collections and archives but they will also need to deal with acquisitions and preservation of digital based photographic images in the form of digital prints and photographic images accompanied by digital files stored on various types of magnetic, optical or other storage media (Rothenberg 1999).

The era of classical or chemical photography started in the beginning of the 19th century with key camera based experiments by Thomas Wedgwood (Davy 1802) and Joseph Nicephore Niepce (Bonnet and Marignier 2003), and with the introduction of the first "practical" photographic process by Daguerre in 1839. Since then more then 150 different photographic processes have been invented, tested and introduced into photographic practice (Wall 1897, Nadeau 1994, Cartier-Bresson 2008). Some of these photographic processes such as Heliography, Chrysotype, Eburneum or 3M Electrocolor processes had only limited practical applications and only rare examples of these processes can be found in museum collections, archives of inventors or deposits of patent submissions.

Some other photographic processes such as Albumen, Tintype, Silver gelatin and Chromogenic C-print processes were used for many decades and millions of photographic images that have been created using these processes can be found in a variety of public and private photographic collections (Naef 1995, Johnson *et al*. 2000, Roberts 2000). A whole range of other photographic processes such as Matte Collodion, Platinotype, and Autochrome were introduced and used for a limited period of time or for a specialized application and later superseded by other more advanced or more convenient photographic processes. A number of old photographic processes were re-introduced after 1960 by photographers interested in alternative photographic processes (James 2002). Figure 3.95 depicts both a date of introduction of the most common photographic processes as well as an approximate time period when the processes were actively used.

What makes the research and care for photographic collections very challenging is the fact that many different photographic images that can be found in collections of 19th and 20th century photography were created using often very diverse chemical, photochemical and physical procedures. This fact is

Fig. 3.95. Timeline of major photographic processes.

responsible for the rather complicated chemical and physical internal structure of photographs and photographic materials of the chemical photography era.

Photographs created using different photographic processes also have very different properties of image stability, different requirements for optimal storage conditions, permissible light exposure during an exhibition, and might require special treatment procedures when being conserved or restored. A correct identification of the photographic process used to create a photograph should also be a mandatory prerequisite before any conservation treatment of any photograph. Only a very limited number of existing photographic collections today can provide detailed registration and conservation records that would include a precise identification of the photographic process used for each photograph in the collection. Many existing registration records of museum collections, historical archives or even auction houses have been created with a rather limited knowledge of photographic processes and process variants and they can often contain a number of identification errors that sometimes made it into exhibition and auction catalogues. Problems of misrepresentation and authentication of photographs will be of growing concern to curators and art dealers due to the constantly increasing monetary value of some vintage photographs. A number of different working methods have been developed and introduced in the past to aid in the identification of photographs (Reilly 1986, Potts and West 2008, Stulik 2008). The majority of these methods are based on the visual examination of photographs or a combination of visual examination and a detailed inspection under a loupe, low power microscope or a stereomicroscope.

3.8.1 Visual and microscopic methods of identification of photographs

Photographic prints can be composed of either simple or very complicated structures made of various inorganic and organic components. Each component of a photographic print has its role in either the formation of a photographic image, as a support of an image layer or as a modifier of the visual, chemical or mechanical properties of a photograph. The complexity of photographic images ranges greatly from the simplest case of two layer photographic structures composed of an imaging metal deposited on a paper support such as salt prints and platinotypes to more complex multiple layer structures of fibre-based baryta coated B&W photographs, modern chromogenic photographs and Polaroid photographs (Fig. 3.96).

The most important part of any photographic print is the image layer that in its lateral distribution of imaging material represents a material record of photographed objects or scenery. The structure of the image layer itself can be very simple or rather complex. A visual examination of photographs followed by a microscopic examination under a microscope or stereomicroscope is usually the first step in the identification of photographs and photographic processes. The visual identification relies on a set of simple visual clues and clue combinations that, when properly interpreted, allow for the reliable identification of a large number of common photographic processes. Both primary and secondary visual clues can be used for the identification of photographs.

The most obvious primary visual clues found on some photographs are the name of a photographer or a printer (signature/stamp), the date of a photograph (negative/positive/both), processing and post processing information, exhibition or photographic competition labels, photographer's or jury notes, photographic material manufacturer's logos or product type information (Agfa, Velox, safety film etc.), tax and postage stamps (Civil war tax stamps 1864–66), etc. The secondary visual clues are more material oriented and can include the size, surface topography (glossy, matt, etc.), tonality and colour of

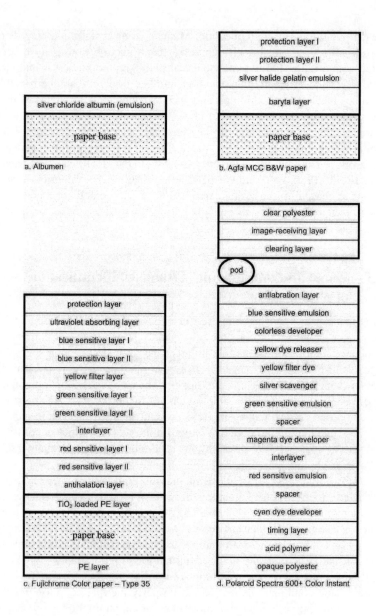

Fig. 3.96. Internal structure of simple and complex photographic images (not to scale).

the image layer, thickness or stiffness of the photograph when handled, etc. The presence or absence of fading of the photographic image as well as the character and extent of deterioration of the photograph can also provide some important clues for identification of a photograph.

The microscopic examination of photographs provides information on the microscopic structure of the image layer and may provide information on the physical structure of a photograph. Paper fibres of photographic paper under the albumen layer are clearly visible when examining albumen photographs and the examination of the edge of silver gelatin photographs may show the presence of a baryta layer between the paper substrate and gelatin emulsion layers (Stulik 2008).

If the visual and microscopic examination of a photograph does not provide enough information to identify a photograph or photographic process, advanced non-destructive instrumental and analytical methods might provide the data needed to finish the identification process.

3.8.2 Identification of photographs using non-contact and non-destructive analytical methods and procedures

Both photographs and photographic negatives are composed of various forms of imaging material in combination with different organic binders on a more or less rigid support that was tailored to provide the required mechanical, chemical and optical properties. The combination of image forming material, organic binders, support material, surface coatings and sometimes the presence of some chemical components left after chemical processing is very often unique for a given photographic process. Knowing the "chemical signature" of various photographic processes and having objective analytical data from the analysis of unknown photographs often allows identification of a photographic process well beyond what can be achieved using visual and/or microscopic observation alone.

Non-invasive (non-destructive and non-contact) analytical and instrumental techniques are always preferred by collection curators and museum directors and these analytical techniques should always serve as the starting point for any analytical investigation of photographs or of any other art object. The most important analytical techniques that can be used for the analytical identification of photographic materials are X-ray Fluorescence Spectrometry (XRF), which is capable of providing information about the presence of inorganic elements and compounds in the analysed photograph and Fourier Transform Infrared Spectrometry (FTIR), which is capable of identifying the most important organic and polymer components of photographic materials.

3.8.3 X-ray Fluorescence Spectroscopy

XRF spectroscopy (Box 2.b) is one of the most widely used and versatile analytical technique for the identification of inorganic elements in art objects (Potts and West 2008). XRF spectroscopy is very sensitive and its major advantage is that the analysis can be conducted without removing a sample (non-destructive analysis) from the analysed photograph and often without any physical contact between the instrument and the object (non-contact analysis). An XRF spectrometer uses primary radiation from an X-ray tube to excite secondary (fluorescence) X-ray emission from an analysed photograph. The detection and energy analysis of this secondary or fluorescent X-ray radiation provides the X-ray spectrum that allows for both qualitative and quantitative analysis of chemical elements present in the analysed photographs. Free standing and portable XRF instruments used widely in art conservation research can detect almost all chemical elements of the Periodic Table of chemical elements with an atomic number greater than sodium (Na). It is very important, when interpreting XRF spectra of photographs, to be aware that even when not present in the recorded XRF spectrum, light elements such as hydrogen, carbon and nitrogen are almost always present in paper-based photographs or plastic substrates of photographic negatives and in complex dyes present in colour photographs.

Most analysed photographs are very thin and are paper or organic polymer based. The primary X-ray beam of the XRF spectrometer penetrates through the entire analysed photograph and in many cases through or deep into the mount of the photograph. The recorded XRF spectrum of a mounted photograph then represents a superposition of elemental compositions of an image layer, baryta layer, paper substrate and of the mounting board or mount. A multiple point analysis of mounted photograph (D-max, D-min, mount and of the back of the photograph and its mount) often provides enough information to assign detected elements to different components and layers of mounted photographs (Fig. 3.97). Table 3.27 shows the inorganic elements found in some of the most important photographic processes of the chemical photography era that can be detected using XRF.

3.8.4 Quantitative XRF analysis of photographs

Silver gelatin black and white (B&W) photographic paper was by far the most commonly used photographic printing material during the 20th century. As a very complex material object, a silver gelatin fibre based baryta coated photograph might harbor some important material clues to support its provenancing and authentication. Scientific investigations conducted at the Getty Conservation Institute have identified a number of chemical and physical

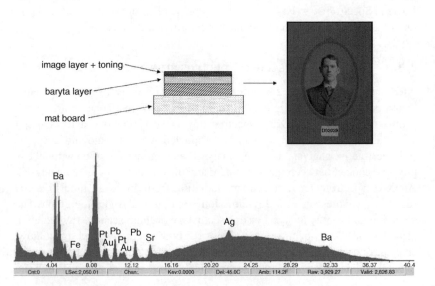

Fig. 3.97. Qualitative analysis of a mounted matt collodion photograph (matt board—Fe, Pb; baryta layer—Ba, Sr; image layer—Ag; toning—Pt, Au).

Table 3.27. Elemental signatures of selected photographic processes

Photographic process	Typical elements	Notes
Daguerreotype	Cu, Ag, Au, (Hg)	low concentration of Hg
Ambrotype	Ag, Ca, K, (Hg)	Ca, K from the glass substrate, Hg from image reduction
Tintype	Fe, Ag	
Salt print	Ag, (Co)	Co from smalt in some paper
Albumen	Ag, (Au)	Au, toning; several elements from the substrate of card mounted images
Collodion	Ag, (Au, Pt), Ba, Sr	Au, Pt single or in combination from toning; Ba, Sr baryta layer; several elements from the substrate of mounted images
Silver gelatin	Ag, Se, (Au, Cu, Fe), Ba, Sr, (Ti)	Se or other metals from toning; Ba, Sr baryta layer; Ti, RC paper
Cyanotypes	Fe	
Platinotype	Pt, Fe, (Pd), Hg	Fe from unwashed developer; Pt + Pd combination print; Pd palladiotype; Pt (Hg) toning
Carbon	No Ag, Cr, possible other elements from pigments	Cr from unwashed dichromate
Pigment	No Ag, Cr, possible other elements from pigments	Cr from unwashed dichromate

markers, or signatures, of baryta-coated B&W photographic papers that could be used in provenancing, authenticating, and in some cases even the dating of photographic materials and photographs.

The XRF analysis of a large number of B&W photographs and photographic papers has shown that 20th century silver gelatin baryta coated photographs contain—in addition to silver—several other chemical elements such as barium, strontium, calcium, and very often chromium. The concentration of these elements in photographs is a function of the photographic paper production technology and composition of raw materials used. The concentration of barium and strontium in the baryta layer of photographic papers is not altered by photographic processing or post treatment of photographs and it varies enough between different manufacturers and production dates to allow for the development of an objective, scientifically based methodology for the provenancing and authentication of photographs. Figure 3.98 shows how the concentration of barium and strontium in a photograph can be used to identify the photographic paper used to print a photograph even when an "unknown" print was selenium toned to mask any visual clues that might help to match the print.

A successful application of the quantitative XRF methodology for provenancing and authentication of photographs depends on the development of the standard operating procedure (SOP) for such an analysis, the availability of high quality quantitative XRF standards for the baryta layer analysis, and the

426 *Materials and case studies*

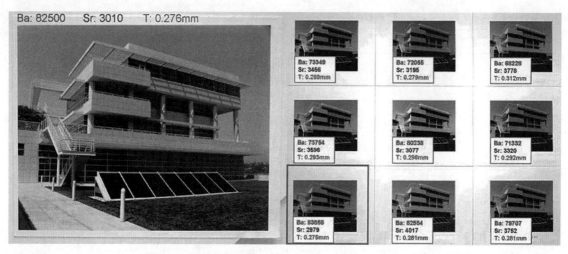

Fig. 3.98. Provenancing and authentication of photographs using the quantitative analysis of the baryta layer of 20th century B&W photographs. The XRF analysis determined that the "unknown" photograph was printed on Forte Elegance, shown in red.

existence of a very comprehensive database of the composition of baryta coated photographic papers. Some more targeted provenancing and authentication studies can also be performed using an individual database of photographic material used by the photographer in question. More details related to the quantitative XRF analysis of photographs and to the scientific methodology of provenancing and authentication of photographs can be found in several published articles (Khanjian and Stulik 2003, Kaplan and Stulik 2008, Stulik and Kaplan 2008).

3.8.5 Fourier Transform Infrared Spectrometry

The most important infrared spectral region (Box 2.e, Table 2.1) for identification of organic material that is typically found in many photographs such as the binder of the image layer, image substrate (paper/polymer) or various coatings and varnishes applied to protect or modify the appearance of photographic images is the mid-IR region of the IR spectrum (Derrick *et al.* 1999).

There are a number of FTIR instruments available on the market that can be used for the identification of various organic materials but most typical FTIR spectrometers work in transmission mode and require the physical removal and special preparation of the sample for FTIR analysis. The preference for non-destructive analysis when analysing photographs utilizes either non-contact reflection analysis or so-called ATR (Attenuated Total Reflection) FTIR analysis. The results of FTIR reflection analysis are more difficult to interpret and some mathematical treatment of the obtained spectra is usually needed in order to obtain FTIR spectrum for interpretation. There are severe limitations when using FTIR reflection spectrometry when searching both commercial and custom made libraries of FTIR spectra. The ATR-FTIR usually represents a better and more practical choice of FTIR analysis when applied to the investigation of photographs. Table 3.28

Table 3.28. Organic components of selected photographic processes

Photographic process	Typical organic components	Notes
Ambrotype	collodion (varnish – shellac)	Might be with or without varnish
Tintype	collodion (varnish)	Might be varnished
Salt print	cellulose (wax)	Wax as a varnish
Albumen	albumin (shellac, collodion, varnish)	Old photographs usually have a thin albumin layer
Collodion	collodion (starch)	Glossy—collodion; matte—collodion and starch
Silver gelatin	gelatin (starch)	Glossy—gelatin; matte—gelatin and starch
Platinotype, Palladiotype, Cyanotype	cellulose	Se or other metals from toning; Ba, Sr baryta layer; Ti, RC paper
Carbon	gelatin	
Pigment	gum arabic	Low concentration

shows the organic components of some of the most common photographic processes.

When recording ATR-FTIR spectra of photographs the side of the photograph to be analysed has to be put into good contact with the ATR crystal of the ATR-FTIR spectrometer. The beam of infrared radiation enters the ATR crystal and most of the infrared beam is reflected from the face of the ATR crystal that is covered by the analysed sample. The infrared beam is not perfectly confined to the inside of the ATR crystal and a small portion of the infrared beam penetrates into the sample surface in contact with the ATR crystal to a depth of the order of the wavelength of the infrared radiation. In the case of photographs, the final ATR-FTIR spectrum might represent a superposition of several FTIR spectra of thin organic layers close to the surface of the analysed photograph. A very illustrative example of this situation is the ATR-FTIR analysis of an albumen photograph coated with a thin layer of beeswax in Fig. 3.99.

The ATR-FTIR spectrum shown in Fig. 3.99 is a superposition of a beeswax coating with a characteristic peak at $1736\,cm^{-1}$ representing the ester bond between long chain aliphatic alcohols and long chain aliphatic acids present in beeswax; a thin layer of albumin has characteristic peaks of Amide I at $1638\,cm^{-1}$ and Amide II at $1533\,cm^{-1}$ peaks of egg albumin and a broad spectral envelope of the cellulosic paper substrate is located around $1030\,cm^{-1}$. The image particles in the photograph are made of elemental silver and gold (toner) that, due to the absence of molecular interactions, cannot be detected by FTIR spectrometry.

The interpretation of the FTIR spectra of photographs and photographic material can be much more complicated and challenging when dealing with FTIR spectra of low intensity, spectra of materials that do not exhibit some

428 *Materials and case studies*

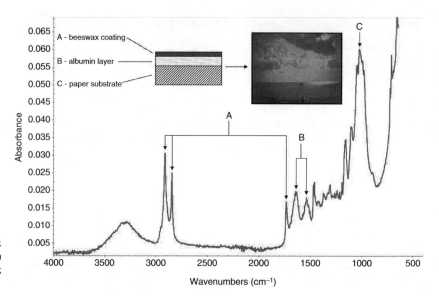

Fig. 3.99. ATR-FTIR analysis of beeswax coated albumen photograph: A—absorption peaks of beeswax; B—albumen peaks; C—spectral envelope of the paper substrate.

unique spectral features or when analysing multiple component samples. FTIR spectrometry is, in comparison with XRF analysis, much less sensitive and this prevents its use in the analysis of minor and trace organic components of photographic images such as sensitizing dyes, dyes of chromogenic colour photographs or plasticizers of negative plastic substrates. More sensitive analytical methods that might require physical sampling may be needed when faced with the analysis of minor organic components in photographs.

3.8.6 Portable laboratory for research of photographs

The Getty Conservation Institute's research project in research and conservation of photographs requires easy access to a variety of photographs and photographic processes that is well beyond what can be found in any single photographic collection or museum. To facilitate access to a variety of photographs and photographic processes and at the same time to comply with current art conservation practice, which calls for restrictions on movement and unnecessary transport of art material, a portable analytical laboratory for research in photographs was assembled (Stulik 2005). Building the portable laboratory was made much easier by recent trends in modern analytical instrumentation. Miniaturization and computerization has made many instruments much smaller and often portable. Using the portable laboratory allows modification of the standard practice of art conservation research whereby the art was brought for analysis to the scientific or conservation laboratory: instead the laboratory goes to the art (Box 2.s). The portable laboratory can travel worldwide, not only for the benefit of the research project but also for the benefit of many collaborating museums, collections and institutions that have important questions about their photographic collections but that lack access to a conservation science laboratory or have limited financial resources to obtain analytical answers using commercial services (Stulik and Wright 2007). Table 3.29 and Fig. 3.100 show primary and secondary analytical and scientific instruments and tools included in the Getty Conservation Institute (GCI) Portable Laboratory.

The portable laboratory needs two operators to work efficiently and can be transported in two standard Pelican cases. It can be checked-in when traveling by air or sent ahead by overnight delivery services. The lab requires only a standard source of electricity (110/220 V) and a desk area of approximately 3×6 feet^2. Experience confirms that it takes about 40 minutes to proceed from locked transportation cases to fully functioning laboratory. Continuous research is actively looking at and testing newly developed and commercially introduced portable instruments that might expand the working capabilities of the Portable Laboratory when analysing and studying photographs and photographic material.

Table 3.29. Components of the GCI portable laboratory

Instrument	Manufacturer	Accessories	Task
S6D stereomicroscope	Leica	Fibre optic light, Pax-it image software, 2× close-up lens, Dell Latitude D620 computer	Microscopic examination and photomicrography
DFC480 camera	Leica		Image capture
DG-3 digital microscope	Scalar		Microscopic examination and photomicrography
Tracer III	Keymaster	Compaq nc6120 computer	Elemental analysis
Travel*IR* ATR-FTIR	Smiths Detection	Omnic FT-IR software, GCI Library of photographic processes, Dell Latitude D610 computer	Organic analysis
Digimatic Caliper	Mitutoyo		Physical measurements
Series 293 micrometer	Mitutoyo		Thickness measurements
Versalume UV lamp	Raytech		Detection of optical brighteners
Spectrocam	Avantes		Densitometry and colorimetry measurements

Fig. 3.100. GCI Portable Laboratory for research in photographs at the Harry Ransom Center at the University of Texas, at Austin, Fall 2003.

3.8.7 Analysis of photographs using invasive methods of analysis

Non-invasive methods of chemical analysis are able to provide the majority of information needed to identify photographs and photographic processes (Stulik and Wright 2007). To answer some other analytical question related to minor and trace elements and components of photographic images or analytical questions related to the detail internal and chemical structure of photographs requires the removal of a physical sample from a photograph and the application of invasive analytical techniques. Similar to the analysis of other types of art objects (paintings, illuminated manuscripts, drawings, etc.), physical sampling is a very ethically sensitive procedure that has to be performed with a great deal of caution and always with the permission of a collection curator or of the photograph's owner. In the majority of cases the sampling of photographs should be avoided. When there is a very compelling need to remove a sample from a photograph the lateral homogeneity of photographs provides a benefit for sampling. The benefit when sampling a photograph is that there is a very high probability that the outside edge of a photographic image might have the same chemical and physical structure as the central part of the photographic image.

Recent developments in micro-sampling techniques provide for the removal of a physical sample for chemical analysis with minimal "damage" to art objects. The "damage" from micro-sampling can be so miniscule that it can be comparable to damage caused to an object by everyday handling. This potentially provides some new opportunities to answer some difficult analytical questions that are beyond the capabilities of strictly non-invasive analytical methods.

Inductively Coupled Plasma Mass Spectrometry (ICP-MS) or its laser ablation modification LA-ICP-MS (Box 2.u) provides enough sensitivity to detect

Fig. 3.101. Cross-section of a Fibre Based Baryta Coated Photographic paper using an ESEM Microscope with an EDX Analyser.

trace concentrations of more than 70 different chemical elements in microscopic samples removed from photographs. The ICP-MS analysis is often used to resolve some interpretation problems of XRF analysis when dealing with spectral interferences (lead–arsenic, etc.), analysis of some very light elements (lithium) or when explaining the presence of some unexpected peaks in the XRF spectra of photographs (rubidium) (Kaplan 2006). Neutron Activation Analysis (NAA, Box 3.a) has been used in the preparation of barium and strontium standards for the XRF quantitative analysis of photographs (Miller 2006). Environmental Scanning Electron Microscopy (ESEM) in combination with Energy Dispersive X-ray analysis allows a detailed investigation of the multilayered internal structure of photographs (Fig. 3.101; Duverne 2006).

Raman spectroscopy is a complementary method of organic analysis to FTIR (Box 2.e). Raman spectroscopy has been used for the identification of inorganic pigments and organic dyes used for the tinting and colouring of early B&W photographs and for dye analysis of various colour photographs (Smith and Clark 2004). The high sensitivity of Gas-Chromatography Mass Spectrometry (GC-MS, Box 3.p) and of Liquid Chromatography Mass Spectrometry (LC-MS, Box 3.o) provides a means for the analysis of minor organic components in photographs and its high selectivity makes it possible to decipher the detailed composition of photographic coatings and varnishes (Schilling, private communication). A monoclonal antibody-based sandwich Enzyme-Linked Immunosorbent Assay (ELISA) and Immunofluorescence Microscopy (IFM) can be used for the detection and differentiation of different protein based components of photographic materials (Box 3.q). These simple techniques are highly sensitive (detection limits below one nanogram) and they can be used to differentiate between different protein types (i.e. gelatin versus. albumin) (Heginbotham *et al*. 2006).

3.8.8 The future of scientific and analytical techniques in research of photographs

There will always be some rare instances when the stated analytical question can not be answered even when using the most advanced analytical instrumentation available. In these instances it is highly advisable to record all of the results of the analytical investigation including unanswered analytical questions and attach it to the conservation record of the object. There is a very good chance that the analytical question that cannot be answered using the analytical methodology and tools today will be able to be answered later using emerging analytical techniques or so far unknown or improved analytical techniques that will be available in the near or more distant future.

Further reading

General books with emphasis on materials

Bowman S.G.E. (ed.) (1991) Science and the past. British Museum Publications, London.

Brothwell D.R., Pollard A.M. (eds) (2001) Handbook of archaeological sciences. John Wiley & Sons, Chichester.

Goffer Z. (2007) Archaeological chemistry. 2nd Edition. Wiley-Interscience, New York.

Henderson J (2000) The science and archaeology of materials. Routledge, London.

Lambert J.B. (1997) Traces of the past. Addison-Wesley, Reading.

Shortland A.J., Freestone I.C. and Rehren T. (eds) (2009) From mine to microscope. Advances in the study of ancient technology. Oxbow Books, Oxford.

Books on specific classes of materials

Lithics, Stones	Kardulias P.N. and Yerkes R.W. (eds) (2003) Written in stone: the multiple dimensions of lithic analysis. Lexington Books, Lexington, MA.
	Rapp G.R. (2009) Archaeomineralogy. 2nd Edition. Springer-Verlag, Berlin.
	Winkler E.M. (1997) Stone in Architecture: Properties, Durability. 3rd revised edition. Springer-Verlag, Berlin.
Ceramics	Barclay K. (2001) Scientific analysis of archaeological ceramics. A handbook of resources. Oxbow Books, Oxford.
	Cuomo di Caprio N. (2007) Ceramica in Archeologia 2. L'Erma di Bretschneider, Roma.
	Rice P.M. (1987) Pottery analysis: A sourcebook. University of Chicago Press, Chicago.
	Velde B. and Druc I.C. (1999) Archaeological ceramic materials: Origin and utilization. Springer-Verlag, Berlin.
Mortars, Cements	Bensted J. and Barnes P. (eds.) (2002) Structure and performance of cements. 2nd Edition. Spon Press, London – New York.
	Hewlett P.C. (ed.) (1998) Lea's Chemistry of cement and concrete. 4th Edition. Butterworth-Heimann, Oxford.

Pigments	Gettens R.J. and Stout G.L. (1966) Painting materials: a short encyclopaedia. Courier Dover Publications.
Zollinger H. (2003) Color chemistry: syntheses, properties, and applications of organic dyes and pigments. 3rd Edition. Verlag Helvetica Chimica Acta, Zürich.
Glass, Faience	Doremus R.H. (1994) Glass science. John Wiley & Sons, New York.
Kingery W.D. and Mccray P. (eds) (1998) The prehistory and history of glassmaking technology. Ceramics and civilization, Vol. VIII. American Ceramic Society, Columbus, OH.
Shelby J.E. (2005) Introduction to glass science and technology. 2nd Edition. The Royal Society of Chemistry, Cambridge.
Tite M. S. and Shortland A. J. (2008) Production technology of faience and related early vitreous materials. Oxford University School of Archaeology, Oxford.
Metals	Porter D.A. and Easterling K.E. (2009) Phase transformations in metals and alloys. 3rd Edition. CRC Press, Boca Raton, FL. p. 520.
Scott D.A. (1991) Metallography and microstructure of ancient and historic metals. The Getty Conservation Institute, Los Angeles.
Tylecote R.F. (1976) A history of metallurgy. The Metals Society, London. Reprinted 1992 The Institute of Materials, London.
Gems	O'Donoghue M. (2006) Gems: their sources, descriptions and identification. 6th Edition. Butterworth-Heinemann, London.
Read P.G. (2005) Gemmology. 3rd Edition. Butterworth-Heinemann, London.
Organics	Grimaldi D.A. (1996) Amber: window to the past. Harry N. Abrams Inc., New York.
Mills J.S., White R. (1999) The organic chemistry of museum objects. 2nd Edition. Butterworth-Heinemann, Oxford.
O'Connor S.A., Brooks M.M., Payton R. and Todd V. (eds) (1990) Archaeological textiles. Proc. Conference on textiles for archaeological conservators, UKIC Archaeological Section, York, April 1988. The United Kingdom Institute for Conservation, London. p. 62.
Rice P.C. (2006) Amber the golden gem of the ages. 4th Edition. AuthorHouse, Bloomington, IN.
Wild J.P. (2003) Textiles in archaeology. 2nd Edition. Shire publications, Aylesbury, UK.
Photography	Lavédrine B. (2003) A guide to the preventive conservation of photograph collections. Getty Conservation Institute, Los Angeles.
Lavédrine B. (2009) Photographs of the past. Process and preservation. Getty Conservation Institute, Los Angeles.

4 Present and future trends: analytical strategies and problems

The application of materials science and scientific methods to the cultural heritage have matured considerably since the starting of archaeometry as a separate field of inquiry, briefly discussed in the introductory section.

Technically, there has been a huge evolution concerning instrumentation, radiation sources, experimental protocols and methods of application of scientific analyses. Some of the recent trends are quite clear and may be tentatively outlined:

1) There is a general tendency towards the use of **microbeams**. Synchrotron sources have the lead, and they will continue to do so because of their intrinsic characteristics. However, there is a growing appreciation of the use of small probes also in the laboratory (micro-FTIR, micro-PIXE, micro-XRF, micro XRD and so on; Janssens and Van Grieken 2004, Creagh 2007), especially concerning the 2D mapping combined with analytical tools (i.e. the merging of microimaging with analytical techniques).

2) As far as large or thick objects are concerned, modern **3D imaging** has added another dimension by the widespread use of tomography and laser surface mapping. Neutrons are fundamentals in this area because of their large and penetrating beams. The present trend is to use neutrons more efficiently, both from spallation sources and reactors, and to expand the capabilities of present experiments by adding the simultaneous detection of several signals from the sample. In the future we ought to see 3D tomographic images simultaneously resolved in density contrast, elemental chemistry and crystallography. Synchrotron sources should do the same at nanometric resolution on small sub-micrometre samples.

3) Another general trend is to make use of **portable instrumentation**. I call it the "labs go to the museum" attitude. At present I am not at all personally satisfied with the way this is being implemented. Many see this development as a wonderful possibility of increasing the number

of analyses in the most diverse situations: archaeological excavations, field surveys, wide screening of museum objects and the like. This is fine, as long as the approximate quality of most data obtained by non-optimized geometries is recognized and used accordingly (Box 2.r). The danger is the proliferation of low-quality or even misleading data that will be hard to fight should they enter the literature.

4) With the widespread use of analytical tools, a desirable trend would be the **standardization of results** and the build up of consistent reference **databases**. There is very little going on in this respect. Each laboratory has its own standards for measurements and local databases of completed analyses. Very limited know-how exchange is being performed, and sometimes the actual measured data are not even published. The only areas where numbers are actually being regularly published are in lead isotope data and Raman spectral identification of pigments. The comparison of data produced by different techniques on the same materials is a problem that is surfacing every once in a while, and it has been sometimes discussed in the literature, especially for metals. It is a persistent problem and one that has to be faced seriously as more labs are coming into the business of cultural heritage analysis.

The latter considerations should be expanded a little.

It is worth remarking that the multi-scale scientific imaging and characterization of complex objects requires innovative ways of storing and retrieve information. Wouldn't it be wonderful to be able to open a database containing, say, a list of archaeological objects, the art works in a collection, or simply a series of objects used in a specific time/place, and by clicking on each object being able to retrieve not only the information on its present location and known history, but all the details concerning the analyses performed on it, and all the past conservation and treatment steps. We may then keep on clicking and get chemical analyses, diffraction spectra, electron micro-images and so on. Then we may retrieve the results or a carefully selected part of the information and compare the data with other objects from the whole database. Should we call it multidimensional analysis?

Of course at present this is science fiction, because so very few analyses are available for the objects of our cultural heritage, and even when data are available they are hardly accessible to anyone, and almost never are they are standardized in such a way that they can be readily compared with freshly measured data. Life for scientific analysis of non-standard materials is stimulating, but very hard...

Virtual reality must come to help scientific analysis. I know it is nice and creative to play with virtual reconstructions, shapes and colours. We are all aware of the quality of modern special movie effects. In extreme conditions one may even tend to be convinced that it is better to see an uncrowded site (archaeological excavation, museum,...) and perfectly lit objects in a virtual form than to see the real thing, as you may enjoy the sight without experiencing the additional trouble of long lines, crowded art galleries, heat-loaded and air-deprived underground burials, or poorly visible distant objects in the dark. Of course I am pushing it a little far: the real thing is the real thing! No one can quietly dismiss the true Indiana Jones experience of re-discovering archaeological sites excavated a hundred years ago and now barely visible under layers of garbage, or the painstaking search among layers of permission documents to analyse published objects long vanished in museum storage basements. Maybe I am slightly biased by a few unhappy personal experiences, but I seriously think that carefully planned systematic digital storage of the information concerning the cultural heritage is a crucial step that must be undertaken to allow proper access to future generations. Serious plans are underway in these directions. The real and virtual reconstruction of the Lascaux II caves and the available virtual tours of many museums are leading us this way.

Talking about trends, there is a rapidly growing tendency, especially in developing countries but not only there, to use the cultural heritage as an asset to be exploited for economic growth and job creation. It is very important to remember that cultural heritage not only helps tourism. Please read Greffe (2004) to rethink the consequences of cultural tourism on economics and on the safeguarding of the cultural heritage itself... However, on the topic of cultural tourism and education, well thought out displays of the current knowledge and enhanced information made possible through virtual reality will certainly be a winning strategy.

I am also thinking about the extreme difficulty in checking databases, cross-checking scientific information, and producing integrated results. The scarcity of funding, but even worse the chronic and general mismanagement of resources (financial and human) and ideas is making it hard to create truly interdisciplinary networks and analytical collaborations. On the one hand, I believe that the scarce available resources should be concentrated in a few high-standard laboratories, which could guarantee proper and well-run instrumentation, quality of results, and continuity in know-how. Too often have I seen doctorate students reaching high scientific levels, and then the whole research field in their research lab collapsing because no jobs are available and there is simply nobody to manage the instruments and the research, or there are mandatory short-term shifts in scientific interest and directions. I do not think cultural heritage should be run on a short-term basis, like the 2–3 years time span of scientific projects. On the other hands scientific laboratories working on cultural heritage must be open to new ideas, external projects and collaborations, in order to be constantly at the forefront of research.

The characterization and conservation of cultural heritage certainly deserves the best technical expertise available. I do not think we are doing our best. Current relationships between archaeology, archaeometry, conservation and management of cultural heritage have a number of shortcomings, some of them related to the presently available curricula at higher education level (Maggetti 2006, Artioli 2009). By outlining the situation and some of the existing problems, I hope that this volume may contribute a little to stimulate positive solutions.

Further reading

Agnew N. and Bridgland J. (2006) Of the past, for the future: Integrating archaeology and conservation. The Getty Conservation Institute, Los Angeles.

Brothwell D.R. and Pollard A.M. (2001) Handbook of archaeological sciences. John Wiley & Sons, Chichester.

Price N.S., Kirby Talley Jr M. and Melucco Vaccaro A. (eds) (1996) Historical and philosophical issues in the conservation of cultural heritage. The Getty Conservation Institute, Los Angeles.

Renfrew C. and Bahn P. (2008) Archaeology: theories, methods and practice. Thames & Hudson, London.

Reference list

Ábelová M. (2008) Schreger pattern analysis of Mammuthus primigenius tusk: analytical approach and utility. Bull. Geosci. 83, 225–232.
Abragram A. (1961) The principles of nuclear magnetism. Clarendon, Oxford.
Abrams E.M. and Freter A. (1996) A Late Classic lime-plaster kiln from the Maya centre of Copan, Honduras. Antiquity 70, 422–428.
Adams P.B. (1984) Glass corrosion A record of the past? A predictor of the future? J. Non-Crystall. Sol. 67, 193–205.
Adams J., Kneller W. and Dollimore D. (1992) Thermal analysis of lime- and gypsum-based medieval mortars. Thermochim. Acta 211, 93–106.
Adams J.L. (2002) Ground stone analysis: A technological approach. The University of Utah Press, Salt Lake City, UT.
Adams F., Janssens K. and Snigirev A. (1998) Microscopic X-ray fluorescence analysis and related methods with laboratory and synchrotron sources. J. Anal. Atomic Spect. 13, 319–331.
Adkins L. and Adkins R.A. (1985) Neolithic axes from Roman sites in Britain. Oxford J. Arch. 4, 69–76.
Adovasio J.M., Soffer O. and Klima B. (1996) Upper Palaeolithic fibre technology: interlaced woven finds from Pavlov I, Czech Republic, c. 26,000 years ago. Antiquity 70, 526–534.
Agnew N. (1997) Conservation of archaeological sites – A holistic philosophy. In: Geiger A. and Eggebrecht A. (eds) World cultural heritage, a global challenge. Documentation on the International Symposium in Hildesheim, Germany, 23 Feb–1 March. pp. 149–155.
Agnew N. (ed.) (1997) Conservation of ancient sites on the Silk Road: Proceedings of an International Conference on the Conservation of Grotto Sites. The Getty Conservation Institute, Los Angeles.
Agnew N. (2003) Sins of omission: Diagnosis, risk assessment and decision. Lessons from three sites. In: Gowing R. and Heritage A. (eds) Conserving the painted past: Developing approaches to wall painting conservation. James & James Science Publishers, London. pp. 75–84.
Agnew N., Bridgland J. (2006) Of the past, for the future: Integrating archaeology and conservation. The Getty Conservation Institute, Los Angeles.
Agostiniani L. (1995) Sui numerali etruschi e la loro rappresentazione grafica. In Silvestri D. (ed.), Atti del Convegno su "Numeri e istanze di numerazione tra preistoria e protostoria linguistica del mondo antico". Napoli, 1–2 December 1995, "AION" 17. pp. 26–30.
Agricola G. (1556) De re metallica. Engl. Transl.: Herbert Clark Hoover and Lou Henry Hoover (1950) Dover Publications Inc., New York.
Ainsworth M.W. (2005) From connoisseurship to technical history: The evolution of the interdisciplinary study of art. The Getty Conservation Institute Newsletter 20, 4–10.
Aitken M.J. (1985) Thermoluminescence dating. Academic Press, London.

Aitken M.J. (1989) Luminescence dating: a guide for non-specialists. Archaeometry 31, 147–159.

Aitken M.J. (1990) Science-based dating in archaeology. Longman, London.

Aitken, M.J. (1998) Introduction to optical dating. Oxford University Press, Oxford.

Aitken M.J., Stringer C.B., Mellars P.A. (eds) (1993) The origin of modern humans and the impact of chronometric dating. Princeton University Press, Princeton.

Alcock J.P. (2006) Food in the ancient world. Greenwood Press, Santa Barbara, CA. pp. 312.

Alessandrini G. (ed.) (1989) Proceedings of Symposium on the Oxalate Films: Origin and Significance in the Conservation of Works of Art. Centro C.N.R. 'Gino Bozza' and Politecnico di Milano, Milano.

Alexander W.R. and Smellie J.A.T. (1998) Maqarin natural analogue project: synthesis report on Phases I, II and III. Nagra Unpublished Project Report, Nagra, Wettingen, Switzerland.

Alexander W.R. and Blaser P.C. (eds) (2002) The use of technical natural analoguers in radioactive waste disposal. Nagra Unpublished Project Report, Nagra, Wettingen, Switzerland.

Al-Naddaf M. (2009) The effect of salts on thermal and hydric dilatation of porous building stone. Archaeometry 51, 495–505.

Alvarez de Buergo M. and Fort Gonzalez R. (2003) Protective patinas applied on stony facades of historical buildings in the past. Constr. Building Materials 17, 83–89.

Ambers J. (1987) Stable carbon isotope ratios and their relevance to the determination of accurate radiocarbon dates for lime mortars. J. Archaeol. Sci. 14, 569–576.

Ambers J. and Freestone I.C. (2005) Introduction. In: Edwards H.G.M. and Chalmers J.M. (eds) Raman spectroscopy in archaeology and art history. The Royal Society of Chemistry, Cambridge.

Ambert P. and Vacquer J. (eds) (2005) La première métallurgie en France et dans les pays limitrophes. Société Préhistorique Française, Carcassonne.

Ambrose W. and Allen C., O'Connor S., Spriggs M., Vasco Oliveira N. and Reepmeyer C. (2009) Possible obsidian sources for artifacts from Timor: narrowing the options using chemical data. J. Archaeol. Sci. 36, 607–615.

Anati E. (ed.) (2000) 40.000 anni di arte contemporanea. L' preistorica d'Europa. Edizioni del Centro Camuno di Studi Preistorici, Capo di Ponte, Brescia.

Anderson D.L. (1989) Theory of the Earth. Blackwell Scientific Publications, Oxford.

Anderson K.B. (1994) The nature and fate of natural resins in the geosphere – IV. Middle and Upper Cretaceous amber from the Taimyr Peninsula, Siberia. Organic Geochem. 21, 209–212.

Anderson K.B. (1996) The nature and fate of natural resins in the geosphers–VII. A Radiocarbon (^{14}C) age scale for description of immature natural resins: an invicton to scientific debate, Organic Geochem. 25, 251–253.

Anderson K.B., Winans R.E. and Botto R.E. (1992) The nature and fate of natural resins in the geospheres – II. Identification, classification and nomenclature of resinites. Organic Geochem. 18, 829–841.

Anderson K.B. and Botto R.E. (1993) The nature and fate of natural resins in the geosphere – III. Re-evaluation of the structure and composition of highgate copalite and glessite. Organic Geochem. 20, 1027–1038.

Anderson K.B. and Crelling J.C. (1995) Amber, resinite, and fossil resins. American Chemical Society, Washington, DC.

Anderson K.B. and La Page B.A. (1995) Analysis of fossil resins from Axel Heiberg Island, Canadian Artic. In: Anderson K.B. and Crelling J.C. (1995) Amber, resinite, and fossil resins. American Chemical Society, Washington, DC. pp. 1–31.

Anderson S.E., Levoy M. (2002) Unwrapping and visualizing cuneiform tablets. IEEE Computer Graphics and Applications. November/December 2002. pp. 82–88.

Andrefsky W. (2005) Lithics: macroscopic approaches to analysis. 2nd revised edition. Cambridge University Press, Cambridge. pp. 301.

Angelini I., Artioli G., Bellintani P., Diella V., Gemmi M., Polla A. and Rossi A. (2004) Chemical analyses of Bronze Age glasses from Frattesina di Rovigo, Northern Italy. J. Arch. Sc. 31, 1175–1184.

Angelini I., Artioli G., Polla A. and De Marinis R.C. (2006) Early Bronze Age faience from North Italy and Slovakia: A comparative archaeometric study. In: Proceedings of the 34th Int. Symposium on Archaeometry, 3–7 May 2004, Zaragoza, Spain. pp. 371–378.

Angelini I. and Bellintani P. (2005) Archaeological ambers from northern Italy: an FTIR–DRIFT study of provenance by comparison with the geological amber database. Archaeometry 47, 441–454.

Angelini I. and Bellintani P. (2006) L'archeometria delle ambre protostoriche: dati acquisiti e problemi aperti. In: "Materie prime e scambi nella protostoria Italiana", Atti della XXXIX Riunione Scientifica dell'Istituto Italiano di Preistoria e Protostoria, Firenze, Italy, 25–27 November 2004. III, 1477–1494.

Anguilano L., Angelini I., Artioli G., Moroni M., Baumgarten B. and Oberrauch H. (2002) Smelting slags from Copper and Bronze Age archaeological sites in Trentino and Alto Adige. In: D'Amico C. (ed.) Atti II Congresso Nazionale di Archeometria. Bologna, 29 January-1 February 2002. Pàtron Editore, Bologna. pp. 627–638.

Aoba T. (2004) Solubility properties of human tooth mineral and pathogenesis of dental caries. Oral Diseases 10, 249–257.

Appolonia L., Giamello M. and Sabatini G. (1996) Caratterizzazione stratigrafica delle pellicole ad ossalati mediante osservazioni in sezione ultrasottile e microdiffrattometria. In: Realini M. and Toniolo L. (eds) Proceedings of the II International Symposium on The oxalate films in the conservation of works of art. Milan, 25–27 March 1996. EDITEAM, Bologna. pp. 360–376.

Arandigoyen M., Bicer-Simsir B., Alvarez J.I. and Lange D.A. (2006) Variation of microstructure with carbonation in lime and blended pastes. Appl. Surf. Sci. 252, 7562–7571.

Arcolao C. and Dal Bò A. (2001) L' delle sostanze proteiche naturali su alcune proprietà degli stucchi. Proc. Convegno di Studi Scienza e Beni Culturali on Lo stucco: Cultura, tecnologia, conoscenza. Bressanone 2001. Edizioni Libreria Progetto, Padova. pp. 527–538.

Arnold J.R. and Libby W.F. (1949) Age determination by radiocarbon content: checks with samples of known age. Science 110, 678–680.

Artioli G. (2004) Caratterizzazione non distruttiva di manufatti archeologici. In: Siano S. (ed.) Proc. Intern Workshop on "Tecnologie e metodologie innovative per lo studio e il restauro di manufatti archeologici", Castiglioncello, 31 Maggio-5 Giugno 2004. Collana "Manufatti Archeologici. Studio e Conservazione". Nardini Editore, Firenze. pp. 478–492.

Artioli G. (2007) Crystallographic texture analysis of archaeological metals: interpretation of manufacturing techniques. Appl. Physics A 89, 899–908.

Artioli G. (2009) Archeometria ed archeologia: il fascino di un amore difficile. Riv. Archaeol., in press.

Artioli G., Dugnani M., Hansen T., Lutterotti L., Pedrotti A. and Sperl G. (2003) Crystallographic texture analysis of the Iceman and coeval copper axes by non-invasive neutron powder diffraction. In: Fleckinger A. (ed.) "La mummia dell'età del rame. 2. Nuove ricerche sull'uomo venuto dal ghiaccio". Collana del Museo Archeologico dell'Alto Adige, Vol. 3, Folio Verlag, Bolzano. pp. 9–22.

Artioli G., Angelini I., Burger E. and Bourgarit D. (2007) Petrographic and chemical investigation of the earliest copper smelting slags in Italy: towards a reconstruction of the beginning of copper metallurgy. Proc. 2nd Intern. Conference "Archaeometallurgy in Europe 2007", Aquileia, 17–21 June 2007. Proceedings on CD.

Artioli G., Angelini I. and Polla A. (2008) Crystals and phase transitions in protohistoric glass materials. Phase Transitions 81, 233–252.

Artioli G., Baumgarten B., Marelli M., Giussani B., Recchia S., Nimis P., Giunti I., Angelini I. and Omenetto P. (2008) Chemical and isotopic tracers in Alpine copper deposits: geochemical links between mines and metal. Geo.Alp 5, 139–148.

Artioli G., Nimis P., Gruppo ARCA, Recchia S., Marelli M. and Giussani B. (2008) Gechemical links between copper mines and ancient metallurgy: the Agordo case study. Rend. Online Soc. Geol. It. 4, 15–18.

Artioli G., Nicola C., Montana G., Angelini I., Nodari L. and Russo U. (2009) The blue enamels in the Baroque decorations of the churches of Palermo, Sicily: Fe^{2+}-coloured glasses from lime kilns. Archaeometry 51, 197–213.

Artioli G. and Angelini I. (2009) Evolution of vitreous materials in Bronze Age Italy. In: Janssens K. (ed.) Modern methods for analysing archaeological and historical glass. John Wiley & Sons, New York. In press.

Asaro F. and Adan-Bayewitz D. (2007) The history of the Lawrence Berkeley National Laboratory instrumental neutron activation analysis programme for archaeological and geological materials. Archaeometry 49, 201–214.

Ashby M.F. (1987) Technology of the 1990s: advanced materials and predictive design. Phil. Trans. Royal Soc. London. Series A, Mathematical and Physical Sciences 322, 393–407.

Ashby M.F. (2001) Drivers for material development in the 21st century. Prog. Mater. Sci. 46, 191–199.

Ashby M.F. (2005) Materials selection in mechanical design. 3rd Edition. Butterworth-Heinemann, Oxford. pp. 603.

Ashby M.F., Gibson L.J., Wegst U. and Olive R. (1995) The mechanical properties of natural materials. I. Material Property Charts. Proc. R. Soc. Lond. A 450, 123–140.

Askeland D.R. (1988) The science and engineering of materials. 2nd Edition. Chapman & Hall, London. pp. 880.

ASTM C-1196 (1991) Standard test method for in-situ compressive stress within solid unit masonry estimated using the flat-jack method. ASTM, Philadelphia.

ASTM C-1197 (1991) Standard test method for in-situ measurement of masonry deformability properties using the flat jack method. ASTM, Philadelphia.

Attanasio D. (2003) Ancient white marbles – Analysis and identification by paramagnetic resonance spectroscopy. L'Erma di Bretschneider, Roma.

Austin J.J., Ross A.J., Smith A.B., Fortey R.A. and Thomas R.H. (2007) problems of reproducibility – does geologically ancient DNA survive in amber-preserved insects? Proc. Roy. Soc. London B 264, 467–474.

Avdelidis N.P., Moropoulou A. (2004) Applications of infrared thermography for the investigation of historic structures. J. Cult. Heritage 5, 119–127.

Avrami E., Mason R. and De la Torre M. (eds) (2000) Values and heritage conservation. Research Report. The Getty Conservation Insitute, Los Angeles.

Avrami E., Guillaud H. and Hardy M. (eds) (2008) Terra literature review. An overview of research in earthen architecture conservation. The Getty Conservation Insitute, Los Angeles.

Bacci M. (2004) Optical spectroscopy and colorimetry. In: Martini M., Milazzo M. and Piacentini M. (eds) (2004) Physics methods in archaeometry. SIF, Bologna – IOS Press, Amsterdam. pp. 1–16.

Bachmann H.G. (1980) Early copper smelting techniques in Sinai and Negev as deduced from slag investigation. In: Craddock P.T. (ed.) Scientific studies of early mining and extractive metallurgy. British Museum Occasional Papers No. 20. British Museum Press, London. pp. 103–134.

Bachmann H.G. (1982) The identification of slags from archaeological sites. Occasional Publication No. 6. The Institute of Archaeology, London.

Baer N.S. and Snethlage R. (eds) (1996) Saving our architectural heritage: The conservation of historic stone structures. John Wiley & Sons, Chichester. pp. 452.

Bailey S.W. (ed.) (1988) Hydrous phyllosilicates. Reviews in Mineralogy, Vol. 19. The Mineralogical Society of America, Washington.

Bailiff I.K. (1997) Retrospective dosimetry with ceramics. Rad. Measurements 27, 923–941.

Baillie M.G. (1995) A slice through time. Dendrochronology and precision dating. B.T. Batsford Ltd, London. pp. 176.

Baillie M.G. and Munro M. (1988) Irish tree rings, Santorini and volcanic dust veils. Nature 332, 344–346.

Baker P. E. (1993) Archaeological stone of Easter Island. Geoarchaeology 8, 127–139.

Bakolas A., Biscontin G., Moropoulou A. and Zendri E. (1998) Characterization of structural Byzantine mortars by thermogravimetric analysis. Thermochimica acta 321, 151–160.

Baksi A.K., Hsu V., McWilliams M.O. and Farrar E. (1992) 40Ar/39Ar dating of the Bruhnes-Matuyama geomagnetic field reversal. Science 256, 356–357.

Balla M. and Gunneweg J. (2007) Archaeological research at the Institute of Nuclear Techniques, Budapest University of Technology and Economics: scholarly achievements of a prosperous long-term collaboration. Archaeometry 49, 373–381.

Bamford C.R. (1977) Colour generation and control in glass. Glass Science and Technology, Vol. 2. Elsevier Scientific Publishing Co., Amsterdam–New York. pp. 225.

Banner J.L. (2004) Radiogenic isotopes: systematics and application to earth surface processes and chemical stratigraphy. Earth Sci. Rev. 65, 141–194.

Barba L., Blancas J., Manzanilla L. R., Ortiz A., Barca D., Crisci G. M., Miriello D. and Pecci A. (2009) Provenance of the limestone used in Teotihuacan (Mexico): A methodological approach. Archaeometry DOI: 10.1111/j .1475–4754.2008.00430.x

Barber D.J. and Freestone I.C. (1990) An investigation of the origin of the colour of the lycurgus cup by analytical transmission electron microscopy. Archaeometry 32, 33–45.

Barber D.J., Freestone I.C. and Moulding K.M. (2009) Ancient copper red glasses: investigation and analysis by microbeam techniques. In: Shortland A.J. and Freestone I.C. and Rehren T. (eds) From mine to microscope. Advances in the study of ancient technology. Oxbow Books, Oxford. pp. 115–127.

Barceló J.A., Forte M. and Sanders D.H. (2000) Virtual Reality in Archaeology. British Archaeological Reports, International Series no. 843. ArcheoPress, Oxford.

Barceló J.A. (2001) Virtual reality and scientific visualization. Working with models and hypothesis. Intern. J. Modern Phys. C, 12, 569–580.

Barceló J.A. (2002) Virtual archaeology and artificial intelligence. In: Virtual archaeology: proceedings of the VAST Euroconference, Arezzo 24–25 November 2000. Archaeopress, pp. 21–28.

Barkoudah Y. and Henderson J. (2006) Plant ashes from Syria and the manufacture of ancient glass: Ethnographic and scientific aspects. J. Glass Studies 48, 297–321.

Barnard H. and Eerkens J.W. (eds) (2007) Theory and practice of archaeological residue analysis. BAR Intern. Series Vol. 1650. pp. 265.

Barthélemy de Saizieu B. and Bouquillon A. (2001) Émergence et evolution des matériaux vitrifiés dans la région de l' du 5e au 3e millénaire (Mehrgarh-Nausharo). Paléorient 26, 93–111.

Bar-Yosef O. (1986) The walls of Jericho: An alternative interpretation. Current Anthropology 27, 157–162.

Barni M., Beraldin J.A., Lahanier C. and Piva A. (2008) Signal Processing in Visual Cultural Heritage. Signal Proc. Mag. 25, 10–12.

Barrett E.P., Joyner L.G. and Halenda P.P. (1951) The determination of pore volume and area distribution in porous substances: I. Computation from nitrogen isotherms. J. Amer. Chem. Soc. 73, 373–380.

Barrett H.H. and Myers K.J. (2004) Foundations of Image Science. John Wiley & Sons.

Barry R.S., Rice S. and Ross J. (2000) Physical chemistry. Oxford University Press, New York.

Barry R.S., Rice S. and Ross J. (2001) Structure of matter: An introduction to quantum mechanics. Oxford University Press, New York.

Baruchel J., Hodeau J.L., Lehmann M.S., Regnard J.R. and Schlenker C. (1993) Neutron and synchrotron radiation for condensed matter studies. HERCULES, Editions de Physique – Springer Verlag.

Baruchel J., Buffière J.Y., Maire E., Merle P. and Peix G. (eds) (2000) X-ray tomography in materials science. Hermes Science Publications, Paris.

Bátora J. (1995) Fayence und Bernstein im nördlichen Karpatenraum während der Frühbronzezeit. In: Hänsel B. (ed.) Handel, Tausch und Verkehr im Bronze- und Früheisenzeitlichen Südosteuropa. PAS 11, München-Berlin.

Bauer M. (1968) Precious stones. Vol. 1 and 2. Dover Publications, New York.

Bayley J. and Rehren T. (2007) Towards a functional and typological reconstruction of crucibles. In: La Niece S., Hook D., Craddock P. (eds) metals and mines. Studies in archaeometallurgy. The British Museum – Archetype Publications, London. pp. 46–55.

Baxter M.J. (1994) Exploratory multivariate analysis in archaeology. Edinburgh University Press, Edinburgh.

Baxter M.J. and Buck C.E. (2000) Data handling and statistical analysis. In: Ciliberto E. and Spoto G. (eds) Modern analytical methods in art and archaeometry. Ch. 20. John Wiley & Sons, New York. pp. 681–746.

Baxter M.J. (2003) Statistics in archaeology. Edward Arnold, London.

Baxter M.J. (2004) Multivariate analysis of archaeometric data. In: Martini M., Milazzo M. and Piacentini M. (eds) Physics methods in archaeometry. SIF, Bologna – IOS Press, Amsterdam. pp. 17–36.

Baxter M.J. (2008) Mathematics, statistics and archaeometry: the past 50 years or so. Archaeometry, 50, 968–982.

Beck C.W. (1971) Amber from Eneolithic Necropolis of Laterza. Origini 5, 301–305.

Beck C.W. (1986) Spectroscopic investigations of amber. Appl. Spectr. Rev., 22, 57–200.

Beck C.W. (1996) Comments on a supposed Clovis 'Mastic'. J. Archaeol. Sci. 23, 459–460.

Beck C.W., Wilbur E. and Meret S. (1964) Infrared spectra and the origin of amber. Nature 201, 256–257.

Beck C.W., Wilbur E., Meret S., Kossove D. and Kermani K. (1965) The infrared spectra of amber and the identification of Baltic amber. Archaeometry 8, 96–109.

Beck C.W., Gerving M. and Wilbur E. (1966) The provenience of archaeological amber. Artifacts. An Annotated Bibliography. Part I, 8th century BC to 1899. Art and archaeology technical abstracts 6, 215–302.

Beck C.W., Gerving M. and Wilbur E. (1967) The provenience of archaeological amber. Artifacts. An Annotated Bibliography. Part 2. Art and archaeology technical abstracts 6, 201–280.

Beck W., Donahue D.J., Jull A.J., Burr G., Broecker W.S., Bonani G., Hajdas I., Malotki E. and Dorn R.I. (1998) Ambiguities in direct dating of rock surfaces using radiocarbon measurements Science 280, 2132 – 2139.

Beck C.W. and Hartnett H.E. (1993) Sicilian amber. In: Proceedings of the Second Conference on Amber in Archaeology, Liblice, 1990. pp. 36–47.

Becker H.W., Bahr M., Berheide M., Borucki L., Buschmann M., Rolfs C., Roters G., Schmidt S., Schulte W.H., Mitchell G.E. and Schweitzer J.S. (1995) Hydrogen depth profiling using 18O beams. Z. Phys. A 351, 453–465.

Bednarik R.G. (1996). Only time will tell: a review of the methodology of direct rock art dating. Archaeometry, 38, 1–13.

Begemann F., Kallas K., Schmitt-Strecker S. and Pernicka E. (1999) Tracing ancient tin via isotope analyses. In: Haptmann A., Pernicka E., Rehren T. and Yalçin Ü. (eds) The beginnings of metallurgy. Der Anschnitt, Beheft 9. Bergbau Museum, Bochum. pp. 277–284.

Bell M., Fowler P.J. and Hillson, S.W. (1996) The experimental earthwork project 1960–1992. Council for British Archaeology, York.

Bell I.M., Clark R.J.H. and Gibbs P. (1997) Raman spectroscopic library of natural and synthetic pigments (pre.-~1850 AD). Spectrochim. Acta Part A 53, 2159–2179.

Bellani V., Giulotto E., Linati L. and Sacchi D. (2005) Origin of the blue fluorescence in Dominican amber. J. Appl. Phys. 97, 16101–16102.

Bellomo R.V. (1993) A methodological approach for identifying archaeological evidence of fire resulting from human activities. Journal of Archaeological Science 20, 525–553.

Benarie M. (1991) The establishment and use of damage functions. In: Baer N.S., Sabbioni C. and Sors A.I. (eds) Science, technology, and European cultural heritage: Proceedings of the European symposium. Bologna, Italy, 13–16 June 1989. Buttherworth-Heinemann Publishers, Guildford. pp. 214–220.

Bender M.M. (1968) Mass spectrometric studies of carbon 13 variations in corn and other grasses. Radiocarbon 10, 468–472.

Benninghoven A., Rüdenauer F.G. and Werner H.W. (1987) Secondary ion mass spectrometry: Basic concepts, instrumental aspects, applications, and trends. J. Wiley & Sons, New York.

Bentley R.A. (2006) Strontium isotopes from the Earth to the archaeological skeleton: A review. J. Archaeol. Meth. Theory 13, 135–187.

Bentur A. (2002) Cementitious materials – Nine millennia and a new century: Past, present, and future. American Society of Civil Engineers, 150th Anniversary Paper. J. Mater. Civ. Engin. JAN/FEB 2002, 2–22.

Berger A. and Loutre M.F. (1991) Insolation values for the climate of the last 10 million years. Quat. Sci. Rev. 10, 297–317.

Berna F., Matthews A. and Weiner S. (2004) Solubility of bone mineral from archaeological sites: the recrystallization window. J. Archaeol. Sci. 31, 867–882.

Berner R.A. and Landis G.P. (1987) Chemical analysis of gaseous bubble inclusions in amber: the composition of ancient air? Amer. J. Sci. 287, 757–762.

Berner R.A. and Landis G.P. (1988) The major gas composition of ancient air: analysis of gas bubble inclusions in fossil amber. Science 239, 1406–1408.

Berns R.S. (2001) The science of digitizing paintings for color-accurate image archives: A review. Journal of Imaging Science and Technology, 45, 305–323.

Berra M., Binda L., Anti L. and Fatticcioni A. (1992) Non destructive evaluation of the efficacy of masonry strengthening by grouting techniques. Proc. International Workshop on Effectiveness of injection techniques for retrofitting of stone and brick masonry walls in seismic areas, March 1992, Polytechnic of Milan, Italy. pp. 63–70.

Berthier S., Boulenguez J., Menu M. and Mottin B. (2008) Butterfly inclusions in Van Schrieck masterpieces. Techniques and optical properties. Appl. Phys. A 92, 51–57.

Betancourt P.P. (2006) The Chrysokamino metallurgy workshop and its territory. Hesperia Suppl. 36. American School of Classical Studies at Athens, Princeton, NJ.

Bhardwaj H.C. (ed.) (1987) Archaeometry of glass. Proc. of the Archaeometrical Session of the XIV Intern. Congr. On Glass, New Delhi 1986. Indian Ceramic Society, Calcutta.

Bicchieri M., Bella M. and Sementilli F. (1999) A quantitative measure of borane tert-butylamine complex effectiveness in carbonyl reduction of aged papers. Restaurator 20, 22–29.

Bicchieri M., Monti M. and Antonelli M.L. (2001) A new low-cost and complete restoration method: a simultaneous non-aqueous treatment of deacidification and reduction. In: Apuente J. (ed.) Proceedings of the 3rd International Conference on Science and Technology for the safeguard of cultural heritage in the Mediterranean basin, Vol. I. Universitad de Alcalà, Alcalà. pp. 276–280.

Bicchieri M., Ronconi S., Romano F.P., Pappalardo L., Corsi M., Cristoforetti G., Legnaioli S., Palleschi V., Salvetti A. and Tognoni E. (2002) Study of foxing stains on paper by chemical methods, infrared spectroscopy, micro-X-Ray Fluorescence spectrometry and Laser Induced Breakdown Spectroscopy. Spectrochim. Acta part B: Atomic Spectroscopy 57, 1233–1246.

Bicchieri M., Sodo A., Piantanida G. and Coluzza C. (2006) Analysis of degraded papers by non-destructive spectroscopic techniques. J. Raman Spectr. 37, 1186–1192.

Bickle P. and Hofmann D. (2007) Moving on: the contribution of isotope studies to the Early Neolithic of Central Europe. Antiquity 81, 1029–1041.

Bilz M. and Grattan D.W. (1996) The ageing of parylene: difficulties with the Arrhenius approach. In: Bridgland J. (ed.) ICOM-CC 11th Meeting, Edinburgh, James and James, London. pp. 925–929.

Bimson M. and Freestone I.C. (1987) Early vitreous materials. British Museum Occasional papers N. 56. British Museum, London.

Binda L. (2005) The importance of investigation for the diagnosis of historic buildings: Application at different scales (centers and single buildings). In: Modena C., Lourenço P.B. and Roca P. (eds) Proc. 4th Int. Seminar on Structural Analysis of Historical Constructions. 10–13 November 2004, Padova, Italy. A.A.Balkema Publishers, Leiden. pp. 29–42.

Binda L., Colla C. and Forde M.C. (1994) Identification of moisture capillarity in masonry using digital impulse radar. J. Constr. Building Mat. 8, 101–107.

Binda L., Gambarotta L., Lagomarsino S. and Modena C. (1999) A multilevel approach to the damage assessment and seismic improvement of masonry buildings in Italy. In: Bernardini (ed.) Seismic damage to masonry buildings. Balkema, Rotterdam. pp. 179–194.

Binda L., Cantini L., Cardani G., Saisi A. and Tiraboschi C. (2007) Use of flat-jack and sonic tests for the qualification of historic masonry. 10NAMC, Proc. of 10th North American Masonry Conference, 3–6 June 2007, St. Louis, Missouri. CD-ROM, ISBN 1-929081-28-6. pp. 791–803.

Biringuccio (Vannoccio Biringuccio) – De la pirotechnia (1540) Eng. Transl: Cyril Stanley Smith and Martha Teach Gnudi (1959) Basic Books, New York.

Black L., Garbev K., Beuchle G., Stemmermann P. and Schild D. (2006) X-ray photoelectron spectroscopic investigation of nanocrystalline calcium silicate hydrates synthesised by reactive milling. Cem. Concr. Res. 36, 1023–1031.

Black L. and Brooker A. (2007) SEM-SCA: combined SEM–Raman spectrometer for analysis of OPC clinker. Ad. Appl. Ceram. 106, 327–334.

Blatt H., Tracy R. and Owens B. (2005) Petrology: igneous, sedimentary, and metamorphic. 3rd Edition. W.H. Freeman, New York. pp. 530.

Bläuer-Böhm C. and Jägers E. (1997) Analysis and recognition of Dolomitic lime mortars. In: Béarat H., Fuchs M., Maggetti M. and Paunier D. (eds) Roman wall painting: materials, techniques, analysis and conservation. Proceedings of the International Workshop, Fribourg, 7–9 March 1996. Institute of Mineralogy and Petrography, Fribourg. pp. 223–235.

Blezard R.G. (1981) Technical aspects of Victorian cement. Chemistry and Industry, 19 Sept 1981, 630–636.

Blezard R.G. (1998) The history of calcareous cements. In: Hewlett P.C. (ed.) Lea's Chemistry of cement and concrete. 4th Edition. Butterworth-Heinemann, Oxford. pp. 1–23.

Bloss F.D. (1999) Optical crystallography. Mineralogical Society of America, Washington.

Blümich B. and Kuhn W. (1992) Magnetic Resonance Microscopy. VCH, Weinheim.

Bogdasarov M.A. (2007) Mineralogy of fossil resins in Northern Eurasia. Geol. Ore Deposits 49, 630–637.

Bohic S., Rey C., Legrand A., Sfihi H., Rohanizadeh R., Martel C., Barbier A. and Daculsi G., (2000) Characterization of the trabecular rat bone mineral: effect of ovariectomy and bisphosphonate treatment. Bone 26, 341–348.

Bonacasa N. and Antonino Buttitta A. (1988) La statua Marmorea di Mozia e la scultura di stile Severo in Sicilia. L'Erma di Bretschneider, Roma. pp. 150.

Bonchin S.L., Zoorob G.K. and Caruso J.A. (2000) Atomic emission, methods and instrumentation. In: Lindon J.C., Tranter G.E. and Holmes J.L. (eds) Encyclopedia of Spectroscopy and Spectrometry, Vol. 1, pp. 42–50. Academic Press, London.

Bonnet M. and Marignier J.-L. (2003) Niepce, correspondence et papiers. Maison Nicephore Niepce, France.

Bonsanti G. and Piccinini F. (eds) (2009) Emozioni in terracotta. Guido Mazzoni – Antonio Begarelli. Sculture del Rinascimento emiliano. Catalogue of the exhibition, Modena 21 March–7 June 2009. Franco Cosimo Panini, Modena.

Boon J.J., Tom A. and Pureveen J. (1993) Microgram scale pyrolysis mass spectrometric and pyrolysis gas chromatographic characterisation of geological and archaeological amber and resin samples. Proc. of the Second Conference on Amber in Archaeology, Liblice, 1990. pp. 9–27.

Booth A.D., Linford N.T., Clark R.A. and Murray T. (2008) Three-dimensional, multi-offset ground-penetrating radar imaging of archaeological targets. Archaeol. Prosp., 15, 93–112.

Bordignon F., Postorino P., Dore P. and Trojsi G. (2007) Raman identification of green and blue pigments in Etruscan polychromes on architectural terracotta panels. J. Raman Spectr. 38, 255–259.

Bøtter-Jensen L., McKeever S.W.S. and Wintle A.G. (2003) Optically stimulated luminescence dosimetry. Elsevier, Amsterdam.

Bouchenaki M. (2003) The interdependency of the tangible and intangible cultural heritage. ICOMOS 14th General Assembly and Scientific Symposium. Victoria Falls. Keynote address. See also: http://www.unesco.org/culture/ich/index.php?lg=EN&pg=home

Bouchenaki M. (ed.) (2004) Defining the intangible cultural heritage. Museum Int., Special Issue 56, 1–197.

Bourgarit D. (2007) Chalcolithic copper smelting. In: La Niece S., Hook D. and Craddock P. (eds) metals and mines. Studies in archaeometallurgy. The British Museum–Archetype Publications, London. pp. 3–14.

Bourgarit D. and Plateau J. (2005) Quand l'aluminium valait de l'or: peut-on reconnaître un aluminium "chimique" d'un aluminium "électrolytique"? ArcheoSciences, Revue d' 29, 95–105.

Bourgarit D., Rostan P., Burger E., Carozza L., Mille B. and Artioli G. (2008) The beginning of copper mass production in the western Alps : the Saint-Véran mining area reconsidered. Historical Metallurgy 42, 1–11.

Boutaine J.L. (2006) The modern museum. In: Bradley D., Creagh D. (eds) Physical techniques in the study of art, archaeology and cultural heritage. Vol. 1. Elsevier, Amsterdam. pp. 1–39.

Bouzek J. (1993) The shift of the amber route. Proc. of the Second Conference on Amber in Archaeology, Liblice, 1990. pp. 141–146.

Bowman S.G.E. (ed.) (1991) Science and the past. British Museum Publications, London.

Bradley D. and Creagh D. (eds) (2006) Physical techniques in the study of art, archaeology and cultural heritage. Vol. 1. Elsevier, Amsterdam.

Bradley D. and Creagh D. (eds) (2007) Physical techniques in the study of art, archaeology and cultural heritage. Vol. 2. Elsevier, Amsterdam.

Brady M. and Coleman D. (2000) Determining the source of felsitic lithic material in Southeastern New England using neodymium isotope ratios. Geoarchaeology, 15, 1–19.

Brandon C., Hohlfelder R.L. and Oleson J.P. (2005) The Roman maritime concrete study (ROMACONS). The Roman harbour of Chersonisos in Crete and its Italian connection. Mediterranée 1.2, 25–29.

Brandon D. and Kaplan W.D. (2008) Microstructural characterization of materials. 2nd Edition. John Wiley & Sons, Chichester. pp. 536.

Brill R.H. and Wampler J.M. (1967) Isotope studies of ancient lead. Am. J. Arch. 71, 63–77.

Brill R.H. and Martin J.H. (eds) (1991) Scientific research in early Chinese glass. Proc. of the Archaeometry of Glass Session of the 1984 Intern. Symp. on Glass, Beijing 7 September 1984. The Corning Museum of Glass, New York.

Brimblecombe P. (2005) Effects of the cultural environment. In: Van Grieken R.and Janssens K. (eds) Cultural heritage conservation and environmental impact assessment by non-destructive testing and micro-analysis. A.A. Balkema Publishers, Leiden. pp. 11–18.

Brody R.H., Edwards H.G.M. and Pollard A.M. (2000) A study of amber and copal samples using FT-Raman spectroscopy. Spectr. Acta A 57, 1325–1338.

Bromblet P., Laboure M. and Orial G. (2003) Diversity of the cleaning procedures including laser for the restoration of carved portals in France over the last 10 years. J. Cult. Herit. 4, Suppl. 1, Lasers in the Conservation of Artworks – LACONA IV. pp. 17–26.

Bronitsky G. and Hamer R. (1986) Experiments in ceramic technology: the effects of various tempering materials on impact and thermal-shock resistance. American Antiquity 51, 89–101.

Bronk H., Rohrs S. and Bjeoumikhov N. (2001) ArtTAX– a new mobile spectrometer for energy-dispersive micro X-ray fluorescence spectrometry on art and archaeological objects. Fresenius Journ. Anal. Chem., 371, 307–316.

Bronk H. and Freestone I.C. (2001) A quasi non-destructive microsampling technique for the analysis of intact glass objects by SEM/EDXA. Archaeometry 43, 517–527.

Brooks D., Bieber A.M. Jr., Harbottle G. and Sayre E.V. (1975) Biblical studies through activation analysis of ancient pottery. In: Beck C.W. (ed.) Archaeological chemistry. Adv. Chem. Ser. 138, 48–80.

Brooks S.O., Glascock M.D. and Giesso M. (1997) Source of volcanic glass for ancient Andean tools. Nature 386, 449–450.

Brothwell D.R. and Higgs E. (eds) (1969) Science in archaeology. Praeger Publishers, New York.

Brothwell D.R. and Pollard A.M. (eds) (2001) Handbook of archaeological sciences. John Wiley & Sons, Chichester.

Brown T.A. (2007) Ancient DNA. In: Brothwell D.R., Pollard A.M. (eds) (2001) Handbook of archaeological sciences. John Wiley & Sons, Chichester. pp. 301–312.

Brownell W.E. (1957) Black coring in structural clay products. J. Amer. Ceram. Soc. 40, 179–187.

Brownell W.E. (1976) Structural clay products. Springer-Verlag, Wien–New York.

Broz M.E., Cook R.E. and Whitney D.L. (2006) Microhardness, toughness, and modulus of Mohs scale minerals. Amer. Mineral. 91, 135–142.

Bruder R., Detalle V. and Coupry C. (2007) An example of the complementarity of laser-induced breakdown spectroscopy and Rama microscopy for wall painting pigments analysis. Journ. Raman Spectr., 38, 909–915.

Brunauer S., Emmett P.H. and Teller E. (1938) Adsorption of gases in multimolecolar layers. J. Amer. Chem. Soc. 60, 309–319.

Brunetti B.G., Matteini M., Miliani C., Pezzati L. and Pinna D. (2007) MOLAB, a mobile laboratory for in situ non-invasive studies in art and archaeology. In: Nimmrichter J., Kautek W. and Schreiner M. (ed.) Lasers in the Conservation of Artworks. Proc. of LACONA VI. Vienna, Austria, 21–25 Sept. 2005. Springer Proceedings in Physics no. 116. Springer-Verlag, Berlin. pp. 453–462.

Buck C.E., Cavanagh W.G. and Litton C.D. (1996) The Bayesian approach to archaeological data interpretation. John Wiley & Sons, Chichester.

Buckle H. (1959) Progress in micro-indentation hardness testing. Metall. Rev. 4, 49–100.

Budd P. (1991) A metallographic investigation of eneolithic arsenical copper artefacts from Mondsee, Austria. J. Hist. Metall. Soc. 25, 99–108.

Budd P., Montgomery J., Evans J. and Chenery C. (2001) Combined Pb-, Sr- and O-isotope analysis of human dental tissue for the reconstruction of archaeological residential mobility. In: Holland J.G. and Tanner S.D. (eds) Plasma source mass spectrometry. Royal Society of Chemistry Special Publication, Cambridge. pp. 311–326.

Budd P., Montgomery J., Evans J. and Trickett M. (2004) Human lead exposure in England from approximately 5500 BP to the 16th century AD. Sci. Total Environ. 318, 45–58.

Bugini R. and Folli L. (2008) Stones used in the Milan architecture. Mat. Constr. 58, 33–50.

Bunge H.J. (2000) Texture analysis. In: Snyder R.L., Fiala J. and Bunge H.J. (eds) Defect and microstructure analysis by diffraction. IUCr Monographs on Crystallography Vol. 10–Oxford University Press, Oxford. pp. 408–531.

Burdukiewicz M.J. (1993) Late Palaeolithic amber in Northern Europe. Proc. of the Second Conference on Amber in Archaeology, Liblice, 1990. pp. 141–146.

Burenhult G. (1993–1994) The illustrated history of humankind. 5 Volumes. HarperCollins, San Francisco.

Burger E. (2008) Métallurgie extractive protohistorique du cuivre: Etude thermodynamique et cinétique des reactions chimique de transformation de minerais de cuivre sulfurés en metal et caractérisation des procédés. Ph.D. Dissertation. Universite Pierre et Marie Curie (Paris 6).

Burger E., Bourgarit D., Wattiaux A. and Fialin M. (2010) The reconstruction of the first copper smelting process in Europe during the 4th and 3rd millennia BC: Where does the oxygen come from? Appl. Phys. A, in press.

Burgoyne T.W. and Hieftje G.M. (1996) An introduction to ion optics for the mass spectrograph. Mass Spectr. Rev. 15, 241–259.

Burleigh R. and Whalley P. (1983) On the relative geological ages of amber and copal. J. Natural History 17, 919–921.

Buzio R., Calvini P, Ferroni A and Valbusa U. (2004) Surface analysis of paper documents damaged by foxing. Appl. Phys. A Materials Science & Processing 79, 383–387.

Caley E.R. (1949) Klaproth as a pioneer in the chemical investigation of antiquities. J. Chem. Education 26, 242–247.

Caley E.R. (1951) Early history and literature of archaeological chemistry. J. Chem. Education 28, 64–66.

Calligaro T. (2005) The origin of ancient gemstones unveiled by PIXE, PIGE and μ-Raman spectrometry. In: Uda M., Demortier G., Nakai I. (eds) X-rays for archaeology. Springer, Berlin. pp. 101–112.

Calligaro T., Mossmann A., Poirot J.-P. and Querre G. (1998) Provenance study of rubies from a Parthian statuette by PIXE analysis. Nucl. Instrum. Meth. Phys. Res. B 136–138, 846–850.

Calligaro T., Dran J.-C. and Salomon J. (2004) Ion beam microanalysis. In: Janssens K. and Van Grieken R. (ed.) (2004) Non-destructive microanalysis of cultural heritage materials. Comprehensive analytical chemistry series. Vol. XLII. Elsevier, Amsterdam. pp. 227–276.

Calligaro T., Coquinot Y., Reiche I., Castaing J., Salomon J., Ferrand G. and Le Fur Y. (2009) Dating study of two rock crystal carvings by surface microtopography and by ion beam analyses of hydrogen. Appl. Phys. A 94, 871–878.

Campbell D.H. (1999) Microscopic examination and interpretation of Portland cement and clinker. 2nd Edition. Portland Cement Association, Skokie, IL. pp. 201.

Campbell D.H. (2007) Geologic origin of Egyptian pyramid blocks and associated structures. Proc. 29th Conference on Cement Microscopy. Quebec City, 20–24 May 2007. International Cement Microscopy Association, CD volume.

Camuffo D. (1998) Microclimate for cultural heritage. Elsevier, Amsterdam.

Camuffo D. (2004) Thermodynamics for cultural heritage. In: Martini M., Milazzo M. and Piacentini M. (eds) Physics methods in archaeometry. SIF, Bologna–IOS Press, Amsterdam. pp. 37–98.

Cano R.J., Poinar H. and Poinar G.O. (1992) Isolation and partial characterisation of DNA from the bee Proplebeia dominicana (Apidae : Hymenoptera) in 25–40 million year old amber. Medical Sci. Res. 20, 249–251.

Capedri S., Venturelli G. and Grandi R. (2002) Trachytes used for paving Roman roads in the Po plain: characterisation by petrographic and chemical parameters and provenance of flagstones. J. Archaeologic. Sci. 30, 491–509.

Capedri S. and Venturelli G. (2003) Trachytes employed for funerary artefacts in the Roman Colonies Regium Lepidi (Reggio Emilia) and Mutina (Modena) (Italy): provenance inferred by petrographic and chemical parameters and by magnetic susceptibility. J. Cult. Herit. 4, 319–328.

Capellini G. (1872) Überdas vorkommen von Bernstein im Bolognesischen und anderen Punkten Italiens. Zeitschrift für Ethnologie, Verhandlungen, 4, 198.

Caple C. (2001) Degradation, investigation, and preservation of archaeological evidence. In: Brothwell D.R. and Pollard A.M. (eds) (2001) Handbook of archaeological sciences. John Wiley & Sons, Chichester. pp. 587–593.

Cappitelli F., Toniolo L., Sansonetti A., Gulotta D., Ranalli G., Zanardini E. and Sorlini C. (2007) Advantages of using microbial technology over traditional chemical technology in removal of black crusts from stone surfaces of historical monuments. Appl. Environ. Microbiol. 73, 5671–5675.

Carriveau G.W. and Han M.C. (1976) Thermoluminescent dating and the monsters of Acambaro American Antiquity 41, 497–500.

Carriveau G.W., Omecinsky D., Wolyniak C., Maccracken L., Dook S., England P. and van Zelst L. (1984) Identification of the Forbes Collection Pigments. In: ICOM Committee for Conservation 7th Triennial meeting preprints, Copenhagen, 1984.

Carrondo M.A and Jeffrey G (1987) Chemical crystallography with pulsed neutrons and synchrotron X-rays. Kluwer Academic Publishers.

Cartechini L., Miliani C., Brunetti B.G., Sgamellotti A., Altavilla C., Ciliberto E. and D'Acapito, F. F. (2008) X-ray absorption investigations of copper resinate blackening in a XV century Italian painting. Appl. Phys. A 92, 243–250.

Carter E.A., Hargreaves M.D., Kononenko N., Graham I., Edwards H.G.M., Swarbrick B. and Torrence R. (2009) Raman spectroscopy applied to understanding prehistoric obsidian trade in the Pacific Region. Vibrational Spect. 50, 116–124.

Cartier-Bresson A. (ed.) (2008) Le vocabulaire technique de la photographie. Marval, Paris.

Cartocci A., Fedi M.E., Taccetti F., Benvenuti M., Chiarantini L. and Guideri S. (2007) Study of a metallurgical site in Tuscany (Italy) by radiocarbon dating. Nucl. Instrum. Meth. Phys. Res. B 259, 384–387.

Casaletto M.P., De Caro T., Ingo G.M. and Riccucci C. (2006) Production of reference "ancient" Cu-based alloys and their accelerated degradation methods. Appl. Phys. A: Mater. Sci. Proces. 83, 617–622.

Casali F. (2006) X-ray and neutron digital radiography and computed tomography for cultural heritage. In: Bradley D. and Creagh D. (eds) Physical techniques in the study of art, archaeology and cultural heritage. Vol. 1. Elsevier, Amsterdam. pp. 41–124.

Caseau B. (2004) The fate of rural temples in late antiquity and the Christianization of the countryside. In: Bowden W., Lavan L. and Machado C. (eds) Recent research on the late antique countryside. Brill, Leiden, The Netheralds.

Cather S. (ed.) (1991) The conservation of wall paintings. Proc. of a Symposium organized by the Courtauld Institute of Art and the Getty Conservation Institute, London, 13–16 July 1987. The Getty Conservation Institute, Los Angeles. pp. 1–12.

Cather S. (2003) Assessing causes and mechanisms of detrimental change to wall paintings. In: Gowing R. and Heritage A. (eds) Conserving the Painted Past: Developing Approaches to Wall Painting Conservation. Post-prints of an English Heritage Conference, 1999. London. pp. 64–74.

Cato (Marcus Porcius Cato) – De Agri Cultura (160 BC) Engl. Transl.: Andrew Dalby (1998) Cato: On Farming. Prospect Books, Totnes.

Cattaneo C., Gelsthorpe K., Phillips P. and Sokol R.J. (1995) Differential survival of albumin in ancient bone. J. Archaeol. Sci. 22, 271–276.

Caubet A. (ed.) (2007) Faiences et matières vitreuses de l' Ancien. Musée du Louvre Éditions, Paris. pp. 309.

Caubet A. and Pierrat-Bonnefois G. (2005) Faïences de l' de l' à l'Iran. Musée du Louvre Éditions, Paris. pp. 206.

Cavarretta G., Gioia P., Mussi V. and Palombo M.R. (eds) (2001) The World of Elephants. Proc. of the 1st International Congress. Consiglio Nazionale delle Ricerche, Rome, Italy.

Cazzato R., Costa G., Dal Farra A., Fornasier M., Toniolo D., Tosato D. and Zanuso C. (2006) Il Progetto Mantegna: storia e risultati. In: Spiazzi A.M., De Nicolò

Salmazo A. and Toniolo D. (eds) Andrea Mantegna. La Cappella Ovetari a Padova. Skira.

Cernych E.N. (1978) Ai Bunar – A Balkan copper mine of the fourth millennium B.C. Proc. Prehist. Soc. 44, 203–218.

Chadefaux C., Vignaud C., Chalmin E., Robles-Camacho J., Arroyo-Cabrales J., Johnson E. and Reiche I. (2009) Color origin and heat evidence of paleontological bones: Case study of blue and gray bones from San Josecito Cave, Mexico. Amer. Mineral. 94, 27–33.

Chahine C. (2000) Changes in hydrothermal stability of leather and parchment with deterioration: a DSC study. Thermochim. Acta 365, 101–110.

Chaimanee Y., Jolly D., Benammi M., Tafforeau P., Duzer D., Moussa I. and Jaeger J.J. (2003) A Middle Miocene hominoid from Thailand and orangutan origins. Nature, 422, 61–65.

Challis K., Priestnall G., Gardner A., Henderson J. and O'Hara S. (2002–2004) Corona remotely-sensed imagery in dryland archaeology: The Islamic city of al-Raqqa, Syria. J. Field Arch., 29, 139 –153.

Chalmin E., Menu M. and Vignaud C. (2003) Analysis of rock art painting and technology of Palaeolithic painters. Meas. Sci. Technol. 14, 1590–1597.

Chandra S. (2002) Properties of concrete with mineral and chemical admixtures. In: Bensted J. and Barnes P. (eds) Structure and performance of cements. 2nd Edition. Spon Press, London – New York. pp. 140–185.

Charles J.A. (1985) Determinative mineralogy and the origins of metallurgy. In: Craddock P.T. and Hughes M.J. (eds) (1985) Furnaces and smelting technology in antiquity. British Museum Occasional Paper No. 48. British Museum Publications, London. pp. 21–28.

Charola A.E. (2000) Salts in the deterioration of porous materials: An overview. J. Amer. Inst. Conserv. 39, 327–343.

Chatterjee A.K. (2002) Special cements. In: Bensted J. and Barnes P. (eds) Structure and performance of cements. 2nd Edition. Spon Press, London – New York. pp. 186–236.

Chen R. and McKeever S.W.S. (1997) Theory of thermoluminescence and related phenomena. World Scientific Publishing Company, Singapore.

Chen T.-H., Calligaro T., Pagès-Camagna S. and Menu M. (2004) Investigation of Chinese archaic jade by PIXE and μRaman spectrometry. Appl. Phys. A 79, 177–180.

Cheng H. and Edwards R.L. (1997) U/Th and U/Pa dating of Nanjing Man. Eos 78, 787.

Cheng H.S., Zhang Z.Q., Xia H.N., Jiang J.C. and Yang F.J. (2002) Non-destructive analysis and appraisal of ancient Chinese porcelain by PIXE. Nucl. Instrum. Meth. Phys. Res. B 190, 488–491.

Chernykh E.N. (1992) Ancient metallurgy in the USSR. Cambridge University Press, Cambridge, UK.

Chiari G. and Lanza R. (1999) Remnant magnetization of mural paintings from the Bibliotheca Apostolica (Vatican, Rome). J. Appl. Geophys. 41, 137–143.

Chiari G. (2008) Saving art in situ. Nature 453, 159.

Chiari G. and Scott D.A. (2004) Pigment analysis: potentialities and problems. Per. Mineral. 73, 227–237.

Chiari G., Giustetto R., Druzik J., Doehne E. and Ricchiardi M.P. (2008) Pre-Columbian nanotechnology: reconciling the mysteries of the Maya blue pigment. Appl. Phys. A 90, 3–7.

Childers B.C. (1994) Long-term lichen-removal experiments and petroglyph conservation: Fremont County, Wyoming, Ranch Petroglyph Site. Rock Art Research 11, 101–112.

Chippindale C. and Taçon P.S.C. (eds) (1998) The archaeology of rock art. Cambridge University Press, Cambridge.

Chisholm J.E. (1992) Powder-diffraction patterns and structural models for palygorskite. Canad. Mineral. 30, 61–73.

Choi S. (2007) Foxing on paper: a literature review. J. Amer. Inst. Cons. 46, 137–152.

Chu V., Regev L., Weiner S. and Boaretto E. (2008) Differentiating between anthropogenic calcite in plaster, ash and natural calcite using infrared spectroscopy: Implications in archaeology. J. Archaeol. Sci. 35, 4905–4911.

Church T. (1994) Lithic resource studies: A sourcebook for archaeologists. Special Publ. Vol. 3. Department of Anthropology, University of Tulsa, Tulsa.

Church T. (2000) Distribution and sources of obsidian in the Rio Grande gravels of New Mexico. Geoarcheol. 15, 649–678.

Cierny J., Marzatico F., Perini R. and Weisgerber G. (2004) Der spätbronzezeitliche Verhüttungsplatz Acqua Fredda am Passo Redebus (Trentino). Alpenkupfer – Rame delle Alpi. Der Aschnitt, Beiheft 17, 155–164.

Ciferri O., Tiano P. and Mastromei G. (2000) Of microbes and art, the role of microbial communities in the degradation and protection of cultural heritage. Springer-Verlag, Berlin.

Ciliberto E. and Spoto G. (eds) (2000) Modern analytical methods in art and archaeometry. John Wiley & Sons, New York.

Clark D.W. (1995) Batza tena; the trail to obsidian. Act. Anthropol. 32, 82–91.

Claro A., Melo M.J., Schäfer S., Melo J.S.S., Pina F., van den Berg K.J. and Burnstock A. (2008) The use of microspectrofluorimetry for the characterization of lake pigments. Talanta 74, 922–929.

Clayton D.D. (1968) Principles of stellar evolution and nucleosynthesis. McGraw-Hill Book Company, New York. pp. 612.

Clementi C., Miliani C., Romani A. and Favaro G. (2006) In situ fluorimetry: A powerful non-invasive diagnostic technique for natural dyes used in artefacts: Part I. Spectral characterization of orcein in solution, on silk and wool laboratory-standards and a fragment of Renaissance tapestry. Spectrochim. Acta Part A: Molecular and Biomolecular Spectroscopy 64, 906–912.

Clifford D.J, Hatcher P.G., Botto R.E., Muntean J.V., Michels B. and Anderson K.B. (1997) The nature and fate of natural resins in the geosphere-VIII.* NMR and Py-GC-MS characterisation of soluble labdanoid polymers, isolated from Holocece class I resins. Organic Geochem. 27, 449–464.

Cloetens P., Ludwig W., Baruchel J., van Dyck D., van Landuyt J., Guigay J.P. and Schlenker M. (1999) Holotomography: Quantitative phase tomography with micrometer resolution using hard synchrotron radiation X-rays. Appl. Phys. Lett. 75, 2912–2914.

Cloetens P., Ludwig W., Guigay J.P., Baruchel J., Schlenker M. and Van Dyck D. (2000) Phase contrast tomography. In: Baruchel J., Buffière J.Y., Maire E., Merle P. and Peix G. (eds) (2000) X-ray tomography in materials science. Hermes Science Publications, Paris. pp. 29–44.

Clottes J. (2002) World rock art. The Getty Conservation Institute, Los Angeles.

Cobean R.H., Vogt J.R., Glascock M.D. and Stocker T.L. (1991) High-precision trace-element characterization of major Mesoamerican obsidian sources and further analyses of artifacts from San Lorenzo Tenochtitlan, Mexico. Latin Amer. Antiquity 2, 69–91.

Coe R.S., Prévot M. and Camps P. (1995) New evidence for extraordinarily rapid change of the geomagnetic field during reversal. Nature 374, 687–692.

Coghlan H.H. (1943) The evolution of the axe from prehistoric to Roman times. J. Royal Anthrop. Inst. Great Britain and Ireland 73, 27–56.

Collepardi M. (1999) Thaumasite formation and the deterioration of historic buildings. Cem. Concr. Composites 21, 147–154.

Collet F., Bart M., Serres L. and Miriel J. (2008) Porous structure and water vapour sorption of hemp-based materials. Constr. Build. Mater. 22, 1271–1280.

Colomban Ph. (2004) Raman spectrometry, a unique tool to analyze and classify ancient ceramics and glasses. Appl. Phys. A 79, 167–170.

Colomban Ph., Tournie A. and Bellot-Gurlet L. (2006) Raman identification of glassy silicates used in ceramics, glass and jewellery: A tentative differentiation guide. Journal of Raman Spectroscopy 37, 841–852.

Colombini M.P., Giachi G., Modugno F., Pallecchi P. and Ribechini E. (2003) The characterization of paints and waterproofing materials from the shipwrecks found at the archaeological site of the Eetruscan and Roman harbour of Pisa (Italy). Archaeometry 45, 659–674.

Colpani F., Marelli M., Giussani B., Recchia S., Angelini I., Baumgarten B. and Artioli G. (2007) Copper isotopic ratio and trace elements spectrometric measurements (ICP-QMS) within Alps and Apennine Cu-ores: Discovering regional geochemical tracers for archaeometrical purposes by advanced chemometric techniques. In: D' C. (ed.) Atti del IV Congr. Naz. AIAr, Pisa, 1–3 February 2006. Pàtron Ed., Bologna. pp. 547–559.

Coluzza C., Bicchieri M., Monti M., Piantanida G. and Sodo A. (2008) Atomic force microscopy application for degradation diagnostics in library heritage. Surf. Interf. Analysis 40, 1248–1253.

Comelli D., D' C., Valentini G., Cubeddu R., Colombo C. and Toniolo L. (2004) Fluorescence lifetime imaging and spectroscopy as tools for nondestructive analysis of works of art. Appl. Opt. 43, 2175–2183.

Compston J.E. (2002) Bone marrow and bone: a functional unit. J. Endocrinology 173, 387–394.

Conyers L.B. (2005) Ground penetrating radar for archaeology. Rowman Altamira.

Cook R.A. and Hover K.C. (1999) Mercury porosimetry of hardened cement pastes. Cem. Concr. Res. 29, 933–943.

Coplen T.B. (1996) New guidelines for reporting stable hydrogen, carbon, and oxygen isotope-ratio data. Geochim. Cosmochim. Acta 17, 3359–3360.

Cornell R.M. and Schwertmann U. (1996) The iron oxides. Structure, properties, reactions, occurrence, and uses. VCH, Weinheim, Germany. pp. 573.

Cotte M., Susini J., Metrich N., Moscato A., Gratziu C., Bertagnini A. and Pagano M. (2006) Blackening of Pompeian cinnabar paintings: X-ray microspectroscopy analysis. Anal. Chem. 78, 7484–7492.

Cotte M., Checroun E., Mazel V., Solé V.A., Richardin P., Taniguchi Y., Walter P., Susini J. (2009) Combination of FTIR and X-Rays Synchrotron-Based Micro-Imaging Techniques for the Study of Ancient Paintings. A Practical Point of View. Review. e-Preservation Science 6, 1–9.

Coulson W. and Wilkie N.C. (1986) Ptolemaic and Roman kilns in the Western Nile delta. Bull. Amer. Schools Oriental Res. 263, 61–75.

Cowen T. (1998) In praise of commercial culture. Harvard University Press, Cambridge.

Cox C. (1992) Satellite imagery, aerial photography and wetland archaeology: an interim report on an application of remote sensing to wetland archaeology: The pilot study in Cumbria, England. World Archaeology, 24, 249–267.

Cox K.G., Bell J.D. and Pankhurst R.J. (1979) The interpretation of igneous rocks. Unwin Hyman, London.

Craddock P.T. (1999) Paradigms of metallurgical innovation in prehistoric Europe. In: Hauptmann A., Pernicka E., Rehren T. and Yalçin Ü. (eds) The beginnings of metallurgy. Der Anschnitt, Beheft 9. Bergbau Museum, Bochum. pp. 175–192.

Craddock P.T. (2007) Scientific investigation of copies, fakes and forgeries. Butterworth-Heinemann, Oxford. pp. 628.

Craddock P.T. and Hughes M.J. (eds) (1985) Furnaces and smelting technology in antiquity. British Museum Occasional Paper No. 48. British Museum Publications, London.

Craddock P.T., Gurney D., Pryor F. and Hughes M.J. (1985) The application of phosphate analysis to the location and interpretation of archaeological sites. Archaeol. J. 142, 361–376.

Craddock P.T. and Bowman S. (1991) Spotting the fakes. In: Bowman S.G.E. (ed) Science and the past. British Museum Publications, London. pp. 141–157.

Craddock P.T. and Giumlia-Mair A. (1993) Hsmn-km, Corinthian bronze, shakudo: black patinated bronze in the ancient world. In: La Niece, S. and Craddock, P.T. (eds) Metal plating and patination Butterworth-Heinemann Publisher, Guildford. pp. 101–127.

Cranmer D. (1987) In: Mark Rothko, 1903–1970. Tate Gallery, London.

Creagh D.C. and Bradley D.A. (ed.) (2000) Radiation in art and archaeometry. Elsevier, Amsterdam.

Creagh D. (2007) Synchrotron radiation and its use in art, archaeometry, and cultural heritage studies. In: Bradley D. and Creagh D. (eds) Physical techniques in the study of art, archaeology and cultural heritage. Vol. 2. Elsevier, Amsterdam.

Cremaschi M. (1996) The rock varnish in the Messak Settafet (Fezzan, Libyan Sahara), age, archaeological context, and palaeo-environmental implication. Geoarchaeology: An Int. Journal 11, 393–421.

Cremaschi M. (1998) Late Quaternary geological evidence for environmental changes in South-Western Fezzan. In: Cremaschi M. and Di Lernia S. (eds) Wadi Teshuinat. Palaeoenvironment and prehistory in South-Western Fezzan (Libya). CNR, Milano. pp. 13–48.

Cremaschi M. (2003) Steps and timing of desertification during Late Antiquity. The case study of the Tanezzuft oasis (Libyan Sahara). In: Liverani M. (ed.) Arid lands in Roman times. Papers from the International Conference, Rome, July 9–10 2001. Arid Zone Archaeology 4, 1–14.

Crisci G.M., Franzini M., Lezzerini M., Mannoni T. and Riccardi M.P. (2004) Ancient mortars and their binder. Per. Mineral. 73, 259–268.

Crowther J.R. (2001) The ELISA Guidebook. Humana Press, New Jersey.

Cunliffe S. (2006) Tourism and cultural risk management. In: Agnew N. and Bridgland J. (eds) Of the past, for the future: Integrating archaeology and conservation. The Getty Conservation Institute, Los Angeles. pp. 194–198.

Currie L.A. (2004) The remarkable metrological history of radiocarbon dating. J. Res. Natl Inst. Stand. Technol. 102, 185–217.

Dalrio G. (1980) Mineralogia del Bolognese. Descrizione e itinerari. Cacciari, Bologna.

Dapiaggi M., Sala M., Artioli G. and Fransen M.J. (2007) Evaluation of phase detection limit on filter-deposited dust particles from Antarctic ice cores. Kristallog. Suppl., 26, 73–78.

D'Armico C. and Starnini E. (2006) Prehistoric polished stone artefacts in Italy: a petrographic and archaeological assessment. Geol. Soc. London, Special Publications 257, 257–272.

D'Arrigo R., Jacoby G., Frank D., Pederson N., Cook E., Buckley B., Nachin B., Mijiddorj R. and Dugarjav C. (2001) 1738 Years of Mongolian Temperature Variability Inferred from a Tree-Ring Width Chronology of Siberian Pine. Geophys. Res. Lett., 28, 543–546.

Davenport T. and Prusak P. (1998) Working knowledge. Harvard Business School Press, Boston.

Davidovits J. (1987) Ancient and modern concretes: what is the difference? Concr. Intern. 9, 23–29.
Davidson T.E. and McKerrell H. (1976) Pottery analysis and Halaf Period trade in the Khabur Headwaters Region. Iraq 38, 45–56.
Davies P. (1992) Is nature mathematical? New Scientist Magazine, 1813, 25.
Davy H. (1802) An account of a method of copying paintings upon glass, and of making profiles by the agency of light upon nitrate of silver. J. Royal Inst. Great Britain 170–174.
Dayagi-Mendels M. (1993) Perfumes and cosmetics of the ancient world. Catalogue. The Israel Museum, Jerusalem. pp. 139.
Deer W.A., Howie R.A. and Zussman J. (1996) An introduction to the rock-forming minerals. 2nd Edition. Prentice Hall. pp. 712.
De Fontenelle B. le Boviers (1686) Entretiens sur la pluralité desmondes. Reprinted by University of California Press, Los Angeles (1990) Conversations on the plurality of worlds.
De Franceschi D. and De Ploëg G. (2003) Origine de l' des faciès sparnaciens (Éocène infèrieur) du Bassin de Paris: le bois de l' producteur. Geodiversitas 25, 633–647.
Degryse P., Henderson J. and Hodgins G. (2009) Isotopes in vitreous materials. (Studies in Archaeological Sciences). Leuven University Press, Leuven. pp. 200.
Degryse P., Schneider J., Kellens N., Waelkens M. and Muchez Ph. (2007) Tracing the resources of iron working at ancient Sagalassos (South-West Turkey): a combined lead and strontium isotope study on iron artefacts and ores. Archaeometry 49, 75–86.
Degryse P. and Schneider J. (2008) Pliny the Elder and Sr-Nd isotopes: tracing the provenance of raw materials for Roman glass production. J. Arch. Sci., 35, 1993–2000.
Deino A.L., Renne P.R. and Swisher C.C. (1998) 40Ar/39Ar dating in paleoanthropology and archaeology. Evol. Anthropolgy, 6, 63–75.
Delalieux F., Cardell-Fernandez C., Torfs K., Vleugels G. and Van Grieken R. (2002) Damage functions and mechanism equations derived from limestone weathering in field exposure. Water, Air, Soil Pollution 139, 75–94.
De la Torre A.G. and Aranda M.A.G. (2003) Accuracy in Rietveld quantitative phase analysis of Portland cements. J. Appl. Cryst. 36, 1169–1176.
De la Torre M. (ed.) (2002) Assessing the values of cultural heritage. Research Report. The Getty Conservation Institute, Los Angeles.
De Marinis R.C. (ed.) (2002) Studi sull' dell' del Bronzo del Lavagnone, Desenzano del Garda. Notiz. Archeol. Bergomensi 10, 1–315.
DeMers M. (2004) Fundamentals of Geographic Information Systems, 3rd edition. John Wiley & Sons, New York.
De Michele V. (1998) The «Libyan Desert Glass» scarab in Tutankhamen's pectoral. Sahara 10, 107–109.
Demortier G. (2004) Ion beam techniques for the analysis of archaeological metallic artefacts. In: Martini M., Milazzo M. and Piacentini M. (eds) (2004) Physics methods in archaeometry. SIF, Bologna – IOS Press, Amsterdam. pp. 99–154.
Demortier G. (2005) Ion beam techniques for the non-destructive analysis of archaeological materials. In: Uda M., Demortier G. and Nakai I. (eds) X-rays for archaeology. Springer, Berlin. pp. 67–100.
Demortier G. and Ruvalcaba-Sil J.L. (1996) Differential PIXE analysis of Mesoamerican jewelry items. Nucl. Instrum. Meth. Phys. Res. B 118, 352–358.
De Navarro J.M. (1925) Prehistoric routes between Northern Europe and Italy defined by the amber trade. Geograph. J. 66, 481–503.

Denker A. and Opitz-Coutureau J. (2004) Proton-induced X-ray emission using 68MeV protons. X-Ray Spectrom. 33, 61–66.

De Nuccio M. and Ungaro L. (2002) I marmi colorati della Roma imperiale. Marsilio Editore, Venezia. pp. 648.

De Paolo D. (1988) Neodymium isotope geochemistry: An introduction. Springer-Verlag, New York.

De Raedt I., Janssens K., Veeckman J., Vincze L., Vekemans B. and Jeffries T.E. (2001) Trace analysis for distinguishing between Venetian and façon-de-Venise vessels of the 16th and 17th century. J. Anal. At. Spectrom. 16, 1012–1017.

Derrick M.R., Stulik D.C. and Landry J.M. (1999) Infrared spectroscopy in conservation science. The Getty Conservation Institute, Los Angeles.

Devia B., Llabres G., Wouters J., Dupont L., Escribano-Bailon M.T., de Pascual-Teresa S., Angenot L. and Tits M. (2002a) New 3-deoxyanthocyanidins from leaves of Arrabidaea chica. Phytochemical Anal. 13, 114–120.

Devia B., Cardale M. and Wouters J. (2002b) The red mantles: *Arrabidaea chica* in archaeological Columbian textiles. In Preprints of the North American Textile Conservation Conference, Philadelphia, 35–46.

De Vries H. (1958) Variation of the concentration of radiocarbon with time and location on Earth. Kon. Ned. Akad. Wetensch. Proc. Ser. B 61, 267–281.

Diamond J. (1999) Guns, germs, and steel. The fates of human societies. W.W. Norton & Company, New York.

Diamond S. (2000) Mercury porosimetry. An inappropriate method for the measurement of the pore size distributions in cement-based materials. Cem Concr. Res. 30, 1517–1525.

Dickin A.P. (2005) Radiogenic isotope geology, 2nd ed. Cambridge: Cambridge University Press.

Dix B. (1982) The manufacture of lime and its uses in the Western Roman provinces. Oxford J. Archaeol. 1, 331–346.

Doehne E. (2002) Salt weathering: A review. In: Siegesmund S., Weiss T., Vollbrecht A. (eds) Natural stone, weathering phenomena, conservation strategies and case studies. Geol. Soc. Special Publ. 205, 51–64.

Donaldson C.H. (1976) An experimental investigation of olivine morphology. Contrib. Mineral. Petrol. 37, 187–213.

Dooryhée E., Menu M. and Susini J. (2006) Synchrotron Radiation in Art and Archaeology. Special Issue of Appl. Phys. A 83, no. 2.

Dooryhée E., Anne M., Bardiès I., Hodeau J.-L., Martinetto P., Rondot S., Salomon J., Vaughan G.B.M. and Walter P. (2005) Non-destructive synchrotron X-ray diffraction mapping of a Roman painting. Appl. Phys. A 81, 663–667.

Doremus R.H. (1994) Glass science. John Wiley & Sons, New York.

Dorn R.I., Jull A.J.T., Donahue D.J., Linick T.W. and Toolin L.J. (1989) Accelerator mass spectrometry radiocarbon dating of rock varnish. Geol. Soc. Am. Bull. 101, 1363–1372.

Dorn R.I. (1994) Rock varnish as evidence of climatic change. In: Abrahams A.D. and Parsons A.J. (eds) Geomorphology of desert environments. Chapman & Hall, London. pp. 539–552.

Dove P.M., De Yoreo J.J. and Weiner S. (eds) (2003) Biomineralization. Rev. Mineral. Geochem., Vol. 54. Mineralogical Society of America-Geochemical Society, Washington, DC. pp. 381.

Dran J.-C., Calligaro T. and Salomon J. (2000) Particle-induced X-ray emission. In: Ciliberto E. and Spoto G. (eds) Modern analytical methods in art and archaeometry. John Wiley & Sons, New York. pp. 135–166.

Dran J.-C., Salomon J., Calligaro T. and Walter P. (2004) Ion beam analysis of art works: 14 years of use in the Louvre. Nucl. Instrum. Meth. Phys. Res. B 219–220, 7–15.

Drennan R.D. (2001) Overview – Numbers, models, maps: computers and archaeology. In: Brothwell D.R. and Pollard A.M. (eds) Handbook of archaeological sciences. John Wiley & Sons, Chichester. pp. 663–670.

Drooker P. and Webster L. (eds) (2000) Beyond cloth and cordage: current approaches to archaeological textile research in the Americas. University of Utah Press, Salt Lake City, UT.

Duller G.A.T. (1996) Recent developments in luminescence dating of Quaternary sediments.Prog. Phys. Geography 20, 127–145.

Dunnell R.C. and Stein J.K. (1989) Theoretical issues in the interpretation of microartifacts. Geoarchaeology: An Int. Journ. 4, 31–42.

Duverne R. (2006) Structural measurements of DOP photographic paper and particle size of baryta coating. Glass Eye Productions, DVD distributed by the Getty Conservation Institute, Los Angeles. [videorecording, 2 DVDs (4h 26min.)]

Duwe S. and Neff H. (2007) Glaze and slip pigment analyses of Pueblo IV period ceramics from East-Central Arizona using time of flight-laser ablation-inductively coupled plasma-mass spectrometry (TOF-LA-ICP-MS). J. Achaeol. Sci. 34, 403–414.

Dykeman D.D., Towner R.H. and Feathers J.K. (2002) Correspondence in tree-ring dating and thermoluminescence dating: A protohistoric Navajo pilot study. American Antiquity, 67, 145–165.

Eckstein D. (2007) Human time in tree rings. Dendrochronologia, 24, 53–60.

Edwards H.G.M. (2000) Art works studied using IR and Raman spectroscopy. In: Lindon J.C., Tranter G.E. and Holmes J.L. (eds) Encyclopedia of Spectroscopy and Spectrometry, Vol. 1, pp. 2–17. Academic Press, London.

Edwards K.J. (2001) Environmental reconstruction. In: Brothwell D.R. and Pollard A.M. (eds) (2001) Handbook of archaeological sciences. John Wiley & Sons, Chichester. pp. 103–109.

Edwards H.G.M. (2005) Case study: prehistoric art. In: Edwards H.G.M. and Chalmers J.M. (eds) (2005) Raman spectroscopy in archaeology and art history. The Royal Society of Chemistry, Cambridge. pp. 84–96.

Edwards H.G.M., Farwell D.W., Holder J.M. and Lawson E.E. (1997) Fourier transform Raman spectra of ivory III: identification of mammalian specimens. Spectrochim. Acta A 53, 2403–2409.

Edwards H.G.M. and de Faria D.L.A. (2004) Infrared, Raman microscopy and fibre-optics Raman spectroscopy (FORS). In: Janssens K. and Van Grieken R. (ed.) Non destructive microanalysis of cultural heritage materials. Comprehensive analytical chemistry series. Vol. XLII. Elsevier, Amsterdam. pp. 359–396.

Edwards H.G.M. and Munshi T. (2005) Diagnostic Raman spectroscopy for the forensic detection of biomaterials and the preservation of cultural heritage. Analyt. Bioanalyt. Chem. 382, 1398–1406.

Edwards H.G.M., Jorge Villar S.E., Hassan N.F.N., Arya N., O'Connor S. and Charlton D.M. (2005) Ancient biodeterioration: an FT–Raman spectroscopic study of mammoth and elephant ivory. Anal. Bioanal. Chem. 383, 713–720.

Edwards H.G.M., Brody R.H., Hassan N.F.N., Farwell D.W. and O'Connor S. (2006) Identification of archaeological ivories using FT-Raman spectroscopy. Anal. Chim. Acta 559, 64–72.

Edwards D.D., Allen G.C., Ball R.J. and El-Turki A. (2006) Pozzolanic properties of glass fines in lime mortars. Adv. Appl. Ceram. 106, 309–313.

Edwards H.G.M., Farwell D.W. and Villar S.E.J. (2007) Raman microspectriscopic studies of amber resins with insect inclusions. Spectr. Acta A 68, 1089–1095.

Eerkens J.W. and Rosenthal J.S. (2004) Are obsidian subsources meaningful units of analysis?: temporal and spatial patterning of subsources in the Coso Volcanic Field, southeastern California. J. Archaeol. Sci. 31, 21–29.

Ehlers E.G. (1972) Intepretation of geological phase diagrams. W.H.Freeman & Co Ltd, San Francisco. pp. 280.

Eighmy J.L. and Sternberg R.S. (eds) (1990) Archaeomagnetic Dating. University of Arizona Press, Tucson.

El-Baz F. (1997) Space age archaeology. Scientific Amer., 277, 60–65.

Ellingham H.J.T. (1944) Reducibility of oxides and sulphides in metallurgical processes. J. Soc. Chem. Ind. London 63, 125–133.

Ellis P.R. (2000) Analysis of mortars (to include historic mortars) by differential thermal analysis. In: Bartos P., Groot C. and Hughes J.J. (eds) Proceedings of the International Rilem workshop on historic mortars: characteristics and tests. Paisley, Scotland, 12–14 May 1999. RILEM, Cachan, France. pp. 133–148.

Eluère C. and Mohen J.-P. (eds) (1991) Découverte du Métal. Picard, Paris.

Emery I. (1980) The primary structures of fabrics – an illustrated classification. The Textile Museum, Washington, DC.

Endlicher G. and Tillmann A. (1997) Lime plaster as an adhesive for hafting Eighteenth-Dynasty flint sickles from Tell el Dab'a, Eastern Nile delta (Egypt). Archaeometry 39, 333–342.

Engelbach R. (1923) The problem of the obelisks. From a study of the unfinished obelisk at Aswan. G.H. Doran Company, New York- T. Fisher Unwin, London.

Ericson J.E. (1982) A potential new chronometric technique: quartz hydration dating. Anthropol. Pap. 23, 299.

Ericson J.E., Dersch O. and Rauch F. (2004) Quartz hydration dating. J. Archaeol. Sci. 31, 883–902.

Espinoza E.O. and Mann M. (1992) Identification guide for ivory and ivory substitutes. 2nd Edition. World Wildlife Fund, Washington, DC.

Evans S.T. (2004) Ancient Mexico and Central America: Archaeology and culture history. London.

Evernden J.A. and Curtis G.H. (1965) Potassium-argon dating of of Late Cenozoic rocks in East Africa and Italy. Current Anthropology 6, 343–364.

Evershed R.P. (2008) Organic residue analysis in archaeology: the archaeological biomarker revolution. Archaeometry 50, 895–924.

Evershed R.P., Dudd S.N., Lockheart M.J., Jim S. (2001) Lipids in archaeology. In: Brothwell D.R. and Pollard A.M. (eds) Handbook of archaeological sciences. John Wiley & Sons, Chichester. pp. 332–349.

Evershed R.P., Berstan R., Grew F., Copley M.S., Charmant A.J.H., Barham E., Mottram H.R. and Brown G. (2004) Archaeology: Formulation of a Roman cosmetic. Nature 432, 35–36.

Farquharson M.J. and Brickley M. (2000) The use of X-ray techniques for bone densitometry in archaeological skeletons. In: Creagh D.C. and Bradley D.A. (eds) Radiation in art and archaeometry. Elsevier, Amsterdam. pp. 151–179.

Faure G. (1998) Principles and Applications of Geochemistry. 2nd edition. Prentice-Hall, Upper Saddle River, N. J.

Feder K.L. (2008) Frauds, myths, and mysteries: Science and pseudoscience in archaeology. 6th Edition. McGraw-Hill, New York. pp. 73–101.

Feller R.L. (1994) Accelerated aging. Photochemical and thermal aspects. The Getty Conservation Institute, Los Angeles.

Feller R.L. and Johnston-Feller R.M. (1978) Use of the International Standards Organization's blue-wool standards for exposure to light. Use as an integrating light monitor for illumination under museum conditions. In: American Institute for Conservation of Historic and Artistic Works. Preprints of papers presented at the 6th Annual Meeting. American Institute for Conservation, Washington. pp. 73–80.

Fermo P., Cariati F., Ballabio D., Consonni V. and Bagnasco Gianni G. (2004) Classification of ancient Etruscan ceramics using statistical multivariate analysis of data. Appl. Phys. A, 79, 299–307.

Ferretti M. (2000) X-ray fluorescence applications for the study and conservation of cultural heritage. In: Creagh D.A. and Bradley D.C. (eds) Radiation in art and archaeometry. Elsevier, Amsterdam.

Figueiredo M.O., Veiga J.P., Silva T.P., Mirão J.P. and Pascarelli S. (2005) Chemistry versus phase constitution of yellow ancient tile glazes: A non-destructive insight through XAS. Nucl. Instrum. Meth. Phys. Res. B 238, 134–137.

Figueiredo M.O., Silva T.P. and Veiga J.P. (2006) A XANES study of the structural role of lead in glazes from decorated tiles, XVI to XVIII century manufacture. Appl. Phys. A 83, 209–211.

Folk R.L. and Valastro S. (1976) Successful technique for dating of lime mortar by Carbon-14. J. Field Archaeol. 3, 203–208.

Forde M.C., Birjandi K.F. and Batchelor A.J. (1985) Fault detection in stone masonry bridges by non-destructive testing. Proc. of the 2nd International Conference Structural Faults and Repair. Engineering Technics Press, Edinburgh. pp. 373–379.

Fornasier M. and Toniolo D. (2005) Fast, robust and efficient 2D pattern recognition for re-assembling fragmented images. Pattern Recognition, 38, 2074–2087.

Fornasier M. (2006) Nonlinear Projection Recovery in Digital Inpainting for Color Image Restoration. J. Math. Imaging Vis., 24, 359–373.

Forte E. and Pipan M. (2008) Integrated seismic tomography and ground-penetrating radar (GPR) for the high-resolution study of burial mounds (tumuli). J. Arch. Science, 35, 2614–2623.

Fowler M.J (2002) Satellite remote sensing and archaeology: a comparative study of satellite imagery of the environs of Figsbury Ring, Wiltshire. Archaeological Prospection 9, 55–69.

Franzini M., Leoni L., Lezzerini M. and Sartori F. (2000) The mortar of the "Leaning Tower" of Pisa: the product of a medieval technique for preparing high-strength mortars. Eur. J. Mineral. 12, 1151–1163.

Fredrickx P., Hélary D., Schryvers D. and Darque-Ceretti E. (2004) A TEM study of nanoparticles in lustre glazes. Appl. Phys. A 79, 283–288.

Freestone I.C. (1994) Chemical analysis of "raw" glass fragments. In: Hurst H.R. (ed.) Excavation at Carthage, Volume II, The circular harbor, North side. Oxford University Press for the British Academ, Oxford. pp. 290.

Freestone I.C. (2001) Post-depositional changes in archaeological ceramics and glasses. In: Brothwell D.R. and Pollard A.M. (eds) Handbook of archaeological sciences. John Wiley & Sons, Chichester. pp. 615–625.

Freestone I.C. (2006) Glass production in Late Antiquity and the Early Islamic period: a geochemical perspective. Geol. Soc. London, Special Publications 2006, 257, 201–216.

Freestone I.C., Gorin Roses Y. and Hughes M.J. (2000) Primary glass from Israel and the production of glass in Late Antiquity and the Early Islamic period. In: Nenna M.D. (ed.) La route du verre. Traveaux de la Maison de l' Méditeranéan, Lyon. pp. 65–83.

Freeth T., Bitsakis Y., Moussas X., Seiradakis J.H., Tselikas A., Mangou H., Zafeiropoulou M., Hadland R., Bate D., Ramsey A., Allen M., Crawley A., Hockley P., Malzbender T., Gelb D., Ambrisco W. and Edminds M.G. (2006) Decoding the ancient

Greek astronomical calculator known as the Antikythera Mechanism. Nature, 444, 587–591.

Frey B.S. (2000) Arts and economics: Analysis and cultural policy. Springer-Verlag, Berlin.

Friedrich M., Remmele S., Kromer B., Hofmann J., Spurk M., Kaiser K.F., Orcel C. and Küppers M. (2004) The 12,460-year Hohenheim oak and pine tree-ring chronology from central Europe — A unique annual record for radiocarbon calibration and paleoenvironment reconstructions. Radiocarbon, 46, 1111–1122.

Friedman I. and Smith R.A. (1960) New method of dating using obsidian; Part 1, the development of the method. Am. Antiquity, 25, 476–522.

Frierman J.D. (1971) Lime burning as the precursor of fired ceramics. Israel Explor. J. 21, 212–216.

Fringeli P. (2000) ATR and reflectance IR spectroscopy, applications. In: Lindon J.C., Tranter G.E., Holmes J.L. (eds) Encyclopedia of spectroscopy and spectrometry. Vol. 1. Academic Press, London. pp. 58–75.

Frischer B., Ryan N.S., Niccolucci F. and Barceló J.A. (2002) From CVR to CVRO: the past, present and future of Cultural Virtual Reality. In: Virtual archaeology: proceedings of the VAST Euroconference, Arezzo 24–25 November 2000. Archaeopress, Pp. 7–18.

Fritts H.C. (2001) Tree rings and climate. The Blackburn Press, Caldwell. Pp. 567.

Frumkin A., Shimron A. and Rosenbaum J. (2003) Radiometri dating of the Siloam Tunnel, Jerusalem. Nature 425, 169–171.

Fuchs M. and Lang A. (2001) OSL dating of coarse-grain fluvial quartz using single-aliquot protocols on sediments from NE Peloponnese, Greece. Quat. Sci. Reviews 20, 783–787.

Fuchs M. and Wagner G.A. (2005) The chronostratigraphy and geoarchaeological significance of an alluvial geoarchive: Comparative OSL and AMS 14C dating from Greece. Archaeometry, 47, 849–860.

Fukushima E. and Roeder S.B.W. (1981) Experimental Pulse NMR. A Nuts and Bolts Approach. Addison-Wesley, Advanced Book Program, Massachusetts.

Furmánek V. and Pieta K. (1985) Počiatky odievania na Slovensku. Bratislava.

Gaetani M.C., Santamaria U. and Seccaroni C. (2004) The use of Egyptian blue and lapis lazuli in the Middle Ages: the wall paintings of the San Saba church in Rome. Studies in Conservation 49, 13–22.

Gaffney C. (2008) Detecting trends in the prediction of the buried past: A review of geophysical techniques in archaeology. Archaeometry, 50, 313–336.

Gaigalas A. and Halas S. (2009) Stable isotopes (H, C, S) and the origin of Baltic amber. Geochronom. 33, 11–14.

Gale N.H. (2009) A response to the paper of A.M. Pollard: What a long, strange trip it's been: lead isotopes and archaeology. In: Shortland A.J., Freestone I.C. and Rehren T. (eds) From mine to microscope. Advances in the study of ancient technology. Oxbow Books, Oxford. pp. 192–196.

Gale N.H., Woodhead A.P., Stos-Gale Z.A., Walder A. and Bowen I. (1999) Natural variations detected in the isotopic composition of copper: possible applications to archaeology and geochemistry. Int. J. Mass Spectr. 184, 1–9.

Gale N.H. and Stos-Gale Z.A. (2000) Lead isotope analyses applied to provenance studies. In: Ciliberto E. and Spoto G. (eds) Modern analytical methods in art and archaeometry. John Wiley & Sons, New York. pp. 503–584.

Galli A., Martini M., Montanari C. and Sibilia E. (2004) Thermally and optically stimulated luminescence of glass mosaic tesserae. Appl. Phys. A 79, 253–256.

Gallucci E., Scrivener K., Groso A., Stampanoni M. and Margaritondo G. (2007) 3D experimental investigation of the microstructure of cement pastes using synchrotron X-ray microtomography. Cem. Concr. Res. 37, 360–368.

Garbassi F. and Mello E. (1984) Surface spectroscopic studies on patinas of ancient metal objects. Studies Conserv. 29, 172–180.

Garbev K., Stemmermann P., Black L., Breen C., Yarwood J. and Gasharova B. (2007) Structural features of C-S-H(I) and its carbonation in air – A Raman spectroscopic study. Part I: Fresh phases. J. Amer. Ceram. Soc. 90, 900–907.

Garrels R.M. and Christ J.L. (1965) Solutions, minerals, and equilibria. Freeman Cooper, San Francisco.

Garrison E.G., McGimsey III C.R. and Zinke O.H. (1978) Alpha-recoil tracks in archeological ceramic dating. Archaeometry 20, 39–46.

Garrison E.G. (2001) Physics and archaeology. Physics Today, October 2001, 32–36.

Garrison E.G. (2003) Techniques in archaeological geology. Springer-Verlag, Berlin–New York. pp. 304.

Garrison T.G., Houston S.D., Golden C., Inomata T., Nelson Z. and Munson J. (2008) Evaluating the use of IKONOS satellite imagery in lowland Maya settlement archaeology. J. Arch. Sci., 35, 2770–2777.

Garstang J. (1936) Jericho: City and necropolis. Annals Archaeol. Anthropol., Liverpool 23, 67–100.

Gartner E.M., Young J.F., Damidot D.A. and Jawed I. (2002) Hydration of Portland cement. In: Bensted J. and Barnes P. (eds) Structure and performance of cements. 2nd Edition. Spon Press, London–New York. pp. 57–113.

Gat J.R. (1996) Oxygen and hydrogen isotopes in the hydrologic cycle. Ann. Rev. Earth Planet. Sci. 24, 225–262.

Gauri L.K., Parks L., Jaynes J. and Atlas R. (1992) Removal of sulphated-crust from marble using sulphate-reducing bacteria. In: Robin G.M. (ed.) Stone cleaning and the nature, soiling and decay mechanisms of stone. Proc. of the International Conference, 14–16 April 1992. Donhead Publishing Ltd., Webster, Edinburgh. pp. 160–165.

Gautier A. (2001) Luminescence dating of archaeometallurgical slag: use of the SAR technique for determination of the burial dose. Quaternary Science Reviews 20, 973–980.

Gaze M.E. and Crammond N.J. (2000) The formation of thaumasite in a cement:lime:sand mortar exposed to cold magnesium and potassium sulfate solutions. Cem. Concr. Composites 22, 209–222.

GCI-AIC Report (2002) Professional development for conservators in the United States. The Getty Conservation Center, Los Angeles – The American Institute for Conservation, Washington.

Geiger S.B. and Weiner S. (1993) Fluoridated carbonatoapatite in the intermediate layer between glass ionomer and dentin. Dental Materials 9, 33–36.

Gelichi S., Piuzzi F. and Cianciosi A. (2008) "Sachuidic presso Forni Superiore". Ricerche archeologiche in un castello della Carnia. Edizioni All' del Giglio sas, Borgo S. Lorenzo, Firenze.

Gershon N. (1994) From perception to visualisation. In: Rosenblum L. *et al.* (eds) Scientific Visualisation. Advances and Challenges. pp. 129–139. New York, Academic Press.

Gerster G. and Trumpler C. (2005) The past from above: Aereial photographs of archaeological sites. J. Paul Getty Museum Series. Getty Trust Publications, Los Angeles.

Gettens R.J. (1962) Maya blue: An unresolved problem in ancient pigments. Ameri. Antiquity 27, 557–564.

Gettens R.J. and Stout G.L. (1942) Painting materials. A short encylopaedia. D. Van Nostrand Company, Princeton, NJ. [reprinted 1966: Dover Publications, New York]

Ghiurca V. and Vavra N. (1990) Occurrence and chemical characterization of fossil resin from Colți (District of Bazu, Romania). Jb. Geol. Paläont. Mh., H5, 283–294.

Giacovazzo C. (ed.) (2002) Fundamentals of crystallography. IUCr Texts on Crystallography, no. 7. Oxford University Press, Oxford.

Gianolla P., Ragazzi E. and Roghi G. (1998) Upper Triassic amber from the Dolomites (Northern Italy). A paleoclimatic indicator? Riv. It. Paleontol. Stratigr. 104, 381–390.

Gibson L.J., Ashby M.F., Karam G.N., Wegst U. and Shercliff H.R. (1995) The mechanical properties of natural materials. II. Microstructures for Mechanical Efficiency. Proc. R. Soc. Lond. A 450, 141–162.

Gilbert M.T.P., Bandelt H.-J., Hofreiter M. and Barnes I. (2005) Assessing ancient DNA studies. Trends Ecol. Evol. 20, 541–544.

Gilboa A., Karasik A., Sharon I. and Smilansky U. (2004) Towards computerized typology and classification of ceramics. J. Archaeol. Sci., 31, 681–694.

Gillies K. and Urch D. (1985) XPS depth profile analysis of the surface decoration on a Pre-Columbian gilded gold copper alloy pectoral. Hist. Metall. 19, 176–185.

Giuffrè A. and Carocci C. (1999) Codice di pratica per la sicurezza e la conservazione del centro storico di Palermo. Laterza, Bari.

Giuli G., Paris E., Pratesi G., Koeberl C. and Cipriani C. (2003) Iron ixidation state in the Fe-rich layer and silica matrix of Libyan Desert Glass: A high-resolution XANES study. Meteor. Planet. Sci. 38, 1181–1186.

Giuliani G., Chaussidon M., Schubnel H.-J., Piat D.H., Rollion-Bard C., France-Lanord C., Giard D., de Narvaez D. and Rondeau B. (2000) Oxygen isotopes and emerald trade routes since antiquity. Science 287, 631–633.

Giuliano M., Mille G., Onoratini G. and Simon P. (2006) Presence of amber in the Upper Cretaceous (Santonian) of La 'Mède' (Martigues, Southeastern France). IRTF characterization. Polavol 5, 851–858.

Giumlia-Mair A. (1996) Das krokodil und Amenemhat III aus el-Faiyum. Antike Welt 4, 313–340.

Giumlia-Mair A. (2005) On surface analysis and archaeometallurgy. Nucl. Instrum. Meth. Phys. Res. B 239, 35–43.

Giuntini L., Lucarelli F., Mandò P.A., Hooper W. and Barker P.H. (1995) Galileo's Writings: Chronology by PIXE. Nucl. Instr. & Meth., B95, 389–392.

Glascock M.D. (1994) New World obsidian: Recent investigation. In: Scott D.A. and Meyers P. (eds) Archaeometry of pre-Columbian sites and artifacts. The Getty Conservation Institute, Los Angeles, CA. pp. 113–134.

Glascock M. (ed) (2002) Geochemical evidence for long-distance exchange. Bergin & Garvey, Greenwood Publishing Group, Westport. pp. 282.

Glascock M.D. and Neff H. (2003) Neutron activation analysis and provenance research in archaeology. Meas. Sci. Technol. 14, 1516–1526.

Glascock M.D., Neff H. and Vaughn K.J. (2004) Instrumental neutron activation analysis and multivariate statistics for pottery provenance. Hyperfine Inter., 154, 95–105.

Glascock M.D., Speakman R.J. and Neff H. (2007) Archaeometry at the University of Missouri Research Reactor and the provenance of obsidian artefacts in North America. Archaeometry 49, 343–358.

Glascock M.D., Speakman R.J. and Burger R.L. (2007) Sources of archeological obsidian in Peru: Descriptions and geochemistry. In: Glascock M.D., Speakman R.J. and

Popelka-Filcoff R.S. (eds) Archaeological chemistry. Analytical techniques and archaeological interpretation. ACS Symposium Series, Vol. 968. American Chemical Society, Washington DC. pp. 522–552.

Gleba M. (2008) Textile production in pre-Roman Italy. Ancient Textile Series Vol. 4. Center for Textile Research–Oxbow Books. pp. 292.

Gnoli R. (1971) Marmora romana. ICCROM, Rome. pp. 253.

Goffer Z. (1980) Archaeological chemistry. John Wiley & Sons, New York. 2nd Edition (2007). Wiley-Interscience, New York.

Goldberg P. and Bar-Yosef O. (1998) Site formation processes in Kebara and Hayonim caves and their significance in Levantine prehistoric caves. In: Akazawa T., Aoki K. and Bar-Yosef O. (eds) Neandertals and modern humans in Western Asia. Springer-Verlag, Berlin–New York. pp. 107–126.

Good I. (2001) Archaeological textiles: A review of current research. Annu. Rev. Anthropol. 30, 209–226.

Goossens R., De Wulf A., Bourgeois J., Gheyle W. and Willems T. (2006) Satellite imagery and archaeology: the example of CORONA in the Altai Mountains. J. Arch. Science, 33, 745–755.

Gosden C. (1994) Social being and time. Cambridge University Press, Cambridge.

Gose W.A. (2000) Palaeomagnetic studies of burned rocks. Journal of Archaeological Science 27, 409–421.

Gosselain O.P. (1992) The bonfire of the enquiries. Pottery firing temperatures: what for? J. Archaeol. Sci. 19, 243–259.

Goudie A.S. and Viles H. (1997) Salt weathering hazards. John Wiley & Sons, Chichester. pp. 241.

Gould S.J. (1990) Time's arrow, time's cycle: Myth and metaphor in the discovery of geological time. Harvard University Press, Harvard.

Gourdin W.H. and Kingery W.D. (1975) The beginning of pyrotechnology: Neolithic and Egyptian limeplaster. J. Field Archaeol. 2, 133–150.

Grabner M., Klein A., Geihofer D., Reschreiter H., Barth F.E., Sormaz T. and Wimmer R. (2007) Bronze age dating of timber from the salt-mine at Hallstatt, Austria. Dendrochronologia, 24, 61–68.

Gramsch B. (1992) Friesack Mesolitic wetlands. In: Coles B. (ed.) The wetland revolution in prehistory. WARP, Exeter-The Prehistoric Society, London. pp. 65–72.

Graves-Brown P., Siân J. and Gamble C. (eds) (1996) Cultural identity and archaeology. Routledge, London.

Greffe X. (2004) Is heritage an asset or a liability? J. Cult. Herit. 5, 301–309.

Greilich S., Glasmacher U.A. and Wagner G.A. (2005) Optical dating of granitic stone surfaces. Archaeometry, 47, 645–665.

Greilich S. and Wagner G.A. (2006) Development of a spatially resolved dating technique using HR-OSL. Radiation Meas. 41, 738–743.

Griffin P.S., Grissom C.A. and Rollefson G.O. (1998) Three late eight millennium plastered faces from 'Ain Ghazal, Jordan. Paléorient 24, 59–70.

Griffiths H. (ed) (1998) Stable isotopes. Integration of biological, ecological and geochemical processes. Bios Scientific Publishers, Oxford.

Grimaldi D.A. (1996) Amber: window to the past. Harry N. Abrams Inc., New York.

Grimaldi D.A., Shedrinsky A., Ross A. and Baer N.S. (1994) Forgeries of fossils in "amber": history, identification, and case studies. Curator 37, 251–275.

Gruen A., Remondino F. and Zhang L. (2002) Reconstruction of the Great Buddha of Bamiyan, Afghanistan. International Archives of Photogrammetry and Remote Sensing, Vol. XXXIV, part 5, pp. 363–368, Corfu, Greece.

Grün R. (1993) Electron spin resonance dating in palaeoanthorpology. Evol. Anthropol., 2, 172–181.

Grün R. and Stringer C.B. (1991) Electron spin resonance dating and the evolution of modern humans. Archaeometry 33, 153–199.

Grupe G. (2001) Archaeological microbiology. In: Brothwell D.R. and Pollard A.M. (eds) Handbook of archaeological sciences. John Wiley & Sons, Chichester. pp. 351–358.

Grzywacz C., Bomin S., Yuquan F. and Wouters J. (2008) Development of identification strategies for Asian organic colorants on historic Chinese wall paintings. ICOM-Committee for Conservation, 15th Triennial Conference, New Delhi. pp. 846–853.

Gualtieri A.F., Viani A. and Montanari C. (2006) Quantitative phase analysis of hydraulic limes using the Rietveld method. Cem. Concr. Res. 36, 401–406.

Guerra M.F. (2004) Fingerprinting ancient gold with proton beams of different energy. Nucl. Instr. Meth. Phys. Res. B 226, 185–198.

Guerra M.F. and Calligaro T. (2004) The analysis of gold: manufacture technologies and provenance of the metal. Meas. Sci. Techn. 14, 1527–1537.

Gunneweg J. and Balla M. (2003) Neutron Activation Analysis: Scroll jars and Common ware. In: Humbert J.-B. and Gunneweg J. (eds) Khirbet Qumrân et 'Aïn Feshkha II, Studies of Anthropology, Physics and Chemistry. Academic Press, Fribourg, Vandenhoeck & Ruprecht, Göttingen. pp. 3–57.

Guo S.L., Liu S.S., Sun F.S., Zhang F., Zhou S.H., Hao X.H., Hu R.Y., Meng W., Zhang P.F. and Liu J.F. (1991) Age and duration of Peking man site by fission track method. Int. J. Rad. Appl. Instrum. Part D. Nucl. Tracks Rad. Meas. 19, 719–724.

Habfast K. (1998) Fractionation correction and multiple collectors in thermal ionisation isotope ratio mass spectrometry. Int. J. Mass Spectr. 176, 133–148.

Habu J. (2004) Ancient Jomon of Japan. Cambridge University Press, Cambridge. pp. 332.

Hain M., Bartl J. and Jacko V. (2003) Multispectral analysis of cultural heritage materials. Meas. Sci. Rev. 3, 9–12.

Hale J., Heinemeier J., Lancaster L., Lindroos A. and Ringbom Å. (2003) Dating ancient mortar. Amer. Sci. 91, 130–137.

Hall C. and Hoff W. (2002) Water transport in brick, stone and concrete. Spon Press, London – New York. pp. 320.

Hammer C.U., Clausen H.B., Friedrich W.L. and Tauber H. (1987) The Minoan eruption of Santorini in Greece dated to 1645 BC? Nature, 328, 517–519.

Hansen E.F., Rodriguez-Navarro C. and Hansen R.D.(1997) Incipient Maya burnt-lime technology: Characterization and chronological variations in Preclassic plaster, stucco and mortar at Nakbe, Guatemala. In: Vandiver P.B., Druzik J.R., Merkel J.F. and Stewart J. (eds) Materials issues in art and archaeology V. Materials Research Society, Pittsburgh. pp. 207–216.

Harbottle G. (1990) Neutron activation analysis in archaeological chemistry. In: Yoshihara K. (ed.) Chemical applications of nuclear probes. Topics Curr. Chem. 157, 57–91.

Harbottle G. and Weigand P.C. (1992) Turquoise in pre-Columbian America. Sci. Amer. 266, 78–85.

Harbottle G. and Holmes L. (2007) The history of the Brookhaven National Laboratory project in archaeological chemistry, and applying nuclear methods to the fine arts. Archaeometry 49, 185–199.

Harland W.B., Armstrong R.L., Cox A.V., Craig L.E., Smith A.G. and Smith, D.G. (1990) A geologic time scale, 1989 edition. Cambridge University Press, Cambridge.

Harrell J.A. and Penrod B.E. (1993) The Great Pyramid debate: evidence from the Lauer sample. J. Geol. Education 41, 358–363.

Harrington C.D and Whitney J.W. (1987) Scanning electron microscope method for rock-varnish dating. Geology 15, 967–970.

Harris E.C. (1979) Principles of archaeological stratigraphy. Academic Press. London.

Harry K.G. (1995) Cation-ratio dating of varnished artefacts: testing the assumptions. Amer. Antiquity 60, 118–130.

Hatcher H., Tite M.S. and Walsh J.N. (1995) A comparison of inductively-coupled plasma emission spectrometry and atomic absorption spectrometry analysis on standard reference silicate materials and ceramics. Archaeometry, 37, 83–94.

Hatherly M. and Hutchison W.B. (1979) An introduction to textures in metals. The Institute of Metals, London.

Hauptmann A. (2007) The archaeometallurgy of copper. Evidence from Faynan, Jordan. Springer, Berlin – New York.

Hauptmann A., Busz R., Klein S., Vettel A. and Werthmann R. (2001) The roots of glazing techniques: Copper metallurgy? Paléorient 26, 113–130.

Hauptmann A. and Yalcin Ü. (2001) Lime plaster, cement and the first puzzolanic reaction. Paléorient 26, 61–68.

Hauptmann A. and Wagner I. (2007) Prehistoric copper production at Timna: thermoluminescence (TL) dating and evidence from the East. In: La Niece S., Hook D. and Craddock P. (eds) metals and mines. Studies in archaeometallurgy. The British Museum – Archetype Publications, London. pp. 67–75.

Haustein M., Roewer G., Krbetschek M.R. and Pernicka E. (2003) Dating archaeometallurgical slags using thermoluminescence. Archaeometry 45, 519–530.

Hawkesworth C.J. and van Calsteren P. (1992) Geological time. In: Brown G.C., Hawkesworth C.J. and Wilson R.C.L. (eds) Understanding the Earth. Cambridge University Press, Cambridge. Chapter 7.

Heaney P.J., Prewitt C.T. and Gibbs G.V. (eds) (1994) Silica. Physical behavior, geochemistry, and materials applications. Rev. Mineral. Vol. 29. The Mineralogical Society of America, Washington, DC.

Hedges R.E.M. (2001) Dating in archaeology; Past, present and future. In: Brothwell D.R. and Pollard A.M. (eds) (2001) Handbook of archaeological sciences. John Wiley & Sons, Chichester. pp. 3–8.

Hedges R.E.M. (2002) Bone diagenesis: An overview of processes. Archaeometry 44 319–328.

Hedges R.E.M. and Reynard L.M. (2007) Nitrogen isotopes and the trophic level of hmans in archaeology. J. Archaeol. Sci. 34, 1240–1251.

Heginbotham A., Millay V. and Quick M. (2006) The use of immunofluorescence microscopy (IFM) and enzyme-linked immunosorbent assay (ELISA) as complementary techniques for protein identification in artists' materials. J. Amer. Inst. Conservation 45, 89–105.

Heide K., Hartmann E., Gerta K. and Wiedemann H.G. (2000) MS-TGA of ancient glasses: an attempt to determine the manufacturing conditions (I). Thermochimica Acta, 365, 147–156.

Heimann R.B. (1986) Nuclear fuel waste management and archaeology: are ancient glasses indicators of long term durability of man made materials? Glass Technol. 27, 96–101.

Heimann R.B. (1989) Assessing the technology of ancient pottery: The use of ceramic phase diagrams. Archaeomaterials 3, 123–148.

Heinemeier J., Jungner H., Lindroos A., Ringbom A., Von Konow T. and Rud N. (1997) AMS 14C dating of lime mortar. Nucl. Inst. Meth. Phys. Res. Section B 123, 487–495.

Henderson J. (1985) The raw materials of early glass production. Oxford J. Archaeol. 4, 267–291.

Henderson J. (1988) Glass production and Bronze Age Europe. Antiquity 62, 435–451.

Henderson J. (1998) Post-Medieval glass: production, characterization and value. In: Kingery W.D. and McCray P. (eds) The prehistory and history of glass making technology. Ceramics and civilization, Vol. VIII. American Ceramic Society, Columbus, OH. pp. 33–59.

Henderson J. (2000) The science and archaeology of materials. Routledge, London.

Henderson J. (2001) Glass and glazes. In: Brothwell D.R. and Pollard A.M. (eds) Handbook of archaeological sciences. John Wiley & Sons, Chichester. pp. 471–482.

Henderson J. (2002) Tradition and experiment in first millennium A.D. glass production. The emergence of Early Islamic glass technology in Late Antiquity. Accounts of Chemical Research 35, 594–602.

Henderson J., Evans J. and Barkoudah Y. (2009) The roots of provenance: glass, plants and isotopes in the Islamic Middle East. Antiquity 83, 414–429.

Henry A. (ed.) (2006) Stone conservation principles and practice. Donhead Publishing Ltd., Shaftesbury, UK. pp. 352.

Hermes A.C., Davies R.J., Greiff S., Kutzke H., Lahlil S., Wyeth P. and Riekel C. (2006) Characterizing the decay of ancient Chinese silk fabrics by microbeam synchrotron radiation diffraction. Biomacromol. 7, 777–783.

Herrmann G. (1968) Lapis lazuli: the early phases of its trade. Iraq 30, 21–57.

Herz N. (1992) Provenance determination of Neolithic to Classical Mediterranean marbles by stable isotopes. Archaeometry 34, 185–194.

Herz N. and Garrison E.G. (1998) Geological methods for archaeology. Oxford University Press, Oxford. pp. 343.

Herz N. and Waelkens M. (1988) Classical marble: geochemistry, technology, trade. Proceedings of the NATO Advanced Research Workshop on Marble in Ancient Greece and Rome: Geology, Quarries, Commerce, Artifacts, Il Ciocco, Lucca, Italy, May 9–13, 1988. NATO ASI Series E, Vol. 153. Dordrecht, London–Boston. pp. 482.

Hill S.J. and Fisher A.S. (2000) Atomic absorption, methods and instrumentation. In: Lindon J.C., Tranter G.E. and Holmes J.L. (eds) Encyclopedia of Spectroscopy and Spectrometry, Vol. 1, pp. 24–32. Academic Press, London.

Hill L.L. (2006) Georeferencing. The MIT Press.

Hillam J. (1998) Dendrochronology: guidelines on producing and interpreting dendrochronological dates. Ancient Monuments Laboratory, Conservation and Technology, English Heritage, London. pp. 35.

Hochleitner B., Desnica V., Mantler M. and Schreiner M. (2003) Historical pigments: a collection analyzed with X-ray diffraction analysis and X-ray fluorescence analysis in order to create a database. Spectrochim. Acta Part B 58, 641–649.

Hodge H.C. and McKay J.H. (1934) The "microhardness" of minerals comprising the Mohs scale. Amer. Mineral. 19, 161–168.

Hodkins G. and Hedges R. (1999) A systematic investigation of the immunological detection of collagen-based adhesives. 6th International Conference on Non-Destructive Testing and Microanalysis for the diagnostics and Conservation of the Cultural and Environmental Heritage, Rome. pp. 1795–1810.

Hoefs J. (2008) Stable isotope geochemistry. 6th Edition. Springer, Berlin.

Holland J.G. and Tanner S.D. (eds) (2001) Plasma source mass spectrometry. Royal Society of Chemistry Special Publication, Cambridge.

Holloway D.G. (1973) The physical properties of glass. Wykeham Pub., London and Winchester. pp. 220.

Hong S., Candelone J.P., Patterson C.C. and Boutron C.F. (1996) History of ancient copper smelting pollution during Roman and Medieval times recorded in Greenland ice. Science, 272, 246–249.

Hopfenberg H.B., Witchey L.C. and Poinar G.O. Jr. (1988) Is the air in amber ancient? Science 241, 717–724.

Höppner B., Bartelheim M., Huijsmans M., Krauss R., Martinek K.-P., Pernicka E. and Schwab R. (2005) Prehistoric copper production in the Inn Valley (Austria), and the earliest copper in Central Europe. Archaeometry 47, 293–315.

Huang K.Z. (2003) Study and conservation of cultural architecture materials. Southeast Cult. 173, 93–96.

Huang Q.-X., Flöy S., Gelfand N., Hofer M. and Pottmann H. (2006) Reassembling fractured objects by geometric matching. ACM Trans. Graph., 25, 569–578.

Hughes M.J., Cowell M.R. and Craddock P.T. (1976) Atomic absorption techniques in archaeometry. Archaeometry, 18, 19–37.

Hughes M.J., Cowell M.R. and Hook D.R. (eds) (1991) Neutron activation and plasma emission spectrometric analysis in archaeology. Techniques and applications. British Museum Occasional Papers 82. British Museum Press, London.

Hughes J.J., Leslie A.B. and Callebaut K. (2001) The petrography of lime inclusions in historic lime based mortars. In: Stamatakis M., Georgali B., Fragoulis D. and Toumbakari E.E. (eds) Proceedings of the 8th Euroseminar on microscopy applied to buildings materials. Athens. pp. 359–364.

Hulbert J.F., Panish P.T. and Prostak K.S. (1996) Chemistry, microstructure, petrology, and diagenetic model of Jurassic dinosaur bones, Dinosaur National Monument, Utah. J. Sedim. Res. 66, 531–547.

Hunt R.W. (1998) Measuring colour. 3rd Edition. Fountain Press, England.

Houston S.D. (ed.) (1998) Function and meaning in classic Maya architecture: a symposium at Dumbarton Oaks, 7–8 October 1994. Published by Dumbarton Oaks. pp. 562.

Huansheng C., Wenquan H., Jiayong T., Fujita Y. and Jianhua W. (1996) PIXE analysis of ancient chinese Qing dynasty porcelain. Nucl. Inst. Meth. Phys. Res. B 118, 377–381.

Hubin A. and Terryn H. (2004) X-ray photoelectron and Auger electron spectroscopies. In: Janssens K. and Van Grieken R. (ed.) (2004) Non-destructive microanalysis of cultural heritage materials. Comprehensive analytical chemistry series. Vol. XLII. Elsevier, Amsterdam. pp. 277–312.

Hurlbut C.S. and Kammerling R.C. (1991) Gemology. John Wiley, New York. pp. 336.

Huysmans J.-K. (1884) À rebours. Charpentier, Paris.

Hyman D.S. (1970) Precolumbian cements: A study of the calcareous cements in Prehispanic Mesoamerican building construction. Ph.D. dissertation, Johns Hopkins University, Baltimore, MD.

Ingram K.D., Daugherty K.E. and Marshall J.L. (1993) The Pyramids – cement or stone. J. Archaeol. Sci. 20, 681–687.

Ingo G.M., Bultrini G., De Caro T. and Del Vais C. (2000) Microchemical study of the black gloss on red- and black-figured Attic vases. Surf. Interf. Anal. 30, 101–105.

Ingo G.M., Angelini E., De Caro T., Bultrini G. and Mezzi A. (2004) Combined use of XPS and SEM + EDS for the study of surface microchemical structure of archaeological bronze Roman mirrors. Surf. Interf. Anal. 36, 871–875.

Ingo G.M., Angelini E., Bultrini G., de Caro T., Pandolfi L. and Mezzi A. (2004) Combined use of surface and micro-analytical techniques for the study of ancient coins. Appl. Phys. A 79, 171–176.

Innes J.L. (1985) Lichenometry. Progress in Physical Geography, 9, 187–245.

Ioannidou E., Bourgarit D., Calligaro T., Dran J.-C., Dubus M., Salomon J. and Walter P. (2000) RBS and NRA with external beams for archaeometric applications. Nucl. Instr. Meth. Phys. Res. B 161–163, 730–736.

Israeli Y. (1991) The invention of blowing. In: Newby M. and Painter K. (eds) Roman glass: two centuries of art and innovation. Soc. Antiq. London Occas. Pap. 13, 46–55.

Ivanovich M.I. (1994) Uranium series disequilibrium: concepts and applications. Radiochimica Acta, 64, 81–94.

Ivanovich M.I. and Harmon S. (1992) Uranium-series disequilibrium: application to Earth, marine, and environmental sciences. Clarendon Press, Oxford.

Ixer R.A. (1999) The role of ore geology and ores in the archaeological provenancing of metals. In: Young S.M.M., Pollard A.M., Budd P. and Ixer R.A. (eds) Metals in antiquity. BAR International Series no. 792. Archaeopress, Oxford, UK. pp. 43–52.

Ixer R.A. and Pattrick R.A.D. (2003) Copper-arsenic ores and Bronze Age mining and metallurgy with special reference to the British Isles. In: Craddock P. and Lang J. (eds) Mining and metal production through the ages. The British Museum Press, London. pp. 9–20.

Jackson M. and Marra F. (2006) Roman stone masonry: Volcanic foundations of the ancient city. Amer. J. Archaeol. 110, 403–436.

Jackson M., Marra F., Deocampo D., Vella A., Kosso C. and Hay R. (2007) Geological observations of excavated sand (*harenae fossiciae*) used as fine aggregate in Roman pozzolanic mortars. J. Roman Archaeol. 20, 25–52.

Jackson C.M. (2005) Glassmaking in Bronze-Age Egypt. Science 308, 1750–1752.

Jackson C.M., Booth C.A. and Smedley J.W. (2005) Glass by design? Raw materials, recipes and compositional data. Archaeometry 47, 781–795.

Jackson C.M. and Wager E. (eds) (2008) Vitreous materials in the Late Bronze Age Aegean: a window to the East Mediterranean world. Sheffield Studies in Aegean Archaeology No. 9, Oxbow Books, Oxford.

Jacobs J.A. (1995) The Earth's magnetic field and reversals. Endeavour, 19, 166–171.

Jakes K.A., Baldia C.M. and Thompson A.J. (2007) Infrared examination of fiber and particulate residues from archaeological textiles. In: Glascock M.D., Speakman R.J. and Popelka-Filcoff R.S. (eds) Archaeological chemistry. Analytical techniques and archaeological interpretation. ACS Symposium Series, Vol. 968. American Chemical Society, Washington DC. pp. 44–77.

James C. (2002) The book of alternative photographic processes. Delmar, Albany.

Jamieson D.N. (2001) New generation of nuclear microprobe systems. Nucl. Instr. Meth. Phys. Res. B 181, 1–11.

Jangl A. and Jangl J. (1989) Ancient legends of gems and jewels. Prisma Press, Coeur d'Alene.

Janssens K. (2004) X-ray based methods of analysis. In: Janssens K. and Van Grieken R. (ed.) Non-destructive microanalysis of cultural heritage materials. Comprehensive analytical chemistry series. Vol. XLII. Elsevier, Amsterdam. pp. 129–226.

Janssens K. and Van Grieken R. (ed.) (2004) Non-destructive microanalysis of cultural heritage materials. Comprehensive analytical chemistry series. Vol. XLII. Elsevier, Amsterdam.

Jenkins R. (1999) X-ray Fluorescence Spectrometry. 2nd Edition. Wiley-Interscience.

Jenkins L.H. and Chung M.F. (1970) Auger electron energies of the outer shell electrons. Surf. Sci. 22, 479–485.

Jerz J.K. and Rimstidt J.D. (2004) Pyrite oxidation in moist air. Geochim. Cosmochim. Acta 68, 701–714.

Jespersen K. (1989) Precipitation of iron-corrosion products on peg-treated wood. In: MacLeod I.D. and Grattan D.W. (eds) Conservation of wet wood and metal. Western Australian museum, Perth. pp. 141–152.

Jianjun M. and Yanxiang L. (2003) Early copper technology in Xinjiang, China: the evidence so far. In: Craddock P. and Lang J. (eds) Mining and metal production through the ages. The British Museum Press, London. pp. 111–121.

Johansen G.A. (2005) Nuclear tomography methods in industry. Nucl. Phys. A, 752, 696c–705c.
Johansson S.A.E. (1989) PIXE : A novel technique for elemental analysis. Endeavour 13, 48–53.
Johansson S.A.E., Campbell J.L. and Malmqvist K.G. (1995) Particle-induced X-ray emission spectrometry (PIXE) Chemical Analysis: A Series of Monographs on Analytical Chemistry and Its Applications. Wiley Interscience. John Wiley & Sons, New York.
Johnsson K. (1997) Chemical dating of bones based on diagenetic changes in bone apatite. Journal of Archaeological Science, 24, 431–437.
Johnson M. and Packard E. (1971) Methods used for the identification of binding media in Italian paintings of the fifteenth and sixteenth centuries. Stud. Conservation 16, 145–164.
Johnson W.S., Rice M. and Williams C. (2000) Photography from 1839 to today: George Eastman House, Rochester, NY. Taschen, New York.
Joichi A., Yomogida K., Awano K. and Ueda Y. (2005) Volatile components of tea-scented modern roses and ancient Chinese roses. Flavour Fragrance J. 20, 152–157.
Jokilehto J. (2002) A history of architectural conservation. Butterworth-Heinemann, Oxford. pp. 386.
Jones A. (2004) Archaeometry and materiality: Materials-based analysis in theory and practice. Archaeometry, 3, 327–338.
Jones M., Craddock P. and Barker N. (eds) (1990) FAKE? The art of deception. British Museum, London.
Jones P.L. (1962) Some observation of methods for identifying proteins in paint media. Stud. Conservation 7, 10–16.
Julia Felix (2000) Operazione Julia Felix. Dal mare al museo. Edizioni della Laguna, Mariano del Friuli, Gorizia. pp. 144.
Juniper A. (2003) Wabi Sabi: The Japanese art of impermanence. Tuttle Publishing, North Clarendon, Vermont.
Junghans S., Sangmeister E. and Schröder M. (1968–1974) Kupfer und bronze in der frühen metallzeit Europas. Studien zu den Anfängen der Metallurgie, 1–3, 4. Berlin.
Junk S.A. (2001) Ancient artefacts and modern analytical techniques – Usefulness of laser ablation ICP-MS demonstrated with ancient gold coins. Nucl. Instr. Meth. Phys. Res. B 181, 723–727.
Junk S.A. and Pernicka E. (2003) An assessment of osmium isotope ratios as a new tool to determine the provenance of gold with platinum-group metal inclusions. Archaeometry 45, 313–331.
Kakoulli I. (2009) Egyptian blue in Greek painting between 2500 and 50 BC. In: Shortland A.J., Freestone I.C. and Rehren T. (eds) From mine to microscope. Advances in the study of ancient technology. Oxbow Books, Oxford. pp. 79–87.
Kantner J. (2000) Realism vs. reality: creating virtual reconstructions of prehistoric architecture. In: Barceló J.A., Forte M. and Sanders D.H. (2000) Virtual reality in archaeology. British Archaeological Reports, International Series no. 843. Archeo-Press, Oxford.
Kaplan A. (2006) Baryta paper musical chairs: where does each element sit? Glass Eye Productions, DVD distributed by the Getty Conservation Institute, Los Angeles. [videorecording, 2 DVDs (4hrs. 26min.)]
Kaplan A. and Stulik D.C. (2008) Expanding the barytome database: the GCI Gevaert collection. In: 15th triennial conference, preprints/ICOM Committee for Conservation, New Delhi, India, September 22–26, 2008. pp. 691–695.
Karatzani A. and Rehren T. (2009) Clothes of gold: metal threads in Bizantine–Greek Orthodox ecclesiastical textiles. In: Moreau J.-F., Auger R., Chabot J. and Herzog A.

(eds) Proc. 36th Intern. Symp. Archaeom. ISA 2006. 2–6 May 2006, Quebec City, Canada. Cahiers d' du CELAT no. 25. Série archéometrie no. 7. CELAT, Université Laval, Québec. pp. 444-9–444-19.

Kardjilov N., Lo Celso F., Donato D.I., Hilger A. and Triolo R. (2007) Applied neutron tomography in modern archaeology. Nuovo Cimento, 30, 79–83.

Kardulias P.N. and Yerkes R.W. (eds) (2003) Written in stone: the multiple dimensions of lithic analysis. Lexington Books, Lexington, MA.

Karkanas P. (2007) Identification of lime plaster in prehistory using petrographic methods: A review and reconsideration of the data on the basis of experimental and case studies. Geoarchaeology: An Intern. J. 22, 775–796.

Karkanas P., Bar-Yosef O., Goldberg P. and Weiner S. (2000) Diagenesis in prehistoric caves: the use of minerals that form in situ to assess the completeness of the archaeological record. J. Archaeol. Sci. 27, 915–929.

Karman M. and Knowles A. (eds) (2002) Past Time, Past Place: GIS for History. ESRI Press.

Kassianidou V. and Knapp A.B. (2005) Archaeometallurgy in the Mediterranean: The social context of mining, technology, and trade. In: Blake E. and Knapp A.B. (eds) The archaeology of Mediterranean prehistory. Wiley-Blackwell, London. pp. 215–251.

Kasztovszky Zs., Visser D., Kockelmann W., Pantos E., Brown A., Blaauw M., Hallebeek P., Veerkamp J., Krook W. and Stuchfield H.M. (2007) Combined prompt gamma activation and neutron diffraction analyses of historic metal objects and limestone samples. Nuovo Cimento, 30C, 67–78.

Kautenburger R., Wannemacher J. and Müller P. (2004) Multi element analysis by X-ray fluorescence: A powerful tool of ivory identification from various origins. J. Radioanal. Nucl. Chem. 260, 399–404.

Kawamura K., and 17 others (2007) Northern Hemisphere forcing of climatic cycles in Antarctica over the past 360,000 years. Nature, 448, 912–916.

Keisch B., Feller R.L., Levine A.S. and Edwards R.R. (1967) Dating and authenticating works of art by measurement of natural alpha emitters. Science 155, 1238–1242.

Kempe D.R.C. and Harvey A.P. (eds) (1983) The petrology of archaeological artefacts. The Clarendon Press, Oxford.

Khanjian H.P. and Stulik D.C. (2003) Infrared spectroscopic studies of photographic material. In: Conservation Science 2002: papers from the conference held in Edinburgh, Scotland, May 22–24, 2008. Edinburgh, Scotland, pp. 195–200.

Khoury H.N., Salameh E., Clark I.D., Fritz P., Bajjali W., Milodowski A.E., Cave M.R. and Alexander W.R. (1992) A natural analogue of high pH cement pore waters from the Maqarin area of northern Jordan. I: introduction to the site. J. Geochem. Explor. 46, 117–132.

Kienlin T.L., Bischoff E. and Opielka H. (2006) Copper and bronze during the Eneolithic and Early Bronze Age: A metallographic examination of axes from the Northalpine region. Archaeometry 48, 453–468.

Kilikoglou V., Vekinis G. and Maniatis Y. (1995) Toughening of ceramic earthenware by quartz inclusions: an ancient art revisited. Acta Metall. Mater. 43, 2959–2965.

Kilikoglou V., Bassiakos Y., Doonan R.C. and Stratis J. (1997) NAA and ICP analysis of obsidian from Central Europe and the Aegean: source characterisation and provenance determination. J. Radioanal. Nuclear Chem. 216, 87–93.

Killick D. (2001) Science, speculation and the origins of extractive metallurgy. In: Brothwell D.R. and Pollard A.M. (eds) Handbook of archaeological sciences. John Wiley & Sons, Chichester. pp. 483–492.

Killick D. (2004) Social constructionist approaches to the study of technology. World. Archaeol. 36, 571–578.

Killick D., van der Merwe N.J., Gordon R.B. and Grébénart D. (1988) Reassessment of the evidence for early metallurgy in Niger, West Africa. J. Archaeol. Sci. 15, 367–394.

King M.E. (1978) Analytical methods and prehistoric textiles. Amer. Antiquity 43, 89–96.

King R.J. (2006) Minerals explained 44, Amber (Part 1). Geology Today 22, 232–237.

Kingery W.D. (1974) A note on the differential thermal analysis of archaeological ceramics. Archaeometry, 16, 109–112.

Kingery W.D. (1981) Plausible inferences from ceramic artifacts. J. Field Archaeol. 8, 457–467.

Kingery W.D. (1987) Microstructure analysis as part of a holistic interpretation of ceramic art and archaeological artifacts. Archaeomaterials 1, 91–99.

Kingery W.D. and Friermann J.D. (1974) The firing temperature of a Karanova sherd and inferences about South-East European Chalcolithic refractory technology. Proc. Prehist. Soc. 40, 204–205.

Kingery W.D., Vandiver P.B. and Prickett M. (1988) The beginnings of pyrotechnology, Part II: Production and use of lime and gypsum plaster in the Pre-Pottery Neolithic near East. J. Field Archaeol. 15, 219–244.

Kingery W.D., Vandiver P.B. and Noy T. (1992) 8,500-year-old sculpted plaster head from Jericho (Israel). Mat. Res. Soc. Bulletin 17, 46–52.

Kinney J.H., Stolken J.S., Smith T.S., Ryaby J.T. and Lane N.E. (2005) An orientation distribution function for trabecular bone. Bone 36, 193–201.

Kirshenblatt-Gimblett B. (2004) Intangible heritage as metacultural production. Museum Intern. 56, 52–65.

Klein S., Lahaye Y., Brey G.P. and Von Kaenel H.-M. (2004) The early Roman imperial aes coinage II: tracing the copper sources by analysis of lead and copper isotopes – copper coins of Augustus and Tiberius. Archaeometry 46, 469–480.

Kleinhanns I.C., Kreissig K., Kamber B.S., Meisel T., Nägler T.F. and Kramers J.D. (2002) Combined chemical separation of Lu, Hf, Sm, Nd, and REEs from a single rock digest: Precise and accurate isotope determination of Lu-Hf and Sm-Nd using multicollector-ICPMS. Anal. Chem. 74, 67–73.

Klemm D.D. and Klemm R. (1991) Mortar evolution in the Old Kingdom of Egypt. In: Pernicka E. and Wagner G.A. (eds) Archaeometry '90. Birkhäuser Verlag, Basel. pp. 445–454.

Klemm D.D. and Klemm R. (2001) The building stones of ancient Egypt – a gift of its geology. African Earth Sci. 33, 631–642.

Klemm D., Schmauder H.P. and Heinze T. (2002) Cellulose. In: Vandamme E.J., De Baets S. and Steinbuchel A. (eds) Biopolymers, Volume 6, Polysaccharides II. Polysaccharides from Eukaryotes. Wiley-VCH, Weinheim.

Klockenkämper R. (1997) Total reflection X-ray analysis. Chemical Analysis Series, Vol. 140. John Wiley and Sons, New York.

Kockelmann W., Pantos E. and Kirfel A. (2000) Neutron and synchrotron radiation studies of archaeological objects. In: Creagh D.C. and Bradley D.A. eds. Radiation in art and archaeometry. Elsevier Science B.V., Amsterdam. pp. 347–377.

Kockelmann W., Chapon L.C., Engels R., Schelten J., Neelmeijer C., Walcha H.M., Artioli G., Shalev S., Perelli-Cippo E., Tardocchi M., Gorini G. and Radaelli P.G. (2006a) Neutrons in cultural heritage research. J. Neutron Res. 14, 37–42.

Kockelmann W., Siano S., Bartoli L., Visser D., Hallebeek P., Traum R., Linke R., Schreiner M. and Kirfel A. (2006b) Applications of TOF neutron diffraction in archaeometry. Appl. Phys. A 83, 175–182.

Kocks U.F., Tomé C.N. and Wenk H.R. (1998) Texture and anisotropy. Cambridge University Press, Cambridge.

Koller D., Trimble J., Najbjerg, Gelfand N. and Levoy M. (2006) Fragments of the City: Stanford's Digital Forma Urbis Romae Project. Proceedings of the Third Williams Symposium on Classical Architecture, Journal of Roman Archaeology Suppl. 61, 237–252.

Koller D. and Levoy M. (2006) Computer-aided Reconstruction and New Matches in the Forma Urbis Romae. Workshop "Formae Urbis Romae - Nuove scoperte" held at the Istituto Archaeologico Germanico in Rome, March 3, 2004 To appear in Bollettino Della Commissione Archeologica Comunale di Roma.

Koon H.E.C., Nicholson R.A. and Collins M.J. (2003) A practical approach to the identification of low temperature heated bone using TEM. J. Archaeol. Sci. 30, 1393–1399.

Körber-Grohne U. (1988) Microscopic methods for identification of plant fibres and animal hairs from the Prince's tomb of Hochdorf, Southwest Germany. J. Archaeol. Sci. 15, 73–82.

Kosmowska-Ceranowicz B. (1984a) Natural environment of amber. In: Amber in nature. Wydawnictwa Geologiczne, Warsaw. pp. 17–23.

Kosmowska-Ceranowicz B. (1984b) Amber in sediment. In Amber in nature. Wydawnictwa Geologiczne, Warsaw, 37–48.

Kosmowska-Ceranowicz B. (1996) Bernstain die lagerstatte und ihre entstehung. In: Slotta and Ganzelewski (eds.) Berstain Tränen der Götter, Katalog der Ausstellung des Deutschen Bergbau Museums. Deutschen Bergbau Museums, Bochum, Germany. 161–168.

Kosmowska-Ceranowicz B. (1997) Amber, treasure of the ancient sea. Muzei Ziemi, Warsaw.

Kosmowska-Ceranowicz B. (1999) Succinite and some other fossil resins in Poland and Europe (deposits, finds, features and differences in IRS). Estudios del Museo de Ciencias Naturales de Alava (Nùm. Espec. 2) 14, 73–117.

Kosmowska-Ceranowicz B. (2003) Amber imitation in the Warsaw amber collection. Acta zoologica cracoviensia, (suppl. Fossil Insects) 46, 411–421.

Kosmowska-Ceranowicz B. (2006) Poland. The story of amber. Muza SA, Warsaw.

Kozlowski S.K. and Kempisty A. (1990) Architecture of the Pre-Pottery Neolithic settlement in Nemrik, Iraq. World Archaeol. 21, 348–362.

Kreimeyer R. (1987) Some notes on the firing colour of clay bricks. Appl. Clay Sci. 2, 175–183.

Kröpelin S. (2002) Damage to natural and cultural heritage by petroleum exploration and desert tourism in the Messak Settafet (central Sahara, southwest Libya). In: Lenssen-Erz T., Tegtmeier U. and Kröpelin S. (eds) Tides of the desert— Contributions to the archaeology and environmental history of Africa in honour of Rudolph Kuper. Heinrich-Barth-Institut, Koln. pp. 405–423.

Krueger H.W. and Sullivan C.H. (1984) Models for carbon isotope fractionation between diet and bone. In: Turnlund J.R. and Johnson P.E. (eds) Stable isotopes in nutrition. American Chemical Society Symposium Series, Washington, DC. Pp. 205–220.

Krzyszkowska O.H. (1990) Ivory and related materials: An illustrate guide. Bulletin of the Institute of Classical Studies Supplement 59. Institute of Classical Studies, London.

Krzyszkowska O.H. (1993) Aegean ivory carving: Towards an evaluation of Late Bronze Age workshop material. In: Lesley Fitton J. (ed.) Ivory in Greece and the Mediterranean from the Bronze Age to the Hellenistic Period. British Museum Occasional Papers 85, 25–35.

Krzyszkowska O.H. and Morkot R. (2000) Ivory and related materials. In: Nicholson P.T. and Shaw I. (eds) Ancient Egyptian materials and technology. Cambridge University Press, Cambridge. pp. 320–331.

Kuhn L.T., Grynpas M.D., Rey C.C., Wu Y., Ackerman J.L. and Glimcher M.J. (2008) A comparison of the physical and chemical differences between cancellous and cortical bovine bone mineral at two ages. Calcified Tissue International 83, 146–154.

Kumar R. and Kumar A.V. (1999) Biodeterioration of stone in tropical environments. The Getty Conservation Institute, Los Angeles. pp. 86.

Kumar S., Snyder D., Duncan D., Cohen J. and Cooper J. (2003) Digital preservation of ancient cuneiform tablets using 3D-scanning. Proceedings of the Fourth International Conference on 3-D Digital Imaging and Modeling. The IEEE Computer society. Document 0-7695-1991-1/03.

Kunz G.F. (1971) The curious lore of precious stones. Dover, New York.

Kuzmin Y.V. (2003) The nature of the transition from the Palaeolithic to the Neolithic in East Asia and the Pacific. Rev. Archaeol. 24, 1–3.

Kuzmin Y.V. (2006) Chronology of the earliest pottery in East Asia: progress and pitfalls. Antiquity 80, 362–371.

Kuzmin Y.V. and Vetrov V.M. (2007) The earliest Neolithic complex in Siberia: the Ust-Karenga 12 site and its significance for the Neolithisation process in Eurasia. Documenta Praehist. 34, 9–20.

Michelangelo: la Cappella Sistina. (1999) Documentazione e interpretazioni. I: rapporto sul restauro del Giudizio Universale. Istituto geografico De Agostini, Novara.

Mortimer C. (1995) Analysis of post-Medieval glass from Old Broad Street, London, with reference to other contemporary glasses from London and Italy. In: Hook D. and Gaimster D.R.M. (eds) Trade and discovery: The scientific study of artefacts from post-Medieval Europe and beyond. British Museum Occasional Paper N. 109. The British Museum, London. pp. 135–144.

Lahanier C. (2004) Information technologies applied to scientific examination of museum collections. In: Martini M., Milazzo M. and Piacentini M. (eds) Physics methods in archaeometry. SIF, Bologna – IOS Press, Amsterdam. pp. 155–178.

Lahlil S., Biron I., Galoisy L. and Morin G. (2008) Rediscovering ancient glass technologies through the examination of opacifier crystals. Appl. Phys. A 92, 109–116.

LaMarche V.C. Jr. and Hirschboeck K.K. (1984) Frost rings in trees as records of major volcanic eruptions. Nature, 307, 121–126.

Lambert J.B., Beck C.W. and Frey J.S. (1988) Analysis of European amber by Carbon-13 nuclear magnetic resonance spectroscopy. Archaeometry 30, 248–263.

Lambert J.B. and Johnson S.C. (1996) Nuclear Magnetic Resonance characterization of cretaceous amber. Archaeometry 38, 325–335.

Lambert J.B. (1997) Traces of the past. Unraveling the secrets of archaeology through chemistry. Addison-Wesley, Reading.

Laming A. (ed.) (1952) La décuverte du passé. Editions A. et J. Picard, Paris.

Langford W.A. (1992) Analysis of hydrogen by nuclear reaction and energy recoil detection. Nucl. Instr. Meth. Phys. Res. B 66, 65–82.

Langenheim J.H. (1969) Amber: a botanical inquiry. Science 163, 1157–1169.

Langenheim J.H. (1995) Biology of amber-producing trees: Focus on case studies of Hymenaea and Agathis. In: Anderson K.B. and Crelling J.C. (eds) Amber, resinite, and fossil resins. American Chemical Society, Washington, DC. pp. 1–31.

Langenheim J.H. (2003) Plant resins: chemistry, evolution, ecology and ethnobotany. Timber Press, Portland, OR.

La Niece S. and Craddock P.T. (eds) (1993) Metal plating and patination : Cultural, technical and historical developments. Butterworth-Heinemann, London.

Larkins F.P. (1977) Semiempirical Auger-electron energies for elements $10 \leq Z \leq 100$. Atomic Data and Nuclear Tables 20, 311–387.

Larson S.G. (1978) Baltic Amber – a Palaeobiological Study. Scandinavian Science Press Ltd., Klampenborg, Denmark.

Larson P.B., Maher K., Ramos F.C., Chang Z., Gaspar M. and Meinert L.D. (2003) Copper isotope ratios in magmatic and hydrothermal ore-forming environments. Chemical Geology 201, 337–350.

Laurenze-Landsberg C., Schmidt C., Mertens L.A. and Schröder-Smeibidl B. (2003) Neutron autoradiography of the painting *Armida abducts the sleeping Rinaldo* (~1637) by Nicolas Poussin. Annual Report 2003. Hahn-Meitner-Institut, Berlin. pp. 36–37.

Lavédrine B. (1996) Prediction of dark stability of color chromogenic films using Arrhenius 'Law and comparison after ten years of natural aging. In: Koch M.S. (ed.) Research techniques in photographic conservation. Conf. Proc., Copenhagen, 14–19 May 1995. Copenhagen. pp. 71–76.

Lavédrine B. (2003) A Guide to the preventive conservation of photograph collections. Getty Conservation Institute, Los Angeles.

Lavédrine B. (2009) Photographs of the past. Process and preservation. Getty Conservation Institute, Los Angeles.

Laws W.D. (1962) An investigation of temple plasters from Tikal, Guatemala, with evidence of the use by the ancient Maya of plant extracts in plaster making. Wrightia 2, 5217–5228.

Lazzarini L. (1982) The discovery of Egyptian Blue in a Roman fresco of the mediaeval period (ninth century A.D.). Studies in Conservation 27, 84–86.

Lazzarini L. (1994) Lapis sarcophagus: an historical and scientific note. In: Conservazione del patrimonio culturale. Contrib. Centro Linceo Interdiscipl. "Beniamino Segre". Vol. 88. Accademia Nazionale dei Lincei, Roma. pp. 103–116.

Lazzarini L. (2004) Archaeometric aspects of white and coloured marbles used in antiquity: the state of the art. Per. Miner. 73, 113–125.

Lazzarini L. (ed.) (2004) Pietre e marmi antichi. Natura, caratterizzazione, origine, storia d'uso, diffusione, collezionismo. CEDAM, Padova.

Lazzarini L. (2006) Poikiloi lithoi, versiculores maculae: I marmi colorati della Grecia antica. Giardini Editori e Stampatori, Pisa. pp. 280.

Lechtman H. (1996) Arsenic bronze: Dirty copper or chosen alloy? A view from the Americas. J. Field Archaeol. 23, 477–514.

Lee J.C., Hsieh H.M., Huang L.H., Kuo Y.C., Wu J.W., Chin S.C., Lee A.H., Linacre A. and Tsai L.C. (2009) Ivory identification by DNA profiling of cytochrome b gene. Int. J. Legal Med. 123, 117–121.

Lee-Thorp J.A., Sealy J.C. and van der Merwe N.J. (1989) Stable carbon isotope ratio diffrerences between bone collagen and bone apatite, and their relationship to diet. J. Archaeol. Sci. 16, 585–599.

Lee-Thorp J.A. (2008) On isotopes and old bones. Archaeometry 50, 925–950.

Lehmann E.H., Vontobel P. and Frei G. (2007) The non-destructive study of museum objects by means of neutron imaging methods and results of investigations. Nuovo Cimento, 30, 93–104.

Leona M. and Lombardi J.R. (2007) Identification of berberine in ancient and historical textiles by surface-enhanced Raman scattering, J. Raman Spectr. 38, 853–858.

Leonardi R. (2005) Nuclear physics and painting. Nucl. Phys. A, 752, 659c–674c.

Leroi-Gourhan A. (1982) The archaeology of Lascaux Cave. Sci. Amer. 246, 80–88.

Leslie A.B. and Hughes J.J. (2002) Binder microstructure in lime mortars: implications for the interpretation of analysis results. Quart. J. Engin. Geol. Hydrogeol. 35, 257–263.

Leute U. (1987) Archaeometry. An introduction to physical methods in archaeology and the history of art. VCH, Weinheim and New York.

Lian O.B. and Roberts R.G. (2006) Dating the Quaternary: progress in luminescence dating of sediments. Quaternary Sci. Rev., 25, 2449–2468.

Libby W.F. (1952) Radiocarbon dating. University of Chicago Press, Chicago.

Liliquist C. (1993) Granulation and glass: Chronological and stylistic investigations at selected sites, ca. 2500–1400 B.C.E. Bull. Amer. Schools Orient. Res. 290–291, 29–94.

Lilyquist C. and Brill R.H. (1993) Studies in early Egyptian glass. The Metropolitan Museum of Art, New York.

Lindroos A., Heinemeier J., Ringbom Å., Brasken M. and Sveinbjörnsdottir Á. (2007) Mortar dating using AMS 14C and sequential dissolution: Examples from medieval, non-hydraulic lime mortars from the Åland Islands, SW Finland. Radiocarbon 49, 47–67.

Linduff K. (2004) Metallurgy in ancient Eastern Eurasia from the Urals to the Yellow River. Edwin Mellen Press, Lampeter.

Littman E.R. (1960) Ancient Mesoamerican mortars, plasters, and stuccos: The use of bark extracts in lime plasters. Amer. Antiquity 25, 4593–4597.

Liverani M. (ed.) (2005) Aghram Nadharif. The Barkat Oasis (Sha' of Ghat, Libyan Sahara) in Garamantian Times. Arid Zone Archaeology, Monographs, Vol. 5. Edizioni All' del Giglio, Firenze.

Livingstone Smith A. (1991) Bonfire II: The return of pottery firing temperatures. J. Archaeol. Sci. 28, 991–1003.

Liz-Marzán L.M. (2004) Nanometals: formation and color. Mat. Today February 2004, 26–31.

Lock G. (ed) (2000) Beyond the map: Archaeology and spatial technologies. IOS Press, Amsterdam.

Lock G. and Harris T. (1997) Analysing change through time within a cultural landscape: conceptual and functional limitations of a GIS approach. In: Sinclair, P. (ed.) Urban origins in Eastern Africa, World Archaeological Congress, One World series.

Loendorf L.L. (1991) Cation-ratio varnish dating and petroglyph chronology in Southeastern Colorado. Antiquity 65, 246–255.

Łojewska J., Missori M., Lubańska A., Grimaldi P., Zięba K., Proniewicz L.M. and Congiu Castellano A. (2007) Carbonyl groups development on degraded cellulose. Correlation between spectroscopic and chemical results. Appl. Phys. A, Materials Science & Processing 89, 883–887.

Long D.G.F., Silveira B. and Julig P. (2001) Chert analysis by infrared spectroscopy. In: Pilon J.L., Kirby M.W. and Thériault C. (eds) A collection of papers presented at the 33rd Annual Meeting of the Canadian Archaeological Association. pp. 255–267.

Long D.A., Edwards H.G.M. and Farwell D.W. (2008) The Goodmanham plane: Raman spectroscopic analysis of a Roman ivory artefact. J. Raman Spectr. 39, 322–330.

Longoni A., Fiorini C., Leutenegger P., Sciuti S., Fronterotta G., Strüder L. and Lechner P. (1998) A portable XRF spectrometer for non-destructive analyses in archaeometry. Nucl. Instr. Meth. Phys. Res. A, 409, 407–409.

Lopez-Molinero A., Castro A., Pino J., Perez-Arantegui J. and Castillo J.R. (2000) Classification of ancient Roman glazed ceramics using the neural network of Self-Organizing Maps. Fresenius J. Anal. Chem., 367, 586–589.

Lorblanchet M., Labeau M., Vernet J.-L., Fitte P., Valladas H., Cachier H. and Arnold M. (1990) Palaeolithic pigments in the Quercy, France. Rock Art Res. 7, 4–20.

Lossau N. and Liebetruth M. (2000) Conservation issues in digital imaging. Spectra, 26, 30–36.

Loy T.H. (1983) Prehistoric blood residues: Detection on tool surfaces and identification of species of origin. Science 220, 1269–1271.

Loy T.H., Jones R., Nelson D.E., Meehan B., Vogel J., Southon J. and Cosgrove R. (1990) Accelerator radiocarbon dating of human blood proteins in pigments from Late Pleistocene art sites in Australia. Antiquity 64, 110–116.

Loy T.H. and Hardy B.G. (1992) Blood residue analysis of 90 000 year old stone tool from Tabun Cave, Israel. Antiquity 66, 24–35.

Loy T.H. and Dixon E.J. (1998) Blood residues on fluted points from Eastern Beringia. Amer. Antiquity 63, 21–46.

Lowell S., Shields J.E. and Thomas M.A., Thommes M. (2004) Characterization of porous solids and powders: surface area, pore size, and density. 4th Edition. Springer-Verlag, Berlin–New York. pp. 347.

Lozano L.F., Peña-Rico M.A., Heredia A., Ocotlán-Flores J., Gómez-Cortés A., Velázquez R., Belío I.A. and Bucio L. (2003) Thermal analysis study of human bone. J. Mater. Science, 38, 4777–4782.

Lucarelli F. and Mandò P.A. (1996) Studying the chronology of Galileo's writings with PIXE. Nuclear Physics News, 6, 24–31.

Lucas A. (1968) Ancient Egyptian materials and industries. 4th Edition. Arnold, London.

Ludwig N. (2004) Thermographic testing on historic building. In: Martini M., Milazzo M. and Piacentini M. (eds) Physics methods in archaeometry. SIF, Bologna – IOS Press, Amsterdam. pp. 481–496.

Ludwig K.R. and Renne P.R. (2000) Geochronology on the palaeoanthropological time scale. Evolutionary Anthropology, 9, 101–110.

MacCurdy G.G. (1912) A new French cavern with Paleolithic mural engravings. Science 36, 269–270.

MacDonald L. (ed.) (2006) Digital heritage. Applying digital imaging to cultural heritage. Elsevier, Amsterdam.

MacGregor A. (1985) Bone, antler, ivory & horn. The technology of skeletal materials since the Roman period. Croom Helm, London & Sidney – Barnes & Noble Books, Totowa, NJ. pp. 245.

Maggetti M. (2006) Archaeometry: Quo vadis? Geol. Soc. London, Special Publications 2006, 257, 1–8.

Maggi R. and Pearce M. (1998) Les mines prehistoriques de Libiola et Monte Loreto (nouvelles fouilles). In: Frère-Sautot M.-Ch. (ed.) Paléométallurgie des cuivres. Monique Mergoil, Montagnac. pp. 89–94.

Magill J., Pfenning G. and Galy J. (2007) The Karlsruhe chart of the nuclides. 7th Edition. European Commission, DG Joint Research Centre, Institute for Transuranium Elements, Karlsruhe. (also available on http://www.nucleonica.net)

Maglioni D., Pancella R., Fruh Y., Canetas J. and Castano V. (1995) Studies on the Mayan mortars technique. In: Vandiver P.B., Druzik J.R., Madrid J.L.G., Freestone I.C. and Wheeler G.S. (eds) Materials issues in art and archaeology IV. Materials Research Society, Pittsburgh. pp. 483–489.

Magne D., Pilet P., Weiss P. and Daculsi G. (2001) Fourier transform infrared microspectroscopic investigation of the maturation of nonstoichiometric apatites in mineralized tissues: a horse dentin study. Bone 29, 547–552.

Mahmid J., Sharir A., Addadi L. and Weiner S. (2008) Amorphous calcium phosphate is a major component of the forming fin bones of zebrafish: Indications for an amorphous precursor phase. Proc. Natl Acad. Sci 105, 12748–12753.

Main P. (1991) Computing and mathematics: Putting two and two together. In: Bowman S.G.E. (ed.) Science and the past. British Museum Publications, London.

Mairinger F. (2004) UV-, IR- and X-ray imaging. In: Janssens K. and Van Grieken R. (ed.) Non destructive microanalysis of cultural heritage materials. Comprehensive analytical chemistry series. Vol. XLII. Elsevier, Amsterdam. pp. 15–72.

Maldague X. (1993) Nondestructive evaluation of materials by infrared thermography. Springer-Verlag, London.

Maldague X. (2001) Theory and practice of infrared technology for non-destructive testing. John Wiley and Sons, New York.

Mancusi-Ungaro C. (1982) In: Yves Klein, 1928–1962. Rice University, Institute for the Arts, Huston.

Mandò P.A. (2007) The Florence accelerator laboratory for Ion Beam Analysis and AMS radiocarbon dating. Il Nuovo Cimento 30C, 85–92.

Maniatis Y. (2004) Scientific techniques and methodologies for the provenance of white marbles. In: Martini M., Milazzo M. and Piacentini M. (eds) Physics methods in archaeometry. SIF, Bologna – IOS Press, Amsterdam. pp. 179–202.

Mann S. (2001) Biomineralization. Principles and concepts in bioinorganic materials chemistry. Oxford University Press, Oxford. pp. 198.

Manniche L. Forman W. (1999) Sacred luxuries: fragrance, aromatherapy & cosmetics in ancient Egypt. Cornell University Press, Ithaca, NY. pp. 160.

Mantler M. and Schreiner M. (2000) X-ray fluorescence spectrometry in art and archaeology. X-Ray Spectrom. 29, 3–17.

Maravelaki-Kalaitzaki P., Zafiropulos V. and Fotakis C. (1999) Excimer laser cleaning of encrustation on Pentelic marble: procedure and evaluation of the effects. Appl. Surf. Sci. 148, 92–104.

Marchant J. (2006) In search of lost time. Nature, 444, 534–538.

Maritan L., Angelini I., Artioli G., Mazzoli C. and Saracino M. (2009) Secondary phosphates in the ceramic materials from Frattesina (Rovigo, North-Eastern Italy). J. Cult. Herit. 50, 144–151.

Marshall A.G., Hendrickson C.L. and Jackson G.S. (1998) Fourier Transform ion cyclotron resonance mass spectrometry: A primer. Mass Spectr. 17, 1–35.

Marshall J. (1990) Glass source book. Chartwell Books Inc., London.

Martín-Gonzáles A., Wierzchos J., Gutiérrez J.C., Alonso J. and Ascaso A. (2009) Double fossilization in eukaryotic microorganisms from Lower Cretaceous amber. BMC Biology 7, 9. doi:10.1186/1741-7007-7-9

Martínez-Richa A., Vera-Graziano R., Rivera A. and Joseph-Nathan P. (2000) A solid-state 13C NMR analysis of ambers. Polymer 41, 743–750.

Martinetto P., Dooryhée E., Anne M., Talabot J., Tsoucaris G. and Walter P. (1999) Cosmetic recipes and make-up manufacturing in ancient Egypt. ESRF Experimental Reports, 32, 10–11.

Martinetto P., Anne M., Dooryhée E., Isnard O. and Walter P. (2003) A non-destructive analysis by neutron diffraction inside make-up containers of ancient Egypt. In: Tsoucaris G. and Lipkowski J. (eds) (2003) Molecular and structural archaeology: cosmetic and therapeutic chemical. NATO Science Series II Mathematics, Physics and Chemistry. Vol. 117. Kluwer Academic Publisher, Dordrecht–Boston–London. pp. 107–111.

Martini M., Milazzo M. and Piacentini M. (eds) (2004) Physics methods in archaeometry. SIF, Bologna – IOS Press, Amsterdam. pp. 503.

Martini M. and Sibilia E. (2001) Radiation in archaeometry: Archaeological dating. Radiation Phys. Chem. 61, 241–246.

Massazza F. (1998) Pozzolana and pozzolanic cements. In: Hewlett P.C. (ed.) Lea's Chemistry of cement and concrete. 4th Edition. Butterworth-Heimann, Oxford. pp. 471–635.

Mathewson C.H. (1915) A metallographic description of some ancient Peruvian bronzes from Machu Picchu. Amer. J. Sci. 40, 525–616.

Matteini M., Moles A. and Giovannoni S. (1994) Calcium oxalate as a protective mineral system for wall paintings: methodology and analyses. In: Fasina V., Ott H. and Zezza F. (eds) La conservazione dei monumenti nel bacino del Mediterraneo. Soprintendenza ai Beni Artistici e Storici di Venezia, Venezia. pp. 155–162.

Matthews J.A. (1994) Lichenometric dating: a review with particular reference to 'Little Ice Age' moraines in southern Norway. In: Beck C. (ed.) Dating in exposed and surface contexts. University of New Mexico Press, Albuquerque. pp. 185–212.

Maurer CA (1980) Electron spin resonance spectroscopy: A potential technique for dating ancient ceramics. Ph.D. Thesis, University of Illinois, Urbana.

Mays S. (1998) The archaeology of human bones. Routledge, London.

Mazurek J., Heginbotham A., Schilling M. and Chiari G. (2008) Antibody assay to characterize binding media in paint. ICOM Committee for Conservation. Vol 2, pp. 678–685.

McCamy C.S. (1992) Munsell value as explicit functions of CIE luminance factor. Color Res. Appl. 17, 205–207.

McConnel D. (1952) The crystal chemistry of carbonated apatite and their relationship to the composition of calcified tissues. J. Dental Res. 31, 53–63.

McConnell D. (1960) The crystal chemistry of dahllite. Am. Mineral. 45, 209–216.

McCown D.E. (1952) Excavations at Nippur, 1948–50. J. Near Eastern Studies 11, 169–176.

McDermott F., Grün R., Stringer C.B. and Hawksworth C.J. (1993) Mass-spectrometric U-series dates for Israeli Neandethal/early modern hominid sites. Nature, 363, 252–255.

McDonald J.K. (1996) House of eternity. The tomb of Nefertari. The Getty Conservation Institute and the J. Paul Getty Museum, Los Angeles).

McDonnell J.G. (2001) Pyrotechnology. In: Brothwell D.R. and Pollard A.M. (eds) Handbook of archaeological sciences. John Wiley & Sons, Chichester. pp. 493–505.

McDougall I. and Harrison T.M. (1999) Geochronology and thermochronology by the 40Ar/39Ar method. 2nd Edition. Oxfrod University Press, Oxford. pp. 261.

McElhinny M.W. and McFadden P.L. (2000) Palaeomagnetism: Continents and oceans. 2nd Edition. Academic Press, San Diego. pp. 386.

McEwan C., Middleton A., Cartwright C. and Stacey R. (2006) Turquoise mosaics from Mexico. The British Museum Press, London. pp. 96.

McGlone C., Mikhail E. and Bethel J. (eds) (2004) Manual of Photogrammetry, 5th Edition. ASPRS, American Society for Photogrammetry and Remote Sensing, Bethesda.

McLafferty F.W. (1980) Interpretation of Mass Spectra. 3rd Edition. University Science Books, Mill Valley.

McLuckey S.A., Busch K.L. and Glish G.L. (1988) Mass spectrometry/mass spectrometry: techniques and applications of tandem mass spectrometry. VCH Publishers, New York.

McNeil M.B. and Little B.J. (1992) Corrosion mechanisms for copper and silver objects in near-surface environments. J. Amer. Inst. Conserv. 31, 355–366.

McNeil M.B., Amos A.L. and Woods T.L. (1993) Adherence of sulfide mineral layers produced by corrosion of copper alloys. Corrosion 49, 755–758.

McNeil M.B. and Selwyn L.S. (2001) Electrochemical processes in metallic corrosion. In: Brothwell D.R. and Pollard A.M. (eds) Handbook of archaeological sciences. John Wiley & Sons, Chichester. pp. 605–614.

Meldrum F.C. and Cölfen H. (2008) Controlling mineral morphologies and structures in biological and synthetic systems. Chem. Rev. 108, 4332–4432.

Melfos V. (2008) Green Thessalian stone: the Byzantine quarries and the use of a unique architectural material from the Larisa area, Greece. Petrographic and geochemical characterization. Oxford J. Archaeol. 27, 387–405.

Melville M.D. and Atkinson G. (1985) Soil colour: its measurement and its designation in models of uniform colour space. Eur. J. Soil Sci. 36, 495–512.

Mercier N., Valladas H., Valladas G., Reyss J.L., Jelinek A., Meignen L. and Joron J.L. (1995) TL dates of burnt flints from Jelinek's excavations at Tabun and their implications. J. Archaeol. Sci. 22, 495–509.

Merkevičius1 A., Bezdicka P., Juškėnas R., Kiuberis J., Senvaitienė J., Pakutinskienė I. and Kareiva A. (2007) XRD and SEM characterization of archaeological findings excavated in Lithuania. Chemija 18, 36–39.

Merwin H.E. (1931) Chemical analysis of pigments. In: Morris E.H., Charlot J. and Morris A.A. (eds) The temple of the warriors at Chichèn Itza, Yucatan. Carnegie Institution of Washington, Public. 406, Washington, DC. pp. 356.

Messiga B. and Riccardi M.P. (2001) A petrological approach to the study of ancient glass. Per. Mineral. 70, 57–70.

Mette B. (2003) Beitrag zur spätbronzezeitlichen kupfermetallurgie im Trentino (Südalpen) im vergleich mit anderen prehistorischen kupferschlaken aus dem Alpenraum. Metalla 10, 1–122.

Michalski S. (2002) Double the life for each five-degree drop, more than double the life for each halving of relative humidity. In: 13th Triennial Meeting. Rio de Janeiro, 22–27 September 2002. Comité de l' pour la Conservation. Preprints, Vol. I. pp. 66–72.

Michelangelo, La Cappella Sistina (1999) Documenti e interpretazioni. I. Rapporto sul restauro del Giudizio Universale. Musei Vaticani, Roma - Istituto geografico de Agostini, Novara.

Middleton A., Freestone I. (eds) (1991) Recent developments in ceramic petrology. British Museum Occasional Paper no. 81. The British Museum, London (reprinted 1997).

Mighall T.M. (2003) Geochemical monitoring of heavy metal pollution and prehistoric mining: Evidence from Copa Hill, Cwmystwyth, and Mount Gabriel, County Cork. In: Craddock P. and Lang J. (eds) Mining and metal production through the ages. The British Museum Press, London. pp. 43–51.

Mighall T.M., Timberlake S., Singh S. and Bateman M. (2007) Records of paleo-pollution from mining and metallurgy as recorded by three ombrotrophic peat bogs in Wales, UK. In: La Niece S., Hook D. and Craddock P. (eds) metals and mines. Studies in archaeometallurgy. The British Museum – Archetype Publications, London. pp. 56–64.

Migliori A., Massi M. and Giuntini L. (2008) Analysis of ancient embroideries by IBA techniques. Surf. Eng. 24, 98–102.

Millard A.R. and Hedges R.E.M. (1996) A diffusion-adsorption model of uranium uptake by archaeological bone. Geochim. Cosmochim. Acta 60, 2139–2152.

Miller D. (1994) Artefacts and the meaning of things. In: Ingold T. (ed.) The companion encyclopedia of anthropology. Routledge, London.

Miller D. (2006) Minor and trace elements in photographic material: analysis and meaning. Glass Eye Productions, DVD distributed by the Getty Conservation Institute, Los Angeles. [videorecording, 2 DVDs (4hrs. 26min.)].

Mills J.S. and White R. (1989) The identity of the resins from the Late Bronze Age shipwreck at Uluburun (Kaş). Archaeometry 31, 37–44.

Mills J.S. and White R. (1999) The Organic Chemistry of Museum Objects. 2nd Edition. Butterworth-Heinemann, Oxford.

Mirguet C., Fredrickx P., Sciau P. and Colomban P. (2008) Origin of the self-organisation of Cu°/Ag° nanoparticles in ancient lustre pottery. A TEM study. Phase Transitions 81, 253–266.

Mirguet C., Dejoie C., Roucau C., De Parseval Ph., Teat S.J. and Sciau Ph. (2009) Nature and microstructure of Gallic imitations of sigillata slips from the La Graufesenque workshop. Archaeometry DOI: 10.1111/j .1475–4754.2008.00452.x

Mirtsou E., Vavelidis M., Ignatiadou D. and Pappa M. (2001) Early Bronze Age faience from Agios Mamas, Chalkidiki: a short note. In: Basiakos Y., Aloupi E. and Facorellis Y. (eds) Archaeometry issues in Greek prehistory and antiquity. Hellenic Society of Archaeometry, Society of Messenian Archaeological Studies, Athens. pp. 309–316.

Missori M., Mondelli C., De Spirito M., Castellano C., Bicchieri M., Schweins R., Arcovito G., Papi M. and Congiu Castellano A. (2006) Modification of the mesoscopic structure of cellulose in paper degradation. Phys. Res. Lett. 97, 238001–238004.

Miziolek A.W., Palleschi V. and Schechter I. (2006) Laser induced breakdown spectroscopy. Cambridge University Press, New York.

Mobilio S. and Vlaic G. (eds) (2003) Synchrotron radiation: Fundamentals, methodologies and applications. SIF Società Italiana di Fisica, Bologna, Atti di Conferenze, Vol. 82.

Modena C. (1995) Criteria for cautious repair of historic buildings. In: Binda L. and Modena C. (ed.) Evaluation and strengthening of existing masonry structures. Proceedings Pro003, RILEM Publications Sarl, The publishing company of RILEM, Bagneux, France. pp. 25–42.

Mohen J.-P. and Walter P. (1994) Le four-creuset, une invention inédite de l' du bronze européen. Techne 1, 103–110.

Molera J., Pradell T. and Vendrell-Saz M. (1998) The colours of Ca-rich ceramicpastes: origin and characterization. Appl. Clay Sci. 13, 187–202.

Molera J., Pradell T., Salvadó N and Vendrell-Saz M. (2009) Lead frits in Islamic and Hispano-Moresque glazed productions. In: Shortland A.J., Freestone I.C. and Rehren T. (eds) From mine to microscope. Advances in the study of ancient technology. Oxbow Books, Oxford. pp. 1–10.

Monaco H.L. and Artioli G. (2002) Experimental methods in X-ray and neutron crystallography. In: C. Giacovazzo (ed.) Fundamentals of Crystallography, Second Edition, Ch. 5. IUCr Texts on Crystallography, Vol. 7. Oxford Science Publications, Oxford. pp. 295–411.

Monna F., Galop D., Carozza L., Tual M., Beyrie A., Marembert F., Château C., Dominik J. and Grousset F.E. (2004) Environmental impact of early Basque mining and smelting recorded in a high ash minerogenic peat deposit. Science of the Total Environment 327, 197–214.

Monod J. (1970) Le hasard et la nécessité. Essai sur la philosophie naturelle de la biologie moderne. Seuil, Paris.

Montana G. and Gagliardo Briuccia V. (1998) I marmi e diaspri del barocco siciliano. Flaccovio Editore, Palermo. pp. 132.

Moorey P.R.S. (1994) Ancient Mesopotamian materials and industries. The archaeolgical evidence. Oxford University Press, Oxford.

Mora P., Mora L. and Philipot P. (1977) La conservation des peintures murales. Editrice Compositori, Bologna. Engl. Transl.: (1984) Conservation of wall paintings. Butterworths, London.

Mordant C. and Mordant D. (1992) Noyen-sur-Seine: a Mesolithic waterside settlement. In: Coles B. (ed.) The wetland revolution in prehistory. WARP, Exeter-The Prehistoric Society, London. pp. 55–64.

Moropoulou A., Bakolas A. and Bisbikou K. (2000) Investigation of the technology of historic mortars. J. Cult. Herit. 1, 45–58.

Morris E.T. (ed.) (1999) The Metropolitan Museum of Art Scents of Time—Perfume from Ancient Egypt to the 21st Century. Bulfinch Press, Boston, MA.

Morris H.R. and Whitmore P.M. (2007) "Virtual fading" of art objects: simulating the future fading of artifacts by visualizing micro-fading test results. Journal of the American Institute for Conservation, 46, 215–228.

Morrison P. and Morrison P. (1994) Powers of ten. Scientific American Library. Revised Edition. W.H. Freeman & Company, San Francisco.

Morton G.R. and Wingrove J. (1969) Slag, cinder and bear. Bull. Hist. Metall. Group 3, 55–61.

Motella De Carlo S. (2009) Tracce di fibre tessili a Chiusa di Pesio. In: Venturino Gambari M. (ed.) Il ripostiglio del Monte Cavanero di Chiusa di Pesio. LineLab.Edizioni, Alessandria. pp. 193–200.

Mott B.W. (1957) Microindentation hardness testing. Butterworths, London.

Muhly J.D. (1977) The copper ox-hide ingots and the Bronze Age Metals trade. Iraq 39, 73–82.

Muhly J.D. (1988) The beginnings of metallurgy in the Old World. In: Maddin R. (ed.) The beginnings of the use of metals and alloys. MIT Press, Cambridge, MA. pp. 2–20.

Muhly J.D. (2007) The first use of metal on Minoan Crete. In: La Niece S., Hook D. and Craddock P. (eds) metals and mines. Studies in archaeometallurgy. The British Museum–Archetype Publications, London. pp. 97–102.

Müller M., Czihak C., Burghammer M. and Riekel C. (2000) Combined X-ray microbeam small-angle scattering and fibre diffraction experiments on single cellulose fibers. J. Appl. Cryst. 33, 817–819.

Müller M., Papiz M.Z., Clarke D.T., Roberts M.A., Murphy B.M., Burghammer M., Riekel C., Pantos E. and Gunneweg J., (2003) Identification of textiles from Khirbet Qumran using microscopy and synchrotron radiation X-ray fibre diffraction. In: Humbert J.-B. and Gunneweg J. (eds) Archaeological Excavations at Khirbet Qumran and Ain Feshka – Studies in Archaeometry and Anthropology, Vol. II, Chapt. XII. Presses Universitaires de Fribourg, Suisse. pp. 177–186.

Müller M., Murphy B., Burghammer M., Riekel C., Roberts M., Papiz M., Clarke D., Gunneweg J. and Pantos E. (2004) Identification of ancient textile fibres from Khirbet Qumran caves using synchrotron radiation microbeam diffraction. Spectrochim. Acta B 59, 1669–1674.

Müller M., Murphy B., Burghammer M., Snigireva I., Riekel C., Gunneweg J. and Pantos E. (2006) Identification of single archaeological textile fibres from the cave of letters using synchrotron radiation microbeam diffraction and microfluorescence. Appl. Phys. A 83, 183–188.

Müller M., Murphy B., Burghammer M., Riekel C., Pantos E. and Gunneweg J. (2007) Ageing of native cellulose fibres under archaeological conditions: textiles from the Dead Sea region studied using synchrotron X-ray microdiffraction. Appl. Phys. A 89, 877–881.

Müller W., Fricke H., Halliday A.N., McCulloch M.T. and Wartho J.A. (2003) Origin and migration of the Alpine Iceman. Science 302, 862–866.

Munro L.E., Longstaffe F.J. and White C.D. (2007) Burning and boiling of modern deer bone: Effects on crystallinity and oxygen isotope composition of bioapatite phosphate. Palaeogeo. Palaeoclim. Palaeoecol. 249, 90–102.

Munsell A.H. (1905) A color notation. G. H. Ellis Co., Boston.

Murad E. and Wagner U. (1998) Clays and clay minerals: The firing process. Hyperfine Interactions 117, 337–356.

Murray A.S. and Olley J.M. (2002) Precision and accuracy in the optically stimulated luminescence dating of sedimentary quartz: a status review. Geochronometria 21, 1–16.

Nadel D., Danin A., Werker E., Schick T., Kislev M.E. and Stewart K. (1994) 19,000-year-old twisted fibers from Ohalo II. Current Anthropol. 35, 451–457.

Nadeau L. (1994) Encyclopedia of printing, photographic, and photomechanical processes. Atelier Luis Nadeau, New Brunswick.

Naef W. (1995) The J. Paul Getty Museum handbook of the photographs collection. The Museum, Malibu.

Narayan C., O'Connor M., Kegel G.H.R., Johnson R., Salmons C. and White C. (1996) PIXE studies on artifacts from Saugus Iron Works. Nucl. Instr. Meth. Phys. Res. B 118, 396–399.

Nassau K. (1983) The physics and chemistry of color. The fifteen causes of color. John Wiley & Sons, New York.

Nassau K. (1984) The early history of gemstone treatments. Gems Gemol. Spring 1984, 22–23.

Nassau K. (1994) Gemstone enhancement. Butterworth, London.

Nassau K. (ed) (1998) Color for science, art, and technology. North Holland-Elsevier, Amsterdam. pp. 510.

Neelmeijer C., Wagner W. and Schramm H.P. (1996) Depth resolved ion beam analysis of objects of art. Nucl. Instr. Meth. Phys. Res. B 118, 338–345.

Neff D. and Dillmann P. (2001) Phosphorus localisation and quantification in archaeological iron artefacts by micro-PIXE analyses. Nucl. Instr. Meth. Phys. Res. B 181, 675–680.

Negroni Catacchio N. and Guerreschi G. (1970) La problematica dell' nella protostoria italiana. Studi Etruschi XXXVIII, 165–183.

Nel A., de Ploëg G., Millet J., Menier J.J. and Waller A. (2004) The French amber: a general conspectus and the Lowermost Eocene amber deposit of Le Quesnoy in the Paris Basin. Geol. Acta 2, 3–8.

Newton R.G. (1966) Some problems in the dating of ancient glass by counting the layers in the weathering crust. Glass Technol. 7, 22–25.

Newton R.G. and Davison S. (1989) Conservation of glass. Butterworths, London.

Newton G.W.A., Bourriau J., French E.B. and Prag A.J.N.W. (2007) INAA of archaeological samples at the University of Manchester. Archaeometry 49, 289–299.

Niccolucci F. (2002) Virtual Archaeology: Proceedings of the VAST Euroconference, Arezzo 24–25 November 2000. European Commission Directorate General for Research. Human Potential Programme. Archaeopress, Oxford.

Nicholson P.T., Jackson C.M. and Trott K.M. (1997) The Ulu Burun glass ingots, cylindrical vessels and Egyptian glass. J. Egyptian Archaeol. 83, 143–153.

Nicholson P.T and Shaw I. (eds) (2000) Ancient Egyptian materials and technology. Cambridge University Press, Cambridge.

Nickerson D. (1976) History of the Munsell color system, company, and foundation. Color Res. Appl. 1, 7–10.

Nielsen-Marsh C.M., Hedges R.E.M., Mann T. and Collins M.J. (2000) A preliminary investigation of the application of differential scanning calorimetry to the study of collagen degradation in archaeological bone. Thermochimica Acta, 365, 129–139.

Niklasson G.A. (1993) Adsorption on fractal surfaces: applications to cement materials. Cem. Concr. Res. 23, 1153–1158.

Nissenbaum A. and Yakir D. (1995) Stable isotope composition of ambers. In: Anderson K.B. and Crelling J.C. (eds) Amber, resinite, and fossil resins. American Chemical Society, Washington, DC. pp. 32–42.

Nobbs J.H. (1985) Kubelka-Munk theory and the prediction of reflectance. Rev. Progr. Color. Related Topics 15, 66–75.

Nociti V. (2007) Dadi e tessere dell' (Tarquinia – Orvieto – Chiusi). Laurea Dissertation. Università degli Studi di Milano.

Northover P., Crossley A., Grazioli C., Zema N., La Rosa S., Lozzi L., Picozzie P. and Paparazzo E. (2008) A multitechnique study of archeological bronzes. Surf. Interface Anal. 40, 464–468.

Northover P. and La Niece S. (2009) New thoughts on niello. In: Shortland A.J., Freestone I.C. and Rehren T. (eds) From mine to microscope. Advances in the study of ancient technology. Oxbow Books, Oxford. pp. 145–154.

Nugent W.R. (1992) Issues in optical disc longevity, from Trithemius to Arrhenius testing. In: Boston G. (ed.) Archiving the audio-visual heritage. Third Joint Technical Symposium. 3–5 May 1990. Canadian Museum of Civilization. Bucks, Ottawa. pp. 89–102.

Oakley K.B. (1980) Relative dating of the fossil hominids of Europe. Bull. Brit. Mus. Nat. Hist. (Geol.), 34, 1–63.

Oates D. (1960) The Excavations at Nimrud (Kal&hbreve;u). Iraq 23, 1–14.

Oates D. (1990) Innovations in mud-bricks: decorative and structural techniques in ancient Mesopotamia. World Archaeol. 21, 388–406.

Oates J.A.H. (1998) Lime and limestone: Chemistry and technology, production and uses. Wiley-VCH.

O'Brien W.F. (1996) Prehistoric copper mining in the British Isles. Shire, Abingdon.

O'Connor S.A., Brooks M.M., Payton R. and Todd V. (ed.) (1990) Archaeological textiles. Proc. Conference on textiles for archaeological conservators, UKIC Archaeological Section, York, April 1988. The United Kingdom Institute for Conservation, London. pp. 62.

Oddy W.A. (1984) In: Thomson J.M.A, Bassett D.A., Duggan A.J., Lewis G.D. and Fenton A. (eds) Manual of curatorship. A guide to museum practice. Butterworth-Heinemann, Oxford. pp. 480–486.

Oddy W.A. and Caroll S. (ed.) (1999) Reversibility – Does it exist? British Museum, Occasional Paper N. 135. The British Museum, London. pp. 180.

Oddy W.A. (2004) Unmasking forged antiquities. How scientific techniques are used to expose fakes in museum collections. In: Martini M., Milazzo M. and Piacentini M. (eds) Physics methods in archaeometry. SIF, Bologna – IOS Press, Amsterdam. pp. 281–297.

Odell G.H. (2004) Lithic analysis. Springer-Verlag, Berlin. pp. 262.

Odler I. (ed.) (2000) Special inorganic cements. Spon Press, London – New York. pp. 395.

Odlyha M. (2000) Thermal analysis. In: Ciliberto E. and Spoto G. (eds) (2000) Modern analytical methods in art and archaeometry. John Wiley & Sons, New York Chapter 11. pp. 279–320.

Odlyha M., Cohen N.S., Foster G.M. and West R.H. (2000) Dosimetry of paintings: determination of the degree of chemical change in museum exposed test paintings (azurite tempera) by thermal and spectroscopic analysis. Thermochimica Acta 365, 53–63.

O'Donoghue M. (2006) Gems: their sources, descriptions and identification. 6th Edition. Butterworth-Heinemann, London. pp. 873.

O'Dubhghaill C., Jones A.H. (2009) Japanese irogane alloys and patination – a study of production and application. In: Bell E. (ed.) The Santa Fe Symposium on Jewelry Manufacturing Technology 2009. Met-Chem research, Albuquerque, NM.

Oguchi H. (1983) Japanese Shakudo. Its history, properties and production from gold-containing alloys. Gold Bull. 16, 125–132.

Oleson J.P., Bottalico L., Brandon C., Cucitore R., Gotti E. and Hohlfelder R.L. (2006) Reproducing a Roman maritime strcture with Vitruvian pozzolanic concrete. J. Roman Archaeol. 19, 31–52.

Olsson A.M.B., Calligaro T., Colinart S., Dran J.C., Loevestam N.E.G., Moignard B. and Salomon J. (2001) Micro-PIXE analysis of an ancient Egyptian papyrus: Identification of pigments used for the "Book of the Dead". Nucl. Instr. Meth. Phys. Res. B 181, 707–714.

Oppenheim L.A. (1965) On royal gardens in Mesopotamia. J. Near Eastern Studies 24, 328–333.

Organ R.M. (1961) The conservation of cuneiform tablets. The British Museum Quarterly 23, 52–58.

Orton C. (1980) Mathematics in archaeology. Cambridge University Press, Cambridge.

Orton C. (2000) Sampling in archaeology. Cambridge University Press, Cambridge.

Orton C., Tyers P. and Vince A.G. (1993) Pottery in archaeology. Cambridge University Press, Cambridge.

Ottaway B.S. and Wang Q. (2004) Casting experiments and microstructure of archeologically relevant bronzes. BAR International Series 1331. Archaeopress, Oxford.

Ou-Yang H., Paschalis E.P., Mayo W.E., Boskey A.L. and Mendelsohn R. (2001) Infrared microscopic imaging of bone: spatial distribution of CO_3-2. J. Bone Mineral Res. 16, 893–900.

Pääbo S., Poinar H., Serre D., Jaenicke-Després V., Hebler J., Rohland N., Kuch M., Krause J., Vigilant L. and Hofreiter M. (2004) Genetic analyses from ancient DNA. Ann. Rev. Genetics, 38, 645–679.

Pagonis V., Kitis G. and Furetta C. (2006) Numerical and practical exercises in thermoluminescence. Springer-Verlag, Berlin.

Pallot-Frossard I., Loisel C., Marie-Victoire E., Textier A. and Detalle V. (2007) From the monument to the microsample. In: Small samples – Big objects. Proc. EU-Artech Seminar, May 2007. Bayerisches Landesamt fuer Denkmalpfledge, Munchen. pp. 80–95.

Pan Y. and Fleet M.E. (2002) Composition of the apatite-group minerals: substitution mechanisms and controlling factors. In: Kohn M.J., Rakovan J. and Hughes J.M. (eds) Phosphates – Geochemical, geobiological, and materials importance. Rev. Mineral. Geochem. Vol. 48. Mineralogical Society of America–Geochemical Society, Washington, DC. pp. 13–49.

Panagiotaki M., Maniatis Y., Kavoussanaki D., Hatton G. and Tite M.S. (2004) The production technology of Aegean Bronze Age vitreous materials. In: Bourriau F.D. and Phillips J. (ed.) Invention and innovation – The social context of technological change II: Egypt, the Aegean and the Near East, 1650–1150 BC. Oxbow Books, Oxford. pp. 155–180.

Paparazzo E. (2006) Recovering the past from surfaces and interfaces: experimental and cultural issues. Surf. Interface Anal. 38, 357–363.

Paparazzo E. and Moretto L. (1999) X-ray photoelectron spectroscopy and scanning Auger microscopy studies of bronzes from the collections of the Vatican Museums. Vacuum 55, 59–70.

Parfitt S.L. *et al.* (2005) The earliest record of human activity in northern Europe. Nature, 438, 1008–1012.

Parsons M.L. (1997) X-ray methods. In: Ewing G.W. (ed.) Analytical instrumentation handbook. Marcel Dekker, New York.

Pasteris J.D., Wopenka B., Freeman J.J., Rogers K., Valsami-Jones E., van der Houwen J.A.M. and Silva M.J. (2004) Lack of OH in nanocrystalline apatite as a function of degree of atomic order: implications for bone and biomaterials. Biomaterials 25, 229–238.

Paynter S. (2009) Links between glazes and glass in mid-2nd millennium BC Mesopotamia and Egypt. In: Shortland A.J., Freestone I.C. and Rehren T. (eds) From mine to microscope. Advances in the study of ancient technology. Oxbow Books, Oxford. pp. 93–108.

Peacock D.P.S. (1980) The Roman millstone trade: A petrological sketch. World Archaeol. 12, 43–53.

Peacock D.P.S. (1997) Charlemagne's black stones: the re-use of Roman columns in early medieval Europe. Antiquity 71, 709–715.

Pearce M. (2000) What this awl means. Understanding the earliest Italian metalworking. In: Ridgway D., Serra Ridgway F.R., Pearce M., Herring E., Whitehouse R.D. and Wilkins J.B. (eds) Ancient Italy in its Mediterranean setting. Studies in honour of Ellen Macnamara. Accordia Specialist Studies on the Mediterranean. Vol. 4. Accordia Research Institute, University of London. pp. 67–73.

Pecchioni E., Fratini F. and Cantisani E. (2008) Le malte antiche e moderne: tra tradizione ed innovazione. Pàtron Editore, Bologna. pp. 238.

Pedrotti A. (2001) L'età del Rame. In: Lanzinger M., Marzatico F. and Pedrotti A. (eds) Storia del trentino. Vol. I. La preistoria e la protostoria. Società Editrice Il Mulino, Bologna. pp. 183–253.

Pedrotti A. (2009) Il riparo Gaban (Trento) e la neolitizzazione della Valle dell'Adige. In: Kruta V., Kruta Poppi L., Lička M. and Magni E. (eds) Antenate di Venere 27,000–4000 a.C. Skira, Geneve-Milano, pp. 39–47.

Peñalver E., Álvarez-Fernández E., Arias P., Delcòs X. and Ontañón R. (2007) Local amber in a Paleolithic context in Cantabrian Spain: the case of La Garma A. Archaeol. Sci. 34, 843–849.

Penniman T.K. (1952) Pictures of ivory and other animal teeth bone and antler. Occasional Papers on Technology, 5. Pitt Rivers Museum, Oxford.

Perlman I. and Asaro F. (1969) Pottery analysis by neutron activation. Archaeometry 11, 21–52.

Pérez-Arantegui J., Molera J., Larrea A., Pradell T., Vendrell-Saz M., Borgia I., Brunetti B.G., Cariati F., Fermo P., Mellini M., Sgamellotti A. and Viti C. (2001) Microstructure and phase equilibria - Luster pottery from the thirteenth century to the sixteenth century: a nanostructured thin metallic film. J. Am. Ceram. Soc. 84, 442–446.

Pérez-Arantegui J., Larrea A., Molera J., Pradell T. and Vendrell-Saz M. (2004) Some aspects of the characterization of decorations on ceramic glazes. Appl. Phys. A: Mater. Sci. Proces. 79, 235–239.

Pernicka E. (1999) Trace element fingerprinting of ancient copper: A guide to technology or provenance? In: Young S.M.M., Pollard A.M., Budd P. and Ixer R.A. (eds) Metals in Antiquity. BAR International Series no. 792. Archaeopress, Oxford, UK. pp. 163–171.

Pernicka E. (2004) Archaeometallurgy: Examples of the application of scientific methods to the provenance of archeological metal objects. In: Martini M., Milazzo M. and Piacentini M. (eds) Physics methods in archaeometry. SIF, Bologna - IOS Press, Oxford. pp. 309–329.

Peschlow-Bindokat A. (1990) Die Steinbruche von Selinunt. Die Cave di Cusa und die Cave di Barone. Mit einem Beitrag von Ulrich Friedrich Hein. Von Zabern, Mainz am Rhein. pp. 66.

Petit J.R. et al. (1999) Climate and atmospheric history of the past 420,000 years from the Vostok ice core, Antarctica. Nature, 399, 429–436.

Petts D. (2002) The reuse of prehistoric standing stones in Western Britain? A critical consideration of an aspect of Early Medieval monument reuse. Oxford J. Archaeol. 21, 195–209.

Philpotts A.R. and Wilson N. (1994) Application of petrofabric and phase equilibria analysis to the study of a potsherd. J. Archaeol. Sci. 21, 607–618.

Pieraccini M. Guidi G. and Atzeni C. (2001) 3D digitizing of cultural heritage. J. Cultur. Heritage, 2, 63–70.

Pigott V.C. and Ciarla R. (2007) On the origin of metallurgy in prehistoric Southeast Asia: the view from Thailand. In: La Niece S., Hook D. and Craddock P. (eds) metals and mines. Studies in archaeometallurgy. The British Museum – Archetype Publications, London. pp. 76–88.

Pigott V.C., Weiss A. and Natapintu S. (1997) Archaeology of copper production: excavations in the Khao Wong Prachan Valley, Central Thailand. In: Ciarla R. and Rispoli F. (eds) South East Asian Archaeology 1992. Proc. 4th Intern. Conf. of the European Association of South-East Asian Archaeologists. Rome, 28 September–4 October 1992. Istituto Italiano per l' e l'Oriente. pp. 119–157.

Pini R. (2002) A high-resolution Late-Glacial – Holocene pollen diagram from Pian di Gembro (Central Alps, Northern Italy). Vegetation History and Archaeobotany, 11, 251–262.

Pinzari F., Pasquariello G. and De Mico A. (2006) Biodeterioration of paper: A SEM study of fungal spoilage reproduced under controlled conditions. Macromol. Symposia 238, 57–66.

Piperno D., Weiss E., Holst I. and Nadel D. (2004) Processing of wild cereal grains in the Upper Palaeolithic revealed by starch grain analysis. Nature 430, 670–673.

Piquè F. (2005) Science for the conservation of wall paintings. Conservation, The GCI Newsletter 20, no. 2. 21–24.

Pittau M. (1996) I dadi da gioco e la questione dei numerali etruschi. Atti del Sodalizio Glottologico Milanese, vol. XXXV–XXXVI, 1994e 1995. pp. 95–105.

Platzner I.T. (1997) Modern isotope ratio mass spectrometry. John Wiley & Sons and London.

Plenderleith H.J., Mora P., Torraca G. and De Guichen G. (1970) Conservation problems in Egypt. UNESCO, Consultant Contract 33, Report 591. International Center for Conservation, ICCROM N. 17820. Rome.

Plinius (Gaius Secundus Plinius) – Naturalis Historia (77 AD) Engl. Transl.: John Bostock (1855) Pliny the Elder – The Natural History. Taylor and Francis, London.

Poinar G.O. Jr. (1992) Life in amber. Stanford University Press, Stanford, CA.

Poinar G.O. Jr., Lambert J.B. and Wu Y. (2004) NMR analysis of amber in the Zubair formation, Khafji oilfield (Saudi Arabia –Kuwait): coal as an oil source rock? J. Petroleum Geol. 27, 207–209.

Polette L.A., Meitzner G., Yacaman M.J. and Chianelli R.R. (2002) Maya blue: application of XAS and HRTEM to materials science in art and archaeology. Microchem. J. 71, 167–174.

Polla A. (2006) Il vetro nella protostoria: indagine archeometrica e caratterizzazione chimico-fisica. Ph.D. Dissertation. Università degli Studi di Milano.

Pollard A.M. (1998) Archaeological reconstruction using stable isotopes. In: H. Griffiths (ed.) Stable Isotopes. Integration of biological, ecological and geochemical processes. βios Scientific Publishers, Oxford. pp. 285–301.

Pollard A.M. (2001) Archaeological science in the biomolecular century. In: Brothwell D.R., Pollard A.M. (eds) Handbook of archaeological sciences. John Wiley & Sons, Chichester. pp. 295–299.

Pollard A.M. (2009) What a long, strange trip it's been: lead isotopes and archaeology. In: Shortland A.J., Freestone I.C. and Rehren T. (eds) From mine to microscope. Advances in the study of ancient technology. Oxbow Books, Oxford. pp. 181–189.

Pollard A.M. and Heron C. (1996) Archaeological chemistry. The Royal Society of Chemistry, Cambridge. 2nd Edition (2008).

Pollard A.M. and Heron C. (2000) Analytical chemistry in archaeology. In: Meyers R.A. (ed.) Encyclopaedia of analytical chemistry. Vol. 15. John Wiley & Sons, Chichester. pp. 13455–13477.

Pollard A.M., Batt C., Stern B. and Young S.M.M. (2007) Analytical chemistry in archaeology. Cambridge Manuals in Archaeology. Cambridge University Press, Cambridge.

Porck H.J. and Teygeler R. (2001) Preservation Science Survey. An Overview of Recent Developments in Research on the Conservation of Selected Analog Library and Archival Materials. European Commission on Preservation and Access, Amsterdam.

Porter D.A., Easterling K.E. and Sherif M. (2009) Phase transformations in metals and alloys. 3rd Edition. CRC Press, Boca Raton, FL. pp. 520.

Postma H., Schillebeeckx P. and Halbertsma R.B. (2004) Neutron resonance capture analysis of some genuine and fake Etruscan copper alloy statuettes. Archaeometry 46, 635–646.

Postma H., Butler J.J., Schillebeeckx P. and van Eijk C.W.E. (2007) Neutron resonance capture applied to some prehistoric bronze axes. Nuovo Cimento 30C, 105–112.

Potgieter-Vermaak S.S., Godoi R.H.M., Van Grieken R., Potgieter J.H., Oujja M. and Castillejo M. (2005) Micro-structural characterization of black crust and laser cleaning of building stones by micro-Raman and SEM techniques. Spectrochim. Acta Part A 61, 2460–2467.

Potts P.J. and West M. (2008) Portable X-ray fluorescence spectrometry: capabilities for in situ analysis. Royal Society of Chemistry, Cambridge.

Pourbaix M. (1974) Atlas of electrochemical equilibria. National Association of Corrosion Engineers, Houston, TX.

Preusser F. (1991) Scientific and technical examination of the tomb of Queen Nefertari at Thebes. In: Cather S. (ed.) The conservation of wall paintings. Proc. of a Symposium organized by the Courtauld Institute of Art and the Getty Conservation Institute, London, 13–16 July 1987. The Getty Conservation Institute, Los Angeles. pp. 1–12.

Price C.A. (1996) Stone conservation. An overview of current research. The Getty Conservation Institute, Los Angeles. pp. 73.

Price N.S., Kirby Talley Jr M. and Melucco Vaccaro A. (eds) (1996) Historical and philosophical issues in the conservation of cultural heritage. The Getty Conservation Institute, Los Angeles.

Promentilla M.A.B., Sugiyama T., Hitomi T. and Takeda N. (2008) Characterizing the 3D pore structure of hardened cement paste with synchrotron microtomgraphy. J. Adv. Concr. Techn. 6, 1–14.

Protzen J.-P. (1985) Inca quarrying and stonecutting. J. Soc. Archit. Histor. 44, 161–182.

Protzen J.-P. and Nair S. (1997) Who taught the Inca stonemasons their skills? A comparison of Tiahuanaco and Inca cut-stone masonry. J. Soc. Archit. Histor. 56, 146–167.

Pryce T.O. (2009) Prehistoric copper production and technological reproduction in the Khao Wong Prachan Valley of Thailand. Ph.D. Dissertation. University College London.

Pryce T.O., Bassiakos Y., Catpotis M. and Doonan R.C. (2007) 'De caerimoniae' technological choices in copper-smelting furnace design at early Bronze Age Chrysokamino, Crete. Archaeometry 49, 543–557.

Pybus D.H. and Sell C.S. (1999) The chemistry of fragrances. Royal Soc. Chem, Cambridge, UK.

Radonjic M., Allen G., Livesey P., Elton N., Farey M., Holmes S. and Allen J. (2001) ESEM characterisation of ancient lime mortars. J. Building Limes Forum 8, 38–49.

Ragazzi E., Roghi G., Giaretta A. and Gianolla P. (2003) Classification of amber based on thermal analysis. Thermochim. Acta, 404, 43–54.

Ramil A., López A.J. and Yáñez A. (2008) Application of artificial neural networks for the rapid classification of archaeological ceramics by means of laser induced breakdown spectroscopy (LIBS). Appl. Phys. A, 92, 197–202.

Ramirez Barat B. and De la Vioa S. (2001) Characterization of proteins in paint media by immunofluorescence: a note on methodological aspects. Stud. Conservation 46, 282–288.

Rampazzi L., Andreotti A., Bonaduce I., Colombini M.P., Colombo C. and Toniolo L. (2004) Analytical investigation of calcium oxalate films on marble monuments. Talanta 63, 967–977.

Ranalli G., Chiavarini M., Guidetti V., Marsala F., Matteini M., Zanardini E. and Sorlini C. (1997) The use of microorganisms for the removal of sulphates on artistic stoneworks. Intern. Biodeter. Biodegr. 40, The III International Symposium on Biodeterioration and Biodegradation, pp. 255–261.

Ranalli D, Scozzafava M. and Tallini M. (2004) Ground penetrating radar investigations for the restoration of historic buildings: the case study of the Collemaggio Basilica (L'Aquila, Italy). J. Cult. Heritage, 5, 91–99.

Rao K.J. (2002) Structural chemistry of glasses. Elsevier, Amsterdam. pp. 568.

Rapp G.R. (2002) Archaeomineralogy. Springer-Verlag, Berlin. pp. 326.

Rapp G.R., Hill C.L. (1998) Geoarchaeology. The Earth-science approach to archaeological interpretation. Yale University Press, New Haven, Connecticut.

Raubenheimer E.J. (1999) Morphological aspects and composition of Africal elephant (Loxodonta africana) ivory. Koedoe 42, 57–64.

Raymo M.E., Lisiecki L. and Nisancioglu K. (2006) Plio–Pleistocene ice volume, Antarctic climate, and the global $\delta 18O$ record. Science, 313, 492–495.

Raymo M.E. and Huybers P. (2008) Unlocking the mysteries of the ice ages. Nature 451, 284–285.

Read P.G. (2005) Gemmology. 3rd Edition. Butterworth-Heinemann, London. pp. 324.

Reade J. (1996) Mesopotamia. British Museum Press, London.

Realini M. and Toniolo L. (eds) (1996) Proceedings of the II International Symposium on The oxalate films in the conservation of works of art. Milano, 25–27 March 1996. EDITEAM, Bologna.

Rebollo N.R., Cohen-Ofri I., Popovitz-Biro P., Bar-Yosef O., Meignen L., Goldberg P., Weiner S. and Boaretto E. (2008) Structural characterization of charcoal exposed to high and low pH: Implications for 14C sample preparation and charcoal preservation. Radiocarbon 50, 289–307.

Rech J.A. (2004) New uses for old laboratory techniques: How radiocarbon dating of mortar and plaster could change the chronology of the ancient Near East. Near Eastern Archaeol. 67, 212–219.

Redman C. (1999) Human impact on ancient environments. University of Arizona Press, Tucson, AZ.

Reedy C.L. (1994) Thin-section petrography in studies of cultural materials. J. Amer. Inst. Conserv. 33, 115–129.

Reed-Hill R.E. and Abbaschian R. (1991) Physical metallurgy principles. The Pws-Kent Series in Engineering. 3rd Edition. CL-Engineering. pp. 944.

Rehder J.E. (1986) Primitive furnaces and the development of metallurgy. J. Hist. Metall. Soc. 20, 87–92.

Rehder J.E. (2000) The mastery and uses of fire in antiquity. McGill-Queen's Press – MQUP, Montreal-Kingston. pp. 216.

Rehr J.J. and Albers R.C. (2000) Theoretical approaches to X-ray absorption fine structure. Rev. Modern Phys. 72, 621–653.

Rehren T. (2000) Rationales in Old World base glass compositions. J. Archaeol. Sci. 27, 1225–1234.

Rehren T. (2008) A review of factors affecting the composition of early Egyptian glasses and faience: alkali and alkali earth oxides. J. Archaeol. Sci. 35, 1345–1354.

Rehren T. and Pernicka E. (2008) Coins, artefacts and isotopes – Archaeometallurgy and archaeometry. Archaeometry 50, 232–248.

Rehren T. and Pusch E.B. (2005) Late Bronze Age glass production at Qantir-Piramesses, Egypt. Science 308, 1756–1758.

Rehren T., Pusch E.B. and Herold A. (2001) Qantir-Piramesses and the organisation of the Egyptian glass industry. In: Shortland A.J. (ed.) The social context of technological change. Egypt and the Near East, 1650–1550 BC. Oxbow Books, Oxford. pp. 223–238.

Reiche I., Vignaud C. and Menu M. (2000) Heat-induced transformation of fossil mastodon ivory into "odontolite." Structural and elemental characterisation. Solid St. Sci. 2, 625–636.

Reiche I., Vignaud C., Champagnon B., Panczer G., Brouder C., Morin G., Solé V.A., Charlet L. and Menu M. (2001) From mastodon ivory to gemstone: The origin of turquoise color in odontolite. Amer. Mineral. 86, 1519–1524.

Reiche I., Morin G., Brouder C., Solé V.A., Petit P.E., Vignaud C., Calligaro T. and Menu M. (2002) Manganese accomodation in fossilised mastodon ivory and heat-induced colour transformation: Evidence by EXAFS. Eur. J. Mineral. 14, 1069–1073.

Reiche I. and Chalmin E. (2008) Synchrotron radiation and cultural heritage: combined XANES/XRF study at Mn K-edge of blue, grey or black coloured palaeontological and archaeological bone material. J. Anal. Atomic Spectrom. 23, 799–806.

Reilly J.M. (1986) Care and identification of 19th-century photographic prints. Eastman Kodak Co., Rochester.

Reimer P.J., Hughen K.A., Guilderson T.P., McCormac F.G., Baillie M.G.L., Bard E., Barratt P., Beck J.W., Brown D.M., Buck C.E., Damon P.E., Friedrich M., Kromer B., Ramsey C.B., Reimer R.W., Remmele S., Southon J.R., Stuiver M. and van der Plicht J. (2002) Preliminary report of the first workshop of the IntCal04 Radiocarbon Calibration/Comparison Working Group. Radiocarbon 44, 653–61.

Reimer P.J., Baillie M.G.L., Bard E., Bayliss A., Beck J. W., Bertrand C.J.H., Blackwell P.G., Buck C.E., Burr G.S., Cutler K.B., Damon P.E., Edwards R.L., Fairbanks R.G., Friedrich M., Guilderson T.P., Hogg A.G., Hughen K., Kromer B., McCormac G., Manning S., Ramsey C.B., Reimer R.W., Remmele S., Southon J.R., Stuiver M., Talamo S., Taylor F.W., van der Plicht J. and Weyhenmeyer C.E. (2004) INTCAL04 Terrestrial Radiocarbon age calibration, 0–26 cal kyr BP. Radiocarbon, 46, 1029–1058.

Remazeilles C., Quillet V., Calligaro T., Dran J.C., Pichon L. and Salomon J. (2001) PIXE elemental mapping on original manuscripts with an external microbeam. Nucl. Instr. Meth. Phys. Res. B 181, 681–687.

Remondino F., El-Hakim S.F., Gruen A. and Zhang L. (2008) Turning images into 3-D models. Signal Proc. Mag., 25, 55–65.

Renfrew C. (1998) Applications of DNA in archaeology: a review of DNA studies of the Ancient Biomolecules Initiative. Ancient Biomol. 2, 107–116.

Renfrew C. and Bahn P. (2008) Archaeology. Theories, methods, and practice. 5th Edition. Thames and Hudson, New York.

Renzulli A., Santi P., Nappi G., Luni M. and Vitali D. (2002) Provenance and trade of volcanic rock millstones from Etruscan–Celtic and Roman archaeological sites in Central Italy. Eur. J. Miner. 14, 175–183.

Resano M., García-Ruiz E. and Vanhaecke F. (2009) Laser ablation-inductively coupled plasma mass spectrometry in archaeometric research. Mass Spectr. Rev. DOI 10.1002/mas .20220.

Rey C., Collins B., Goehl T., Dickson I.R. and Glimcher M.J. (1989) The carbonate environment in bone mineral: A resolution-enhanced Fourier Transform Infrared Spectroscopy study. Calcif. Tissue Int. 45,157–164.

Rey C., Shimizu M., Collins B. and Glimcher M.J. (1990) Resolution-enhanced Fourier Transform Infrared Spectroscopy study of the environment of phosphate ions in the early deposits of solid phase of calcium phosphate in bone and enamel, and their evolution with age. I: Investigations in the $\nu 4$ PO4 domain. Calcif. Tissue Int. 46, 384–394.

Rey C., Shimizu M., Collins B. and Glimcher M.J. (1991) Resolution-enhanced Fourier Transform Infrared Spectroscopy study of the environment of phosphate ions in the early deposits of solid phase of calcium phosphate in bone and enamel, and their evolution with age. II: Investigations in the $\nu 3$ PO4 domain. Calcif. Tissue Int. 49, 383–388.

Rey C., Miquel J.L., Facchini L., Legrand A.P. and Glimcher M.J. (1995) Hydroxyl Group in bone mineral. Bone 16, 583–586.

Reyes-Valerio C. (1993) De Bonampak al Templo mayor. El azul maya en Mesoamerica. Coleccion América Nuestra, Vol. 40. Siglo Veintiuno Editores, Mexico – Agroasemex, Madrid.

Reynard L.M. and Hedges R.E.M. (2008) Stable hydrogen isotopes of bone collagen in palaeodietary and palaeoenvironmental reconstruction. J. Archaeol. Sci. 35, 1934–1942.

Riccardi M.P., Messiga B. and Duminuco P. (1999) An approach to the dynamics of clay firing. Appl. Clay Sci. 15, 393–409.

Rice P.C. (2006) Amber the golden gem of the ages. 4th Edition. AuthorHouse, Bloomington, IN.

Rice P.M. (1987) Pottery analysis. A sourcebook. The University of Chicago Press, Chicago.

Rice P.M. and Kingery W.D. (eds) (1997) The prehistory & history of ceramic kilns. Ceramic and Civilization, Volume VII. The American Ceramic Society, Westerville, OH.

Richards W.J., Gibbons M.R. and Shields K.C. (2004) Neutron tomography developments and applications. Appl. Rad. Isotopes, 61, 551–559.

Richardson F.D. and Jeffes J.H.E. (1948) The thermodynamics of substances of interest in iron and steel making from 0 °C to 2400 °C: I-Oxides. J. Iron Steel Inst. 160, 261–270.

Richardson I.G. (2002) Electron microscopy of cements. In: Bensted J. and Barnes P. (eds) Structure and performance of cements. 2nd Edition. Spon Press, London–New York. pp. 500–556.

RILEM Paper Lum-D2 (1991) – In-situ stress tests on masonry based on the flat jack. RILEM Recommendations for the testing and use constructions materials. Published by E. & F.N. Spon, London, 1994. pp. 503–505.

RILEM Paper Lum-D3 (1991)- In-situ strength/elasticity tests on masonry based on the flat jack. RILEM Recommendations for the testing and use constructions materials. Published by E. & F.N. Spon, London, 1994. pp. 506–508.

RILEM Recommendations, TC 127-MS. Recomm. MS.D .1 (1996) Measurement of mechanical pulse velocity for masonry. Mater. Struct. 29, 463–466.

Rimstidt J.D. and Vaughan D.J. (2003) Pyrite oxidation: a state-of-the-art assessment of the reaction mechanism. Geochim. Cosmochim. Acta 67, 873–880.

Ripley B.D. (1996) Pattern recognition and neural networks. Cambridge University Press, Cambridge.

Riva G., Bettio C. and Modena C. (1997) The use of sonic wave technique for estimating the efficiency of masonry consolidation by injection. Proc. 11th International Brick/Block Masonry Conference, Shanghai, China, October 1997. pp. 28–39.

Robens E., Benzler B., Büchel G., Reichert H. and Schumacher K. (2002) Investigation of characterizing methods for the microstructure of cement. Cem. Concr. Res. 32, 87–90.

Roberts P. (2000) PhotoHistorica: landmarks in photography: rare images from the collection of the Royal Photographic Society. Artisan, New York.

Roberts R.G. (1997) Luminescence dating in archaeology: from origins to optical. Radiation Measur., 27, 819–892.

Roberts R.G., Walsh G., Murray A., Olley J., Jones R., Morwood M., Tuniz C., Lawson E., Macphail M., Bowdery D. and Naumann I. (1997) Luminescence dating of rock art and past environments using mud-wasp nests in northern Australia. Nature 387, 696–699.

Robinson N., Evershed R.P., Higgs W.J., Jerman K. and Eglinton G. (1987) Proof of a pine wood origin for pitch from Tudor (Mary Rose) and Etruscan shipwrecks: application of analytical organic chemistry in archaeology. Analyst 112, 637–643.

Rodriguez-Navarro C., Doehne E. and Sebastian E. (2000) Influencing crystallization damage in porous materials through the use of surfactants: Experimental results using sodium dodecyl sulfate and cetyldimethylbenzylammonium chloride. Langmuir 16, 947–954.

Rogers A.K. (2008) Obsidian hydration dating: accuracy and resolution limitations imposed by intrinsic water variability. J. Archaeol. Sc., 35, 2009–2016.

Roghi G., Ragazzi E. and Gianolla P. (2006) Triassic Amber of the Southern Alps (Italy). Polaris 21, 143–154.

Rollefson G.O. and Köhler-Rollefson I. (1992) Early Neolithic exploitation patterns in the Levant: Cultural impact on the environment. Popul. Environm.: J. Interdisc. Stud. 13, 243–254.

Ronca P., Tiraboschi C. and Binda L. (1997) In-situ flat-jack tests matching new mechanical interpretations. In: 2nd Int. Brick/Block Masonry Conf., Shanghai, China. Vol. 1. pp. 357–366.

Rossi P.P. (1982) Analysis of mechanical characteristic of brick masonry tested by means of in-situ tests. In 6th IBMaC, Roma, Italy.

Rostan P. and Rossi M. (2002) Approche économique et industrielle du complexe minier et métallurgique de Saint-Véran (Hautes-Alpes) dans le contexte de l'Age du Bronze dans les Alpes du Sud. Bull. Etudes Préhist. Archéol. Alpines, 13, 77–96.

Rostoker W., Pigott V.C. and Dvorak J.R. (1989) Direct reduction to copper metal by oxide–sulfide mineral interaction. Archaeomaterials 3, 69–87.

Rothenberg B. (1999) Archaeo-metallurgical researches in the Southern Arabah 1959–1990, part I: Late Pottery Neolithic to Early Bronze Age IV. Palestine Explor. Quart. 131, 68–89.

Rothenberg B., Tylecote R.F. and Boydell P.J. (ed.) (1979) Chalcolithic copper smelting: Excavations and experiments. Monograph No. 1. Institute for Archaeometallurgical Studies, London.

Rothenberg B. and Merkel J.F. (1995) Late Neolithic Copper Smelting in the Arabah. Inst. Archaeo-Metall. Studies Newsletter 15, 1–8.

Rothenberg J. (1995) Ensuring the longevity of digital documents. Scientific American, 272, 42–47.

Rothenberg J. (1999) Avoiding technological quicksand: finding a viable technical foundation for digital preservation: a report to the Council on Library and Information Resources. Council on Library and Information Resources, Washington D.C.

Rowan Y.M. and Ebeling J.R. (eds) (2008) New approaches to old stones. Recent studies of ground stone artifacts. Equinox Publishing Ltd., London–Oakville.

Rowlands M.J. (1971) The archaeological interpretation of prehistoric metalworking. World Archaeol. 3, 210–224.

Ruskin J. (1849) The Seven Lamps of Architecture. Second edition: G. Allen, Sunnyside, Kent (1880). Reprinted by Dover Publications, New York (1989).

Russ J., Kaluarachchi W.D., Drummond L. and Edwards H.G.M. (1999) The nature of a whewellite-rich rock crust associated with pictographs in Southwestern Texas. Studies Conserv. 44, 91–103.

Rutten F.J.M., Roe M.J., Henderson J. and Briggs D. (2006) Surface analysis of ancient glass artefacts with ToF-SIMS: A novel tool for provenancing? Appl. Surf. Sci. 252, 7124–7127.

Rutter N.W. and Blackwell B. (1995) Amino acid racemization dating. In: Rutter N.W., Catto N.R. (eds) Dating methods for Quaternary deposits. Geological Association of Canada, Geotext 2, 125–167.

Ruvalcaba-Sil J.L., Manzanilla L., Melgar M. and Lozano Santa Cruz R. (2008) PIXE and ionoluminescence for Mesoamerican jadeite characterization. X-Ray Spectrom. 37, 96–99.

Ryder M.L. (1964) Fleece evolution in domestic sheep. Nature 204, 555–559.

Ryder M.L. (1965) Report of textiles from Çatal Hüyük. Anatolian Studies 15, 175–176.

Ryder M.L. (1972) Wool of the 14th Century BC from Tell el-Amarna, Egypt. Nature 240, 355–356.

Ryder M.L. (1973) Ancient Scythian wool from Crimea. Nature 242, 480–481.

Ryder M.L. (1983) A re-assessment of Bronze Age wool. J. Archaeol. Sci. 10, 327–331.

Ryder M.L. (1987) The evolution of fleece. Sci. Amer. 256, 112–119.

Ryder M.L. (1990) Wool remains from Scythian burials in Siberia. Oxford J. Archaeol. 9, 313–321.

Ryder M.L. (1990) Skin, and wool-textile remains from Hallstatt, Austria. Oxford J. Archaeol. 9, 37–49.

Plinius (Gaius Secundus Plinius) – Naturalis Historia (77 AD) Engl. Transl.: John Bostock (1855) Pliny the Elder – The Natural History. Taylor and Francis, London.

Sahagún (Bernardo de Sahagún) – Florentine Codex: General history of the things of New Spain. Engl. Transl.: Arthur J.O. Anderson and Charles E. Dibble (1950–1982) 12 Books in 13 Volumes. University of Utah Press, Salt Lake City.

Salamon M., Tuross N., Arensburg B. and Weiner S. (2005) Relatively well preserved DNA is present in the crystal aggregates of fossil bones. Proc. Natl Acad. Science 102, 13783–13788.

Salomon J., Dran J.-C., Guillou T., Moignard B., Pichon L., Walter P. and Mathis F. (2008) Ion-beam analysis for cultural heritage on the AGLAE facility: impact of PIXE/RBS combination. Applied Physics A 92, 43–50.

Saltzman B. (2001) Dynamical Paleoclimatology: Generalized Theory of Global Climate Change. Academic Press, San Diego.

Sánchez del Río M., Martinetto P., Reyes-Valerio C., Doorhyée E. and Suárez M. (2006) Synthesis and acid resistance of Maya blue pigment. Archaeometry 48, 115–130.

Sánchez del Río M., Gutiérrez-León A., Castro G.R., Rubio-Zuazo J., Solís C., Sánchez-Hernández R., Robles-Camacho J. and Rojas-Gaytán J. (2008) Synchrotron powder diffraction on Aztec blue pigments. Appl. Phys. A 90, 55–60.

Sánchez del Río M., Suárez M. and Garcia-Romero E. (2009) The occurrence of palygorskite in the Yucatan peninsula: ethno-historic and archaeological contexts. Archaeometry 51, 214–230.

Sanders D.H. (2001) Persuade or perish: moving virtual heritage beyond pretty pictures of the past. Proceedings of the Seventh International Conference on Virtual Systems and Multimedia, Berkeley, USA, 25–27 October 2001. IEEE, pp. 236–246.

Sandford M.K. (ed.) (1993) Investigations of ancient human tissue. Gordon and Breach publishers, Langhorne. pp. 456.

Santucci L. and Plossi Zappalà M. (2001) Cellulose viscometric oxidometry. Restaurator 22, 51–65.

Saturno W.A., Sever T.L., Irwin D.E., Howell B.F. and Garrison T.G. (2007) Putting us on the map: remote sensing investigation of the ancient Maya landscape. In: Wiseman J.R., El-Baz F. (Eds.), Remote Sensing in Archaeology. Springer, New York, pp. 137–160.

Sauerbier M., Schrotter G., Lambers K. and Eisenbeiss H. (2006) Multi-resolution image-based visualization of archaeological landscapes in Palpa, Peru. In: Campana S., Forte M. (eds) From Space to Place. Proc. of the 2nd International Conference on Remote Sensing in Archaeology, CNR, Rome, Italy, December 4–7, 2006. BAR International Series 1568. Archaeopress, Oxford. pp. 353–359.

Sauter F., Jordis U. and Hayek E. (1992) Chemisiche untersunchungen der Kittschäftungs-materialien. In: Höpfel H., Platzer W. and Spindler K. (eds) Der Mann im Eis, Band 1, Bericht über das Internationale Symposium 1992. Eigenverlang der Universität Innsbruck, Innsbruck. pp. 435–441.

Sax M., Walsh J.M., Freestone I.C., Rankin A.H. and Meeks N.D. (2008) The origin of two purportedly pre-Columbian Mexican crystal skulls. J. Archaeol. Sci. 35, 2751–2760.

Sayre E.V. and Smith R.W. (1961) Compositional categories of ancient glass. Science 133, 1824–1826.

Sayre E.V. (1963) The intentional use of antimony and manganese in ancient glasses. Advances in Glass Technology, Part 2. Plenum Press, New York. pp. 263–282.

Scarre C. and Healy F. (eds) (1993) Trade and exchange in prehistoric Europe. Oxbow Books, Oxford.

Scarre C. (2005) The human past. World prehistory and the development of human societies. Thames & Hudson, London. pp. 784.

Scherer G.W. (2004) Stress from crystallization of salt. Cem. Concr. Res. 34, 1613–1624.

Schmidt A.R., Ragazzi E., Coppellotti O. and Roghi G. (2006) A microword in Triassic amber. Nature Brief Comm. 444, 835.

Schoenberg R., Nägler T.F. and Kramers J.D. (2000) Precise Os isotope ratio and Re-Os isotope dilution measurements down to the picogram level using multicollector inductively coupled plasma mass spectrometry. Int. J. Mass Spectr. 197, 85–94.

Schoeninger M.J. (1995) Stable isotope studies in human evolution. Evol. Anthrop. 4, 83–98.

Scholz C.A., Johnson T.C., Cohen A.S., King J.W., Peck J.A., Overpeck J.T., Talbot M.R., Brown E.T., Kalindekafe L., Amoako P.Y.O., Lyons R.P., Shanahan T.M., Castaneda I.S., Heil C.W., Forman S.L., McHargue L.R., Beuning K.R., Gomez J. and Pierson J. (2007) East African megadroughts between 135 and 75 thousand years ago and bearing on early-modern human origins. Proc. Natl. Acad. Sci., 104, 16416–16421.

Schreiner M. (1991) Glass of the past: The degradation and deterioration of medieval glass artifacts. Microchim. Acta 104, 255–264.

Schreiner M. (2004) Corrosion of historic glass and enamels. In: Janssens K. and Van Grieken R. (ed.) Non-destructive microanalysis of cultural heritage materials. Comprehensive analytical chemistry series. Vol. XLII. Elsevier, Amsterdam. pp. 713–754.

Schreiner M., Grasserbauer M. and March P. (1988) Quantitative NRA and SIMS depth profiling of hydrogen in naturally weathered medieval glass. Fresenius J. Analyt. Chem. 331, 428–432.

Schuller M., Berra M., Fatticcioni A., Atkinson R. and Binda L. (1994) Use of tomography for diagnosis and control of masonry repairs. Proc. 10th International Brick/Block Masonry Conference. Calgary, Canada, July 1994. pp. 438–447.

Schumann W. (2006) Gemstones of the world. 3rd Edition. Sterling Pub Co Inc., New York. pp. 272.

Schwab R., Heger D., Höppner B. and Pernicka E. (2006) The provenance of iron artefacts from Manching: a multi-technique approach. Archaeometry 48, 433–452.

Schwarcz H.P. (1991) Some theoretical aspects of isotope paleodiet studies. J. Archaeol. Sci. 18, 261–275.

Schwarcz H.P. (1992) Uranium series dating in palaeoanthropology. Evol. Anthropol., 1, 56–62.

Schwarcz H.P. (2002) Chronometric dating in archaeology: A review. Acct. Chem. Res. 35, 637–643.

Schwartz J. (1962) The pernicious influence of mathematics on science. In: Nagel E., Suppes P. and Tarski A. (eds) Logic, Methodology and Philosophy of Science. Proc. of the 1960 International Congress. Stanford University Press, Stanford.

Schwartz A.J., Kumar M. and Adams B.L. (2000) Electron Backscatter Diffraction in Materials Science. Springer Verlag, Heidelberg.

Schweitzer M.H, Suo Z., Avci R., Asara J.M., Allen M.A., Arce F.T. and Horner J.R. (2007) Analyses of soft tissue from Tyrannosaurus rex suggest the presence of protein. Science 316, 277–280.

Sciau P., Relaix S., Roucau C., Kihn Y. and Chabanne D. (2006) Microstructural and microchemical characterization of Roman period terra sigillate slips from archeological sites in Southern France. J. Amer. Ceram. Soc. 89, 1053–1058.

Scollar I., Tabbagh A., Hesse A., Herzog I. (1990) Archaeological prospecting and remote sensing. Topics in remote sensing, 2. Cambridge University Press, Cambridge and New York. pp. 674.

Scott D.A. (1991) Metallography and microstructure of ancient and historic metals. The Getty Conservation Institute, Los Angeles.

Scott D.A. (1995) Goldwork of pre-Columbian Costa Rica and Panama: a technical study. Mat. Res. Soc. Symp. Proc. 352, 499–526.

Scott D.A. (2000) A review of copper chlorides and related salts in bronze corrosion and as painting pigments. Studies Conserv. 45, 39–53.

Scott D.A. (2001) The application of scanning X-ray fluorescence microanalysis in the examination of cultural materials. Archaeometry, 43, 475–482.

Scott D.A. (2002) Copper and bronze in art. Corrosion, colorants, conservation. The Getty Conservation Institute, Los Angeles.

Scott D.A., Newman M., Schilling M., Derrick M.R. and Khanjian H.P. (1996) Blood as a binding medium in a Chumash Indian pigment cake. Archaeometry 38, 103–112.

Scrivener K.L., Füllman T., Gallucci E., Walenta G. and Bermejo E. (2004) Quantitative study of Portland cement hydration by X-ray diffraction/Rietveld analysis and independent methods. Cem. Concr. Res. 34, 1541–1547.

Sebera D.K. (1994) Isoperms: An environmental management tool. Commission on Preservation and Access, Washington.

Seferiades M. (1995) La route neolithique des spondyles de la Mediterranée a la Manche. In: Otte M. (ed.) Nature et culture. Colloque de Liège, 13–17 Décembre 1993. E.R.A.U.L. 68, Liège. pp. 291–358.

Selwitz C. and Doehne E. (2002) The evaluation of crystallization modifiers for controlling salt damage to limestone. J. Cult. Herit. 3, 205–216.

Serpico M. and White R. (2000) Resins, amber and bitumen. In: Nicholson P.T. and Shaw I. (eds) Ancient egyptian materials and technology. Cambridge University Press, Cambridge, UK.

Serruys Y., Tirira J. and Trocellier P. (eds) (1996) Forward recoil spectrometry: Applications to hydrogen determination in solids. Springer-Verlag, Berlin.

Sever T.L. and Irwin D.E. (2003) Landscape archaeology: Remote-sensing investigation of the ancient Maya in the Peten rainforest of northern Guatemala. Ancient Mesoamerica, 14, 113–122.

Shackley M.S. (1998) Gamma rays, X-rays and stone tools: Some recent advances in archaeological geochemistry. J. Arch. Sci., 25, 259–270.

Shackley M.S. (ed.) (1998) Archaeological obsidian studies: Method and theory. Advances in Archaeological and Museum Science, Volume 3. Plenum Press in cooperation with the Society for Archaeological Sciences, New York.

Shackley M.S. (2008) Archaeological petrology and the archaeometry of lithic materials. Archaeometry 50, 194–215.

Shahack-Gross R., Bar-Yosef O. and Weiner S. (1997) Black coloured bones in Hayonim Cave, Israel: Differentiating between burning and oxide staining. J. Archaeol. Sci. 24, 439–436.

Shalev S., Kahanov Y. and Doherty C. (1999) Nails from a 2,400 years old shipwreck: a study of copper in a marine archaeological environment. JOM February 1999, 14–18.

Shelby J.E. (2005) Introduction to glass science and technology. 2nd Edition. The Royal Society of Chemistry, Cambridge. pp. 291.

Shennan S. (1988) Quantifying archaeology. Edinburgh University Press, Edinburgh, and Academic Press, San Diego. 2nd Edition Reprinted (1997).

Shepard A.O. (1962) Maya blue: alternative hypothesis. Amer. Antiquity 27, 565–566.

Shimoyama M., Ninomiya T. and Ozaki Y. (2003) Nondestructive discrimination of ivories and prediction of their specific gravity by Fourier-transform Raman spectroscopy and chemometrics. The Analyst 128, 950–953.

Shimoyama M., Morimoto S. and Ozaki Y. (2004) Non-destructive analysis of the two subspecies of African elephants, mammoth, hippopotamus, and sperm whale ivories by visible and short-wave near infrared spectroscopy and chemometrics. The Analyst 129, 559–563.

Shishlina N.I., Orfinskaya O.V. and Golikov V.P. (2002) Textile from the Bronze Age North Caucasus. In: Piotrovsky Yu.Yu. (ed.) Eurasian steppe in the Prehistory and the Middle Ages. Papers in the Honour of M.P. Gryaznov (St-Petersburg). pp. 253–257. (In Russian.)

Shishlina N.I., Orfinskaia O.V. and Golikov V.P. (2003) Bonze Age textiles from the North Caucasus: new evidence of fourth millennium BC fibres and fabrics. Oxford J. Achaeol. 22, 331–344.

Shortland A.J. (2000) Vitreous materials at Amarna: the production of glass and faience in 18th dynasty Egypt. British Archaeological Reports International Series S827. Archaeopress, Oxford.

Shortland A.J. (2002) The use and origin of antimonate colorants in early Egyptian glass. Archaeometry 44, 517–530.

Shortland A.J. (2004) Evaporites of the Wadi Natrun: seasonal and annual variation and its implication for ancient exploitation. Archaeometry 46, 497–516.

Shortland A.J., Schachner L., Freestone I. and Tite M. (2006) Natron as flux in the early vitreous materials industry: sources, beginnings, and reasons for decline. J. Archaeol. Sci. 33, 521–530.

Shortland A.J., Tite M.S. and Ewart I. (2006) Ancent exploitation and use of the cobalt alums from the Western Oases of Egypt. Archaeometry 48, 153–168.

Shortland A.J., Shishlina N. and Egorkov A. (2007) Origins and production of faience beads in the North Caucasus and the Northwest Caspian Sea region in the Bronze Age. In: Lyonnet B. (ed.) Les cultures du Caucase: leur relations avec le Proche-Orient. Éditions Recherche sur les Civilisations, CNRS Éditions, Paris. pp. 269–283.

Shortland A.J. (2009) The fish's tale: a foreign glassworker at Amarna? In: Shortland A.J., Freestone I.C. and Rehren T. (eds) From mine to microscope. Advances in the study of ancient technology. Oxbow Books, Oxford. pp. 109–114.

Shugar A. and Rehren Th. (2002) Formation and composition of glass as a function of temperature. Glass Technol. 43C, 145–150.

Siano S., Margheri F., Pini R., Mazzinghi P. and Salimbeni R. (1997) Cleaning processes of encrusted marbles by Nd:YAG lasers operating in free-running and Q-switching regimes. Appl. Opt. 36, 7073–7079.

Siano S., Bartoli L., Zoppi M., Kockelmann W., Daymond M., Dann J.A., Garagnani M.G. and Miccio M. (2003) Proc. Archaeometallurgy in Europe, Associazione Italiana di Metallurgia, Milano. Vol 2, 319.

Siart C., Eitel B. and Panagiotopoulos D. (2008) Investigation of past archaeological landscapes using remote sensing and GIS: a multi-method case study from Mount Ida, Crete. J. Arch. Science, 35, 2918–2926.

Sickels L.-B. (1981) Organic additives in mortars. Edinburgh Architecture Research, Vol. 8. Dept. of Architecture, University of Edinbourgh Publ., Edinbourgh. pp. 7–20.

Silvestri A., Molin G. and Salviulo G. (2006) Sand for Roman glass production: an experimental and philological study on source of supply. Archaeometry 48, 415–432.

Silvestri A., Molin G. and Salviulo G. (2008) The colourless glass of Iulia Felix. J. Archaeol. Sci. 35, 331–341.

Silvestri A., Molin G. and Salviulo G. (2008) The coloured glass of Iulia Felix. J. Archaeol. Sci. 35, 1489–1501.

Simon S. and Utz R. (2007) Experiences with large collections and big monuments – how to approach the analytical challenger. In: Small samples – Big objects. Proc. EU-Artech Seminar, May 2007. Bayerisches Landesamt fuer Denkmalpfledge, Munchen. pp. 105–127.

Sinopoli C.M. (1991) Approaches to archaeological ceramics. Plenum Press, New York and London.

Skalski A.W. and Veggiani A. (1990) Fossil resin in Sicily and in the Northern Apennines: Geology and organic content. Prace Muzeum Ziemi 41, 37–49.

Skibo J.M., Schiffer M.B. and Reid K.C. (1989) Organic-tempered pottery: an experimental study. American Antiquity 54, 122–146.

Skibsted J. and Hall C. (2008) Characterization of cement minerals, cements and their reaction products at the atomic and nano scale. Cem. Concr. Res. 38, 205–225.

Slater C., Preston T. and Weaver L.T. (2001) Stable isotopes and the international system of units. Rapid Commun. Mass Spectrom. 15, 1270–1273.

Smith B.N. and Epstein S. (1971) Two categories of 13C/12C ratios for higher plants. Plant Physiology 47, 380–384.

Smith D.C., Edwards H.G.M., Bouchard M., Brody R., Rull-Perez F., Withnall R. and Coupry C., (2000) Mobile Raman microscopy (MRM): a powerful non-destructive polyvalent in situ archæometric tool for microspectrometrical analysis of cultural heritage in the next millennium (ARCHAEORAMAN): geomaterials, biomaterials and pigments. In: Guarino A. (ed.) Proc. of the Second International Congress on 'Science and Technology for the Safeguard of Cultural Heritage in the Mediterranean Basin, vol. 2. Nanterre, Paris, 5–9 July 1999. Elsevier, Amsterdam, pp. 1373–1375.

Smith D.C. (2006) A review of the non-destructive identification of diverse geomaterials in the cultural heritage using different configurations of Raman spectroscopy. Geol. Soc. London, Special Public., 257, 9–32.

Smith G.D. and Clark R.J.H. (2004) Raman microscopy in archaeological science. J. Archaeol. Sci., 32, 1137–1160.

Smith C.I., Craig O.E., Prigodich R.V., Nielsen-Marsh C.M., Jans M.M.E., Vermeer C. and Collins M.J. (2005) Diagenesis and survival of osteocalcin in archaeological bone. J. Archaeol. Sci. 32, 105–113.

Smith P.R. and Wilson M.T. (2001) Blood residues in archaeology. In: Brothwell D.R. and Pollard A.M. (eds) (2001) Handbook of archaeological sciences. John Wiley & Sons, Chichester. pp. 313–322.

Soffer O., Vandiver P.B., Klima B. and Svoboda J. (1993) The pyrotechnology of performance art: Moravian Venuses and Wolverines. In: Knecht H., Pike-Tay A. and White R. (eds) Before Lascaux: The complex record of the Early Upper Paleolithic. CRC Press, Baton Rouge, LA. pp. 259–275.

Sonck-Koota P., Lindroos A., Lill J.-O., Rajander J., Viitanen E.-M., Marra F., Pehkonen M.H., Suksi J. and Heselius S.-J. (2008) External-beam PIXE characterization of volcanic material used in ancient Roman mortars. Nucl. Inst. Meth. Phys. Res. B 266, 2367–2370.

Song Y.X. and KaiWu T.G. (1982) The exploitation of the works of Nature. First written in 1587. Times and Culture Publishing Co., Taibei.

Sowers J.M. (2000) Rock varnish chronometry. In: Noller J.S., Sowers J.M. and Lettis W.R. (eds.), Quaternary Geochronology: Methods and Applications. American Geophysical Union Reference Shelf 4, Washington. pp. 241–260.

Speakman R.J. and Glascock M.D. (2007) Acknowledging fifty years of neutron activation analysis in archaeology. Archaeometry 49, 179–183.

Spindler K. (1993) The man in the Ice. Weidenfeld and Nicholson, London.

Spoto G. (2000) Secondary ion mass spectrometry in art and archaeology. Thermochim. Acta 365, 157–166.

Spragg R.A. (2000) IR spectroscopy sample preparation methods. In: Lindon J.C., Tranter G.E. and Holmes J.L. (eds) Encyclopedia of spectroscopy and spectrometry. Vol. 2. Academic Press, London. pp. 1058–1071.

Stadtarchäologie Wien (2004) Enter the Past: The E-way Into the Four Dimensions of Cultural Heritage: CAA 2003, Computer Applications and Quantitative Methods in Archaeology : Proceedings of the 31st Conference, Vienna, Austria, April 2003. British Archaeological Reports, Archaeopress, Oxford.

Stauffer A. (2002) Tessuti. In: Von Eles P. (ed.) Guerriero e sacerdote. Autorità e comunità nell' del Ferro a Verucchio. La tomba del trono. Quaderni di Archeologia dell' Romagna. Vol. 6, Firenze. pp. 192–234.

St. Clair L.L. and Seaward M.R.D. (2004) Biodeterioration of stone surfaces: lichens and biofilms as weathering agents of rocks and cultural heritage Springer-Verlag, Berlin.

Stech Wheeler T., Maddin R. and Muhly J.D. (1979) Ancient metallurgy: Materials and techniques. J. Metals 31, 16–18.

Steiger M. (2005) Crystal growth in porous materials – II Influence of crystal size on the crystallization pressure. J. Cryst. Growth 282, 470–481.

Stern B., Heron C., Tellefsen T. and Serpico M. (2008) New investigations into the Uluburun resin cargo. J. Archaeol. Sci. 35, 2188–2203.

Sternberg R.S. (2008) Archaeomagnetism in archaeometry – A semi-centennial review. Archaeometry, 50, 983–998.

Stevenson C.M., Abdelrehim I.M. and Novak S.W. (2001) Infra-red photoacoustic and secondary ion mass spectrometry measurements of obsidian hydration rims. Journ. Arch. Sci., 28, 109–115.

Stos Z.A. (2009) Across the wine dark seas... sailor tinkers and royal cargoes in the Late Bronze Age Eastern Mediterranean. In: Shortland A.J., Freestone I.C. and Rehren T. (eds) From mine to microscope. Advances in the study of ancient technology. Oxbow Books, Oxford. pp. 163–180.

Stos-Gale Z.A. (1995) Isotope archaeology – a review. In: Beavis J. and Barker K. (eds) Science and site. Bournemouth University School of Conservation Sciences, Poole, UK. pp. 12–28.

Stout E.C., Beck C.W. and Kosmowska-Ceranowicz B. (1995) Gedanite and gedanosuccinite. In: Anderson K.B. and Crelling J.C. (eds) Amber, resinite, and fossil resins. American Chemical Society, Washington, DC. pp. 130–148.

Stout E.C., Beck C.W., Anderson K.B. (2000) Identification of rumanite (Romanian amber) as thermally altered succinite (Baltic amber). Physics and Chemistry of Minerals, 27, 665–678.

Stulik D.C. (2005) Getty Conservation Institute portable analytical laboratory for photograph conservation: the first three years. In: 14th triennial meeting: preprints (ICOM Committee for Conservation), The Hague, September 12–16, 2005. The Hague. pp. 556–564.

Stulik D.C. (2008) Une Nouvelle Methodologie Pour L' des Photographies. In: Cartier-Bresson A. (ed.) Le Vocabulaire Technique de la Photographie. Marval, Paris. pp. 436–451.

Stulik D.C. and Wright P. (2007) Challenging the museum's collections, the research project between the Getty Conservation Institute and the National Media Museum. Archive February 2007, 28–32.

Stulik D.C. and Kaplan A. (2008) A new scientific methodology for provenancing and authentication of 20th century photographs: nondestructive approach. Available at: http://www.ndt.net/article/art2008/papers/050Stulik.pdf

Sturgeon T. (1950) The dreaming jewels. Greenberg. pp. 217.

Taft W.S., Mayer J.W., Aderhold H.C., Keller M. and Rizzo G. (1992) Neutron induced autoradiography and PIXE analysis. In: Vandiver P.B., Druzik J.R., Wheeler G.S. and Freestone I.C. (eds) Proc. Symposium on Materials issues in art and archaeology III. 27 April–1 May 1992, San Francisco. Materials Research Society, Pittsburgh. pp. 319–323.

Taniguchi Y., Hirao Y., Shimadzu Y. and Tsuneki A. (2002) The first fake? Imitation turquoise beads recovered from a Syrian neolithic site, Tell El-Kerkh. Stud. Conserv. 47, 175–183.

Targowski P., Rouba B., Góra M., Tymińska-Widmer L., Marczak J. and Kowalczyk A. (2008) Optical coherence tomography in art diagnostics and restoration. Appl. Phys. A, 92, 1–9.

Taylor E.W. (1949) Correlation of the Mohs's scale of hardness with the Vickers's hardness numbers. Mineral. Mag. 28, 718–721.

Taylor H.F.W. (1997) Cement chemistry. 2nd Edition. Thomas Thelford, London.

Taylor R.E. (2000) Fifty years of radiocarbon dating. Am. Sci., 88, 60–67.

Taylor R.E. and Aitken M.J. (1997) Chronometric dating in archaeology. Plenum, New York.

Technè (1999) Couleur et perception. Centre de Recherche et de Restauration des Musées de France. N. 9–10. pp. 176.

Technè (2007) La couleur des peintres. Centre de Recherche et de Restauration des Musées de France. N. 26. pp. 130.

Thickett D., Odlyha M. and Ling D. (2002) An improved firing treatment for cuneiform tablets. Studies in Conservation 47, 1–11.

Thornton C.P. (2007) Of brass and bronze in prehistoric South west Asia. In: La Niece S., Hook D. and Craddock P. (eds) metals and mines. Studies in archaeometallurgy. The British Museum – Archetype Publications, London. pp. 123–135.

Tilley C. (1999) Metaphor and material culture. Blackwell, Oxford.

Timberlake S. (2007) The use of experimental archaeology/archaeometallurgy for the understanding and reconstruction of Early Bronze Age Mining and Smelting. In: La Niece, Hook D. and Craddock P.T. (eds) Metals and Mines. Archetype Publications, London. p. 37–45.

Tite M.S. (1972) Methods of physical examination in archaeology. Seminar Press, London.

Tite M.S. (1991) Archaeological science – past achievements and future prospects. Archaeometry, 31, 139–151.

Tite M.S. (1996) In defence of lead isotope analysis. Antiquity 70, 959–962.

Tite M.S. (2001) Overview – Materials study in archaeology. In: Brothwell D.R. and Pollard A.M. (eds) Handbook of archaeological sciences. John Wiley & Sons, Chichester. pp. 443–448.

Tite M.S. and Bimson M. (1989) Glazed steatite: an investigation of the methods of glazing used in ancient Egypt. World Archaeol. 21, 87–100.

Tite M.S., Freestone I., Mason R., Molera J., Vendrell-Saz M. and Wood N. (1998) Lead glazes in antiquity – Methods of production and reasons for use. Archaeometry 40, 241–260.

Tite M.S. and Shortland A.J. (2008) Production technology of faience and related vitreous materials. Monograph 72. Oxford University School of Archaeology, Oxford. pp. 232.

Tite M.S., Pradell T. and Shortland A.J. (2008) Discovery, production and use of tin-based opacifiers in glasses, enamels, and glazes from the Late Iron Age onwards: a reassessment. Archaeometry 50, 67–84.

Todd J.M (1993) The continuity of amber artefacts in ancient Palestine from the Bronze Age to the Byzantines. Proc. of the Second Conference on Amber in Archaeology, Liblice, 1990. pp. 236–248.

Tonidandel L., Ragazzi E., Roghi G. and Traldi P. (2008) Mass Spectrometry in the characterization of ambers. I. Studies of amber samples of different origin and ages by laser desorption ionization, atmospheric pressure chemical ionization and atmospheric pressure photoionization mass spectrometry. Rapid Commun. Mass Spectr. 22, 630–638.

Towle A., Henderson J., Bellintani P. and Gambacurta G. (2001) Frattesina and Adria: report of scientific analyses of early glass from the Veneto. Padusa 37, 7–68.

Tratebas A.M. and Chapman F. (1996) Ethical and conservation issues in removing lichens from petroglyphs. Rock Art Research, 13, 129–133.

Trevisani E., Papazzoni C.A., Ragazzi E., Roghi G. (2005) Early Eocene amber from the "Pesciara di Bolca" (Lessini Mountains, Northern Italy). Palaeo, 223, 260–274.

Trueman C.N.G., Behrensmeyer A.K., Tuross N. and Weiner S. (2004) Mineralogical and compositional changes in bones exposed on soil surfaces in Amboseli National Park, Kenya: Diagenetic mechanisms and the role of sediment pore fluids. J. Archaeol. Sci. 31, 721–739.

Trueman C.N.G., Privat K. and Field J. (2008) Why do crystallinity values fail to predict the extent of diagenetic alteration of bone mineral? Palaeog. Palaeoclim. Palaeoecol. 266, 160–167.

Tsoucaris G., Martinetto P., Walter P. and Leveque J.L. (2001) Chemistry and cosmetic materials in ancient civilizations. Annales Pharmaceutiques Francaises 59, 415–422.

Tsoucaris G. and Lipkowski J. (eds) (2003) Molecular and structural archaeology: cosmetic and therapeutic chemical. NATO Science Series II Mathematics, Physics and Chemistry. Vol. 117. Kluwer Academic Publisher, Dordrecht–Boston–London. pp. 272.

Tykot R.H. (2004) Scientific methods and applications to archaeological provenance studies. In: Martini M., Milazzo M. and Piacentini M. (eds) (2004) Physics methods in archaeometry. SIF, Bologna – IOS Press, Amsterdam. pp. 407–432.

Tylecote R.F. (1976) A history of metallurgy. The Metals Society, London.

Tylecote R.F. (1985) The apparent tinning of bronze axes and other artifacts. J. Hist. Metal. Soc. 19, 169–175.

Tylecote R.F. (1987) The early history of metallurgy in Europe. Longman, London. pp. 391.

Tylecote R.F. (1992) A history of metallurgy. 2nd Edition. The Institute of Materials, London. P. 205. (1st edition 1976).

Tuniz C. (2001) Accelerator mass spectrometry: ultra-sensitive analysis for global science. Rad. Phys. Chem., 61, 317–322.

Turner W.E.S. (1956) Studies in ancient glasses and glass-making processes. Part V. Raw materials and melting processes. J. Soc. Glass Techn. 40, 276T–300T.

Turrell G. and Corset J. (eds) (1996) Raman microscopy: Developments and applications. Academic Press, London.

Uhlig H.H. and Revie R.W. (1985) Corrosion and corrosion products. John Wiley & Sons, New York.

Ungar T., Martinetto P., Ribarik G., Dooryhee E., Walter P., Anne M. (2002) Revealing the powdering methods of black makeup in Ancient Egypt by fitting microstructure based Fourier coefficients to the whole x-ray diffraction profiles of galena. J. Appl. Phys. 91, 2455–2465.

Urey H.C. (1947) The thermodynamic properties of isotopic substances. J. Chem. Soc. 562–581.

Valaczkai T. and Ghiurca V. (1997) Amber from Romania. Sonderheft Metalla 63–66.

Valladas H. (1992) Thermoluminescence dating of flint. Quat. Sci. Rev. 11, 1–5.

Valladas H. (2003) Direct radiocarbon dating of prehistoric cave paintings by accelerator mass spectrometry. Measur. Sci. Technology 14, 1487–1492.

Valle S., Zanzi L., Binda L., Saisi A. and Lenzi G. (1998) Tomography for NDT applied to masonry structures: Sonic and/or EM methods. In: Sinopoli A. (ed.) Arch bridges. Balkema, Rotterdam, pp. 243–252.

Valluzzi M.R. (2007) On the vulnerability of historical masonry structures: analysis and mitigation. RILEM Materials and Structures 40, 723–743.

Valluzzi M.R., da Porto F. and Modena C. (2004) Behavior and modeling of strengthened three-leaf stone masonry walls. RILEM Materials and Structures, MS 267, Vol. 37, April 2004, pp. 184–192.

Van Balen K. (2005) Carbonation reaction of lime, kinetics at ambient temperature. Cem. Concr. Res. 35, 647– 657.

Vandamme E.J., De Baets S. and Steinbuchel A. (2002) Biopolymers, Volume 6, Polysaccharides II: Polysaccharides from Eukaryotes. Wiley-VCH, Weinheim.

Van Deman E.B. (1912a) Methods of determining the date of Roman concrete monuments I. Amer. J. Archaeol. 16, 230–251.

Van Deman E.B. (1912b) Methods of determining the date of Roman concrete monuments II. Amer. J. Archaeol. 16, 387–432.

Vandenabeele P. (2004) Raman spectroscopy in art and archaeology. J. Raman Spectr. 35, 607–609.

Vandenabeele P. and Moens L. (2005) A study of artists' materials based on Raman spectroscopy and total-reflection X-ray fluorescence analysis (TXRF). In: Van

Grieken R. and Janssens K. (eds) Cultural heritage conservation and environmental impact assessment by non-destructive testing and micro-analysis. Taylor and Francis Group, London. pp. 27–35.

Van den Haute P. and De Corte F. (1998) Advances in fission-track geochronology. Kluwer Academic Publishers, Dordrecht. pp. 331.

Van der Plicht J., Van der Sanden W.A.B., Aerts A.T. and Streurman H.J. (2004) Dating bog bodies by means of 14C-AMS. J. Archaeol. Sci. 31, 471–491.

Van der Snickt G., De Nolf W., Vekemans B. and Janssens K. (2008) μ-XRF/ μ-RS vs. SR μ-XRD for pigment identification in illuminated manuscripts. Appl. Phys. A 92, 59–68.

Van der Snickt G., Dik J., Cotte M., Janssens K., Jaroszewicz J., De Nolf W., Groenewegen J. and Van der Loeff L. (2009) Characterization of a degraded cadmium yellow (CdS) pigment in an oil painting by means of synchrotron radiation based X-ray techniques. Anal. Chem. 81, 2600–2610.

Vandiver P.B., Soffer O., Klima B. and Svoboda J. (1989) The Origins of Ceramic Technology at Dolni Vestonice, Czechoslovakia. Science 246, 1002–1008.

Vandiver P.B. and Vasil' S.A. (2002) A 16,000 year-old ceramic human-figurine from Maina, Russia. Mat. Res. Soc. Symp. Proc. 712, II6.9.1- II6.9.11.

Van Gool L., Waelkens M., Mueller P., Vereenooghe T. and Vergauwen M. (2004) Total recall: a plea for realism in models of the past. Proceeding of XXth ISPRS Congress, 12–23 July 2004 Istanbul, Turkey. Commisson 5. Pp. 332–343.

Van Strydonck M., Dupas M., Dauchot-Dehon M., Pachiaudi C. and Marechal J. (1986) The influence to contaminating (fossil) carbonate and the variations of $\delta 13C$ in mortar dating. Radiocarbon 28, 702–710.

Van Strydonck M., Vanderborg K., De Jong A. and Keppens E. (1992) Radiocarbon dating of lime fractions and organic material from buildings. Radiocarbon 34, 873–879.

Vavra N. (1993) Chemical characterization of Fossil Resins ("Amber") – A critical review of methods, problems and possibilities: Determination of mineral species, botanical sources and geographical attribution. Proc. of a Symposium held in Neukirchen am Grobvenediger (Salzburg/Austria), September 1990. Volker Höck – Friedrich Koller editors. pp. 147–157.

Veiga J.P. and Figueiredo M.O. (2008) Calcium in ancient glazes and glasses: a XAFS study. Appl. Phys. A 92, 229–233.

Veiga J.P. and Figueiredo M.O. (2008) A XANES study on the structural role of zinc in ancient tile glazes of Portuguese origin. X-Ray Spectrom. 37, 458–461.

Velde B. (1992) Introduction to clay minerals. Chapman & Hall, London.

Velde B. and Druc I.C. (1999) Archaeological ceramic materials: Origin and utilization. Springer-Verlag, Berlin.

Verhoeven J.D. (1975) Fundamentals of physical metallurgy. John Wiley & Sons, New York. pp. 592.

Viles H.A. (2001) Scale issue in weathering studies. Geomorphol. 41, 63–72.

Vitruvius (Marcus Vitruvius Pollio) (70 BC–15 AD) De Architectura. Engl. Transl.: Rowland I.D., Howe T.H. (eds) (1999) Vitruvius: Ten Books on Architecture, Cambridge University Press, Cambridge.

Vlaardingerbroek M.T. and den Boer J.A. (1999) Magnetic resonance imaging. Springer, New York.

Vrba E.S., Denton G.H., Partridge T.C. and Burckle L.H. (eds) (1996) Paleoclimate and evolution, with emphasis on human origins. Yale University Press, New Haven, Connecticut.

Wagner G.A. (1998) Age determination of young rocks and artifacts. Physical and chemical clocks in Quaternary geology and archaeology. Springer, Berlin. pp. 466.

Wagner G.A. and Van den Haute P. (1992) Fission-track-dating. Ferdinand Enke, Stuttgart – Kluwer Academic Publishing, Dordrecht. pp. 285.

Wagner U., Gebhard R., Grosse G., Hutzelmann T., Murad E., Riederer J., Shimad I. and Wagner F.E. (1998) Clay: An important raw material for prehistoric man. Hyperfine Interactions 117, 323–335.

Walderhaug O. and Walderhaug E.M. (1998) Weathering of Norwegian rock art – a critical review. Norwegian Archaeological Review, 31, 119–139.

Walker C.B.F. (1987) Cuneiform. University of California Press, Los Angeles.

Wall E.J. (1897) Dictionary of photography for the amateur and professional photographer. Hazell, Watson and Viney, London.

Walsh J. (2008) Legends of the crystal skulls. Archaeology 61, 36–40.

Walter P., Menu M. and Dran J.-C. (1992) Dating of archaeological flints by fluorine depth profiling: new insights int the mechanism of fluorine uptake. Nucl. Instrum. Meth. Phys. Res. B 64, 494–498.

Walter P., Martinetto P., Tsoucaris G., Breniaux R., Lefebvre M.A., Richard G. and Talabot J. Dooryhee E. (1999) Making make-up in Ancient Egypt. Nature 397, 483–484.

Walter R.C. (1994) Age of Lucy and the first family. Geology, 22, 6–10.

Walter R.C. and Aronson J.L. (1982) Revisions of K/Ar ages for the Hadar hominid site, Ethiopia. Nature, 296, 122–127.

Ward-Perkins J.B. (1972) Quarrying in antiquity: Technology, tradition and social change. Oxford University Press, Oxford.

Warren B.E. (1969) X-ray diffraction. Addison-Wesley, Reading, MA.

Watchman A.L. (1991) Age and composition of oxalate-rich crusts in the Northern Territory, Australia. Studies Conserv. 36, 24–32.

Watchman A.L. (2000) A review of the history of dating rock varnishes. Earth Sci. Rev. 49, 261–277.

Waterbolk H.T. and Butler J.J. (1965) Comments on the use of metallurgical analysis in prehistoric studies. Helinium, 5, 227–251.

Wayman M. and Craddock P.T. (1993) Wu tong: a neglected Chinese decorative technology. In: La Niece S. and Craddock P.T. (eds) (1993) Metal plating and patination: Cultural, technical and historical developments. Butterworth-Heinemann, London. pp. 128–134.

Wedepohl K.H. (1995) The composition of the continental crust. Geochim. Cosmochim. Acta 59, 1217–1232.

Wehmiller J.F. and Miller G.H. (2000) Aminostratigraphic dating methods in Quaternary geology. In Noller J.S., Sowers J.M. and Lettis W.R. (eds) Quaternary geochronology: methods and applications. American Geophysical Union, Washington D.C. pp. 187–222.

Weiner S. and Bar-Yosef O. (1990) States of preservation of bones from prehistoric sites in the Near East: a survey. J. Archaeol. Sci. 17, 187–196.

Weiner S., Kustanovich Z., Gil-Av E. and Traub W. (1980) Dead-sea scroll parchments – unfolding of the collagen molecules and racemization of aspartic-acid. Nature 287, 820–823.

Weiner S. and Price P.A. (1986) Disaggregation of bone into crystals. Calcif. Tissue Intern. 39, 365–375.

Weiner S. and Wagner H.D. (1998) The material bone: Structure–mechanical function relations. Ann. Rev. Mater. Sci. 28, 271–298.

Weisgerber G. (2003) Spatial organization of mining and smelting at Feinan, Jordan: Mining archaeology beyond the history of technology. In: Craddock P., Lang J. (eds) Mining and metal production through the ages. The British Museum Press, London. pp. 76–89.

Weisgerber G. and Goldenberg G. (eds) (2004) Alpenkupfer – Rame delle Alpi. Der Anschnitt, Beiheft 17. Deutsches Bergbau-Museum, Bochum.

Weisler M.I. and Clague D.A. (1998) Characterization of archaeological volcanic glass from Oceania: the utility of three techniques. In: Shackely M.S. (ed.) Archaeological obsidian studies: Method and theory. Advances in Archaeological and Museum Science, Volume 3. Plenum Press in cooperation with the Society for Archaeological Sciences, New York. pp. 103–128.

Wenk H.R. (2002) Texture and anisotropy. In: Karato S.-I. and Wenk H.R. (eds) Plastic deformation in minerals and rocks. Rev. Mineral. Geochem. Vol. 51. The Mineralogical Society of America, Washington, DC. pp. 291–329.

Wenk H.R. and Van Houtte P. (2004) Texture and anisotropy. Rep. Prog. Phys. 67, 1367–1428.

Wenk H.R. (2006) Neutron scattering in Earth Sciences. Reviews in Mineralogy and Geochemistry. Vol. 63. The Mineralogical Society of America, Washington.

Wernecke D.C. (2008) A burning question: Maya lime technology and the Maya forest. J. Ethnobiol. 28, 200–210.

Werner R.A. and Brand W.A. (2001) Referencing strategies and techniques in stable isotope ratio analysis. Rapid Commun. Mass Spectrom. 15, 501–519.

Wertime T.A. (1964) Man's first encounters with metallurgy. Science 146, 1257–1267.

Wertime T.A. and Wertime S.F. (eds) (1982) Early pyrotechnology: The evolution of the first fire-using industries. Smithsonian Institution Press, Washington.

Wertime T.A. (1983) The furnace versus the goat: The pyrotechnologic industries and Mediterranean deforestation in antiquity. J. Field Archaeol. 10, 445–452.

Wess T., Alberts I., Hiller J., Drakopoulos M., Chamberlain A.T. and Collins M. (2001). Microfocus Small Angle X-ray Scattering reveals structural features in archaeological bone samples: detection of changes in bone mineral habit and size. Calcif. Tissue Int. 70, 103–110.

West R.G. (1971) Studying the past by pollen analysis. Oxford University Press, Oxford.

Weyl W.A. (1951) Coloured glass. Society of Glass Technology, Sheffield, UK [Reprinted 1999].

Wheatley D., Gillings M. (2002) Spatial Technology and Archaeology: The Archaeological Applications of GIS. CRC Press.

White W.M., Albarède F. and Télouk P. (2000) High precision analysis of Pb isotope ratios by multi-collector ICP-MS. Chem. Geol. 167, 257–270.

Whitmore P.M. (2002) Pursuing the fugitive: Direct measurement of light sensitivity with micro-fading tests. In: Stratis H.K. and Salvesen B. (eds) The broad spectrum: Studies in the materials, techniques, and conservation of color on paper. Archetype Publications, London. Pp. 241–244.

Willerslev E. and Cooper L. (2005) Ancient DNA. Proc. R. Soc. B 272, 3–16.

Williams A. and Edge D. (2007) The metallurgy of some Indian swords. Gladius 27, 149–176.

Williams R. (2004) Lime kilns and lime burning. Osprey Publishing, Oxford. pp. 52.

Williams-Thorpe O. (1995) Obsidian in the Mediterranean and the Near East: a provenancing success story. Archaeometry 37, 217–248.

Williams-Thorpe O. (2008) A thousand and one columns: Observations on the Roman granite trade in the Mediterranean area. Oxford J. Archaeol. 27, 73–89.

Williams-Thorpe O. and Thorpe R.S. (1993) Geochemistry and trade of eastern Mediterranean millstones from the Neolithic to Roman periods. J. Archaeol. Sci. 20, 263–309.

Williams-Thorpe O., Jones M.C., Potts P.J. and Webb P.C. (2006) Preseli dolerite bluestones: axe-heads, Stonehenge monoliths, and outcrop sources. Oxford J. Arch. 25, 29–46.

Willis A.R. and Cooper D.B. (2008) Computational reconstruction of ancient artifacts. IEEE Signal Processing Magazine, 25, 65–83.

Wilson L. and Pollard A.M. (2001) The provenance hypothesis. In: Brothwell D.R. and Pollard A.M. (eds) Handbook of archaeological sciences. John Wiley & Sons, Chichester. pp. 507–517.

Wilson M.A., Carter M.A., Hall C., Hoff W.D., Ince C., Savage S.D., McKay B. and Betts I.M. (2009) Dating fired-clay ceramics using long-term power law rehydroxylation kinetics. Proc. Roy. Soc. A., in press.

Winkler W., Kirchner E.Ch., Asenbaum A. and Musso M. (2001) A Raman spectroscopic approach to the maturation process of fossil resins. *J. Raman Spectr.* **32**, 59–63.

Winter J. (2005) In: Proc. of the Workshop on "Scientific examination of art. Modern techniques in conservation and analysis". Washington, DC, March 19–21, 2003. The National Academies Press, Washington.

Wintle A.G. (1996) Archaeologically-relevant dating techniques for the next century. Small, hot and identified by acronyms. J. Archaeological Science, 23, 123–138.

Wintle A.G. (1998) Luminescence dating: laboratory procedures and protocols. Radiat. Meas. 27, 769–818.

Wintle A.G. (2008) Fifty years of luminescence dating. Archaeometry 50, 276–312.

Wintle A.G., Murray A.S. (2006) A review of quartz optically stimulated luminescence characteristics and their relevance in single-aliquot regeneration dating protocols. Radiation Meas. 41, 369–391.

Wolfe A.P., Tappert R., Muehlenbachs K., Boudreau M., McKellar R., Basinger F.J. and Garret A. (2009) A new proposal concernine the botanical origin of Baltic amber. Proc. Roy. Soc B, in press. doi: 10.1098/rspb .2009.0806.

Woll A.R., Bilderback D.H., Gruner S., Gao N., Huang R., Bisulca C. and Mass J. (2005) Confocal X-ray fluorescence (XRF) microscopy: A new technique for the nondestructive compositional depth profiling of paintings. Mater. Res. Soc. Symp. Proc. 852, OO2.5.1–OO2.5.10.

Wong H.S., Head M.K. and Buenfeld N.R. (2006) Pore segmentation of cement-based materials from backscattered electron images. Cem. Concr. Res. 36, 1083–1090.

Wood N. (1999) Chinese glazes: their chemistry, origins and recreation. A&C Black, London.

Wood N. (2009) Some implications of the use of wood ash in Chinese stoneware glazes of the 9th-12th centuries. In: Shortland A.J., Freestone I.C. and Rehren T. (eds) From mine to microscope. Advances in the study of ancient technology. Oxbow Books, Oxford. pp. 51–59.

Worsley P. (1990) Lichenometry. In: Goudie A. (ed.) Geomorphological techniques. 2nd Edition. Unwin Hyman, London. pp. 442–428.

Wouters J. (1987) Analyse des colorants des tapisseries brugeoises des XVIe et XVIIe siècles. In "Bruges et la tapisserie". Mouscron, Bruges. pp. 515–526.

Wouters J. (1993) Kleurstofanalyse van Koptisch textiel (Dye analysis of Coptic textiles). In: De Moor A. (ed.) "Koptisch Textiel" (Coptic Textiles). Publicaties van het Provinciaal Archeologisch museum van Zuid-Oost-Vlaanderen, site Velzeke. pp. 53–64.

Wouters J. (2001) The dye of Rubia peregrina. Dyes Hist. Archaeol. 16/17, 145–157.

Wouters J. (2008) Protecting cultural heritage: reflections on the position of science in multidisciplinary approaches. Chem. Intern. 30, 4–7.

Wouters J. and Rosario-Chirinos N. (1992) Dyestuff analysis of Precolumbian Peruvian textiles by high performance liquid chromatography and diode-array detection. J. Amer. Inst. Conservation 31, 237–255.

Wouters J. and Verhecken A. (1989) The Coccid insect dyes. HPLC and computerized diode-array analysis of dyed yarns. Stud. Conservation 34, 189–200.

Wouters J., Vanden Berghe I., Richard G., Breniaux R. and Cardon D. (2008) Dye analyses of selected textiles from three Roman sites in the Eastern desert of Egypt a hypothesis on the dyeing technology in Roman and Coptic Egypt. Dyes Hist. Archaeol. 21, 1–16.

Wright L.E. and Schwarcz H.P. (1996) Infrared and isotopic evidence for diagenesis of bone apatite at Dos Pilas, Guatemala: Palaeodietary implications. J. Archaeol. Sci. 23, 933–944.

Yalçin Ü. (2000) Anfänge der metallverwendung in Anatolien. In: Anatolian metal I. Der Anschnitt Beihefte 13. Deutsche Bergbaum-Museum, Bochum. pp. 17–30.

Yamamoto S., Otto A., Krumbiegel G. and Simonet B.R.T. (2006) The natural product biomarkers in succinite, glessite and stantienite ambers from Bitterfeld, Germany. Rev. Palaeobot. Palynol. 140, 27–49.

Yamauchi K., Taniguchi Y. and Uno T. (2007) International Symposium on the Conservation and Restoration of Cultural Property: Mural paintings of the Silk Road : cultural exchanges between East and West : proceedings of the 29th Annual International Symposium on the Conservation and Restoration of Cultural Property, National Research Institute for Cultural Properties, Tokyo, January 2006. Japan Center for International Cooperation in Conservation, and Tokyo Bunkazai Kenkyujo. Archetype, London.

Yang F.W., Zhang B.J., Pan C.C. and Zeng Y.Y. (2009) Traditional mortar represented by sticky rice lime mortar – One of the great inventions in ancient China. Science in China, Series E: Technological Sciences 52, 1641–1647.

Yates E., Valdor J., Haslam S., Morris H., Dell A., Mackie W. and Knox J. (1996) Characterization of carbohydrate structural features recognized by anti-arabinogalactan-protein monoclonal antibodies. Glycobiology 6, 131–139.

Yellin J. and Maeir A.M. (2007) Four decades of Instrumental Neutron Activation Analysis and its contribution to the archaeology of the ancient Land of Israel. Isr. J. Earth Sci. 56, 123–132.

Ynsa M.D., Chamón J., Gutiérrez P.C., Gomez-Morilla I., Enguita O., Pardo A.I., Arroyo M., Barrio J., Ferretti M. and Climent-Font A. (2008) Study of ancient Islamic gilded pieces combining PIXE-RBS on external microprobe with sem images. Appl. Phys. A: Mater. Sci. Proces. 92, 235–241.

Young S.M.M., Pollard A.M., Budd P. and Ixer R.A. (eds) (1999) Metals in antiquity. BAR International Series, Vol. N. 792. Archaeopress, Oxford.

Yu H.Y., Chen D. (2004) Protection and development of Qiantan River's dyke constructed in Ming and Qing dynasty as a tourism resource. Zhejiang Hydrotech. 134, 9–10.

Yubao L., Klein C.P.A.T., Zhang X.D. and de Groot K. (1993) Relationship between the colour change of hydroxyapatite and the trace element manganese. Biomat. 14, 969–972.

Zacharias N., Christodoulos M., Philaniotou-Hadjianastasiou O., Hein A. and Bassiakos Y. (2006) Fine-grain TL dating of archaeometallurgical furnace walls. J. Cult. Heritage 7, 23–29.

Zachariasen W.H. (1932) The atomic arrangement in glass. J. Am. Chem. Soc. 54, 3841–3851.

Zeng Y.Y., Zhang B.J. and Liang X. (2008) A case study and mechanism investigation of typical mortars used on ancient architecture in China. Thermochimica Acta 473, 1–6.

Zerboni A. (2008) Holocene rock varnish on the Messak plateau (Libyan Sahara): Chronology of weathering processes. Geomorphology 102, 640–651.

Zhang J., Harbottle G., Wang C. and Kong Z. (1999) Oldest playable musical instruments found at Jiahu early Neolithic site in China. Nature 401, 366–368.

Zhang J., Xiao X. and Lee Y.K. (2004) The early development of music. Analysis of the Jiahu bone flutes. Antiquity 78, 769–778.

Zhao C. and Wu X. (2000) The dating of Chinese early pottery and a discussion of some related problems. Documenta Praehist. 27, 233–239.

Zhu X.K., O' R.K., Guo Y., Belshaw N.S. and Rickard D. (2000) Determination of natural Cu-isotope variation by plasma source mass spectrometry: implications for use as geochemical tracers. Chem. Geol. 163, 139–149.

Ziegler J.F., Wu C.P., Williams P., White C.W., Terreault B., Scherzer B.M.U., Schulte R.L., Schneid E.J., Magee C.W., Ligeon E., 'Ecuyer J.L., Lanford W.A., Kuehne F.J., Kamykowski E.A., Hofer W.O., Guivarc'h A., Filleux C.H., Deline V.R., Evans C.A. Jr., Cohen B.L., Clark G.J., Chu W.K., Brassard C., Blewer R.S., Behrisch R., Appleton B.R. and Allred D.D. (1978) Profiling hydrogen in materials using ion beams. Nucl. Instr. Meth. 149, 19–39.

Zimmer C. (2001) How old is it? Solving the riddle of ages. Natl. Geographic, 200, 78–101.

Zimmerman D.W. and Huxtable J. (1971) Thermoluminescent dating of Upper Palaeolithic fired clay from Dolní Vêstonice. Archaeometry 13, 53–57.

Zouridakis N., Salliege J.F., Person J.F. and Filippakis S. (1987) Radiocarbon dating of mortars from ancient Greek palaces. Archaeometry 29, 60–68.

Zucchiatti A., Bouquillon A., Lanterna G., Lucarelli F., Mandò P.A., Prati P., Salomon J. and Vaccari M.G. (2002) PIXE and µ-PIXE analysis of glazes from terracotta sculptures of the della Robbia workshop. Nucl. Instr. Meth. Phys. Res. B 189, 358–363.

Zucchiatti A., Bouquillon A. and Katona I. and D'Alessandro A. (2006) The 'Della Robbia blue': a case study for the use of cobalt pigments in ceramics during the Italian Renaissance. Archaeometry 48, 131–152.

Permission list

All reasonable effort has been made to contact the holders of copyright in materials reproduced in this book. Any omissions will be rectified in future printings if notice is given to the publisher.

All websites quoted in this book have been visited before July 15th, 2009.

Fig. 1.4 With permission of the Ministero per i Beni e le Attività Culturali, Archivio Fotografico SBSAE di Modena e Reggio Emilia.

Fig. 1.5 With permission of Royal Ontario Museum, © ROM.

Fig. 1.6 Compagnia Generale Riprese Aeree di Parma, S.M.A. authorization no. 248, 19/03/1987.

Fig. 1.7 Includes material © [2006]GeoEye, Telespazio for Italy, all rights reserved.

Fig. 2.4.8 © The J. Paul Getty Trust 1991, all rights reserved.

Fig. 2.6 N.A. Sharp, NOAO/NSO/Kitt Peak FTS/AURA/NSF.

Fig. 2.7 © 2009 GILBERTO ARTIOLI. Permission is granted to copy, distribute and/or modify this document under the terms of the GNU Free Documentation License, Version 1.2 or any later version published by the Free Software Foundation; with no Invariant Sections, no Front-Cover Texts, and no Back-Cover Texts. A copy of the licence is included in the section entitled "GNU Free Documentation License".

Fig. 2.9 Reprinted from *Nuclear Instruments and Methods in Physics Research Section B: Beam Interactions with Materials and Atoms*, Vol. 181, No. 1–4, C. Remazeille, V. Quillet, T. Calligaro, J. C. Dran, L. Pichon and J. Salomon, "PIXE elemental mapping on original manuscripts with an external microbeam. Application to manuscripts damaged by iron-gall ink corrosion", pp. 681–687, © 2001, with permission of Elsevier.

Fig. 2.10 Reprinted from *Nuclear Instruments and Methods in Physics Research Section B: Beam Interactions with Materials and Atoms*, Vol. 181, No. 1–4, A. -M. B. Olsson, T. Calligaro, S. Colinart, J.C. Dran, N. E. G. Lövestam, B. Moignard and J. Solomon, "Micro-PIXE analysis of an ancient Egyptian papyrus: Identification of pigments used for the *Book of the Dead*", pp. 707–714 © 2001, with permission of Elsevier.

Fig. 2.11 Reprinted by permission from Macmillan Publishers Ltd: *Nature*, Vol. 422, No. 6927, Y. Chaimanee, D. Jolly, M. Benammi, P. Tafforeau, D. Duzer *et al.*, "A Middle Miocene hominoid from Thailand and orangutan origins", pp. 61–65, © Nature 2003.

Fig. 2.14 With permission of Museo Gaetano Chierici di Paletnologia.

Fig. 2.15 With permission of NARDINI EDITORE © Nardini Press srl.

Fig. 2.17 © The J. Paul Getty Trust 1991, all rights reserved.

Fig. 2.19 With permission of SCANCO MEDICAL AG.

Fig. 2.20 With permission of NARDINI EDITORE © Nardini Press srl.

Fig. 2.21 With permission of NARDINI EDITORE © Nardini Press srl.

Fig. 2.22 With permission of Museo Archeologico dell'Alto-Adige, ©Museo Archeologico dell'Alto-Adige, http://www.iceman.it.

Fig. 2.25 With permission of NOAA, data sources: Berger and Loutre 1991 and Kawamura *et al.* 2007.

Fig. 2.b.2 Modified and reproduced with permission of John Wiley & Sons Inc., from Jenkins R., 1999, *Fluorescence Spectrometry*, 2nd Edition, Wiley-Interscience, © John Wiley & Sons, Inc. (1999).

Fig. 2.e.1 © The J. Paul Getty Trust 1991, all rights reserved.

Fig 2.h.1 Reprinted from *II Mathematics, Physics and Chemistry, NATO science series*, Vol. 117, 2003, Page No. 107, "Molecular and structural archaeology: cosmetic and therapeutic chemical", Martinetto P., Anne M., Dooryhée E., Isnard O. and Walter P. In: G. Tsoucaris and J. Lipkowski (eds), with kind permission of Springer Science and Business Media.

Fig. 2.h.2 With permission of AIM, Associazione Italiana Metallurgia and Salvatore Siano, CNR, Istituto di Fisica Applicata "Nello Carrara".

Fig. 2.i.3 Reprinted from *Journal of the American Ceramic Society*, Vol. 84, Pérez-Arantegui J., Molera J., Larrea A., Pradell T., Vendrell-Saz M., Borgia I., Brunetti B.G., Cariati F., Fermo P., Mellini M., Sgamellotti A. and Viti C., "Microstructure and phase equilibria. Lustre pottery from 13th to the 16th century: a nanostructured thin metallic film", pp. 442–446, © 2001, with permission of Wiley-Blackwell.

Fig. 2.l.2 Reprinted from "Crystals and phase transitions in protohistoric glass materials", Artioli G., Angelini I. and Polla A., *Phase Transitions,* 2008, Taylor & Francis, by permission of the publisher (Taylor & Francis Group, http://www.informaworld.com).

Fig. 2.m.1 With permission of C2RMF, Centre de Recherche et de Restauration des Musées de France.

Fig. 2.m.2 © Ospedale Regionale di Bolzano/Museo Archeologico dell'Alto-Adige.

Fig. 2.m.3 *Il Nuovo Cimento C*, vol. 30, 2007, pp. 93–104, Lehmann E.H., Vontobel P. and Frei G., 2007, "The non-destructive study of museum objects by means of neutron imaging methods and results of investigations", with kind permission of Società Italiana di Fisica.

Fig. 2.n.1 Modified and reprinted from *Nuclear Physics A*, vol. 752, Johansen G.A., "Nuclear tomography methods in industry", pp. 696c–705c, © 2005, with permission of Elsevier.

Fig. 2.n.6 Reprinted from "Crystals and phase transitions in protohistoric glass materials", Artioli G., Angelini I. and Polla A., *Phase Transitions,* 2008, Taylor & Francis, by permission of the publisher (Taylor & Francis Ltd, http://www.tandf.co.uk/journals).

Fig. 2.o.1 Reprinted from *Nuclear Physics A*, vol. 752, R. Leonardi "Nuclear physics and painting", pp. 659c–674c, © 2005, with permission of Elsevier.

Fig. 2.o.2 Reprinted from *Nuclear Physics A*, vol. 752, R. Leonardi "Nuclear physics and painting", pp. 659c–674c, © 2005, with permission of Elsevier.

Fig. 2.o.3 © All rights reserved. Reproduced with the permission of the Canadian Conservation Institute of the Department of Canadian Heritage, 2009.

Fig. 2.p.1 Permission is granted to copy, distribute and/or modify this document under the terms of the GNU Free Documentation License, Version 1.2 or any later version published by the Free Software Foundation; with no Invariant Sections, no Front-Cover Texts, and no Back-Cover Texts. A copy of the licence is included in the section entitled "GNU Free Documentation License".

Fig. 2.p.4 Includes material © (2006) GeoEye, Telespazio for Italy, all rights reserved.

Fig. 2.q.1 Alinari, Fratelli Archivi Alinari © Archivio Alinari, Firenze.

Fig. 2.q.2 Reprinted from *Journal of Mathematical Imaging and Vision*, vol. 24, 2006, pp. 359–373, "Nonlinear Projection Recovery in Digital Inpainting for Color Image Restoration", Fornasier M., with kind permission of Springer Science and Business Media.

Fig. 2.q.3 Huang Q.-X., Flöy S., Gelfand N., Hofer M. and Pottmann H. "Reassembling fractured objects by geometric matching", ACM Trans. Graph., vol. 25, pp. 569–578, © 2006 Association for Computing Machinery, Inc. Reprinted by permission. http://doi.acm.org/213215.

Fig. 2.q.4 Huang Q.-X., Flöy S., Gelfand N., Hofer M. and Pottmann H. "Reassembling fractured objects by geometric matching", ACM Trans. Graph., vol. 25, pp. 569–578, © 2006 Association for Computing Machinery, Inc. Reprinted by permission. http://doi.acm.org/213215.

Fig. 2.u.1 With permission of the University of Arizona.

Fig. 2.u.2 With permission of the University of Arizona.

Fig. 3.1 Ashby M.F., "Technology of the 1990s: advanced materials and predictive design", *Philosophical Transaction of the Royal Society. Series A, Mathematical and Physical Sciences*, vol. 322, pp. 393–407, 1987, Royal Society.

Fig. 3.6 With permission of the Archivio della Soprintendenza per i Beni Librari Archivistici ed Archeologici della Provincia Autonoma di Trento. Photo: Renato Perini.

Fig. 3.7 With permission of the Ufficio dei Beni Culturali, Sezione dello Sviluppo Territoriale, Divisione dello Sviluppo Territoriale e della Mobilità, Dipartimento del Territorio, Repubblica e Canton Ticino.

Fig. 3.17 With permission of Galleria Regionale della Sicilia – Palermo.

Fig. 3.18 With permission of Regione Siciliana Assessorato dei Beni Culturali e Ambientali e della Pubblica Istruzione - Dipartimento dei Beni Culturali e Ambientali, della Educazione Permanente e dell'Architetura e dell'Arte Contemporanea - Soprintendenza per i Beni Culturali ed Ambientali, Servizio per i Beni Archeologici – Trapani.

Fig. 3.21 Permission is granted to copy, distribute and/or modify this document under the terms of the GNU Free Documentation License, Version 1.2 or any later version published by the Free Software Foundation; with no Invariant Sections, no Front-Cover Texts, and no Back-Cover Texts. A copy of the licence is included in the section entitled "GNU Free Documentation License".

Fig. 3.34 © Field Museum, #A105154c_241509.

Fig. 3.36(a) With permission of INAH – Instituto Nacional de Antropología e Historia.

Fig. 3.38 Permission is granted to copy, distribute and/or modify this document under the terms of the GNU Free Documentation License, Version 1.2 or any later version published by the Free Software Foundation; with no Invariant Sections, no Front-Cover Texts, and no Back-Cover Texts. A copy of the licence is included in the section entitled "GNU Free Documentation License".

Fig. 3.42(a, b, c) Images © the Metropolitan Museum of Art. Reproduction of any kind is prohibited without express written permission in advance from The Metropolitan Museum of Art.

Fig. 3.43 With permission of Rijksmuseum van Oudheden.

Fig. 3.45 With permission of Ministero per i Beni e le Attività Culturali – Soprintendenza per i Beni Archeologici della Sardegna. Reproduction of any kind is prohibited without express written permission in advance from the Ministero per i Beni e le Attività Culturali – Soprintendenza per i Beni Archeologici della Sardegna.

Fig. 3.46 With permission of Ministero per i Beni e le Attività Culturali – Archivio della Soprintendenza per i Beni Archeologici del Piemonte. Reproduction of any kind is prohibited without express written permission in advance from the Ministero per i Beni e le Attività Culturali – Archivio della Soprintendenza per i Beni Archeologici del Piemonte.

Fig. 3.49 © The Trustees of the British Museum. All rights reserved

Fig. 3.50 With permission of Musei Civici di Pavia.

Fig. 3.51 With permission of Museo Civico Gaetano Filangieri.

Fig. 3.57 With permission of The J. Paul Getty Museum.

Fig. 3.75 With permission of Ministero per i Beni e le Attività Culturali – Dipartimento per i Beni Culturali e Paesaggistici – Direzione Regionale per i Beni Culturali e Paesaggistici del Lazio - Soprintendenza per i Beni Archeologici dell'Etruria meridionale. Reproduction and distribution of any kind are prohibited without express written permission in advance from the Ministero per i Beni e le Attività Culturali – Dipartimento per i Beni Culturali e Paesaggistici – Direzione Regionale per i Beni Culturali e Paesaggistici del Lazio - Soprintendenza per i Beni Archeologici dell'Etruria meridionale.

Fig. 3.91 © The J. Paul Getty Trust 1991, all rights reserved.

Fig. 3.92 © The J. Paul Getty Trust 1991, all rights reserved.

Fig. 3.d.7 © The J. Paul Getty Trust 1991, all rights reserved.

GNU Free Documentation License

Version 1.3, 3 November 2008; copyright (C) 2000, 2001, 2002, 2007, 2008 Free Software Foundation, Inc. <http://fsf.org/>; everyone is permitted to copy and distribute verbatim copies of this license document, but changing it is not allowed.

0. Preamble

The purpose of this License is to make a manual, textbook, or other functional and useful document "free" in the sense of freedom: to assure everyone the effective freedom to copy and redistribute it, with or without modifying it, either commercially or noncommercially. Secondarily, this License preserves for the author and publisher a way to get credit for their work, while not being considered responsible for modifications made by others.

This License is a kind of "copyleft", which means that derivative works of the document must themselves be free in the same sense. It complements the GNU General Public License, which is a copyleft license designed for free software.

We have designed this License in order to use it for manuals for free software, because free software needs free documentation: a free program should come with manuals providing the same freedoms that the software does. But this License is not limited to software manuals; it can be used for any textual work, regardless of subject matter or whether it is published as a printed book. We recommend this License principally for works whose purpose is instruction or reference.

1. Applicability and Definitions

This License applies to any manual or other work, in any medium, that contains a notice placed by the copyright holder saying it can be distributed under the terms of this License. Such a notice grants a world-wide, royalty-free license, unlimited in duration, to use that work under the conditions stated herein. The "Document", below, refers to any such manual or work. Any member of the public is a licensee, and is addressed as "you". You accept the license if you copy, modify or distribute the work in a way requiring permission under copyright law.

A "Modified Version" of the Document means any work containing the Document or a portion of it, either copied verbatim, or with modifications and/or translated into another language.

A "Secondary Section" is a named appendix or a front-matter section of the Document that deals exclusively with the relationship of the publishers or authors of the Document to the Document's overall subject (or to related matters) and contains nothing that could fall directly within that overall subject. (Thus, if the Document is in part a textbook of mathematics, a Secondary Section may not explain any mathematics.) The relationship could be a matter of historical connection with the subject or with related matters, or of legal, commercial, philosophical, ethical or political position regarding them.

The "Invariant Sections" are certain Secondary Sections whose titles are designated, as being those of Invariant Sections, in the notice that says that the Document is released under this License. If a section does not fit the above definition of Secondary then it is not allowed to be designated as Invariant. The Document may contain zero Invariant Sections. If the Document does not identify any Invariant Sections then there are none.

The "Cover Texts" are certain short passages of text that are listed, as Front-Cover Texts or Back-Cover Texts, in the notice that says that the Document is released under this License. A Front-Cover Text may be at most 5 words, and a Back-Cover Text may be at most 25 words.

A "Transparent" copy of the Document means a machine-readable copy, represented in a format whose specification is available to the general public, that is suitable for revising the document straightforwardly with generic text editors or (for images composed of pixels) generic paint programs or (for drawings) some widely available drawing editor, and that is suitable for input to text formatters or for automatic translation to a variety of formats suitable for input to text formatters. A copy made in an otherwise Transparent file format whose markup, or absence of markup, has been arranged to thwart or discourage subsequent modification by readers is not Transparent. An image format is not Transparent if used for any substantial amount of text. A copy that is not "Transparent" is called "Opaque".

Examples of suitable formats for Transparent copies include plain ASCII without markup, Texinfo input format, LaTeX input format, SGML or XML using a publicly available DTD, and standard-conforming simple HTML, PostScript or PDF designed for human modification. Examples of transparent image formats include PNG, XCF and JPG. Opaque formats include proprietary formats that can be read and edited only by proprietary word processors, SGML or XML for which the DTD and/or processing tools are not generally available, and the machine-generated HTML, PostScript or PDF produced by some word processors for output purposes only.

The "Title Page" means, for a printed book, the title page itself, plus such following pages as are needed to hold, legibly, the material this License requires to appear in the title page. For works in formats which do not have any title page as such, "Title Page" means the text near the most prominent appearance of the work's title, preceding the beginning of the body of the text.

The "publisher" means any person or entity that distributes copies of the Document to the public.

A section "Entitled XYZ" means a named subunit of the Document whose title either is precisely XYZ or contains XYZ in parentheses following text that translates XYZ in another language. (Here XYZ stands for a specific section name mentioned below, such as "Acknowledgements", "Dedications", "Endorsements", or "History".) To "Preserve the Title" of such a section when you modify the Document means that it remains a section "Entitled XYZ" according to this definition.

The Document may include Warranty Disclaimers next to the notice which states that this License applies to the Document. These Warranty Disclaimers are considered to be included by reference in this License, but only as regards disclaiming warranties: any other implication that these Warranty Disclaimers may have is void and has no effect on the meaning of this License.

2. Verbatim Copying

You may copy and distribute the Document in any medium, either commercially or noncommercially, provided that this License, the copyright notices, and the license notice saying this License applies to the Document are reproduced in all copies, and that you add no other conditions whatsoever to those of this License. You may not use technical measures to obstruct or control the reading or further copying of the copies you make or distribute. However, you may accept compensation in exchange for copies. If you distribute a large enough number of copies you must also follow the conditions in section 3.

You may also lend copies, under the same conditions stated above, and you may publicly display copies.

3. Copying in Quantity

If you publish printed copies (or copies in media that commonly have printed covers) of the Document, numbering more than 100, and the Document's license notice requires Cover Texts, you must enclose the copies in covers that carry, clearly and legibly, all these Cover Texts: Front-Cover Texts on the front cover, and Back-Cover Texts on the back cover. Both covers must also clearly and legibly identify you as the publisher of these copies. The front cover must present the full title with all words of the title equally prominent and visible. You may add other material on the covers in addition. Copying with changes limited to the covers, as long as they preserve the title of the Document and satisfy these conditions, can be treated as verbatim copying in other respects.

If the required texts for either cover are too voluminous to fit legibly, you should put the first ones listed (as many as fit reasonably) on the actual cover, and continue the rest onto adjacent pages.

If you publish or distribute Opaque copies of the Document numbering more than 100, you must either include a machine-readable Transparent copy along with each Opaque copy, or state in or with each Opaque copy a computer-network location from which the general network-using public has access to download using public-standard network protocols a complete Transparent copy of the Document, free of added material. If you use the latter option, you must take reasonably prudent steps, when you begin distribution of Opaque copies in quantity, to ensure that this Transparent copy will remain thus accessible at the stated location until at least one year after the last time you distribute an Opaque copy (directly or through your agents or retailers) of that edition to the public.

It is requested, but not required, that you contact the authors of the Document well before redistributing any large number of copies, to give them a chance to provide you with an updated version of the Document.

4. Modifications

You may copy and distribute a Modified Version of the Document under the conditions of sections 2 and 3 above, provided that you release the Modified Version under precisely this License, with the Modified Version filling the role of the Document, thus licensing distribution and modification of the Modified Version to whoever possesses a copy of it. In addition, you must do these things in the Modified Version:

A. Use in the Title Page (and on the covers, if any) a title distinct from that of the Document, and from those of previous versions (which should, if there were any, be listed in the History section of the Document). You may use the same title as a previous version if the original publisher of that version gives permission.
B. List on the Title Page, as authors, one or more persons or entities responsible for authorship of the modifications in the Modified Version, together with at least five of the principal authors of the Document (all of its principal authors, if it has fewer than five), unless they release you from this requirement.
C. State on the Title page the name of the publisher of the Modified Version, as the publisher.

D. Preserve all the copyright notices of the Document.
E. Add an appropriate copyright notice for your modifications adjacent to the other copyright notices.
F. Include, immediately after the copyright notices, a license notice giving the public permission to use the Modified Version under the terms of this License, in the form shown in the Addendum below.
G. Preserve in that license notice the full lists of Invariant Sections and required Cover Texts given in the Document's license notice.
H. Include an unaltered copy of this License.
I. Preserve the section Entitled "History", Preserve its Title, and add to it an item stating at least the title, year, new authors, and publisher of the Modified Version as given on the Title Page. If there is no section Entitled "History" in the Document, create one stating the title, year, authors, and publisher of the Document as given on its Title Page, then add an item describing the Modified Version as stated in the previous sentence.
J. Preserve the network location, if any, given in the Document for public access to a Transparent copy of the Document, and likewise the network locations given in the Document for previous versions it was based on. These may be placed in the "History" section. You may omit a network location for a work that was published at least four years before the Document itself, or if the original publisher of the version it refers to gives permission.
K. For any section Entitled "Acknowledgements" or "Dedications", Preserve the Title of the section, and preserve in the section all the substance and tone of each of the contributor acknowledgements and/or dedications given therein.
L. Preserve all the Invariant Sections of the Document, unaltered in their text and in their titles. Section numbers or the equivalent are not considered part of the section titles.
M. Delete any section Entitled "Endorsements". Such a section may not be included in the Modified version.
N. Do not retitle any existing section to be Entitled "Endorsements" or to conflict in title with any Invariant Section.
O. Preserve any Warranty Disclaimers.

If the Modified Version includes new front-matter sections or appendices that qualify as Secondary Sections and contain no material copied from the Document, you may at your option designate some or all of these sections as invariant. To do this, add their titles to the list of Invariant Sections in the Modified Version's license notice. These titles must be distinct from any other section titles.

You may add a section Entitled "Endorsements", provided it contains nothing but endorsements of your Modified Version by various parties—for example, statements of peer review or that the text has been approved by an organization as the authoritative definition of a standard.

You may add a passage of up to five words as a Front-Cover Text, and a passage of up to 25 words as a Back-Cover Text, to the end of the list of Cover Texts in the Modified Version. Only one passage of Front-Cover Text and one of Back-Cover Text may be added by (or through arrangements made by) any one entity. If the Document already includes a cover text for the same cover, previously added by you or by arrangement made by the same entity you are acting on behalf of, you may not add another; but you may replace the old one, on explicit permission from the previous publisher that added the old one.

The author(s) and publisher(s) of the Document do not by this License give permission to use their names for publicity for or to assert or imply endorsement of any Modified Version.

5. Combining Documents

You may combine the Document with other documents released under this License, under the terms defined in section 4 above for modified versions, provided that you include in the combination all of the Invariant Sections of all of the original documents, unmodified, and list them all as Invariant Sections of your combined work in its license notice, and that you preserve all their Warranty Disclaimers.

The combined work need only contain one copy of this License, and multiple identical Invariant Sections may be replaced with a single copy. If there are multiple Invariant Sections with the same name but different contents, make the title of each such section unique by adding at the end of it, in parentheses, the name of the original author or publisher of that section if known, or else a unique number. Make the same adjustment to the section titles in the list of Invariant Sections in the license notice of the combined work.

In the combination, you must combine any sections Entitled "History" in the various original documents, forming one section Entitled "History"; likewise combine any sections Entitled "Acknowledgements", and any sections Entitled "Dedications". You must delete all sections Entitled "Endorsements".

6. Collections of Documents

You may make a collection consisting of the Document and other documents released under this License, and replace the individual copies of this License in the various documents with a single copy that is included in the collection, provided that you follow the rules of this License for verbatim copying of each of the documents in all other respects.

You may extract a single document from such a collection, and distribute it individually under this License, provided you insert a copy of this License into the extracted document, and follow this License in all other respects regarding verbatim copying of that document.

7. Aggregation with Independent Works

A compilation of the Document or its derivatives with other separate and independent documents or works, in or on a volume of a storage or distribution medium, is called an "aggregate" if the copyright resulting from the compilation is not used to limit the legal rights of the compilation's users beyond what the individual works permit. When the Document is included in an aggregate, this License does not apply to the other works in the aggregate which are not themselves derivative works of the Document.

If the Cover Text requirement of section 3 is applicable to these copies of the Document, then if the Document is less than one half of the entire aggregate, the Document's Cover Texts may be placed on covers that bracket the Document within the aggregate, or the electronic equivalent of covers if the Document is in electronic form. Otherwise they must appear on printed covers that bracket the whole aggregate.

8. Translation

Translation is considered a kind of modification, so you may distribute translations of the Document under the terms of section 4. Replacing Invariant Sections with translations requires special permission from their copyright holders, but you may include translations of some or all Invariant Sections in addition to the original versions of these Invariant Sections. You may include a translation of this License, and all the license notices in the Document, and any Warranty Disclaimers, provided that you also include the original English version of this License and the original versions of those notices and disclaimers. In case of a disagreement between the translation and the original version of this License or a notice or disclaimer, the original version will prevail.

If a section in the Document is Entitled "Acknowledgements", "Dedications", or "History", the requirement (section 4) to Preserve its Title (section 1) will typically require changing the actual title.

9. Termination

You may not copy, modify, sublicense, or distribute the Document except as expressly provided under this License. Any attempt otherwise to copy, modify, sublicense, or distribute it is void, and will automatically terminate your rights under this License.

However, if you cease all violation of this License, then your license from a particular copyright holder is reinstated (a) provisionally, unless and until the copyright holder explicitly and finally terminates your license, and (b) permanently, if the copyright holder fails to notify you of the violation by some reasonable means prior to 60 days after the cessation.

Moreover, your license from a particular copyright holder is reinstated permanently if the copyright holder notifies you of the violation by some reasonable means, this is the first time you have received notice of violation of this License (for any work) from that copyright holder, and you cure the violation prior to 30 days after your receipt of the notice.

Termination of your rights under this section does not terminate the licenses of parties who have received copies or rights from you under this License. If your rights have been terminated and not permanently reinstated, receipt of a copy of some or all of the same material does not give you any rights to use it.

10. Future Revisions of this License

The Free Software Foundation may publish new, revised versions of the GNU Free Documentation License from time to time. Such new versions will be similar in spirit to the present version, but may differ in detail to address new problems or concerns. See http://www.gnu.org/copyleft/.

Each version of the License is given a distinguishing version number. If the Document specifies that a particular numbered version of this License "or any later version" applies to it, you have the option of following the terms and conditions either of that specified version or of any later version that has been published (not as a draft) by the Free Software Foundation. If the Document does not specify a version number of this License, you may choose any version ever published (not as a draft) by the Free Software Foundation. If the Document specifies that a proxy can decide which future versions of this License can be used, that proxy's public statement of acceptance of a version permanently authorizes you to choose that version for the Document.

11. Relicensing

"Massive Multiauthor Collaboration Site" (or "MMC Site") means any World Wide Web server that publishes copyrightable works and also provides prominent facilities for anybody to edit those works. A public wiki that anybody can edit is an example of such a server. A "Massive Multiauthor Collaboration" (or "MMC") contained in the site means any set of copyrightable works thus published on the MMC site.

"CC-BY-SA" means the Creative Commons Attribution-Share Alike 3.0 license published by Creative Commons Corporation, a not-for-profit corporation with a principal place of business in San Francisco, California, as well as future copyleft versions of that license published by that same organization.

"Incorporate" means to publish or republish a Document, in whole or in part, as part of another Document.

An MMC is "eligible for relicensing" if it is licensed under this License, and if all works that were first published under this License somewhere other than this MMC, and subsequently incorporated in whole or in part into the MMC, (1) had no cover texts or invariant sections, and (2) were thus incorporated prior to November 1, 2008.

The operator of an MMC Site may republish an MMC contained in the site under CC-BY-SA on the same site at any time before August 1, 2009, provided the MMC is eligible for relicensing.

How to use this License for your documents

To use this License in a document you have written, include a copy of the License in the document and put the following copyright and license notices just after the title page:

> Copyright (c) YEAR YOUR NAME.
> Permission is granted to copy, distribute and/or modify this document
> under the terms of the GNU Free Documentation License, Version 1.3
> or any later version published by the Free Software Foundation;
> with no Invariant Sections, no Front-Cover Texts, and no Back-Cover Texts.
> A copy of the license is included in the section entitled
> "GNU Free Documentation License".

If you have Invariant Sections, Front-Cover Texts and Back-Cover Texts, replace the "with... Texts." line with this:

> with the Invariant Sections being LIST THEIR TITLES, with the
> Front-Cover Texts being LIST, and with the Back-Cover Texts being LIST.

If you have Invariant Sections without Cover Texts, or some other combination of the three, merge those two alternatives to suit the situation.

If your document contains nontrivial examples of program code, we recommend releasing these examples in parallel under your choice of free software license, such as the GNU General Public License, to permit their use in free software.

Image acknowledgements

The following persons and institutions are acknowledged for making pictures available for publication.

Agnew Neville, GCI, Getty Conservation Institute, Los Angeles, CA, USA

Anguilano Lorna, ETC, Experimental Technique Centre, Brunel University West London, Uxbridge, Middlesex, UK

Ashby Michael F., EDC, Engineering Design Centre, Department of Engineering, University of Cambridge, Cambridge, UK

Bleuet Pierre, ESRF, European Synchrotron Radiation Facility, Grenoble, France

Bodsworth Jon, http://www.egyptarchive.co.uk/index.htm

Boyle Ninian, Venturescope, Emsworth, United Kingdom

Bourgarit David, C2RMF, Centre de Recherche et de Restauration des Musées de France - CNRS, Paris, France

Breger Dee, Department of Materials Science and Engineering, Drexel University, Philadelphia, PA, USA

Brunetti Brunetto Giovanni, Laboratorio di Chimica Generale, Dipartimento di Chimica, Università degli Studi di Perugia, Perugia, Italy

Calligaro Thomas, C2RMF, Centre de Recherche et de Restauration des Musées de France – CNRS

Cardale de Schrimpf Marianne

Cardarelli Andrea, Dipartimento di Scienze Storiche, Archeologiche e Antropologiche dell'Antichità, Università degli Studi di Roma "La Sapienza", Roma, Italy

Cardon Dominique, CNRS, Lyon, France

Carter Andrew, Thermocronometry Research Laboratory, UCL and Birkbeck Earth Science, University College, London, United Kingdom

Chaimanee Yaowalak, Paleontology Section, Department of Mineral Resources, Bangkok, Thailand

Chistè Paolo, Università degli Studi di Trento, Trento, Italy

Cremaschi Mauro, Dipartimento di Scienze della Terra "A. Desio", Università degli Studi di Milano, Milano, Italy

De Marinis Raffaele C., Dipartimento di Scienze dell'Antichità - sezione di Archeologia, Università degli Studi di Milano

Fornasier Massimo, RICAM, Johann Radon Institute for Computational and Applied Mathematics, Linz, Austria

Flöry Simon, Geometric Modeling and Industrial Geometry, Institute of Discrete Mathematics and Geometry, Vienna University of Technology, Vienna, Austria

Fujimoto Jane, GCI, Getty Conservation Institute

Furmanek Vaclav, Archaeological Institute Slovak Academy of Sciences, Bratislava, Slovak Republic

Gallo Filomena, Dipartimento di Geoscienze, Università degli Studi di Padova, Padova, Italy

Gelfand Natasha, Nokia Research Center, Palo Alto, CA, USA

Gordon Jacoby, Tree-Ring Laboratory, Lamont-Doherty Earth Observatory, Columbia University, New York, NY, USA

Grissino-Mayer Henry D., Laboratory of Three-Ring Science, University of Tennessee, Knoxville, TN, USA

Gruen Armin, Institute of Geodesy and Photogrammetry, ETH Zurich (Swiss Federal Institute of Technology), Zurich, Switzerland

Grzywacz Cecily M., CGI, Getty Conservation Institute

Heginbotham Arlen

Hofer Michael, Geometric Modeling and Industrial Geometry, Institute of Discrete Mathematics and Geometry, Vienna University of Technology, Vienna, Austria

Howard Louisa, Electron Microscope Facility, Dartmouth College, Hanover, NH, USA

Huang Qi-Xing, Geometric Computing Group, Stanford University, Stanford, CA

Keller Robert, National Institute of Standards and Technology, Gaithersburg, MD, USA

Khanjian Herant, GCI, The Getty Conservation Institute

Lehmann Eberhard, PSI, Paul Sherrer Institute, Villigen, Switzerland

Leonardi Renzo, Dipartimento di Fisica, Università degli Studi di Trento, Trento, Italy

Ludwig Nicola G., Istituto di Fisica Generale Applicata, Università degli Studi di Milano

Maritan Lara, Dipartimento di Geoscienze, Università degli Studi di Padova

Mazzoli Claudio, Dipartimento di Geoscienze, Università degli Studi di Padova

Martinetto Pauline, C2RMF, Laboratoire de Recherche des Musées de France – CNRS

Müller Martin, GKSS Forschungszentrum, Geesthacht, Germany

Pedrotti Annaluisa, Dipartimento di Filosofia, Storia e Beni Culturali, Università degli Studi di Trento, Trento, Italy

Perez-Arantegui Josefina, Departamento de Química Analítica, Universidad de Zaragoza, Zaragoza, Spain

Perini Renato

Pini Roberta, CNR-IDPA, Laboratorio di Palinologia e Paleoecologia, Università degli Studi di Milano Bicocca, Milano, Italy

Piovesan Rebecca, Dipartimento di Geoscienze, Università degli Studi di Padova

Polla Angela, Dipartimento di Geoscienze, Università degli Studi di Padova

Pottmann Helmut, Geometric Modeling and Industrial Geometry, Institute of Discrete Mathematics and Geometry, Vienna University of Technology

Ravazzi Cesare, CNR, Istituto per la Dinamica dei Processi Ambientali, Dalmine, Italy, http://www.idpa.cnr.it

Recchia Sandro, Dipartimento di Scienze Chimiche ed Ambientali, Università degli Studi dell'Insubria, Como, Italy

Rehren Thilo, Institute of Archaeology, University College of London

Rémazeilles Céline, Laboratoire d'Etude des Matériaux en Milieux Agressifs, Département Genie Biologique, Université de La Rochelle, La Rochelle, France

Remondino Fabio, Institute of Geodesy and Photogrammetry, ETH Zurich (Swiss Federal Institute of Technology)

Salvioni Davide, Laboratorio Analisi, MAPEI S.p.A., Milano

Sedova Tatiana, Institute of Philosophy Slovak Academy of Sciences

Sharp Nigel A., NOAO, National Optical Astronomy Observatory, Association of Universities for Research in Astronomy, National Science Foundation, Tucson, AZ, USA

Siano Salvatore, I.F.A.C., Istituto di Fisica Applicata "Nello Carrara", CNR, Firenze, Italy

Sibilia Emanuela, Laboratorio di Archeometria, Dipartimento di Scienza dei Materiali, Università degli Studi di Milano Bicocca, http://dating.mater.unimib.it

Tafforeau Paul, ESRF, European Synchrotron Radiation Facility

Tawastenhielm FranceAnne-Marie (previously Olsson)

Tirabassi Iames, Musei Civici di Reggio Emilia, Italy

Tissue Brian M., Department of Chemistry, Virginia Polytechnic Institute and State University, Blacksburg, VA, USA

Toniolo Domenico, Università degli Studi di Padova

Walter Philippe, C2RMF, Centre de Recherche et de Restauration des Musées de France– CNRS

Wong Lori, GCI, Getty Conservation Institute
Michel Zabé

AIM, Associazione Italiana Metallurgia

CCI, Canadian Conservation Institute of the Department of Canadian Heritage, Ottawa, Canada

Columbia University, Department of Astronomy, New York, NY, USA

C2RMF, Centre de Recherche et de Restauration des Musées de France, Paris, France

ESRF, European Synchrotron Radiation Facility, Grenoble, France

GCI, The Getty Conservation Institute, Los Angeles, CA, USA

INAH- Instituto Nacional de Antropología e Historia, Cuauhtémoc, México

LBL, Advanced Light Source, Lawrence Berkeley National Laboratory, University of California, Berkeley, CA, USA

LRMH, Laboratoire de Recherche de Monuments Historiques, Champs-sur-Marne, France

MAPEI S.p.A., Milano, Italy

MTSN, Museo Tridentino di Scienze Naturali, Trento, Italy

Musée d'art de Joliette, Joliette, Quebec, Canada

NOAA, National Oceanic and Atmospheric Administration, Washington, DC

NOAO, National Optical Astronomy Observatory, Association of Universities for Research in Astronomy, National Science Foundation, Tucson, AZ, USA

ORNL, Oak Ridge National Laboratory, Oak Ridge, TN, USA

SCANCO Medical AG, Brüttisellen, Switzerland

SERC, the Science Education Resource Center at Carlton College, http://serc.carleton.edu/index.html

UCL, University College London, London, United Kingdom

University of Arizona, Tucson, Arizona, USA

University of Calgary, Calgary, AB, Canada

USGS, U.S. Geological Survey, Denver, CO, USA

Index

Note: Illustrations are indicated by italic page numbers, tables are indicated by bold page numbers.

AAS **20**, 40
absolute time
   dating methods 166–198
   scale 130–136
absolute zero 141
absorbance 31, 44
absorbed dose 189
absorption **20**, 30, 52
   coefficient 69, 303–304
   contrast 69
   edge 303, *303*
   imaging 31
   of light 150
   spectroscopy 30, 40–47
   measurement 116
   micro-tomography 62
   resonant 408
   spin resonance 193
Abu Matar, Israel 330
accelerated test 123, 139, 142–145, *146*, 277
accelerator mass spectrometry 173, **177**, 179, 181
accuracy 111, 128–129
acetic acid 148
acetonitrile 411
acid rain 224
Acqua Fredda, Redebus, Italy *208*
activation energy 141, **142**
acylation 415
admixtures 258
adobe 230
aerial photography *11*, 86
aerophotogrammetry 86
AES **20**, 37, 301
AFm phases 257
AFM **60**, 393–394
AFt phases 257
Afunfun, Niger 335
agate 353
age 24
   conventional radiocarbon 181
   calibrated radiocarbon 181

ageing
   artificial 139
   natural 137, 140, *141*
   of materials 137–157
   photochemical 149–153
   thermal 141–148
Agordo, Belluno, Italy 329
AGP 417
Agricola, Georgius 205, 320
Ai Bunar, Bulgaria 332
AIDS 416
'Ain Ghazal, Jordan 245
Ajjûl, Mesopotamia 287
Akhenaten *287*
alabaster 314
albumen 419–428, **425–426**
   FTIR spectrum *428*
Al Claus, France 330
Alexander the Great 288
Alexandria, Egypt 288
alite 252, **252**, 256
   /belite ratio 256–257
alizarin **268**, 406–407
alkali elements 280, 290
alkaline
   earths 281, 291
   environment 264
alloys **308**, 309, 327
   hardness 341, **342**
Al Mihdar Mosque, Tarim, Yemen 230
Almizaraque, Spain 330
alpha-recoil tracks 173, 183
Alpine
   area 322, 329, 330, 332
   copper mines 306
   smelting slags 334–336, *334–336*
alteration
   of glass 295–298
   sample 99, 138–140
alumina-rich materials 232, *233*
aluminium
   electrochemically produced 314
   metal 314

Amarna, Egypt 287, *287*, 398
amber 372–381
   archaeological 380–381
   Baltic *115*
   characterization 383–384
   chemical properties 375, 381
   definition **368–369**
   density 113
   deposits 375–379
   formation 372–373
   geology 375–379
   inclusions in 379–380
   IR analysis 115, *115*, 382–383, *382*
   nomenclature **368–369**, 373–377
   physical properties **374**
ambergris 368, 373
Ambohitralanana, Madagascar *322*
aminoacid 359
   racemization method 163–164
amorphous solids 278–279
AMS 173, **177**, 179
Amur River, Russia 229
amylopectin 248
analytical
   accuracy 111, 128–129
   cost 28
   methods 5
   precision 101, 111, 128–129
   projects 112
   protocol 400
   questions 103, **104**, 107
   results 102
   strategies 29, 106–111, 123
Anatolia 330
ancient glass types and compositions **286**
Andean cultures 406
anglesite **324**, 337
Ångstrom, units 18
angular momentum 408
anhydrite 250–251
   solubility 250

animal
　breeding 398
　glue 416–418
　hair **396**, 398
annealing temperature **281**
anodic reaction 317
anorthite **237**, 239–240
anoxic conditions 318, 396
antibody 415–418
　fluorescent 416
　secondary 416
antigen 417
Antikythera Mechanism 75
antimonate opacifiers 285, 295
antimony
　as decolourant 288, 293
antler 272, *273*, 323, 360
Antwerp, Belgium 290
apatite *116*, 162–163, 356–360, **357**, *357*
　annealing temperature 184
　blue fossil 362
　group minerals **357**
　fission tracks 184, *184*
　isotope data 187
Apliki mine, Cyprus 332
aquamarine *350*, **351**
Aquileia, Veneto, Italy 288
Arabah, Near East 322
arabinogalactan 416–417
aragonite-calcite transformation 176
archaeological
　glass composition 285–295, **286**
　organic materials 356
　prospection 84
　science 1, 10
　site conservation 123
　textiles 395, **396**
archaeology 7, 9–12
archaeomagnetism 164–166, *165*
archaeometallurgy 306, 322
archaeometry 1–5
Archimedes' principle 260
architecture
　conservation 224–225, 253–256
　decorations 223, *223–224*
　earthen 230
　heritage preservation 253
Arikamedu, India 288
Arrabidea chica 406, 406
Arrhenius' law **144**, 145–149
arrow points *212*
Arslan Tepe, Anatolia 330
Art Nouveau **282**, 283, 298
artist 7
artist's materials 7–9
Asia, South-Western metallurgy 331

Asian Organic Colourants project 404–405
Aşıklı Höyük, Turkey 246
Aspdin, Joseph 252
Aswan red granite 217, *220*
atacamite 318
atmospheric ^{14}C curve 169, 181, *182*
atomic
　absorption spectroscopy 40–42
　emission spectroscopy 37
　force microscopy 23, **60**, 393–394
　mass 170
　number 33, 170
ATR 45, 115, 426
ATR-FTIR 45, 426
attapulgite 270
attenuated total reflectance 45, 426
Auger
　effect 301
　electron *301*
　electron spectroscopy 301–303
Aurora Borealis 32
austenite 312, *312*
authenticity 79, 108, 111, 266, 274
　ceramics 190–192
　modern paintings 136, 274
automatic reassembling 90–93
autoradiography 70, 215
Aztecs 270, 348
azurite **56**, 319, 323

Babylon, Mesopotamia 230, 287, 314, 349
bacteria
　sulphur-reducing 225
Badarian graves, Egypt 287
bainite 312, *312*
Balkans 330
balsam 369
　container 287
Baltic amber *115*, 368, 378
　deposits 373–374
　FTIR spectra *115, 382*
Bamiyan Valley, Hazarajat, Afghanistan 85–86
Baratti, Tuscany, Italy 332
Baroque decorations 293
baryta paper 107, 421–426, *426, 431*
baryte 108, **268**
Baschenis, Simone *263*
bassanite **243**, 250–251
　dissolution 251
bast 398
bear 337
Becke lines 65

beeswax 427
　AMS amount **177**
　FTIR spectrum *428*
belite 252, **252**, 256
bellow 333
Belus River, Israel 289
BET 260
binders 242–265, 417
　characterization techniques 265, 417
biodeterioration 356
biological materials 355
biomineralization 353, 355
　processes 355
biotite *247*
birefringence 65
Biringuccio, Vannuccio 205, 320
birthstone 348
BJH 260
black
　copper 312
　core 240
　crusts 225, 274
　gloss 241, 300
　pigments **268–269**
blast furnace 331
bloodstone 353
bloomery process 331
blue
　colourants 153
　glass 292–295, *294, 300*
　pigments **268–269**
Blue Wool Lightfastness Standards 153, *154*
bluestone 210
Bohr magneton 193
Boltzmann statistics *261*
bone 356–360, *358*
　alteration 364–367
　AMS amount **177**
　astragalus 36
　blue-stained 362, *362*
　chemical changes in 162–163, 365
　density 113, 115–116
　fluorination 163, 365
　fossil 365
　FTIR spectra 365–366
　hierarchical organization 357–359, *358*
　isotopes 186–187
　regeneration process 364
　talus 363
bornite **324**, *325*, 329
botallakite 318
Bovier de Fontenelle, Bernard 139
BP 181
Bragg's law 51
brass **308**, 331

breccia, coarse-grained *225*
bricks 238
  sun-dried 230
brightness 275, 277
Britain 330
Brixen, South-Tyrol, Italy *335*
Brixlegg-Mariahilfbergl, Tyrol, Austria 323, 330
brochantite **324**, 326
bronze
  chemical analysis 112, 328
  Corinthian 312
  corrosion 318–319
  desease 318
  hardness **342**
  in China 331
  tin 310–311, *311*, 331
Bronze Age
  Aegean 332
  faience *284*, 287
  glass 283–295, *287*, *294*, *297*
  mine *325*, 332
brucite **243**, 244
Bruges tapestry 407
Buddha
  Bamiyan statues 85, *86*
building diagnostics 59, 86–87
building stones
  Alpine stones in Milan 217
  ancient Egypt 217
  Baroque decorations in Palermo 217
  degradation 225
  Roman stones 217
  provenance 217
  resistance to compression 223
  re-use 222, *223*
Burma 314, 349
butterfly wing 272
Byzantine mosaics 248, 285

^{14}C method 177, 181–183
C2S **252**
C3A **252**
C3S **252**
C4AF **252**
Cabezo Juré, Spain 330
Cabrières, France 323, 330, 332
cactus 248
cadmium
  yellow 274
  orange 120–122
  sulphate 274
Caesarea, Israel 247
Calceranica, Valsugana, Italy 332

calcite 108, 176, 211, 225, 243, 257
  EPR signal *194*
  lime-derived 245
  magnesian 244
  microcrystalline 244
calcium
  aluminates 232, *233*, **252**
  aluminosolicates **243**
  antimonate 293, 295
  bicarbonate 263
  carbonate 211, 234, 242–244
  fluoride 295
  hydroxide 243–244
  ions 248
  oxalate 225
  oxide 242–244
  phosphate 295, 356–360, **357**
  sulphate anhydrous 250–251
  sulphate emihydrate 250–251
  sulphate dihydrate 225, **243**, 250–251, 257
calendar 166
  dates 181
calibration
  curve, AAS 41
  of colour 93
  protocols 83, 128
  radiocarbon procedures 181–183
  standards for MS 185, **186**, 414
Campeche, Mexico *271*
capillary column 413
carbon
  absorption in iron 331
  dioxide 205, *206–207*, **237**, 243, 257
  in the food chain 187
  -iron system 312–313, *312*, 331
  monoxide 205, *206–207*
  stable isotopes 185, **185**
carbonate
  biomineralization 353
  clay-rich 249
  dating **174**, 175–177
  decomposition **237**, 243–244
  fine-grained 223
  -hydroxylapatite 357, **357**
  magnesian 244
  marine record 181
  minerals 236, *237*
  oxalate patina 225
  shells 291
  surface alteration 225
  surface cleaning by bacteria 225
carbonation reaction 243, **243**, 257
carbonic acid 263
carburized iron 312, 323, 331
carnelian 353
carrier gas 413

casein 416–417
cast iron 331
Castel-Grande, Bellinzona, Switzerland *212*
Castelrotto, Bolzano, Italy *347*, 348
Catal Hüyük, Turkey 397–398
cathodic reaction 317
cathodoluminescence 66
cation-oxygen bond 279
Cato 244
Caucasus 287, 398
Cave of Letters, Nahal Hever, Israel 399
Çayönü Tepesi, Turkey 330
CCMP 38
cellulose 108, *385*, 384–388, 396, **396**
  acetate 148–149
  degree of crystallinity 386
  fibre microdiffraction *399*
cement *233*, 242–265, **259**
  admixtures 258
  composites 257
  compressional resistance 257–258
  degradation 259–265, 318
  hydration process 257
  lime-based 242–250
  paste 242
  Portland 242
  quick-setting 252
  set control 258
  setting 257
  steel reinforced 258
  types **259**
cementite 312, *312*
ceramics 229–242
  analysis by NAA 216
  Biblical times 216
  black core 240
  characterization of 239–242
  coarse 238
  coating 240–241
  dating 241
  drying 234–238
  fakes 190–192
  fine 238
  firing 234–241
  firing temperatures 50
  Greek and Cypriot 112
  impermeability 238
  inclusions 98
  materials 200
  mechanical strength 237, **238**
  observation by OM *65*
  perforated cylinders 333
  pipes 238
  pore formation 237–238
  provenancing 241–241
  Qumran 216

ceramics (*cont.*)
 temper 234
 thermoluminescence 190–192, *191*
 typology 241
Čerenkov radiation 32, 301
cerussite **324**, 337
chaîne opératoire 7
chalcedony 350–353
chalcocite 318, *318*, **324**
Chalcolithic Age
 axes 341, 347–348, *347*
 copper metallurgy 322–323
 crucible smelting 323
 mine *325*, 332
chalcophile elements 328
chalcopyrite **324**, 326, 329, 335, *336*
chamotte 234
charcoal 267
 AMS amount **177**
Charlemagne 222
charred tissues 364
chemical
 aggression 154–156, 162
 analysis 33, **34**, 101, 239
 bonds 33, 278–279
 changes in materials 162–163
 composition of clay products 232–233
 compound identification 43, 48
 degradation 138–140
 fingerprints 49
 immiscibility 340
 information 24, 33
 maps 61, 68, *68*, 76, 313
 photography 419
 processes 13
 reactions 56, 123, **238**
 shift 408
 signature of photographs 423–426, **425**
 stratigraphy 135
 time scales 136–157
 tracers 49
chert **211**, *212*, 214, 353
Chichén Itza, Yucatàn, Mexico 270
China
 HLHB glass **286**, 288
 lime mortars 248
 tin-bronze 331
Chiusa di Pesio hoard, Cuneo, Italy 396–398
chloride 318, *318*, 320
chlorine 262, 317–318, *318*
chlorophyll *47*, 410
chromatic component 275
 antagonistic 277–278
chromaticity
 coordinates *276*
 diagram 276, *276*

chromatograph 403
chromatography 410, 413
 gas 413–415
 liquid 410–413
chromium oxide **206**, 308
chromophore ions
 in pigments **268–269**
 in glass 292–295, **293**
chrysocolla **324**, 335
Chrysokamino, Crete 332
Chumash Painted Cave, California 227
CIE 276
 L*a*b space 277
 standard diagram 276–278
CIELAB 277–278
 coordinate system *277*
cinder 337
cinematographic film deterioration 148–149
cinnabar 274
CITES 361
Civitavecchia, Italy 249
clay
 based materials **211**, 229–242
 figurines 229
 fired 165, 190, 234–238
 glazing 287
 lining 230
 minerals 210, 232–233
 -rich carbonates 249
 tablets 92, 230–232, *231*
Climate Notebook 147
climatic
 chamber 142
 changes 262
clinker *233*, 242–265, **252**
 based materials 251–265
CMY colour system 278, *278*
coal 205, **211**
cobalt
 alums 293
 ions in glass 293, **293**
cocciopesto 246
cochineal 402
collagen 56, 163, 357–360, **396**, 416
collodion **425–426**
colloidal behaviour 233
colorimetry 23, 275
Colosseum, Rome, Italy 248
colour 23, 25, 46, 266–274
 charts 278
 components 276
 desktop 278
 fading 93, 153–154, 277
 matching functions 276
 measurement 274–278
 mechanisms of 46, 266, 275

perception 275
retinal stimuli 278
saturation 275
science 266
systems 278
colour photographs
 thermal ageing *144*
coltan 320
columbite-tantalite 320
combustion 205–208
Commission Internatioale de l'Eclairage, CIE 276
composite materials 199, 355, 419–432
compositional zoning 240
computed tomography 71–75
computer
 graphics 58
 science 127
conchoidal fracture 283
concrete 99, **243**, 257
 degradation 262, 318
 high performance 258
 lightweight 248
 Portland 257
condensation 262
Congo 320
conical button *284*, 287
conservation 1, 6, 12–13, 59
 costs 124
 clay tablets 230–232
 diagnostics 100
 organic materials 356
 strategies 264
 treatment 262
conservation science 1, 13, 123
Convention on International Trade in Endangered Species 361
cooling rate 335
coordination geometry 33, 304
copal 368, 373
copper
 alpha-phase 310, *310–311*
 awl 323
 axe *94, 118*, 197, 341
 black 312, 326
 casting texture *344*, 347
 chemical analysis 112, 328
 chloride 318, *319*
 corrosion 318–319
 deposits 323
 early production 330–331
 hardness 340–341, **342**
 ingots 73
 ions in glass 292–294, **293**, *294*
 isotopes 329, **329**
 -lead system 309–310, *310*

metal particles 293
minerals **324**
minor elements in 101
native 329, 341
oxhide ingots 332
oxidation-reduction 294, 317–318
phosphate 319
reduction 335
resinate 305
slags 332–333, *333*
smelting 208, 326, 330, 334–339
sulphide 310, 318, *318*
-tin system 310, *311*
-water system 317–318, *318*
XPS spectrum *300*
Copper Age 322
axes 347
coral 353
Corinthian bronze 312
corrosion 317–320
cosmetics 266, 269, 356
cosmic rays 177, 188
cosmogenic nuclides 173
cotectic line 291
cotton **396**, 397, *406*
covalent bond character 279
"cristallo" glass **286**, 290
cristobalite **237**, 239
melting point 280
critical
angle 349
temperature 279
cross-dating 197
crucible 290, 333
-furnace 333
smelting 323, 330, 338, *339*
crysotile asbestos **396**
crystal
dendritic *344*
domains 345
lattice 278–279
lattice planes 307
orientation 345, *345*
phase identification 50
phase maps 62
structure 202
texture analysis 345–348
crystallinity
of gypsum plaster 250
crystallization pressure 262
crystallochemical substitutions 357
crystallographic texture analysis 343–348
C-S-H **243**, 256–257
CT 71
CTA 343–348
cultural tourism 124, 226, 436

cuneiform
digital library initiative 232
inscriptions 92, 230–232
cupellation 334, 337
cuprite *73*, 293, *294*, 317, *318, 320*, 323, 329
cuprorivaite 287
Curie temperature 84, 165
Cusa quarries 217, *219*
Cuzco, Peru 210
cyan-magenta-yellow colour system 278, *278*
Cyprus 287, 331–332

D_{65} 276, *276*
Daguerre, Louis Jacques Mandé 419
dahllite 357–360, **357**, 364–367
recrystallization 365
damage functions 123
Damascus steel **308**, 331
data
consistency 112
modelling 79
normalized 102
processing 128
quality 28–29
significance 113
treatment 125–128
databases 82–83, 435
of ceramics NAA data 216
dating
cross- 197
electron traps methods 188–196, **188**
fission track method 183
interval of application 132, **134**
methods 130–198
optical 190–191
pitfalls 196–197
radiometric methods 175–177, **174**
sediments 196
Davy, Humphry 201
daylight
average illumination 276
exposure 190
DCP 38
De Broglie 17
Debye cones 345
decolourant in glass **282**, 288
deforestation 208
deformation 307
degradation process 261
dehydroxilation 234, **237**
delafossite **324**, 338, *339*
della Robbia, Luca 8, 289, *289*
delta notation 185

dendritic crystals *344*
dendrochronology 167–169
densitometry 115
density 113
depth profiling 62, 75, 180, 314–315
derivatization 415
desert
glass 280, *280–281*
varnish 226
detection limit 103, **104**
deterioration 138
devitrification 279
dew point 262
DEXA 16
DFJ 254
diagenesis 357, 362, 364
diagnosis 123
diamagnetic compounds 193
diamond 279
diatomaceous earth 289
dice 362–364
number combinations on 363
dichroic effect 295
differential
scanning calorimetry 55–56
thermal analysis 55–56
diffraction 23, 25, 47–48
angle of 51
diagrams *117, 120–122*
electron 53–55
grazing incidence 241
high-temperature 50
neutron 52–53
rings *117, 118, 120*
X-ray 50–51, 117
diffractometer 51
diffuse reflectance Fourier transform spectroscopy 45
diffusion kinetics 279
diopside **237**
dioptase 335
dislocations 307
Djenne Mosque, Mali 230
djurleite 319
DNA 364
studies 356, 361, 398
Dodson, Richard W. 214
Dolní Věstonice, Czech Republic 229
Dolnoslav, Bulgaria 330
dolomite 244
EPR signal *194*
dominant wavelength *276*, 277
Douglass, Andrew 169
DP 389
DRIFT 45, 115

dry hydration 244
DSC 55
DTA 55
DTG 55
ductile behaviour 307, 323
DUETTO instrument 119–122
durability of materials 131, 137
dyes 266, 399–407

Earth
　age 131
　crust 307
　magnetic field 164–166
　Sciences 47–50
earthen architecture 230
earthenware 238, **238**
Easter Island 210
EBSD 54, 343–345
ED 53
Eddystone lighthouse 249
EDS 36
ED-XRF 36
EELS **20**, 301
efflorescence 264
Egypt
　Aswan red granite 217, *220*
　Ptolemaic 351
　pyramid controversy 251
　Western Desert 280, *280*
　Western Oases 293
Egyptian
　blue pigment 283, 287
　cosmetics 269
　early plaster 246
　faience production 287
　glass 287, *294*
　glass beads 287
　glass-making techniques 287
　green pigment 283, 287
　polychrome glass 287, *287*
　textiles 407, *407*
E_h 318, *318*
E_h-pH diagram 317–319, *318*
Eifel, Germany 247
Elba Island, Tuscany, Italy 332
electrical resistance survey 84
electrochemical reaction 317
electro-spray ionization 180
electromagnetic
　beam penetration **96**
　radiation 17–21
　spectrum *18*, **19**
　visible spectrum 275
　wave *17*

electron 17
　backscatter diffraction 54, **60**, 343
　backscattered *19*, 66–68
　beam 18, *19*
　configuration 33, 202
　defects in solids 188–195
　diffraction 53–55
　energy loss spectroscopy 301–303
　-hole pair 185, 350
　microscope 55–57, 66–68
　mobility in metals 307
　paramagnetic resonance
　　spectroscopy 192–195
　probe micro-analyser 37
　secondary *19*, 66–68
　spectroscopy for chemical analysis 299
　spin resonance 192–195
　spin states 192
　traps 188–195, **188**
electronegative character 280
elemental
　analysis 33, 34, 101
　composition 202
　concentration 101
Eleonor of Aragon *224*
Elephas
　maximus 360
　primigenius 360
ELISA 416, *418*
Ellingham diagrams 205, *206–207*, 308
ELSD 413
elution 411, 413, 417
embroidery 342
emerald 314, 349, **351**
emission 30, 32
　centers 188
　characteristic lines 34, 38
　spectroscopy 30
enamel 238, 285, 295
endothermic reaction 56, 205
energy 16, 17–24, 31
　activation 141
　dispersive spectrometry 36
　energy scale *18*
　energy levels 17, **19**, 23, 30, 35
　kinetic 142
　of photons 17, 193
　of reaction 205
　of the probe beam 59–60
　quantum 17, 29
　recoil detection 63, 316
　transfer 23, 29
engobe 239
Ensor, James 274
enthalpy 56, 279
entropy 137, 279

environmental
　reconstruction 133
　scanning electron microscope 68, *431*
enzyme-linked immunosorbent assay 416–418
EPMA **20**, 37
EPR **20**, 193
ERD 63, 316
ERDA 316
Eremitani Church, Padova, Italy 90–92
Erlitou-Erligang cultures 331
ESCA 299
e.s.d. 103, 129
ESEM 68, *431*
ESI-MS 180
ESR 192
Este 240
estimated standard deviation 103, 112, 129
Etruria, Southern 363
Etruscan
　alphabet 363, *364*
　dice 362–364, *362, 364*
　necropolis 332, 398
ettringite 257, *258*
　secondary 255, 262
EU-ARTECH project 267
eutectic 289, 312, 331, *337*
evaporation 262
evaporative light scatterer 413
evaporite **211**, 290
evolved gas analysis 55
EXAFS 304
excited state 30
exothermic reaction 56, 205
experimental protocols 14, 28
　smelting reconstruction 333
Eyring-Polany equation 146

faience 278–299, 284, *284*
　blue-turquoise colour 292–293
　tomographic image *73*
falherz 328
falhore 328
Faraday, Michael 201
favrile glass *282*, 283
fayalite 332, 334–337, *335–336*
　density 338
Faynan, Jordan 330
　project 306, 322
　slags 335
feldspar 233, *233*, 234, *247*
　luminescence 190–191
Ferriere, Genova, Italy 331
ferrite (alpha-Fe) 312, *312*
ferrite (C4AF) **252**

ferromagnetism 165
ferrum noricum 331
fibre 395–399, *397*
   analysis 389–390, 396
   identification by microbeams 399
   natural 396, **396**
fibre optics
   fluorescence spectroscopy 272
   reflectance spectroscopy 46, 78
fibrils 357–360
fire-based processes 204–208, 234–238
fire-clay **238**
firefly 32
firing technology 208
fission tracks 173, 183, *183, 184*
   induced 183–184
flash set 257
flat-jack tests 254
flax 396–399, **396**
FLIm 272
flint 210, **211**, 323
   hardness 323
fluid
   in porous media 259–265, 409
fluorescence 32, 34, 61, 150
   life-time imaging 272
fluorine
   substitution in bone 162–163, 365
   uptake in flint 163
fluxing agent 233–234, 280–281, 334–335
fly ash 258, **259**
FOFS 272
Forbes, Edward Waldo 270
forensic science 82
forest glass 289
forging 313
Forma Urbis Romae 92
FORS 46
fossil resins 368, 373
four-creuset 333
Fourier transform
   infrared spectroscopy 44
   mass spectrometry 179
   Raman spectroscopy 45
foxing 386, *387*
francolite **357**, 365
Frattesina, Rovigo, Italy *234–235*, 288, *292, 294*, 295
free energy 279
   of reaction 205, *206–207*
frequency perturbation 408
Friesack, Germany 397
frit 239, 283
FT-ICR-MS 179
FTIR 44, 115, 249, 426
FTRS 45

fuel 205, 244, 332
FUN test 163
furnace 165–166, 196, 204–208, 240, 244, 330, 333
   blast 331
   history of technology 205
   Late Bronze Age *208*
   reconstruction 208
   smelting *339*

g factor 193–194, *193*
Gaban, Trento, Italy 272, *273*, 330, *334*
galena **324**, 337
Galileo 125, 136
Gara Ouda, Fezzan, Libya *217–218*
Gardenia augusta, pigment *405*
garnet 314
gas chromatography
   instrument configuration *414*
   mass spectrometry 179, 413–415
   scan mode 415
   selected ion monitoring 415
gas constant 148
GC-MS 179, 413–415
Geary, Frank 9, 209
gehlenite 236, **237**, 239–240
gemstones 314, 348–355, *350*, **351–352**
   authentication 349
   irradiation 350
   synthetic 349
   thermally treated 350
   trade 349
Genista tinctoria 413
genistein *412*, 413
geochemistry 211, 327
geoglyphs 87
geographical information systems 83, 89
geological
   age of metals 326
   materials 210–225, **211**
   time scale 131, *132*
geomagnetic polarity reversals 164–166
geomaterials 210, **211**
   composing objects **213**
   investigation 214
geometric representation 79
geophysical techniques 84–85
geopolymers 251
georeferencing 89
Ghat, Fezzan, Libya *231*
Giara di Gesturi, Sardinia *212*
Giovanetto di Mozia *225*
GIS 83, 89
GIXRD 241, 342

glaciations 135
glass 67, *95*, *233*, 278–299
   alteration *95*, 291, 295–298
   ancient compositions 285–289, **286**
   blowing 288
   chalcogenides 280
   characterization 298–299
   chemical analysis 102–103
   chemical components **282**
   colourants 281, 292–295
   decolourant **282**, 288
   dissolution 296
   forming process 283
   homogeneous 283, *284*
   hydration layer 159–161, *161*, 296–298
   impact 280, *280–281*
   iridescent *282*, 283
   making 289–295
   matrix 284–285
   opacifiers 284, 294–295
   resistance to leaching 281, 291
   science 280
   silica 279–283, **282**
   transformation region 279
   transition temperature 279
   transparency 292
   viscosity 281, **281**, 291–292, *292*
   working temperature **281**, *292*
   Zachariasen model 279
glassy faience 284, *284*, 287
glaze *233*, 238, 285, 289, 295
   Della Robbia 289
   Pb-rich 285, 289
glazed
   Chinese stoneware 289
   pottery 238, 287, 289, 305
   stones 285
   terracotta 289
global positioning systems 89
Goddess Ishtar 314
Goethe, Wolfgang 293
gold
   -copper particles 295
   in mosaic tesserae 285
Goldschmidt, Victor Moritz 279
gossan 323
Gossypium hirsutum **396**
GPR 84
GPS 89
Grado, Veneto, Italy 288
green
   pigments **268–269**
greenstone 214, 222
   polished axes 218
grindstones 218
grog 234, *235*

Groppello Cairoli, Pavia, Italy 288
Grotthus-Draper's law 32, 150
ground
    penetrating radar 84, 255
    state 30
    stones 220, *221*
Guane, Colombia *406*
gunmetal **308**, 331
gypsum **211**, 225, 242, 298
    based materials 250–251
    dehydration 250–251, 263
    plaster 230, 242–251, **243**

haemoglobin identification 222
Halaf, Syria 230
half-life *171*
halloysite 233
Hallstatt salt mines, Austria 398
Hallstatt culture 331
halophytic plants 290
hammer scales 333, 337
Hamoukar, Syria 287
Han Dynasty 288–289, 331
hardness 323, **342**
    measurement 340–341
harenae fossiciae 246
Hayonim Cave, Israel 245
heat 55, 205
heliotrope 353
Hellenistic glass 288
hematite 164, 240, 267
    Curie point of 166
heritage ecosystem 124
Hertz, units 18
Hertz and Hallwachs experiment 299
High Medieval HKEG glass **286**, 288
high performance liquid
        chromatography 410–413
high temperature
    processing 232
    resistance 222
Hippopotamus amphibius 360
HMG glass **286**
Hochdorf Prince burial, Stuttgart, Germany 398
holotomography 74
hominid *80*, 164
Hooke's law 34
hornstone 353
HPLC 400, 410
    chromatogram *412*
HR-ED 55
human eye 275
humidity 145–148, 261–262
Huysmans, Joris-Karl 354–355

HV, units 341, **342**
hydration
    dating 159–161, 296–297
    dry process 244
    layer 159–161, *161*, 296–298, 316
    process in cement 257
    rim 160
hydraulic mortar 247
hydrogen profiles 316, *316*
hydrolysis 138, **142**, 296, 385, 414
hydronium ions 296
hydrophobic strength 412
hydroxyl group 43–44
    in glass 296
    substitution in bone 162–163
hydroxylapatite 162, 357–364, **357**
hyperalkaline water 256
hyperfine
    coupling 194
    splitting *193*
    structure 194

ICC 278
Iceman *118*, 135, 142, *346*
ICOMOS Charters 253, 264
ICP 38–39, 180
ICP-MS 180
ICP-OES 39
identity 12
    cultural 123
Ijen volcano, Java 320
Ikawaten, Niger 335
Ilakaka, Madagascar *321*
illite 233
illuminance 151
image
    3D 72, 79, 434
    analysis 285
    backscattered electron 67
    digital 66, 79
    high resolution EM 68
    recording 58
    reproduction 57
    resolution 72
    scale 58
    secondary electron 67
    storage 419
Image Permanence Institute 147–148
imaging 23, 25, 79–83
    objects 59
    phase contrast 74
    pixel 60
    satellite 86–89
    science 57, 434

immiscible system 309
immunology 416–417
impactite 280
in situ analysis 75, 97
Inca 248
    walls 210
incense 369
Indian HAG glass **286**, 288
indigo molecule 270
Indigo suffruticosa 270
indigotin **269**, 413
inductively coupled plasma 38, 180
industrial revolution 200
inert component 257
information
    complementary 111
    content 29, 58, 118–125
    digital storage 419, 436
    supports 13, 419
    technology 79, 127
    visual 57
infrared
    spectroscopy 42–46, 249, 426–428
    thermography 86, 255
ink *61*, 152, *394*
    stratigraphy 136
INS **20**
instrumental noise 103
intangible cultural heritage 123, 201
INTCAL04 181, *182*
INTCAL98 181, *182*
International Color Consortium 278
International Commission on
        Illumination 276
invasiveness 60, **63**
ion
    beam 63, 179, 313
    beam analysis 313–317
    cyclotron resonance 179
    exchange 296
    selected monitoring 415
    source 414
ionization **20**
IR 42, 78, 115
irogane 311
iron
    -carbon system 312–313, *312*
    carburized 312, 323, 331
    cast 331
    EPR signal *194*
    forging 313, 337
    hammer scales 333, 337
    hardness **342**
    metal 312–313
    minerals **324**
    oxides 84, 164, 234, 238–240, 267, 334, 337

oxidation state 240, 267, 300, 335
phosphates *235*, 319
slags 313, 332, 337
smelting 334, 337
smithing 334, 337
sulphate 319
wrought **342**
irradiance 151–152
IRSF 366
IRT 86
Ishtar gate 287
Ishtar, Goddess 314
Islamic HMEIG glass **286**, 288
isochron line 327
isocratic elution 411
isotopes 170
    certified standards **186**
    fractionation 187, 326
    lead 185, **186**
    measurement 178–180
    radioactive 170, **174**
    stable 170, 184–187, **185–186**
isotopic
    abundances 103
    ratios 178, 185
    ratio measurement 180
    signature of artwork 136
    tracers 49
ivory 45, 360–364
    AMS amount **177**
    blue fossil 362
    Schreger lines 361
    texture patterns 361
    tusk 360

jade 314
Jaina Island, Campeche, Mexico *271*
Japanese sword 311
Jarmo, Kurdistan 397
jasper 353
Jericho, Israel 230, 245, 397
Jomon 229
Julia Felix shipwreck 288

K-Ar method 175
Kachiqhat red granite 210
kaolinite 233, 239
Kermes vermilio 406
kermesic acid 406
Khao Wong Prachan Valley, Thailand 331, 333
Kikuchi lines 54, 345
kiln 204–208, 240

lime- 243–244, 293
rotary 256
shaft 256
kinetics 13, 131, 145–149
    decay curve *261*
    of degradation 259, *261*
    of transformation 279
King Herod 247
King Offa of Mercia 222
King's Valley, Egypt 263
Kirchhoff and Bunsen experiment 30
Klaproth, Martin Heinrich 201
Klein, Yves 9
knowledge management 90
knucklebone 363
Kollman, Bolzano, Italy 348
Krakatoa eruption *169*
Kubelka-Munch theory 275

LA 180
La Ceñuela, Spain 330
LaDAR 85
lake 266, 406
    kermes 406
    madder **268**, 406
Lambert-Beer's law 31, *31*, 69
Lampyris noctiluca 32
Landsat Thematic Mapper 88
lapis sarcophagus 222
lapislazuli **269**, 349, **352**
large scale facilities 21–23, 343
Larmor frequency 409
Lascaux caves, France 228
laser
    beam ablation 96, 180
    cleaning of surfaces 225
    induced breakdown spectroscopy 40
    print 143
    scan of objects 79, 230–232
    lattice 278–279
Late Antiquity HIMT glass **286**, 288
Laurana, Francesco 224
Lavagnone, Brescia, Italy *284*
LC-MS 179
lead
    antimonate 293, 295
    arsenate 295
    -copper system 309–310, *310*
    containing pigments 274
    cupellation 334
    glaze 289
    isochron line 327
    isotope method 326–327
    isotope ratios 186, **186**, 326–327

oxide 337
slagging 337
smelting 334
stannate 295
Leaning Tower, Pisa, Italy 248
leather *56*
Le Corbusier 258
Leonardo 8
Leopol'dovič Rostropovič, Mstislav 139
leucite *247*
Levantine glass **286**, 288
Levi, Primo 320
libethenite 319
Libiola, Liguria, Italy 323, 332
LIBS 40
Libyan
    desert glass 280, *280–281*
    Sahara *217, 218, 221*
lichen 161–162
    growth curve 161–162
lichenometry 161–162, *162*
LiDAR 85
life cycle of objects 201
life expectancy
    of cultural heritage 137, 149
    prediction 145–149
lifetime 130–131, 139
light 18, 25, 30
    ageing tests 150–152
    annual exposure limits 153
    detection and ranging 85
    dosis 151
    exposure 93
    influence on materials 149–153
    interference 25, 297
    polarized 64
    reflection 275
    -sensitive materials 93, 153, 200
    source temperature 276, *276*
    stability assessment 153
    visible spectrum 275
lightfastness 93
lignosulphonate 265
Ligurian Eneolithic copper mines *221*, 323, *325*, 332
lime 230, *233*, 242–244
    hydraulic 249
    -kiln 243–244, 293
    in glass 291
    lumps 245, 249
    mortar **243**, 244, *245*
    plaster 242–244, *245*
    production 242–244
    putty 243, 244
    slaking **243**, 244
lime-based binders 242–250

limestone 211, **211**, 242, **243**, 251
   argillaceous 249, 251
   blue Liassic 249
   calcination 242–244, **243**
   of Selinunte temples 217, *219*
   Portland 252
limits of detection 16, 101–104, **104**
linen 396–399, **396**
Linum bienne 397
Linum usitatissimum **396**
LIPS 40
liquid chromatography
   instrumental configuration *411*
   mass spectrometry 179
litharge **324**, 337
lithics 210–225, **213**, 322
   small-size artefacts 219
   surface microwear 219
LMG glass **286**
LMHK glass **286**
LOAEL 138
local pollen zones 159
London, UK 290
long term behaviour of cultural
    heritage 136–157
Los Millares, Spain 330
lowest observed adverse effect level 138
Loxodonta africana 360
LPZ 159
Lucretius 204
luminescence 32, 78, 188, 189–192
lustre-ware 8, *54*, *68*, 241, 285, 295, 301
luteolin *412*, 413
lux, units 151
Lycurgus cup 295, 353

Madagascar *322, 323*
magnesian putty 244
magnesium
   carbonate 244
magnetic field
   Earth's 164–166
   gradient survey 84
   polar wander 164, *165*
   polarity 164
   static 408
magnetic moment 408
magnetic resonance imaging 408
magnetite 164, **324**, 335
   Curie point of 166
magnetization 84
magnetostratigraphy 164
magnification 66, 69

Mahaiza, Madagascar *321*
Maina, Russia 229
Majkop culture 398
majolica 238, **238**
major elements 101–102
malachite *56*, *73*, 319, 323, **324**, 329, 334
MALDI 180
Ma'Mikhael shipwreck 319
mammoth 360
   hair 398
Mammut sp 360
manganese
   as decolourant 288, 291
   EPR signal in marble *194*
   ions in ivory 362
   oxides 267
Mantegna 90–92
Maqarin, Jordan 252, 256
marble 108, 211, **211**
   EPR signal *194*, 195
   Greek classical 217
   high quality quarries 224
   provenancing 185
   maximum grain size analysis 218
   Mediterranean white marble trade
    217–218
   sulphated surfaces 225
marcasite 319, **324**
Marghera, Venezia, Italy *262*
marl 240, 249, 256
martensite 312, *312*
masonry 253–256
mass
   analyser 178–180, 414
   /charge ratio 178
   fractionation 178, 185
   selective detector 178–180, 414
   spectra 178–180, 414
   spectra NIST database 414
   spectrometer 178, *178*
mass spectrometry 178–180
   accelerator 173, **177**, 179
   gas.chromatography 179, 413–415
   Fourier transform 179
   inductively coupled plasma 180
   isotope measurement by 173
   liquid chromatography 179
   multicollector 180
   quadrupole 179
   radiocarbon measurement by 177
   secondary ion 179
   sector 178
   spark source 180
   tandem 178
   thermal ionization 180

time-of-flight 179
master layer sequence 158
mastic 369, 372
mastodon 360–361
material culture 9–12
materiality 12
materials
   compatibility 255, 264
   composite 199, 419
   development 201
   durability 131, 137
   light-weight resistant 200
   measurable properties **203**
   nature of 202, *202*
   organic 355–407
   paradigm 204
   refractory 232
   reversibility 264
   scientific investigation of 201
   structural 209–265, **211**
   technical properties 201–204
   technical selection 209–210
   use by man 199–204, *200*
materials properties
   electrical *204*
   mechanical 202, **203**, *204*
   physico-chemical 202, **203**
matrix-assisted laser desorption-
    ionization 180
matte 313, *339*
Mavrovouni mine, Cyprus 332
Maxwell-Boltzmann distribution 142
Maya
   blue pigment 269–270, *271*
   Classical period 248
   wall paintings 270, *271–272*
Mayan Lowlands 244
Mazzoni, Guido 8
MC 180
MC-ICP-MS 180
MDT 253
Medieval stained glass 289, 298
Megiddo, Israel 287
Meitner, Lise 301
melilite 240, 335
melting
   point 234
   temperature 279, **281**, 291, *292*
Mercurago, Piemonte, Italy *284*
mercury
   intrusion porosimetry 260
   picnometry 260
Mesoamerica 248, 348
   rock crystal skulls 316–317
Mesopotamia 230, 287

metacinnabar 274
metallic bond 307
metallography 64, *65, 307*, 343–348
metallurgical
　casting moulds 222–223
　pollution 334
　processing 309, 321–322
　sites 331–334
　slags dating 190, 196
　smelting slags 208, 334–339
metallurgy 305
　extractive 323
metals 305–348
　alloy **308**, 309
　characterization 339–348
　chemical analysis 102, 314, 339–340
　coated embroidery 342
　corrosion 317–320
　extraction 320–334
　forgeries 319–320
　hardness 323, 340–341, **342**
　historical use *200*
　inclusions 98, 340
　manufacturing 343–348
　microstructure 340–341, 343–348
　nanoparticles *54, 68*, 267, 295
　native 322
　object investigation 339–342
　oxides 205
　physical properties 340–341
　-polluted soils 334
　provenancing 326–331
　science 306, 307–313
　segregation *98*, 340
　smelting 322, 334–339
　surface alteration 95, 300, 317–320
　surface analysis 342
　observation by OM *65*
　technology 341
　texture 343–348
　thermomechanical history 343–344
　used in the past **308**
metastable solids 279
meteoritic impact *280*
methacrylic polymers 265
Mexico, Central 248, 348
Miaoyan, China 229
Michelangelo 7, 9
Michelson interferometer 44, *44*
micro-artefacts 210
micro-beam techniques 98, 340, 399, 434
　external 313
microbial activity 318, 356

micro-fading test 93, 153, *154*, 277
micro-indentation test 340–341, **342**
micro-mapping 62
micro-sampling 94, 113–114, 340, 430
microscopy 25
　dark field 65
　electron 55, 66–68
　optical 47, 59
　polarized light 64
microstructure
　of metals 340–341, 343–348
micro-textural evolution of copper *344*
microtomography 74, 260
microwave heating 17
Middle Age
　bone turquoise 362
　gem belief 348
　mortars 248
　plasters *245*
Middle East 230, 245
　plasters in 242, 251
Milland, South-Tyrol, Italy 330, *334*
millstones 218
mine 320–339
mineral 64, 210–225
　aluminosilicate 233
　composing objects **213**
　for metal extraction **324**
　gem 348–355, **351–352**
　Mohs scale 342
　ore 323, **324**, 327
　phases 232
　secondary 327, 329
mineralogical
　analyses 232, 239
　information 24
mineralogy 47–48, 64, 210
Minoan Crete 248, 287, 332
minor
　elements 101–102, 328–329
　destructive tests 253
MIP 38
mixed alkali source 290–291
mixing valve 411
moai 210
Mogao, China *100, 264*, 404–405, *404*
Mohs scale 323, 341, **342**
moldavite *280*
molecular
　vibrations 33
monochromator 39, 44
Monod, Jacques
Monodon monoceros 360
monosulphoaluminate 257
Montale terramara, Modena, Italy *362*

Monte Loreto, Liguria, Italy 323, *325*, 332
monticellite 335
montmorillonite 233
Moreau, Gustave 354
morganite *350*, **351**
mortar 99, 242–265
　analytical protocol for 249
　dating 250
　hydraulic 247
　organic molecules in 248
　Portland 257
　water resistant 246–247
mosaic 285
　glasses 288
　tesserae 285
Moseley diagrams *36*
Mössbauer spectroscopy **20**, 239–240, 267
Mothya, Sicily *225*
MRI 408
MS 178
MSD 414
mullite *233*, **237**, 239–240
multidimensional analysis 435
multivariate analysis 126
Munsell
　colour system 278
　tables 278
Murgul, Anatolia 330
Mycenean
　Greece 287
　Mediterranean 361

NAA **20**, 52, 214, 339
nacre 353
nantokite 318, *319*
naphthalene sulphonate 265
native
　copper 329, 335, 341
　metals 322, **324**
natron **286**, 288–291
natural
　ageing 137, 140, *141*
　dose 188–189
　dose rate 188–189
　fibres **396**
　glow curve 189
　organic dyes 399–407
　organic pigments 399–407
　polymers 200
Nazca, Peru 87
NCP 364
ND 52, 188
NDR 188

NDT 253
Nefertari tomb 263
Nemrik, Iraq 230
neodymium 105
Neolithic
  changes 229
  pre-pottery 230
  Venus of Gaban 272, *273*
Nernst glower 44
network
  forming elements 279, **282**
  modifiers 280, **282**
neutron
  activation analysis 52, 214–216, 339
  based analysis 214–216
  capture resonant analysis 214–216
  diffraction 52–53
  irradiation 70, 183–184
  radiography 69–71, 76
  resonance capture analysis 52
  sources 21
  tomography 71–75
New World 331
NEXAFS 304
niello **308**, 310–311, 319
Niepce, Joseph Nicephore 419
Niger 331
Nimrud, Iraq 230
NIP 119
Nippur, Iraq 230
NIST library of mass spectra 414
nitrogen
  depletion in bone 163
Nižná Myšľa, Slovakia *287*
NMR 408
  parameters for nuclei **409**
non-destructive tests 253
non-invasive
  analysis 52, 71–75, 94–97, 266, 343, 391–395
  diffraction 117
  neutron resonance analysis 215
  portable techniques 119–122
Noric steel **308**, 331
Norsun Tepe, Anatolia 330
North Pole *165*
Noyen, France 397
NRA 63, 316
NRCA 52, 215
NRRA 316
nuclear
  fission 171, 183–184
  magnetic resonance 408–410
  reaction 170, **171**
  reaction analysis 63, 316

nuclide chart *176*
number of samples 97
nuraghe 210, *212*
Nuzi, Iraq 287

Oakley, Kenneth 163
observation
  qualitative 27
  quantitative 27, 31
obsidian 45, 210, *212*, 280
  hardness 323
  hydration 159–161, *161*
  provenancing 218–219
ochre 166, 272, *273*
OCP 360
octacalcium phosphate 360
ODF 346
Odobenus rosmarus 360
odontolite 362
OES **20**, 37
Olduvaian fossils 175
olivine, fayalitic 334–337
  chain 335, *335–336*
  cooling rate 335
  feathered 335, *336*
  hopper 335
  skeletal 335, *335–336*
  spinifex-type *336*
  texture in slags *335–336*
Ollantaytambo, Peru 210
OM 47, 64
onyx 353
opacifiers in glass 284–285, 294–295
opal 296, 353
opalescent glass 296, *298*
OPC 257
Oppenheimer, Julius Robert 214
optical
  coherence tomography 74
  emission spectroscopy 37–40
  microscopy 47, 64–66, *234–235*, 239, *245*, *247*, 249, *335*, 343–344, *343*
optically stimulated luminescence **188,** 189–192, 196
  regenerative-dose protocols 191, 196
  single aliquot 190–191, 196
optoluminescence 32
opus caementicium 248
orange
  pigments **268–269**
ordinary Portland cement 257
ore minerals 320–339, **324**
  oxidic 323, **324**

sulphidic 323, **324**
organic
  binders in pigments 272–273
  complexes in dyes 266
  components in photography **427**
  dyes 399–407
  materials 355–407
  molecules in binders 258, 265
  pigments 399–407
  residues 355–356
  residues on stone tools 220–222, 356
organo-clay complex 270
orientation
  distribution function 346–347
  maps 345, *345*
OSL **20**, 189
osmium isotopes 329, **329**
osteoclastogenesis 364
ovalbumin 416–418
Ovetari Chapel, Padova, Italy 90–92
oxidation 137, 205–207, 386
  reactions *206–207*
oxidation state maps 62
oxides
  free energy of formation *206*
  glass-forming 278–279
oxygen
  partial pressure 205
  stable isotopes 186–187, **186**
ozone 152

painting 75–78, 99, 399
  binding medium 272–273, 399–400, 417
  exfoliation *264*
  stratigraphy *76, 100, 405*
Pakefield, England 164
palaeoclimatic reconstruction 159, *160*, 169
palaeodiet 187
palaeomagnetism 164–166
Palermo, Sicily 293
Palpa, Peru 87
palygorskite 270
palynology 158–159, *160*
palynomorphs 158
Pantheon, Rome, Italy 248
paper
  AMS amount **177**
  carbonylic groups 389
  carboxylic groups 389
  deacidification 154, 388
  degradation 154, 195, 384–388
  degree of polymerization 389
  destructive analysis 388–390

EPR signal *194*, 195
FTIR spectra *391*
foxing 386, *387*
hydrolysis 385
non-destructive analysis 391–395
oxidation 386
pH 388–389
Raman spectra *392*
RC *151*
paramagnetic
  defects in solids 193, *194*
  ions 193, 195
paramagnetism 165, 193
paratacamite 318, 335
parchment *56*
Parker, James 251
particle
  accelerators 21
  beam penetration **96**
  induced gamma-ray emission 313
  stopping power 96
Paschen-Runge polychromator *39*
passive layer 317
pathological mineralization 356
patina
  calcium oxalate 225
  in metals 318–319
Pauling, Linus Carl 279
Pb-isotope method 326–328
PDA 401, 413
PDA-MS 413
pearl 353
pearlite 312, *312*
peat 159
penetration depth 76, 95–97, **96**
perfume 356
periclase **243**, 244
periodic table 33
permanence index 147
petroglyphs 162, 226, *227–228*
petrographic analysis 239
petrology 47, 64, 210
pewter **308**
PGAA **20**, 52, 215
phase
  contrast mode 116
  crystalline 50
  diagrams 56, 239
  identification 42, 48, 50, 53, 64, 78
  mobile 411, 413
  partitioning 413
  quantification 49–50
  stationary 410, 413
Phoenician glass *283*
phosgenite **324**, 337

phosphate 319, 356–364, **357**
  group 357, *357*
  octacalcium 360
phosphorescence 32
photochemical
  activation 32
  ageing 149–153
  damage 138
  reaction 150
photodiode array 401, 413
photoelectric effect 299
photogrammetry 85
photography 419–432
  identification of type 420–428
  invasive analysis 430–432
  layer structure 421–423, *422*
  main processes 419–420, *420*
  non-invasive analysis 423–429
  portable laboratory 428–430, **429**, *430*
photo-oxidation 150
photo-reduction 150
Physeter macrocephalus 360
physical information 24
phytoliths 246
PI 149
picnometry 260
pietra ollare 222
Pieve di S. Siro, Cemmo, Italy *223*
PIGE 313
pigment 45, *62*, 78, 266–274
  archaeomagnetic dating of 166
  binders 272–273
  black 267, **269**
  blue **269**
  collections of 270–271
  degradation 273–274
  for cool surfaces 275
  green **269**
  identification 266–274
  in glaze 239
  non-invasive analysis 266
  orange **268**
  organic 399–407
  red 166, 267, **268**
  yellow *43*, **268–269**
  thermogravimetry *56*
  violet **269**
  white **268**
Piltdown forgery 163
Pisa, Tuscany, Italy 248
pitch 369
PIXE **20**, 313–317
pixel mapping 60, 88
Plank's constant 17, 30

plant ash 246, 288
  soda-rich 290
plant gum 414, 416–417
plasma 353
plaster 99, 230, 242–251
  dolomitic **243**
  gypsum 250–251
  lime 242–244, **243**
  of Paris 242, **243**, 250
plastic component 233–234
plasticizer 258, 265
Plato 204
plattenschlake 335, *336*
Pletz von Mozze, Luserna, Italy 332
Plinius 205, 222, 246, 289, 291, 320
poise, unit **281**
polarizing microscope *64*
pole figure 346–348, *347*
  inverse 346, *347*
pollen *67*, 158–159, *160*
  AMS amount **177**
pollution
  chambers *155*
  influence on ageing 154–156
polychromator 39
polycrystalline aggregates 50
polyethylene 150
polypeptide chain *358*, 359
polysulphides ore 323, 328
polyvinylic alcohol 265
population dynamics 135, 187
Populonia, Tuscany, Italy 332
porcelain 238, **238**, 289, 314
pore size distribution 410
porosimetry 260
porosity 260–261
  in binders 261
  in ceramics 237
  in materials 260, 409
  measurement 260
portable
  instrumentation 97, 119–122, 339, 434
  laboratory for photography 428–430, **429**, *430*
Portland
  cement 242, **243**, 252–265
  clinker *233*, **252**
  composition *233*, **252**
portlandite 243–244, 248, 257, *258*
Portuguese glazed tiles 305
potassium
  -argon method 175
  sulphate 251
pottery 238
Pourbaix diagram 317–320, *318*
Poviglio, Reggio Emilia, Italy *11*

pozzolan **211**, *233*, **243**, 246–249, 258, **259**
Pozzolane Rosse, ignimbrite 246
pozzolanic
  components 258
  minerals *247*
  mortar 246, **247**, 314
  reaction 246
Pozzuoli, Neaples, Italy 246
ppm, units 101, **104**
prase 353
PRAXIS instrument 267
precious stones 353
precision 101, 111, 128–129
pre-pottery Neolithic 230, 245, 251
Preseli dolerite 210
preservation index 149
probe 17, 59
prognosis 123
prompt-gamma activation analysis 52, 215
protein
  binding 416–417
  collagen 357, 396
  in artworks 416–418
  non-collagenous 364
  osteocalcin 364
proton 77, 313, 408
  backscattered 313
  induced X-ray emission **20**, 313–317
provenancing 48, 185
  building stones 217
  ceramics 241–242
  metals 326–331
pumice 248, 280
purple-red light region *276*, *277*
purpurin 406–407
Py-GC/MS 415
pyrite 319
  disease in wood 319
pyrolisis 415
  gas chromatography 415
pyrotechnology 204–208, 234–238
pyroxene 335

Qantir-Piramesses, Egypt 287, *294*
Qau, Egypt 287
Qiantan river, China 248
Qing Dynasty 314
QMS 179
quantum
  energy 17, 29
  mechanics 17
  theory 30, 32, 150
  vibrational states 33, 141
  yield 150

quartz 114, *212*, 232, 283, 334
  alpha-beta transition 236, *237*
  density 338
  fission tracks 184
  hydration 159–161, 316
  melting point 280
  sand 244–246, 289
  skulls 161, 316–317
  thermoluminescence 190
quartzite **211**, 289–290
Quaternary Period 132, *133*, 173
quicklime 242, **243**, 252
  early production 245
Qumran, Israel 216, 399

rabbit skin glue *43*
racemic mixture 163
racemization process 163
radiation
  Čerenkov 32, 302
  characteristic spectrum 34
  dose 150, 188, 189–192
  dosimetry 184, 188
  induced defects 183
  scattered 18, *19*
radicals, free 150, 193
radio frequency 409
radioactive decay
  constant 170, **174**
  long-lived 340
  parent-daughter evolution 171–172
  processes 170–177, **174**
  rate 171
radioactivity 170
  delayed 76
  induced 53, 71
  measurement 173
radiocarbon
  calibrated age 181
  conventional age 181
  method 177, 181–183
radiogenic species 171
radiography 31, **60**, 69–71
  neutron 70–71
  X-ray 69–70
radiometric dating
  ^{14}C method **174**, 177, 181–183
  K-Ar system **174**, 175
  U series **174**, 175–176, *176*
Raman
  scattering 42
  spectroscopy 42–46, 78, 240, 249, 266–267
  surface enhanced 45

random network theory 279
Rano Raraku tuffs 210
Rapanui 210
rare earth elements 281, 329, **329**
rate constants
  dependence on temperature 149
  of chemical reactions 123
Ravenna, Romagna, Italy 248
RBS 313
RC paper *151*
reciprocity
  failure 152–153
  principle 150, 152, 155
red
  glass 293–295, *294*
  pigments **268–269**
red-green-blue colour system 278, *278*
redox
  conditions 240, 323
  reactions 273
REE 329, **329**
referenced geographical systems 83, 89
reflectance spectrum 276
reflected light optical microscopy 64
reflectometry 61, 75–78
  total- 349
refractive index 65, 292, 349
refractory 232, *233*
relative time
  dating methods 157–166
  scale 131–136
remnant magnetization 84, 164
remote sensing 84–89
reproduction
  images 57
research infrastructures 114
resin 415, **368**, 370–372
  characterization 372, 383–384
  composition 371
  definition **368–369**
  fossilization 372–373
  physical properties 373–374
  plant source of 370–372
resistance to compression 257
resolution
  annual 167
  spatial 48, 72, 88
  spectral 88
resonance
  centre 188
  energy 316
restoration
  computer assisted 90
retention time 411, 414
reversibility 264
RF 409

RGB colour system 278, *278*
RH, relative humidity 147–148
Rhodes 287
risk
   assessment 123–124
   of damage 96
RL-OM 64
roasting 335
rock art 162, 227–228
   dating 226
   deterioration 226
   documentation 228–229
   engravings *227–228*
   preservation 226–229
rocks 64, 210–225, **211**
   composing objects **213**
   monomineralic 211
   polymineralic 211
   surface 162, 225
   thermal properties 222–223
ROMACONS project 247
Roman
   cement 251
   glass **286**, 285–291, *288, 298*
   harbors 247
   lime plaster 242–246
   mortar 246
   sarcophagi 222
Romanesque decorations *223–224*
Ross Island, Ireland 332
Rothenberg, Beno 306
Rothko, Mark 8
RS 42
Rubia peregrina 407
Rubia tinctorum **268**, 406
ruby 314, 349, **351**
rumanite 373, **377**
Rumiqolqa andesite 210
Ruskin, John 138
rust 317
Rutherford backscattering 313

Sacuidic, Carnia, Italy *245*
saffron 43
Sahagùn 348
Saint Véran, Queyras, France 323, *325*, 332
Salsola sp 290
salt
   efflorescence 264
   crystallization 255, 262, 298
   soluble 262
   weathering 262
SAM 302

sample
   homogenization 98
   position 99
   radio-activation 340
   representativity 27, 99–100
   surface layer 99
   volume 97–99, 113–114
sampling 27, 94–100
   strategies 97, 100
sandstones 223
San Juan River, Utah, rock engravings *228*
San Pietro di Zuri, Sardinia *224*
Santorini, Greece 246
Santu Antine, Sardinia *212*
San Vigilio, Pinzolo, Italy *263*
SAR 191, 196
sard 353
Sardinia 210, *212, 221*, 353
   glass bead *284*
sardonyx 353
satellite
   CORONA 88, **88**
   GeoEye 88, **88**
   IKONOS *11*, 88, **88**
   imagery *11*, 86–89
   Landsat 88, *88*, **88**
SAXS 398–399
scale
   energy *18*
   macroscopic 25, 58, 202
   mesoscopic 25, 59, 202
   microscopic 25, 59, 202
   nanoscopic 25, 202
   space 24–27, 58
   time
Scandinavia 322
scanning
   Auger microscope 302
   electron microscope 54, **60**, 66–68
   tunneling microscope 23, **60**
scattering
   anelastic **20**, 23
   elastic **20**, 23
Schio theater, Vicenza, Italy *260, 263*
Schottky-Frenkel defects 188
Schreger lines 361
science in archaeology 1
scientist 14–15
SCLF 175
Scythian Crimea 398
seal stone 353
Sebera permanence index 149
secondary ion mass spectrometry 179
   time of flight 180
sector MS instrumentation 178
secular equilibrium 172

security of context 131, 197
selection
   criteria of structural materials 209–210
   rules in spectroscopy 42
Selinunte, Sicily 217, *219*
SEM 55, 66
SEM-EDS **20**, 67
separation
   columns 411–412
   techniques 410
sepiolite 270
septariae 252
Serra Pelada, Brazil 320
SERS 45
SEXAFS 304
SFJ 254
shakudo **308**, 311, 319
shape acquisition systems 91
shell 353
Shiqmim, Israel 330
shrinkage 234–237, 257
Siberia 361, 398
Sicily 217, *219*, 293
siderite 319
Sidon-Saïda, Lebanon 288
silanol groups 412
silica 232, *233*, 334, 337, *337*, 412
   glass 278–279
   -lime reactions 246, 252
   microcrystalline 353
   sand 289
silk **396**, 403
   yarn 413
Silk Road 361
silver **324**, 337
   gelatin 419–426, **425–426**
   photographic print *151*, 419–426
   reduction 337
   sulphide 310
SIM 415
simetite **377**
SIMS 179
single-aliquot regenerative-dose protocol 191
sintering 234–237, *236*
Skouriotissa mine, Cyprus 332, *333*
slag
   blast-furnace *233*, 258, **259**
   classification *339*
   copper smelting 332–333, *333*
   dating 338
   density 338
   immature 338
   iron smelting 337
   smelting 334–339
   tapped 337
   viscosity 313

slagging process 334–339
slaked lime **243**, 244–246
slaking 242–246, **243**
slip
  ceramics 239
  systems in metals 307, *307*, 343–344, *344*
Slovakia 287
SLS glass **286**
slurry 242–244
small-angle X-ray scattering 398–399, *399*
smalt 285
Smeaton, John 249, 251
smectite 233
smelting process 334–339
  early activities 330
  elemental fractionation 328
  experimental 333
  mineral charge 323
  slags 334–339
Smith, Cyril Stanley 306
smithing 337
soda-lime silica glass 285–289, **286**
sodium
  carbonate 290
  phosphate 295
soil
  colour of 278
  erosion 196
solar cycles *135*, 169
solid
  solution 309
  state detectors 215
sonic pulse velocity 253
space 24–27
spectrometry 29, 61
  radioactivity measurement 173
spectrophotometry 23, 46, 275, 417
spectroscopy 23, 25, 29–34
  reflectance 46
  vibrational 33, 42–46
Sphinx, Giza, Egypt 251
spin
  -echo experiment 409–410
  lattice relaxation 409
  transition 193
Spondylus sp 353
SRC 258
SRIXE 35
stability diagram 317–320, *318*
stable isotopes 184–187, **185–186**
standardization 435
statistics 125–128
steatite 285

steel
  corrosion 262
  Damascus **308**, 331
  hardness **342**
  Noric **308**
  oxidation 262
  phase diagram *312*
  reinforced concrete 258
  wootz **308**, 331, **342**
Steno, Nicholas 157
Sticky Rice Bridge, China 248
sticky rice solution 248
stone hammers 220, *221*
Stonehenge, England 210
stoneware 238, **238**
Stradivarius 139
stratigraphy 133, 157–158
  chemical 135
strontium
  substitution in bones 163
  stable isotopes 186, **186**
structural
  clay products **211**, 229–242, *232*
  materials 209–265, **211**
structure diagnostics 84
structure-processing-properties relationship 204
stucco decoration *260*
Sturgeon, Theodore 355
succinite 368, 373, 376
  FTIR spectra *115*, 382
  paleobotanical origin 380
sulphate 236, *237*, 262
  -reducing bacteria 318
sulphide
  co-smelting 323
  inclusion in slag *336*
  inclusion in metal *343*
  matte 326
  poly- 323, 328
  primary 327, 329
  reactions *207*, *318*
sulphur 318
sulphuric acid 319
superposition law 157
surface
  contamination 299, 339
  reflectance 267
  techniques 299–303, 342
survey
  intersite 84
  intrasite 84
  large scale 10, *11*, 59, 84–89
Sweden 331
SXRF 37

sylilation 415
synchrotron radiation 21, 62, 303, 399
  diffraction *118*
  induced X-ray emission 35
  tomography 74, 80, *81*
syngenite 298

tandem MS 178
tar 369, 372
Tarim, Yemen 230
Tarquinia, Tuscany, Italy *362*, 363
technical art history 111
techniques 17–93
  bulk 59
  choice of 106–111
  dating 24, **134**
  geophysical 84–85
  incident probe **20**
  information provided 24, **34, 109**
  instrumental design 19
  laser-based 61
  micro-beam 98
  nature of the interaction **20, 34**
  non-invasive 95, 343
  optimization of 114–118
  portable 97
  selection of 106–111
  sensitivity of 115
  separation 410
  surface 59
technological processes 12
teeth 80, *81*, 187, 356–364
  AMS amount **177**
tektite 280, *280*
Tell Atchana, Turkey 287
Tell Braq, Syria 230, 287
Telloh, Iraq *231*
TEM 55, 66
temper 233–234, 239
temperature
  annealing 184, **281**
  critical 279
  Curie 84, 165
  firing 50
  glass transition 279
  glass working **281**, *292*
  light source 276
  melting 234, 279, **281**, 291, *292*
Temple of Venus, Pompei, Italy **247**
Templo Rojo, Cacaxtla, Mexico *271–272*
tenacity 323
tennantite **324**, 328
ternary diagrams 232, *233*

terra sigillata 302
terramara *11, 234–235*
tetrahedrite **324**, 328
Tevere river, Italy 246
textiles 395–399
   archaeological 395, **396**
   Coptic Egyptian 407
   dyes 400–407, *406–407*
   pre-Columbian 407
   Roman Egyptian 407
texture
   analysis 25, **60**, 343–348
   crystallographic 54, 202, 343–348
   in mortar 249
   maps 54
   microcrystalline 309
TGA 55
thaumasite 255, 262
Thera 247
   eruption 158
theriomorphic vessel *225*
thermal
   ageing 141–148
   agitation 141
   analysis 55–57
   annealing *343–344*
   degradation 138
   insulation 238
   ionization 180
thermodynamic
   equilibrium 261
   of combustion 205–207
   stability 131
thermographic testing 85–87, 255
thermogravimetry 55
thermo-hygrometric changes 147
thermoluminescence 32, 189–192, 196, 241
   curves *191*
   spectrally resolved *192*
thermomechanical history 343–344
Thessalian ophicalcites 218
thin sections 48, 64, 98, 214, *234–235*, 239
Thomsen, Christian Jürgensen 305
"Three Age" system 305
Ticino river, Italy 290
tidelines 149
Tiffany, Louis Comfort 283
Tiffany glass *282*
Tikal, Guatemala *88*
time 13, 28, 130–198
   resolution 136, 167
   sequence 110, 130
time scale
   absolute 130–136
   chemical 136–157

   cross-referencing 133
   geologic 131, *132*
   relative 131–136
time-of-flight 53, 179
time-temperature path 232, 238, 335
time weighted preservation index 149
Timna, Israel 330
   project 306, 322
   slags 335
TIMS 173, 180
tin
   -copper system 310, *311*
   dioxide 295
   isotopes 329, **329**
   oxide opacifier 289, 295
   surface enrichment 319–320
titanite
   annealing temperature 184
   fission tracks 184
titanium
   use in architecture 9, 209
TL **20**, 189
Tlaxcala, Mexico *271–272*
TL-OM 64
TOF 52
TOF-SIMS 180
tomography 31, **60**, 62, 71–75
tourmaline *350*, **352**
trace elements 101, 242, 327, **329**
   in copper 328–329, **329**
   in gold 328, **329**
tracers 49, **329**
trachyte **211**, 222, *224*
transition elements 281
transmission electron microscope 55, **60**, 66–68
transmitted light optical microscopy 64
Trass 247
tree rings 167–169, *168*
   sequences 167–169, 181
tristimulus values 276
trophic level 187
TRXRF 37
trydimite 280
Tswett, Mikhail 410
Tuc d'Audoubert, Ariège, France 229
tumbaga **308**, 309, 342
Turkey 287
turquoise 348–349, **352**
   fossil 362
tusk 360
   AMS amount **177**
Tutankhamun *280*
tuyere 333
TWPI 149
Tylecote, Ronald Frank 306

ultraviolet *17*, 46
   spectroscopy 46–47
Uluburun shipwreck 287, 361
undercooled liquid 279
Upper Cretaceous chalk 252
uranium
   decay series 175–176, *176*
   disequilibrium 175–176
   series method 175–176, *176*
   spontaneous decay 183–184, *184*
   substitution in bone 163
uranium-thorium method 175–176
Ust-Karenga 12 site, Siberia, Russia 229
U-Th method 175–176
UV *17*, 46, 78, *412*

Val Camonica rock engravings *227*
valence-sensitive spectroscopies 273
Van de Graaff generator 313
varnish *76*, 78
   desert 226, 228
vegetation 159
Venetian glass **286**, 290
   "façon de Venise" 290
Verucchio, Rimini, Italy 398
vibrational
   modes 34, 42
   states 42
Vicat, Joseph 252
Vicat, Louis 252
Vickers
   indentation test 340–341, *341*
   micro-hardness values **342**
vinegar syndrome 148
violet
   pigments **268–269**
Viollet le Duc, Eugène 138
virtual
   access 79
   archaeology 79
   cultural heritage 82
   preservation 230–232
   reality 79, 82, 90–93, 435
   reconstruction 73, 79–82, 90–93
   rendering 79
Vis *17*, 46, 78, *412*
viscosity of glass **281**, 291–292, *292*
visible
   spectroscopy 46–47
visualization 58, 79–83
vitrification 238, 252
Vitruvius 246–247
vivianite *235*, 319, 362

volcanic
  ash 246–248
  glass **211**, 280
  particles *247*
  tuffs 246–248
Volturno river, Campania, Italy 289
volume
  of material analysed 113–114
  probed volume 27
voxel 72
VPDB 185

Wabi-Sabi 137
WARA 228
water absorption 260–262
  coefficient 249
  porosimetry 260
water molecule *34*, 43, 259–265, 296
  chemisorbed 262
  cycle 186
  diffusion in silica 160–161, 296
  in bone 357–360
  in clays 233–237
  loss during firing **237**
  physiosorbed 262
wave
  electromagnetic *17*
  evanescent 45
  frequency 17, *18*, **19**, 30, 44
wave-particle duality 17
wavelength 17, 30, 51
  complementary 277
  dominant *276*, 277
  dispersive spectrometry 36
  wavelength scale *18*, **19**
wavenumbers 44
WDS 36

WD-XRF 37
weathering
  glass 296–298
  stone *156*
  tests 155–156, *156*
weddellite 225
Wedgwood, Thomas 419
Wertime, Theodore A. 306
whewellite 225
white
  pigments **268–269**
wide area survey 84–89
Wissler, Clark 169
wollastonite 240, 287, 295
wood 167–169, 205, 244, 319
  AMS amount **177**
  fossil 217, *217–218*
  preservation 319
wool 396–398, **396**, *396*
  fleece type 398
  yarn 403
wootz steel **308**, 331, **342**
working temperature **281**
World Archives of Rock Art 228
wüstite 337, *337–339*
wu tong 311, 319

XAFS 304
XANES 303
XAS **20**, 303
XPS **20**, 299
X-ray
  absorptiometry 116
  absorption fine structure 304
  absorption spectroscopy 303–305
  absorption near-edge structure 303
  beam 35

Booklet 301
characteristic 34–35, 66
diffraction 50–51, 117, 119–122, 249
fluorescence spectroscopy 34–37, 423–426
photoelectron spectroscopy 299–301
powder diffraction 50–51, 114, 117, 239
radiography 69–71, 77, 116
small-angle scattering 398–399, *399*
tomography 71–75, 116
XRD **20**, 47–48, 50–51, 249, 266–267
  crystallinity index in bone 366
  portable 119–122
XRF **20**, 34
  portable 119–122
  scanning 37
  total reflection 37

Yarmouk river, Jordan 256
yarn 399–400
yellow
  dye *412*, 413
  pigments **268–269**
Younger Dryas 135
Yucatàn, Mexico 270, 348
Yuchanyan, China 229

Zachariasen model of glass *279*
Zeeman interaction 192–195
zeolite **243**, 246–247, *247*
  -rich tuffs 246
Zhoukoudian caves, China 184
zinc **308**, 313
  oxide **206**, 308
zircon
  annealing temperature 184

Printed and bound by CPI Group (UK) Ltd, Croydon, CR0 4YY
24/11/2024
01793623-0001